T0336190

Handbook of Digital Twins

Over the last two decades, Digital Twins (DTs) have become the intelligent representation of future development in industrial production and daily life. Consisting of over 50 chapters by more than 100 contributors, this comprehensive handbook explains the concept, architecture, design specification and application scenarios of DTs.

As a virtual model of a process, product or service to pair the virtual and physical worlds, DTs allow data analysis and system monitoring by using simulations. The fast-growing technology has been widely studied and developed in recent years. Featured with centralization, integrity and dynamics, it is cost-effective to drive innovation and performance. Many fields saw the adaptation and implementation across industrial production, healthcare, smart city, transportation and logistics. World-famous enterprises such as Siemens, Tesla, ANSYS and General Electric have built smart factories and pioneered digital production, heading towards Industry 4.0.

This book aims to provide an in-depth understanding and reference of DTs to technical personnel in the field, students and scholars of related majors, and general readers interested in intelligent industrial manufacturing.

Dr Zhihan Lyu is an Associate Professor at the Department of Game Design, Uppsala University, Sweden. He is also IEEE Senior Member, British Computer Society Fellow, ACM Distinguished Speaker, Career-Long Scientific Influence Rankings of Stanford's Top 2% Scientists, Marie Skłodowska-Curie Fellow, Clarivate Highly Cited Researcher and Elsevier Highly Cited Chinese Researcher. He has contributed 300 papers including more than 90 papers on IEEE/ACM Transactions. He is the Editor-in-Chief of Internet of Things and Cyber-Physical Systems (KeAi), an Associate Editor of a few journals including *ACM TOMM, IEEE TITS, IEEE TNSM, IEEE TCSS, IEEE TNSE* and *IEEE CEM*. He has reviewed 400 papers. He has received more than 20 awards from China, Europe and IEEE. He has given more than 80 invited talks for universities and companies in Europe and China. He has given 20 keynote speeches at international conferences.

Handbook of Digital Twins

Edited by
Zhihan Lyu

CRC Press is an imprint of the
Taylor & Francis Group, an **informa** business

Front cover image: Gorodenkoff/Shutterstock

First edition published 2024
by CRC Press
2385 NW Executive Center Drive, Suite 320, Boca Raton FL 33431

and by CRC Press
4 Park Square, Milton Park, Abingdon, Oxon, OX14 4RN

CRC Press is an imprint of Taylor & Francis Group, LLC

ISBN: 978-1-032-54607-0 (hbk)
ISBN: 978-1-032-54608-7 (pbk)
ISBN: 978-1-003-42572-4 (ebk)

DOI: 10.1201/9781003425724

Typeset in Palatino
by codeMantra

Contents

Part 5 Digital Twins in Management

Part 6 Digital Twins in Industry

Part 7 Digital Twins in Building

Part 10 Digital Twins in Medicine and Life

Contributors

Päivi Aaltonen
MORE SIM Research Platform, LUT School of Business and Administration
LUT University
Lappeenranta, Finland

Islam Asem Salah Abusohyon
Università degli studi di Palermo
Palermo, Italy

Sofia Agostinelli
CITERA Research Centre
Sapienza University of Rome
Rome, Italy

Ashwin Agrawal
Civil and Environmental Engineering
Stanford University
Stanford, CA

Giuseppe Aiello
Università degli studi di Palermo
Palermo, Italy

George Arampatzis
School of Production Engineering and Management
Technical University of Crete
Chania, Greece

Rebeca Arista
Industrial System Digital Continuity Specialist at Airbus SAS
Leiden, the Netherlands

Zeynep Baysal
Ostim Technical University
OSTIM, Turkey

Marcelo Behar
PricewaterhouseCoopers LLP
New York, New York

Abdeljalil Beniiche
Optical Zeitgeist Laboratory
Institut national de la recherche scientifique
Quebec, Canada

Antonio Bono
Department of Computer Science, Modeling, Electronics and Systems Engineering
University of Calabria
Rende, Italy

Hui Cai
Department of Electrical Engineering and Information Technology
Ilmenau University of Technology
Ilmenau, Germany

Serdar Çelik
Ostim Technical University
Ostim, Turkey

Marianna Charitonidou
Faculty of Art Theory and History
Athens School of Fine Arts
Athens, Greece

Dawei Chen
InfoTech Labs
Toyota Motor North America
Plano, Texas

Long Chen
School of Architecture, Building and Civil Engineering
Loughborough University
Loughborough, England

YangQuan Chen
University of California Merced
Merced, California

Kai Cheng
Brunel University London
Uxbridge, England

David Christopher
Brunel University London
Uxbridge, England

Chiara Cimino
Associate Professor at University of Turin
Department of Management, Economics, and Industrial Engineering
Politecnico di Milano
Milan, Lombardia, Italy

Marianna Crognale
Department of Structural and Geotechnical Engineering
Sapienza University of Rome
Rome, Italy

Paul M D'Alessandro
Customer Transformation
PricewaterhouseCoopers LLP
New York, New York

Luigi D'Alfonso
Department of Computer Science, Modeling, Electronics and Systems Engineering (DIMES)
University of Calabria
Rende, Italy

João Miguel da Costa Sousa
IDMEC, Instituto Superior Técnico
Universidade de Lisboa
Lisbon, Portugal

Susana Margarida da Silva Vieira
IDMEC, Instituto Superior Técnico
Universidade de Lisboa
Lisbon, Portugal

Yanning Dai
School of Instrumentation and Optoelectronic Engineering
Beihang University
Beijing, China

Istvan David
Université de Montréal
Montreal, Canada

Melissa De Iuliis
Department of Structural and Geotechnical Engineering
Sapienza University of Rome
Rome, Italy

José Ferreira de Rezende
Federal University of Rio de Janeiro (UFRJ)
Rio de Janeiro, Brazil

Marianne T DeWitt
Customer Transformation
PricewaterhouseCoopers LLP
New York, New York

Pedro Henrique Diniz
Federal University of Rio de Janeiro (UFRJ)
Rio de Janeiro, Brazil

Leiting Dong
School of Aeronautic Science and Engineering
Beihang University
Beijing, China

Yongquan Dong
Chongqing Jiaotong University
Chongqing, China

Ivan Dychka
Faculty of Applied Mathematics
National Technical University of Ukraine
Kyiv, Ukraine

Pavlos Eirinakis
Department of Industrial Management and Technology
University of Piraeus
Piraeus, Greece

Georgios Falekas
Department of Electrical and Computer Engineering
Democritus University of Thrace
Komotini, Greece

Giuseppe Fedele
Department of Informatics, Modeling, Electronics and Systems Engineering (DIMES)
University of Calabria
Rende, Italy

André Filipe Simões Ferreira
Hovione Farmaciência S.A.
Loures, Portugal

Gianni Ferretti
Automatic Control
Cremona campus of the Politecnico di Milano
Cremona, Italy

Anselmo Filice
Department of Environmental Engineering, Afference to Department of Informatics, Modeling, Electronics and Systems Engineering (DIMES)
University of Calabria
Rende, Italy

Martin Fischer
Civil and Environmental Engineering
Stanford University
Stanford, California

Nikolai Fomin
V. A. Trapeznikov Institute of Control Sciences of Russian Academy of Sciences
Moscow, Russia

Marco Francesco Funari
Department of Civil and Environmental Engineering
University of Surrey
Guildford, England

Shuo Gao
School of Instrumentation and Optoelectronic Engineering
Beihang University
Beijing, China

Gabriele Garnero
Interuniversity Department of Regional and Urban Studies and Planning
Università degli Studi di Torino
Turin, Italy

Vincenzo Gattulli
Department of Structural and Geotechnical Engineering
Sapienza University of Rome
Rome, Italy

Behnam Ghalamchi
Mechanical Engineering
California Polytechnique State University
San Luis Obispo, California

Ning Gou
Brunel University London
Uxbridge, England

Jascha Grübel
Cognitive Science
ETH Zurich
Zurich, Switzerland

Furkan Guc
University of California Merced
Merced, California

Dayalan R. Gunasegaram
CSIRO Manufacturing
Geelong, Australia

Atul Gupta
Merative
Ann Arbor, Michigan

Mohammed Adel Hamzaoui
Lab-STICC
Université Bretagne Sud Lorient
Lorient, France

Daguang Han
School of Civil Engineering
Southeast University
Nanjing, China

Zhu Han
Department of Electrical and Computer Engineering
University of Houston
Houston, Texas

Richard Heininger
Business Informatics-Communications Engineering
Johannes Kepler University
Linz, Austria

Anca-Simona Horvath
Research Laboratory for Art and Technology
Aalborg University
Aalborg, Denmark

Kaixin Hu
Smart City and Sustainable Development Academy
Chongqing Jiaotong University
Chongqing, China

Thomas Ernst Jost
Business Informatics-Communications Engineering
Johannes Kepler University
Linz, Austria

Nathalie Julien
Lab-STICC
Université Bretagne Sud Lorient
Lorient, France

Eric Guiffo Kaigom
Computer Science and Engineering
Frankfurt University of Applied Sciences
Frankfurt, Germany

Kostas Kalaboukas
Gruppo Maggioli
Santarcangelo di Romagna, Greece

Vivek Kant
Human Factors and Sociotechnical Systems Studios
IDC School of Design
Indian Institute of Technology Bombay
Mumbai, India

Athanasios Karlis
Department of Electrical and Computer Engineering
Democritus University of Thrace
Komotini, Greece

Dimitris Kiritsis
Sustainable Manufacturing
Ecole Polytechnique Federale de Lausanne (EPFL)
Lausanne, Switzerland

Zafeirios Kolidakis
Department of Electrical and Computer Engineering
Democritus University of Thrace
Komotini, Greece

Mariusz Kostrzewski
Warsaw University of Technology
Faculty of Transport
Warszawa, Poland

Serge P. Kovalyov
Institute of Control Sciences of Russian Academy of Sciences
Moscow, Russia

Esra Kumaş
Ostim Technical University
Ostim, Turkey

Emil Kurvinen
Materials and Mechanical Engineering Research Unit, Machine and Vehicle Design
University of Oulu
Oulu, Finland

Antero Kutvonen
LUT School of Engineering Science
LUT University
Lappeenranta, Finland

Cecilia Lee
Royal College of Art
London, United Kingdom

Alberto Leva
Automatic Control at Politecnico di Milano
Milan, Italy

Shangkuan Liu
Brunel University London
Uxbridge, England

Xiaocheng Liu
School of Computer Science and Technology
Qingdao University
Qingdao, China

Yanhui Liu
Southwest Jiaotong University
Chengdu, China

Stavros Lounis
ELTRUN E-Business Research Center, Department of Management Science and Technology
Athens University of Economics and Business
Athens, Greece

Paulo B. Lourenço
Department of Civil Engineering
University of Minho
Minho, Portugal

Jinzhi Lu
Ecole Polytechnique Federale de Lausanne (EPFL)
Lausanne, Switzerland

Zhihan Lyu
Department of Game Design
Uppsala University
Uppsala, Sweden

Martin Maier
Optical Zeitgeist Laboratory
Institut national de la recherche scientifique
Quebec, Canada

Utpal Mangla
Telco Industry & EDGE Clouds
IBM
Toronto, Canada

Giulia Marcon
University of Palermo
Palermo, Italy

Roman V. Meshcheryakov
V. A. Trapeznikov Institute of Control Sciences of Russian Academy of Sciences
Moscow, Russia

Andrey Mozokhin
Department of Automated Systems of Process Control of SMGMA Group
Moscow, Russia

Petra Müller-Csernetzky
Design Management and Innovation
Lucerne School of Engineering and Architecture
Lucerne, Switzerland

Ahsan Muneer
School of Business and Management
LUT University
Lappeenranta, Finland

Andre Nemeh
Strategy and Innovation
Rennes School of Business
Rennes, France

Tobias Osterloh
RWTH Aachen University
Aachen, Germany

Busra Ozen
Department of Civil Engineering
Aydin Adnan Menderes University
Aydin, Turkey

Gozde Basak Ozturk
Department of Civil Engineering
Aydin Adnan Menderes University
Aydin, Turkey

Hamide Özyürek
Department of Business Administration
Ostim Technical University
Ostim, Turkey

Ilias Palaiologou
Department of Electrical and Computer Engineering
Democritus University of Thrace
Komotini, Greece

Andreas Pester
Faculty of Informatics and Computer Science
The British University in Egypt
Cairo, Egypt

Heli Ponto
Forum Virium Helsinki Oy
Helsinki, Finland

Panagiota Pouliou
CITA – Center of Information Technology and Architecture
KADK
Copenhagen, Denmark

Salvatore Quaranta
Università degli studi di Palermo
Palermo, Italy

Laavanya Ramaul
School of Business and Management
LUT University
Lappeenranta, Finland

Guoqian Ren
College of Architecture and Urban Planning
Tongji University
Shangai, China

Stefano Rinaldi
Department of Information Engineering
University of Brescia
Brescia, Italy

Jürgen Roßmann
Electrical Engineering
RWTH Aachen University
Aachen, Germany

Christian Esteve Rothenberg
University of Campinas
Campinas, Brazil

Jože Martin Rožanec
Information and Communication Technologies
Jožef Stefan International Postgraduate School
Ljubljana, Slovenia

Timo Ruohomäki
Forum Virium Helsinki Oy
Helsinki, Finland

Jussi Salakka
Mechanical Engineering
Oulu University
Oulu, Finland

Ville Santala
Forum Virium Helsinki Oy
Helsinki, Finland

João Afonso Ménagé Santos
Hovione Farmaciência S.A.; IDME, Instituto Superior Técnico
Universidade de Lisboa
Lisbon, Portugal

Jyrki Savolainen
School of Business and Management
LUT University
Lappeenranta, Finland

Philip Scarf
Cardiff Business School
Cardiff University
Cardiff, Wales

Oleg Schekochikhin
Department of Information Security
Kostroma State University
Kostroma, Russia

Elena F. Sheka
Institute of Physical Researches and Technology of the Peoples' Friendship University of Russia
Moscow, Russia

Muhammad Shoaib
Information Systems Department, King Saud University
Politecnico di Milano
Milan, Italy

Valeria Shvedenko
LLC T-Innovatic
St. Petersburg, Russia

Vladimir Shvedenko
Federal Agency for Technical Regulation and Metrology ROSSTANDART
The Russian Institute of Scientific and Technical Information of the Russian Academy of Sciences (VINITI RAS)
Moscow, Russia

Seppo Sierla
Aalto University
Espoo, Finland

Xinya Song
Department of Electrical Engineering and Information Technology
Ilmenau University of Technology
Ilmenau, Germany

Christian Stary
Business Informatics-Communications Engineering
Johannes Kepler University
Linz, Austria

Nenad Stojanović
Nissatech Innovation Centre
Germany

Oliver Stoll
Lucerne School of Engineering and Architecture
Lucerne, Switzerland

Jayasurya Salem Sudakaran
Human Factors and Sociotechnical Systems Studios, IDC School of Design
Indian Institute of Technology Bombay
Mumbai, India

Olga Sulema
Computer Systems Software Department
National Technical University of Ukraine
Kyiv, Ukraine

Yevgeniya Sulema
Computer Systems Software Department
National Technical University of Ukraine
Kyiv, Ukraine

Eugene Syriani
Department of Computer Science and Operations Research
Université de Montréal
Montreal, Canada

Lavinia Tagliabue
University of Turin
Turin, Italy

Chenyu Tang
Department of Engineering
University of Cambridge
Cambridge, England

Gloria Tarantino
Università degli Studi di Torino |
UNITO · Dipartimento
Interateneo di Scienze, Progetto e
Politiche Del Territorio
Politecnico di Torino
Turin, Italy

Frits van Rooij
IDE Americas Inc.
Carlsbad, California
Salford Business School
University of Salford
Salford, England

Sai Phanindra Venkatapurapu
Customer Transformation
PricewaterhouseCoopers LLP
New York, New York

Jairo Viola
University of California Merced
Merced, California

Juho-Pekka Virtanen
Forum Virium Helsinki Oy
Helsinki, Finland

Annalaura Vuoto
Department of Civil Engineering
University of Minho
Minho, Portugal

Dan Wang
Department of Computing
The Hong Kong Polytechnic
University
Hong Kong, China

Jiaqi Wang
School of Instrumentation and
Optoelectronic Engineering
Beihang University
Beijing, China

Jie Wang
Beijing Institute of Nanoenergy and
Nanosystems
Chinese Academy of Sciences
Beijing, China
School of Nanoscience and
Technology
University of Chinese Academy of
Sciences
China

Bianca Weber-Lewerenz
Faculty of Civil Engineering
RWTH Aachen University
Aachen, Germany

Shaun West
Lucerne School of Engineering and
Architecture
Lucerne University of Applied
Sciences and Arts
Lucerne, Switzerland

Dirk Westermann
Department of Electrical
Engineering and Information
Technology
Ilmenau University of Technology
Ilmenau, Germany

Chunli Ying
School of Architecture, Building and
Civil Engineering
Loughborough University
Loughborough, England

Yatong Yuan
China Construction Fifth Engineering Bureau
Guangdong, China

Jiayue Zhang
Department of Mechanical Engineering
State Key Laboratory of Tribology
Tsinghua University
Shenyang Architectural and Civil Engineering Institute
Tsinghua University
Beijing, China

Xiaochen Zheng
Sustainable Manufacturing (ICT4SM)
Ecole Polytechnique Fédérale de Lausanne (EPFL)
Lausanne, Switzerland

Yu Zhang
Shenyang Jianzhu University
Shenyang, China

Xuan Zhou
School of Aeronautic Science and Engineering
Beihang University
Beijing, China

Yifei Zhu
Shanghai Jiao Tong University
Shanghai, China

Part 1

Introduction

1

Overview of Digital Twins

Zhihan Lyu
Uppsala University

Xiaocheng Liu
Qingdao University

1.1 Introduction

This book consists of 50 chapters contributed by 129 authors. This chapter is a general introduction to each chapter of the book. From the second chapter, the concept of digital twinning, architecture description, design specification, and application scenarios are introduced. Section 2 introduces the concept and development of digital twins. Section 3 introduces the key technologies to promote the development of digital twins. Section 4 introduces some general frameworks and construction methods of digital twins. Section 5 introduces the application of digital twins in management and operation. Section 6 introduces the application of digital twins in industry. Section 7 introduces the application of digital twins in building construction. Section 8 introduces the application of digital twins in transportation. Section 9 introduces the application of digital twins in the energy industry. Section 10 introduces the application of digital twins in health and life.

1.2 Thinking about Digital Twins

Ashwin Agrawal and Martin Fischer designed a framework to enable users to find suitable Digital Twins applications, to help practitioners systematically think about the basic factors that affect successful Digital Twins deployment in Chapter 2. Realizing these factors in practice can improve the probability of success and accelerate the application of Digital Twins in the industry.

Pedro Henrique Diniz examines the application of the Digital Twins paradigm to the field of computer networks in Chapter 3. At present,

DOI: 10.1201/9781003425724-2

only industrial tools that deal with life-cycle network management through intention-based network automation and closed-loop automation can be effectively classified as Network Digital Twins, mainly because they maintain two-way communication between physical and virtual environments.

Xiaochen Zheng et al. introduced the concept of cognitive Digital Twins, which reveals a promising development of the current twins paradigm toward a more intelligent, comprehensive, and full life-cycle representation of complex systems in Chapter 4. Compared with the current Digital Twins concept, cognitive Digital Twins enhances cognitive ability and autonomy. This chapter first introduces the evolution process of cognitive Digital Twins concept.

Marco Francesco Funari et al. outline the concept of Digital Twins in the Architecture, Engineering, Construction, and Operation domain in Chapter 5. Then, some applications in the integrity protection of architectural heritage structures are critically discussed. The Digital Twins concept prototype of heritage building structural integrity protection is proposed.

1.3 Digital Twins Technology

Serge P. Kovalyov provides an overview of Digital Twins model–specific technology in Chapter 6. Integrated physical models and simulations, statistical machine learning models, and knowledge-based models play a central role.

Mohammed Hamzaoui and Nathalie Julien aim to introduce the general deployment method of Digital Twins in Chapter 7. Considering the position that Digital Twins may occupy in various fields as key technologies of digital transformation, we emphasized the key requirements of this method.

Istvan David and Eugene Syriani outlined how to use machine learning to automatically build simulators in Digital Twins in Chapter 8. The methods discussed in this chapter are particularly useful in systems that are difficult to model because of their complexity.

Dan Wang et al. introduced how to apply the Digital Twins technology to simulate physical/end side with limited resources and use rich resources on virtual/computing side in Chapter 9. The concept of Digital Twins is applied to the federal distribution analysis problem, and the global data distribution is obtained by aggregating partial observation data of different users.

Esra Kumaş et al. proposed a model that combines blockchain technology with Digital Twins in Chapter 10, because it provides benefits for decentralized data storage, data invariance, and data security. The integration of Digital Twins and blockchain ensures the security and integrity of data accumulation generated by the Internet of Things from relevant stakeholders of the system by verifying transactions in the blockchain ledger.

Emil Kurvinen et al. believe that real-time physics can accurately study the machine operated by humans in Chapter 11, so that human actions can be better integrated into the machine. For high-tech products, the use of physical-based Digital Twins can explore design options and their impact on the overall performance, such as the dynamic behavior of machines.

1.4 Digital Twins Design and Standard

Andreas Pester et al. provide the classification and analysis of Digital Twins types based on recent research in this scientific area in Chapter 12.

Richard Heininger et al. designed and ran the abstraction layer required by Digital Twins as part of the Cyber-Physics System in Chapter 13. A layered Digital Twins modeling method is proposed, which promotes the use of coarse granularity abstraction in the composition of Cyber-Physics System, while retaining the controllability of Cyber-Physics System for operational purposes.

In Chapter 14, Vivek Kant and Jayasurya Salem Sudakaran believe that the human-centered design of Digital Twins and their interfaces is crucial to ensure the effective use of this technology and provide the highest possible benefits for human users. It solves all kinds of problems, from the larger theme designed for human beings to the specific details of Human Machine Interaction design, to achieve interactivity and visualization.

Chiara Cimino et al. propose the specification of a tool that can help ensure consistency among such a heterogeneous set of Digital Twins in Chapter 15, making consistent the set of models and data that are processed during the design phase. The aim is to create a knowledge base of the system, which will serve the design and be useful to analyze the system throughout its life cycle.

Shaun West et al. introduced a human-centered approach to developing Digital Twins in Chapter 16, which can create new value propositions in intelligent service systems. When creating and designing Digital Twins, a people-centered, system-based design lens can support value co-creation and gain multiple perspectives of value within the system.

In Chapter 17, Abdeljalil Beniiche and Martin Maier first introduced the evolution of mobile networks and the Internet, then briefly discussed 6G vision, and elaborated various blockchain technologies. They borrow ideas from the biological superorganism with brain-like cognitive abilities observed in colonies of social insects for realizing internal communications via feedback loops, whose integrity is essential to the welfare of Society 5.0, the next evolutionary step of Industry 4.0.

Timo Ruohomäki and others distinguish urban Digital Twins from industrial Digital Twins in Chapter 18. Urban Digital Twins should be based on complex and scalable information models to maintain the key artifacts of social structure. The urban Digital Twins is about a large organism of a city, a complex urban system.

In Chapter 19, Tobias Osterloh et al. believe that the combination of Digital Twins and modern simulation technology provides significant benefits for the development and operation of robot systems in challenging environments. In the future, integrating big data into concepts will provide new possibilities for predictive maintenance and further match simulation data with available operational data.

1.5 Digital Twins in Management

Vladimir Shvedenko et al. considered managing complex systems through interactive Digital Twins, and described the realization of the principle of process function management of multi-structure system elements in Chapter 20. The main advantage of the proposed method is that the management system is built as open to its improvement, function expansion, and interaction with other systems.

In Chapter 21, Gozde Basak OZTURK and Busra OZEN introduce the integration of artificial intelligence and building information modeling to create a Digital Twins that improves the knowledge management process in the architectural, engineering, operation, and facility management. The progress of information and communication technologies and AI technology has improved the ability of building information modeling to transform static BIM model into dynamic Digital Twins.

Frits van Rooij and Phil Scarf discussed the application of Digital Twins in the context of engineering asset management in Chapter 22. Special attention is paid to the design principles of the maintenance plan Digital Twins. These principles are introduced as a framework, and a real case is used to illustrate how to use this framework to design Digital Twins.

In Chapter 23, Päivi Aaltonen et al. believe that the organizational barriers and facilitation factors to achieve Digital Twins maturity have not been widely discussed, but they are similar to the barriers and facilitation factors to achieve AI maturity. The author discusses the concept of AI and Digital Twins maturity and their relationship.

Petra Müller-Csernetzky and others described the innovation process, prototype stages, and relevant business models of five selected intelligent service projects and tried to apply Digital Twins to these links in Chapter 24. It can be learned from practice that when designing Digital Twins, it is important to be able to scale up and down in the time dimension, because doing so will outline the system dynamics and the main inputs and outputs.

Sofia Agostinelli summarizes the existing definition and specific use, complexity level, and system architecture of Digital Twins in Chapter 25. Lessons can be learned and applied to architecture, engineering, construction, and operation.

1.6 Digital Twins in Industry

Seppo Sierla analyzed recent work on Digital Twins in the process industry in Chapter 26. It shows different types of processes and different use cases of Digital Twins.

In Chapter 27, Dayalan R. Gunasegaram points out that Digital Twins offers an ideal method by which operations can be autonomously controlled and optimized in the highly connected smart factories of the Industry 4.0 era. Digital Twins can also optimize the various operations within factories for improved profitability, sustainability, and safety.

Jože Martin Rožanec et al. shared their experience in the methods we followed when implementing and deploying cognitive Digital Twins in Chapter 28. This concludes by describing how specific components address specific challenges involving three use cases that correspond to crude oil distillation, metal forming processes, and the textile industry.

In Chapter 29, Ning Gou et al. 's innovative concept of ultra-precision machining based on digital twin and its realization and application prospects are proposed. It may provide new insights for the future development of ultraprecision machining tools or processing systems in the Industry 4.0 era.

Giulia Marcon and Giuseppe Aiello research and solve the conceptualization, design and development of the Digital Twins of the logistics system in the shipbuilding industry in Chapter 30, in which the material handling operation is planned and managed in the space of the virtual shipyard, and the autonomous mobile robot and cooperative robot technology are used to improve the safety and efficiency of the operation.

George Falekas et al. introduce the concept of Digital Twins under the scope of electrical machine diagnostics and provide a Digital Twins framework of electrical machine predictive maintenance in Chapter 31.

In Chapter 32, Ahsan Muneer and Jyrki Savolainen discuss the applicability of Digital Twins in the board industry, and identified several practical problems in model building, data availability, and the use of unstructured data. The key issues of building and implementing Digital Twins are related to data availability and how to effectively use data, especially in the case of unstructured datasets that are traditionally utilized only by the human operators for high-level decisions.

Jascha Grübel believes that Digital Twins have a lot of untapped potential in Chapter 33, especially when they are combined with rigorous practices from experiments. The author shows the possibility of the combination of Digital Twins and disease algorithm codes.

Jairo Viola et al. proposed a new development framework in Chapter 34, which uses Digital Twins to make control and predictive maintenance intelligent. The case shows the ability of Digital Twins in the intelligent control of temperature uniformity control system and intelligent predictive maintenance of mechatronics test bench system.

1.7 Digital Twins in Building

In Chapter 35, Gabriele Garnero and Gloria Tarantino give a general overview of the current application fields of 3D urban models, to classify 3D data requirements into specific applications and clarify which types of 3D models with specific characteristics are suitable for this purpose. Then, a practical application example is shown in the Swedish environment, and a 3D building model was developed for Vaxholm City, Stockholm County.

Muhammad Shoaib et al. proposed a green Digital Twins framework based on case studies in Chapter 36. It can be concluded that the process of sustainability assessment through Digital Twins is highly dependent on building information modeling and other input data. The sustainability parameters assessment is quite efficient, fast, and transparent through Digital Twins.

Bianca Weber Lewerenz believes that Digital Twins is the most effective method in Chapter 37, which can start to ride the waves in the wave of digital transformation of the construction industry, take advantage of various opportunities, master unique challenges, and set new standards in the digital era.

Marianna Crognale et al. implemented a general data platform for vibration data visualization in Chapter 38. The work develops an approach that integrates a 3D information model and IoT systems to generate a detailed BIM, which is then used for structural simulation via finite element analysis.

Anca-Simona Horvath and Panagiota Pouliou drew and summarized the current situation of Digital Twins art in architecture in Chapter 39. Digital Twins should take the data they use seriously and consider the need for data storage and processing infrastructure in their entire life cycle, because this ultimately constitutes a sustainability issue.

Chunli Ying et al. proposed a method based on Digital Twins in Chapter 40, which is used to control the processing accuracy and installation quality of structural steel rigid frame (SSRS) bridges. It can provide accurate three-dimensional dimensions, eliminate human interference to the measured data, and use more flexible and systematic data processing algorithms to greatly improve the speed and quality of data.

Marianna Charitonidou introduced how the digital twin of city size can promote the sustainable development goals in Chapter 41. In the context of the current data-driven society, urban digital twins are often used to test scenarios related to sustainable environmental design.

1.8 Digital Twins in Transportation

Mariusz Kostrzewski briefly summarized the application of most Digital Twins in the transportation branch in Chapter 42.

In Chapter 43, Yuan Zhou and Leiting Dong established a Digital Twins drive framework to realize damage diagnosis and predict fatigue crack growth. In the example of a cyclic helicopter component, the uncertainty of the Digital Twins is significantly reduced, and the evolution of structural damage can be well predicted. The proposed method, using the ability of Digital Twins, will help to achieve condition-based maintenance.

Antonio Bono et al. proposed an integrated strategy for managing and checking infrastructure using drones and Digital Twins in Chapter 44. This strategy can provide the real-time status of buildings and perfectly process location information.

1.9 Digital Twins in Energy

Nikolai Fomin and Roman V. Meshcheryakov discuss the Digital Twins security of network physical water supply systems in Chapter 45. By using the safety assessment method based on Digital Twins, the safety system of the water supply company is improved.

In Chapter 46, Dirk Westermann understands Digital Twins as a real-time digital representation of physical components based on measurement data and analysis knowledge. It enables power suppliers to transform their operations through actionable insight to achieve better business decisions. In other words, grid operators can improve operations, reduce unplanned outages, and manage fluctuations in market conditions, fuel costs, and weather conditions.

In Chapter 47, Elena F. Sheka believes that with the increasing amount of modeling data, it is inevitable that the concept of Digital Twins will change from ordinary modeling. This chapter takes the material science of high-tech graphene materials as an example to introduce an example of this concept reflection.

Triboelectric nanogenerator (TENG) is a technology that transforms the changes of the physical world into electrical signals. Jiayue Zhang and Jie Wang introduce the mechanism of common TENG, common self-powered sensors based on TENG, and various scenarios of TENG in Digital Twins applications in Chapter 48. In addition, this section also discusses the future application potential of TENG in Digital Twins.

1.10 Digital Twins in Medicine and Life

In Chapter 49, João A. M. Santos et al. introduced the current paradigm of Digital Twins applied in the pharmaceutical industry, studied the Digital

Twins applied in the pharmaceutical supply chain and pharmaceutical management more deeply, and proposed the future research direction.

Chenyu Tang et al. introduced the development status of human Digital Twins in Chapter 50. The success of Digital Twins technology in industrial application makes people more confident in building human Digital Twins models.

In Chapter 51, Sai Phanindra Venkatapurapu et al. describe the opportunities and challenges of Digital Twins for Proactive and Personalized Healthcare.

Part 2

Thinking about Digital Twins

2

What Is Digital and What Are We Twinning?: A Conceptual Model to Make Sense of Digital Twins

Ashwin Agrawal and Martin Fischer
Stanford University

2.1 Motivation

We often find ourselves having some version of the following conversation when we meet a friend or colleague that we have not seen for a while:

What are you working on these days?
Digital Twins.
Interesting! I have heard a lot about it. How do you define a Digital Twin?

A brief explanation is then followed by:

I fail to make sense of Digital Twins. Some say it is a digital representation and others include Artificial Intelligence and Machine Learning. I honestly think it is a new buzzword people have created.

This confusion among researchers and practitioners regarding the Digital Twin (DT) concept motivates this work. We start by providing more context about the problem, then review the many ways the term "Digital Twin" has been used in the literature before proposing a conceptual model that unifies these existing perspectives and thus alleviates the confusion.

DOI: 10.1201/9781003425724-4

2.2 Introduction

In digital transformation, DTs have emerged as a key concept because they provide a bi-directional coordination between the virtual (digital) and the physical worlds. It provides a multi-dimensional view of asset performance through simulation, prediction, and decision-making based on real-world data (Autodesk 2021). According to a recent Gartner report, over two-thirds of the companies that have implemented sensor technology plan to deploy at least one DT in production by 2022 (Gartner 2019).

DT, however, is still ambiguous and has varied interpretations because of the wide spectrum of technologies and market needs. It is unable to distinguish itself from general computing and simulation models (Khajavi et al. 2019). The boundary between different digitalization approaches such as Building Information Modelling (BIM), Cyber Physical Systems (CPS), Industry 4.0, and DT is extremely blurry in practice (Agrawal et al. 2022). As a result of this confusion, practitioners may hold unrealistic hopes about DT, misalign their strategies, misallocate resources, and ultimately reject DT as hype (Agrawal, Fischer, and Singh 2022).

Researchers and practitioners have attempted to alleviate the confusion by defining DT in different ways. However, it appears that each definition has grown directly from the set of examples and applications that the definer had in mind. There are some people who believe that DT means a complex real-time updated digital model with prescriptive and predictive capabilities (Gabor et al. 2016; Glaessgen and Stargel 2012), while others think it means a simple digital representation (Canedo 2016; Schroeder et al. 2016). Further, Boje et al. (2020) note that there has been little research in this field, resulting in rebranding and reusing emerging technology capabilities like Prediction, Simulation, Artificial Intelligence (AI), and Machine Learning (ML) as necessary constituents of DTs, forcing the practitioners to adapt these technologies without really understanding their purported benefits (Love, Matthews, and Zhou 2020).

Further evidence of the practical importance of the problem is provided by Agrawal, Singh, and Fischer (2022), who analyse Google search trends for the keyword "Digital Twin". They use a web tool to scrape the data for the most common user queries. Based on the analysis, the authors found very strong evidence of the existing confusion regarding DTs. For instance, some of the most searched queries about DTs are: "How is DT different from cyber physical systems?", "How is DT different from the Information model?", and "How to use AI in a DT?", further confirming our observed problem and motivation for this work.

To alleviate the confusion, this work proposes a conceptual framework that can help researchers and practitioners make sense of DTs. To do so, the framework asks two fundamental questions (see Figure 2.1): In a DT, (1) What is digital? and (2) What are we twinning? The triviality of these questions often makes them go unnoticed. However, we argue that for researchers and

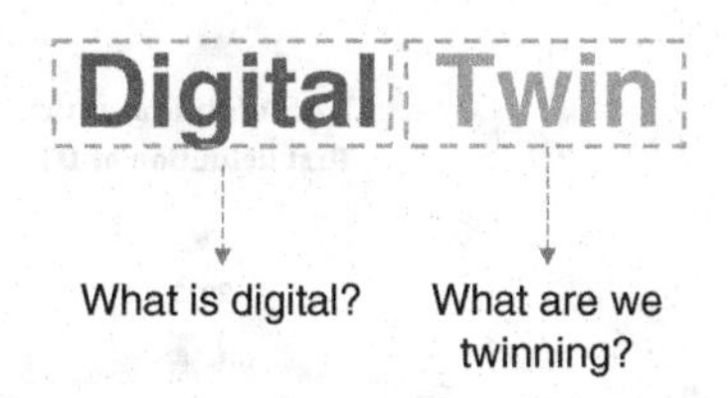

FIGURE 2.1
Motivating questions for our conceptual framework.

practitioners to make sense of DTs, they need to address these two fundamental questions for every DT application because the meaning of 'Digital' and 'Twin' varies depending on the situation, use case, and context. Our proposed framework would aid them in answering these questions.

This work is organized as follows: A brief background about DTs is presented in Section 2.3. Our conceptual framework is described in Section 2.4, before its application is demonstrated in Section 2.5. Finally, the chapter is concluded in Section 2.6.

2.3 Background

This section is divided into three parts. The first section describes a brief history about DTs. In the second section, we use illustrative definitions of DTs from literature and examine them through the two fundamental questions we outlined earlier: What is digital and What are we twinning? This will help readers gain a better understanding of the variety present in DT definitions, and thus appreciate the need to ask our two fundamental questions. Finally, in the third section, we summarize the gap in the literature.

2.3.1 History of DT

Let us have a quick look at the development timeline of DTs as demonstrated in Figure 2.2. As a part of NASA's Apollo programme in the 1960s, at least two identical spacecrafts were constructed, and one remained on earth called the twin (Schleich et al. 2017). The twin was used extensively for training during flight preparation, run simulations, and assist astronauts in critical conditions. In this sense, every prototype that mimics real-world operating conditions can be seen as a twin.

However, constructing a physical twin of every asset/entity, like NASA did of its aircraft, is almost impractical (and costly). Therefore, the idea of "twinning" was extended to create a twin digitally, making a twin practical and cost effective. In this regard, Michael Grieves first introduced the concept of virtually twinning a physical object as a part of a university course

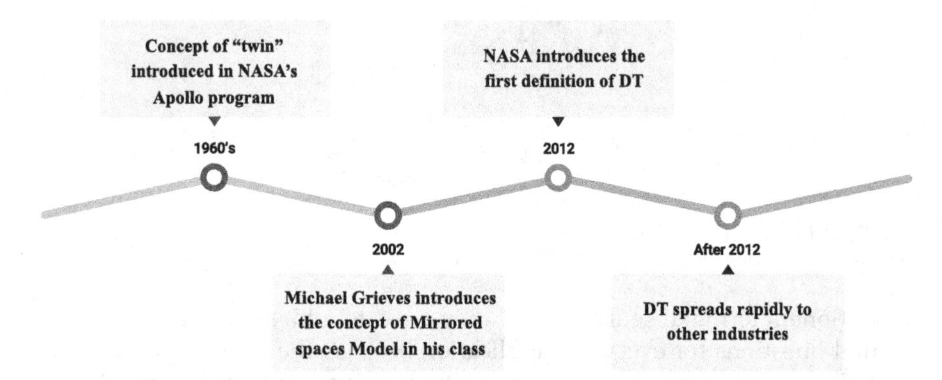

FIGURE 2.2
Timeline for digital twin development.

on Product Lifecycle Management in 2002/03 (Grieves and Vickers 2017). As of yet, the term DT was not coined, and therefore Grieves called it as the Mirrored Spaces Model, a digital information construct of a physical system, linked with the physical system in question. Broadly speaking, Grieves described the concept to be consisting of three parts: (1) the Physical entity, (2) the Digital entity, and (3) the data flow between them. Change in the state of the physical object leads to a change in the state of the digital object and vice versa.

The term "Digital Twin" was attached to the concept defined by Grieves in 2011/12 (Grieves and Vickers 2017). NASA later coined the first definition of DT in 2012 as (Glaessgen and Stargel 2012): "an integrated multi-physics, multi-scale, probabilistic simulation of a vehicle or system that uses the best available physical models, sensor updates, fleet history, etc., to mirror the life of its flying twin".

Thereafter, owing to its usefulness, the concept has not been limited to the aerospace industry and has spread across a spectrum of industries such as manufacturing, and building and construction. It has been used for a variety of applications including optimization of the product lifecycle, production planning and control, and predicting and managing maintenance.

2.3.2 Usage of DT in the Literature

To orient ourselves, let us examine and compare some of the existing definitions of DT in the literature through the lenses: What is digital and What are we twinning?

1. **NASA's DT** (Glaessgen and Stargel 2012): "A Digital Twin is an integrated multi-physics, multi-scale, probabilistic simulation of a vehicle or system that uses the best available physical models, sensor updates, fleet history, etc., to mirror the life of its flying twin".

What does digital mean for NASA's definition?: Having multi-physics, multi-scale, probabilistic simulation capabilities.

What is being twinned in NASA's definition?: Physical product or system (e.g., the aircraft).

2. **Rios's DT** (Rios et al. 2015): "Product digital counterpart of a physical product".

   What does digital mean for Rios's definition?: Digital representation.

   What is being twinned in Rios's definition?: Physical product.

3. **buildingSMART's DT** (buildingSMART International 2021):

   A digital twin (DT) – also referred to as digital shadow, digital replica or digital mirror – is a digital representation of a physical asset. Linked to each other, the physical and digital twin regularly exchange data throughout the PBOD lifecycle and use phase. Technology like AI, ML, sensors, and Internet of Things (IoT) allow for dynamic data gathering and right-time data exchange to take place.

   What does digital mean for buildingSMART's definition?: Digital representation supplemented with technologies such as AI, ML, sensors, and IoT.

   What is being twinned in buildingSMART's definition?: Physical product.

4. **IBM's DT** (IBM 2021): "A digital twin is a virtual representation of an object or system that spans its lifecycle, is updated from real-time data, and uses simulation, machine learning and reasoning to help decision-making".

   What does digital mean for IBM's definition?: Real-time digital representation spanning over life cycle which uses simulations, ML, and reasoning.

   What is being twinned in IBM's definition?: Object or System.

5. **GE's DT** (GE 2021): "Software representation of a physical asset, system or process designed to detect, prevent, predict, and optimize through real time analytics to deliver business value".

   What does digital mean for GE's definition?: Software representation to detect, prevent, predict, and optimize.

   What is being twinned in GE's definition?: Physical asset, system or process.

6. **PwC's DT** (PricewaterhouseCoopers 2021): "Digital Twin captures a virtual model of an organization and helps accelerate strategy. The model can identify elements that are hindering or enabling strategy execution and suggests specific recommendations based on embedded pattern recognition".

   What does digital mean for PwC's definition?: Virtual representation with pattern recognition.

   What is being twinned in PwC's definition?: Organization.

7. **Gartner's DT** (Gartner 2021): "A digital twin is a digital representation of a real-world entity or system. The implementation of a digital twin is an encapsulated software object or model that mirrors a unique physical object, process, organization, person or other abstraction".

   What does digital mean for Gartner's definition?: Digital representation.

   What is being twinned in Gartner's definition?: Physical object, process, organization, person, or other abstraction.

To summarize, by asking the question – What is digital? – we find a wide range of digital capabilities that have been envisioned for DT most notably including monitoring, prediction, and simulation. For example, Rios et al. (2015) suggest that DT is a digital representation of the real-world physical object, Glaessgen and Stargel (2012) emphasize the need to have predictive and simulation capabilities, and buildingSMART International (2021) recognizes the importance of AI and optimization capabilities in a DT. Similarly, Digital Twin Consortium (2021) reasons that a DT should be synchronized at a specified frequency and fidelity. On the other hand, Autodesk (2021) and IBM (2021) state that DT should be updated with real-time data and be always kept up-to-date.

On the twinning front, when we ask the question – What are we twinning? – we find that most of the literature has focused on DT as a digital replica of tangible physical assets. Figure 2.3 shows the top 100 words

FIGURE 2.3
Word cloud for the definitions of digital twin in literature.

of a word cloud constructed by compiling the different definitions for DT available in the literature review performed by Negri, Fumagalli, and Macchi (2017). The top occurring words are 'physical', and 'product', reflecting a bias towards a product-focused DT. However, there are also some other existing definitions that believe that the promise of DT should not be just limited to the physical tangible product, and be extended to processes, systems, and the organization. For example, according to PwC, DT captures the virtual model of an organization (PricewaterhouseCoopers 2021). Gartner proposes one of the broadest definitions of DT, which states that DT can be a digital representation of a physical object, process, organization, person, or other abstractions (Gartner 2021). Parmar, Leiponen, and Thomas (2020) argue that a DT can be a living digital simulation model of the organization that updates and changes as the organization evolves and can allow scenarios to be thoroughly tested to predict the performance of potential tactics and strategies. Papacharalampopoulos and Stavropoulos (2019) showcase the use of process DT for optimization of thermal processes in manufacturing.

2.3.3 Gap in the Literature

The review of the literature reveals that researchers and practitioners have offered a variety of definitions and applications of DT, each hoping to explain their use of the word 'Digital Twin'. Some people think of DTs as a complex real-time updated digital model with predictive and prescriptive capabilities, while others see it as a simple digital representation. This existing confusion regarding the DT concept risks it for being rejected as just hype by the people due to its vagueness (Wright and Davidson 2020). Therefore, to aid practitioners in understanding DTs, a framework that ensures that practitioners are making sense of DTs considering their use case must be developed, which is described in the following section.

2.4 Framework to Make Sense of DTs

The framework is an integration of the two fundamental questions – What is digital and What are we twinning? – that we outlined earlier in the paper (see Figure 2.4). The following sections detail these two perspectives.

Digital Perspective (What Is Digital?): "Digital" is a fuzzy word and can mean anything ranging from as simple as a digital picture to increasingly complex models with predictive and prescriptive capabilities. In the absence of a framework, a default strategy that practitioners use when selecting digital capabilities in DTs is to select the ones that are "best-suited" according to them without exhaustively evaluating other possibilities (Parasuraman, Sheridan, and Wickens 2000).

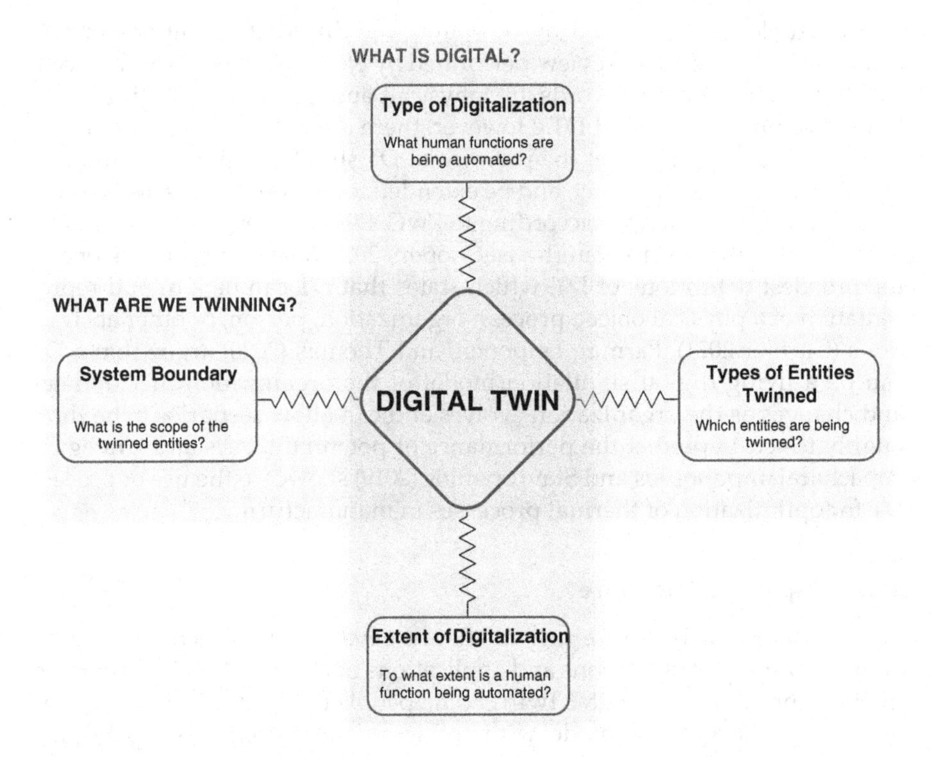

FIGURE 2.4
Spring model: four forces shaping a digital twin.

While going with practitioner's "best-suited option", without evaluating other possibilities, can be a viable option in the cases where the practitioner happens to be creative and visionary, in most cases, it risks: (1) selecting a sub-optimal or an overly optimistic solution that might later have to be altered or disposed-off completely (Fjeld 2020) and (2) misaligning with other project stakeholders who might have a totally different conceptualization about the digital capabilities a DT would deliver (Agrawal, Fischer, and Singh 2022). Therefore, to alleviate the above challenge, our framework forces the practitioners to ask themselves before deploying DTs: What does digital mean for their use case?

Twinning Perspective (What Are We twinning?): Even if the ambition in many cases is to create a so-called "most comprehensive DT", where all entities across the lifecycle are fully digital, the reality is that most work systems today and in the foreseeable future will only be partially digital. Feng et al. (2020) note that the idea to have a DT that is a comprehensive virtual replica of all physical and functional activities can result in huge costs (West and Blackburn 2017) and make it difficult for manufacturers to adopt, invest, and implement DT. Therefore, to alleviate this challenge, our framework asks the practitioners a second fundamental question: What are they planning to twin for their use case?

The digital perspective aids practitioners in selecting the technological capabilities in a DT, while the twinning perspective forces them to identify the right scope of digitally twinned entities. The digital perspective anchors on the elements: (1) type of digitalization and (2) extent of digitalization. Together, these two elements can help practitioners understand what digital means to them. Similarly, the twinning perspective anchors on the elements: (1) system boundary and (2) types of entities twinned. The following paragraphs describe these elements in detail:

Type of Digitalization: All digital capabilities are not created equal. While some provide more benefits, they are difficult for the practitioners to adapt, invest, and deploy (Agrawal et al. 2022). On the other hand, relatively basic capabilities may provide fewer benefits, but reduce the necessary efforts for adoption (Lu et al. 2020). Therefore, it becomes important for practitioners to select the right type of digitalization for their use case.

To choose the right type of digitalization, it is important to first understand the types of digitalization available, and then select the one that meets the needs of practitioners. But what are the types of digitalization available? Types of digitalization, a four-step hierarchy by Gartner (2013), provide this categorization for digitalization. The hierarchy is as follows: Description ("What is happening") → Diagnostic ("Why is it happening?") → Prediction ("What will happen?") → Prescription ("What should be done?"). To complete any task, humans start by gathering the required information and understanding what is happening (i.e., description), explain why it might be happening (i.e., diagnostic), predict what may happen next based on this understanding (i.e., prediction), and finally decide what to do next (i.e., prescription).

The IoT, wireless monitoring, augmented reality/virtual reality, drones, visualizations, and 3D scans are a few examples of current technologies that are essentially trying to improve the description level. The diagnostic level includes technologies and techniques such as confirmatory data analysis, statistical tests, and causal inference, as they help to diagnose why certain things are happening. Design simulations, AI methods, and predictive analytics fall in the prediction category.

Extent of Digitalization: Following the type of digitalization, practitioners must decide on how much to digitalize. Any digitalization type is not all or none, but varies across a continuum of levels, from no automation to full automation. At a higher level, DT would have more autonomy and would be able to handle more difficult tasks. While this would provide more benefits, it would also be challenging to achieve. On the other hand, at a lower level, most of the work would still be completed by the human operator with some assistance from DTs. Therefore, the extent of digitalization forces practitioners to consider which scenarios should be handled by humans and which situations by DT.

System Boundary: Many DT deployment errors emanate from people failing to define the exact boundaries for DTs and expecting anything and everything from it. Some people argue that DT should be made for the whole lifecycle of an asset to produce maximum value. Even if this is true, practitioners might not have direct control over many parts of the asset lifecycle due to contractual or legal structures. As a result, practitioners should consider the following two factors when defining the appropriate system boundaries for DT deployment:

1. *Local versus global optimization:* Although a global optimization (end-to-end optimization) of an asset or an entity would be ideal, in most real-world scenarios, this is not possible due to practical, legal, and financial considerations. Due to this, practitioners do not have control over the entire lifecycle, thus emphasizing the need to define the appropriate system boundaries for twinning before creating a DT.

   For example, in a building construction project, an ideal situation would be to use DTs to produce a building that is optimized together for design, construction, operations, maintenance, and sustainability. As these are interdependent functions, like design affects construction and maintenance, an ideal or globally optimized DT would require all disciplines to work together from the start of the project. But this rarely occurs in real life, and usually different disciplines are handled by independent contractors. Therefore, a global DT for the whole lifecycle of the asset is out of question. In the best-case scenario, each contractor would locally use their own DTs to optimize the function that they are responsible for without thinking of the other disciplines (or global optimization).
2. *Level of detail of insights:* DTs can provide insights at different scales, from fine details to large systems. Therefore, practitioners should determine the scale of insights they want and define the system boundaries accordingly because this will impact the data required for DTs and algorithms they use. Defining the scale too broadly obscures any relevant operational insights and defining the scale too narrowly makes the problem less worthwhile for solving. For example, let us consider a DT that analyzes productivity on a project. If DT provides the insights just on an aggregate level, like productivity was down 10% last month, then practitioners cannot understand where to improve. More detailed insights reflecting the change on a component level would have been better.

Types of Entities Twinned: Typically, all production activities can be viewed as delivering a 'Product' (or service) by an 'Organization' following a certain 'Process' to make the delivery possible. Therefore, the aspects of a project that the project delivery team or a company can, at least to some extent, control, fall into three categories: Product, Organization, and Process (POP) (Fischer et al. 2017). Hence, each element of POP can be improved digitally or, in a way, digitally twinned. As discussed in Section 2.3, limiting the twinning to only one element (most people restrict it to product) can limit the benefits of DT.

For example, on a manufacturing project, a practitioner makes a DT that predicts whether an equipment needs to be shut down for maintenance. While DT could detect the need for repair if the equipment (or product) is only twinned, it would not be intelligent enough to know that shutting down for repairs will cause production delays and bottlenecks because it does not understand the process or the organization in which the equipment functions. Therefore, the decision to shut down would still have to be made by the engineering team.

A practitioner must therefore decide which elements they would like to twin. Of course, the natural instinct is to start with the product, but we recommend thinking further ahead like twinning the process in the manner of (Papacharalampopoulos and Stavropoulos 2019) who optimizes processes using DT, as well as (Parmar, Leiponen, and Thomas 2020) who optimizes organization with DTs.

Why a Spring Representation: In Figure 2.4, we use a spring representation to connect the four elements to show that 'Digital' and 'Twin' components of a DT are greatly intertwined with one another, and a change in one affects the other. Further, the model emphasizes that finding the appropriate DT is an iterative process, where the practitioner answers one question, observes the effects on other parts, and repeats the process.

For example, if a practitioner decides to implement "prediction" type of digitalization on a construction project, they will have to reduce the system boundaries if they did not have appropriate data for the whole project. Similarly, it would be difficult to make accurate predictions if the practitioner only twins the asset (or the building) without considering the process or the organization behind it as discussed above. Agrawal, Fischer, and Singh (2022) also observe the impact of the digital element on the twinning element in a highway maintenance case study.

2.5 Case Study

We demonstrate the framework's application in a real-life case study to illustrate how practitioners find it useful in practice. Below we first describe the situation, and then detail how the framework was used in this case study.

2.5.1 Situation Description

Firm X maintains and manages the pavements and the highways. Recently, it got a contract to manage a part of one of the fastest highways in the country, stretching just over 60 miles. Although the project is very charismatic, it has very strict and hefty fines for non-compliance with the following rules:

1. Cracks and potholes on the road
2. Traffic signs and road markings in good condition
3. Maintenance of roadside trees and spaces

Bob, a manager on this project, has been quite nervous lately, and is looking to create a DT to manage and maintain the health condition of the road surface and the surrounding areas.

He consults his various colleagues for ideas. Mike, the operator of the roadway, suggests that DT should be able to acquire the data for the current road conditions, and thus act as a real-time virtual replica of the roadway. On the other hand, Jay, a colleague of Bob, suggests that DT should be also able to comprehend the acquired data, and act as a decision support system by making predictions about the future road conditions. Tom, a friend of Bob, jumps in by suggesting that DT in addition to making the predictions should be able to adapt itself in uncertain situations like the COVID-19 pandemic.

The issue that Bob faces is this: Given the different perspectives that people have about DTs, how does he conduct a structured brainstorming to make sense of what DTs mean for their use case? Because if different stakeholders have widely varying interpretations of DT, it may result in false hopes and misalignments between stakeholders, with little consensus about the deliverables to expect and resources to allocate.

2.5.2 Use of the Framework

2.5.2.1 Deciding on What Does Digital Mean

Practitioners used a two-step process: (1) brainstorming in a structured way using the framework to develop many ideas for different types and extents of digitalization, and (2) selecting the appropriate one based on the value to be gained and the technology available.

1. Description – Helps to better understand what is happening on the road/pavement.
    a. *No digitalization:* Operators go for patrolling and try to manually detect cracks.
    b. *Low extent of digitalization:* One operator drives a car with a camera mounted on it, and then another operator (sitting in office) looks through the video to detect cracks/potholes.

c. *High extent of digitalization:* Drones automatically take pictures of the road, and computer vision algorithm detects the cracks (no operator required).

No digitalization method is very inefficient because cracks are hard to detect at 70 mph. Introducing some digitalization may solve this problem, but most of the work still must be done by humans. In the highest level of digitalization, operators are less needed, and they just need to supervise the system.

2. Diagnostic – Helps to understand why something might be happening.
   a. *No digitalization:* Practitioners take an experienced guess.
   b. *Low/High extent of digitalization:* DT can help to understand the reasons behind the cracks by analysing weather patterns and daily traffic conditions.
3. Prediction – Helps to understand what might happen in the future.
   c. *No digitalization:* Practitioners use an empirical approach to come up with a prediction for crack propagation.
   d. *Low extent of digitalization:* The DT can predict the crack propagation in normal pre-determined circumstances like with usual traffic and weather conditions.
   e. *High extent of digitalization:* The DT's prediction can adapt itself based on the changing environment, e.g., traffic reduced due to COVID-19 pandemic or extreme weather circumstances.

   A lack of digitalization forces practitioners to use existing empirical approaches to predict the crack propagation. The introduction of some digitalization helps to solve this problem by making DTs take care of normal circumstances and humans taking care of anomalies and unusual situations. In the highest extent of digitalization, the DT can learn from itself based on the data gathered without the practitioner having to adjust the algorithm and therefore take care of unusual situations as well.
4. Prescription – Helps to decide what should be done in the future.
   a. *No digitalization:* Operators decide based on experience whether to repair the crack.
   b. *Low extent of digitalization:* DT can handle easy or trivial cases, like a dangerous pothole or a big crack but not the cases which require an intelligent evaluation.
   c. *High extent of digitalization:* To decide whether to perform preventive maintenance or not, the algorithm can intelligently evaluate what the additional cost would be if the cracks were not repaired now.

The above brainstorming sheet illustrates how practitioners have used the digital perspective of the framework to conduct a structured brainstorming session. Each of the above ideas (or even all of them) could have been implemented in this DT. In the absence of this structured and exhaustive brainstorming, many different definitions of what digital means may have arisen, leading to the confusion. Based on their technological resources and careful evaluation of alternatives/value, the team decided that they would like to have a high extent of digitalization in description, i.e., use drones to automatically take pictures of the road, and computer vision algorithm to detect the road defects. So, this is what digital would mean in the context of this case study.

2.5.2.2 Determining What to Twin

To answer the question what we are twinning, practitioners needed to decide on two things: System boundaries and the entities to be twinned.

The project team started by selecting the scope of the project (system boundaries). It was decided to focus on defects caused by damaged road markings despite there being many defects to choose from, such as defects in traffic markings, vegetation situation around the road, potholes, and guard rails. This was selected based on two criteria: (1) importance to the life safety of users and (2) frequency of occurrence as determined by analysing the defects log from 21 quarters, starting from 2013. It was thus concluded that damaged road markings are one of the most frequently reported defects in roads and pose the maximum risk to life safety of users. Clearly defining the scope of the project at the very onset gave clear and actionable directions to the project team, thus alleviating any existing confusion and aligning the expectations of all the stakeholders.

Next, it was decided that for this project, twinning only the road/pavement (or the product) would be necessary and sufficient. It would, however, be necessary to twin the process of defect repair and the organization performing it in the future when the team is predicting the cost and time to repair the defects and prescribing a solution using DT (since different processes and organizations would have impact on time and cost).

2.6 Conclusion

In our motivation, we gave an example of a conversation with a friend where the main question that arose was: "What is an ideal DT? or How do we define a DT that industry needs?" We do not believe that there is one universal answer for this question. Each organization is unique in its market position and technological capabilities. Hence, DT to be deployed in practice must be evaluated individually for each organization and project – a "one-size-fits-all" approach would not work.

The second author of the chapter, who has guided over 2000 DT implementations since 2008 quotes: "In my experience, I rarely see people fully evaluating a DT to decide if it is suitable for their use case. It is always the first thought option of DT that practitioners want to implement without really thinking of other better possibilities". Fjeld (2020) interviews 15 practitioners and reports a similar finding. Practitioners getting lured by technology and not exhaustively evaluating other viable options is a well-documented problem in broader technology evaluation and selection literature as well (Love, Matthews, and Zhou 2020). A lack of this understanding can result in confusion about DTs among practitioners, unrealistic hopes from the technology, and ultimately a rejection of DT as hype.

The proposed framework helps alleviate the above problem by proposing two fundamental questions that every practitioner should ask before implementing DTs: (1) What is digital? and (2) What are we twinning? By answering these questions for each DT application, practitioners can alleviate confusion and ensure a clear understanding of the specific use case and context. Awareness of these factors in practice can increase the likelihood of success and accelerate the adoption of DT in the industry.

References

Agrawal, Ashwin, Martin Fischer, and Vishal Singh. 2022. "Digital Twin: From Concept to Practice." *Journal of Management in Engineering* 38 (3): 06022001. https://doi.org/10.1061/(ASCE)ME.1943-5479.0001034.

Agrawal, Ashwin, Vishal Singh, and Martin Fischer. 2022. "A New Perspective on Digital Twins: Imparting Intelligence and Agency to Entities." arXiv. https://doi.org/10.48550/arXiv.2210.05350.

Agrawal, Ashwin, Vishal Singh, Robert Thiel, Michael Pillsbury, Harrison Knoll, Jay Puckett, and Martin Fischer. 2022. "Digital Twin in Practice: Emergent Insights from an Ethnographic-Action Research Study." In *Construction Research Congress 2022*, 1253–60. Arlington, VA: ASCE. https://doi.org/10.1061/9780784483961.131.

Autodesk. 2021. "Digital Twins in Construction, Engineering, & Architecture." Autodesk. 2021. https://www.autodesk.com/solutions/digital-twin/architecture-engineering-construction.

Boje, Calin, Annie Guerriero, Sylvain Kubicki, and Yacine Rezgui. 2020. "Towards a Semantic Construction Digital Twin: Directions for Future Research." *Automation in Construction* 114 (June): 103179. https://doi.org/10.1016/j.autcon.2020.103179.

buildingSMART International. 2021. "Digital Twins." *BuildingSMART International* (blog). 2021. https://www.buildingsmart.org/digital-twins/.

Canedo, Arquimedes. 2016. "Industrial IoT Lifecycle via Digital Twins." In *2016 International Conference on Hardware/Software Codesign and System Synthesis (CODES+ISSS)*, 1. Pittsburg, PA: IEEE. https://doi.org/10.1145/2968456.2974007.

Digital Twin Consortium. 2021. "Digital Twin." 2021. https://www.digitaltwinconsortium.org/initiatives/the-definition-of-a-digital-twin.htm.

Feng, B., S. Kim, S. Lazarova-Molnar, Z. Zheng, T. Roeder, and R. Thiesing. 2020. "A Case Study of Digital Twin for Manufacturing Process Invoving Human Interaction." 2020. https://www.semanticscholar.org/paper/A-CASE-STUDY-OF-DIGITAL-TWIN-FOR-A-MANUFACTURING-Feng-Kim/50f3cd21e4860470e8d762fb591193868d9fe878.

Fischer, Martin, Howard W. Ashcraft, Dean Reed, and Atul Khanzode. 2017. *Integrating Project Delivery*. John Wiley & Sons.

Fjeld, Tord Martin Bere. 2020. "Digital Twin - Towards a Joint Understanding within the AEC/FM Sector." https://ntnuopen.ntnu.no/ntnu-xmlui/handle/11250/2779306.

Gabor, Thomas, Lenz Belzner, Marie Kiermeier, Michael Till Beck, and Alexander Neitz. 2016. "A Simulation-Based Architecture for Smart Cyber-Physical Systems." In *2016 IEEE International Conference on Autonomic Computing (ICAC)*, 374–79. Wuerzburg: IEEE. https://doi.org/10.1109/ICAC.2016.29.

Gartner. 2013. "Extend Your Portfolio of Analytics Capabilities." *Gartner*. 2013. https://www.gartner.com/en/documents/2594822/extend-your-portfolio-of-analytics-capabilities.

Gartner. 2019. "Gartner Survey Reveals Digital Twins Are Entering Mainstream Use." *Gartner*. 2019. https://www.gartner.com/en/newsroom/press-releases/2019-02-20-gartner-survey-reveals-digital-twins-are-entering-mai.

Gartner. 2021. "Digital Twin." Gartner. 2021. https://www.gartner.com/en/information-technology/glossary/digital-twin.

GE. 2021. "Digital Twin." 2021. https://www.ge.com/digital/applications/digital-twin.

Glaessgen, Edward, and David Stargel. 2012. "The Digital Twin Paradigm for Future NASA and U.S. Air Force Vehicles." In *53rd AIAA/ASME/ASCE/AHS/ASC Structures, Structural Dynamics and Materials Conference 20th AIAA/ASME/AHS Adaptive Structures Conference 14th AIAA, 1818*. Honolulu, Hawaii: American Institute of Aeronautics and Astronautics. https://doi.org/10.2514/6.2012-1818.

Grieves, Michael, and John Vickers. 2017. "Digital Twin: Mitigating Unpredictable, Undesirable Emergent Behavior in Complex Systems." In *Transdisciplinary Perspectives on Complex Systems: New Findings and Approaches*, 85–113. Cham: Springer International Publishing. https://doi.org/10.1007/978-3-319-38756-7_4.

IBM. 2021. "What Is a Digital Twin?" 2021. https://www.ibm.com/topics/what-is-a-digital-twin.

Khajavi, Siavash H., Naser Hossein Motlagh, Alireza Jaribion, Liss C. Werner, and Jan Holmström. 2019. "Digital Twin: Vision, Benefits, Boundaries, and Creation for Buildings." *IEEE Access* 7: 147406–19. https://doi.org/10.1109/ACCESS.2019.2946515.

Love, Peter E. D., Jane Matthews, and Jingyang Zhou. 2020. "Is It Just Too Good to Be True? Unearthing the Benefits of Disruptive Technology." *International Journal of Information Management* 52 (June): 102096. https://doi.org/10.1016/j.ijinfomgt.2020.102096.

Lu, Yuqian, Chao Liu, Kevin I-Kai Wang, Huiyue Huang, and Xun Xu. 2020. "Digital Twin-Driven Smart Manufacturing: Connotation, Reference Model, Applications and Research Issues." *Robotics and Computer-Integrated Manufacturing* 61 (February): 101837. https://doi.org/10.1016/j.rcim.2019.101837.

Negri, Elisa, Luca Fumagalli, and Marco Macchi. 2017. "A Review of the Roles of Digital Twin in CPS-Based Production Systems." In *Procedia Manufacturing, 27th International Conference on Flexible Automation and Intelligent Manufacturing,* FAIM2017, 27–30 June 2017, Modena, Italy, 11 (January): 939–48. https://doi.org/10.1016/j.promfg.2017.07.198.

Papacharalampopoulos, Alexios, and Panagiotis Stavropoulos. 2019. "Towards a Digital Twin for Thermal Processes: Control-Centric Approach." *Procedia CIRP, 7th CIRP Global Web Conference - Towards Shifted Production Value Stream Patterns through Inference of Data, Models, and Technology (CIRPe* 2019*)*, 86 (January): 110–15. https://doi.org/10.1016/j.procir.2020.01.015.

Parasuraman, R., T. B. Sheridan, and C. D. Wickens. 2000. "A Model for Types and Levels of Human Interaction with Automation." *IEEE Transactions on Systems, Man, and Cybernetics. Part A, Systems and Humans* 30 (3): 286–97. https://doi.org/10.1109/3468.844354.

Parmar, Rashik, Aija Leiponen, and Llewellyn D. W. Thomas. 2020. "Building an Organizational Digital Twin." *Business Horizons* 63 (6): 725–36. https://doi.org/10.1016/j.bushor.2020.08.001.

PricewaterhouseCoopers. 2021. "Digital Twin." *PwC*. 2021. https://www.pwc.com/gx/en/issues/transformation/digital-twin.html.

Rios, José, Juan Carlos Hernandez, Manuel Oliva, and Fernando Mas. 2015. *Product Avatar as Digital Counterpart of a Physical Individual Product: Literature Review and Implications in an Aircraft. Transdisciplinary Lifecycle Analysis of Systems*, 657–66. Philadelphia, PA: IOP Press. https://doi.org/10.3233/978-1-61499-544-9-657.

Schleich, Benjamin, Nabil Anwer, Luc Mathieu, and Sandro Wartzack. 2017. "Shaping the Digital Twin for Design and Production Engineering." *CIRP Annals* 66 (1): 141–44. https://doi.org/10.1016/j.cirp.2017.04.040.

Schroeder, Greyce N., Charles Steinmetz, Carlos E. Pereira, and Danubia B. Espindola. 2016. "Digital Twin Data Modeling with AutomationML and a Communication Methodology for Data Exchange." *IFAC-PapersOnLine, 4th IFAC Symposium on Telematics Applications TA 2016* 49 (30): 12–17. https://doi.org/10.1016/j.ifacol.2016.11.115.

West, Timothy D., and Mark Blackburn. 2017. "Is Digital Thread/Digital Twin Affordable? A Systemic Assessment of the Cost of DoD's Latest Manhattan Project." *Procedia Computer Science, Complex Adaptive Systems Conference with Theme: Engineering Cyber Physical Systems, CAS*, October 30–November 1, 2017, Chicago, IL, 114 (January): 47–56. https://doi.org/10.1016/j.procs.2017.09.003.

Wright, Louise, and Stuart Davidson. 2020. "How to Tell the Difference between a Model and a Digital Twin." *Advanced Modeling and Simulation in Engineering Sciences* 7 (1): 13. https://doi.org/10.1186/s40323-020-00147-4.

3

When Digital Twin Meets Network Engineering and Operations

Pedro Henrique Diniz
Federal University do Rio de Janeiro (UFRJ)

Christian Esteve Rothenberg
University of Campinas

José Ferreira de Rezende
Federal University do Rio de Janeiro (UFRJ)

3.1 Introduction

Over the past few years, the emergence of several new network applications, such as Industry 4.0, IoT, augmented reality, and 5G mobile networks, has brought a series of complex requirements that are hardly met by traditional human-centric network management approaches. The operation and management of networks has become more difficult due to the complexity of managed networks and the services delivered by them. Developing, validating, and deploying new network technologies, both at the hardware and software levels, will become increasingly complex due to the risk of interfering with the services in operation.

The concept of Network Digital Twin (NDT) has recently gained attention from Internet standards bodies, both with the aim of defining what a NDT is (Zhou et al. 2022) and identifying use cases in which the adoption of NDT can bring great benefits (ITU Telecommunication Standardization Sector 2020). NDT can be defined as a virtual representation of the physical network that can be used to emulate, analyze, diagnose, and control the physical network (Zhou et al. 2022). To this end, there must be a communication relationship between the physical network and the virtual network, represented by the logical objects associated with each of the physical objects of the network, which can be switches, routers, firewalls, hosts, links, and so on.

By exchanging data between physical and logical devices, NDT helps achieve full network management and automation, simplifying daily tasks.

 DOI: 10.1201/9781003425724-5

NDT adds value to network management by optimizing resilience and verifying network changes before deployment, using "what-if" scenarios in a virtual environment, for example.

This chapter presents a characterization of the NDT based on its main properties and a systemic analysis of NDT through different software technologies and a review of publications related to the NDT are also carried out, producing a categorized review of recent works based on their main properties and application scenarios. This work provides an overview of how different technological strategies for the realization of NDT can be categorized and what are some of the main challenges to adopt in the context of computer network operation.

3.2 Network Digital Twins: Background, Definitions, and Systemic View

Several recent works present definitions of the concept of the Digital Twin (DT) (Tao et al. 2019, Minerva, Lee and Crespi 2020, Fuller et al. 2020, Lim, Zheng and Chen 2019, Rasheed, San and Kvamsdal 2020, Barricelli, Casiraghi and Fogli 2019, Jones et al. 2020). However, as pointed out by Fuller et al., both academia and industry have not sufficiently distinguished the difference between DT and generic computer models and simulations. Thus, there is still no definitive definition of DT in the literature. In this way, we identified that the considerations made by Fuller et al., when applied to the concept of NDT, support the identification of the existing differences between the various software technologies that introduced the application of the DT concept over time.

3.2.1 Foundational Properties of Digital Twins

DT is composed of two main entities, physical and logical, as converged in the main literature. The term refers to both the physical object and its logical counterpart, as well as their coupling, rather than exclusively to the logical part. Physical objects can be devices, products, or software/hardware resources. So, DT encompasses both physical and logical components and their relationship.

The physical entity, called Physical Object (PO), refers exclusively to the physical aspects of the DT. In contrast, the logical entity refers to virtualizing the physical object's characteristics generally implemented via software. The physical entity can also be referred to as an object, physical object, artifact, or product, while the logical entity, called Logical Object (LO), has as synonyms digital object, clone, counterpart, duplicate, among others. The DT couples the two different entities, a real (physical) that is relevant to the physical world, like a router or a network switch in the networking domain, and a virtual (softwarized) embodiment such as an emulated network equipment.

In order to describe the concepts of DT, Minerva et al. (2020) present a set of fundamental properties of DTs. This set of properties characterizes a DT and identifies a set of functionalities associated with it, thus being able to apply them to different contexts, situations, and domains. In this chapter, we seek to elucidate how these properties can be restricted and applied in the case of an NDT toward a valuable definition, characterization, and categorization of related work. Next, key properties of an NDT are introduced.

a. *Representativeness and contextualization:* generally speaking, the LO has to be as credible as the original PO. However, representing the PO in all its facets and implications can be extremely difficult and costly. Thus, DT must be supported by a model designed and implemented with a set of objectives and purposes specific to its application domain and refer to the context in which it operates. The LO must represent at least the properties, characteristics, and behaviors that are necessary and sufficient to qualify an LO.
b. *Reflection:* refers to the ability of a PO to be accurately measured and represented in relation to the application's objectives. A PO is described by an LO as a set of values related to status, attributes, and behaviors that can change over time. The ability of reflection suggests that each relevant PO value is uniquely represented in the LO.
c. *Entanglement:* means the communication relationship between the PO and the LO, a key concept of a DT to represent the link between a PO and its LO, where all information necessary to describe the PO must be passed to the LO.
d. *Replication:* refers to the ability of replicating a PO in different environments. A PO can be virtualized and replicated multiple times in a virtualized space.
e. *Persistence:* alludes to the DT's ability to be persistent over time. The PO may have real-world limitations that restrict its functioning over time. So, the LO is the main factor that enables the persistence of the DT since it must be able to mitigate these limitations to support the constant availability of the DT.
f. *Memorization:* is the ability to store and represent all present and past data relevant to the DT.
g. *Composability:* focuses on the ability to group multiple objects into a single composite object and then observe and control the behavior of the composite object as well as individual components.
h. *Accountability/Management:* refers to the ability to completely and accurately manage the DT, interacting between different LOs, in

order to compose them into larger aggregates to enable the construction of the DT.

i. *Servitization:* means the ability to offer the DT consumer the association of a product/device with services, functionalities, processes, and data access of a PO through capabilities, tools, and software interfaces guaranteed by one or more LOs of a DT.
j. *Predictability:* revolves around the ability to incorporate an LO in a specific environment and simulate its behavior and interactions with other objects in the future or during a specific period to support the determination of the behavior of the PO in the context of its operation.
k. *Programmability:* is the ability to provide Application Programming Interfaces (API) to program DT functions.

In short, the presented properties support the definition of what a DT is and which system characteristics support its operation. Some of the properties presented here are fundamental for a system to be characterized as a DT and can be considered as a basic/minimal implementation of the DT where the LO fully represents and acts like the PO within an operational context exhibiting the properties of: (a) *Representativeness and Contextualization,* (b) *Reflection,* and (c) *Entanglement.* Meanwhile, other properties, here called Additional Properties, extend and increase the intrinsic value of the DT relationship by offering additional features that may or may not be important to the system depending on its application context, such as Replication, Persistence, Memorization, Composability, Management, Servitization, Predictability, and Programmability. The definition of these properties is important to characterize what can be defined as an NDT and to understand the functionalities and characteristics covered by implementations of this concept, thus allowing its categorization.

3.2.2 Network Digital Twin Definitions and Misconceptions

Based on works that seek to standardize the NDT concept (Sun et al. 2021, Zhou et al. 2022), we now revisit the generic properties of DT presented by Minerva et al. (2020) in the context of computer networks. Next, we present how different works (Fuller et al. 2020, Wu, Zhang and Zhang 2021) categorize DT implementations, based on the execution of DT properties. At the end of this chapter, we perform the categorization of NDT works, based mainly on these definitions.

As stated by Zhou et al. (2022), NDT can be defined as a virtual representation of the physical network that can be used to analyze, diagnose and control the physical network. For this to occur, there must be an entanglement relationship between the physical network and the virtual network, represented by the logical objects associated with each of the physical objects

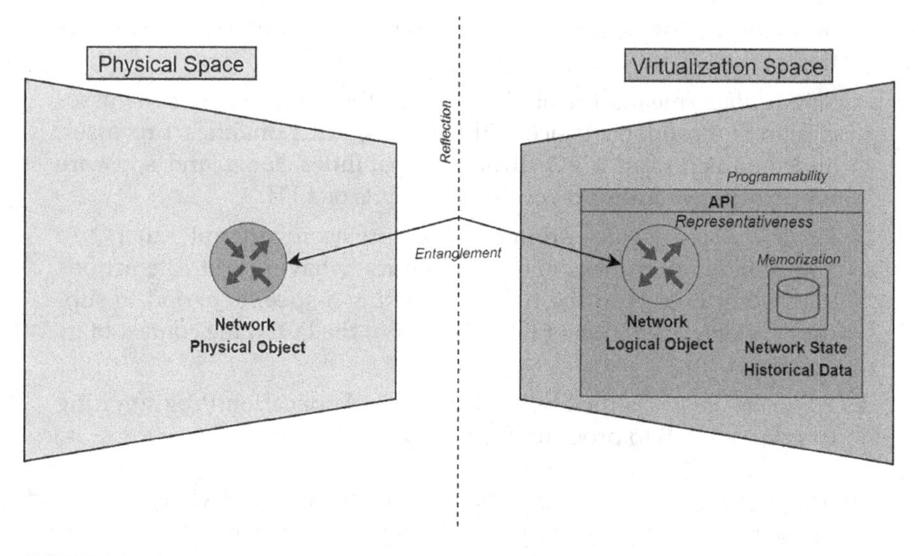

FIGURE 3.1
Representation of network digital twin main properties. Adapted from Minerva, Lee and Crespi 2020.

of the network, which can be switches, routers, firewalls, hosts, links, or even a set of all these entities forming a network. These properties, shown in Figure 3.1, represent the NDT concept and the entire system that supports it. The NDT system must be able to represent the physical objects through logical objects together with their Entanglement and Reflection properties to control and manage the physical objects. On the other hand, the properties of Memorization and Programmability are extremely relevant in the current context of computer networks, as they allow us to answer questions of great importance in the scenario of network operations, for example, to identify "who-what-when" caused an error and enable network automation, respectively.

As presented by Fuller et al. (2020) and Wu et al. (2021), several concepts and applications similar to the DT emerged over time, and several misconceptions regarding its definition could be seen. In this chapter, we present a brief description of these concepts, classified into three different types, as presented by Fuller et al. (2020): (1) Digital Model; (2) Digital Shadow; and (3) DT. All these definitions are mainly associated with the Entanglement property and its respective association characteristic, which represents the communication direction between PO and LO.

i. *Digital model:* is a digital version of a physical object, where there is no automatic data exchange between the physical object and the digital model. In this case, there is no way of automatically exchanging data between the physical object and the logical object. Once the logical object is created, a change made to the physical object has no impact on the logical object and vice versa.

ii. *Digital shadow:* is a digital representation of a physical object that has a single-direction information exchange flow from the physical object to the logical object. A change in the state of the physical object leads to a change in the state of the logical object, but the opposite is not true.

iii. *Digital twin:* represents when the flow of data between a physical object and a logical object occurs in both directions and is fully integrated. This means that a change in the physical object automatically leads to a change in the logical object and vice versa.

The presentation of these different definitions, in addition to helping us to identify the misconceptions found in the literature, helps us to identify how different approaches to network twinning emerged over time, culminating in several recent works, described in Section 3.3, which represent the definition of DT in its entirety.

3.2.3 Network Digital Twin – A Systemic View

It is challenging to have a meaningful discussion on how to build a DT even further when considering its applicability to computer networks. Its specificities depend not only on the application area where it will be used but also on the specific use case within that particular area. So, modeling the DT through a black box approach is very useful, as it can help to define the best way to implement the DT according to the intended results.

In computing and engineering, a black box is a system that can be viewed solely in terms of its inputs and outputs. An NDT can be thought of as a black box whose inputs include the network configuration and other factors such as queuing techniques and routing protocols, while the outputs represent the resulting network performance. This performance is determined by the behavior of the network's components, such as routers, switches, hosts, and links, under a given workload. Measuring network performance requires different metrics, including network paths, packet forwarding behaviors, delay, packet loss, and link capacity utilization. By analyzing these metrics, NDTs can be used to model and optimize network behavior, making them an important tool for network engineers and operators.

Figure 3.2 presents a systemic view for characterizing NDT based on a black box modeling approach. Its constitutive properties can be classified according to its application context, its inputs and outputs, and the intended result of its application.

Next, we use qualitative attributes for the key properties of an NDT for classifying an NDT implementation. The proposed approach aims to support the categorization of related work discussed later in this chapter as well

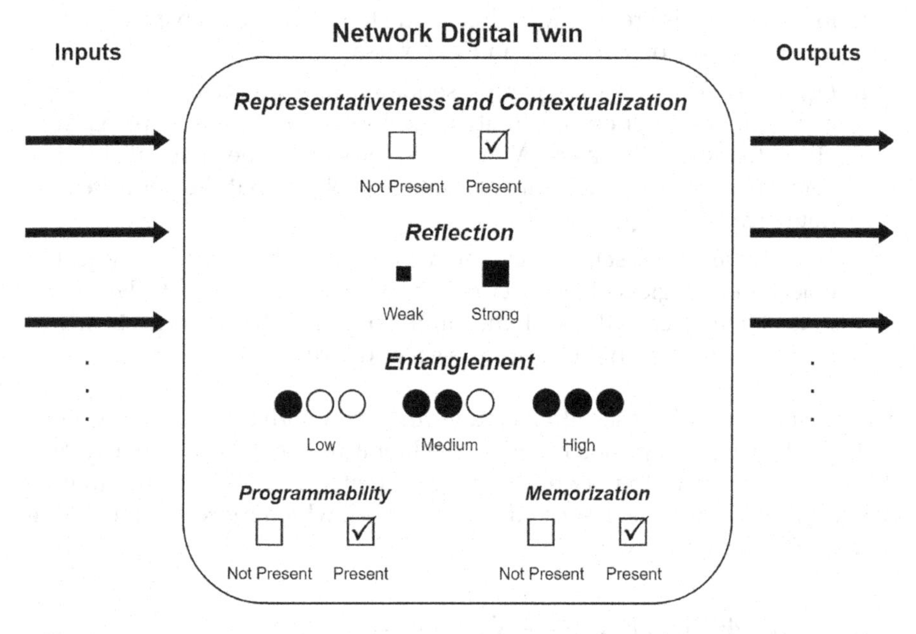

FIGURE 3.2
Network digital twin systemic view based on its properties.

as future developments. The basic classification mechanism for each of the properties is briefly described as follows:

- *Representativeness and contextualization:* classified into two levels, being Present or Not Present. That is, the DT is supported by a model designed with a set of objectives specific to its application domain, yes or no.
- *Reflection:* represents the ability of POs to be accurately measured and represented in relation to the objectives of their application. When LOs are represented through discrete math, stochastic math, or data models, it means that abstractions are performed regarding the classification, processing, and forwarding processes of data packets. In this case, we rate the Reflection property as Weak. Otherwise, when, for example, LOs are represented through emulated objects that implement exactly the software available on commercial equipment, then Reflection is classified as Strong.
- *Entanglement:* indicates the communication relationship between POs and LOs. This relationship was classified into three different levels associated with the automatic exchange of data between them:
    - *Low:* there is no automatic data exchange between the physical asset and the digital model.

    - *Medium:* communication in the PO to LO direction occurs automatically throughout the entire DT lifecycle. A change in PO leads to a change in LO. However, the opposite is not true.
    - *High:* if data between PO and LO automatically flow between them in both directions, what indicates full integration between them?
- Memorization: classified in two levels, being Present or Not Present. That is, the DT is able or not to store and represent all relevant present and past data in the context of its application.
- Programmability: classified into only two levels, as Present and Not Present. DT provides a set of APIs that allow programming some of its features, yes or no, respectively.

The model presented is an important contribution of the proposed methodology to dissect the evolving landscape of related works around NDT that differ in their technical approach and the delivered DT features, or even diverge in the foundational view of an NDT. The application of this model to related works is explored next and used for comparison in Table 3.1.

3.3 Enabling Technologies and Related Work

Depending on the expected result of the target DT, different implementation approaches can be adopted. In the specific case of NDTs, four main approaches can be identified: (1) simulation; (2) emulation; (3) analytical models; and (4) machine intelligence. Each approach has been used over time to model a computer network environment in different contexts with different goals. In the context of NDTs, their main function is to perform the modeling of the physical environment in a virtual setting. More recently, industry-driven solutions have emerged following different viewpoints to realize the digital environment. They are based on new enabling technologies coupled to network lifecycle management tasks and therefore delivering solutions characterizable as NDT platforms. Next, we seek to present an overview of the main software-based techniques used over time to model computer networks and how they fit into the different types of DT definitions. At the end of this section, we present a temporal view of the emergence of different works related to the different technologies presented and how they can be categorized.

3.3.1 Simulation

Network simulators have been used for a long time to model computer network behavior. They work by using discrete events to model each packet's

TABLE 3.1

Network Digital Twin Related Works Categorized and Classified Based on Its Main DT Properties

Related Work	Application Domain	Problem Scope	Technology Base	DT Category	DT Properties				
					R&C	Rf	E	M	P
NEAT (Fantom et al. 2022)	Network Verification	Network testing automation system for network DevOps via enhancing CI/CD workflows	**Emulation**	**Digital Model**	☑	■	●●○	☑	☑
FlowDT (Ferriol-Galmés, Cheng, et al. 2022)	Network Performance Analysis	Performance evaluation of computer networks	**Machine Intelligent**	**Digital Model**	☑	■	●○○	☐	☑
RouteNet-E (Ferriol-Galmés, Rusek, et al. 2022)	Network Performance Analysis	Performance evaluation of computer networks	**Machine Intelligent**	**Digital Model**	☑	■	●○○	☐	☑
Batfish (Fogel et al. 2015, Beckett et al. 2017)	Network Verification	Network configuration analysis to detect configuration errors and run "what-if" scenarios	**Machine Intelligent**	**Digital Shadow**	☑	■	●●○	☑	☑
Optical Failure Localization (Mayer et al. 2022)	Optical Network Performance	ML-assisted soft-failure localization with partial telemetry	**Machine Intelligent**	**Digital Shadow**	☑	■	●●○	☑	☑
Cisco DNA Center (Szigeti et al. 2018)	Automated network lifecycle management	Network controller and intent-based management dashboard	**Machine Intelligent**	**Digital Twin**	☑	■	●●●	☑	☑
NetBrain PDAs (NetBrain Technologies Inc. 2022)	Network verification and Intent-based automation	Intent-based hybrid network automation and visibility platform	**Machine Intelligent**	**Digital Twin**	☑	■	●●●	☑	☐
NetGraph (Hong et al. 2021)	Automated network lifecycle management	Operational lifecycle management for data center networks based on vendor-agnostic intent-based automation through data models	**Machine Intelligent**	**Digital Twin**	☑	■	●●●	☑	☑
Nokia FSS (Nokia 2022)	Automated network lifecycle management	Operational lifecycle management for data center networks, based on intent-based automation through advanced telemetry	**Emulation**	**Digital Twin**	☑	■	●●●	☐	☑

R&C, Representativeness and Contextualization; Rf, Reflection; E, Entanglement; M, Memory; P, Programmability.

behavior, from its generation to its consumption, while also incorporating mathematical models to simplify more complex behaviors like signal propagation and packet loss. Simulators are software solutions and different types are available for different applications (Pan and Jain 2008). Different types of simulators are available for various applications, such as testing protocol characteristics within a network, and for educational and research purposes. Network simulation tools have played an important role in the implementation of the Internet and the TCP/IP reference model. REAL (Keshav 1988), NEST (Bacon et al. 1988), Network Simulation (NS) (Chaudhary et al. 2012, NS-2 1996, Henderson, Lacage and Riley 2008), and OMNET++ (Varga 1999) are just few examples of the most relevant simulation tools available in academia in the past 40 years.

3.3.2 Emulation

Simulators use mathematical models to represent network behavior, and track each packet's flow, while emulators replicate network behavior and can functionally replace networks. Although simulation tools can also be used for emulation, virtualization technologies like virtual machines and containers offer software components that mirror those used in real networks, but without hardware acceleration. Nevertheless, virtualization guarantees more reliable behavior of virtual networks than physical networks, especially in terms of data forwarding.

Over time, several emulators have been developed, both with a focus on academia or research, such as EMPOWER (Zheng and Ni 2003), IMMUNES (Zec and Mikuc 2004), or Mininet (Lantz, Heller and Mckeown 2010), as well as aimed at industry, such as GNS3 (Neumann 2015), EVE-ng (Dzerkals 2022), or CORE (Ahrenholz 2010). More recently, emulation mechanisms with the ability to model heterogeneous network devices with high fidelity have emerged. These alternatives are based on a series of hardware virtualization techniques such as Virtual Machines (VM) or containers, which guarantee support for pre-existing network systems, such as Cisco, Juniper, and Nokia Service Router Linux (SR Linux) router images. Examples of such cases are Containerlab (Dodin 2021) and NEAT (Fantom et al. 2022).

3.3.3 Analytic Models

In addition to simulation and emulation strategies, there are also options for analytical models. Currently, the Queuing theory is the most popular analytical tool used for modeling computer networks (Ferriol-Galmés, Rusek, et al. 2022). It allows us to model the network as a series of interconnected queues that are evaluated analytically. Each of these queues can be modeled by a series of mathematical equations (Cooper 1981). Since the beginning of the Internet, analytical models based on queuing theories have been used, mainly to perform network performance analysis (Samari and Schneider 1980,

Tipper and Sundareshan 1990). More recently, several works have been published seeking to evaluate the performance of Software-Defined Networks (SDN) and OpenFlow through analytical models (Goto et al. 2019, Xiong et al. 2016).

3.3.4 Machine Intelligent Approaches

Recent works leveraging machine learning techniques such as neural networks seek to analyze the performance of computer networks, such as the case of RouteNet-E (Ferriol-Galmés, Rusek, et al. 2022) and FlowDT (Ferriol-Galmés, Cheng, et al. 2022). Such approaches share the same objectives as the analytical models, being able to model a network of queues, with different sizes and scheduling policies, and seek to identify network performance through estimates of network metrics such as packet loss, jitter, and delay. However, similar to analytical models, data-driven approaches rely on manual (expert-based) abstractions to model system input, such as input configurations, queue scheduling, input traffic, and so on.

All these works of simulators, emulators, analytical models, and neural networks mentioned so far can be categorized as Digital Models instead of DTs, when used as software technologies to model physical objects in a virtual environment. This type of category is mainly characterized by the method of mirroring the physical object. As mentioned earlier, the digital model does not feature an automatic exchange of data between the physical object and the logical object. Once it is created, changes to the physical object are not reflected in the logical object. This fact characterizes the absence of a high entanglement ratio in the approaches presented, leading to their categorization as Digital Models.

In recent developments, SDN and Network Analytics (NA) are being explored under the paradigm of Knowledge-Defined Networking (KDN) (Mestres et al. 2017) to leverage AI techniques in the context of network operation and control.

Approaches based on standardized network protocols and data models (e.g., NETCONF/YANG) seek to generate a logical model of the control plane that represents the network configuration, topology, and the mechanism that routers execute to compute the data plane. This type of approach requires that the tools that adopt it have a complete view of the various device configurations and support southbound interfaces responsible for obtaining configuration data automatically along with high-volume, fine-granular telemetry data efficiently at scale (e.g., streaming telemetry and in-band telemetry). They also seek to generate a data plane model that considers the environment (e.g., link status and prefixes advertised between neighbors) together with the control plane model. A good example of this type of application is Batfish (Beckett et al. 2017, Fogel et al. 2015), whose main function is to analyze and verify network configurations. As a result, network administrators can get indications of where errors are on the network, but the tool does not implement techniques to apply proposed changes automatically. This characteristic results in classifying Batfish into a Digital Shadow, since an automatic communication relationship from LO to PO is not established.

In recent years, new tools and ML/AI techniques have been proposed by multiple approaches with different objectives to build DTs of target network domains. For example, the "ML-assisted Soft-Failure Localization Based on Network Digital Twins" (Mayer et al. 2022) adopts NETCONF/YANG-based data models to perform telemetry data collection and store it in a database for the construction of LOs from them. These data are consumed by a fault detection and localization pipeline using Machine Learning (ML) techniques, based on Artificial Neural Networks (ANN). In this case, communication between POs and LOs (PO $\rightarrow$ LO) is established automatically, but it is not established in the opposite direction (PO $\leftarrow$ LO), given that correction actions are proposed, but not applied directly. Therefore, the work can be categorized as Digital Shadow.

3.3.5 Industry-Driven Solutions

Recently, the networking industry has been evolving their network management systems and even categorizing them as NDTs in their entirety by leveraging techniques such as Closed Control Loops (CCL) and Intent-Based Network (IBN) Automation to ensure network management throughout its lifecycle, from Day 0 (project) and Day 1 (implementation) to Day 2 (monitoring, analysis, and change management). Examples of these cases are Cisco Digital Network Architecture Center (DNA Center) (Szigeti et al. 2018), Nokia Fabric Services System (FSS) (Nokia 2022), and Huawei NetGraph (Hong et al. 2021) which are systems for network management (enterprise and data center).

One of the main differences between them is the fact that both Nokia FSS and Cisco DNA Center are intended for the management of in-house equipment, while NetGraph implements a vendor-agnostic device data model that allows it to manage equipment from any manufacturer. Cisco DNA Center implements a proprietary model of its devices that virtualizes the logs of traditional equipment into a management platform, for later manual analysis or by additional ML techniques. Unlike NetGraph and DNA Center, Nokia FSS adopts a Digital Sandbox (virtual), which emulates the physical network made of SR Linux devices through containers, to validate changes. In addition to these systems, NetBrain Technologies Inc. (2022) has made available since 2004 its system called NetBrain Problem Diagnosis Automation System (PDAs), which carries out the verification of network changes on a virtual environment based on proprietary data models and performs Intent-based Network Automation. As all these systems implement a network lifecycle management mechanism, validating changes and applying them to the physical environment automatically, they can then be fully categorized as NDTs.

3.3.6 Wrap-Up of NDT Approaches

Figure 3.3 presents the timeline of the different approaches to NDT and how each one of them can be categorized according to the different definitions of DT, as presented by Fuller et al. (2020) and Wu et al. (2021). Each of these categories has had several examples of application over time and may or may not be classified as DT in its entirety. They could sometimes be better characterized as Digital Shadow or Digital Model, rather than DT. As seen, each of these definitions is mainly associated with the communication relationship existing between PO and LO, defined here as the Entanglement property.

3.4 When DT Meets Network Engineering and Operations

Adopting NDT platforms offers several advantages over traditional network management tools. One of the main differences is the continuous mapping between the real and virtual environment, which enables the optimization of network operation and management processes. In addition, NDT can be applied in various contexts, such as engineering and network design. The most recent works in the field of NDT can be classified according to their properties and categorized according to their definitions. This section will explore these different scenarios and categories in more detail.

3.4.1 Network Engineering: Projects and Planning

Using NDT in network engineering improves the reliability of the network and workflow from design to daily operation. It can aid in the validation of hardware and software architecture, including interoperability tests with deployed protocols and configurations, and offer a high-fidelity environment for planning and testing that ensures improved reliability.

3.4.2 Network Operation: Autonomic Network Management

Validating the compliance of network changes before their actual deployment is critical to ensure risk mitigation during the network Operation, Administration, and Maintenance (OAM) process (Yahui et al. 2019). Human intervention is open to errors throughout the full process, which are extremely common (Meza et al. 2018, Tong et al. 2018, Zeng et al. 2012), and network operators need tools to avoid them and thus ensure high availability and network resilience.

New technologies, such as Closed Control Loop Automation (CCLA) together with IBN, provide autonomous management of networks in highly complex environments, e.g., scaling datacenter networks on demand.

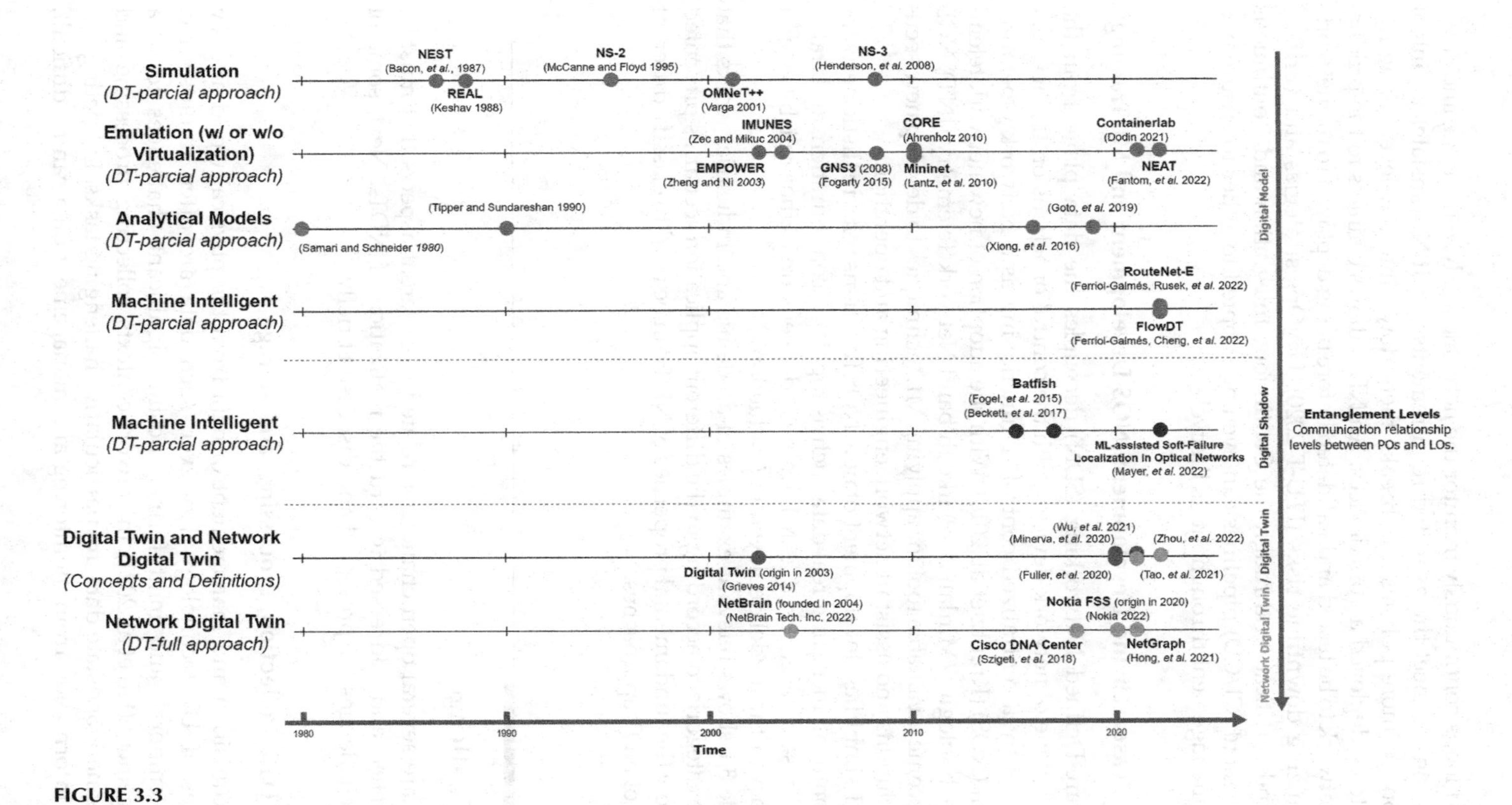

FIGURE 3.3
Network digital twins timely concepts, definitions, and software technology approaches.

Closed loops continuously monitor the network, analyze the data, and carry out actions to meet the expected requirements. An IBN translates an intent into one or more policies and breaks them down into low-level configurations to be deployed across devices. An NDT allows changes to the production network to be tested and validated before being deployed to the network, minimizing downtime risks (ITU-T 2020). DevOps strategies and verification and certification through the Continuous Integration and Continuous Deployment (CI/CD) pipelines can even be adopted together to ensure network management throughout its lifecycle.

3.4.3 Reasearch and Development: NOS Development and ML Training

Software-Defined Networking (SDN) decouples the data plane from the control plane of network devices, with the control of actions on the network managed by a centralized controller functioning as a Network Operating System (NOS) (Kreutz et al. 2014). With the adoption of new network telemetry technologies (Minlan 2019) like In-band Network Telemetry (INT), NOS has become a valuable tool for applying ML techniques to develop new security solutions and assist in network engineering and operation.

NDT simplifies development of new NOS functions with realistic network environment, including real data such as topology, configuration, and traffic load. These datasets used as ML training data ensure higher reliability and accuracy of the model deployed in production.

Table 3.1 shows relevant examples of academic and industrial works that can be categorized according to the different application scenarios presented and how the fundamental properties of NDT can be used to classify different types of NDT approaches.

3.5 Challenges

There are several open challenges related to different aspects that must be researched and addressed toward the realization of NDTs. Next, some of these challenges are presented and discussed briefly.

3.5.1 Data Collection, Processing, and Storage

The acquisition and management of data from the physical world are key features of DT technology. However, such data are often multi-source, multi-temporal, and multi-format, posing significant challenges to data management (Hu et al. 2021). In network contexts, collecting, processing, and storing network state data are particularly challenging tasks. Extracting relevant information from monitoring and management data can be difficult, and standardizing interfaces for different monitoring sources is crucial to ensure agile data collection and processing. Furthermore, network telemetry techniques and standardized interfaces are essential to ensure the correct representation of the network state in a timely, fine-grained, and standardized manner. However, storing network state data and network traffic load can be resource-intensive, and reducing data size through sampling or compression techniques is vital to mitigate this issue (Almasan, Ferriol-Galmés and Paillisse, et al. 2022a, Almasan, Ferriol-Galmés and Suárez-Varela, et al. 2022b).

3.5.2 Modeling Objects

Building accurate and functional data models of devices throughout their lifecycle is challenging due to the diversity and variability of the physical environment. In the context of networks, a unified mapping of data models from network equipment used by NDT for vendor-specific heterogeneous devices is difficult to perform. In addition, software implementation aspects of distributed systems in a network environment impact NDT performance, making modeling accurate mechanisms such as congestion control algorithms, queuing policies, and routing schemes cumbersome. Balancing requirements such as heterogeneity support, accuracy, flexibility, and scalability further increase complexity. Emulated resources offer a solution, but can become computationally costly as the environment scales. Depending on the use case, using analytical modeling technologies and neural networks can be more adequate (Hong et al. 2021, Hu et al. 2021, Hui et al. 2022, Zhou et al. 2022).

3.5.3 Entanglement and Real-Time Requirements

Entanglement, which involves bidirectional communication between the physical and virtual environments, is a critical aspect of DT. It ensures that actions taken in the real world are informed by real-time data obtained from physical objects in the virtual environment. Implementing this process is crucial for defining the DT in a relevant context, but it can be extremely complex, with changes in the virtual environment leading to corresponding changes in the physical environment (Fuller et al. 2020, Hu et al. 2021, Zhou et al. 2022). In many use cases, automating and simplifying the modeling and processing of the models is essential (Zhou et al. 2022) to reduce the time and computational resources required for tasks in the virtual environment, while ensuring that they match the hardware and software resources of the NDT. However, it can be challenging to achieve real-time requirements for the DT, depending on the amount of data required and the time taken to analyze it. In certain scenarios, such as cybersecurity, real-time data analysis is critical to maintaining the network's proper functioning. In other scenarios, however, real-time requirements may not be as stringent, allowing analyses to be performed over a longer period. Identifying the real-time requirements of the application scenario is essential for designing the NDT system's hardware and software to ensure its effective use.

3.5.4 Standardization

The diversity of protocols, functionalities, and services implemented through heterogeneous technologies by different network manufacturers is a significant challenge for establishing a unified NDT system. Even more, when it is intended to apply them in different scenarios and standardize them in terms of technologies across different layers (Hu et al. 2021, Zhou et al. 2022). Thus, it is necessary to propose a unified NDT architecture with well-defined components and functionalities so that follow-up adjustments can be made depending on the approach adopted for the different application scenarios. Afterward, the interfaces between the different layers of the architecture can be identified and defined by adopting existing technologies or not (Zhou et al. 2022).

3.5.5 Scalability of NDT Systems

The size of the network and the amount of data needed to model these environments can cause several problems in ensuring network scalability. Large volumes of network data can help to achieve more accurate network models. However, the acquisition, modeling, and massive storage of these data are problems that arise within the context of scalability (Almasan, Ferriol-Galmés and Paillisse, et al. 2022a, Almasan, Ferriol-Galmés and Suárez-Varela, et al. 2022b, Zhou et al. 2022). They are, in general, caused by the high cost of communication and massive data storage caused by the excessive use of capacity by the adoption of NDT (Zhou et al. 2022). Therefore, adopting efficient data collection tools and compression techniques is required to minimize the data transferred on the network and the space occupied by them. Furthermore, while a DTN system is used by a network environment that grows over time, its performance degrades in terms of accuracy or execution time, concluding that the choice of the technological approach may not have been the most appropriate (Almasan, Ferriol-Galmés and Suárez-Varela, et al. 2022b). For example, when an NDT model based on neural networks takes too long to execute, turning its adoption in a production environment unfeasible, adopting alternative approaches, e.g., based on mathematical abstractions, such as simulations or analytical models, may be more suitable. This is another factor that should guide the choice of the right technological approach depending on the scenario in which the NDT will be applied.

3.6 Conclusions

This chapter examines the application of the DT paradigm to the field of computer networks, i.e., NDT. An NDT system maintains constant synchronization between a real network environment and a twin virtual environment, ensuring that actions in the real world are performed based on the analysis of real-time data on the virtual environment. Several practical implementations from industry and academia have been proposed in recent years. However, many do not implement key NDT features in their entirety, which complicates attempts to define and categorize NDT approaches. We observe that network simulation approaches have been adopted since the beginning of the Internet and could since then be categorized as a type of NDT, the Network Digital Models. We were also able to identify that only industrial tools dealing with full lifecycle network management through intent-based network automation and closed-loop automation can be effectively classified today as NDT in their full design, mainly because they maintain two-way communication between the physical and virtual environment.

We highlight that, in addition to maintaining bidirectional communication between the physical and virtual worlds, further challenges are also of great relevance today, such as the difficulty of maintaining software and hardware requirements to achieve key objectives of target application scenarios. The specific results expected by using an NDT will likely end up guiding the choice of the technological approach. For example, depending on the size of the network to be modeled, approaches with a higher level of abstraction may prove more efficient. However, new practical implementations of the different approaches are necessary to ensure the maturity of the models and the collection of sufficient data from the production environment. So far, there is still a lack of a unified definition of an NDT, a reference architecture, and common implementation artifacts (e.g., open data models, APIs, standard interfaces, etc.), potentially compromising the future interoperability of NDT systems.

References

Ahrenholz, Jeff. "Comparison of CORE Network Emulation Platforms." *2010-Milcom 2010 Military Communications Conference. IEEE*, 2010: 166–171.

Almasan, Paul, et al. "Digital Twin Network: Opportunities and Challenges." 2022a.

Almasan, Paul, et al. "Network Digital Twin: Context, Enabling Technologies and Opportunities." IEEE Communications Magazine *60*.11 (2022): 22–27.

Bacon, David F., Alexander Dupuy, Jed Schwartz, and Yechiam Yemini. "NEST: A Network Simulation and Prototyping Tool." USENIX *Winter. 1988*: 17–78.

Barricelli, Barbara Rita, Elena Casiraghi, and Daniela Fogli. "A Survey on Digital Twin: Definitions, Characteristics, Applications, and Design Implications." *IEEE Access*, 7 (2019): 167653–167671.

Beckett, Ryan, Gupta Aarti, Ratul Mahajan, and David Walker. "A General Approach to Network Configuration Verification." *2017 Conference of the ACM Special Interest Group on Data Communication (SIGCOMM'17)*. New York: Association for Computing Machinery, Inc, 2017. 155–168.

Chaudhary, Rachna, Shweta Sethi, Rita Keshari, and Goel Saksi. "A Study of Comparison of Network Simulator-3 and Network Simulator-2." *International Journal of Computer Science and Information Technologies*, 3 (2012): 3085–3092.

Cooper, Robert B. "Queueing Theory: A 90 Minute Tutorial." *Proceedings of the ACM'81 Conference*, 1981: 119–122.

Dodin, Roman. "Containerlab...and How People Use It." *NLNOG*. Amsterdam: Nokia, 2021.

Dzerkals, Uldis. *EVE-NG Professional*. EVE-NG LTD, 2022.

Fantom, Will, Paul Alcock, Ben Summs, Charalampos Rotsos, and Nicholas Race. "A NEAT Way to Test-Driven Network Management." *NOMS 2022-2022 IEEE/IFIP Network Operations and Management Symposium*. Budapest, Hungary: IEEE, 2022: 1–5.

Ferriol-Galmés, Miquel, et al. "RouteNet-Erlang: A Graph Neural Network for Network Performance Evaluation." *IEEE INFOCOM 2022 - IEEE Conference on Computer Communications*, 2022: 2018–2027.

Ferriol-Galmés, Miquel, Xiangle Cheng, Xiang Shi, Shihan Xiao, Pere Barlet-Ros, e Albert Cabellos-Aparicio. "FlowDT: A Flow-Aware Digital Twin for Computer Networks." *ICASSP, IEEE International Conference on Acoustics, Speech and Signal Processing - Proceedings*. Singapore: IEEE, 2022. 8907–8911.

Fogel, Ari, et al. "A General Approach to Network Configuration Analysis." *12th USENIX Symposium on Networked Systems Design and Implementation (NSDI'15)*. Oakland, CA: USENIX Association, 2015. 469–483.

Fuller, Aidan, Zhong Fan, Charles Day, and Chris Barlow. "Digital Twin: Enabling Technologies, Challenges and Open Research." *IEEE Access* 8 (2020): 108952–108971.

Goto, Yuki, Bryan Ng, Winston K. G. Seah, and Yutaka Takahashi. "Queueing Analysis of Software Defined Network with Realistic OpenFlow-Based Switch Model." *Computer Networks*, 164 (2019): 106892.

Henderson, Thomas R., Mathieu Lacage, and George F. Riley. "Network Simulations with the NS-3 Simulator." *SIGCOMM Demonstration*, 14 (2008): 527.

Hong, Hanshu, et al. "NetGraph: An Intelligent Operated Digital Twin Platform for Data Center Networks." *Proceedings of the ACM SIGCOMM 2021 Workshop on Network-Application Integration (NAI'21)*. Proceedings of the ACM SIGCOMM 2021 Workshop on Network-Application Integration, 2021. 26–32.

Hu, Weifei, Tangzhou Zhang, Xiaoyu Deng, Zhenyu Liu, e Jianrong Tan. "Digital Twin: A State-of-the-Art Review of Its Enabling Technologies, Applications and Challenges." *Journal of Intelligent Manufacturing and Special Equipment*, 2021: 1–34.

Hui, Linbo, Mowei Wang, Liang Zhang, Lu Lu, and Yong Cui. "Digital Twin for Networking: A Data-driven Performance Modeling Perspective." *arXiv preprint arXiv:2206.00310*, 2022.

ITU Telecommunication Standardization Sector. *Network* 2030 *Architecture Framework*. Technical Specification, ITU, 2020.

Jones, David, Chris Snider, Aydin Nassehi, Jason Yon, and Ben Hicks. "Characterising the Digital Twin: A Systematic Literature Review." *CIRP Journal of Manufacturing Science and Technology* 29 (2020): 36–52.

Keshav, Srinivasan. *REAL: A Network Simulator.* Computer Science Department, University of California, California: Univ. of California at Berkeley, 1988, 1–16.

Kreutz, Diego, Fernando Ramos, Paulo Verissimo, Christian Rothenberg, Slamak Azodolmolky, and Steve Uhlig. "Software-Defined Networking: A Comprehensive Survey." *Proceedings of the IEEE,* 2014: 14–76.

Lantz, Bob, Brandon Heller, and Nick Mckeown. "A Network in a Laptop: Rapid Prototyping for Software-Defined Networks." *Ninth ACM SIGCOMM Workshop on Hot Topics in Networks - Hotnets '10*. Monterey, CA: Association for Computing Machinery, 2010. 1–6.

Lim, Kendrik Yan Long, Pal Zheng, and Chun-Hslen Chen. "A State-of-the-Art Survey of Digital Twin: Techniques, Engineering Product Lifecycle Management and Business Innovation Perspectives." *Journal of Intelligent Manufacturing*, 31 (2019): 1313–1337.

Mayer, Kayol Soares, et al. "Demonstration of ML-assisted Soft-Failure Localization Based on Network Digital Twins." *Journal of Lightwave Technology*, 40 (2022): 4514–4520.

Mestres, Albert, et al. "Knowledge-Defined Networking." *ACM SIGCOMM Computer Communication Review*, 47 (2017): 1–10.

Meza, Justin, Tianyin Xu, Kaushik Veeraraghavan, and Onur Mutlu. "A Large Scale Study of Data Center Network Reliability." *Proceedings of the Internet Measurement Conference 2018 (IMC'18)*. Boston, MA: ACM, 2018. 393–407.

Minerva, Roberto, Gyu Myoung Lee, and Noel Crespi. "Digital Twin in the IoT Context: A Survey on Technical Features, Scenarios and Architectural Models." *Proceedings of the IEEE*, 108 (2020): 1785–1824.

Minlan, Yu. "Network Telemetry: Towards a Top-Down Approach." *ACM SIGCOMM Computer Communication Review*, 49 (2019): 11–17.

NetBrain Technologies Inc. "NetBrain: Problem Diagnosis Automation System." Whitepaper, 2022.

Neumann, Jason C. *The Book of GNS3: Build Virtual Network Labs Using Cisco, Juniper, and More.* San Francisco, CA: No Starch Press, 2015.

Nokia. "Nokia Fabric Services System Release 22." Data sheet, 2022.

NS-2. 1996.

Pan, Jianli, and Raj Jain. *A Survey of Network Simulation Tools: Current Status and Future Developments*. Project Report, Computer Science and Engineering, Washington University in St. Louis, St. Louis MI: Washington University, 2008.

Rasheed, Adil, Omer San, and Trond Kvamsdal. "Digital Twin: Values, Challenges and Enablers from a Modeling Perspective." *IEEE Access*, 8 (2020): 21980–22012.

Samari, N. K., and G. Michael Schneider. "A Queueing Theory-Based Analytic Model of a Distributed Computer Network." *IEEE Transactions on Computers*, C-29 (1980): 994–1001.

Sun, Tao, et al. "Digital Twin Network (DTN): Concepts, Architecture, and Key Technologies." *Acta Automatica Sinica*, 47 (2021). doi:10.16383/j.aas.c210097.

Szigeti, T., D. Zacks, M. Falkner, e S. Arena. *Cisco Digital Network Architecture: Intent-based Networking for the Enterprise*. Indianapolis, IN: Cisco Press, 2018.

Tao, Fei, He Zhang, Liu Ang, and A. Y. C. Nee. "Digital Twin in Industry: State-of-the-Art." *IEEE Transactions on Industrial Informatics*, 15 (2019): 2405–2415.

Tipper, David, and Malur K. Sundareshan. "Numerical Methods for Modeling Computer Networks under Nonstationary Conditions." *IEEE Journal on Selected Areas in Communications*, 8 (1990): 1682–1695.

Tong, Van, Hai Anh Tran, Sami Souihi, and Abdelhamid Mellouk. "Network Troubleshooting: Survey, Taxonomy and Challenges." *2018 International Conference on Smart Communications in Network Technologies, SaCoNeT 2018. El Oued*, Algeria: IEEE, 2018. 165–170.

Varga, András. "Using the OMNET++ Discrete Event Simulation System in Education." IEEE Transactions on Education 42.4 (1999): 11..

Wu, Yiwen, Ke Zhang, and Yan Zhang. "Digital Twin Networks: A Survey." *IEEE Internet of Things Journal*, 8 (2021): 13789–13804.

Xiong, Bing, Kun Yang, Jinyuan Zhao, Wei Li, and Keqin Li. "Performance Evaluation of OpenFlow-Based Software-Defined Networks Based on Queueing Model." *Computer Networks*, 102 (2016): 172–185.

Yahui, Li, et al. "A Survey on Network Verification and Testing with Formal Methods: Approaches and Challenges." *IEEE Communications Surveys and Tutorials*, 21 (2019): 940–969.

Zec, Marko, and Miljenko Mikuc. "Operating System Support for Integrated Network Emulation in IMUNES." *1st Workshop on Operating System and Architectural Support for the on demand IT InfraStructure (OASIS)*. 2004: 3–12.

Zeng, Hongyi, Peyman Kazemian, George Varghese, and Nick Mckeown. *A Survey on Network Troubleshooting*. Stanford HPNG Technical Report TR12-HPNG-061012, Stanford, CA: Stanford University, 2012.

Zheng, Pei, and Lionel M. Ni. "EMPOWER: A Network Emulator for Wireline and Wireless Networks." *IEEE INFOCOM 2003. Twenty-second Annual Joint Conference of the IEEE Computer and Communications Societies (IEEE Cat. No. 03CH37428). IEEE, 2003*, 3: 1933–1942.

Zhou, Cheng, et al. "Digital Twin Network: Concepts and Reference Architecture." *Internet-draft draft-zhou-nmrg-digitaltwin-network-concepts*, edited by IETF. March 7, 2022.

4

Cognitive Digital Twins

Xiaochen Zheng and Jinzhi Lu
Ecole Polytechnique Fédérale de Lausanne (EPFL)

Rebeca Arista
Airbus SAS

Jože Martin Rožanec
Jožef Stefan International Postgraduate School

Stavros Lounis
Athens University of Economics and Business

Kostas Kalaboukas
Gruppo Maggioli

Dimitris Kiritsis
Ecole Polytechnique Federale de Lausanne (EPFL)

4.1 Introduction

Digital Twin (DT) is one of the key enabling technologies for the Industry 4.0 revolution. It represents a comprehensive physical and virtual description of a product or a system that captures relevant attributes and behaviors of a component, product, or system [1,2]. After a decade of rapid development, DT has been applied to almost all industrial sectors covering different lifecycle phases such as concept, design, production, and maintenance, among others [3]. Although the benefit of DT is undoubted, it is challenging to deal with specific complex industrial systems, which requires the integration of multiple relevant DTs corresponding to different system levels and lifecycle phases.

Although the concept of DTs was initially defined for the manufacturing sector, it has been expanded to other industrial sectors [4]. Manufacturing industries have experienced a considerable need and transformation toward "agility" and "resilience" to cope with the risks and disruptions imposed

DOI: 10.1201/9781003425724-6

by COVID. Many supply chains had to be re-engineered and manufacturing processes had to be more "configurable." The use of DTs with cognition services on top of them allows industries to simulate various "what-if" scenarios, understand their impact, and finally decide on the optimal solution.

Modern industrial systems are getting increasingly complex due to the continuous adoption of advanced technologies. The complexity management of such modern systems is challenged particularly by the Product Lifecycle Management (PLM). Moreover, a complex system usually contains multiple subsystems and components which may have their own DTs. Various stakeholders might create these DTs based on different protocols and standards whose data structures are usually heterogeneous in terms of syntax, schema, and semantics. Such heterogeneity makes the integration of DT models a challenging task. In order to cope with these challenges, recent research has attempted to enhance the current DT paradigm with cognitive and autonomous capabilities enabled by advanced technologies such as semantic modeling, cognitive computing, and Model-based Systems Engineering, among others. These efforts led to a novel concept named Cognitive Digital Twin (CDT), or Cognitive Twin (CT), representing a promising evolution trend for the next generation of DTs.

The term *cognitive* in CDT is inspired by cognitive science, which studies the circulation and treatment of information in the human brain [5]. The classic definition of cognition includes *all the processes by which the sensory input is transformed, reduced, elaborated, stored, recovered and used* [6,7]. A typical cognitive process of human beings includes the following steps: first, sensing and obtaining information from ambient physical environments with perceptive sense organs such as skin, eyes, and ears; secondly, transmitting and processing information in the brain through nerves; and finally sending analysis results to certain body parts through nerves and activating appropriate behavior response [5]. In addition, some fundamental aspects of cognitive capabilities can be extracted from the above process, such as attention (selective focus), perception (forming useful precepts from raw sensory data), memory (encoding and retrieval of knowledge), reasoning (drawing inferences from observations, beliefs, and models), learning (from experiences, observations, and teachers), problem-solving (achieving goals), or knowledge representation [8]. The target of CDT is to enable these cognitive capabilities to some extent based on various advanced technologies such as semantic engineering, knowledge graph, cognitive computing, and systems engineering, among others.

4.2 CDT Definitions

The first appearance of the *Cognitive Digital Twins* concept can be traced back to an industry workshop in 2016 when Ahmed [9] discussed the cognitive evolution of IoT technologies and proposed the CDT concept, which was defined as *a digital representation, augmentation and intelligent companion of its physical twin as a whole, including its subsystems across all of its life cycles and evolution phases.* This definition emphasizes a CDT's key characteristics, including intelligence, cross-system levels, and cross-full lifecycle.

A similar concept was presented later by Fariz [10] during a workshop in 2017. Some specific functions of CDT are identified in this new definition, i.e., *CDT uses real-time data from IoT sensors and other sources to enable learning, reasoning and automatically adjusting for improved decision making.* This definition approached the CDT concept from the perspectives of cognitive computing and artificial intelligence and is used by IBM Watson as an example to demonstrate cognitive engineering scenarios.

In addition to the efforts from the industry sector, researchers from academia are also promoting the CDT concept, although the term *cognition* or *cognitive* is not always used. Boschert et al. [1] prospected the paradigm of next-generation DT (nexDT): *A description of a component, product, system or process by a set of well-aligned, descriptive and executable models* which *is the semantically linked collection of the relevant digital artifacts including design and engineering data, operational data and behavioral descriptions* and *evolves with the real system along the whole life cycle and integrates the currently available and commonly required data and knowledge.* Its primary motivation is that the currently isolated DT models cannot fulfill all purposes and tasks across the entire lifecycle. The integration of multiple DT models for different business objectives is needed. Semantic technologies like knowledge graphs are promising tools to connect PLM systems, cloud solutions, and other data artifacts and devices.

Fernandez et al. [11] explored the functions of hybrid human–machine cognitive systems and specified the Symbiotic Autonomous Systems (SAS). The authors considered CDT as *a digital expert or copilot, which can learn and evolve, and that integrates different sources of information for the considered purpose.* It focuses on the cooperation and convergence of human and machine augmentation, with increasing intelligence and consciousness, leading toward a continuous symbiosis of humans and machines. Based on the SAS context, the authors created an Associative Cognitive DT (AC-DT) framework to enhance the applications of CDTs. AC-DT is an augmented contextual description of an entity that aims at a specific cognitive purpose and includes the relevant associated connections with other entities.

Lu et al. [12] proposed a formal definition of CTs as "DTs with augmented semantic capabilities for identifying the dynamics of virtual model evolution, promoting the understanding of interrelationships between virtual models and enhancing the decision-making". A framework based on a knowledge graph was created to support the development of the CDTs. To facilitate the application of CDTs, they provided a tool-chain consisting of multiple existing software and platforms to empower different components of the CDT models. In a following study [13], the authors further explored the CDT concept using MBSE approaches and formally defined it based on the ISO/IEC/IEEE 42010 standard.

The Enhanced Cognitive Twin (ECT) concept [14] was proposed by Eirinakis et al. by introducing advanced cognitive capabilities to the DT artifact that enable supporting decisions, with the end goal to enable DTs to react to inner or outer stimuli. The ECT can be deployed at different hierarchical levels of the production process, i.e., at sensor-, machine-, process-, employee or even factory-level, aggregated to allow horizontal and vertical interplay.

This definition specifies the core cognitive elements of CDT and emphasizes that it should cross hierarchical levels during implementations from a process-oriented perspective.

Abburu et al. [2] reviewed existing concepts related to DTs and provided a three-layer approach to clarify different "twins" into DTs (isolated models of physical systems), hybrid twins (interconnected models capable of integrative prediction of unusual behaviors), and CTs (extended with expert and problem-solving knowledge capable of dealing with unknown situations). According to this categorization, CDT is defined as *An extension of Hybrid Digital Twins incorporating cognitive features that enable sensing complex and unpredicted behavior and reason about dynamic strategies for process optimization, leading to a system that continuously evolves its digital structure as well as its behavior.* A five-layer implementation architecture is also proposed in this study containing a software toolbox, which is applied to several relevant use cases in the process industry to verify its possible applicability.

Aiming at creating smart manufacturing systems, Ali et al. [15] envisioned CDT as "an extension of existing digital twins with additional capabilities of communication, analytics, and intelligence in three layers: access, analytics, and cognition". The cognitive layer is enabled by edge computing, domain expertise, and global knowledge bases. It supports the integration of multiple DTs by building customized communication networks, thus performing autonomous decision-making. It also depicted an ecosystem of CDTs by connecting many CDTs of different systems and domains.

The CDT for manufacturing systems was proposed [8] based on advances in cognitive science, artificial intelligence technologies, and machine learning techniques. According to the fundamental aspects of cognition, they define CDT as a DT with additional cognitive capabilities, including perception, attention, memory, reasoning, problem-solving and learning.

Based on the review of the existing CDT definitions, Zheng et al. [16] summarized the core characteristics of CDT:

- *DT-based:* meaning a CDT is an extended version of DT. It contains essential elements of DT, including the physical entity, the digital entity, and the connections between them. In addition, CDT usually contains multiple DT models corresponding to different system levels and lifecycle phases under a unified semantics topology definition.

- *Cognition capability:* meaning a CDT should enable human-like intelligent activities such as attention, perception, comprehension, memory, reasoning, prediction, decision-making, or problem-solving.
- *Full-lifecycle management:* meaning a CDT should consist of digital models covering different phases across the system's entire lifecycle, and be capable of integrating all available data, information, and knowledge.
- *Autonomy capability:* A CDT should conduct autonomous activities without human assistance or a minimum level of human intervention.
- *Continuous evolving:* A CDT should be able to evolve along the entire system lifecycle.

On the basis of existing CDT definitions, the CDT is defined as [16]: *a digital representation of a physical system that is augmented with certain cognitive capabilities and support to execute autonomous activities; comprises a set of semantically interlinked digital models related to different lifecycle phases of the physical system, including its subsystems and components, and evolves continuously with the physical system across the entire lifecycle.*

The main differences between them are their structural complexity and cognitive capability. First, a CDT is usually more complex than a DT in terms of architecture and the number of lifecycle phases involved. Most DTs correspond to a single system (or product, subsystem, or component) and focus on one of the lifecycle phases. In contrast, a CDT should consist of multiple digital models corresponding to different subsystems and components of a complex system and focus on multiple lifecycle phases of the system. A CDT can be constructed in many cases by integrating multiple related DTs using ontology definition, semantic modeling, and lifecycle management technologies. These DTs may correspond to different subsystems and components mapped to different lifecycle phases, and each evolves along with the system lifecycle.

Second, cognitive capabilities are essential for CDTs, whereas DTs do not necessarily possess such capabilities. Most existing DTs are applied for visibility, analytics, and predictability services, such as condition monitoring, functional simulation, dynamic scheduling, abnormal detection, or predictive maintenance. These services are usually enabled by data-based and model-based algorithms using the data collected from the physical entities. However, a cognitive capability is required to reach a higher level of automation and intelligence, for example, to enable sensing complex and unpredicted behaviors and to generate dynamic strategies autonomously. To achieve this target, the data-based and model-based algorithms cannot integrate the complex data and models from different systems and lifecycle phases with heterogeneous specifications and standards. Addressing this challenge requires more technologies such as semantic modeling, systems engineering, and PLM. For instance, a unified ontology can represent physical entities, virtual entities, and the topology between them which is the basis for realizing cognitive capabilities. Top-level ontologies can be used to integrate different ontologies synchronized with virtual entities across the lifecycle to support the reasoning for cognitive decision-making.

4.3 CDT Architectures

In addition to the definitions, different architectures of CDT have been proposed, including conceptual frameworks for formalizing the CDT concept and reference architectures to support CDT implementations. These architectures are critical for CDT development and applications. This section reviews some of the existing CDT architectures from the literature.

A CDT reference framework was proposed [9] that specifies the key architectural building blocks of CDT in its early development phase. According to this framework, the digital representations of the physical entities in the virtual space are based on the Cognitive Digital Twin Core (CDTC), which contains metadata, self-defense mechanisms, and governing rules to empower the CDT functions. The CDTC is composed of six layers including anchors (data workers), surrogates (knowledge workers), bots (makers), perspectives (interfaces), self-management (administrators), and defense systems (guardians). This architecture framework summarizes the key functions that a CDT should support and the enabling mechanisms.

Lu et al. [13] designed a knowledge-graph-centric framework for CDT according to the ISO/IEC/IEEE 42010 standard [17]. It consists of five main components covering multiple domains: industrial system dynamic process modeling, ontology-based cross-domain knowledge graphs, CDT construction for dynamic process simulation, CDT-based analysis for process optimization, and service-oriented interface for data interoperability. This framework is designed to support decision-making during IoT system development. It takes inputs from business domains and provides outputs to asset domains.

Abburu et al. [2] proposed an architectural blueprint for CDT that consists of five layers providing a set of model-driven and data-driven services. The five layers include the data ingestion and preparation layer, model management layer, service management layer, twin management layer, and a user interaction layer. Moreover, to better standardize the proposed architecture, its layers and components are mapped to the Big Data Value Association (BDVA) reference models and the Artificial Intelligence Public Private Partnership (AI PPP). Several use cases are also introduced to demonstrate the ability and versatility of the proposed architecture (Figure 4.1).

The recently published standard *ISO 23247, Automation systems and integration — Digital Twin framework for manufacturing* [18] defines the principles and

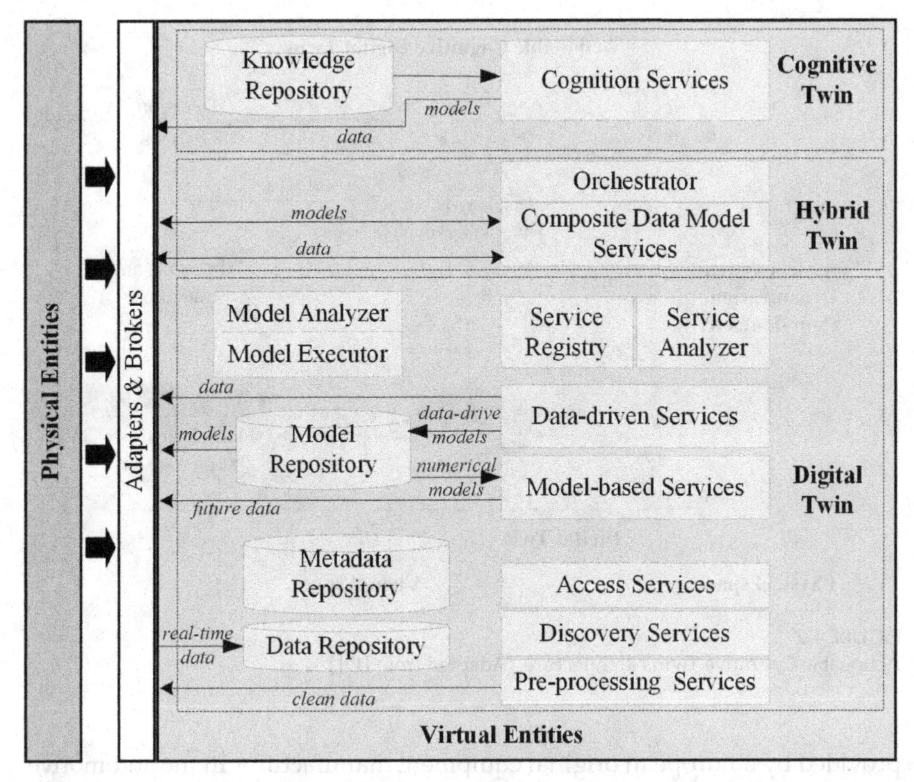

FIGURE 4.1
Cognitive Twin Toolbox conceptual architecture. (Adapted from [2,16]. ©: CC BY-NC-ND 4.0.)

requirements for developing DTs in the manufacturing domain and provides a framework to support the creation of DTs of observable manufacturing elements (e.g., personnel, equipment, materials, or facilities). This standard represents the trend of standardization for DT in different domains. Although mainly focused on the manufacturing domain, it provides a valuable reference for DT standardization in other domains and cross domains. Eirinakis et al. [4] proposed a conceptual architecture for implementing CDTs in resilient production. It is based upon the entities and interconnections in the framework proposed in the ISO 23247 standard.

Rožanec et al. [19] proposed an actionable CT enabled by a knowledge graph to support reasoning, binding different knowledge processes and types of knowledge and providing an actionable dimension. Furthermore, in their work, the authors describe how a cognition-first DT can be created by discerning how the different types of cognition can be realized in the digital domain, and the resulting knowledge leveraged to solve specific problems. The concept was described along with scenarios concerning demand forecasting and production planning. The use cases were developed with data

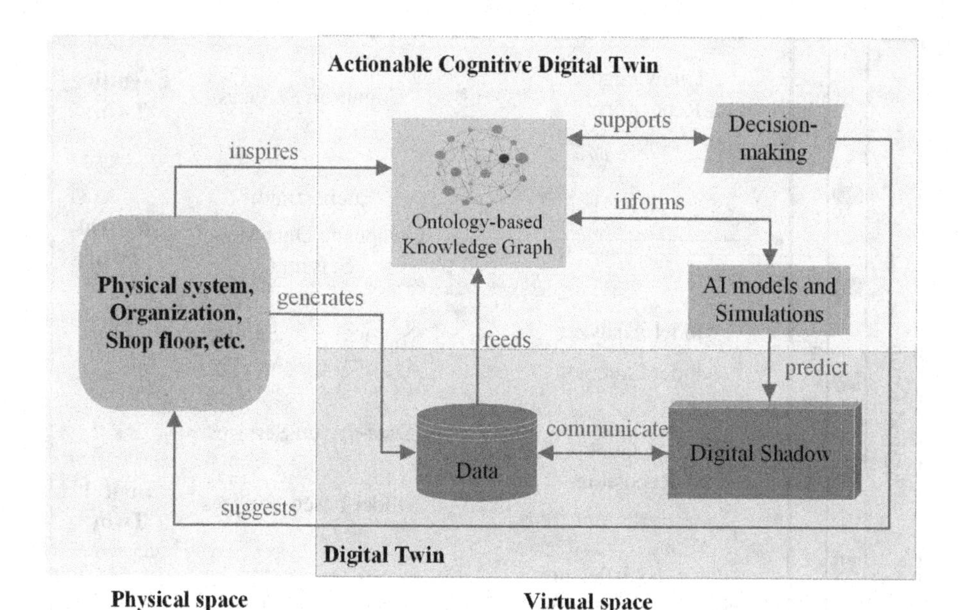

FIGURE 4.2
Actionable Cognitive Twins architecture. (Adapted from [19].)

provided by a European original equipment manufacturer in the automotive industry. The qualitative and quantitative analysis of two use cases was used to verify the advantages of the proposed CDT approach. In Figure 4.2, we provide an overview of the actionable CT concept.

Based on these existing architectures, Zheng et al. [16] designed a comprehensive reference architecture for CDT to cover all its key elements and characteristics. When developing the reference architecture, the RAMI4.0 [20] is used to improve interoperability. RAMI4.0 is a widely adopted reference architecture for industry 4.0 digitization. It provides a three-dimensional framework covering the most important aspects of Industry 4.0, including the lifecycle value stream, six common business scenarios, and the communication layers corresponding to a smart factory hierarchy. These three dimensions overlap some of the key elements of the CDT definition, especially the lifecycle value stream dimension and the multi-layer approach for structuring physical systems and communication levels. Therefore, based on the RAMI4.0 and considering the existing CDT architectures, a three-dimensional CDT reference architecture covering the key elements of the CDT definition is designed, as shown in Figure 4.3.

The three dimensions of this reference architecture include full lifecycle phases, system hierarchy levels, and six functional layers. The elements of each dimension are explained below:

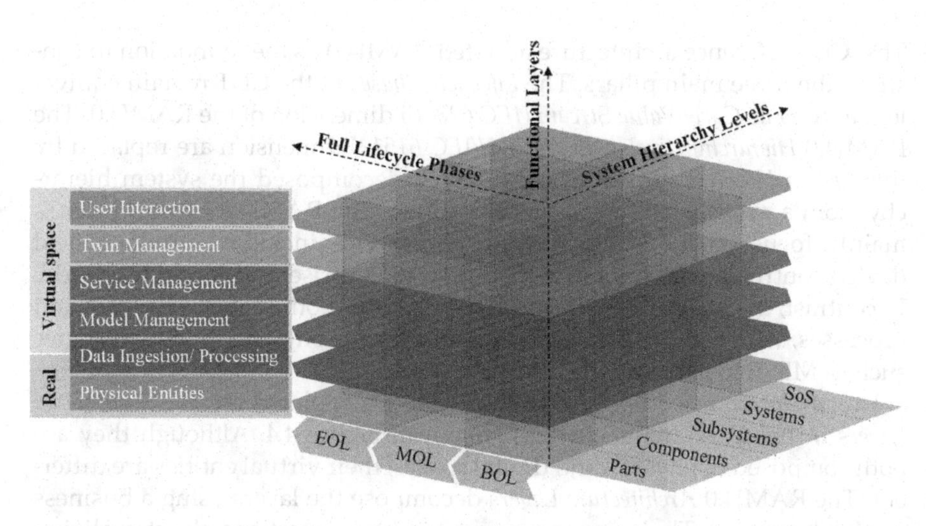

FIGURE 4.3
Cognitive Digital Twin reference architecture based on RAMI4.0 [16]. (©: CC BY-NC-ND 4.0).

- *Full lifecycle phases:* This dimension focuses on the full lifecycle management and continuous evolving capabilities of the CDT definition. During the entire lifecycle of a system, many digital models are created to support different lifecycle phases along this dimension. For example, a typical product lifecycle includes production design, simulation, manufacturing process planning, production, maintenance, and recycling. Each of these phases may have multiple related digital models.
- *System hierarchy levels:* This dimension provides a hierarchical approach to specify the structure and boundary of a CDT. Modern industrial systems are usually highly complex system-of-systems (SoS). Therefore, it is crucial to define the scope of a CDT properly. When developing CDTs, it might be challenging to create the models separately simply according to the physical hierarchy of a factory as specified in RAMI4.0. Therefore, we adopted a systems engineering methodology to define the structure of a complex into SoS, system, subsystem, components, and parts [21].
- *Functional layers:* This dimension specifies the different functions provided by a CDT. It is designed based on the architectural blueprint of the COGNITWIN Toolbox (CTT) proposed [2]. The Physical Entities layer represents the system in the physical space, and the other five layers represent different functions of CDT in the digital space, including Data Ingestion and Processing, Model Management, Service Management, Twin Management, and User Interaction.

This CDT reference architecture adopted RAMI4.0 as the foundation to construct the three main pillars. The *Lifecycle Phases* of the CDT remain equivalent to the *Life Cycle Value Stream (IEC 62890)* dimension of the RAMI4.0. The RAMI4.0 *Hierarchy Levels (IEC 62264//IEC 61512)* dimension are replaced by the *System Hierarchy Levels* in CDT, which decomposed the system hierarchy from a systems engineering perspective. The RAMI4.0 *Hierarchy Levels* mainly focus on the physical system hierarchy, including product, field device, control device, station, work centers, enterprise, and connected world. In contrast, the CDT *System Hierarchy Levels* cover both physical systems and processes, which allow the application of systems engineering technologies such as MBSE to facilitate CDT development.

The *Architecture Layers* of the RAMI4.0 are replaced by the *Functional Layers* in the CDT architecture as shown in Figure 4.4. Although they are both composed of physical and digital spaces, their virtual entities are different. The RAMI4.0 *Architecture Layers* decompose the layers using a business oriented strategy. They aim to answer the basic questions about realizing a business idea by specifying organization and business processes, functions of the asset, necessary data, access to information, and integration of

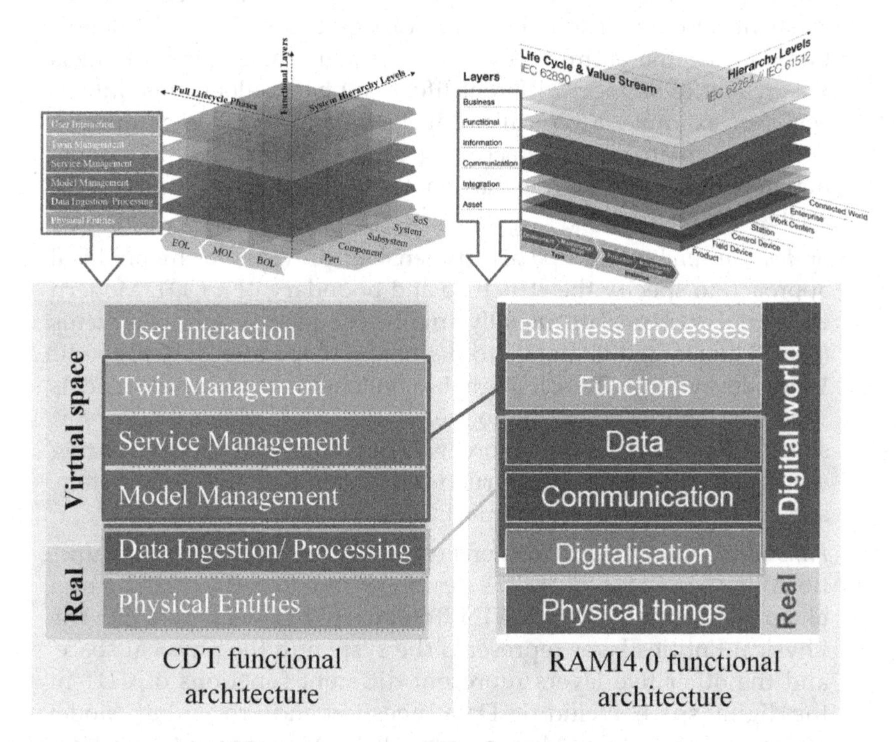

FIGURE 4.4
Comparison of functional architecture between Cognitive Digital Twin and RAMI4.0 [16]. (©: CC BY-NC-ND 4.0).

assets into the real world through the digitalization layer [20]. In contrast, the CDT *Functional Layers* focus more on the functions of the twins in the virtual space, which include Model Management, Service Management, and Twin Management layers. They are integrated with the physical entities through the Data Ingestion/Processing layer and interact with users through the User Interaction layer on the top.

4.4 CDT Applications

As an emerging concept, CDT has not yet been widely implemented and verified in the industry. Therefore, most published studies either explore the theoretical perspectives of CDT or focus on the CDT vision. However, several ongoing studies and projects aim to verify the CDT feasibility by applying it to different industry scenarios.

Zhou et al. [22] integrated dynamic knowledge bases with DT models to enable knowledge-based intelligent services for autonomous manufacturing. The knowledge-driven DT supports intelligent perceiving, simulating, understanding, predicting, optimizing, and controlling strategy. This approach was applied to three use cases: manufacturing process planning, production scheduling, and production process analysis. Moreover, Yitmen et al. [23] use CDTs to support building information management. Through CDTs, process optimization and decision-making are mainly supported based on knowledge graph modeling and reasoning across the entire building lifecycle.

The EU project COGNITWIN (Cognitive plants through proactive self-learning hybrid DTs)[1] is a dedicated project aiming to enhance the cognitive capabilities of existing process control systems, thus enabling self-organizing and offering solutions to unpredicted behaviors. The previously mentioned conceptual architecture, as shown in Figure 4.1, was applied to several use cases from the process industry [2].

The FACTLOG (Energy-aware Factory Analytics for Process Industries)[2] project aims to improve the cognition capabilities of complex process systems by combining data-driven and model-driven DTs. CDT is one of the main enabling technologies in this project. An application case of this project has been presented [24,25], where an actionable CT is applied to support demand forecasting and production planning in a European original equipment manufacturer in the automotive industry.

[1] https://www.sintef.no/projectweb/cognitwin/.

[2] https://www.factlog.eu/.

In another EU project QU4LITY (Digital Reality in Zero Defect Manufacturing),[3] CDT has been used as a semantically enhanced version of a DT to support autonomous quality. In order to improve product quality, a data model named RMPFQ (Resource, Material, Process, Feature/Function, Quality) [26,27] was adopted to support the ontology development. A CDT model is created for an aircraft assembly system enabling trade-offs for assembling schedules, data flow across model-based systems engineering, semantic models, system architectures, and co-simulation for verification [28,29].

4.5 Discussion

As an emerging concept, CDT is still in a very early stage of its development. Consequently, there are many challenges to be resolved in order to realize its vision fully. Some primary challenges for CDT development are summarized as follows, indicating potential opportunities for future studies:

1. *Knowledge management:* A functional and comprehensive knowledge base is the core of the cognitive capabilities of a CDT. The main challenges for CDT knowledge management can be categorized into knowledge representation, acquisition, and knowledge update [30].
2. *Integration of DT models:* Multiple DT models involved in a CDT implementation might be created separately by different stakeholders of a complex system corresponding to its different subsystems or components across the entire lifecycle. They need to be integrated and appropriately orchestrated into the CDT architecture enabled by the twin management layer. Moreover, stakeholders might adopt different standards, protocols, and structures for their DT models. This heterogeneity leads to the challenges of interoperability issues at the data level. Moreover, this brings the DT models with more complex features and services for integration.
3. *Standardization:* The standardization of DTs is the basis for enabling the CDT paradigm. Several Standards Developing Organizations (SDOs) are developing DT standards such as the ISO, W3C-WoT, the Industrial Internet Consortium (IIC), and the Plattform Industrie 4.0 [31]. It is not easy to unify and align the relevant DT standards developed by different SDOs. Some existing standards and protocols can be considered a substitute. For example, Plattform Industrie 4.0 provides the Asset Administration Shell (AAS) as a part of the

[3] https://qu4lity-project.eu/.

RAMI4.0 [20]. ETSI Industry Specification Group (ISG) proposes the Next Generation Service Interfaces-Linked Data (NGSI-LD) APIs [32]. The W3C-WoT working group proposes the WoT Thing Description (WoT TD), which is an official W3C Recommendation [33]. Such standards and APIs are discussed in previous studies [2,31].

4. *Implementation:* a CDT might include multiple physical systems covering several lifecycle phases with stakeholders from different enterprises. The implementation of CDTs thus requires both intra- and inter-organization collaborations. In addition to the above mentioned interoperability issue, it brings new challenges in project management, data privacy/security concerns, and intellectual property protection. Furthermore, the lack of successful demonstrators of CDT implementation further increases its risks. From a managerial perspective, the CDT development and implementation should be organized by experienced experts and follow proper project management methodologies.

Acknowledgement

The work presented in this paper has been partially supported by the EU H2020 project FACTLOG (869951) – Energy-aware Factory Analytics for Process Industries, and EU H2020 project QU4LITY (825030) – Digital Reality in Zero Defect Manufacturing.

Bibliography

[1] Stefan Boschert, Christoph Heinrich, and Roland Rosen. Next generation digital twin. In Proc. tmce, pages 209–218. Las Palmas de Gran Canaria, Spain, 2018.

[2] Sailesh Abburu, Arne J Berre, Michael Jacoby, Dumitru Roman, Ljiljana Stojanovic, and Nenad Stojanovic. Cognitwin-hybrid and cognitive digital twins for the process industry. In 2020 IEEE International Conference on Engineering, Technology and Innovation (ICE/ITMC). IEEE, 2020: 1–8.

[3] Fei Tao, He Zhang, Ang Liu, and AYC Nee. Digital Twin in industry: State-of-the-art. *IEEE Transactions on Industrial Informatics*, 15(4):2405–2415, 2019.

[4] Pavlos Eirinakis, Stavros Lounis, Stathis Plitsos, George Arampatzis, Kostas Kalaboukas, Klemen Kenda, Jinzhi Lu, Joze M Rozanec, and Nenad Stojanovic. Cognitive digital twins for resilience in production: A conceptual framework. *Information*, 13(1):33, 2022.

[5] Min Chen, Francisco Herrera, and Kai Hwang. Cognitive computing: Architecture, technologies and intelligent applications. *IEEE Access*, 6:19774–19783, 2018.

[6] RL Solso, MK MacLin, and OH MacLin. *Cognitive Psychology*. Pearson Education, New Zealand, 2005.

[7] Liqiao Xia, Pai Zheng, Xinyu Li, Robert X Gao, and Lihui Wang. Toward cognitive predictive maintenance: A survey of graph-based approaches. *Journal of Manufacturing Systems*, 64:107–120, 2022.

[8] Mohammad Abdullah Al Faruque, Deepan Muthirayan, Shih-Yuan Yu, and Pramod P Khargonekar. Cognitive digital twin for manufacturing systems. In 2021 Design, Automation & Test in Europe Conference & Exhibition (DATE). IEEE, 2021: 440–445.

[9] Ahmed El Adl. The cognitive digital twins: Vision, architecture framework and categories. Technical report, 2016. https://www.slideshare.net/slideshow/embed_code/key/JB60Xqcn.

[10] Fariz Saracevic. Cognitive digital twin. Technical report. https://www.slideshare. net/BosniaAgile/cognitive-digital-twin, 2017.

[11] Felipe Fernandez, Angel Sanchez, Jose F Velez, and A Belen Moreno. Symbiotic autonomous systems with consciousness using digital twins. From Bioinspired Systems and Biomedical Applications to Machine Learning: 8th International Work-Conference on the Interplay Between Natural and Artificial Computation, IWINAC 2019, Almería, Spain, June 3–7, 2019, Proceedings, Part II 8. Springer International Publishing, 2019: 23–32.

[12] Jinzhi Lu, Xiaochen Zheng, Ali Gharaei, Kostas Kalaboukas, and Dimitris Kiritsis. Proceedings of 5th International Conference on the Industry 4.0 Model for Advanced Manufacturing: AMP 2020. Springer International Publishing, 2020: 105–115.

[13] Lu Jinzhi, Yang Zhaorui, Zheng Xiaochen, Wang Jian, and Kiritsis Dimitris. Exploring the concept of cognitive digital twin from model-based systems engineering perspective. *The International Journal of Advanced Manufacturing Technology*, 2022, 121(9-10): 5835–5854.

[14] Pavlos Eirinakis, Kostas Kalaboukas, Stavros Lounis, Ioannis Mourtos, Joze M Rozanec, Nenad Stojanovic, and Georgios Zois. Enhancing cognition for digital twins. 2020 IEEE International Conference on Engineering, Technology and Innovation (ICE/ITMC). IEEE, 2020: 1–7.

[15] Muhammad Intizar Ali, Pankesh Patel, John G Breslin, Ramy Harik, and Amit Sheth. Cognitive digital twins for smart manufacturing. In *IEEE Intelligent Systems*, 2021, 36(2): 96–100.

[16] Xiaochen Zheng, Jinzhi Lu, and Dimitris Kiritsis. The emergence of cognitive digital twin: Vision, challenges and opportunities. *International Journal of Production Research*, 2022, 60(24): 7610–7632.

[17] Jean Duprez. Approach to structure, formalize and map MBSE metamodels and semantic rules. *INCOSE International Symposium*, 29(1):22–36, 2019.

[18] ISO 23247-1. ISO 23247-1:2021 automation systems and integration - digital twin framework for manufacturing – Part 1: Overview and general principles. Technical report, 2021. ISO 23247-1:2021(en), Automation systems and integration — Digital twin framework for manufacturing — Part 1: Overview and general principles

[19] Joze M Rozanec, Jinzhi Lu, Jan Rupnik, Maja Skrjanc, Dunja Mladenic, Blaz Fortuna, Xiaochen Zheng, and Dimitris Kiritsis. Actionable cognitive twins for decision making in manufacturing. *International Journal of Production Research*, 60(2):452–478, 2022.

[20] Karsten Schweichhart. *Reference Architectural Model Industrie 4.0 (rami 4.0). An Introduction*. 2016. Available online: http://www.plattform-i40.de/I40/Navigation/EN/InPractice/Online-Library/online-library.html.

[21] Alexander Kossia*koff, William N Sweet, et al. Systems Engineering: Principles and Practices. John Wiley & Sons, 2020.*

[22] Guanghui Zhou, Chao Zhang, Zhi Li, Kai Ding, and Chuang Wang. Knowledge-driven digital twin manufacturing cell towards intelligent manufacturing. *International Journal of Production Research*, 58(4):1034–1051, 2020.

[23] Ibrahim Yitmen, Sepehr Alizadehsalehi, İlknur Akıner, and Muhammed Ernur Akıner. An adapted model of cognitive digital twins for building lifecycle management. *Applied Sciences*, 2021, 11(9): 4276.

[24] Joze M Rozanec, Jinzhi Lu, Jan Rupnik, Majà Skrjanc, Dunja Mladenic, Blaz Fortuna, Xiaochen Zheng, and Dimitris Kiritsis. Actionable cognitive twins for decision making in manufacturing. arXiv 2021." arXiv preprint arXiv:2103.12854..

[25] Joze M Rozanec and Lu Jinzhi. Towards actionable cognitive digital twins for manufacturing. In *International Workshop on Semantic Digital Twins Co-located with ESWC. SeDiT@ ESWC*, 2020, 2615: 1–12.

[26] Xiaochen Zheng, Foivos Psarommatis, Pierluigi Petrali, Claudio Turrin, Jinzhi Lu, and Dimitris Kiritsis. A quality-oriented digital twin modelling method for manufacturing processes based on a multi-agent architecture. *Procedia Manufacturing*, 51:309–315, 2020.

[27] Xiaochen Zheng, Pierluigi Petrali, Jinzhi Lu, Claudio Turrin, and Dimitris Kiritsis. RMPFQ: A quality-oriented knowledge modelling method for manufacturing systems towards cognitive digital twins. *Frontiers in Manufacturing Techno*logy, 2022, 2: 901364.

[28] Xiaochen Zheng, Xiaodu Hu, Rebeca Arista, Jinzhi Lu, Jyri Sorvari, Joachim Lentes, Fernando Ubis, and Dimitris Kiritsis. A semantic-driven tradespace framework to accelerate aircraft manufacturing system design. *Journal of Intelligent Manufactu*ring, 2022: 1–24.

[29] Rebeca Arista, Xiaochen Zheng, Jinzhi Lu, and Fernando Mas. An ontology-based engineering system to support aircraft manufacturing system design. *Journal of Manufacturing Systems*, 68:270–288, 2023.

[30] Sailesh Abburu, Arne J Berre, Michael Jacoby, Dumitru Roman, Ljiljana Stojanovic, and Nenad Stojanovic. Cognitive digital twins for the process industry. In *Proceedings of the The Twelfth International Conference on Advanced Cognitive Technologies and Applications (COGNITIVE 2020)*, Nice, France. 2020: 25–29.

[31] Michael Jacoby and Thomas Uslander. Digital twin and internet of things-current standards landscape. *Applied Sciences*, 10(18):6519, 2020.

[32] Alliance, Open Mobile. Next Generation Service Interfaces Architecture. Tech. rep. May, 2010.

[33] W3C-WoT. Web of things (wot) thing description - w3c recommendation 9 April 2020. Technical report, 2020. Web of Things (WoT) Thing Description (w3.org).

5

Structural Integrity Preservation of Built Cultural Heritage: How Can Digital Twins Help?

Annalaura Vuoto
University of Minho Campus de Azurém

Marco Francesco Funari
University of Surrey

Paulo B. Lourenço
University of Minho Campus de Azurém

5.1 Introduction

Digital transformation is spreading in several areas, ranging from the industrial production of mechanical components to infrastructure monitoring and maintenance. Nevertheless, the Architecture, Engineering, Construction and Operation (AECO) industry is one of the most reluctant industries to introduce a comprehensive digitalisation process (European Commission 2021). Figure 5.1 reports a helpful classification of the leading digital technologies currently adopted in the construction sector, performed by the European Construction Sector Observatory (ESCO) (European Commission 2021).

The technologies are organised into three categories: data acquisition, automating processes and digital information and analysis. Data acquisition has become a relatively simple task thanks to the rapid development of related technologies. The actual use of digital data represents the future of digitalisation which should be performed by employing state-of-the-art automatic processes. Furthermore, by automating certain activities, the final quality of the project increases, workers are also less exposed to risks, and new materials and techniques can be deployed. Hence, the digital information and analysis category is paramount in connecting innovative technologies and processing available data. In this framework, Digital Twin (DT) concept is

 DOI: 10.1201/9781003425724-7

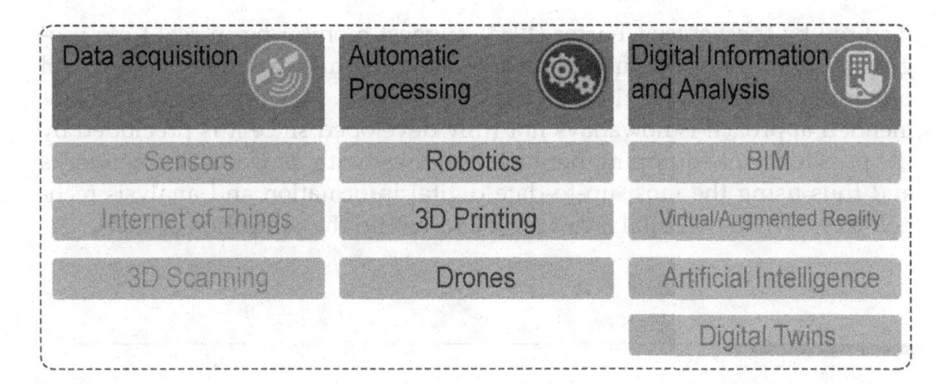

FIGURE 5.1
Three categories of digital technologies in construction. Adapted from the European Commission (2021).

intended as the highest expression of the digital information and analysis management processes.

However, the DT paradigm is often erroneously understood as a simple geometric representation of a specific asset resulting in a tendentious concept overlapping with Computer Aided Design (CAD) and Building Information Modeling (BIM) models. One should note how, while CAD models are static representations of shapes, the ability to include dynamic data, i.e., forces and loads, temperature variations, and rate-dependent phenomena, is typical of a DT (Grieves and Vickers 2016). This concept is relevant for a correct understanding of the DT, which must be understood primarily as a computational model to perform analyses and predictions related to the behaviour of physical assets when subject to boundary conditions changes rather than as a simple geometrical representation.

Researchers and practitioners agree about the potential key role that a correct implementation of the DT paradigm in the AECO domain could play, particularly in enhancing the development of sustainable management of built assets. In fact, the building industry is responsible for a tremendous amount of carbon emissions, causing serious environmental issues. Moreover, such a trend will be boosted by the expected increase of the world population within the next decades, resulting in exceptional housing and energy demand. Therefore, extending the life of existing structures and infrastructures will materially reduce carbon-consuming activities. In this perspective, Built Cultural Heritage (BCH) has to be considered a resource for achieving new sustainable solutions since it offers an immense potential to support the transition towards a healthier, greener, fairer society and economy (Europa Nostra 2021). In this context, preservation of their structural integrity is a key element for the heritage buildings' re-use as it ensures both the safety of the occupiers and the preservation of the historical and cultural values. In order to preserve heritage buildings, state-of-the-art recommendations, inspired by the Venice Charter (1964), suggest monitoring in real-time progressive damage of existing structures, avoiding massive interventions and providing immediate action in case of a disaster. However, such a recommended approach is nowadays not fully developed since it is precluded by the possibility of equipping heritage buildings with dense sensor networks and thus using the most up-to-date digital information and analysis technologies, Which potential is not exploited within the protection of BCH.

5.2 DT Definition in the AECO Domain

Although DT is increasingly gaining attention and relevance in civil engineering, the available literature is still insufficient to acquire appropriate knowledge of such technology (Camposano, Smolander, and Ruippo 2021). One should note that, unlike other domains, the implementation of the DT paradigm in the civil engineering field has been conditioned by the pre-existence of the BIM paradigm, which has gained massive acceptance over the last two decades, especially for the Operational and Maintenance (O&M) phase of modern buildings. In fact, the literature analysis underlines how DT and BIM are technologies frequently and mistakenly overlapped (Daniotti et al. 2022). The majority of the analysed review papers regarding DT implementation in the civil engineering domain intend the DT concept as a BIM model that is progressively implemented during the building's lifecycle (see Figure 5.2). Moreover, the lack of scientific studies that appropriately address the DT concept, despite the extensive presence of the keyword 'Digital Twin' is an issue.

Camposano et al. (2021) reported opinions from the prominent representatives of the AECO industry within the scope to conceptualise the DT paradigm. DT was identified as a digital model (could be CAD or BIM). Such a visual representation is enhanced with real-time data to monitor the asset state by integrating statistics or artificial intelligence (AI) models to enable intelligent predictive capabilities. DT could ideally replace human intervention and autonomously determine action on behalf of the physical asset. They concluded that CAD or BIM digital models are generally considered descriptive, whereas DT is expected to be more dynamic, representing geometry and structural behaviour evolution under specific boundary conditions changes.

Deng et al. (2021) proposed a five-level ladder taxonomy reflecting the evolution concept from BIM to DT. Figure 5.3 represents their discretisation: Level 1 – BIM, Level 2 – BIM-supported simulations, Level 3 – BIM integrated with IoT, Level 4 – BIM integrated with AI techniques for predictions and Level 5 – ideal DTs. Each level is to be understood as an enhancement of the basic digitalisation technology, i.e., BIM, which allows for providing

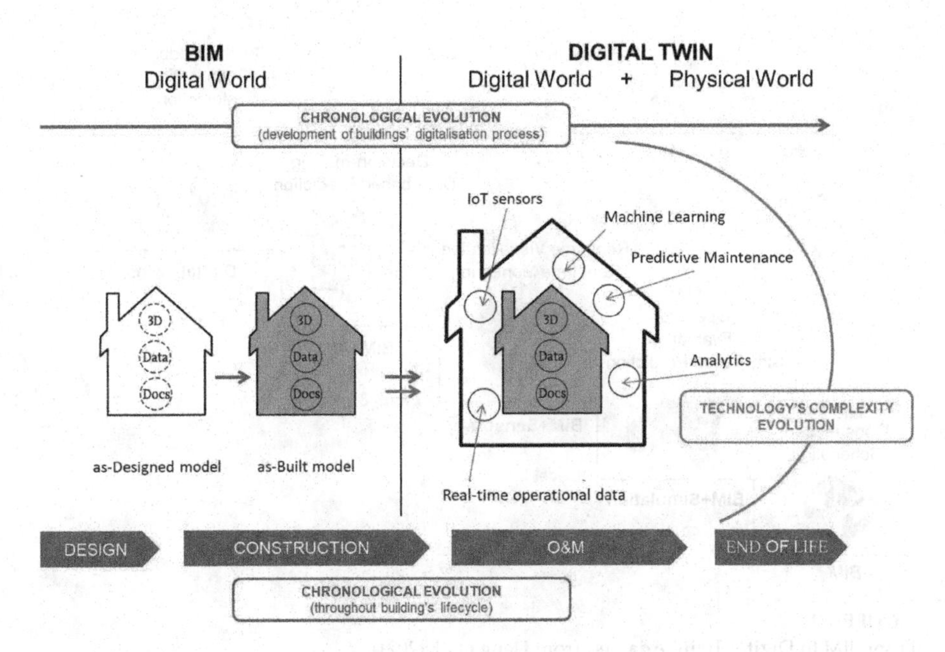

FIGURE 5.2
Digital Twin as an evolution of BIM.

ever-increasing services and benefits. According to the authors, an ideal DT contains real-time monitoring and prediction capabilities and enables automated feedback control for the adjustment of building parameters when necessary. Deng et al. (2021) also identified specific implementations of DTs in building projects based on both the area of application, i.e., research domain, and the level of complexity of the employed technology, according to the levels shown in Figure 5.3. As regards BIM models used as a support for simulations (Level 2), they found implementations for: (1) Construction Process Simulations, (2) Energy Performance Evaluation and (3) Thermal Environment Evaluation. In addition, applications for assessing embodied carbon, lighting, and acoustics were also referenced. Moreover, BIM models integrated with IoT techniques (Level 3) are used for: (1) Construction Process Monitoring, (2) Energy Performance Management, (3) Indoor Envi ronment Monitoring, (4) Indoor Thermal Comfort, (5) Space Management, (6) Hazards Monitoring and (7) Community Monitoring. Lastly, when real-time predictions are added to BIM models and IoT sensors (Level 4), applications are found for: (1) Construction Process Prediction, (2) Thermal Comfort Prediction, (3) Management of indoor systems, (4) Life Cycle Cost (LCC) of buildings' facilities, and (5) Smart Cities DT.

Other authors correctly treated DT and BIM as independent concepts, underlining differences in terms of final utility for the users. While BIM is suitable for integrating cost estimation and time schedule data to enhance

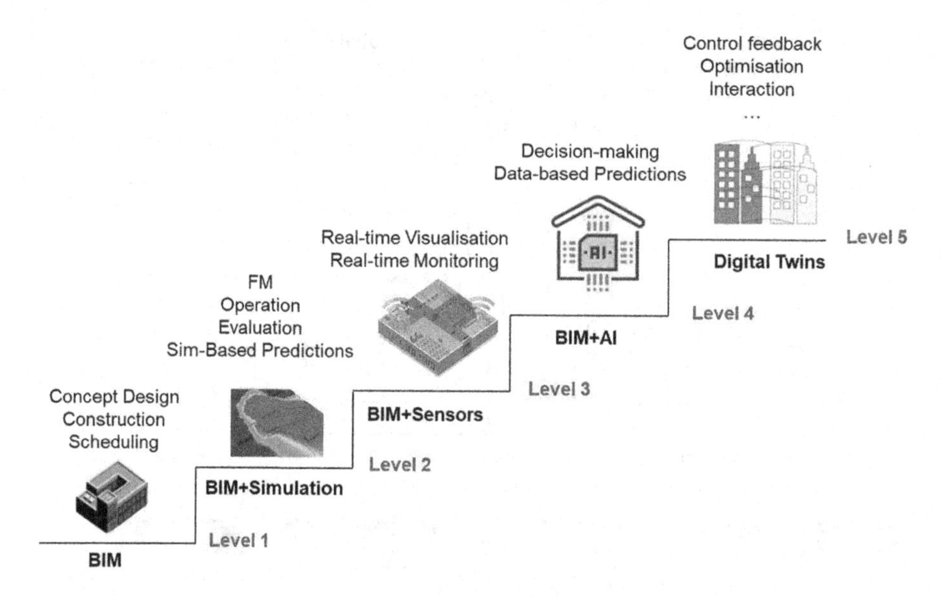

FIGURE 5.3
From BIM to Digital Twin. Adapted from Deng et al. (2021).

the efficiency of a construction project, DT is designed to integrate real-time sensor or data readings and data analysis tools resulting in a boosted interaction between the building and the environment and users.

Sacks et al. (2020) stated that the DT should not be viewed simply as a logical progression from BIM or as an extension of BIM tools integrated with sensing and monitoring technologies. Instead, DT is a holistic model that prioritises closing the control loops by basing management decisions on reliable, accurate, thorough and timeous information. A further consideration was made by the authors, who noted how BIM tools provide excellent product design representations using object-oriented vector graphics, which is not ideal for incorporating the raster graphics of point clouds acquired through the scanning physical assets. This is particularly inconvenient for the implementation of DT concept for existing buildings. In fact, an object-based approach is often not the most suitable for their representation, whereas a reality-based approach realised by remote sensor acquisitions could be more effective.

5.2.1 DT Implementations for the Structural Integrity Preservation of Heritage Buildings

The analysis of the existing literature related to DT paradigm implementation in the AECO domain revealed that limited attempts had been developed so far in the context of heritage buildings. Indeed, most of the existing applications were conceived for built infrastructures rather than heritage buildings. In addition, most techniques within the DT paradigm were tested only across specific areas of interest, mainly related to Facility Management. Extending the search to the structural engineering framework, Chiachío et al. (2022) recently reviewed relevant contributions in this field. From 2018 to 2022, only 22 relevant studies were found, most of them addressing structural monitoring and maintenance issues of infrastructures (mainly bridges). According to their analysis, most of the considered research papers mainly focus on the DT model generation phase, consisting of the definition of a model that provides a geometrical representation of the physical assets (Lu and Brilakis 2019).

Few studies address key features of the DT paradigm, such as real-time updating and learning of the virtual models by exploiting the data gathered from the physical counterpart or automated decision-making. Specifically, Jiang et al. (2022) presented a DT-driven framework for fatigue lifecycle management of steel bridges. They proposed a probabilistic multiscale fatigue deterioration model to predict the entire fatigue process, derive maintenance and timely update inspection/monitoring planning.

However, the main DT concept's gap within the structural engineering community is the lack of paradigm definition. Only two review articles defined a proper "structural DT" framework. Chiachío et al. (2022) provided a functional DT technology integration framework within the structural O&M context, suitable for simulations, learning and management. This framework is further validated in a proof of concept using a laboratory-scale test structure monitored using IoT sensors and actuators.

Regarding the implementation of the DT paradigm for BCH structural integrity preservation, no studies deal with the definition of a suitable framework. Some refer to the DT paradigm but only address specific and limited aspects of the technology's holistic character (Funari et al. 2021a, Funari et al. 2021b). The most recurrent applications of the DT paradigm in the field of BCH conservation found in the literature are summarised in Table 5.1 and discussed as follows.

Several works address the application of DT principles using heritage building information model(HBIM) models as a digital replica. Some authors refer

TABLE. 5.1

Application of DT paradigm for BCH conservation

Application		Type of model
Graphical representation		HBIM model
Information storage and data structuring		HBIM model
Management planning for preventive conservation		HBIM model
Digital fruition of BCH and museums		3D model /Augmented Reality
Behaviour simulation	Environmental conditions	Numerical model
	Structural conditions	
Structural Health Monitoring (SHM)		Numerical model Surrogate model

to DT technology as an HBIM model used essentially for graphical representation and storing information and data to increase knowledge and record all the happening changes, which are crucial tasks for BCH conservation. For example, Mol et al. (2020) constructed HBIM models of heritage timber structures to collect geometrical information and data from non-destructive testing (NDT) for real-time assessment of eventual changes in structures' conditions. Patrucco et al. (2022) structured an information system based on an HBIM model of a modern architecture heritage building to store information from NDT tests and on-site investigations. These informed models support preventive conservation policies and effective decision-making within the maintenance programming of BCH.

A further area where DT implementation is explored concerns the digital fruition of BCH and museums. Gabellone (2022) developed a distance visit approach to an underground historical oil mill in Gallipoli (Puglia). According to the DT perspective, Covid-19 recommends the development of immersive platforms to provide virtual tours assisted by real remote wizards and make it possible by reconstructing stereoscopic scenes..

The creation and visualisation of virtual models and data storage are key features of a DT. However, the applications found in the literature did not extensively investigate structural simulation aspects. Zhang et al. (2022) established an underground heritage site preservation mechanism through a dynamic ventilation system based on DT technology to control the relative humidity. They used an HBIM model to develop a web-based platform coupled with IoT sensors for real-time and dynamic control of the overall relative humidity level. It provides a preservation mechanism through a dynamic ventilation system based on DT technology for daily environmental control, operation and maintenance of ventilation equipment and emergency response.

As regards structural integrity preservation of heritage buildings, Angjeliu et al. (2020) developed a DT application for the Duomo of Milan. Their workflow involves three steps: (1) inputs (data collection and classification); (2) simulation model generation; and (3) simulation model calibration and validation.

For the sake of clarity, their DT model is a FE model implemented by following the hierarchy concept based on construction phases sequence and load path. Structural analysis was performed to predict both the dynamic response and damage under certain conditions. Although following the well-known steps of structural assessment through numerical methods, this work first attempted to frame them into a DT paradigm by investigating which are the key features of a DT simulation model and developing a tailored workflow for BCH structures.

A similar study was developed by Funari et al. (2021a, 2021b), which proposed a visual program algorithm for the DT generation of heritage buildings. Such an approach has been adopted to model a Portuguese monument for which a FE model has been generated and validated by implementing Operational Modal Analysis (OMA).

Other studies addressed the importance of including SHM as part of DT implementation since it can concretely address the real-time connection between physical and virtual assets (Kita et al. 2021).

5.3 DT Concept Prototyping for the Heritage Buildings' Structural Integrity Preservation

The literature analysis highlights the practical possibilities and potentials of the DT paradigm for structural integrity preservation but also reports difficulties and concerns for achieving its full implementation. As mentioned earlier, it is paramount to identify a unified definition for the DT concept attesting to its validity in a cross-domain panorama. However, it emerges how DT definition is strongly domain-dependent, meaning that what has been proven effectively working in other engineering-based fields could fail within the BCH structural integrity preservation. As well known, the DT paradigm was initially developed in aerospace (Glaessgen and Stargel 2012) and manufacturing domains, characterised by a high level of knowledge and control of the physical assets. In these specific fields, the validated procedures for design and testing allow a high level of asset knowledge, as well as of the individual components and their assembling. A real-time bi-directional correspondence between the physical asset and its digital counterpart can be meaningful in these cases. Such ideal conditions cannot occur in the AECO industry and even less in the BCH conservation subdomain. Indeed, structural engineers working within the architectural heritage's conservation must accept the condition to perform safety evaluations having limited knowledge of the assets and thus handling a considerable number of uncertainties arising from material characterisation, geometrical definition, understanding of the boundary and loading conditions and interaction with the environment and humans. It is because heritage buildings were built following the so-called rule of thumb and thus without responding to any specific codes and standards. Even though one would like to reach a higher knowledge level of the built assets, it could not be economical because of the considerable cost of hi-tech surveying and testing approaches.

A further relevant difference between industrial artefacts and BCH is the intensity of the structural response when subjected to specific loading conditions. While the industrial artefacts (as well as modern buildings) are designed with the scope to generate ductile systems, historical buildings were typically designed to resist gravity loads.

One should note that damage and degradation in historical buildings are long-time processes that occur over centuries; instead, if we consider exceptional actions, i.e., earthquakes or hurricanes, they are characterised by a very low probability of happening and dramatic consequences for our economy and society. Bearing in mind the concepts mentioned above, the main consequence is that real-time data exchange between a physical asset and its digital replica might be useless since the physical asset probably might not have a good ductility level. It would result in the incapability of the digital counterpart to simulate the ongoing damage and thus make a decision before the near collapse phase approaches. Furthermore, one should note how such an approach would not be economically sustainable on a large scale since the massive presence of heritage buildings populating our cities, particularly in Europe. Therefore, in our opinion, the most rational application of the DT paradigm would be to develop digital tools and numerical models able to simulate several scenarios ensuring the capability of performing a large number of computations (in a short time), taking into account the variability of some parameters. Thus, computational results must be stored and analysed using statistical tools to develop engineering charts and/or decision-making tools that would drive stakeholders to a rational approach. Ideally, such predictive tools must be integrated with NDT testing, performed with increasing levels of accuracy and detail in order to optimise economical resources for the structure's testing.

The following subsection discusses some insights about imperative components that must be included in a DT project: geometrical, computational modelling and predictive tools.

5.3.1 DT Creation: Geometrical Model Generation and Computational Strategies

The geometrical definition of the digital model is the basis of the DT concept implementation. Creating such models is often expensive and time-consuming; hence, several authors are committed to developing tools for the automation of 3D modelling, starting from raw geometrical data acquisition (Fortunato et al. 2017; Castellazzi et al. 2022).

Despite considerable advances in computational engineering, one of the main difficulties associated with numerical modelling is the laborious compatibility between 3D geometries modelled in CAD systems and numerical simulators. It is estimated that 80% of the total time spent on numerical simulations goes into cleaning the geometries and pre-processing, and only the remaining 20% is employed for actual simulation and analysis (Rasheed, San, and Kvamsdal 2020). Thanks to the evolution of geomatics methodologies, several solutions are available nowadays for the generation of refined models of real-world structures, exploiting either automatic or semi-automatic meshing of the point clouds and resorting to manual or parametric modelling approaches.

A second imperative component is the computational modelling strategy. One should note how such a choice depends on the specific scenario one would simulate. The taxonomies proposed by D'altri et al. (2020) are here taken into account. Four main approaches to numerical modelling masonry

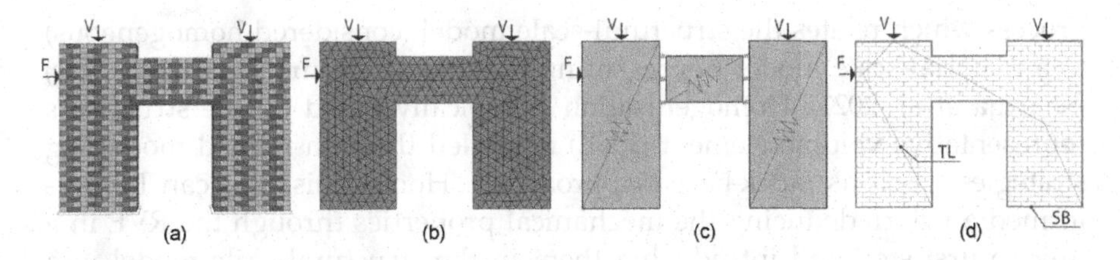

FIGURE 5.4
Numerical modelling strategies: (a) block-based model; (b) continuum model; (c) macroelement model; and (d) geometry-based model.

structures were identified: (1) block-based models, (2) continuum models, (3) macroelement models and (4) geometry-based models.

Block-based models can account for the actual masonry texture since masonry is modelled following the unit-by-unit representation (see Figure 5.4a). Block-based approaches consider rigid or deformable blocks interacting according to a frictional or cohesive-frictional contact modelling (Ravi Prakash et al. 2020). Three main strategies can be grouped within this sub-class: (1) distinct element method (DEM) (Sarhosis et al. 2016), which was introduced by Cundall and Strack (1979); (2) discontinuous deformation analysis (DDA) (Shi 1992), which takes into account the deformability of blocks and fulfils the assumption of no tension between blocks and no penetration of one block into another; and (3) non-smooth contact dynamics (NSCD) method (Moreau 1988), characterised by a direct contact formulation in its non-smooth form, implicit integrations schemes and energy dissipation due to blocks impact. In conclusion, block-based models provide the most accurate modelling strategy for masonry structures, representing actual masonry bonding and constructive details. For this reason, they are computationally demanding and thus more appropriate for small models and localised analysis (Lourenco 2019).

Continuum models consider masonry as a deformable continuum material. This approach, defined as macro-modelling in Lourenco (2019) and Roca et al. (2010), is largely employed for structural analysis of masonry structures (see Figure 5.4b). It is suitable for the global modelling of complex geometries involving massive elements like piers and walls combined with vaults and arches. Indeed, it does not require significant computational effort compared with the discontinued approach because the mesh discretisation does not have to describe the actual block-by-block discretisation (Degli Abbati et al. 2019; Szabó et al. 2021). The most relevant issue in this approach is related to the mechanical modelling of the material properties, which may be performed with: (1) direct approaches and (2) homogenisation procedures. Direct approaches are based on constitutive continuum laws that can represent masonry's overall behaviour. Such laws can be calibrated by means of experimental testing or available data (Milani et al. 2012). In homogenisation procedures, the constitutive law of the material is derived from a homogenisation process which relates the structural-scale model (considered homogeneous) to a material-scale model (representing the main masonry heterogeneities) (Sharma et al. 2021). Homogenisation is typically based on the structure's representative volume element (RVE) modelled through refined modelling strategies (such as block-based approaches). Homogenisation can be performed a priori, deducing the mechanical properties through the RVE in a unique first step and introducing them in the structural-scale model in a second step. Alternatively, multiscale approaches (Funari et al. 2022) allow for a stronger coupling between the material-scale and the structural-scale models in a step-by-step fashion.

In macroelement models, the structure is idealised into rigid or deformable panel-scale structural components. Typically, two main structural components may be identified: piers and spandrels (see Figure 5.4c). Macroelement models are the most widely diffused strategies for the seismic assessment of masonry structures, especially among practitioners, due to their simplified implementation and computation. On the other hand, they present some drawbacks. Firstly, decoupling in-plane and out-of-plane failure could lead to a conventional estimate of the seismic capacity, as in reality, out-of-plane and in-plane damages can simultaneously arise. Finally, the structure needs to be a priori idealised in piers and spandrels, and this could lead to the definition of a mechanical system that does not well represent the actual one, especially in the presence of very irregular opening layouts (Dolatshahi and Yekrangnia 2015; Pantò et al. 2017). Alternative formulations discretise the structure with macro distinct elements (Malomo and DeJong 2021).

Geometry-based models assume the structure's geometry as the only input, in addition to the loading condition (see Figure 5.4d). The structure is modelled as a rigid body whose structural equilibrium is assessed through limit analysis theorems, i.e., static or kinematic. Static theorem-based computational approaches can provide very useful outcomes for the investigation of the equilibrium states in masonry vaulted structures and also appear especially suitable for predicting the collapse mechanism (and the collapse multiplier) in complex masonry structures (Block, Ciblac, and Ochsendorf 2006). Kinematic theorem-based limit analysis approaches have been widely used in the last decades to effectively assess existing masonry buildings (Giuffrè 1991) since they allow a fast seismic assessment of masonry buildings through the analysis of local collapse mechanisms (Funari et al. 2020, 2021a, 2021b; Colombo et al. 2022).

All the above-discussed approaches have pros and cons. The choice to use a numerical modelling strategy rather than another depends on the type of analysis, data quality, time availability, etc. A DT, intended as a computational tool, should enable simulating different phenomena, and this will be achievable by seeking to integrate modelling strategies having a different level of complexity, from super simple and quick analytical models to super-complex multiscale or micro-modelling strategies (Lourenço et al. 2022).

5.3.2 Predictive Models for the Assessment of the Structural Behaviour of Heritage Buildings

One of the DT paradigm's peculiarities is the possibility of storing a large amount of information and processing it using technologies with high computational potential and sophistication. This allows for dealing with the uncertainty of the parameters that affect the structural performance of the physical assets. Such a feature of the DT paradigm appears worth investigating for the structural assessment of historical masonry buildings. Masonry components, namely, units and joints, show wide variability, depending on environmental factors and raw material availability on the construction site. In addition, construction technology and masons' skills affect the structures' local failure mechanisms and loading-bearing capacity (Vadalà et al. 2022; Szabó, Funari, and Lourenço 2023). Due to masonry's inherently non-linear and scattered material properties (Gonen et al. 2021), addressing all types of failure modes and providing accurate predictions of the structural response is challenging.

Due to their variability, masonry properties show a probabilistic nature. Therefore, probabilistic structural models that account for uncertainties and allow to analyse their effects on the response can be a reliable tool for problems associated with random variables and have been recently explored by a few authors (Pulatsu et al. 2022; Saloustros et al. 2019). Furthermore, hybrid models, coupling numerical and probabilistic analysis, can be generated to assess the structural response of masonry structures, including the effect of uncertainties. In addressing this type of structural engineering problem, a significant improvement could come from employing proper statistical or machine learning methods since they allow to enrich and relax the sets of assumptions that models make on data, enabling the creation of models over more complex and less understood scenarios.

5.4 Conclusions and Future Development

This chapter provides a synthesis of the potential innovation that the DT paradigm might bring within the structural integrity preservation of BCH. The imperative components that must be included for applying DT paradigm, i.e., geometrical model generation, suitable computational modelling strategies and the development of predictive tools implementing statistical and machine learning methods, were surveyed. Future research and innovation opportunities for a meaningful implementation of DTs in the context of BCH structural preservation can be identified as follows: (1) address a systematic comparison between the concepts of DT and HBIM to clearly define differences and interconnections among the two concepts; (2) define a methodological framework for the implementation of a structural DT; (3) develop

DT as a computational model for simulations and predictive purposes; and (4) increase practical implementations within the structural engineering field and conservation of BCH.

References

Angjeliu, Grigor, Dario Coronelli, and Giuliana Cardani. 2020. "Development of the Simulation Model for Digital Twin Applications in Historical Masonry Buildings: The Integration between Numerical and Experimental Reality." *Computers and Structures* 238 (October): 106282. https://doi.org/10.1016/j.compstruc.2020.106282.

Block, P, T Ciblac, and J Ochsendorf. 2006. "Real-Time Limit Analysis of Vaulted Masonry Buildings." *Computer and Structures* 84 (29–30): 1841–52.

Camposano, José Carlos, Kari Smolander, and Tuomas Ruippo. 2021. "Seven Metaphors to Understand Digital Twins of Built Assets." *IEEE Access* 9: 27167–81. https://doi.org/10.1109/ACCESS.2021.3058009.

Castellazzi, Giovanni, Nicolò Lo Presti, Antonio Maria D'Altri, and Stefano de Miranda. 2022. "Cloud2FEM: A Finite Element Mesh Generator Based on Point Clouds of Existing/Historical Structures." *SoftwareX* 18: 101099.

Chiachío, Manuel, María Megía, Juan Chiachío, Juan Fernandez, and María L. Jalón. 2022. "Structural Digital Twin Framework: Formulation and Technology Integration." *Automation in Construction* 140 (August): 104333. https://doi.org/10.1016/J.AUTCON.2022.104333.

Colombo, Carla, Nathanaël Savalle, Anjali Mehrotra, Marco Francesco Funari, and Paulo B Lourenço. 2022. "Experimental, Numerical and Analytical Investigations of Masonry Corners: Influence of the Horizontal Pseudo-Static Load Orientation." *Construction and Building Materials* 344: 127969. https://doi.org/10.1016/j.conbuildmat.2022.127969.

Cundall, P. A. and O. D. Strack. 1979. "A Discrete Numerical Model for Granular Assemblies." *Geotechnique* 29 (1): 47–65.

D'altri, Antonio Maria, Vasilis Sarhosis, Gabriele Milani, Jan Rots, Serena Cattari, Sergio Lagomarsino, Elio Sacco, Antonio Tralli, Giovanni Castellazzi, and Stefano De Miranda. 2020. "Modeling Strategies for the Computational Analysis of Unreinforced Masonry Structures: Review and Classification" *Archives of Computational Methods in Engineering* 27: 1153–85. https://doi.org/10.1007/s11831-019-09351-x.

Daniotti, Bruno, Alberto Pavan, Cecilia Bolognesi, Claudio Mirarchi, and Martina Signorini. 2022. "Digital Transformation in the Construction Sector: From BIM to Digital Twin." In *Digital Transformation*, edited by Antonella Petrillo, Fabio De Felice, Monica Violeta Achim and Nawazish Mirza, 13. Intech Open. https://doi.org/10.5772/intechopen.103726.

Degli Abbati, Stefania, Antonio Maria D'Altri, Daria Ottonelli, Giovanni Castellazzi, Serena Cattari, Stefano de Miranda, and Sergio Lagomarsino. 2019. "Seismic Assessment of Interacting Structural Units in Complex Historic Masonry Constructions by Nonlinear Static Analyses." *Computers & Structures* 213: 51–71.

Deng, Min, Carol C. Menassa, and Vineet R. Kamat. 2021. "From BIM to Digital Twins: A Systematic Review of the Evolution of Intelligent Building Representations in the AEC-FM Industry." *Journal of Information Technology in Construction* 26 (November 2020): 58–83. https://doi.org/10.36680/J.ITCON.2021.005.

Dolatshahi, K. M. and M. Yekrangnia. 2015. "Out-of-Plane Strength Reduction of Unreinforced Masonry Walls because of in-Plane Damages." *Earthquake Engineering & Structural Dynamics* 44 (13): 2157–76.

Europa Nostra. 2021. "European Cultural Heritage Green Paper: Executive Summary," no. March.

European Commission. 2021. "Digitalisation in the Construction Sector - Analytical Report European Construction Sector Observatory," no. April: 1–159.

Fortunato, Giuseppe, Marco Francesco Funari, and Paolo Lonetti. 2017. "Survey and Seismic Vulnerability Assessment of the Baptistery of San Giovanni in Tumba (Italy)." *Journal of Cultural Heritage* 26: 64–78. https://doi.org/10.1016/j.culher.2017.01.010.

Funari, Marco F., Luís C. Silva, Nathanael Savalle, and Paulo B. Lourenço. 2022. "A Concurrent Micro/Macro FE-Model Optimized with a Limit Analysis Tool for the Assessment of Dry-Joint Masonry Structures." *International Journal for Multiscale Computational Engineering* 26: 65–85. https://doi.org/10.1615/IntJMultCompEng.2021040212.

Funari, Marco Francesco, Ameer Emad Hajjat, Maria Giovanna Masciotta, Daniel V. Oliveira, and Paulo B. Lourenço. 2021. "A Parametric Scan-to-FEM Framework for the Digital Twin Generation of Historic Masonry Structures." *Sustainability* 2021a, 13 (19): 11088. https://doi.org/10.3390/SU131911088.

Funari, Marco Francesco, Anjali Mehrotra, and Paulo B. Lourenço. 2021b. "A Tool for the Rapid Seismic Assessment of Historic Masonry Structures Based on Limit Analysis Optimisation and Rocking Dynamics." *Applied Sciences* 11 (3): 942. https://doi.org/10.3390/app11030942.

Funari, Marco Francesco, Saverio Spadea, Paolo Lonetti, Francesco Fabbrocino, and Raimondo Luciano. 2020. "Visual Programming for Structural Assessment of Out-of-Plane Mechanisms in Historic Masonry Structures." *Journal of Building Engineering* 31: 101425. https://doi.org/10.1016/j.jobe.2020.101425.

Gabellone, Francesco. 2022. "Digital Twin : A New Perspective for Cultural Heritage Management and Fruition." *Acta IMEKO* 11 (1): 1–7.

Giuffrè, A. 1991. *Letture Sulla Meccanica Delle Murature Storiche*. Kappa.

Glaessgen, E. H., and D Stargel. 2012. "The Digital Twin Paradigm for Future NASA and US Air Force Vehicles." In *AAIA 53rd Structures, Structural Dynamics, and Materials Conference*, Honolulu, Hawaii.

Gonen, Semih, Bora Pulatsu, Serdar Soyoz, and Ece Erdogmus. 2021. "Stochastic Discontinuum Analysis of Unreinforced Masonry Walls: Lateral Capacity and Performance Assessments." *Engineering Structures* 238 (March): 112175. https://doi.org/10.1016/j.engstruct.2021.112175.

Grieves, Michael, and John Vickers. 2016. *Digital Twin: Mitigating Unpredictable, Undesirable Emergent Behavior in Complex Systems. Transdisciplinary Perspectives on Complex Systems: New Findings and Approaches*. Springer. https://doi.org/10.1007/978-3-319-38756-7.

Jiang, Fei, Youliang Ding, Yongsheng Song, Fangfang Geng, and Zhiwen Wang. 2022. "Digital Twin-Driven Framework for Fatigue Lifecycle Management of Steel Bridges Digital Twin-Driven Framework for Fatigue Lifecycle Management of Steel Bridges." *Structure and Infrastructure Engineering* 19 (12): 1826–46. https://doi.org/10.1080/15732479.2022.2058563.

Kita, Alban, Nicola Cavalagli, Ilaria Venanzi, and Filippo Ubertini. 2021. "A New Method for Earthquake-Induced Damage Identification in Historic Masonry Towers Combining OMA and IDA." *Bulletin of Earthquake Engineering* 19 (12): 5307–37.

Lourenco, Paulo B. 2019. "Computations on Historic Masonry Structures," no. July. https://doi.org/10.1002/pse.120.

Lourenço, Paulo B., M. F. Funari, and L. C. Silva. 2022. "Building Resilience and Masonry Structures: How Can Computational Modelling Help?" *Computational Modelling of Concrete and Concrete Structures*, edited by Günther Meschke, Bernhard Pichler, and Jan G. Rots, 30–37. CRC Press.

Lu, Ruodan, and Ioannis Brilakis. 2019. "Digital Twinning of Existing Reinforced Concrete Bridges from Labelled Point Clusters." *Automation in Construction* 105 (September): 102837. https://doi.org/10.1016/J.AUTCON.2019.102837.

Malomo, Daniele and Matthew J. DeJong. 2021. "A Macro-Distinct Element Model (M-DEM) for Simulating the in-Plane Cyclic Behavior of URM Structures." *Engineering Structures* 227 (April 2020): 111428. https://doi.org/10.1016/j.engstruct.2020.111428.

Milani, Gabriele, Siro Casolo, Andrea Naliato, and Antonio Tralli. 2012. "Seismic Assessment of a Medieval Masonry Tower in Northern Italy by Limit, Nonlinear Static, and Full Dynamic Analyses." *International Journal of Architectural Heritage* 6 (5): 489–524. https://doi.org/10.1080/15583058.2011.588987.

Mol, Alvaro, Manuel Cabaleiro, Hélder S. Sousa, and Jorge M. Branco. 2020. "HBIM for Storing Life-Cycle Data Regarding Decay and Damage in Existing Timber Structures." *Automation in Construction* 117 (September): 103262. https://doi.org/10.1016/J.AUTCON.2020.103262.

Moreau, JJ. 1988. "Unilateral Contact and Dry Friction in Finite Freedom Dynamics." *Nonsmooth Mechanics and Applications*, edited by J. J. Moreau and P. D. Panagiotopoulos, 1–82. Springer.

Pantò, Bartolomeo, Linda Giresini, Mauro Sassu, and Ivo Caliò. 2017. "Non-Linear Modeling of Masonry Churches through a Discrete Macro-Element Approach." *Earthquake and Structures* 12 (2): 223–36. https://doi.org/10.12989/eas.2017.12.2.223.

Patrucco, G, S Perri, G Sammartano, E Fillia, I Matteini, E Lenticchia, R Ceravolo, Structural Health, and Concrete Assessment. 2022. "3D Models and Non-Destructive Investigations: Towards a Meeting in Digital Twins." *The International Archives of the Photogrammetry, Remote Sensing and Spatial Information Sciences* XLIII (June): 6–11.

Pulatsu, Bora, Semih Gonen, Fulvio Parisi, Ece Erdogmus, Kagan Tuncay, Marco Francesco Funari, and Paulo B Lourenço. 2022. "Probabilistic Approach to Assess URM Walls with Openings Using Discrete Rigid Block Analysis (D-RBA)." *Journal of Building Engineering* 61: 105269. https://doi.org/10.1016/j.jobe.2022.105269.

Rasheed, Adil, Omer San, and Trond Kvamsdal. 2020. "Digital Twin: Values, Challenges and Enablers From a Modeling Perspective." *IEEE Access* 8: 21980–2012. https://doi.org/10.1109/ACCESS.2020.2970143.

Ravi Prakash, P., Bora Pulatsu, Paulo B. Lourenço, Miguel Azenha, and João M. Pereira. 2020. "A Meso-Scale Discrete Element Method Framework to Simulate Thermo-Mechanical Failure of Concrete Subjected to Elevated Temperatures." *Engineering Fracture Mechanics* 239 (August): 107269. https://doi.org/10.1016/j.engfracmech.2020.107269.

Roca, Pere, Miguel Cervera, Giuseppe Gariup, Luca Pela', P Roca,·M Cervera,·G Gariup, M Cervera, G Gariup, and L Pela'. 2010. "Structural Analysis of Masonry Historical Constructions. Classical and Advanced Approaches." *Archives of Computational Methods in Engineering* 17: 299–325. https://doi.org/10.1007/s11831-010-9046-1.

Sacks, Rafael, Ioannis Brilakis, Ergo Pikas, Haiyan Sally Xie, and Mark Girolami. 2020. "Construction with Digital Twin Information Systems." *Data-Centric Engineering* 1 (6). https://doi.org/10.1017/DCE.2020.16.

Saloustros, Savvas, Luca Pelà, Francesca R. Contrafatto, Pere Roca, and Ioannis Petromichelakis. 2019. "Analytical Derivation of Seismic Fragility Curves for Historical Masonry Structures Based on Stochastic Analysis of Uncertain Material Parameters." *International Journal of Architectural Heritage* 13 (7): 1142–64. https://doi.org/10.1080/15583058.2019.1638992.

Sarhosis, V., K. Bagi, JV. Lemos, and G. Milani. 2016. *Computational Modeling of Masonry Structures Using the Discrete Element Method*. IGI Global.

Sharma, S., L. C. Silva, F. Graziotti, G. Magenes, and G. Milani. 2021. "Modelling the Experimental Seismic Out-of-Plane Two-Way Bending Response of Unreinforced Periodic Masonry Panels Using a Non-Linear Discrete Homogenized Strategy." *Engineering Structures* 242 (September): 112524. https://doi.org/10.1016/J.ENGSTRUCT.2021.112524.

Shi, G-H. 1992. "Discontinuous Deformation Analysis: A New Numerical Model for the Statics and Dynamics of Deformable Block Structures." *Engineering Computations* 9 (2): 157–68.

Szabó, Simon, Marco Francesco Funari, and Paulo B Lourenço. 2023. "Masonry Patterns' Influence on the Damage Assessment of URM Walls: Current and Future Trends." *Developments in the Built Environment* 13: 100119. https://doi.org/10.1016/j.dibe.2023.100119.

Szabó, Simon, Andrea Kövesdi, Zsolt Vasáros, Ágnes Csicsely, and Dezső Hegyi. 2021. "The Cause of Damage and Failure of the Mud-Brick Vault of the Khan in New-Gourna." *Engineering Failure Analysis* 128 (October): 105567. https://doi.org/10.1016/J.ENGFAILANAL.2021.105567.

Vadalà, Federica, Valeria Cusmano, Marco Francesco Funari, Ivo Caliò, and Paulo B. Lourenço. 2022. "On the Use of a Mesoscale Masonry Pattern Representation in Discrete Macro-Element Approach." *Journal of Building Engineering* 50: 104182. https://doi.org/10.1016/j.jobe.2022.104182.

Venice Charter. 1964. "International Charter for the Conservation and Restoration of Monuments and Sites." The Getty Converstaion Institute, Venice, Italy.

Zhang, Jiaying, Helen H. L. Kwok, Han Luo, Jimmy C. K. Tong, and Jack C. P. Cheng. 2022. "Automatic Relative Humidity Optimization in Underground Heritage Sites through Ventilation System Based on Digital Twins." *Building and Environment* 216: 108999. https://doi.org/10.1016/j.buildenv.2022.108999.

Part 3

Digital Twins Technology

6

Key Technologies of Digital Twins: A Model-Based Perspective

Serge P. Kovalyov
Russian Academy of Sciences

6.1 Introduction

Recall that a digital twin (DT) is a virtual copy of a physical asset that promptly reproduces and alters the state and behavior of the asset [1]. DT intensely employs Industry 4.0 data-driven artificial intelligence technologies and simulations for adaptive monitoring, forecasting, optimization, and controlling asset performance [2]. The most accurate is the DT initiated at the beginning of the asset lifecycle, developed incrementally together with the asset, and put into operation synchronously with it [3]. Computer-aided design (CAD) tools apply to create a DT at the design stage, long before the asset comes into existence. Such an early DT is useful as a virtual testbed for evaluating alternative design decisions and choosing the best one.

Moreover, the initial DT is synthesized fully automatically when advanced artificial intelligence technologies, known as generative design [4], are used. The generative design process consists of the following steps: deriving a multi-objective optimization problem from stakeholder requirements for the asset, the automatic search for the best solution to this problem, and the subsequent physical implementation of the solution through automatic manufacturing. The solution determines the initial DT, so the latter accurately reflects the physical asset at the beginning of their synchronous operation. Later in the lifecycle, the generative design technology can be invoked again to automate planning of an asset upgrade or renovation, with the design space narrowed to a suitable "neighborhood" of the up-to-date DT. For example, generative design has been successfully applied to dependable parts in mechanical engineering, using topological optimization and 3D printing.

We will employ a much more complex asset as a running example: a power energy system. DT-based energy control applications show promising results in load/generation forecasting, power flow analysis, optimal energy storage scheduling, fault diagnosis, electrical machine health monitoring, and other

DOI: 10.1201/9781003425724-9

relevant areas [5]. Yet, these applications are difficult to scale to real-world power grids and cost-effectively adapt to everyday operation and maintenance. A full-fledged grid DT requires high-end IT resources and a lot of skilled labor to set up: it has to promptly follow long-running multi-step processes involving numerous widely distributed disparate facilities and affected by hidden technical, social, and economic factors. For similar reasons, generative design technologies are much less productive in the power industry than in mechanical engineering [6].

Major difficulties stem from the complexity and diversity of models that have to be included in the DT to describe the complex asset properly [7]. Models differ from each other both horizontally, by parts/units they describe, and vertically, by concerns/viewpoints/aspects they represent. The integration and coherent operation of numerous disparate models tend to consume a huge amount of computer and human resources. In essence, the DT composition process has to reproduce the asset construction process digitally. Thus, DT development activities belong to the problem domain of model-based systems engineering (MBSE). Any technology should be largely model-based to apply for DT. Figure 6.1 shows relations between DT and MBSE [8].

To date, MBSE is a mature methodology supported by an advanced theoretical basis, a convincing success story, and powerful technologies and tools.

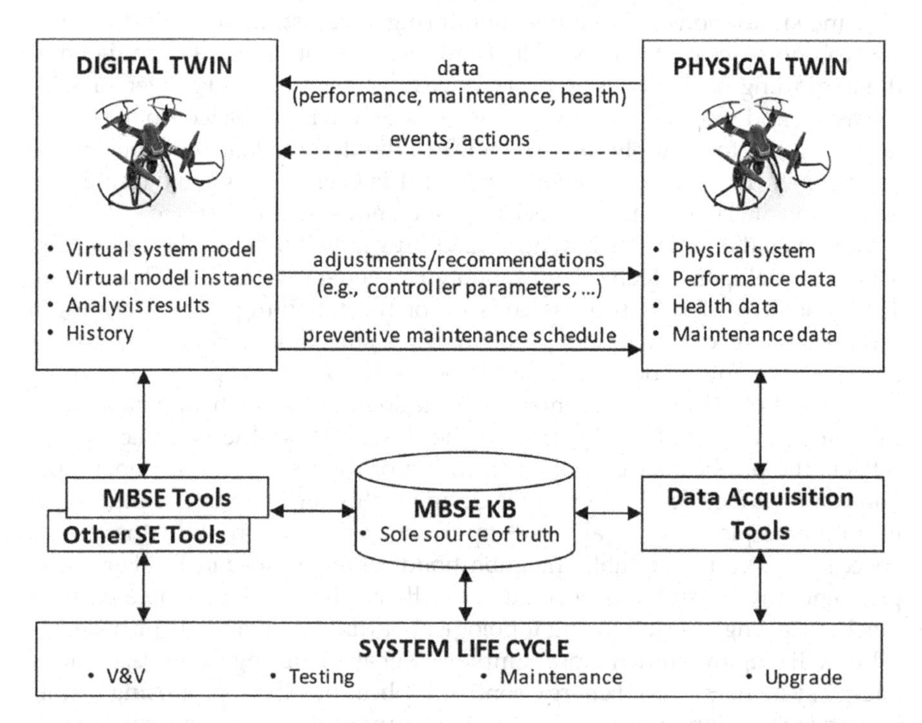

FIGURE 6.1
Relations between digital twin and model-based systems engineering [11].

To fight complexity, the heterogeneous domain is considered as a whole only at the first (architectural design) and the last (integration) phases of a typical development process. At the intermediate phases, models are developed in parallel using different languages and tools, oblivious of each other, yet ready for smooth integration via semantic mediation [9]. To facilitate rigorous description and verification of model composition and integration procedures, MBSE has recently obtained a formal foundation built upon the mathematical framework of category theory [10,11]. Indeed, category theory is a branch of higher algebra specifically aimed at the unified representation of objects of different nature and relations between them [12].

This chapter presents the key technologies that facilitate model-based design and development of the DT. We show that innovative digital technologies empowered by advanced algebraic techniques make DT cost-effective to compose and operate while maintaining sufficient completeness, accuracy, and usability for practical purposes.

This chapter is organized as follows. Section 6.2 presents the model-based DT architecture. Sections 6.3–6.8 provide overviews of the following key technologies: mathematical modeling, Internet of Things (IoT), interactive schematics and augmented reality (AR), electronic document management, master data management, and ontological modeling, respectively. Section 6.9 introduces an algebraic approach to DT composition based on category theory. Some conclusions and directions for further research are outlined in Section 6.10.

6.2 The Model-Based Digital Twin Architecture

Different approaches to the DT architectural decomposition exist [1]. For example, a layered architecture is proposed, with the physical asset on the lowest layer, intelligent control services on the upper layer, and various mediating components in-between. Alternatively, the functional decomposition applies to break DT down to data storage, information exchange, calculation, visualization, and other modules. In contrast, the MBSE perspective suggests decomposing by models that differ in kind and viewpoint, highlighting the specifics of DT among other digital systems. The typical DT consists of models of the following kinds [3]:

- ontology model;
- digital diagrams and visuals;
- electronic documentation;
- information model;
- real-time data;
- mathematical and simulation models.

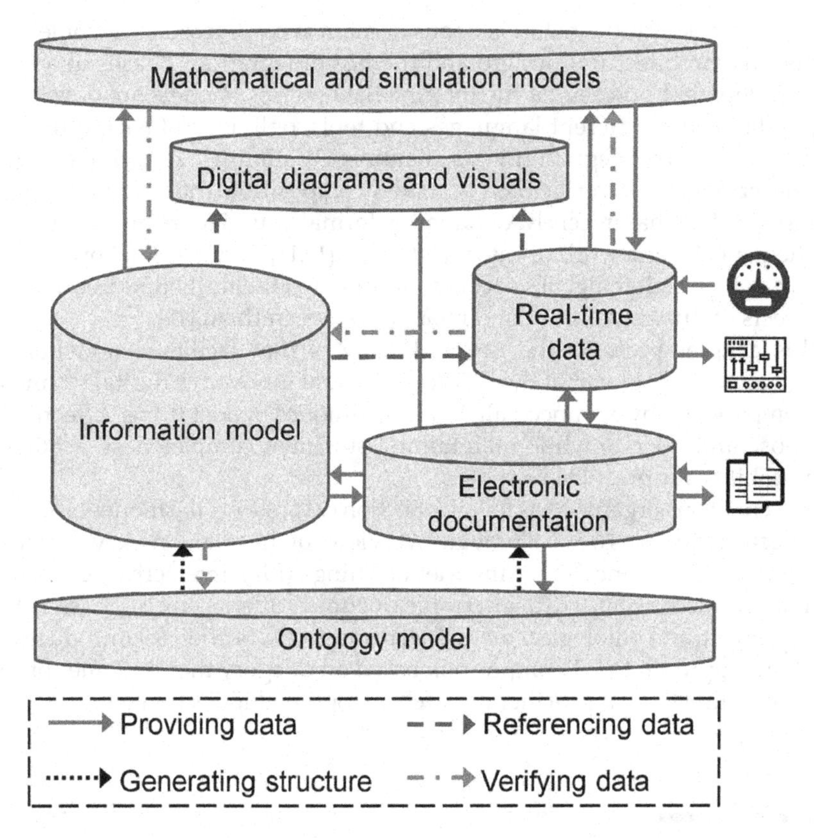

FIGURE 6.2
The model-based digital twin architecture.

To provide convenient access to the models, they are often designed as (micro) services [13]: the service-oriented architecture hides model implementation details from the user and provides flexibility in orchestrating the interaction between models. During the asset operation, the models interact intensively with each other and with the environment. Model interaction use cases include data exchange, data referencing, cross-model data verification, and data structure generation, as shown in Figure 6.2 and described in detail below.

6.3 Mathematical Modeling and Simulation

Mathematical (in a broad sense) models of an asset directly implement the purpose of its DT: analyze the asset operational state and determine control commands and/or recommendations necessary to bring it to a (sub)optimal state.

The analysis includes evaluation of unknown or missing state variables and forecasting variable values for some period into the future. The control actions are then determined by solving an optimization problem, targeted to maximize performance and/or minimize losses, over the asset state space.

For a more comprehensive analysis and control, models apply that are capable of simulating (i.e. imitating in a digital environment) the asset behavior in different operational scenarios. The extensive use of simulation is one of the most significant achievements of MBSE. At the design stage, simulation facilitates comparing alternative solutions for a particular design problem. Displayed as an interactive graphical animation, the simulation is useful as an instructional tool to train the asset personnel. When twinning complex multi-step processes, the DT employs simulation of a particular kind called discrete-event [14]. The discrete-event simulation represents the asset behavior as a discrete sequence of typed events that indicate elementary changes in the asset state and environment.

Many tools support composition and execution of mathematical models and simulations, ranging in scale from a smartphone app to a supercomputer suite. Small-scale models may be used for rapid prototyping of a DT or to twin a simple asset. Large-scale assets, like power systems, require more powerful tools.

Currently, there are three fundamentally different types of the models:

- "First-principles" (physical, biological, economical, and so on) models based on the numerical solution of equations and optimization problems that describe phenomena involving the asset;
- statistical models based on machine learning on the historical asset state and behavior data, including neural networks;
- asset expert knowledge and decision-making models, such as rule-based and multi-agent systems.

Numerical models based on solving discretized algebraic, differential, and stochastic equations have a long development history in computer-aided engineering (CAE) software. CAE tools include solvers and optimizers for mechanical and multiphysical stress analysis (using finite element method (FEM) or so-called mesh-free methods), computational fracture mechanics, multibody dynamics and kinematics, computational fluid dynamics, computational electromagnetics, and many other areas. Traditionally, CAE tools were intended for use in a design office to evaluate engineers' decisions, processing manually entered data under very weak time constraints. Recently, widely available computers reach performance sufficient to execute CAE algorithms in soft real time along with the asset operation, processing data that arrive from sensors. This allows using CAE in DT during the asset operation stage.

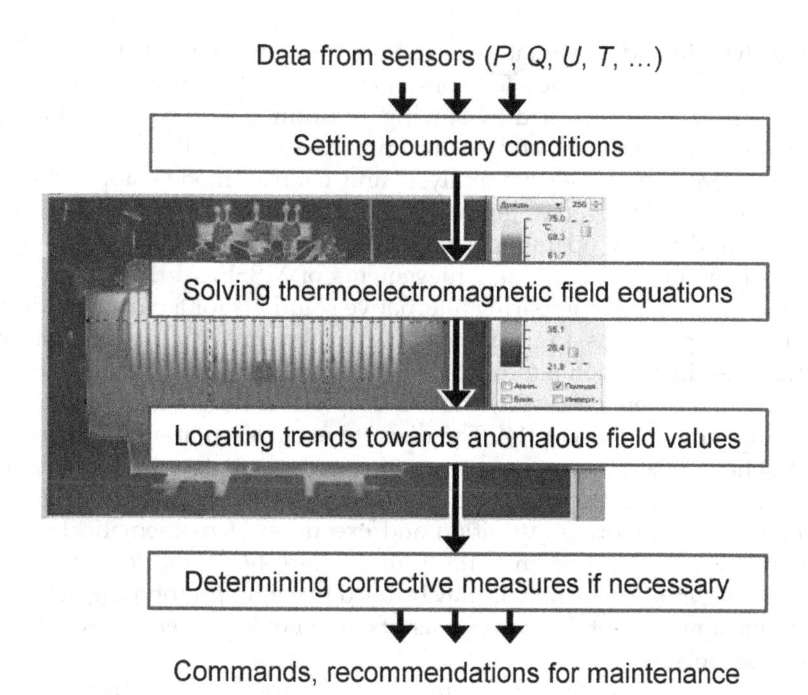

FIGURE 6.3
The power transformer model.

For example, Figure 6.3 shows a thermoelectromagnetic model of a power transformer [15]. The model combines high accuracy (more than 95%) with real-time calculation speed. Input data for the model include power, voltage, temperature, and properties of transformer oil, continuously measured by sensors installed on the transformer. The latter is approximated as a nonlinear magnetic circuit with dielectric placeholders imitating technological gaps. At any moment of time, the collected input data determine the boundary conditions and coefficients for field equations. The equations are then solved using FEM to calculate magnetic field and temperature distribution across the transformer body. Then, the post-processor compares the current field distribution with ones computed earlier, locates trends toward anomalous values, and determines (sub)optimal corrective actions if necessary: partial or total unloading of the transformer (to be done automatically) and/or maintenance procedures. The power transformer DT based on this model extends the transformer's useful life, increases safety, and reduces maintenance costs.

When the processes to be digitally twinned are governed by unknown rules (usually going beyond physics) and/or contain hidden volatile patterns, machine learning–based models apply. Such models make predictions or decisions, without being explicitly programmed, extrapolating from a training sample, viz. a set of asset behavior scenarios observed with different values of influencing factors (features). The most widely known machine learning models are based on artificial neural networks; other models include decision trees, Bayesian networks, and genetic algorithms. The accuracy of the model mainly depends on the recognition quality of the only relevant feature and the statistical quality of the sample..

For example, deep (multilayer) artificial neural networks apply in smart grid DT for energy load/generation/price forecasting, power machine health assessment and prediction, faults diagnosis, and other control procedures [16]. The neural network with an architecture tailored for time-series prediction, such as LSTM (Long Short-Term Memory), can forecast by only looking at the energy profiles, agnostic of the physics of generation and consumption processes. To improve forecasting accuracy and reduce uncertainty, the data sets are augmented with ancillary features such as energy market demand or weather conditions. To identify relevant features and generate rich training samples, high-end CAD-class software for offline grid simulation is useful. Good samples allow training the model to reach 95% or better accuracy.

The DT can contain models of yet another type based on expert knowledge presented in a machine-readable format. The large corpus of knowledge pertinent to monitoring and control is expressible as rules of the form "IF *condition* THEN *action*". The DT continuously evaluates such rules and immediately issues prescribed actions as soon as real-time data satisfy their conditions [17]. More complex decision-making scenarios involve multiple reasoning parties motivated by different and even conflicting intentions yet targeted to reach the consensus. Such parties are modeled as communicating software agents. For example, an agent represents each distributed energy resource (DER) in a microgrid targeting to maximize the DER economic effect, and the optimal grid operation mode is determined by consensus between the agents under the reliability and electrical stability constraints [18].

The above overview justifies equipping DT with several fundamentally different models of the same phenomena to calibrate and verify each other. Moreover, very deep intertwining of different models to increase accuracy is gaining popularity. For example, consider the so-called physics-informed neural network: its loss function, to be minimized by training, includes not only a purely statistical error but also some measure of violation of relevant physical laws. For a power system DT, such network can be trained to take a topology similar to the twinned grid. The network is capable of solving the power flow problem with acceptable accuracy even if electrical mode data are only available for a limited subset of grid nodes [19].

The main use cases for the mathematical models, hence for the DT as a whole, are:

- multicriteria evaluation and forecasting the asset operation and performance indicators;
- determining, optimizing, testing, and applying operation modes, commands, and control devices settings;

- predictive equipment health monitoring, assessment of failure rate, and maintenance needs;
- calibration and verification of models and control algorithms;
- validation and evaluation of design decisions;
- comprehensive virtual training of the asset operating personnel.

6.4 Internet of Things

The IoT feeds a DT with data that describe the actual state of an asset and environment in soft real time. Physical or virtual sensors produce the data: meters, controllers, audio and video recorders, human–machine interfaces, and all kinds of information systems. Accordingly, the data include time series of measured or calculated physical quantities, event logs, multimedia streams, economic indicators, and so on. For example, a power system DT collects electrical mode parameters (active and reactive power, current, voltage, phase, frequency) in the grid nodes, equipment state and health variables, telesignals, weather conditions, and energy market indicators.

Each collected data value is annotated with metadata, including the data type, the timestamp, and an identifier of the source (a reference to the appropriate information model element). Upon arrival at the DT, the real-time data are placed to the fast storage such as a time-series database (TSDB) and then delivered to other components for processing as needed.

IoT also closes the DT-based control loop, transporting commands and settings determined by mathematical models to asset controllers and actuators. Thus, IoT has no alternative as a handy technology for real-time bidirectional interaction between the asset and its DT. Yet, IoT has certain shortcomings as such, major of which are associated with security, complexity, and connectivity [20]. Indeed, IoT devices are attractive to hackers: they share factual data over the internet and are vulnerable since price limitations discourage investment in firmware-level security. At the same time, honest programmers have troubles composing a code to align, collate, and pre-process high dimensional irregular data arriving from numerous poorly interoperable sources.

Anyway, the DT employs real-time data according to the following major use cases:

- assessing the state of the asset as a whole as well as of its parts;
- detecting out-of-range values of the state variables, especially emergency indicators;
- feeding mathematical and simulation models with input data;
- collecting the observational history of the asset operation in different conditions;

- search for patterns in the asset's behavior;
- structural and parametric identification and verification of the information model;
- verification of mathematical and simulation models, including retro-forecasting;
- generation and verification of operation reports;
- visualization of state variables in tables, graphs, diagrams, and histograms.

6.5 Interactive Diagrams and Augmented Reality

In two dimensions, a diagram symbolically represents relationships between certain objects: spatial (a map or a layout), causal (a flow chart), taxonomical (a class inheritance hierarchy), partonomical (a product structure graph), teleological (a goal diagram), and any blend of these. In 3D, a geometric model displays a physical asset virtually as if it were real. A diagram in a digital format can be very large yet graspable due to zooming, browsing, rotating, and search capabilities. Numerous tools support the creation and displaying diagrams and visuals in a plethora of formats, from a smartphone diagram editor app to a CAD suite to a planet-scale geographic information system (GIS). To facilitate diagrammatic description of a complex asset in every aspect, MBSE promotes such tools as systems modeling language (SysML).

In an information-rich environment, a diagram can be made interactive by associating symbols with data that describe their denotata. Further, AR technologies display symbols and data right near the relevant parts of a real physical asset viewed through a digital camera. For example, diagrams and visuals of several types represent a power system: a single line diagram, an electrical plan of a building, a power transmission lines map, an equipment maintenance process flow chart, etc. When the diagrams are interactive, they display the current values of key mode parameters in grid nodes, highlighting the abnormal ones. AR improves the equipment maintenance and repair: consider the power transformer DT, described in Section 6.3, capable of highlighting places with an anomalous magnetic field or temperature by bright colors when a worker examines the transformer housing through a smartphone or a headset (cf. a background picture in Figure 6.3). The DT can also guide maintenance procedures necessary to fix anomalies [21].

The following use cases define how DT employs diagrams and visuals:

- compact user-friendly visual presentation of the asset structure and state as a symbolical image;

- (soft) real-time indication of places where important events occur;
- search for information about asset parts by topological criteria with respect to displayed relations.

6.6 Electronic Document Management

A document is a uniquely identified self-contained unit of information readable by a human or a machine. Each document has metadata associated with it: title, author(s), keywords, and so on. Various standards apply to define the metadata structure and semantics, preferably with references to the domain ontology [22]. For permanent storage, documents annotated with metadata are placed into an archive organized in a hierarchy of folders according to document subjects. The archive often has a paper backup, into which hard copies of electronic documents are printed and from which incoming paperwork is scanned. The paper backup helps to avoid losses and unauthorized changes. Yet, nowadays, another approach to prevent tampering with documents is available: the hash of the document is stored in a blockchain upon creation and verified upon access.

The following use cases are relevant for electronic documentation as a DT part:

- presentation of information about the asset in a self-contained semi-structured human-readable form;
- identification and verification of reasons and permissions to execute control actions;
- extraction of ontological concepts, relationships, and axioms;
- designing digital diagrams and visuals;
- updating and verification of the information model;
- updating and verification of real-time data;
- search for documents by context and metadata;
- collecting the documentation archive.

6.7 Master Data Management

Master data describe an asset structure and static properties, providing context for asset business transactions and operation (including parameters for mathematical models). Structured in the machine-readable form of entities, attributes, and relationships, the master data constitute the asset information model. Master data can be imported into the information model from the asset documentation as well as exported from the model to documents.

The most convenient storage for master data is a relational database. It provides a number of advantages, including data accuracy, integrity, flexibility, scalability, and security [23]. For example, consider a power system information model. Its core consists of a comprehensive database of installed power and control equipment: generators, transformers, electrical power consuming equipment, switches, controllers, telecommunication channels, and so on. The database tables form an inheritance hierarchy by equipment taxonomy. Each table has numerous class-specific attributes that describe the equipment in every aspect: identification, performance, operating conditions, ownership/rent, cost, control, maintenance, etc. Ancillary tables store information to which equipment database records refer: about energy facilities, grid topology, measurement, and control channels.

The main use cases for the information model as a DT component are:

- supplying mathematical and simulation models with master data;
- fine-grained binding of real-time data to the asset parts;
- assigning unique identifiers to parts;
- fine-grained part-wise data access control;
- collecting the history of asset changes during its lifecycle;
- verifying master data by comparing with the factual asset structure;
- attribute-based search for information about parts;
- generation and verification of the asset design and construction documentation;
- visualization of the asset structure and properties of parts in tables and hierarchy trees.

6.8 Ontological Modeling

A domain ontology is a widely recognized tool to improve interoperability between disparate models. An ontology formally describes the logical structure of the domain concepts, including their definitions, properties, and relationships with each other. When designing any model presented in the previous sections, the ontology applies as a semantically solid source of domain terms and knowledge.

The ontology is represented as a taxonomy tree of concepts that are endowed with properties of various types, associated by various relationships (equivalence, proximity, partonomy, etc.), constrained by various axioms, and instantiated in various individuals. It is possible to automatically infer nontrivial knowledge from axioms by applying deductive rules of the so-called descriptive logic. In a machine-readable form, the ontology is written on the OWL (Ontology Web Language) using various tools such as the interactive editor Protégé.

In ontologies useful as the semantic basis of a DT, the central role is played by the taxonomy of *Processes* to be twinned: monitoring, steering, maintenance, and so on. Taxonomies of other top-level concepts relate to them: *Events* trigger processes, *Actors* initiate processes and participate in them, equipment and other *Resources* are used/affected, and *Models* formally represent everything [24]. Axioms express physical laws, design rules, and control objectives.

For example, ontology modeling has a long history in the electric power industry [25]. A reference source of the domain concepts is the Common Information Model (CIM) specified in the standards IEC 61968, 61970, and 62325 for interoperability across electricity-related information systems. However, CIM doesn't describe energy processes, hence is insufficient for a DT. Some missing concepts are specified in the Smart Appliances REFerence (SAREF) family of ontologies designed for interaction with IoT devices, the Domain Analysis–Based Global Energy Ontology (DABGEO) that claims to be a complete semantic description of energy management scenarios, and other relevant sources.

The ontology applies as the semantic basis of a DT in the following use cases:

- information model structure generation;
- reference data maintenance;
- communication protocols design;
- development of quality models of services;
- development of document templates and user interface forms;
- axiomatic inference and verification of design decisions;
- learning the subject domain.

6.9 Algebraic Approach to Digital Twin Composition

To automate the DT composition, especially when employing the generative design, a rigorous formal description and verification technique is much desired. As stated in the Introduction, such a technique is feasible to build upon the mathematical framework of category theory. Category theory allows to separate compositionality from other aspects of models by shifting the representation focus from their internal contents to behavior with respect to each other (a systems engineer's "black box" point of view).

Recall that a category is a class of abstract objects pairwise related by morphisms (arrows). The precise definition takes just a few lines: a category C consists of a class of objects `Ob` C and a class of morphisms `Mor` C endowed with the following operations. Firstly, two objects are associated with each morphism f: a domain (arrow source) $\mathtt{dom}\,f$ and a codomain (target) $\mathtt{codom}\,f$, so that the pair of equalities $\mathrm{dom}\,f = A$ and $\mathtt{codom}\,f = B$ is depicted as an arrow $f : A \to B$. Secondly, for any pair of morphisms f, g that satisfies the condition $\mathrm{codom}\,f = \mathtt{dom}\,g$, the composition of morphisms $g \circ f : \mathtt{dom}\,f \to \mathtt{codom}\,g$ is defined, and it is associative: for any three morphisms f, g, h such that codom $f = \mathtt{dom}\,g$ and $\mathtt{codom}\,g = \mathtt{dom}\,h$, the equality $h \circ (g \circ f) = (h \circ g) \circ f$ holds. Thirdly, any object A has an associated identity morphism $1_A : A \to A$ such that $f \circ 1_A = 1_B \circ f = f$ for any morphism $f : A \to B$. A classic example of a category is **Set** that consists of all sets and all maps between sets: the composition law for maps is defined by the standard substitution, and the identity map of a set sends each element to itself.

In the MBSE context, we denote by C a category used as a virtual "model catalog". Objects of such a catalog category are all possible models of a certain kind. Its morphisms describe all actions that could be performed with models when composing systems from components. Depending on the kind of models, actions include variable substitution, reference resolution, embedding of a submodel, quotienting, etc. It is easy to see that C is indeed a category since it has composition of morphisms (sequential execution of actions) and identity morphisms (idle "doing nothing" action with any model). A configuration of a complex system composed of multiple models, such as a DT, is represented as a diagram in C.

For example, consider the categorical representation of an information model [26]. Here, objects are sets of typed data that populate database tables, and morphisms are generated by foreign keys as maps between tables that combine them into data sets with a complex structure. Hence, all possible information models pertaining to a certain domain constitute a subcategory of **Set**. A diagram in such a category shows how to compose a database.

Mathematical models, being the most important among DT components, are the most technically difficult to represent categorically. For example, a categorical representation of DERs as finite automata with states labeled by power demand regions is proposed to formalize DER aggregation and solve a power flow problem [27]. As a less involved yet apt example, consider discrete-event simulation. Recall from Section 6.3 that the simulation is a discrete sequence of typed events that represents some operational scenario. Algebraically, a scenario is represented by a set of events partially ordered by causal dependencies and labeled by event types; scenario assembling actions are defined as maps that preserve the order and the labeling (i.e. homomorphisms) [28]. All scenarios and assembly actions constitute a category denoted as **Pomset**.

In an arbitrary category C, the universal construction called colimit is rigorously defined. Given a diagram, its colimit, if exists, transforms it into a holistic complex model whose structure the diagram represents. Conceptually, a colimit expresses in algebraic terms the common-sense view of a system as a "container" that includes all parts, respecting their structural interconnections, and nothing else.

Consider the simplest nontrivial colimit, called a pushout, for a span-shaped diagram $f : P \leftarrow G \rightarrow S : g$. This diagram can be interpreted as the structure of a system composed of parts P and S joined by a "glue" G. Various mediators that integrate disparate data volumes, models, or whole systems are representable as glues. In information model design, a "many-to-many" relation between two tables, e.g., the ownership/rent relation between actors (legal entities) and equipment units, can act as glue. In discrete-event simulation, glue, for example, represents a controller that establishes a (sub)optimal coherent operation of two equipment units, such as a wind turbine and an electric energy storage device. In general, the target integral model constructed via gluing should include both parts and respect the glue in the sense that tracing the glue inclusion through either one of two parts amounts to the same action. Moreover, the target model should contain nothing except two glued parts, i.e., it should be unambiguously identified within any arbitrary model that contains the parts and respects the glue. Figure 6.4 presents these criteria as a commutative diagram where the vertex R denotes the target model, the edges $p : P \rightarrow R$ and $s : S \rightarrow R$ denote inclusions of P and S into it, respectively, and T denotes an arbitrary enclosing model. It is easy to verify that the object R, provided that it exists, is defined uniquely up to an isomorphism. Under mild technical conditions, a colimit of any finite diagram can be calculated by a sequence of pushouts.

When formalizing a generative design technology, a suitable category of diagrams represents a design space. Requirements for an asset, translated

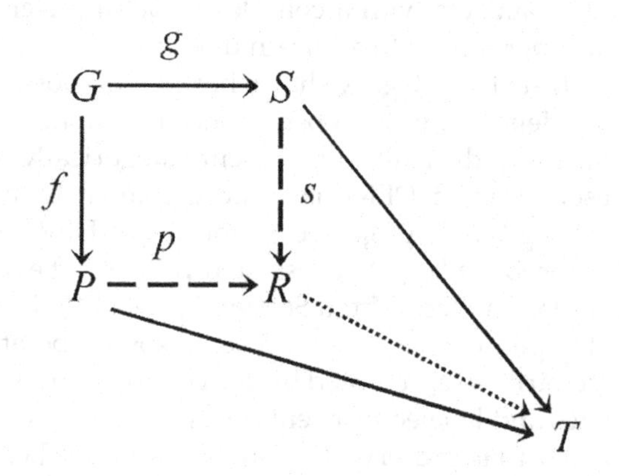

FIGURE 6.4
The pushout.

into the language of models that constitute the category C, bound the design space through checking whether a diagram represents an admissible candidate asset configuration. For example, consider the simple case where requirements completely specify the target asset and prescribe to build it by gluing together the two specified parts, yet with an unspecified glue. Borrowing the notation from the pushout diagram above, assume that the integral model R, the part models P and S, and the inclusion actions $p : P \to R$ and $s : S \to R$ are known and fixed. The generative design procedure amounts to the search for an optimal (with respect to some objective function) glue model G and actions that form a span-shaped diagram whose pushout produces the model R. In many practical cases, where C consists of sets with some structure (including information modeling and simulation), the gluing design space is constructed using the following theorem derived from Proposition 2 of [29].

Theorem 1

Let P, S, R be sets, $p : P \to R \leftarrow S : s$ be a pair of maps, $[p, s] : P \coprod S \to R$ be the canonical map that acts as p on P and as s on S. A span-shaped diagram that has a pushout with edges p and s exists if and only if for each $r \in R$, whenever either one of the sets $p^{-1}(r)$ or $s^{-1}(r)$ is empty, there exists a unique $t \in P \coprod S$ such that $[p, s](t) = r$.

Applied to the information model design, this theorem paves the way to reconstruct a minimal (in terms of a number of records) "many-to-many" relation between two tables, given a data set that specifies how to relate their records; consider for example an archive of documents that establish the equipment ownership/rent relations as mentioned above. In discrete-event simulation, this construction produces a minimal (in terms of a number of events) scenario of a controller that executes a given process including the coordinated operation of two equipment units.

More involved generative design problems may be solved by navigating the design space along diagram morphisms according to some optimization strategy, e.g., a gradient descent. The navigation path can be calculated by means of computer algebra. To represent the design space of a complex heterogeneous asset, such as a power system, advanced categorical constructions apply, such as a multicomma [30].

6.10 Conclusion

The key model-based DT technologies and algebraic techniques aim to make the DT cost-effective to compose and operate while maintaining sufficient completeness, accuracy, and usability. We have successfully tested the technologies on a DT of a small DER prosumer facility [3]. Now we are scaling them up [31] and applying for other kinds of assets [30], which require further research in a number of directions.

References

[1] G. Steindl et al, "Generic digital twin architecture for industrial energy systems," *Applied Sciences*, vol. 10, p. 8903, 2020.

[2] A. Rasheed, O. San, and T. Kvamsdal, "Digital twin: Values, challenges and enablers from a modeling perspective," *IEEE Access*, vol. 8, pp. 21980–22012, 2020.

[3] S. K. Andryushkevich, S. P. Kovalyov, and E. Nefedov, "Composition and application of power system digital twins based on ontological modeling," In *Proceedings of the 17th IEEE International Conference on Industrial Informatics INDIN'19*, pp. 1536–1542, 2019.

[4] H. Sun and L. Ma, "Generative design by using exploration approaches of reinforcement learning in density-based structural topology optimization," *Designs*, vol. 4(2), p. 10, 2020.

[5] P. Palensky et al, "Digital twins and their use in future power systems," *Digital Twin*, vol. 1, p. 4, 2021. doi:10.12688/digitaltwin.17435.1.

[6] S. P. Kovalyov, "An approach to develop a generative design technology for power systems," In *Proceedings of the VIth International Workshop IWCI'2019, Advances in Intelligent Systems Research Series*, vol. 169, pp. 79–82, 2019.

[7] C. Brosinsky, D. Westermann, and R. Krebs, "Recent and prospective developments in power system control centers: Adapting the digital twin technology for application in power system control centers," In *Proceedings of the IEEE International Energy Conference, ENERGYCON*, pp. 1–6, 2018.

[8] A. M. Madni, C. C. Madni, and S. D. Lucero, "Leveraging digital twin technology in model-based systems engineering," *Systems*, vol. 7(1), p. 7, 2019.

[9] H. Broodney, U. Shani, and A. Sela, "Model integration - Extracting value from MBSE," *INCOSE International Symposium*, vol. 23(1), pp. 1174–1186, 2014.

[10] M. A. Mabrok and M. J. Ryan, "Category theory as a formal mathematical foundation for model-based systems engineering," *Applied Mathematics and Information Sciences*, vol. 11(1), pp. 43–51, 2017.

[11] S. P. Kovalyov, "Category theory as a mathematical pragmatics of model-based systems engineering," *Informatics and Applications*, vol. 12(1), pp. 95–104, 2018.

[12] S. Mac Lane, *Categories for the Working Mathematician*. 2nd ed. Springer, Cham, 1998, 314 p.

[13] M. Redeker et al, "A digital twin platform for Industrie 4.0," In *Data Spaces*, Springer, Cham, 2022, pp. 173–201.

[14] G. Tsinarakis, N. Sarantinoudis, and G. Arampatzis, "A discrete process modelling and simulation methodology for industrial systems within the concept of digital twins," *Applied Sciences*, vol. 12, p. 870, 2022.

[15] A. I. Tikhonov et al, "Development of technology for creating digital twins of power transformers based on chain and 2D magnetic field models," *South-Siberian Scientific Bulletin*, vol. 29, pp. 76–82, 2020. [In Russian]
[16] Z. Shen, F. Arraño-Vargas, and G. Konstantinou, "Artificial intelligence and digital twins in power systems: Trends, synergies and opportunities," *Digital Twin*, vol. 2, p. 11, 2023. doi:10.12688/digitaltwin.17632.2.
[17] W. Kuehn, "Digital twins for decision making in complex production and logistic enterprises," *International Journal of Design & Nature and Ecodynamics*, vol. 13(3), pp. 260–271, 2018.
[18] F. Z. Harmouch et al, "An optimal energy management system for real-time operation of multiagent-based microgrids using a T-cell algorithm," *Energies*, vol. 12, p. 3004, 2019.
[19] L. Pagnier and M. Chertkov, "Physics-informed graphical neural network for parameter & state estimations in power systems," arXiv, 2021. https://arxiv.org/abs/2102.06349. Accessed April 15, 2023.
[20] A. Sharma et al., "Digital twins: State of the art theory and practice, challenges, and open research questions," arXiv, 2020. https://arxiv.org/abs/2011.02833. Accessed April 15, 2023.
[21] D. Mourtzis, V. Zogopoulos, and F. Xanthi, "Augmented reality application to support the assembly of highly customized products and to adapt to production re-scheduling," *International Journal of Advanced Manufacturing Technologies*, vol. 105, pp. 3899–3910, 2019.
[22] M.-A. Sicilia (Ed.), *Handbook of Metadata, Semantics and Ontologies*. World Scientific, 2013, 580 p.
[23] P. Pedamkar, Relational Database Advantages. *EDUCBA*, 2020. https://www.educba.com/relational-database-advantages/. Accessed April 15, 2023.
[24] S. P. Kovalyov and O. V. Lukinova, "Integrated heat and electric energy ontology for digital twins of active distribution grids," In *Proc. RSES 2021, AIP Conference Proceedings*, vol. 2552, pp. 080005-1–080005-8, 2023.
[25] Z. Ma et al, "The application of ontologies in multi-agent systems in the energy sector: A scoping review," *Energies*, vol. 12, p. 3200, 2019.
[26] D. Spivak and R. Kent, "Ologs: A categorical framework for knowledge representation," *PLoS One*, vol. 7(1), p. e24274, 2012.
[27] J. S. Nolan et al., "Compositional models for power systems," *In Proceedings on Applied Category Theory 2019, EPTCS*, vol. 323, pp. 149–160, 2020.
[28] V. R. Pratt, "Modeling concurrency with partial orders," *International Journal of Parallel Programming*, vol. 15(1), pp. 33–71, 1986.
[29] S. P. Kovalyov, "Leveraging category theory in model based enterprise," *Advances in Systems Science and Applications*, vol. 20(1), pp. 50–65, 2020.
[30] S. P. Kovalyov, "Algebraic means of heterogeneous cyber-physical systems design," In *Cyber-Physical Systems: Modelling and Industrial Application, Studies in Systems, Decision and Control Series*, vol. 418, Springer, Cham, pp. 3–13, 2022.
[31] S. P. Kovalyov, "Design and development of a power system digital twin: A model-based approach," *In 2021 3rd International Conference on Control Systems, Mathematical Modeling, Automation and Energy Efficiency (SUMMA)*, pp. 843–848, 2021.

7

A Generic Deployment Methodology for Digital Twins – First Building Blocks

Mohammed Adel Hamzaoui and Nathalie Julien
Université Bretagne Sud Lorient

7.1 Introduction

Digital Twin (DT) technology gained prominence across various disciplines and engineering fields, peaking in the Gartner Hype Cycle for Internet Technology (IT) in GNU Compiler Collection (GCC) in 2019 and ranking third in trending technologies for 2020 (IEEE Computer Society 2019). It has been applied across diverse domains:

Healthcare: DT is well-suited for healthcare 4.0, with applications in digital patients, pharmaceuticals, hospitals, and wearable technologies (Erol, Mendi, and Dogan 2020). Projects include the Human Brain Project, Brain Initiative, and Blue Brain Cell Atlas project (Erö et al. 2018), along with patient-based simulation models (Sim and Cure 2022), heart DTs (van Houten 2018), virtual humans, hospital department DTs (Siemens Healthineers 2018), and wireless devices for patient observation (Kellner 2015).

Smart cities and buildings: Cities have developed into smarter cities over the last years as a result of the widespread information and communications technology, utilizing data-driven models and Artificial Intelligence(AI) solutions (White et al. 2021). DTs of cities include Dublin's Docklands (White et al. 2021), Virtual Singapore (Farsi et al. 2020), and Zurich (Shahat et al. 2021). Intelligent buildings use IoT and microsensors for control and efficiency, with projects like the West Cambridge site (Lu et al. 2020) and structural health monitoring (Shu et al. 2019). Building Information Modeling(BIM) has also been used for building DTs given the advantages offered by this tool for the development of digital models in the building field (Kaewunruen et al. 2018).

DOI: 10.1201/9781003425724-10

Transportation and logistics: DTs have been used in transportation and processing scheduling (Yan et al. 2021), digital shopfloor management systems (Brenner and Hummel 2017), and Automated Guided Vehicle(AGV) transportation (Martínez-Gutiérrez et al. 2021). Applications extend to elevators (Gonzalez et al. 2020), smart electric vehicles (Bhatti et al. 2021), pipeline DTs (Sleiti et al. 2022), traffic control systems (Wang et al. 2021; Saifutdinov et al. 2020), and Adaptive Traffic Control Systems (Dasgupta et al. 2021).

Energy 4.0: DTs provide disruptive advantages in energy production, transit, storage, and consumption (Karanjkar et al. 2018; Li et al. 2022). Examples include energy consumption optimization for robotic cells (Vatankhah Barenji et al. 2021), energy cyber-physical systems (Saad et al. 2020), energy storage systems for electric vehicles (Vandana et al. 2021), and smart electrical installations (Fathy et al. 2021; O'Dwyer et al. 2020).

Smart manufacturing: DT applications in industry can be categorized into four phases: design, production, service, and retirement (Liu et al. 2021). Applications range from iterative optimization (Xiang et al. 2019), virtual validation (Howard 2019), automated flow-shop production systems (Liu et al. 2019), autonomous manufacturing (Zhang et al. 2019), Manufacturing Execution Systems (Cimino et al. 2019), start-up phases acceleration (Zhu et al. 2021), optimization of time-based maintenance strategies (Savolainen and Urbani 2021), Remaining Useful Life (RUL) assessment (Werner et al. 2019), and applications in the retirement phase such as disassembly, reusage, disposal, retrospect, and upgrade (Falekas and Karlis 2021).

In summary, DT technology has been applied across various domains, including healthcare, smart cities and buildings, transportation and logistics, energy management, and smart manufacturing. The wide range of applications and ongoing research showcase the adaptability and potential of DT technology in addressing a multitude of concerns and improving the efficiency and effectiveness of systems across these sectors.

7.2 The Digital Twin, between the Quest for Diversity and the Need for Standardization

The DT represents a multifaceted technology, encompassing a range of tools and addressing various concerns and application areas. To strike a balance between adaptability and structural robustness, it must employ different descriptive models, adhere to well-defined standards, and utilize efficient development and deployment methods. In the following sections, we will discuss the key models and standards that offer a comprehensive and synergistic understanding of the DT concept. In order to cover all these models, a generic and standardized approach is necessary to be able to define, implement, and bring to life the different DTs specific to the various applications; this is precisely the object of this paper.

7.2.1 The Reference Architecture Model Industrie 4.0 (RAMI 4.0)

This reference architecture model was developed by the German Electrical and Electronic Manufacturers' Association (ZVEI) to support Industry 4.0 initiatives, and is defined by its creators as follows: "consists of a three-dimensional coordinate system that describes all crucial aspects of Industrie 4.0. In this way, complex interrelations can be broken down into smaller and simpler clusters" (Hankel and Rexroth 2015).

The first axis, "Hierarchy Levels" from IEC 62264, encapsulates various factory functionalities. The second, "Life Cycle & Value Stream," is grounded in IEC 62890 for lifecycle management. The third axis, "Layers," deconstructs a machine into organized attributes. Although this model significantly contributes to Industry 4.0 organization, its generic nature doesn't specifically address DTs as a unique technology. For effective DT development and deployment, employing more specialized methods may be necessary.

7.2.2 The Five-Dimension Digital Twin Model

Tao et al. (2019) suggested a paradigm in their book to explain the DT technology. This model is made up of five main parts that each reflect one of the key characteristics of a DT. The proposed model is an expansion of the Three-Dimensional DT model that Grieves initially introduced in 2003 in his product lifecycle management (PLM) course at the University of Michigan (first mentioned in the literature in Grieves (2014)).

In addition to the three dimensions of Grieves' model – physical element, virtual element, and data and information – two more dimensions were added by Tao et al. (2019), as indicated in expression (7.1):

$$M_{DT} = (PE, VE, S_S, DD, CN) \tag{7.1}$$

The Physical Entity (PE) represents the real-world object, while the Virtual Entity (VE) consists of an assortment of models (Tao et al. 2019). Services (S_S) cater to both entities, and Data (DD) comprise the information that drives the DT. Connections (CN) link the various DT components. This model effectively embraces the DT's wider aspects. S_S standardizes and packages DT functions, facilitating convenient and on-demand utilization.

7.2.3 ISO 23247 – Digital Twin Framework for Manufacturing

A general development framework that may be instantiated for use in case-specific implementations of DTs in manufacturing is what this ISO standard series attempts to provide (International Organization for Standardization 2021). The standard is divided into four sections: (1) overview and general principles, (2) reference architecture, (3) digital representation, and (4) information exchange (Shao 2021). These four sections include instructions on how to define goals and scope, examine modeling needs, encourage the use of common terminologies, construct a general reference architecture, assist information modeling of OMEs (Observable Manufacturing Elements), and synchronize information between a DT and its OME (Shao 2021).

7.2.4 The Digital Twin Capabilities Periodic Table

The Digital Twin Capabilities Periodic Table (CPT) focuses on use case capability needs rather than technology (Pieter 2022). Developed by the Digital Twin Consortium (DTC), a global ecosystem, it clusters capabilities around similar characteristics and addresses overall requirements, platforms.

The CPT proposal, while promising, is incomplete as it doesn't provide organization guidance for DT capabilities, relying on external expertise for DT development. A structured proposal with a clear process and hierarchy for capability implementation is necessary.

These models and standards allow a good formalization of the DT concept, and set the basis for a good handling of this technology as well as its integration in the migration process toward Industry 4.0 and digital revolution in general. However, it is still necessary to develop a generic methodology for the development and deployment of this technology whatever the application domain and the uses dedicated to it. It is this concern that we are interested in, since we propose through this chapter a generic methodology of development and deployment of the DT, articulated on several successive levels of abstraction.

7.3 A Generic Development and Deployment Methodology

There are a plurality of definitions of what should be a DT, and that varies along with the service, the granularity, the application domain, the technical focus and the actors (Boschert, Heinrich, and Rosen 2018; Kritzinger et al. 2018). The DT frequently has a form-defined simulation model as its only option. The real interests and difficulties of DT, however, lie in the identification and management of data throughout the lifecycle of the represented object, the development of simulation models in different information systems, the management of big data with numerical consistency and security, and the transparency of the management to various users (Lu et al. 2020; Negri et al. 2017; Uhlenkamp et al. 2019).

7.3.1 Main Attributes

The DT is not just a collection of data or models; it is also an information and meta-information-based data structure that enables the combination of models to give the necessary usages for its application. The DT attributes must change together with the necessary services over the lifespan of the object. It can be characterized by the following attributes:

- A continuous digital thread between real and virtual spaces, *at an appropriate rate of synchronization* (International Organization for Standardization 2021).
- The ability to manage relevant, reliable, and secure data with suitable accuracy.
- A DT architecture allowing to permanently increment, validate, and enhance models on-line and off-line.
- Partial or total control of the object through a decision loop that is formalized.
- Learning ability.
- Prediction ability in a delay compatible with decision support.

7.3.2 Deployment Flow

In order to implement our deployment approach, we rely on the well-known 5C Cyber-Physical Systems(CPS) Architecture (Figure 7.1) suggested in (Lee et al. 2015). Thus, we have five deployment processes, the first two of which, at the Configuration and Cognition levels, are independent of any technology concerns, while they appear in the last three stages gradually. Delaying the technological choices allows the development of a generic architecture and a higher modularity, facilitating reusing of basic building blocks in order to make the development of DTs easier. Our goal is to provide support for the deployment of DTs in contexts where there are not necessarily experts in this field. Actually, most of the interest in the DT comes from experts in specific application areas. Therefore, the implementation of generic methodologies guiding in the expression of needs and the dimensioning of the DT with respect to the use and not the technology is a necessity to extend the use of this technology by making it accessible to non-specialists

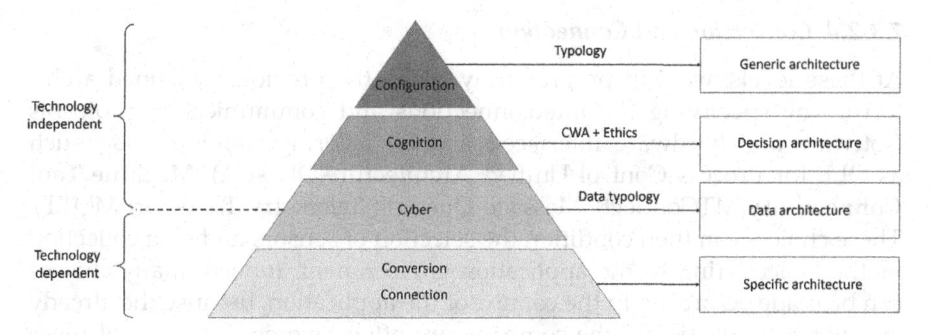

FIGURE 7.1
Deployment methodology based on the 5C CPS architecture.

7.3.2.1 Configuration

At the Configuration Level, we describe the key attributes of the DT architecture for the desired application. To do so, we propose a "Typology" that intends to effectively guide DT designers through this stage in order to create a first generic architecture. The latter is made up of 13 criteria that characterize the DT and its interactions as thoroughly as possible, and will be explained in detail in Section 7.4.

7.3.2.2 Cognition

The DT evolves toward autonomous systems, raising questions about human integration. This leads to developing cyber-physical-social systems where the human operator is a fully realized entity interacting with the system (Perez et al. 2022). A cooperative strategy is needed, considering both human and DT influences on the physical counterpart (Kamoise et al. 2022). This involves studying interaction modalities, task allocation, decision-making levels, and identifying situations with ethical issues.

7.3.2.3 Cyber

Regarding this level of deployment, it will be necessary to establish the first sketches of the data architecture that the DT will have, as well as the identification of the life cycle of the data within the DT. It is at this level that the first technological choices begin to be made, or even the identification of the needs from which these technological choices will stem (decentralization, cybersecurity, system's reactivity, etc.). Data flow graphs have been elaborated to represent data lifecycle and enhance to adjust the sizing of the DT architecture according to the data volume adapted to the needs.

7.3.2.4 Conversion and Connection

At these levels, we will progressively refine the previously defined architecture by specifying the interconnections and communication protocols (software and hardware interfaces) and by favoring standard tools such as OLE for Process Control Unified Architecture(OPC-UA), **Machine Tool Connectivity**(MTConnect), Message Queuing Telemetry Transport(MQTT). These choices can then condition the selection of sensors and data collection methods according to the application environment. Indeed, many choices can be made according to the context of the application, because the already existing technologies in the company are often favored. The goal of these steps is to produce metrics that will allow the comparison of various technological choices in order to select the architecture that best meets the needs in terms of cost, performance, agility, security, etc.

7.4 Typology

In the sequel, we outline the 13 characteristics that make up our typology for the DT. These criteria aim at establishing a general architecture of the DT by defining the basic and fundamental requirements (user's needs) to which the DT will have to provide an efficient response (and this still without making any technological choices on the DT components).

7.4.1 DT Entity

The ISO/DIS 23247 standard defines the entity as "thing (physical or non-physical) having a distinct existence" together with the OME as an *item that as an observable physical presence or operation in manufacturing.* OMEs include equipment, material, personnel, process, facility, environment, product, and supporting document. However, as our typology is not restricted to manufacturing applications, we choose to qualify this object as DT entity.

7.4.2 Type

According to Grieves' classification of DT in Grieves and Vickers (2017), there are three distinct types: DT aggregate, DT prototype, and DT instance. The prototype is an item that is utilized mostly during the design process. The instance describes the specific piece of production-related equipment. The aggregate consists of all DT instances deployed in the exploitation or maintenance phases for the same entity.

7.4.3 Level

A DT level is defined with four different levels:

- Component, elementary level, a part of equipment or system.
- Equipment, composed of components, a part of system or system of systems.
- Devices can be part of the system.
- System of system, composed of systems.

Using an AGV as an example, you can either perceive this DT entity as a component of an equipment since it interacts with the production line but without internal information, or you can see it as an equipment that is a part of the system that is the production line and is made up of a battery, an engine, wheels, etc., on which we can perform monitoring and maintenance.

7.4.4 Maturity

We have categorized digital representations according to their maturity, as shown in Figure 7.2, in order to distinguish between the digital shadow, the digital mirror, and the DT. The digital mirror is a collection of mathematical models that describe the thing and how it behaves, much like a 3D simulation, but without any actual physical or digital interaction.

All of the numerical activity of the physical item is represented by the digital shadow, which also enables dynamic representation by following all of the alterations of its physical counterpart in a one-way thread. In the third step, both physical and virtual parts are merged in what is called a Cyber-Physical System, where the DT represents the numerical part of this system. A partial or complete control feedback from the DT to the physical object is also possible in real-time communication in addition to communication from the DT

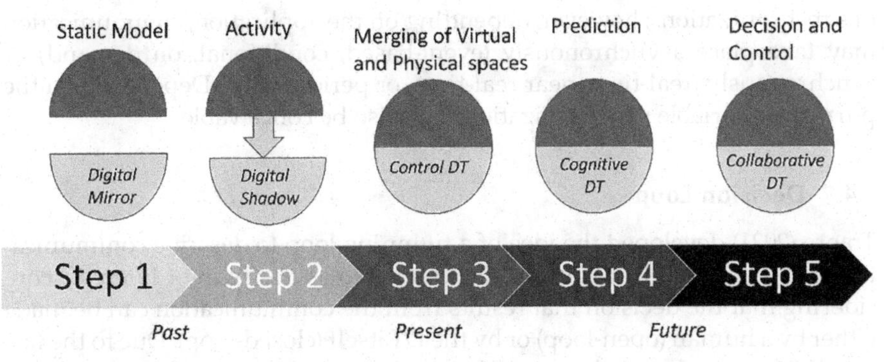

FIGURE 7.2
Numerical maturity (Julien and Martin 2021).

to the physical object. By incorporating predictive capability into this Control DT, it can evolve into Cognitive DT. And, even into a Collaborative DT with more decision and control on its counterpart as well as human interaction and collaboration.

7.4.5 Topology

It clarifies how the observable manufacturing element and the digital counterpart are related. We extended the proposal of Schroeder et al. (2021) to the four following topologies:

- Disconnected, there is no direct connection between the DT and its entity, the information comes from external components or users. This configuration, also called pre-DT, actually foreshadows the actual DT in its early design phase to operate validation and tests before connecting the physical and virtual entities.
- Connected, the DT is directly connected to its entity:
  - Embedded, the DT is included in its entity, the communication may be facilitated but the computational capabilities will be limited.
  - Disjoint, even if the DT is linked to its OME, it remains physically distant from it.
  - Joint, this represents an in-between of the two previously mentioned topologies, as one part of the DT is embedded in the OME while the other part remains physically distant from it.

7.4.6 Synchronization

Now that the DT and its entity are linked, they can speak to one another directly or indirectly. According to the ISO/DIS Standard, the next challenge is to ensure that this communication is carried out "at an appropriate rate of synchronization"; however, depending on the application, communication may take place asynchronously (event-based, conditional, on demand) or synchronously (real-time, near real-time, or periodically). Depending on the purpose, a variable synchronization may also be conceivable.

7.4.7 Decision Loop

Traore (2021) developed the idea of a twinning loop to describe communication between the DT and its entity (referred to as the Twin of Interest), considering that the decision that results from the communication can be made either by a human (open-loop) or by the DT itself (closed-loop). Due to the fact that DT frequently only has partial control over its object, we have improved this representation by adding a third option, the mixed loop. In addition, we may consider that the decision circuit may change depending on use. It should be noted that if there is no direct or indirect loop from the virtual representation to the entity, this is not a DT but a digital mirror or shadow.

7.4.8 Users

International Organization for Standardization (2021) states that suitable interfaces must be made available to various DT users of different natures, such as:

- Human
- Device
- Software/Application
- Another DT

7.4.9 Usages

Julien and Martin (2021) presented seven usages that can be combined to obtain services and applications:

- *Analysis:* control, monitoring, regulation...,
- *Simulation:* design, performance improvement, management...,
- *Optimization:* for a product, a process, a workstation...,
- *Cognition:* to predict future situations or risks,
- *Collaboration:* with human, system, other DTs or factories...,
- *Comparison:* between predicted and actual situation, for decision support and adaptation,
- *Conceptualization:* translation of visual information into a symbolic one to communicate with human.

The hierarchy of usages according to the models is shown in Figure 7.3. Analysis and simulation will be performed using the models and the gathered data first. At this level, the visibility and traceability of products and processes are the Control DT's main contributions. The creation of optimizations and predictions will therefore be made possible by these results in combination with artificial intelligence algorithms, allowing the DT models to acquire higher anticipation and deliver a Cognition DT. Finally, the DT will progress toward autonomy through collaboration with its physical or virtual surroundings; this Collaborative DT is able to offer both immersive environments and dashboards for analysis.

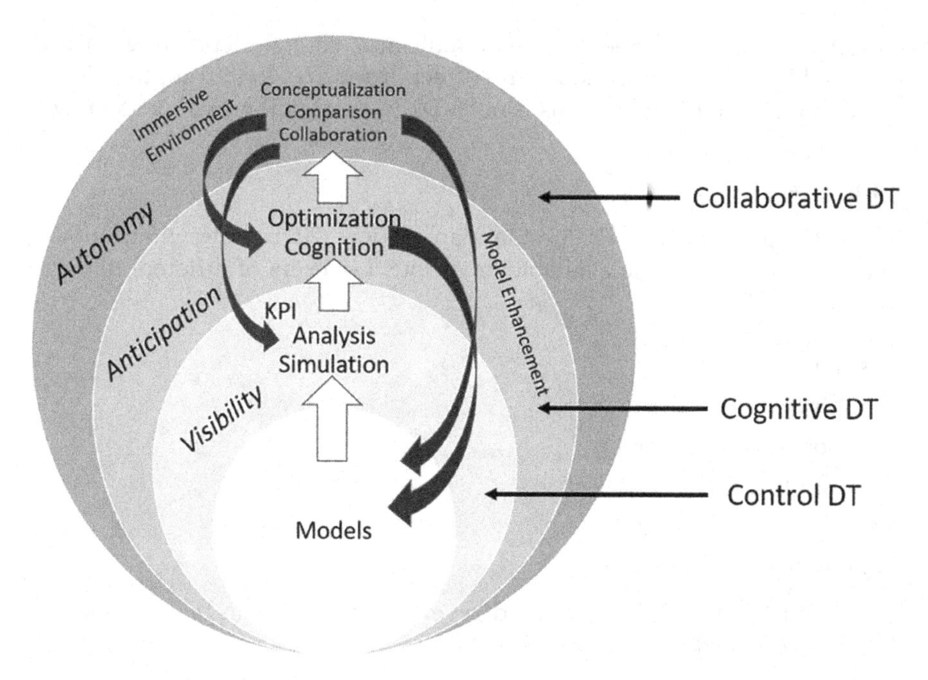

FIGURE 7.3
Usage hierarchy.

7.4.10 Applications

DT applications, as per the ISO/DIS Standard:

- Real-time control
- Off-line analytics
- Predictive maintenance
- Health check
- Engineering design

DT development will increasingly impact business operations, especially production management and predictive maintenance. Currently, we have found three additional typical DT applications:

- On-line reconfiguration
- Diagnostic support
- Operator training

Initially, our typology had ten elements, but after applying it to industrial projects for validation, three new criteria were added for a better representation of DT concept and its evolution: "version," "accuracy," and "models."

7.4.11 Version

This refers to the different versions that the DT can have during its life cycle. For example, there may be a first DT internal version (at the developer or supplier of the DT) of a particular equipment (IV0). This version can be used later to develop two other internal versions, one in Prototype (IV1) and the other in Aggregate (IV2). Moreover, these versions may be used to develop new versions dedicated to the client (EV0) whose features will be adapted to the customer needs.

7.4.12 Accuracy

Accuracy is associated to the main metric for DT validation, in other words, to what level of precision the DT is supposed to respond in order to provide a solution to clearly expressed operational needs. This can be done in a "fixed" way (one and the same precision is kept whatever the situations encountered), "variable" (the DT adapts to the situations it is confronted with, and adjusts its precision according to the need, it thus has a certain level of autonomy), or "configurable" (the precision changes according to pre-determined situations, the DT is thus parameterized in advance in order to adjust its precision when it is in front of pre-determined cases). An example of Dynamic Accuracy Management for DT is discussed in Julien and Hamzaoui (2022).

7.4.13 Models

Modeling needs for digital replication of physical objects can be classified into three criteria: realism, precision, and generality (Levins 1966). Realism encompasses underlying processes, precision measures exactness, and generality refers to the model's applicability in various contexts (Dickey-Collas et al. 2014). Levins (1966) noted that model development involves favoring two criteria over the third, resulting in empirical, mechanistic, and analytical models.

Another classification identifies "probabilistic" (stochastic) and "deterministic" models. Deterministic models compute future occurrences precisely without randomness, while stochastic models handle input uncertainty, resulting in varying outcomes with identical initial conditions. Sometimes, a mixed mode is required with both types of models.

7.5 Typology's Comprehensiveness

In order to evaluate to what extent the proposed typology describes the DT in a sufficiently exhaustive and comprehensive way, we decided to confront it with the most robust model available in the literature, namely, the 5D model. And, this is in order to see how far it is consistent with theoretical notions formalizing the concept of DT. We can see via Figure 7.4 that all the elements of the typology fit perfectly into the 5D representation of the DT. Moreover, we identify the direct environments interacting with the DT, while highlighting the different borders and edges of each concept (the DT, the cyber-physical system, and the social cyber-physical system).

By extending the 5D model, we hope to take a first step toward identifying the DT's immediate environment, followed by its internal dynamics, its evolution as a result of potential interactions, and finally a precise description of the ecosystem in which the DT operates.

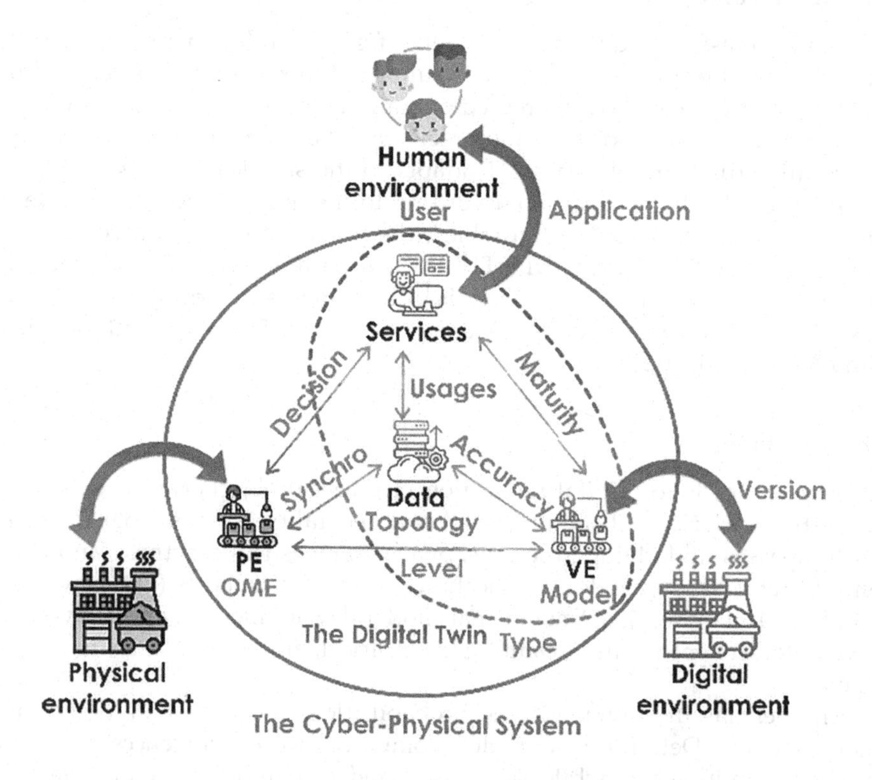

FIGURE 7.4
Extended 5D model.

7.6 Use Cases and Applications

We showcase applications of the presented typology to demonstrate its advantages in identifying DT needs and dimensioning functionalities. Within our team we are testing our approach on different applications and projects: Assistive Devices, Additive Manufacturing, Maintenance 4.0, Battery Design, Civil Engineering, and Data Center Reliability. Table 7.1 summarizes the typology of three DTs.

TABLE 7.1

Typology Applied to Different Use Cases

Applications Criteria	Lithium-Ion Battery Design	AMR Fleet	Predictive Maintenance
DT entity (OME)	Battery	Robot/Fleet	Machine
Type	Instance	Instance/Aggregate	Instance
Version	Internal version IV0	Internal "instance" version IV0 Internal "aggregate" version IV1	Internal version IV0
Level	Equipment	Equipment/System	Equipment
Maturity	Control DT	Collaborative DT	Cognitive DT
Topology	Disconnected	Connected embedded/ Connected disjoint	Connected
Decision loop	Open loop	Mixed loop	Open loop/Mixed
User	Human agent: Design Engineer	Human agent/Fleet DT	Production and maintenance manager and operator
Usages	Analysis/Comparison	Simulation/ Optimization/ Collaboration	Analysis/Cognition
Application	Off-line analytics	Real-time control/ Trajectory optimization	Health diagnosis and predictive maintenance
Synchronization	Synchronous	Synchronous/ Asynchronous	Synchronous
Accuracy	Fixed (for the metrics: State of Health and State of Charge)	Fixed (for the metric: Throughput)	Configurable
Models	Empirical (Electrical Equivalent Circuit Models) → Deterministic	Analytical (Motion models) → Deterministic	Analytical (Markov chains) → Probabilistic

7.6.1 Lithium-Ion Battery Design

The project involves developing a DT with an electrical model to simulate and analyze battery performance, including charge–discharge cycles, thermal management, and capacity degradation. The DT can be further enhanced with electrochemical or multi-physics models to improve accuracy.

7.6.2 AMR Fleet

The project requires development of individual "instance" DTs for each robot, simulating kinematics, dynamics, and control algorithms. The "aggregate" twin integrates these instance twins to optimize fleet coordination, routing, and scheduling. Techniques like model predictive control, collision avoidance algorithms, and multi-agent systems can be employed.

7.6.3 Predictive Maintenance

The DT employs a probabilistic analytical model, such as Markov chains, to predict machine failure likelihood and maintenance requirements. Other models, such as neural networks, finite state machines, or partial differential equations, can be combined to improve prediction accuracy. Metrics like temperature, humidity, and vibration can be monitored to adapt the model based on environmental conditions.

7.7 Conclusion

Through this chapter, we aimed at presenting a generic deployment methodology for DT. We highlighted the crucial need for such a methodology, considering the place that the DT is likely to take in various domains as a key technology of the digital transition. The method we propose is declined according to a 5C hierarchy in order to ensure a deployment process as generic and comprehensive as possible. The typology presented allows the establishment of a generic architecture ensuring the adequacy of the solution (DT) to the expressed needs and requirements. The latter has been validated on different industrial applications (industrial maintenance, control and monitoring of lithium batteries, quality management of food products) in order to best refine its attributes. The following steps allow to address the particularity of the case study in a gradual way. The "Cognition" part is still being worked on by our team, and first results have been published. The first outlines of the "Cyber," "Conversion," and "Connection" stages are gradually being drawn up. All this in the red thread of the establishment of a generic methodology that allows, whatever the field of application, to develop, adjust, and dimension the right DT that corresponds to the actual needs and requirements.

References

Bhatti, Ghanishtha, Harshit Mohan, and R. Raja Singh. 2021. "Towards the Future of Smart Electric Vehicles: Digital Twin Technology." *Renewable and Sustainable Energy Reviews* 141 (May): 110801.

Boschert, Stefan, Christoph Heinrich, and Roland Rosen. 2018. "Next Generation Digital Twin." *Proc. tmce. Las Palmas de Gran Canaria, Spain*, 2018, 2018: 7–11.

Brenner, Beate, and Vera Hummel. 2017. "Digital Twin as Enabler for an Innovative Digital Shopfloor Management System in the ESB Logistics Learning Factory at Reutlingen - University." *Procedia Manufacturing* 9: 198–205.

Cimino, Chiara, Elisa Negri, and Luca Fumagalli. 2019. "Review of Digital Twin Applications in Manufacturing." *Computers in Industry* 113 (December): 103130.

Dasgupta, Sagar, Mizanur Rahman, Abhay D. Lidbe, Weike Lu, and Steven Jones. 2021. "A Transportation Digital-Twin Approach for Adaptive Traffic Control Systems." arXiv preprint arXiv:2109.10863, 2021.

Dickey-Collas, Mark, Mark R. Payne, Verena M. Trenkel, and Richard D. M. Nash. 2014. "Hazard Warning: Model Misuse Ahead." *ICES Journal of Marine Science* 71 (8): 2300–2306.

Erö, Csaba, Marc-Oliver Gewaltig, Daniel Keller, and Henry Markram. 2018. "A Cell Atlas for the Mouse Brain." *Frontiers in Neuroinformatics* 12 (November): 84.

Erol, Tolga, Arif Furkan Mendi, and Dilara Dogan. 2020. "The Digital Twin Revolution in Healthcare." *In 2020 4th International Symposium on Multidisciplinary Studies and Innovative Technologies (ISMSIT)*, 1–7. Istanbul, Turkey: IEEE.

Falekas, Georgios, and Athanasios Karlis. 2021. "Digital Twin in Electrical Machine Control and Predictive Maintenance: State-of-the-Art and Future Prospects." *Energies* 14 (18): 5933.

Farsi, Maryam, Alireza Daneshkhah, Amin Hosseinian-Far, and Hamid Jahankhani, eds. 2020. *Digital Twin Technologies and Smart Cities. Internet of Things*. Cham: Springer International Publishing.

Fathy, Yasmin, Mona Jaber, and Zunaira Nadeem. 2021. "Digital Twin-Driven Decision Making and Planning for Energy Consumption." *Journal of Sensor and Actuator Networks* 10 (2): 37.

https://www.gartner.com/en/newsroom/press-releases/2019-10-14-gartner-s-2019-hype-cycle-for-it-in-gcc-indicates-pub#:~:text=The%202019%20Gartner%2C%20Inc.,to%20translate%20into%20mainstream%20adoption.

Gonzalez, Mikel, Oscar Salgado, Jan Croes, Bert Pluymers, and Wim Desmet. 2020. "A Digital Twin for Operational Evaluation of Vertical Transportation Systems." *IEEE Access* 8: 114389–114400.

Grieves, Michael. 2014. "Digital Twin: Manufacturing Excellence through Virtual Factory Replication." *Global Journal of Engineering Science and Researches NC-Rase* 18: 6–15.

Grieves, Michael, and John Vickers. 2017. "Digital Twin: Mitigating Unpredictable, Undesirable Emergent Behavior in Complex Systems." In *Transdisciplinary Perspectives on Complex Systems*, edited by Franz-Josef Kahlen, Shannon Flumerfelt, and Anabela Alves, 85–113. Cham: Springer International Publishing.

Hankel, Martin, and Bosch Rexroth. 2015. "The Reference Architectural Model Industrie (RAMI 4.0)." *Zvei* 2 2: 4–9.

Howard, Dwight. 2019. "The Digital Twin: Virtual Validation in Electronics Development and Design." In *2019 Pan Pacific Microelectronics Symposium (Pan Pacific)*, 1–9. Kauai, HI: IEEE.

IEEE Computer Society. 2019. "IEEE Computer Society's Top 12 Technology Trends for 2020." IEEE Computer Society. 2019. https://www.computer.org/press-room/2019-news/ieee-computer-societys-top-12-technology-trends-for-2020.

International Organization for Standardization. 2021. "Automation Systems and Integration - Digital Twin Framework for Manufacturing (ISO No. 23247)." https://www.iso.org/obp/ui/#iso:std:iso:23247:-1:ed-1:v1:en.

Julien, N., and E. Martin. 2021. "How to Characterize a Digital Twin: A Usage-Driven Classification." *IFAC-PapersOnLine* 54 (1): 894–899.

Julien, Nathalie, and Mohammed Adel Hamzaoui. 2022. "Integrating Lean Data and Digital Sobriety in Digital Twins through Dynamic Accuracy Management." In *International Workshop on Service Orientation in Holonic and Multi-Agent Manufacturing*, 107–112. Bucharest, Romania: Springer.

Julien, Nathalie, and Eric Martin. 2021. "Typology of Manufacturing Digital Twins: A First Step towards A Deployment Methodology." *In Service Oriented, Holonic and Multi-agent Manufacturing Systems for Industry of the Future. SOHOMA 2021. Studies in Computational Intelligence*, vol. 1034, 161–172. Cham: Springer.

Kaewunruen, Sakdirat, Panrawee Rungskunroch, and Joshua Welsh. 2018. "A Digital-Twin Evaluation of Net Zero Energy Building for Existing Buildings." *Sustainability* 11 (1): 159.

Kamoise, N., C. Guerin, M. Hamzaoui, & N. Julien. 2022. Using Cognitive Work Analysis to Deploy Collaborative Digital Twin. In European Safety and *Reliability Conference*, August 2022. Dublin, Ireland: Singapore.

Karanjkar, Neha, Ashish Joglekar, Sampad Mohanty, Venkatesh Prabhu, D. Raghunath, and Rajesh Sundaresan. 2018. "Digital Twin for Energy Optimization in an SMT-PCB Assembly Line." In *2018 IEEE International Conference on Internet of Things and Intelligence System (IOTAIS)*, 85–89. Bali: IEEE.

Kellner, Tomas. 2015. "Meet Your Digital Twin: Internet for the Body Is Coming and These Engineers Are Building It." https://www.ge.com/news/reports/these-engineers-are-building-the-industrial-internet-for-the-body.

Kritzinger, Werner, Matthias Karner, Georg Traar, Jan Henjes, and Wilfried Sihn. 2018. "Digital Twin in Manufacturing: A Categorical Literature Review and Classification." *IFAC-PapersOnLine* 51 (11): 1016–1022.

Lee, Jay, Behrad Bagheri, and Hung-An Kao. 2015. "A Cyber-Physical Systems Architecture for Industry 4.0-Based Manufacturing Systems." *Manufacturing Letters* 3 (January): 18–23.

Levins, Richard. 1966. "The Strategy of Model Building in Population Biology." American *Scientist* 54: 421–431.

Li, Hongcheng, Dan Yang, Huajun Cao, Weiwei Ge, Erheng Chen, Xuanhao Wen, and Chongbo Li. 2022. "Data-Driven Hybrid Petri-Net Based Energy Consumption Behaviour Modelling for Digital Twin of Energy-Efficient Manufacturing System." *Energy* 239 (January): 122178.

Liu, Mengnan, Shuiliang Fang, Huiyue Dong, and Cunzhi Xu. 2021. "Review of Digital Twin about Concepts, Technologies, and Industrial Applications." *Journal of Manufacturing Systems* 58 (January): 346–361.

Liu, Qiang, Hao Zhang, Jiewu Leng, and Xin Chen. 2019. "Digital Twin-Driven Rapid Individualised Designing of Automated Flow-Shop Manufacturing System." *International Journal of Production Research* 57 (12): 3903–3919.

Lu, Qiuchen, Ajith Kumar Parlikad, Philip Woodall, Gishan Don Ranasinghe, Xiang Xie, Zhenglin Liang, Eirini Konstantinou, James Heaton, and Jennifer Schooling. 2020. "Developing a Digital Twin at Building and City Levels: Case Study of West Cambridge Campus." *Journal of Management in Engineering* 36 (3): 05020004.

Lu, Yuqian, Chao Liu, Kevin I-Kai Wang, Huiyue Huang, and Xun Xu. 2020. "Digital Twin-Driven Smart Manufacturing: Connotation, Reference Model, Applications and Research Issues." *Robotics and Computer-Integrated Manufacturing* 61 (February): 101837. https://doi.org/10.1016/j.rcim.2019.101837.

Martínez-Gutiérrez, Alberto, Javier Díez-González, Rubén Ferrero-Guillén, Paula Verde, Rubén Álvarez, and Hilde Perez. 2021. "Digital Twin for Automatic Transportation in Industry 4.0." *Sensors* 21 (10): 3344.

Negri, Elisa, Luca Fumagalli, and Marco Macchi. 2017. "A Review of the Roles of Digital Twin in CPS-Based Production Systems." *Procedia Manufacturing* 11: 939–948.

O'Dwyer, Edward, Indranil Pan, Richard Charlesworth, Sarah Butler, and Nilay Shah. 2020. "Integration of an Energy Management Tool and Digital Twin for Coordination and Control of Multi-Vector Smart Energy Systems." *Sustainable Cities and Society* 62 (November): 102412.

Perez, M. J., Meza, S. M., Bravo, F. A., Trentesaux, D., & Jimenez, J. F. (2022). Evolution of the Human Digital Representation in Manufacturing Production Systems. In *International Workshop on Service Orientation in Holonic and Multi-Agent Manufacturing*, 201–211. Cham: Springer.

Saad, Ahmed, Samy Faddel, and Osama Mohammed. 2020. "IoT-Based Digital Twin for Energy Cyber-Physical Systems: Design and Implementation." *Energies* 13 (18): 4762.

Saifutdinov, Farid, Ilya Jackson, Jurijs Tolujevs, and Tatjana Zmanovska. 2020. "Digital Twin as a Decision Support Tool for Airport Traffic Control." In *2020 61st International Scientific Conference on Information Technology and Management Science of Riga Technical University (ITMS)*, 1–5. Riga, Latvia: IEEE.

Savolainen, Jyrki, and Michele Urbani. 2021. "Maintenance Optimization for a Multi-Unit System with Digital Twin Simulation: Example from the Mining Industry." *Journal of Intelligent Manufacturing* 32 (7): 1953–1973.

Schroeder, Greyce, Charles Steinmetz, Ricardo Rodrigues, Achim Rettberg, and Carlos Eduardo Pereira. 2021. "Digital Twin Connectivity Topologies." *IFAC-PapersOnLine* 54(1): 737–742.

Shahat, Ehab, Chang T. Hyun, and Chunho Yeom. 2021. "City Digital Twin Potentials: A Review and Research Agenda." *Sustainability* 13 (6): 3386.

Shao, Guodong. 2021. "Use Case Scenarios for Digital Twin Implementation Based on ISO 23247." National Institute of Standards and Technology. https://doi.org/10.6028/NIST.AMS.400-2.

Shu, Jiangpeng, Kamyab Zandi, Tanay Topac, Ruiqi Chen, and Chun Fan. 2019. "Automated Generation of FE Model for Digital Twin of Concrete Structures from Segmented 3D Point Cloud." In *Structural Health Monitoring 2019, 2019.* DEStech Publications, Inc.

Siemens Healthineers. 2018. "From Digital Twin to Improved Patient Experience." 2018. https://www.siemens-healthineers.com/perspectives/mso-digital-twin-mater.html.

Sim & Cure. n.d. "Improving Patient Care in Brain Aneurysm Treatment." Accessed July 5, 2022. https://sim-and-cure.com/.

Sleiti, Ahmad K., Wahib A. Al-Ammari, Ladislav Vesely, and Jayanta S. Kapat. 2022. "Carbon Dioxide Transport Pipeline Systems: Overview of Technical Characteristics, Safety, Integrity and Cost, and Potential Application of Digital Twin." *Journal of Energy Resources Technology* 144 (9): 092106.

Tao, Fei, Meng Zhang, and Andrew Yeh Chris Nee. 2019. *Digital Twin Driven Smart Manufacturing*. London: Academic press.

Traore, Mamadou Kaba. 2021. "Unifying Digital Twin Framework: Simulation-Based Proof-Of-Concept," *IFAC-PapersOnLine*, 2021, 54(1): 886–893.

Uhlenkamp, Jan-Frederik, Karl Hribernik, Stefan Wellsandt, and Klaus-Dieter Thoben. 2019. "Digital Twin Applications : A First Systemization of Their Dimensions." In *2019 IEEE International Conference on Engineering, Technology and Innovation (ICE/ITMC)*, 1–8. Valbonne Sophia-Antipolis, France: IEEE.

van Houten, Henk. 2018. "How a Virtual Heart Could Save Your Real One." 2018. https://www.philips.com/a-w/about/news/archive/blogs/innovation-matters/20181112-how-a-virtual-heart-could-save-your-real-one.html.

Vandana, Akhil Garg, and Bijaya Ketan Panigrahi. 2021. "Multi-Dimensional Digital Twin of Energy Storage System for Electric Vehicles: A Brief Review." *Energy Storage* 3 (6): e242.

Vatankhah Barenji, Ali, Xinlai Liu, Hanyang Guo, and Zhi Li. 2021. "A Digital Twin-Driven Approach towards Smart Manufacturing: Reduced Energy Consumption for a Robotic Cell." *International Journal of Computer Integrated Manufacturing* 34 (7–8): 844–859.

Wang, Songchun, Fa Zhang, and Ting Qin. 2021. "Research on the Construction of Highway Traffic Digital Twin System Based on 3D GIS Technology." *Journal of Physics: Conference Series* 1802 (4): 042045.

Werner, Andreas, Nikolas Zimmermann, and Joachim Lentes. 2019. "Approach for a Holistic Predictive Maintenance Strategy by Incorporating a Digital Twin." *Procedia nManufacturing* 39: 1743–1751.

White, Gary, Anna Zink, Lara Codecá, and Siobhán Clarke. 2021. "A Digital Twin Smart City for Citizen Feedback." *Cities* 110 (March): 103064.

Xiang, Feng, Zhi Zhang, Ying Zuo, and Fei Tao. 2019. "Digital Twin Driven Green Material Optimal-Selection towards Sustainable Manufacturing." *Procedia CIRP* 81: 1290–1294.

Yan, Jun, Zhifeng Liu, Caixia Zhang, Tao Zhang, Yueze Zhang, and Congbin Yang. 2021. "Research on Flexible Job Shop Scheduling under Finite Transportation Conditions for Digital Twin Workshop." *Robotics and Computer-Integrated Manufacturing* 72 (December): 102198.

Zhang, Chao, Guanghui Zhou, Jun He, Zhi Li, and Wei Cheng. 2019. "A Data- and Knowledge-Driven Framework for Digital Twin Manufacturing Cell." *Procedia CIRP* 83: 345–350.

Zhu, Zexuan, Xiaolin Xi, Xun Xu, and Yonglin Cai. 2021. "Digital Twin-Driven Machining Process for Thin-Walled Part Manufacturing." *Journal of Manufacturing Systems* 59 (April): 453–466.

8

Automated Inference of Simulators in Digital Twins

Istvan David and Eugene Syriani
Université de Montréal

8.1 Introduction

Digital Twins are real-time and high-fidelity virtual representations of physical assets [1]. The intent of a Digital Twin is to provide data-intensive software applications with a proxy interface for the underlying asset. This is achieved by mirroring the prevalent state of the underlying system through the continuous processing of real-time data originating from the sensors of the physical asset. This one-directional dependent virtual representation is often referred to as a *Digital Shadow*. Digital Twins are composed of possibly multiple Digital Shadows and additional facilities that enable exerting control over the physical asset through its actuators.

This bi-directional coupling of physical and digital counterparts enables advanced engineering scenarios, such as automated optimization, real-time reconfiguration, and intelligent adaptation of physical systems. Such scenarios are typically enabled by simulators [2]. Simulators are programs that encode a probabilistic mechanism on a computer and enact its changes over a sufficiently long period of time [3]. Statistical methods are readily available to process the results of simulations, known as simulation *traces*.

However, the inherent complexity of systems subject to digital twinning renders the construction of these simulators an error-prone, time-consuming, and costly endeavor [4]. The complexity of constructing precise models tends to increase rapidly with the growth of the underlying system. Spiegel et al. [5] show that identifying assumptions even in simple models such as Newton's second law of motion $\left(\vec{p} = m\vec{v}\right)$ might be infeasible. This is particularly concerning when considering the size and heterogeneity of systems subject to digital twinning. The inherent uncertainty and stochastic features of some domains—such as bio-physical systems [6]—further exacerbate this problem.

Therefore, managing the efforts and costs associated with simulator construction is a matter of paramount importance. Traditionally, reusability

DOI: 10.1201/9781003425724-11

and composability of simulator components have been seen as an enabler to scaling up simulation engineering. However, these techniques are severely hindered by vertical challenges (stemming from inappropriate abstraction mechanisms), horizontal challenges (stemming from different points of views), and increased search friction due to the abundance of information [7]. In addition, Malak and Paredis [8] point out that the ability to reuse a model does not necessarily mean that it should be reused. However, model reuse often results in the segmentation of the knowledge required for validation across models of possibly different domains. Substantial efforts have been made to enable sound reuse and composition of simulator components. For instance, one can port the results of component-based software engineering to simulator engineering [9]. Also, one can rely on validity frames [10] to capture the context in which assumptions of a model are considered valid. However, these techniques still fall short of appropriately scaling up the engineering of simulators.

This chapter introduces the reader to an alternative approach to automating simulator construction by machine learning. Machine learning is a particularly appropriate technique for the inference and configuration of simulators because of the high volume of the data generated in systems subject to digital twinning. Specifically, in this chapter, we rely on the Discrete Event System Specification (DEVS) [11] as the formalism of simulation foundations; and we choose reinforcement learning [12] as the machine learning method for inference. The approach infers DEVS models by observing the system to be modeled. As the system traverses its state space, the reinforcement learning agent learns to react to these states by adapting the DEVS model to produce the same behavior as the system to be modeled. The versatility of DEVS vastly increases the reusability of the inferred knowledge.

8.1.1 Structure

The rest of this chapter is structured as follows. Section 8.2 provides a general introduction to the foundations of Digital Twin simulators (Section 8.2.1), the DEVS formalism (Section 8.2.2), and reinforcement learning (Section 8.2.3). Based on these preliminaries, Section 8.3 presents an approach for the automated inference of Digital Twin simulators by reinforcement learning. Section 8.4 provides a general discussion of the approach and contextualizes it within the broader topic of Digital Twins. Finally, Section 8.5 concludes the chapter and outlines potential directions for future work.

8.2 Foundations

This section provides the reader with an overview of selected foundational topics relevant to the automated inference of simulators in Digital Twins.

8.2.1 Simulators in Digital Twins

The complexity of engineered systems has been at a steady pace for decades. Not only is the number of system components increasing—as illustrated by Moore's law [13] and its variants—but the heterogeneity of the components is on the rise as well [14]. To cope with this ever-increasing complexity, the engineering of intricate systems, such as Digital Twins, is best approached through modeling and simulation.

The typical role of simulation has shifted from the design phase of complex systems to their operational phase. Boschert and Rosen [15] identify four significant phases of this evolution, shown in Figure 8.1. In the early 1960s, simulation was used in isolated cases of select domains employing the few experts in simulation, such as mechanical engineering and military intelligence. About two decades later, the first wave of simulation tools arrived, with a particular focus on engineering problems. With the advent of multi-disciplinary systems engineering—highlighted by domains such as mechatronics and early incarnations of today's cyber-physical systems—the role of simulation has been firmly positioned in the design phase of systems engineering. Simulation has become a mainstay of engineering complex systems. Finally, the emergence of Digital Twins shifted the usage of simulators toward the operational phase of systems.

Today, simulators are first-class components of Digital Twins and enablers of the sophisticated features and services Digital Twins provide. As shown in Figure 8.2, the complexity of these features and services is a result of two factors. First, data in Digital Twins tend to be more dynamic than in traditional software applications. Often, models cannot be materialized in the memory to support reasoning because data are dispersed across the temporal dimension. Streams of sensor readings are prime examples of this challenge. Second, and related to the previous point, simulation scenarios in Digital Twins are often of a continuous or live nature. This is contrasted with one-shot scenarios in which a simulator is instantiated on demand, used to

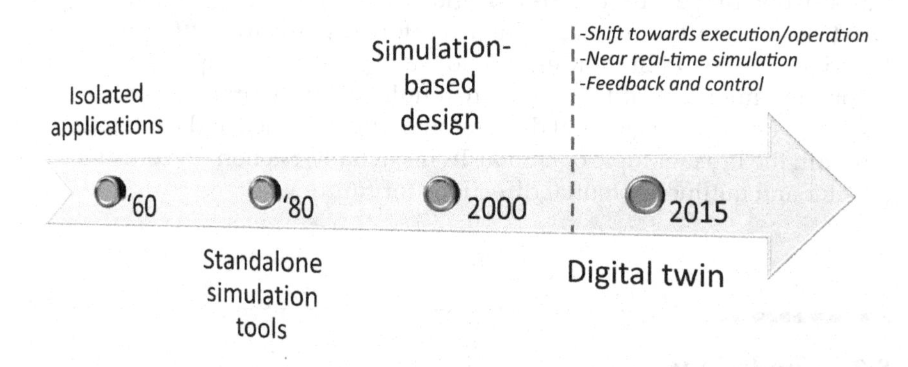

FIGURE 8.1
The role of simulation in the engineering of complex systems. Adopted from Boschert and Rosen [15].

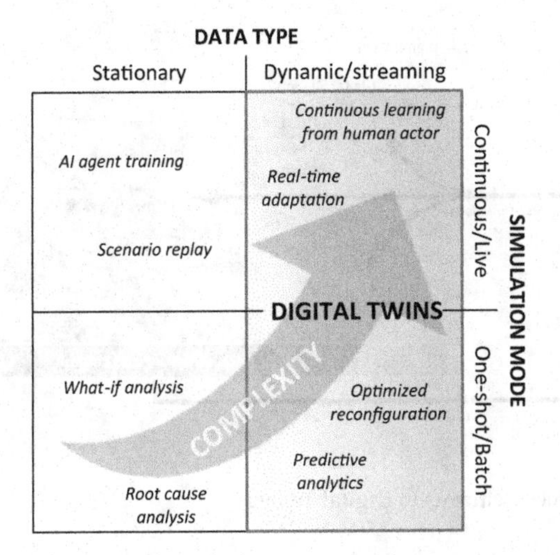

FIGURE 8.2
Characteristic Digital Twin services and the source of their complexity.

simulate a quantitative property, and then disposed. Pertinent examples of these complex services include:

- real-time adaptation of the system to a changing environmental context;
- optimized reconfiguration to align with changing business goals;
- predictive analytics for the early identification of threats, such as failures and errors;
- what-if analysis to evaluate human decisions to be made;
- providing a learning environment for training purposes of human and computer agents.

At the core of the simulator, the physical asset is represented by a formal model, from which complex algorithms calculate the metrics of interest. This model has to capture the specificities of the physical asset in appropriate detail to consider the results of the simulation representative of the physical asset. Typically, simulation models reuse or rely on models already present in the Digital Twin, e.g., the models defined by Digital Shadows (Figure 8.3).

A model [16] is an abstraction of reality and allows its users to focus on the essential aspects of the problem at hand while still providing a basis for rigorous methods to analyze, verify, and operationalize models. How to model the problem at hand depends on the specific use case. For example, a safety analysis might rely on discrete and abstract models, such as a Petri net; or rely on acausal continuous models, such as bond graphs or differential equations. The following section introduces the reader to a versatile modeling formalism with exceptionally high utility and potential in the simulation of complex systems, such as Digital Twins.

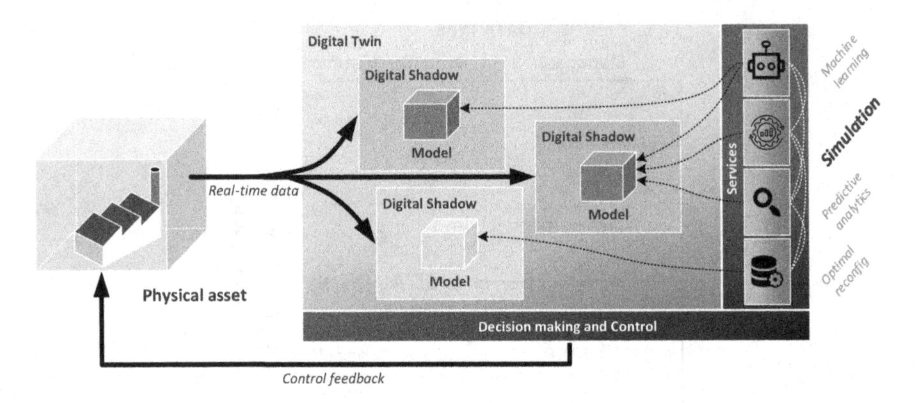

FIGURE 8.3
Simulators are first-class citizens in Digital Twins.

8.2.2 Discrete Event System Specification

Investing resources into the learning process only pays off if the learned simulation model is representative of a class of problems rather than a specific case. For example, after learning how to construct the Digital Twin of a robot arm, one would like to use this knowledge to construct the Digital Twin of a robot leg. This generalization can be achieved at the level of the simulation model by choosing the appropriate simulation formalism. The DEVS formalism [11] is a prime candidate for this purpose, as the versatility of DEVS allows modeling the reactive behavior of real systems in great detail, including timing and interactions with the environment—two aspects predominantly present in the physical assets of Digital Twins [17].

DEVS is a formalism for modeling and simulation of systems. DEVS is a hierarchically compositional formalism. That is, it allows for constructing system specifications of arbitrary complexity through the sound composition of more primitive DEVS models. DEVS is closed under composition, i.e., composing two DEVS models results in a new DEVS model. Although discrete by nature, DEVS also supports modeling and simulating continuous state systems and hybrid continuous-discrete state systems. As Vangheluwe [18] points out, DEVS can serve as an effective assembly language between different simulation formalisms. Due to its ability to tackle the complexity and heterogeneity of the modeled system, DEVS is especially suitable for building simulators for Digital Twins.

The discrete event abstraction of DEVS allows computing the prevalent state of the modeled system based on its previous state and the latest observed event. The state of a DEVS model is changed only upon observing events but remains constant in between. Events in DEVS are timed, i.e., they allow omitting intermediate points in time when no interesting events occur. This is in contrast with simulation formalisms with a fixed time step, where time increments with a fixed value. Flexibility in choosing the appropriate time granularity of DEVS events is especially useful in complex simulation and co-simulation scenarios of Digital Twins [19].

Events can originate from the DEVS model itself or from other DEVS models. This allows DEVS models to be reactive to other components' changes and to events generated by the modeled system. Being able to ingest real-time data is a distinguishing requirement against simulators of Digital Twins, and DEVS is an appropriate choice in this respect as well.

Due to its remarkable expressiveness, DEVS has been a particularly popular choice in academic and commercial simulation tools, such as Python-based Discrete Event Simulation (Python-PDEVS)[1] and MathWorks' SimEvents[2].

8.2.2.1 Running Example

To illustrate the capabilities of DEVS modeling, we use the running example of a Digital Twin of a biological production room. The production room is isolated from the external environment, and every physical property is artificial. The lighting is turned on and off periodically, simulating 12 hours of daylight and 12 hours of night. Simultaneously, a heating, ventilation, and air conditioning (HVAC) system takes care of the proper temperature and humidity in the production room. The HVAC system is either in a heating state or in a cooling state, and its behavior follows the cycles of the lighting system. When the lights turn on, the HVAC system switches to heating mode to increase the temperature of the air. When the lights turn off, the HVAC system switches to cooling mode to decrease the temperature of the air. For safety reasons, the HVAC system is allowed 2 hours to prepare for the new operation mode after the switch is requested.

The Digital Twin supports two operation modes: observation and simulation. In observation mode, the Digital Twin provides human and machine agents with the prevalent state of the physical room in real time. In simulation mode, different performance indicators of the room are calculated based on scenarios enacted by a simulator. For example, one might calculate the overall energy consumption of the room for an extended period of time by simulating the behavior of the lighting and HVAC components.

In this chapter, we use DEVS to construct models serving as the basis of simulation.

8.2.2.2 Atomic DEVS

Atomic DEVS models are the simplest DEVS models. They describe the autonomous behavior of a system or its component. Figure 8.4 represents two atomic DEVS models of the running example as timed automata. The model

---

[1] https://msdl.uantwerpen.be/documentation/PythonPDEVS.

[2] https://www.mathworks.com/products/simevents.html.

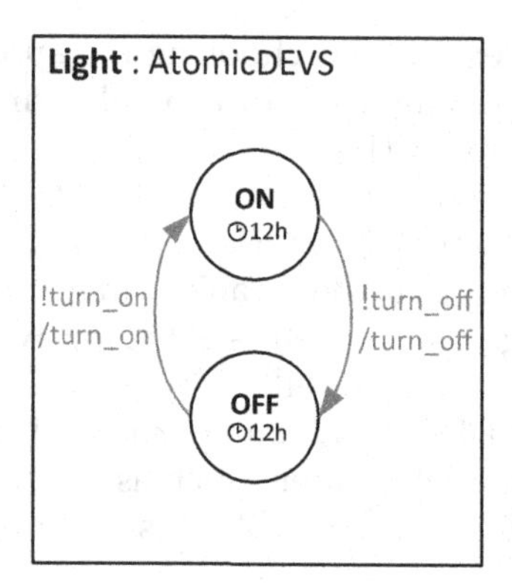

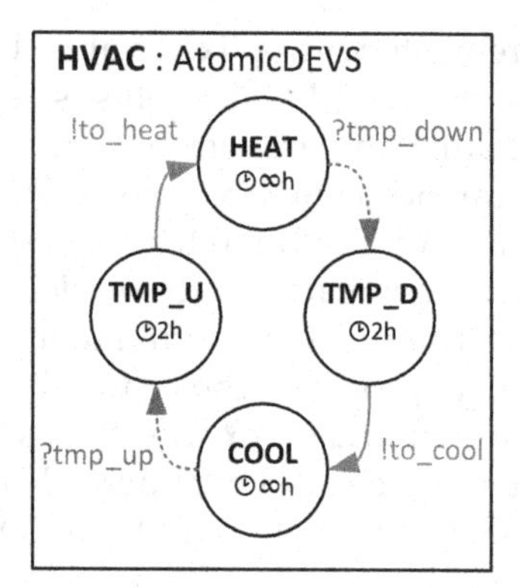

FIGURE 8.4
Atomic DEVS models from the Digital Twin of a biological production room.

of the lighting system in Figure 8.4a shows that the lights are turned on and off periodically, every 12 hours. Timed transitions take the lighting system from one state to another. These transitions are independent of the surroundings of the lighting system; hence, they are referred to as internal transitions. In contrast, the model of the HVAC system in Figure 8.4b relies on external transitions. Once the HVAC system is in the heating (HEAT) or cooling (COOL) state, the system cannot advance to the next state in a timed fashion. Transitions occur upon receiving an event, or the system remains in its current state. By convention, this behavior is expressed by a time advance set to infinity (∞) for each state. Upon receiving the tmp_up event in the COOL state, the system transitions into the TMP_U state to prepare for the HEAT state. The system then proceeds to the TMP_U state for 2 hours before transitioning to the HEAT state in an automatic fashion. The transition from HEAT to COOL via the TMP_D follows the same logic.
We use the following definition of atomic DEVS to provide formal grounds.

Definition 8.1: Atomic DEVS

An atomic DEVS model is defined as the eight-tuple $M = \langle X, Y, S, q_{\text{init}}, \Delta_{\text{int}}, \Delta_{\text{ext}}, \lambda, ta \rangle$.

S is the set of states in the DEVS model. In the *Light* model in Figure 8.4a, S = {ON, OFF}. Δ_{int}: $S \rightarrow S$ is the internal transition function that maps the current state to the next state. In the *Light* model, Δ_{int} = {OFF → ON, ON → OFF}. Internal transitions capture the autonomous behavioral dynamics of a DEVS model as they are driven by internal factors and are independent of external factors. The internal factor determining *when* state transitions occur is the $ta : S \rightarrow \mathbb{R}^{+}_{0,+\infty}$ time advance function. In the *Light* model, ta = {OFF → 12h, ON → 12h}. That is, after exhibiting the OFF state for 12 hours, the system transitions to the next state. According to Δ_{int}, this next state will be the ON state. The model is initialized in the $q_{init} \in Q$ initial state, with $Q = \{(s,e) | s \in S, 0 \leq e \leq ta(s)\}$ being the set of total states, i.e., states with the time elapsed in them. In the *Light* model, q_{init} = {OFF, 0.0}. That is, the room is initialized with the OFF state, with 0.0 elapsed time at the beginning of the simulation.

X denotes the set of input events the model can react to. $\Delta_{ext} : Q \times X \rightarrow S$ is the external transition function to the next state when an event $x \in X$ occurs. In Figure 8.4, these events are prefixed with a question mark (?). For example, in the *HVAC* DEVS model, X = tmp_up, tmp_down. These events are provided by external event sources, for example, other DEVS models. To this end, Y denotes the set of output events of the DEVS model that other models can react to. In Figure 8.4, these events are prefixed with an exclamation point (!). For example, in the *HVAC* model, Y = {to_heat, to_cool}. Finally, the $\lambda : S \rightarrow Y$ output function specifies how output events are generated based on the state of the model. Output events are generated before the new state is reached. That is, the output function associates an output event with the state that is being left. For example, in the *HVAC* model, λ = {TMP_U → to_heat, TMP_D → to_cool}.

8.2.2.3 Coupled DEVS

DEVS supports the hierarchical composition of models into more complex DEVS models. Such complex DEVS models are called coupled DEVS models. As discussed previously, the lighting and HVAC components of the running example can be modeled as atomic DEVS models. However, the tmp_up and tmp_down events in the *HVAC* model need to be supported by the transitions in the *Light* model. The coupled DEVS formalism provides the means to model this connection.

Figure 8.5 shows a DEVS model coupling the two atomic DEVS models previously seen in Figure 8.4. In this coupled DEVS model, the *HVAC* model reacts to state transitions in the *Light* model by observing its output events arriving to its command (COM) port from the output (OUT) port of the *Light* model. As the *Light* model is turned on, the turn_on event notifies the *HVAC* model to activate the transition from the COOL state to the TMP_U state.

We use the following definition of coupled DEVS to provide formal grounds.

Definition 8.2

(Coupled DEVS). An coupled DEVS model is defined as the seven-tuple $C = \langle X_{self}, Y_{self}, D, \{M\}, \{I\}, \{Z\}, select \rangle$.

D is the set of model instances that are included in the coupled model. In the coupled *System* model, D={light, hvac}. {M} is the set of model

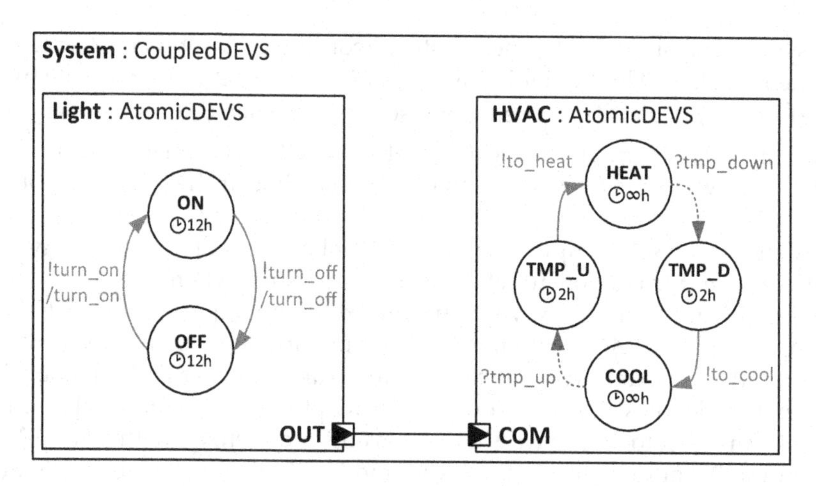

FIGURE 8.5
Coupled DEVS model, integrating the atomic DEVS models in Figure 8.4.

specifications such that $\forall d \in D \exists M_d : M_d = \langle X, Y, S, q_{init}, \Delta_{\text{int}}, \Delta_{ext}, \lambda, ta \rangle_d$, as defined in Definition 1. In the coupled *System* model, M_{light} is the model in Figure 8.4a, and M_{hvac} is the model in Figure 8.4b. To connect model instances within a coupled model, {*I*} defines the set of model influences such that $\forall d \in D \exists I_d : d \rightarrow D' \subseteq D \setminus \{d\}$. In the *System* model, it is the *Light* model influencing the *HVAC* model. That is, I_{light} = {hvac} and $I_{hvac} = \varnothing$. Similar to atomic DEVS models, coupled DEVS models can define input and output events to interface with other models. X_{self} and Y_{self} denote the input and output events, respectively. In the example Figure 8.5, the couple d DEVS model does not define its own events. Thus, $X = \varnothing$ and $Y = \varnothing$. Due to the independence of models $d \in D$ within the coupled DEVS model, events within the coupled models might happen in parallel. To retain ambiguous execution semantics, the *select*: $2^D \rightarrow D$ tie-breaking function defines which model to treat with priority in case of conflicting models. In the *System* DEVS model, we choose to prioritize the *Light* component over the *HVAC* component, i.e., *select* = {{light, hvac} → light, {light} → light, {hvac} → hvac}. Finally, to foster the reuse of DEVS models, the set of {Z} translation functions allow translating the output events of model $d_1 \in D$ onto the input events of $d_2 \in D$, potentially modifying their content. In the coupled *System* DEVS model, the turn_on output event of the *Light* model should trigger the transition from the COOL to the TMP_U state in the HVAC model. That is, the turn_on output event should be translated to the tmp_up input event. Similarly, turn_off should be translated to tmp_down. That is, Z={turn_on → tmp_up, turn_off → tmp_down}.

8.2.3 Reinforcement Learning

Reinforcement learning [12] is a machine learning paradigm and a family of algorithms implementing goal-directed learning from interactions

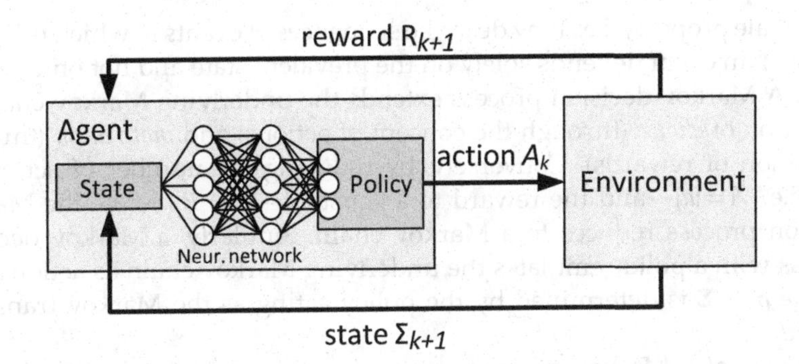

FIGURE 8.6
Deep reinforcement learning – conceptual overview.

with the environment. Reinforcement learning focuses on learning how to react to specific states of the environment—that is, how to map situations to actions—–in order to maximize a reward signal.

Figure 8.6 shows the conceptual overview of reinforcement learning. In a typical reinforcement learning setting, the *agent* explores its *environment* sequentially and exploits it to achieve a result. At each step, the agent carries out an *action*, observes the new *state* Σ_{k+1} of the environment, and receives a *reward* R_{k+1}. Through a repeated sequence of actions, the agent learns the best actions to perform in the different states. The product of the reinforcement learning process is the *policy* π that maps situations the agent can encounter to actions the agent can execute in those situations. Thus, the policy is defined as $\pi(a \in A | \sigma \in \Sigma)$ to choose action a given the current state σ. The policy can be represented in a simple lookup table or can be as intricate as a neural network. The latter case is known as *deep reinforcement learning*, where the eventual decision about the next action is encoded in the output layer of a neural network.

Markov decision processes [20] and Markov chains [21] are particularly apt choices to formalize the process of reinforcement learning and the subsequent behavior of the trained agent.

Definition 8.3: (Markov Decision Process)

A Markov decision process is defined as four-tuple $\langle \Sigma, A, P_A, R_A \rangle$.

Here, Σ denotes the set of states of the environment the agent can observe. A denotes the set of actions the agent can perform on the environment. $P_A : \langle (\sigma \in \Sigma, a \in A) \mapsto \Pr(\sigma' | \sigma, a) \rangle$ is the policy, that is, the probability of the 'agent executing action a in state σ, which will result in the new state σ'. $R_A : (\sigma \in \Sigma, a \in A) \mapsto \mathbb{R}$ is the numeric reward received for choosing action a in state σ.

Markov chains [21] are the formal structures underpinning Markov decision processes. A Markov chain is a stochastic state transition system with a martingale property, i.e., they describe sequences of events in which the probability of an event depends solely on the prevalent state and not on previous states. A Markov decision process extends the underlying Markov chain by the notion of *choice* (through the concept of actions) and *motivation* (through the notion of rewards). Conversely, by reducing the number of actions to one—i.e., $A = \{a\}$—and the reward to a constant—e.g., $R_A \mapsto 1$—the Markov decision process reduces to a Markov chain. Similarly, a Markov decision process with a policy emulates the underlying Markov chain as action $a \in A$ in state σ *in* Σ is determined by the policy acting as the Markov transition matrix.

The aim of a Markov decision process is, therefore, to identify the best policy P_A that maximizes the reward (in policy-based methods) or a more long-term, accumulative value (in value-based methods). The behavior of a reinforcement learning agent can be formally captured in this compact formal structure. This formalization helps define how reinforcement learning can be programmed to learn simulators, and in particular, simulation models conforming to DEVS.

8.2.3.1 Various Flavors of Reinforcement Learning

While the reward function quantifies the immediate reward associated with an action, it is often more useful to optimize the agent's behavior for a long-term *value* function. The value function V is the expected sum of rewards along a trajectory of actions. *Value-based* reinforcement learning methods aim to maximize this value function. In contrast, *policy-based* methods aim to maximize the reward function in every state. *Actor-critic* methods provide a trade-off between value-based and policy-based methods, in which the actor implements a policy-based strategy, and the critic criticizes the actions made by the actor based on a value function. Different methods offer different benefits and perform with different utility in specific learning problems. Value-based approaches are deterministic because they select actions greedily when maximizing the value function; however, this might lead to under-exploration. In contrast, policy-based methods explore the state space more thoroughly but produce more noisy estimates, potentially resulting in unstable learning processes. Recently, non-cumulative reward functions with a martingale property—that is, $\mathbb{E}(V|a_1,...a_n) = R(s_n, a_n)$—have been successfully utilized to address these problems [22].

The above methods belong to the *active* type of reinforcement learning methods, in which the goal is to learn an optimal policy. In contrast, *passive* reinforcement learning methods aim to evaluate how good a policy is. Finally, in *inverse* reinforcement learning settings, the goal is to learn the reward function under a fixed policy or value function and state space.

Specifically, this chapter illustrates an approach for learning DEVS models by a policy optimization method called Proximal Policy Optimization (PPO) [23]. PPO is a policy gradient method that improves upon its predecessor,

Trust Region Policy Optimization (TRPO) [24], by eliminating performance drops while simplifying implementation. Due to its performance, PPO has become a popular choice recently for reinforcement learning tasks.

Structurally, PPO is implemented in an actor-critic fashion, where the actor optimizes the policy by immediate rewards, and the critic drives the decisions of the actor by considering the value function of long-term rewards. The value function maintained by the critic is used in actor updates, thereby allowing the critic to influence the actor. PPO alternates between interacting with the environment and optimizing an objective function by stochastic gradient ascent or descent. The agent updates the actor and critic in a batch fashion, allowing for multiple gradient updates in each epoch (i.e., one forward and one backward pass of the training examples in the batch). At the end of each batch, the agent updates its actor and critic networks.

8.2.3.2 Reinforcement Learning in Simulator Engineering

Reinforcement learning has been successfully employed for various problems of simulator engineering before. Imitation learning has been shown to be a feasible approach to infer DEVS models by Floyd and Wainer [25]. In an imitation learning setting, the agent learns from another agent or human by observing their reactions to a sequence of inputs. Such techniques show an immense utility when eliciting the tacit knowledge of an expert is a particularly challenging task and expert demonstrations are easier to carry out. A substantial challenge in imitation learning is the low number of training samples, and therefore, the quality of demonstrations is paramount for success. For example, demonstrations must sufficiently cover the characteristic cases and edge-cases of problem-solving process. The amount of learning input is not a problem when learning DEVS models of autonomous systems, and therefore, the techniques of reinforcement learning can be fully leveraged.

A less interactive way to learn from examples has been provided by Otero et al. [26], who use inductive learning to infer general knowledge about the system in the form of DEVS models. The approach relies on a behavior description language by which the system has to be described in appropriate detail, serving as the training input to the learning algorithm.

Deep reinforcement learning has been used by Sapronov et al. [27] to learn and store the parameterization of discrete event models. The eventual policy encodes an optimal control strategy for the parameters of the simulator. Such a fusion of a reinforcement learning trained neural network with discrete event simulations offers better customizability of the simulator and the ability to optimize its configuration from an ingested data stream by updating the neural network.

A special form of classic reinforcement learning, contextual bandits [28] have been shown to be an appropriate choice in the runtime adaptation and reconfiguration of Digital Twins with continuous control [29,30].

Contextual bandits reduce the environment to a one-state Markov decision process and focus solely on mapping action-reward combinations. This simplification is admissible when the results of an action can be assessed immediately. Continuous control problems of Digital Twins are prime examples of such settings. Digital Twin settings typically entail high-performance simulators for the automated optimization of the physical asset. The tuning of such simulation models is crucial in achieving appropriate precision and performance.

8.3 Inference of Simulators by Reinforcement Learning

In our approach, simulators are inferred by observing the states of the physical system. As shown in Figure 8.7, Digital Twin settings typically provide advanced facilities to observe the system to build a simulator for. Particularly, the models of the corresponding Digital Shadow can be used to gather real-time information about the physical asset. In many cases, interacting with the environment is not feasible or raises safety concerns. In such cases, simulated environments can be used to infer simulators [31]. The advantage of simulated environments is that the learning agent can learn long trajectories (sequences of actions) in a short time as the responses of the environment are simulated. In addition, machine learning capabilities are already present in a typical Digital Twin setting, further simplifying the task of inferring simulators.

Reinforcement learning excels in solving sequential decision-making problems based on an explicitly modeled set of actions. Thus, reinforcement learning is a particularly appropriate choice in engineering problems where the set of engineering actions is finite and effectively enumerable, such as DEVS model construction. As a learning approach of the unsupervised

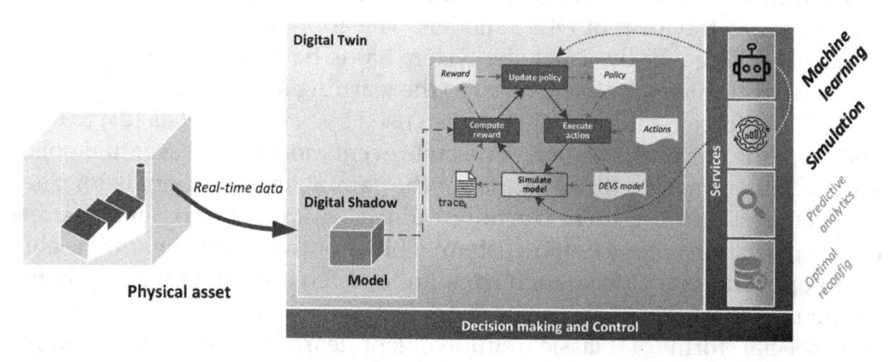

FIGURE 8.7
Overview of the approach.

kind, reinforcement learning does not require a priori labeled data. Instead, the learning agent is provided with a set of engineering actions, and by trial-and-error, the agent learns to organize these actions into sequences of high utility. The learning agent iteratively chooses actions and executes them, thereby updating the candidate DEVS model. The candidate DEVS model is simulated with an appropriate simulator. The learning environment can be equipped with a standalone DEVS simulator for this purpose, or the Digital Twin can provide it as a service (as shown in Figure 8.7). The output of the simulation in iteration k is trace$_k$. The trace stores behavioral information about the simulation, which provides means for comparing the behavior of the simulator with that of the real system. The reward is computed by comparing the trace of the simulation and the trace (history) of the real system. This cycle is repeated multiple times, ensuring enough time for the agent to infer a well-performing policy by continuously updating it.

This training session results in two important artifacts. First, the final DEVS model which will be used in the simulator service of the Digital Twin. Second, the inferred intelligence persisted in the policy of the agent. This intelligence approximates the tacit knowledge of the domain expert and is able to solve not only the specific problem at hand but other congruent problems as well. Thus, this intelligence can be reused across Digital Twins built for different physical assets.

In the rest of this section, we outline a formal framework for integrating DEVS and reinforcement learning, and whenever necessary, we illustrate concepts through the illustrative example introduced in Section 8.2.2.1.

8.3.1 Formal Framework

To utilize reinforcement learning in the inference of DEVS models, DEVS construction operations are needed to be translated to reinforcement learning terms. In the following, we briefly review such a translation, provided by David and Syriani [32] and David et al. [33].

Given a Markov Decision Process $MDP = \langle \Sigma, A, P_A, R_A \rangle$ and the atomic DEVS specification $M = \langle X, Y, S, q_{\text{init}}, \Delta_{\text{int}}, \Delta_{\text{ext}}, \lambda, ta \rangle$ (Definition 1), the mapping MDP $\rightarrow M$ is briefly given as follows.

8.3.1.1 States

Every state $\sigma \in \Sigma$ of the Markov Decision Process encodes one DEVS configuration M. That is, $\forall \sigma \in \Sigma : \sigma \mapsto M$. We use the notation σ_M to denote the state that encodes M.

8.3.1.2 Actions

Intuitively, actions in the Markov Decision Process that learns to construct DEVS models are DEVS operators. Every action is applied to a DEVS model M, resulting in a new model M'. Subsequently, a new state encoding M', $\sigma_{M'}$ is added to the states of the MDP. That is, $\forall a \in A : a(M) \mapsto M', \Sigma \mapsto \Sigma \cup \{\sigma_{M'}\}$. In accordance with the Markov property of the underlying search process, in each state, the probability of executing action $a \in A$ is determined by policy P_A.

Here, we only show one example used later in this chapter. Setting the time advance function to a specific value is defined as:

$$\text{update } Ta\left(M, s, t' \in \mathbb{R}^{+}_{0,+\infty}\right) : M' = M \,|\, ta(s) = t'. \tag{8.1}$$

The detailed list of atomic and coupled DEVS actions is provided in Ref. [32].

8.3.1.3 Reward Function

To guide the learning agent toward learning the right DEVS model, the reward function R_A maps each state to a metric. That is, $R_A : \Sigma \to \mathbb{R}$. The reward is calculated by simulating the candidate model and evaluating how similar its behavior is to the behavior of the modeled system.

The behavior of a simulation is reflected in its trace. Roughly speaking, $t(\mathcal{S}(M)) = \{(\tau_1, y_1), (\tau_2, y_2), \ldots, (\tau_n, y_n), \ldots\}$, so that each event $y_i \in M.Y$. ($M.Y$ denotes the set of output events – see Definition 1).

To measure the similarity of traces or the lack thereof, we need an appropriate distance metric. Roughly, a distance metric on the set of traces T is a non-negative real-valued function $d : T \times T \to \mathbb{R}$, such that $d(t_1, t_2) = 0 \Leftrightarrow t_1 = t_2$ (identity); and $d(t_1, t_2) = d(t_2, t_1)$ (commutativity).

Based on the above, we are ready to define the general reward function for learning DEVS models as $R = -\left|d\left(t(\mathrm{M}), t(\mathcal{S})\right)\right|$. That is, the reward is inversely proportional to the distance between the trace of the model under construction M and the trace of the system to be modeled S.

8.3.2 Agent Setup

Every reinforcement learning approach requires a specific setup. In this section, we show how a PPO agent (see Section 8.2.3.1) can be used to implement a reinforcement learning mechanism. Table 8.1 shows an example parameterization of the PPO agent implementation of Tensorforce, the reinforcement learning library of TensorFlow.[3] These settings have been shown to be successful in the inference of DEVS models previously [32].

The number of updates per batch is defined by the *multi_step* parameter. The gradient ascent or descent update is performed on batches of action trajectories defined by the *batch size*. A crucial element of policy update is the update step. If the policy is updated in too large steps, the policy performance

[3] https://www.tensorflow.org/.

TABLE 8.1
Example Agent Settings

Parameter	Value
agent	PPO
multi_step	10
batch_size	10
likelihood_ratio_clipping	0.2
discount	1.0
Exploration	0.01
subsampling_fraction	0.33
learning rate	1e-3
l2_regularization	0.0
entropy_regularization	0.0
States	type = 'float' shape = (4,) min_value = 0 max_value = 24
Actions	type = 'int' shape = (4,) num_ values = 5

can deteriorate. A conservative *likelihood ratio clipping* of 0.2 removes the incentive for changing the objective function beyond the [0.8, 1.2] interval. Setting the *discount* factor of the reward function to 1.0 allows the agent to collect trajectories up to the horizon without degrading the weight of later decisions. Setting the *exploration* parameter to 0.01 allows the agent some flexibility to randomly explore. That is, the agent can deviate from the policy in 1% of the cases. The *subsampling fraction* parameter controls the ratio of steps taken in an episode used for computing back-propagation. *L2 regularization* adds random noise to the loss to prevent overfitting. This parameter should be set appropriately based on some manual experimentation of the agent's tuning. *Entropy regularization* would help balance the probabilities of the policy to keep the entropy at maximum, but in the context of this book chapter, we can safely assume equally likely actions at the beginning of the training. The neural network storing the policy is composed of seven layers: four dense layers with three register layers in between them, as suggested for Tensorforce settings with multiple actions.

8.3.3 States

The state of the agent has to store the four time advance values. This information can be captured in a 4×1 tensor of decimal numbers (floats) as follows: $\left[ta_{\text{off}}, ta_{\text{on}}, ta_{tmp\_u}, ta_{tmp\_d} \right]$. In the running example, the Digital Twin has to simulate cycles of days. Therefore, the following reasonable constraints can be specified: $0 \leq ta_{on}, ta_{off}, ta_{tmp\_u}, ta_{tmp\_d} \leq 24$.

8.3.4 Actions

To allow the agent to manipulate the time advance values of the DEVS model under construction, actions should be specified in accordance with Eq. (1.1). Below, we show five typical actions.

Macro-increase *Inc*(*ta*): $ta \mapsto ta + 0.1$

Micro-increase *inc*(*ta*): $ta \mapsto ta + 0.01$

No change *noc*(ta): $ta \mapsto ta$

Micro-decrease *dec*(*ta*): $ta \mapsto ta - 0.01$

Macro-decrease Dec(*ta*): $ta \mapsto ta - 0.1$

Macro-steps improve the convergence of the agent to a rudimentary solution, while micro-steps allow the agent to refine the solution within a solution area. To speed up the learning process, the agent should be able to change each of the four time advance values encoded in the state in each step. Therefore, it is a good idea to encode the five actions operating on the four time advance values as a 4×5 tensor.

The response to a chosen action is then obtained, as shown in Listing 1.

Here, we only show the update of the first state variable, i.e., ta_{off}. The response mechanism of the agent expects the chosen actions on its input (action). The first element of the action tensor (action[0]) corresponds to the action chosen for the first element of the state, i.e., ta_{off}. The chosen action is represented by an integer. The switch case in Lines 10–21 interprets the value under action[0] and reacts accordingly. Changes to the rest of the state variables are calculated in a similar way. Finally, the new state is returned in Line 30.

Tensorforce provides means for specifying action masks which allow for controlling which actions are enabled in specific states. The action mask in Listing 2 ensures that the agent's actions cannot bring the time advance values outside the [0, 24] interval. Element [*i*, *j*] in the action mask tensor corresponds to element [*i*, *j*] of the action tensor and element [*i*] of the state tensor. For example, mask element [1,1] `(Self.ta[elf.tatensor.` determines that the first action in the action tensor (*Inc* (*ta*)) action can be executed on the first value of the state (ta_{off}) only if $ta_{off} \leq 23.9$. Clearly, this means that in each timestep *k* (see Figure 8.7), $ta_{off}(k) \leq 24$.

8.3.5 Reward

As per Section 8.3.1, the reward function is formulated as the distance between the trace of the candidate model and the reference trace.

Following the notations of Figure 8.7, the reward in step k is defined as $r_k = -|d(\text{trace}_k, \text{trace}_0)|$. The goal of the agent is to maximize the expected cumulative reward, i.e., $\max\left\{E\left\{\left[\sum_{k=0}^{H} a_k r_k\right]\right\}\right\}$, where H denotes the horizon of the action trajectory, a_k denotes the kth action, and r_k denotes the kth reward. To ensure faster convergence to the learning objective, a more aggressively increasing penalty can be used, e.g., $-r^2$. The distance metric is the Euclidean distance between the traces. Accordingly, the two2-norm (or Euclidean norm) of the trace vectors can be used, i.e., $r_k = \left(\sum_{n=1}^{|\text{trace}_k|} |x_n|^2\right)^{\frac{1}{2}} - \left(\sum_{n=1}^{|\text{trace}_0|} |x_m|^2\right)^{\frac{1}{2}}$, where $|\cdot|$ denotes length with vector arguments $trace_k$ and $trace_0$, and absolute with scalar arguments x_n and x_m.

```
1   def response(self, action):
2   on_action = action[0]
3   off_action = action[1]
4   tmp_u_action = action[2]
5   tmp_d_action = action[3]
6   update = np.array([0, 0, 0, 0])
7
8   #Compute ON action
9   if on_action == 0: update += np.array([0.1, 0, 0, 0])
10  elif on_action == 1: update += np.array([0.01, 0, 0, 0])
11  elif on_action == 2: update += np.array([0, 0, 0, 0])
12  elif on_action == 3: update += np.array([-0.01, 0, 0, 0])
13  elif on_action == 4: update += np.array([-0.1, 0, 0, 0])
14  else: raise Exception('Unknown action')
15
16  #Compute OFF action
17  ...
18  #Compute TMP_U action
19  ...
20  #Compute TMP_D action
21  ...
22  return self.ta+update
```

Listing 8.1: Action mask controlling the applicability of actions

8.4 Discussion

We discuss the advantages and limitations of the presented approach, as well as outline some challenges.

```
1  np.array([
2  [self.ta['off'] <= 23.9, self.ta['off'] <= 23.99,
3  True, self.ta['off'] >= 0.01, self.ta['off'] >= 0.1],
4  [self.ta['on'] <= 23.9, self.ta['on'] <= 23.99,
5  True, self.ta['on'] >= 0.01, self.ta['on'] >= 0.1],
6  [self.ta['tmp_u'] <= 23.9, self.ta['tmp_u'] <= 23.99,
7  True, self.ta['tmp_u'] >= 0.01, self.ta['tmp_u'] >= 0.1],
8  [self.ta['tmp_d'] <= 23.9, self.ta['tmp_d'] <= 23.99,
9  True, self.ta['tmp_d'] >= 0.01, self.ta['tmp_d'] >= 0.1],
10 ])
```

Listing 8.2: Action mask controlling the applicability of actions

8.4.1 Advantages of This Approach

As we have shown, reinforcement learning aligns well with the construction of simulators. As reported by previous work [32], the illustrative setup in Section 8.3 performs well on DEVS construction problems. The state encoding and the reward function in are general enough to accommodate simulation formalisms other than DEVS. Basically, it is the set of actions (Section 8.3.4) that make the presented approach specific to DEVS. By the separation of formalism-specific concerns, the approach promises high reusability across various formalisms used in Digital Twins.

Our approach illustrates that the complexity of simulator construction reduces to the essential complexity of setting up and tuning reinforcement learning agents. Accidental complexity is largely mitigated, mostly due to the high automation and minimized need for human intervention. The engineering of Digital Twins is challenged by the enormous complexity of the systems subject to twinning, and therefore, approaches such as the one outlined in this book chapter project to be key enablers in the next generation of Digital Twins.

With proper automation, these approaches reduce the costs of developing simulators. As a consequence, the development costs of Digital Twin services relying on simulation will decrease as well. Reduced costs, in turn, enable more sophisticated services and features. For example, training machine learning agents in the simulated worlds of Digital Twins allows the agent to explore potentially unsafe or hazardous settings without actual physical risk.

As emphasized previously, DEVS is a versatile simulation formalism. Consequently, it is not surprising to see numerous Digital Twin use cases relying on DEVS. Such use cases with substantial upside include the optimization of manufacturing processes through simulation [34], explicitly modeled interactions between services and with data Markov Decision Process(APIs) that allow for better validation and verification of interaction protocols [35], and explicit models at runtime driven by programmed graph transformations [36].

8.4.2 Limitations

While the eventually inferred model behaves similarly to the system to be modeled, this approach faces limitations.

It is not always obvious whether learning a similarly behaving model is sufficient or structural constraints should be respected as well. To overcome this limit, one needs to ensure that behavioral congruence is sufficient for the problem at hand. In addition, the trace information must be rich enough to allow reasoning about the property on which the similarity between the underlying asset and the simulation model is based on. Alternative distance measures can be considered to further improve the reward function, such as dynamic time warping and various kernel methods [37,38]. In many practical settings, ensuring behavioral congruence is sufficient, and this threat to validity may be negligible.

While opting for DEVS makes the outlined approach fairly generic, the approach is reusable only if appropriate congruence is ensured between applications. As demonstrated in Ref. [32], it is feasible to parameterize the outlined approach in a way that the agent and the reward function are reusable for a class of models. However, the generalizability of the policy might be limited to structurally similar models. Transposing the originally inferred policy to other problems without retraining the agent is a challenge to be investigated. One promising direction to explore is the ability of validity frames [10] to capture the contextual information of the physical asset and its environment. The conditions under which the policy is transferable to other problems could be expressed in a formal way.

8.4.3 Open Challenges

Below, we outline some of the challenges and opportunities we find the most important currently.

8.4.3.1 Human–Machine and Machine–Machine Learning

We see opportunities in learning settings with the human or other machine learning agents involved. While humans perform poorly in specifying optimal models, they perform considerably well in specifying reasonable ones [39]. As a consequence, rudimentary models or model templates provided by domain experts would increase the performance of the approach. Furthermore, domain experts can be employed in the oversight of the learning process as well. In such interactive settings, the human expert can be queried by the machine at the points of the exploration where the search algorithm indicates tough choices for the machine. Learning from the human expert can be enabled to further improve the agent's performance. Despite the high costs of human reward function in naive reinforcement learning approaches [40], learning from the human has been shown to be feasible in numerous settings [41,42].

Alternative techniques, such as learning from demonstration [43] and active learning [44], should be considered to augment the approach with. Once reinforcement agents become experienced enough with different instances and flavors of the same class of Digital Twins, new agents can be trained by experienced agents. Such settings would allow removing the human from the loop and replacing the manually assembled default policies with the ones the agents themselves inferred from previous cases.

8.4.3.2 Policy Reuse in Congruent Problems

One of the main benefits of building inference methods specifically for DEVS is that the wide range of problems that can be mapped onto DEVS can effectively leverage the approach presented in this book chapter. This facilitates the much sought-after reuse of simulators and simulator components [7,8]. Reusing the policy of the agent is different from simulator reuse. When a policy is reused, previously learned patterns of simulator inference are used to solve slightly different simulator inference problems.

Policy reuse is typically achieved by transfer learning [45], in which the output layer of the neural network encoding the policy is removed, and the rest of the network is retrained on the new problem. This, in effect, associates the previously learned solution patterns with the tokens of the problem at hand. The effectiveness of transfer learning in simulator inference is determined by the similarity of the original problem and the new one.

8.4.3.3 Ensembles of Inverse Learning Methods

The approach presented in this book chapter uses a generic reward function based on behavioral traces. However, some problems might necessitate more specialized reward functions. As discussed in Section 4.4, appropriately capturing the reward function becomes challenging as the complexity of the underlying system increases. Similarly, we used a set of generic DEVS model engineering operations, but domain-specific approaches might prefer to leverage a more refined set of actions. For example, the exact value by which the time advance is increased or decreased in Section 4.3.4 was an arbitrary choice at our end.

To approach the estimation of these parameters in a sound way, inverse reinforcement learning methods [46] can be employed. Inverse reinforcement learning aims to learn the reward function under a fixed policy. Similarly, the action space can be inferred under a fixed policy and reward function. We foresee advanced learning settings in which the reward function, value function, and actions are inferred in an iterative fashion by an ensemble of inverse reinforcement learning algorithms. This will enable simulator inference processes that are better tailored to the problem at hand.

8.4.3.4 Co-design of Digital and Physical Capabilities

The feasibility of using machine learning methods is heavily influenced by the amount and quality of data that can be used to train the machine learning agent. The same holds for the services provided by Digital Twins. In the context of the approach presented in this book chapter, data quality and quantity can be improved by considering data concerns in the early design phases of Digital Twins. We anticipate research directions focusing on the co-design of Digital Twin services and the instrumentation of the physical asset (such as the sensor and actuator architecture, data management infrastructure, and appropriate network configuration). Such integrated approaches will enable more efficient harvesting of data on the physical asset, give rise to more sophisticated real-time data processing capabilities [47], and improve the costs associated with the engineering of Digital Twins.

8.4.3.5 The Need for Unified Digital Twin Frameworks

The lack of unified Digital Twin frameworks adversely affects the engineering of their simulators [48], even if automated as discussed in this chapter. Limitations of reports on Digital Twin have been recently identified as a factor that hinders understanding and development of Digital Twin tools and techniques [49]. Experience reports often omit important characteristics of the Digital Twin, such as the scope of system-under-study, the time-scale of processing, and life-cycle stages. Standards such as the Reference Architectural Model Industry 4.0 (RAMI 4.0) [50] and various academic frameworks [51,52] are the first steps toward alleviating this issue, although they only address select aspects of Digital Twins.

The benefits of ontology-based semantic knowledge management have been demonstrated in the design of complex heterogeneous systems [53]. The success of ontology-based techniques in the realm of cyber-physical systems [54] sets promising future directions in their application in the design of Digital Twins.

8.5 Conclusion

As the complexity of services provided by Digital Twins keeps increasing, the importance of advanced simulation capabilities to support those services grows as well. Therefore, proper automation is paramount in enabling the construction of simulators.

This chapter outlined how machine learning can be used to automate the construction of simulators in Digital Twins. We illustrated the concepts of such an approach based on DEVS and reinforcement learning. Depending on the problem at hand, other simulation formalisms and other machine learning approaches might be appropriate choices as well. However, DEVS projects to be an appropriate choice for a wide range of problems as it is commonly considered the assembly language of simulation formalisms [18]. Reinforcement learning is a natural fit for problems in which previously labeled data are not available, but model operations can be exhaustively listed. The formal foundations of DEVS provide a good starting point for the latter, allowing reinforcement learning to be leveraged for inferring DEVS models.

The approach discussed in this chapter is of particular utility in the context of systems that are challenging to model due to their complexity, such as Digital Twins. We anticipate numerous lines of research on this topic unfolding in the upcoming years.

Bibliography

[1] Adil Rasheed, Omer San, and Trond Kvamsdal. Digital twin: Values, challenges and enablers from a modeling perspective. *IEEE Access*, 8:21980–22012, 2020.

[2] Benoit Combemale et al. A hitchhiker's guide to model-driven engineering for data-centric systems. *IEEE Software*, 2020, 38(4): 71–84.

[3] Sheldon M Ross. *Simulation (Statistical Modeling and Decision Science)* (5th ed.). Academic Press, 2012. ISBN 978-0-12-415825-2.

[4] Francis Bordeleau, Benoit Combemale, Romina Eramo, Mark van den Brand, and Manuel Wimmer. Towards model-driven digital twin engineering: Current opportunities and future challenges. In Systems Modelling and Management: First International Conference, ICSMM 2020, Bergen, Norway, June 25–26, 2020, Proceedings 1. Springer International Publishing, 2020: 43–54.

[5] Michael Spiegel, Paul F Reynolds Jr., and David C Brogan. A case study of model context for simulation composability and reusability. In Proceedings of the Winter Simulation Conference, 2005. IEEE, 2005: 8 pp.

[6] Thijs Defraeye, Chandrima Shrivastava, Tarl Berry, Pieter Verboven, Daniel Onwude, Seraina Schudel, Andreas Buhlmann, Paul Cronje, and Rene M Rossi. Digital twins are coming: Will we need them in supply chains of fresh horticultural produce? *Trends in Food Science & Technology*, 109:245–258, 2021. DOI: 10.1016/j.tifs.2021.01.025.

[7] Ernest H Page and Jeffrey M Opper. Observations on the complexity of composable simulation. In Proceedings of the 31st conference on Winter simulation: Simulation---a bridge to the future-Volume 1. 1999: 553–560.

[8] Richard J Malak and Chris J J Paredis. Foundations of validating reusable behavioral models in engineering design problems. In Proceedings of the 2004 Winter Simulation Conference, 2004. IEEE, 2004, 1.

[9] Robert G Bartholet, David C Brogan, Paul F Reynolds, and Joseph C Carnahan. In search of the philosopher's stone: Simulation composability versus component-based software design. In Proceedings of the Fall Simulation Interoperability Workshop. 2004.

[10] Simon Van Mierlo, Bentley James Oakes, Bert Van Acker, Raheleh Eslampanah, Joachim Denil, and Hans Vangheluwe. Exploring validity frames in practice. In *International Conference on Systems Modelling and Management*. Cham: Springer International Publishing, 2020: 131–148.
[11] Bernard P Zeigler, Alexandre Muzy, and Ernesto Kofman. *Theory of Modeling and Simulation: Discrete Event & Iterative System Computational Foundations*. Academic Press, 2018.
[12] Richard S Sutton and Andrew G Barto. *Reinforcement Learning: An Introduction*. Robotica, 1999, 17(2): 229–235.
[13] Robert R Schaller. Moore's law: Past, present and future. *IEEE Spectrum*, 34(6):52–59, 1997.
[14] Magnus Persson, Martin Torngren, Ahsan Qamar, Jonas Westman, Matthias Biehl, Stavros Tripakis, Hans Vangheluwe, and Joachim Denil. A characterization of integrated multi-view modeling in the context of embedded and cyber-physical systems. In *Proceedings of the International Conference on Embedded Software, EMSOFT 2013, Montreal, QC, Canada, September 29–October 4, 2013*, pages 10:1–10:10. IEEE, 2013. DOI: 10.1109/EMSOFT.2013.6658588.
[15] Stefan Boschert and Roland Rosen. Digital twin-the simulation aspect. In Mechatronic futures: Challenges and solutions for mechatronic systems and their designers, 2016: 59–74.
[16] Marvin Minsky. *Matter, Mind and Models*. 1965.
[17] Saurabh Mittal et al. Digital twin modeling, co-simulation and cyber use-case inclusion methodology for iot systems. In 2019 Winter Simulation Conference (WSC). IEEE, 2019: 2653–2664.
[18] Hans Vangheluwe. DEVS as a common denominator for multi-formalism hybrid systems modelling. In Cacsd. conference proceedings. IEEE international symposium on computer-aided control system design (cat. no. 00th8537). IEEE, 2000: 129–134.
[19] Claudio Gomes, Casper Thule, David Broman, Peter Gorm Larsen, and Hans Vangheluwe. Co-simulation: A survey. *ACM Computing Surveys*, 51(3):49:1–49:33, 2018. DOI: 10.1145/3179993.
[20] Martin L Puterman. Markov decision processes. *Handbooks in Operations Research and Management Science*, 2:331–434, 1990.
[21] James R Norris and James Robert Norris. *Markov Chains. Number 2*. Cambridge University Press, 1998.
[22] Nelson Vadori et al. Risk-sensitive reinforcement learning: a martingale approach to reward uncertainty. In Proceedings of the First ACM International Conference on AI in Finance. 2020: 1–9.
[23] John Schulman, Filip Wolski, Prafulla Dhariwal, Alec Radford, and Oleg Klimov. Proximal policy optimization algorithms. arXiv preprint arXiv:1707.06347, 2017.
[24] John Schulman, Sergey Levine, Pieter Abbeel, Michael I Jordan, and Philipp Moritz. Trust region policy optimization. In Proceedings of the 32nd Inter*national Conference on Machine Learning, ICML 2015, Lille, France, 6–11 July 2015, Volume 37 of JMLR Workshop and Conference Proceedings*, pages 1889–1897. JMLR.org, 2015.
[25] Michael W Floyd and Gabriel A Wainer. Creation of devs models using imitation learning. In Proceedings of the 2010 Summer Computer Simulation Conference. 2010: 334–341.

[26] Ramon P Otero, David Lorenzo, and Pedro Cabalar. Automatic induction of devs structures. Computer Aided Systems Theory—EUROCAST'95: A Selection of Papers from the Fifth International Workshop on Computer Aided Systems Theory Innsbruck, Austria, May 22–25, 1995 Proceedings 5. Springer Berlin Heidelberg, 1996: 305–313.

[27] Andrey Sapronov, Vladislav Belavin, Kenenbek Arzymatov, Maksim Kar pov, Andrey Nevolin, and Andrey Ustyuzhanin. Tuning hybrid distributed storage system digital twins by reinforcement learning. *Advances in Systems Science and Applications*, 18(4):1–12, 2018.

[28] Wei Chu, Lihong Li, Lev Reyzin, and Robert E Schapire. Contextual bandits with linear payoff functions. Computer Aided Systems Theory—EUROCAST'95: A Selection of Papers from the Fifth International Workshop on Computer Aided Systems Theory Innsbruck, Austria, May 22–25, 1995 Proceedings 5. Springer Berlin Heidelberg, 1996: 305–313.

[29] Constantin Cronrath, Abolfazl R Aderiani, and Bengt Lennartson. Enhancing digital twins through reinforcement learning. 2019 IEEE 15th International conference on automation science and engineering (CASE). IEEE, 2019: 293–298.

[30] Nikita Tomin et al. Development of digital twin for load center on the example of distribution network of an urban district. E3S Web of Conferences. EDP Sciences, 2020, 209.

[31] Marius Matulis and Carlo Harvey. A robot arm digital twin utilising reinforcement learning. *Computers & Graphics*, 95:106–114, 2021.

[32] Istvan David and Eugene Syriani. DEVS model construction as a reinforcement learning problem. 2022 Annual Modeling and Simulation Conference (ANNSIM). IEEE, 2022: 30–41.

[33] Istvan David, Jessie Galasso, and Eugene Syriani. Inference of simulation models in digital twins by reinforcement learning. 2021 ACM/IEEE International Conference on Model Driven Engineering Languages and Systems Companion (MODELS-C). IEEE, 2021: 221–224.

[34] Istvan David, Hans Vangheluwe, and Yentl Van Tendeloo. Translating engineering workflow models to devs for performance evaluation. 2018 Winter Simulation Conference (WSC). IEEE, 2018: 616–627.

[35] Simon Van Mierlo, Yentl Van Tendeloo, Istvan David, Bart Meyers, Addis Gebremichael, and Hans Vangheluwe. A multi-paradigm approach for modelling service interactions in model-driven engineering processes. *Proceedings of the Model-Driven Approaches for Simulation Engineering Symposium, SpringSim (Mod4Sim) 2018, Baltimore, MD, USA, April 15–18, 2018*, pages 6:1–6:12. ACM, 2018.

[36] Eugene Syriani and Hans Vangheluwe. DEVS as a semantic domain for programmed graph transformation. *Discrete-Event Modeling and Simulation: Theory and Applications*, pages 3–28. CRC Press, Boca Raton, FL, 2010. https://www.crcpress.com/Discrete-Event-Modeling-and-Simulation-Theory-and-Applications/Wainer-Mosterman/9781420072334.

[37] Donald J Berndt and James Clifford. Using dynamic time warping to find patterns in time series. Usama M Fayyad and Ramasamy Uthurusamy, editors, *Knowledge Discovery in Databases: Papers from the 1994 AAAI Workshop, Seattle, Washington, USA, July 1994. Technical Report WS- 94-03*, pages 359–370. AAAI Press, 1994.

[38] Bernhard Scholkopf. The kernel trick for distances. *Advances in Neural Information Processing Systems 13, Papers from Neural Information Processing Systems (NIPS) 2000, Denver, CO, USA*, pages 301–307. MIT Press, 2000.

[39] Marco A Wiering and Martijn Van Otterlo. Reinforcement learning. *Adaptation, Learning, and Optimization*, 2012, 12(3): 729.

[40] Paul Christiano, Jan Leike, Tom B Brown, Miljan Martic, Shane Legg, and Dario Amodei. Deep reinforcement learning from human preferences. *arXiv preprint arXiv:1706.03741*, 2017, 30.

[41] W Bradley Knox and Peter Stone. Reinforcement learning from simultaneous human and MDP reward. *AAMAS*, pages 475–482, 2012.

[42] Robert Loftin et al. Learning behaviors via human-delivered discrete feedback: Modeling implicit feedback strategies to speed up learning. *Autonomous Agents and Multi-Agent Systems*, 30(1):30–59, 2016.

[43] Brenna D Argall, Sonia Chernova, Manuela Veloso, and Brett Browning. A survey of robot learning from demonstration. *Robotics and Autonomous Systems*, 57(5):469–483, 2009.

[44] Burr Settles. Active learning. *Synthesis Lectures on Artificial Intelligence and Machine Learning*, 6(1):1–114, 2012.

[45] Lisa Torrey and Jude Shavlik. Transfer learning. *Handbook of Research on Machine Learning Applications and Trends: Algorithms, Methods, and Techniques*, pages 242–264. IGI Global, 2010.

[46] Andrew Y Ng and Stuart Russell. Algorithms for inverse reinforcement learning. *Proceedings of the Seventeenth International Conference on Machine* Learning *(ICML 2000), Stanford University, Stanford, CA, USA, June 29–July 2, 2000*, pages 663–670. Morgan Kaufmann, 2000.

[47] Istvan David, Istvan Rath, and Daniel Varro. Foundations for streaming model transformations by complex event processing. *Software and Systems Modeling*, 17(1):135–162, 2018. DOI: 10.1007/s10270-016-0533-1.

[48] Fei Tao, He Zhang, Ang Liu, and Andrew Y C Nee. Digital twin in industry: State-of-the-art. *IEEE Transactions on Industrial Informatics*, 15(4):2405–2415, 2019. DOI: 10.1109/TII.2018.2873186.

[49] Bentley James Oakes, Ali Parsai, Bart Meyers, Istvan David, Simon Van Mierlo, Serge Demeyer, Joachim Denil, Paul De Meuleneare, and Hans Vangheluwe. *Improved Reporting on Digital Twins: An Illustrative Example and Mapping to Asset Administration Shell*. Springer, 2023. To appear.

[50] Martin Hankel and Bosch Rexroth. The reference architectural model industrie 4.0 (rami 4.0). Zvei, 2015, 2(2): 4–9.

[51] Yuqian Lu, Chao Liu, Kevin I-Kai Wang, Huiyue Huang, and Xun Xu. Digital twin-driven smart manufacturing: Connotation, reference model, applications and research issues. *Robotics and Computer-Integrated Manufacturing*, 61:101837, 2020. DOI: 10.1016/j.rcim.2019.101837.

[52] `Shohin Aheleroff, Xun Xu, Ray Y Zhong, and Yuqian Lu. Digital twin as a service (dtaas) in industry 4.0: An architecture reference model. *Advanced Engineering Informatics*, 47:101225, 2021. DOI: 10.1016/j.aei.2020.101225.

[53] Ken Vanherpen, Joachim Denil, Istvan David, Paul De Meulenaere, Pieter J Mosterman, Martin Torngren, Ahsan Qamar, and Hans Vangheluwe. Ontological reasoning for consistency in the design of cyber-physical systems.

In *1st International Workshop on Cyber-Physical Production Systems, CPPS@ CPSWeek 2016, Vienna, Austria, April 12, 2016*, pages 1–8. IEEE Computer Society, 2016. DOI: 10.1109/CPPS.2016.7483922.

[54] Zihang Li, Guoxin Wang, Jinzhi Lu, Didem Gurdur Broo, Dimitris Kiritsis, and Yan Yan. Bibliometric analysis of model-based systems engineering: Past, current, and future. *IEEE Transactions on Engineering Management*, pages 1–18, 2022. DOI: 10.1109/TEM.2022.3186637.

9

Digital Twin for Federated Analytics Applications

Dan Wang
The Hong Kong Polytechnic University

Dawei Chen
Toyota Motor

Yifei Zhu
Shanghai Jiao Tong University

Zhu Han
University of Houston

9.1 Introduction

As we are stepping into the era of the Internet of Everything (IoE), the number of devices connecting to the Internet is growing rapidly, which will achieve 41.6 billion in 2025, generating 79.4 zettabytes (ZB) of data according to the forecast of International Data Corporation (IDC) [1]. Such a tremendous amount of data poses great challenges to systems concerning scalability, mobility, and availability. One of the key challenges is to closely connect physical devices, such as sensors, Industrial Internet of Things (IIoTs), and Internet of Vehicles (IoVs), with virtual computing resources, such as edge server, base station, roadside units (RSUs), and remote cloud center. Thanks to the flourishing developments of data acquisition and processing, high-performance computing, low-latency communication, and other techniques, the digital twin is proposed to be an effective tool for the cyber-physical association, which provides advanced analytics by utilizing real-time data rendered by physical devices [2,3]. The obtained analytics results can be used to optimize the existing device's performance or assist decision-making. For example, in industrial applications, the digital twin can realize diverse purposes, such as virtual evaluation, asset management, process evaluation and optimization, predictive maintenance, and fault detection and

DOI: 10.1201/9781003425724-12

diagnosis, which still rely on big data processing and cannot be performed on resource-constrained devices [4].

Although the techniques of hardware designs are in prosperous progress, on-device computation-intensive tasks execution is still challenging regarding big data [5]. To address this issue, multi-access edge computing (MEC) can be a solution, which geographically distributively deploys edge nodes within the network to undertake computing tasks. Besides, since edge nodes possess computation ability to a certain extent, in this way, the transmission distance can be reduced and tasks are executed almost locally [6,7]. Consequently, the computing resources limitation problem can be solved, and simultaneously the overall latency is reduced and quality of service (QoS) can be remarkably improved.

Practically, in the majority of cases, obtaining data from one single device is not sufficient. For example, in automotive scenarios, the digital twin is able to mimic the virtual model of a connected vehicle through its behavioral and operational data. To evaluate the safety for planned passing, gathering surrounding vehicles, and creating corresponding models are also necessary. Therefore, federated analytics can be an effective way for collaborative decision-making. *Federated analytics* is proposed by the Google AI team [8], which allows data scientists to derive analytical insights among distributed datasets without collecting the data in a centralized server. Actually, federated analytics follows the framework of federated learning, obtaining global aggregated results from uploads of individuals. The difference is that federated analysis only focuses on basic data science needs, such as model evaluation, data quality assessment, and heavyweight discoverers, and there is no learning part. [9]. Intuitively, there are several reasons to apply federated analytics for digital twins applications.

- Firstly, the digital twin is made up of two parts: physical objects or process that generates data and a virtual server that performs computation and makes decisions. Therefore, regarding a federated analytics problem, individual devices can be regarded as physical objects in digital twin and aggregator can be regarded as virtual server of the digital twin.
- Secondly, in federated analytics, what is transmitted within digital twins is the device's individual analytical result instead of raw data. As a result, the communication latency can be decreased further.
- Thirdly, devices may be privacy-sensitive, especially for healthcare and personal applications. In federated analytics, since data will not be shared within digital twins, the users' privacy can be guaranteed as well.

From the perspective of data science, obtaining the distribution of the whole dataset based on evidence is important. However, because of the limitation of sensor range and mobility of connected devices, the generated individual datasets are only observed from a part of the entire data, i.e., individual datasets are not independent and identically distributed (non-i.i.d.), which makes it hard to estimate the global distribution with the data from one device [10]. Intuitively, the distribution of the dataset can be characterized by distribution parameters, like mean and variance for Gaussian distribution. Therefore, once the parameters can be evaluated, the global data distribution can be obtained accordingly. Fortunately, Bayesian approaches have been proved to be an effective method to tackle global parameter and posterior estimation-related problems. The Bayesian approach estimates an unknown parameter by minimizing the posterior probability from all observed evidence. One typical method is the Monte Carlo method, which evaluates and approximates target distributions with random sampling techniques. The drawback of the Monte Carlo method is that the scheme heavily relies on the large volume repeating sampling. Therefore, how to conduct efficient sampling and guarantee the quality of samples at the same time becomes challenging.

To overcome this issue, we proposed a federated Markov chain Monte Carlo with delayed rejection (FMCMC-DR), which employs the convergence and detailed balance properties of Markov chain, rejection–acceptance sampling, and delayed rejection to assist the Monte Carlo method in realizing efficient sampling and global state space exploration.

The rest of this chapter is organized as follows. Section 9.2 introduces the specific scenario, analyzes mathematical model for federated analytics, and formulates the corresponding parameter estimation problem based on edge-assisted digital twin. In Section 9.3, related Markov Chain Monte Carlo (MCMC) methods and definitions are given and the proposed FMCMC-DR algorithm is introduced in detail. Section 9.4 describes the conducted numerical simulation and discusses the results accordingly. Finally, a conclusion is drawn in Section 9.5.

9.2 Modeling of Digital Twin

In this section, the scenario of an edge-assisted digital twin for federated analytics is described. Also, we model the physical side of the digital twin in Section 9.2.1 and the virtual side of digital twin in Section 9.2.2, respectively. Finally, the corresponding federated analytics problem formulation is given.

9.2.1 Physical Side of Digital Twin

We consider there are multiple users $N=1.....i...,N$ taking part in a federated analytics task. Because of the limitation of sensor range and mobility, each user i can only have partial observations among the whole evidence or dataset. Here, the collection of individual dataset of each user i is denoted as R_i. And $R = \{R_1, \dots R_i, \dots R_N\}$ denotes the entire dataset. Due to privacy concerns, individual datasets cannot be shared within the network. Therefore, for each user, the on-device private data analysis will be performed based on R_i locally. Accordingly, the individual analyzed output or individual data insights can be obtained by:

$$x_i = f_i(R_i), \tag{9.1}$$

where f_i denotes the local data distribution among observations of R_i. Here, x_i is the representative parameters that can describe f_i, such as mean and variance.

9.2.2 Virtual Side of Digital Twin

After individual evaluation is done, all the participants will upload their dataset features to the aggregator, which is acted by an edge server covering this area. Intuitively, this aggregation process can be written as:

$$y = g(x), \tag{9.2}$$

where $x = (x_1 \dots, x_i, x_N)$ is the locally obtained data feature vector. Private data cannot be shared while sampling technique requires true observations. In federated analytics, raw data cannot be transmitted through network due to privacy concerns.

Therefore, by using x, the distribution of individuals can be evaluated so that the virtual observations can be constructed, which is denoted as $e = e_1 \dots, e_i \dots, e_N$. Since the global distribution is obtained from the aggregated individual distribution, $g(e)$ can be described in a federated way, i.e., a linearly weighted sum of N kernel functions, which is given by:

$$\begin{aligned} g(e) &= \phi(e)^T \omega \\ &= \sum_{i=1}^{N} \omega_i \phi_i(e_i) \end{aligned} \tag{9.3}$$

where $\varphi(x) = (\varphi_1, \dots \varphi_N)^T$ is the kernel vector, the corresponding parameters can be denoted as $\theta = \{\theta_1, \dots \theta_i, \dots \theta_N\}$, and $\omega = (\omega_1, \dots \omega_N)^T$ is the associated weight vector.

In total, the summation of all the weights equals to one, i.e., $\sum_{i=1}^{N} \omega_i = 1$, where N is finite and known. Essentially, the kernel functions belong to a parametric family. Since individual datasets are non-i.i.d. due to different behavior characteristics, the kernel functions are chosen as Gaussian functions, as is suggested in Ref. [11]. Gaussian mixtures have proved to be an effective way for modeling heterogeneous populations [12]. Intuitively, Eq. (9.3) can be rewritten in the following form:

$$g(e|\omega,\theta)=\sum_{i=1}^{N}\omega_i f_i(e_i|\theta_i) \tag{9.4}$$

In the particular case in which the $f(e_i\ \theta_i)$ are all normal distributions, $\theta=(\mu_1, \sigma_1, \ldots, \mu_n, \sigma_n)$ represents each client's belief. $f(e_i\ \theta_i)$ is the density of a Gaussian distribution with mean μ_i and variance σ_i. θ denotes the unknown mean and variance for global distribution. The purpose is to find the parameter ω and θ so that the global data distribution can be obtained. Since the parameters are unknown, from the Bayesian perspective, Eq. (9.4) can be interpreted as to infer the posterior probability distribution of the model parameters given the measurement or observations e. Then, the estimated distribution parameter will be broadcast to all the clients. In this way, the client can be aware of the difference between a local distribution and global distribution. The overall scenario is illustrated in Figure 9.1.

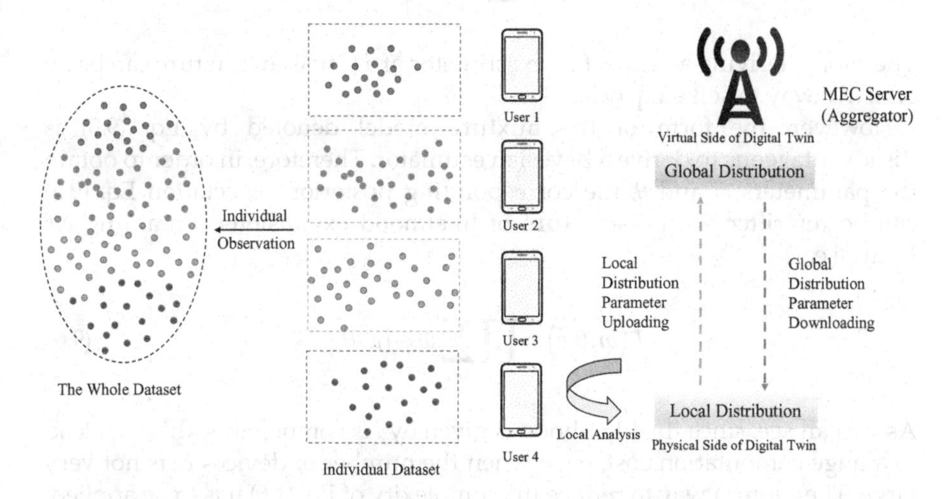

FIGURE 9.1
Scenario description: we consider there are multiple users taking part in a federated analytics task to obtain the whole dataset distribution. Each user can only have partial observations among the whole dataset due to the limitation of sensing and mobility range. For each user, the on-device private data analysis will be performed locally because of privacy concern. After individual evaluation is done, all the participants will upload their dataset features to the aggregator, which is done by an edge server covering this area. By doing aggregation, the estimation of distribution regarding the whole dataset can be obtained. Then, the estimated distribution will be broadcast to all the clients to make them aware of the difference between a local distribution and global distribution. This process will be performed iteratively to keep the global distribution updated with more samples observed from the user side.

9.3 Methodology

In this section, the concept of the missing data approach is introduced in Section 9.3.1. Then, the MCMC and standard MH is introduced in detail in Section 9.3.2. Finally, the delayed rejection technique is elaborated in Section 9.3.3, and the proposed FMCMC-DR algorithm is given accordingly.

9.3.1 Missing Data Approach

As is discussed in the previous section, the main challenge to obtain the global distribution is to estimate and calculate the unknown parameters, i.e., ω and θ. Assume the kernel functions are convex, the expectation of Eq. (9.3) is also a convex combination of expectations, i.e.,

$$\begin{aligned} \mathrm{E}[e] &= \sum_{i=1}^{N} \omega_i \mathrm{E}^{\phi_i}[e] \\ &= \sum_{i=1}^{N} \omega_i \mathrm{E}[e_i|\theta_i] \end{aligned} \tag{9.5}$$

Therefore, deriving an expectation estimator of a Gaussian mixture can be an effective way to solve Eq. (9.3).

However, the form of the mixture model denoted by Eq. (9.3) is disadvantageous to derive a Bayesian estimator. Therefore, in order to obtain the parameters ω and θ, the corresponding posterior expectation Eq. (9.5) can be rewritten as the structure of likelihood expansions containing N^K terms, i.e.,

$$L(\omega,\theta|e) = \prod_{j=1}^{K} \sum_{i=1}^{N} \omega_i \phi_i (e_j|\theta_i). \tag{9.6}$$

As we can see, since the likelihood is given by N^K components, this will lead to a huge computation cost, even when the number of devices N is not very large. Therefore, a way to reduce the complexity of Eq. (1.6) has to be applied, where one of common but effective approach is missing data approach.

One interpretation of the mixture model is that the observation happens among a dataset that is made up of several strata, while the allocation of observation to a specific group is lost. Therefore, for each evidence e_i, it has a probability of ω_i to be allocating into sub-group φ_i. Because the structure is not clear or lost, an auxiliary variable can be introduced, which aims at identifying the specific sub-group or mixture component each observation is generated from. This is also the reason why this method is named as the missing data approach. Here, we denote this auxiliary variable as z_j.

To construct z_j, the association of observations e from N mixture components and z_j is

$$e_j | Z_j = i \sim \phi(e_i|\theta_i) \tag{9.7}$$

and

$$p(Z_j = i) = \omega_i \tag{9.8}$$

This auxiliary variable z_j identifies to which component the observation x_j belongs to [13]. In other words, for each sample of data $e = (e_1 \ldots, e_N)$, we assume a missing dataset $z = (z_1, \ldots, z_N)$, rendering the corresponding labels to indicate the specific mixture sub-population from which the observation has been generated.

By utilizing the constructed auxiliary random variable, the likelihood in Eq. (9.6) can be simplified into:

$$\begin{aligned} L(\omega, \theta | e, Z) &= \prod_{j=1}^{K} \omega_{zj} \phi(e_j | \theta_{zj}) \\ &= \prod_{i=1}^{N} \omega_i^{k_i} \left[\prod_{j:z_j=i} \phi(e_j|\theta_i) \right] \end{aligned} \tag{9.9}$$

where k_i equals to the number of $z_j = i$ and $\sum_i k_i = K$. Then, the posterior probability that the observation e_j belongs to the i-th mixture component is:

$$p(z_j = i | e_j, \omega, \theta) = \frac{\omega_i \phi(e_j|\theta_i)}{\sum_{i=1}^{N} \omega_i \phi(e_j|\theta_i)} \tag{9.10}$$

Since we assume the kernel functions are Gaussian, the corresponding estimated likelihood can be further written as the following:

$$L(\omega, \mu, \sigma | e, z) \propto \prod_{i=1}^{N} (\omega_i \sigma_i)^{k_i} \exp\left(-\frac{\sigma_i}{2} \sum_{j:z_j=i} (e_j - \mu_i)^2 \right) \tag{9.11}$$

Accordingly, the posterior probability is:

$$P(z_j = i | e_j, \omega, \mu, \sigma) = \frac{\omega_i \sigma_i \exp\left(-\frac{\sigma_i}{2} (e_j - \mu_i)^2 \right)}{\sum_{i=1}^{N} \omega_i \sigma_i \left(\exp\left(-\frac{\sigma_i}{2} (x_j - \mu_i)^2 \right) \right)} \tag{9.12}$$

9.3.2 Markov Chain Monte Carlo Method

In order to obtain the posterior probability, we firstly introduce the MCMC method, which utilizes the properties of the Monte Carlo method and Markov chain. Generally, parameter distribution cannot be obtained analytically. Therefore, the Monte Carlo aims to conduct repeated random sampling and use the expectation of sampling to estimate or approximate parameters. Therefore, how to perform high-volume sampling efficiently is challenging, which motivates the utilization of the Markov chain. The definition of a Markov chain is as follows.

Definition 9.1

A Markov chain is a sequence of random variables Q_1, Q_2, Q_3,..., where the probability of moving to the next state is dependent on the current state while independent on the previous states:

$$\begin{aligned} &P(Q_{t+1} = q \mid Q_1 = q_1, Q_2 = q_2, \dots, Q_n = q_n) \\ &= P(Q_{t+1} = q \mid Q_n = q_n). \end{aligned} \tag{9.13}$$

By this definition, for each Markov chain, there exists a state transition matrix that describes the probabilities of the transition for any two states in this chain, which can be written as:

$$T = \begin{bmatrix} P_{11} & P_{12} & \dots & p_{1n} \\ P_{21} & P_{22} & \dots & P_{2n} \\ \vdots & \vdots & \vdots & \vdots \\ P_{n1} & P_{n2} & \dots & P_{nn} \end{bmatrix} \tag{9.14}$$

With regard to matrix T, there are several properties.

Theorem 9.1

For an irreducible, aperiodic Markov chain, with finite or infinite state space, the existence of a stationary distribution π ensures that the Markov chain will converge to π as t approaches ∞, i.e.,

$$\lim_{t \to \infty}, P_{ij}^t = \pi(j) \tag{9.15}$$

where the stationary distribution can be described as $\pi = [\pi_1, \pi_2, \dots \pi_n]$ and the summation of all the elements equals to one, i.e., $\sum_{j=1}^{n} \pi(j) = 1$.

Another important property of the Markov chain is the detailed balance, which is defined as follows.

Definition 9.2

Let $Q_1, Q_2, \ldots$ be a Markov chain with stationary distribution π. The chain is said to be reversible with respect to π or statisfy detailed balance with respect to π if

$$\pi_i P_{ij} = \pi_j P_{ji}, \forall i, j \tag{9.16}$$

Intuitively, if the stationary distribution of the Markov chain is exactly the posterior probability distribution, the problem can be solved. The Monte Carlo method utilizes the property of the Markov chain for sampling. Suppose a random sample q_0 is given. According to the convergence property of the Markov chain, after t times sampling, the probability distribution converges to a stationary state $\pi(q)$.

Then, suppose sampling starts from $t+1$, the newly obtained π_{t+1} needs to be equal to π_t. With the detailed balance property, we can derive $\pi T = \pi$. If we can obtain the matrix T, the sampling can be proceeded through random walk. Now the problem is how to get T, where the Metropolis-Hastings algorithm is the effective method.

The purpose is to construct transition probability matrix T. In the Metropolis-Hastings algorithm, the acceptance rate is employed to evaluate the feasibility to adopt the newly obtained sample from $Q(q^*|q_t)$. Therefore, to interpret this, we can consider these two parts contributes to form T together, i.e., $T(q^*|q_t) \rightarrow \mathbb{Q}(q^*|q_t) \times \min\left(1, \dfrac{\pi(q^*)\mathbb{Q}(q_1|q^*)}{\pi(q_1)\mathbb{Q}(q^* q_1)}\right)$. When the acceptance rate equals to 1, u is smaller than 1, which indicates that $Q(q^*|q_t)$ is the value of T. On the contrary, when the acceptance is less than one, the value $\dfrac{\pi(q^*)\mathbb{Q}(q_1|q^*)}{\pi(q_1)\mathbb{Q}(q^*|q_1)}$ is the required T [14].

However, for the mixture model, there are multiple parameters to estimate, where a Federated MCMC algorithm is proposed to solve this problem. In order to model parameters, ω, μ and σ, we assume conjugate prior and posterior pairs as the following:

1. $\left[\omega \sim D(\delta_1, \ldots, \delta_n), \omega \mid e, z \sim D(\delta 1*, \ldots, \delta_n *)\right]$,
2. $\left[\sigma_i \sim G\left(\dfrac{\alpha}{2}, \dfrac{\beta}{2}\right), \sigma_i \mid e, z \sim G\left(\dfrac{\alpha^*}{2}, \dfrac{\beta^*}{2}\right)\right]$,
3. $\left[\mu_i \mid \sigma_i \sim N\left(m_i, \dfrac{1}{a_i \sigma_i}\right), \mu_i \mid e, z, \sigma_i \sim N\left(m_i^*, \dfrac{1}{a_i^* a_i}\right)\right]$

where $\delta_i^* = \delta_i + k_i, \alpha_i^* = \alpha + k_i, b_i^* = b + \sum_{j:z_j=i}(e_j - \mu_i)^2, a_i^* = a_i + k_i, m^*_i = \frac{a_i m_i + k_i \bar{e}_i}{a_i + k_i}$, and $\bar{x}_i = \frac{1}{k_i}\sum_{j:z_j=i} e_j$ D denotes the Dirichlet distribution, G denotes the Gamma distribution, and N denotes the normal distribution.

A basic feature of a mixture model is that it is invariant under the permutation of the indices of the components. This implies that the component parameters θ_i are not identifiable marginally: we cannot distinguish component 1 (or θ_1) from component 2 (or θ_2) from the likelihood, because they are exchangeable. Therefore, for identifiability reasons, we assume that $\mu_1 < \ldots < \mu_n$.

Note that $D(\delta_1, \ldots, \delta_n)$ follows the Dirichlet distribution with density

$$h(\omega_1, \ldots \omega_k) \sim \prod_{i=1}^{n} \omega_i^{\delta_i - 1} \tag{9.17}$$

The usual prior choice is to take $(\delta_1, \ldots, \delta_n) = (1, \ldots, 1)$ to impose a uniform prior over the mixture weights. Note that this prior choice is equivalent to use the following reparameterization:

$$\begin{gathered} \omega_1 = \eta_1, \\ \ldots \\ \omega_i = (1-\eta_1)\ldots(1-\eta_{i-1})\eta_i, \end{gathered} \tag{9.18}$$

where $\eta_i \sim B(1, n-i+1)$ and B stands for the Bernoulli distribution.

9.3.3 Federated Markov Chain Monte Carlo with Delayed Rejection

Suppose now the Markov chain arrives at a stationary state at time t. A new sample q^* will be proposed from distribution $Q(q^*\ q_1)$, where the corresponding acceptance probability is given by:

$$\begin{aligned} A_1 &= \min\left(1, \frac{\pi(q^*)Q(q_1|q^*)}{\pi(q_1)Q(q^*|q_1)}\right) \\ &= \min\left(1, \frac{\lambda_1}{\varepsilon_1}\right), \end{aligned} \tag{9.19}$$

where $\lambda_1 = \pi(q^*)Q(q_1|q^*)$ and $\varepsilon_1 = \pi(q_1)Q(q^*|q_1)$ for the convenience of illustration. If the new sample is accepted indicating a good fitting, while if it is rejected, a bad local fitting occurs and a new q^* (i.e. q_2) needs to be generated. However, for the purpose of preserving current stationary distribution, the acceptance probability of q_2 should be carefully designed and computed. One feasible way is to utilize detailed balance property at each stage separately and obtain the acceptance rate that can preserve stationary status [15], i.e., Eq. (9.20).

$$\begin{aligned} A_2 &= \min\left(1, \frac{\pi(q_2)\mathbb{Q}(q_2|q_1)\mathbb{Q}(q_2,q_1|q^*)\left[1-A_1(q_2,q_1)\right]}{\pi(q^*)\mathbb{Q}(q^*|q_1)\mathbb{Q}(q^*|q_1,q_2)\left[1-A_1(q^*,q_1)\right]}\right) \\ &= \min\left(1, \frac{\lambda_2}{\varepsilon_2}\right) \end{aligned} \tag{9.20}$$

Here $\pi(q_2)\mathbb{Q}(q_2|q_1)\mathbb{Q}(q_2,q_1|q^*)\left[1-A_1(q_2,q_1)\right]$ is denoted as $\lambda_2 \cdot \pi(q^*)\mathbb{Q}(q^*|q_1)$ $\mathbb{Q}(q^*|q_1,q_2)\left[1-A_t(q^*,q_1)\right]$ which is denoted as ε_2 for the convenience of illustration. Since the procedure comes to $t=2$ stage, which means at $t=1$ stage we have $\lambda_1 < \varepsilon_1$.

Therefore, in Eq. (9.20), A_1 can be replaced with $\frac{\lambda_1}{\varepsilon_1}$ in denominator, which is Eq. (9.21).

$$\begin{aligned} A_2 &= \min\left(1, \frac{\lambda_{t+1}}{\mathbb{Q}(q^*|q_1,q_2)\left[\pi(q^*)\mathbb{Q}(q^*|q_1)-\pi(q_1)\mathbb{Q}(q_1|q^*)\right]}\right) \\ &= \min\left(1, \frac{\lambda_2}{\mathbb{Q}(q^*|q_1,q_2)\left[\varepsilon_1-\lambda_1\right]}\right) \end{aligned} \tag{9.21}$$

$$\begin{aligned} A_t &= \min\left(1, \left\{\frac{\pi(q_t)\hat{\mathbb{Q}}\left[1-A_1(q_t,q_{t-1})\right]\cdots\left[1-A_{t-1}(q_t,\ldots,q_1)\right]}{\pi(q^*)\mathbb{Q}(q^*|q_i)\ldots\mathbb{Q}(q^*|q_1,\ldots,q_t)}\right\}\right) \\ &= \min\left(1, \frac{\lambda_t}{\varepsilon_t}\right) \end{aligned} \tag{9.22}$$

Iteratively, we can derive a general delayed rejection procedure as the following. At the t-th stage, if the sample of the previous stage is rejected, the acceptance rate can be written as Eq. (9.22), where $\hat{\mathbb{Q}} = \mathbb{Q}(q_t \mid q_{t-1})\ldots\mathbb{Q}(q_t,q_{t-1},\ldots q^*)$.

Similarly, at the t-th stage, we have $\lambda_m < \varepsilon_m$ for $\forall m = 1,\ldots,t-1$ so that A_m can always be written in the form of $\frac{\lambda_m}{\varepsilon_m}$, for $\forall m = 1,\ldots,t-1$. Intuitively, the recursive equation

$$\varepsilon_t = \mathbb{Q}(q^* \mid q1,\ldots,q_t)(\varepsilon_{t-1} - \lambda_{t-1}) \tag{9.23}$$

can be derived, which utilizes the iteratively detailed balance at each stage [16]. Note that the path from q_m to q^* is the reverse of the path from q^* to q_m, which can be interpreted as: propose q_{m-1} from q_m, if it is rejected, propose q_{m-2} from q_{m-1} again, and repeating this process until q^* is accepted. This procedure is able to give a reversible Markov chain invariant stationary distribution π. In addition, the advantage to have this back-forward path going through to q_1 is that, since q_1 is rejected at stage $t = 1$, its indicated $A_1(q^*, q_1)$ is small as well. This provides a good reason to believe $A_1(q_2, q_1)$ will be small. However, as we can see in Eq. (9.22), the term 1 $A_1(q_2, q_1)$ is the numerator component of A_2, resulting in a high acceptance probability for candidates at stage two, such that it can be advantageous for exploring the whole state space [17].

In order to obtain the parameter of joint global distribution, the Metropolis-Hastings algorithm can be an effective tool. However, as is discussed above, delayed rejection is able to enhance the performance by exploring the entire state space [18]. Therefore, to collaborate the Metropolis-Hastings algorithm with delayed rejection, we propose a method named as FMCMC-DR. For MCMC-DR, like MH method, initial values for unknown parameters will be generated from the conjugate prior distributions. Then, the rejection–acceptance sampling method is adopted as the updating scheme. Inside this, we need to compare the values of reference and acceptance rate. In order to utilize the advantage of backward exploration as delayed rejection, the calculation of acceptance rate proceeds as described in Eq. (9.22).

9.4 Results

In this section, the number of mixture distribution N is set as 4. The default number of parameters is set as 100. The default number of iteration is set as 10,000. These default values may be changed in specific cases when comparison analysis is conducted. As for the parameters of conjugate priors, we adopt the suggestions in Ref. [19]. For the prior $\omega \sim D(\delta_1, \ldots, \delta_n)$, we take $(\delta_1, \ldots, \delta_n) = (1,..,1)$ so as to impose a uniform prior over the mixture weights. For the prior $\sigma \sim G\left(\frac{\alpha}{2}, \frac{\beta}{2}\right)$, the hyperparameters α and β are both set as 0.02. For the prior μ_i, the hyperparameter m_i is set as 0 and a_i is set as 0.2. Except for the proposed FMCMC-DR method, we also adopt RW-MCMC and standard MH algorithms as baseline methods. The sampling technique of RW-MCMC is based on random walk, which means at each iteration the sampling moves to a neighboring state randomly. Also, all the obtained new samples will be accepted instead of probability-based acceptance. We compare the performance by calculating the relative distances of the chain points from the origin and count the points that are inside given probability limits.

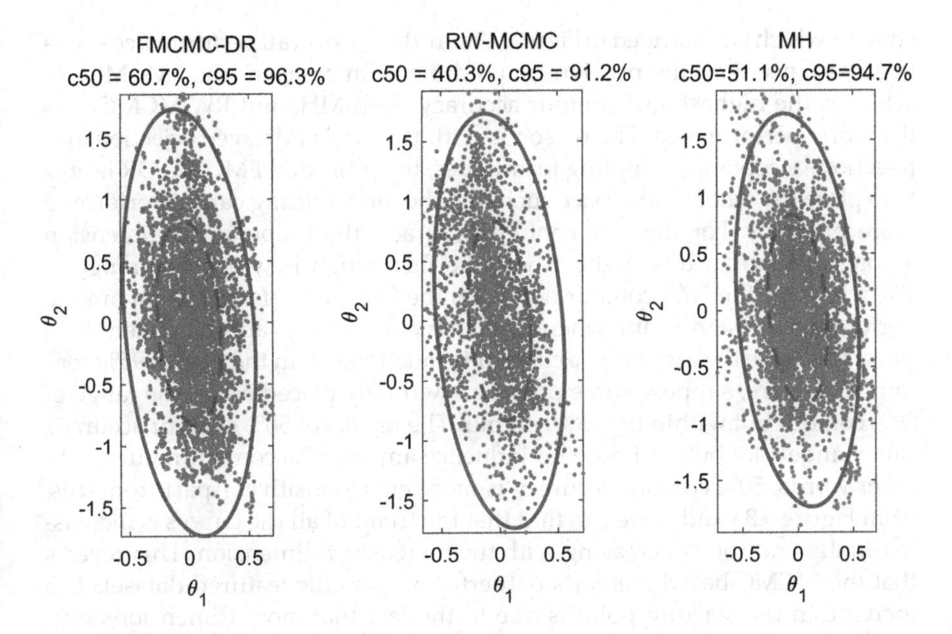

FIGURE 9.2
50% and 95% contour correctness results of three algorithms.

Since we utilize the Gaussian mixture model, which allows for an exact computation, for example, 50% and 95% probability regions, the correctness of MCMC-based methods can be easily verified [20]. In addition, for the visualization purpose, the first two-dimensional parameters are illustrated together with the correct probability contours. The results for proposed FMCMC-DR, RW-MCMC, and MH are illustrated in Figure 9.2a–c, respectively.

These three figures exhibit the sampled points as well as 50% and 95% probability regions. Obviously, no matter from 50% contour or 95% contour accuracy perspective, the proposed FMCMC-DR achieves the highest precision, which is 60.7% and 96.3%, respectively, which indicates that the delayed rejection technique can indeed improve the performance of MCMC because of the ability of the intact state space exploration. However, for RW-MCMC, it has 40.3% accuracy for 50% contour measurement, which is 0.8% lower than the standard MH method. In addition, the accuracy of 95% contour measurement for RW-MCMC is 91.2%, which is lower than the standard MH method as well, where the corresponding value is 94.7%. This is understandable because, for the sampling technique, the standard MH method adopts rejection–acceptance sampling so that the sample that is bad for local fitting will be discarded. On the contrary, RW-MCMC utilized the random walk for sampling so that the bad samples will also be included, resulting in the worst performance among these three techniques.

For more reliable statics and demonstration, we change the number of parameters dimension from 10 to 100 by step of 10 to see the performance change, which is illustrated in Figure 9.3a and b. The overall performance keeps in accordance with the previous fixed dimension case, i.e., the FMCMC-DR achieves the highest 95% contour accuracy, then MH, and RW-MCMC gets the worst performance. The reason is with the help of delayed rejection and rejection–acceptance sampling techniques, the proposed FMCMC-DR is able to explore the whole state space such that the model fitting can be performed more precisely. For the 95% contour accuracy, the influence of dimension among performance is slight, where the fluctuation is within the range of 4%, while for the 50% contour accuracy, the influence of dimension among performance is significant, where the fluctuation scale is approximately 10%. This is because the area of 50% contour is much less than the area of 95% contour. Therefore, suppose some points are wrongly placed out of the range of 50% contour but within the 95% contour. The results of 50% contour accuracy can change a lot but will pose no influence among 95% contour accuracy. In other words, 50% contour accuracy is more error-sensitive. Apart from this, from Figure 9.3a and b, we can find that the trend of all the curves is increasing firstly and then decreasing with the increase of dimension. This reveals that the MCMC-based methods outperform a specific featured dataset. The increase in the starting point is due to the fact that more dimensions give more features to the posterior approximation.. But with dimension continuing to increase, the curse of dimension happens, leading to the decrease of accuracy. Therefore, for a given algorithm or machine learning model, there will be the best data dimension that suits. For example, in this case, the most suitable data are those with 50-dimensional or 60-dimensional parameters.

In addition, we also discuss the convergence performance of the three algorithms. In this sub-task, the total number of iteration is set as 200,000. The number of mixture distributions is set as 10. The number of parameter is set as 50. The convergence curves are shown in Figure 9.4, which shows the convergence tendency of mean values. Obviously, all of the three curves converge to a near-zero value. The proposed FMCMC-DR method converges the fastest, which is approximately around 70,000 iterations. However, for MH and RW-MCMC, the convergence curves fluctuate through the process with the amplitude decreasing. Basically, the turning point for them appears in about 100,000 iterations. The reason for the quick convergence of FMCMC-DR is the utilization of the rejection–acceptance sampling method and delayed rejection. Rejection–acceptance sampling examines the acceptance of the candidate and determines whether to accept the proposed sample according to the comparison result with the calculated acceptance rate. Further, the delayed rejection helps to find out better or more accurate sample candidates via backward exploration. Therefore, the proposed FMCMC-DR obtained the best convergence rate. Also, the endpoints of the three curves have gaps to the true mean value, i.e., zero, which is because the accuracy cannot achieve 100%, as is reflected in previous results. Besides, the difference of gaps demonstrates the different accuracy levels achieved by the three methods. Obviously, FMCMC-DR gets the least gap, and the second one is

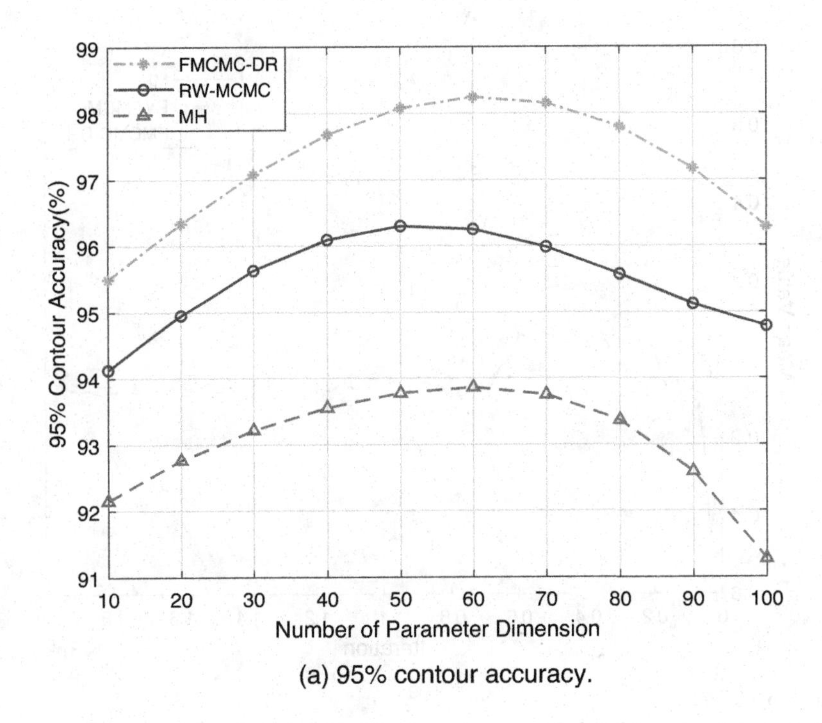

(a) 95% contour accuracy.

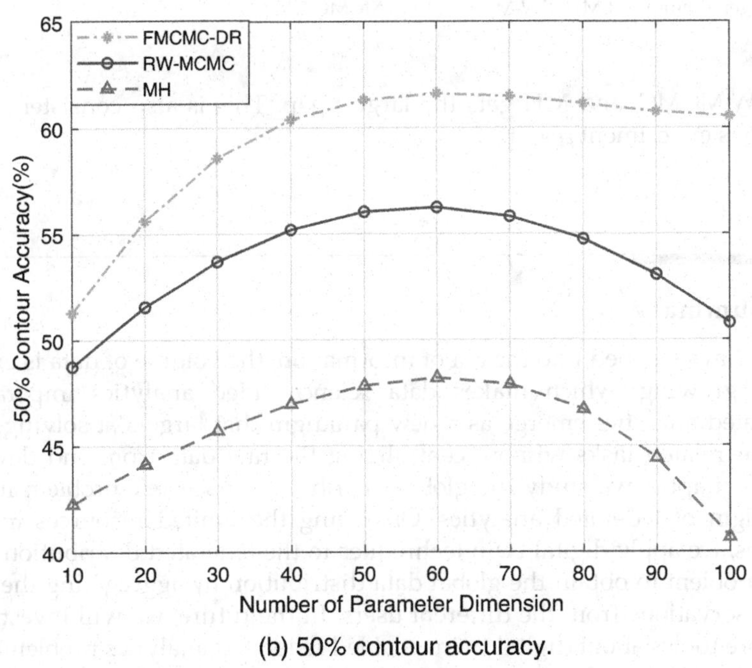

(b) 50% contour accuracy.

FIGURE 9.3
50% and 95% contour accuracy with different parameter dimensions.

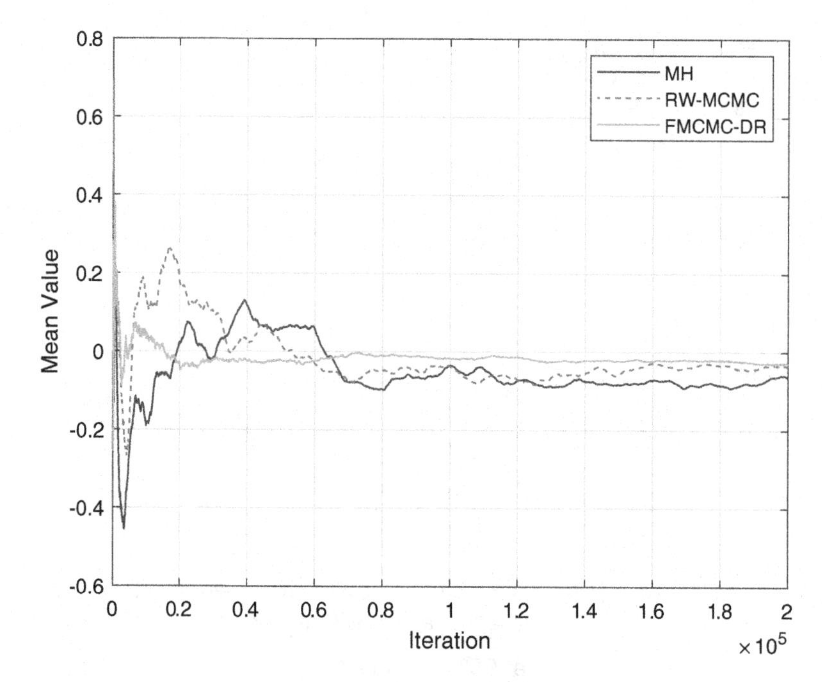

FIGURE 9.4
Convergence curves of MH, RW-MCMC, and FMCMC-DR.

the RW-MCMC, and MH gets the largest gap. This is also consistent with previous experiment result.

9.5 Summary

As we have stepped into the era of information, the volume of data is explosively growing, which makes data science–aided analytics imperative. Federated analytics emerge as a new paradigm that targets at solving data science–related tasks without centralizing the raw data from end devices. In this chapter, we study the global distribution discovery problem in the paradigm of federated analytics. Observing the limited resources in end devices, we apply digital twin techniques to the federated distribution analytic problem to obtain the global data distribution by aggregating the partial observations from the different users. In the future, we will investigate on more topics about digital twin–enabled federated analytics problems, for example, the digital twin–assisted federated recommendations. On-device search queries are directly related to suggest relevant contents to users.

On the digital twin of the physical side, the analytics method can be applied to extract the recommendation features. Then, the features can be uploaded to the virtual side of the digital twin to evaluate the popularity from all of the users and rank the items to be recommended from the current context. In addition, more topics about digital twin–enabled federated analytics problems can be explored such as federated histograms over closed sets, federated data popularity discovery over open sets, federated density of vector space, as well as privacy and security management for secure federated aggregation.

Bibliography

[1] Carrie MacGillivray, Marcus Torchia, Ashutosh Bisht, Nigel Wallis, Andrea Siviero, Gabriele Roberti, Svetlana Khimina, Yuta Torisu, Roberto Membrila, Krishna Chinta, and Jonathan Leung. Worldwide internet of things forecast, IDC, 2020–2024, August 2020.

[2] Dawei Chen, Dan Wang, Yifei Zhu, and Zhu Han. Digital twin for federated analytics using a Bayesian approach. *IEEE Internet of Things Journal*, 8(22):16301–16312, 2021.

[3] Fei Tao, He Zhang, Ang Liu, and Andrew YC Nee. Digital twin in industry: State-of-the-art. *IEEE Transactions on Industrial Informatics*, 15(4):2405–2415, April 2019.

[4] Yu Zheng, Sen Yang, and Huanchong Cheng. An application framework of digital twin and its case study. *Journal of Ambient Intelligence and Humanized Computing*, 10(3):1141–1153, March 2019.

[5] Quan Yu, Chao Han, Lin Bai, Jingchao Wang, Jinho Choi, and Xuemin Shen. Multiuser selection criteria for mimo-noma systems with different detectors. *IEEE Transactions on Vehicular Technology*, 69(2):1777–1791, February 2020.

[6] Dawei Chen, Yin-Chen Liu, BaekGyu Kim, Jiang Xie, Choong Seon Hong, and Zhu Han. Edge computing resources reservation in vehicular networks: A meta-learning approach. *IEEE Transactions on Vehicular Technology*, 69(5):5634–5646, 2020.

[7] Dawei Chen, Xiaoqin Zhang, Li Wang, and Zhu Han. Prediction of cloud resources demand based on hierarchical pythagorean fuzzy deep neural network. *IEEE Transactions on Services Computing*, 14(6):1890–1901, November 2021.

[8] Daniel Ramage and Stefano Mazzocchi. Federated analytics: Collaborative data science without data collection,*Software Engineer, Google Research,* May 2020.

[9] Peter Kairouz. Federated learning and analytics at Google and beyond, *Research Scientist, Google,* September 2020.

[10] Longbing Cao. Data science: a comprehensive overview. *ACM Computing Surveys (CSUR)*, 50(3):1–42, June 2017.

[11] Eugenia Koblents and Joaquın Mıguez. A population Monte Carlo scheme with transformed weights and its application to stochastic kinetic models. *Statistics and Computing*, 25(2):407–425, March 2015.

[12] Shahin Boluki, Mohammad Shahrokh Esfahani, Xiaoning Qian, and Edward R Dougherty. Constructing pathway-based priors within a Gaussian mixture model for Bayesian regression and classification. *IEEE/ACM Transactions on Computational Biology and Bioinformatics*, 16(2):524–537, November 2017.
[13] Michael J Daniels and Joseph W Hogan. *Missing Data in Longitudinal Studies: Strategies for Bayesian Modeling and Sensitivity Analysis*. CRC Press, 2008.
[14] Martyn Plummer, Nicky Best, Kate Cowles, and Karen Vines. Coda: Convergence diagnosis and output analysis for MCMC. *R News*, 6(1):7–11, March 2006.
[15] Luke Tierney and Antonietta Mira. Some adaptive Monte Carlo methods for Bayesian inference. *Statistics in Medicine*, 18(17–18):2507–2515, August 1999.
[16] Peter J Green and Antonietta Mira. Delayed rejection in reversible jump metropolis-hastings. *Biometrika*, 88(4):1035–1053, December 2001.
[17] Antonietta Mira et al. On metropolis-hastings algorithms with delayed rejection. *Metron*, 59(3–4):231–241, April 2001.
[18] Heikki Haario, Marko Laine, Antonietta Mira, and Eero Saksman. Dram: Efficient adaptive MCMC. *Statistics and Computing*, 16(4):339–354, December 2006.
[19] Jose G Dias and Michel Wedel. An empirical comparison of EM, SEM and MCMC performance for problematic Gaussian mixture likelihoods. *Statistics and Computing*, 14(4):323–332, April 2004.
[20] Heikki Haario, Eero Saksman, and Johanna Tamminen. Componentwise adaptation for high dimensional MCMC. *Computational Statistics*, 20(2):265–273, June 2005.

10

Blockchain-Based Digital Twin Design

Esra Kumaş, Hamide Özyürek, Serdar Çelik, and Zeynep Baysal
OSTIM Technical University

10.1 Introduction

Advancements in computer technologies have increased human abilities millionfold in terms of collecting, processing and storing information. When combined with network technologies, such technologies have also accelerated the transfer of information from one place to another via networks a millionfold (Tonta and Küçük 2005). Invention of the transistor in 1947 has brought great changes to the field of communication and electronics followed by developments in computer technology which have made data processors smaller, faster and cheaper. Data processors, which are getting smaller, faster and cheaper, have also increased the number of products using computer technology (Brynjolfsson and McAffe 2014). Exponential development has the same exponential effect not only on the technology itself but also on the entire process, such as the speed, efficiency and cost-effectiveness of the process. This speed will then give birth to the second level of exponential growth, in other words, the speed of exponential growth will grow exponentially (Kurzweil 2016). New technologies such as cyber-physical systems, smart factories, cloud technologies, 3D printing technologies, IoT, cyber security, big data, learning robots, and virtual reality (VR) which are considered as components of Industry 4.0 (Hermann et al. 2015) are other elements of pressure that change the way companies do business.

Digital Twins (DT) are a virtual image of a physical object and/or operational process that allows the product and system environment to be monitored based on real data. Combined with the concept of IoT introduced by Kevin Ashton in 1999 to describe objects that can perceive and communicate with the environment (Ashton 2009), it has paved the way for the creation of a system that can understand the increasing complexity and respond quickly. The ability to code and monitor objects has enabled companies to be more efficient, reduce error, speed up processes, prevent losses, and use flexible and complex organizational systems owing to IoT (Ferguson 2002). The IoT's ability to connect everything to everyone on an integrated global network

DOI: 10.1201/9781003425724-13

has made DT possible. As DTs are used for the collection and monitoring of instant data, they analyse the data they collect, extract, optimize the process, give advice to users and even make their own decisions and implement them. During these processes, problems such as managing the increasing number of data, preventing data loss and ensuring the security and transparency of the data may arise. Blockchain is a type of distributed ledger technology (DLT). Since there is no single central authority controlling the ledger, it is a highly secure and transparent way of storing information in an immutable and chronological record. In this respect, blockchain has the potential to be an effective technology to ensure the trust, transparency and reliability of the large amount of data used and produced by DTs.

The following section provides a definition of blockchain, its background, stages, structure, benefits, consensus algorithms, network types and application areas. In the second part, definition and background of DTs, their stages, concept and maturity model are explained and the literature on DT and blockchain technologies was examined. Finally, a model is proposed for an integrated use of DT and blockchain technologies.

10.2 Blockchain

10.2.1 Definition and Background of the Blockchain

Blockchain is the underlying technology of the cryptocurrency called Bitcoin that was popularized by Satoshi Nakamoto in 2008 (Cong and He 2019). Each block connected through cryptography contains the previous block's cryptographic karma, timestamp and transaction data (Chang et al. 2020).

Bitcoin allows each user on the network (blockchain) to anonymously broadcast all transactions across the network in a way that all other users on the network are aware of all transactions (Pimentel and Boulianne 2020). In daily life, money transfers can take place between two parties without the need for informing or an intervention from a trusted third party. This transaction is called a peer-to-peer (P2P) network. What blockchain does is that it enables this transaction to be transferred to an online platform (Nakamoto 2008).

Transparency of the blockchain data, accountability of transactions, unmediated nature of transactions and data symmetry are only a few of the qualities of blockchain technology that make it immensely essential for almost all sectors. Blockchain is a distributed database which offers an unprecedented technology due to its decentralized structure in which all records on the network are copied to each user's computer. Each individual user on the blockchain network is called a "node", and the digital ledger is defined as a "chain" consisting of individual data blocks (Hughes et al. 2019). When new data are added to the network in a chronological order, a new "block" is created and then added to the "chain". Each additional block updates all previous nodes on the blockchain while also containing all previous data blocks which provides the key reason why the blockchain technology is considered to be secure. Majority of the users on the network are required to verify and confirm the new data for a block to be added to the blockchain network which creates a challenge more difficult than a single user making an alteration on an independent network. This requisite for verifying all transactions prevents double-spending of cryptocurrencies while also prohibiting fraud.

10.2.2 Structure of the Blockchain

Cryptocurrencies work based on cryptographic evidence by using a "hash" (consisting of a summary and a series of letters and numbers) with a chain of digital signatures to connect each transaction to the previous one. This collection of data and hashes is called a "block", and these blocks contain timestamps and are linked to each other to create a secure chain. As instant data are added to the chain with specific timestamps, all inputs are recorded by users without any missing data (Yli-Huumo et al. 2016).

Moreover, as each addition to the chain contains a unique identifier code based on the previous block, it ensures accuracy and security in data tracking (Nam et al. 2021). Users trust the data held on a blockchain without having to know and trust each other or rely on a central authority such as a bank, credit company or government and this is why blockchain is called the "Trust Machine" (Miraz and Ali 2018). The chain-like structure chronologically links data blocks and prevents any tampering or copying of the distributed ledger. As Bahga and Madisetti (2016) demonstrate, each link on this chain-like structure or the so-called block contains the previous hash, the nonce (a number used only once), the merkle root (containing all transaction hash on the block in a tree-like structure) and a timestamp.

10.2.3 Advantages of the Blockchain Technology

Blockchain reinforces data collection (Suri 2021) and data management (Lin et al. 2017; Kamilaris et al. 2019) and allows transactions to be recorded and verified independently (Graglia and Mellon 2018). Due to this feature, land registry records can be easily kept by using blockchain technology (Veuger 2020). Personal documents including birth certificates, driver's licences, social security numbers and passports, can be stored on the blockchain as it provides a solution to the identity theft problem with digital identities (Gupta et al. 2017; Dogru et al. 2018). It can also be used worldwide for issuing and verifying educational certificates such as degrees, transcripts as well as documents showing students' qualifications, achievements and professional talents (Alammary et al. 2019) or as in IBM's Food Trust project, for monitoring the supply chain (Pathak 2021). It enables the provision of correct product origin information to the consumers (Nikolakis et al. 2018). The MediLedger project supported by CryptoSeal, Genentech and Pfizer and the BlockVerify project which authenticates the products offer solutions to problems in this area (Radochia 2019).

Blockchain applications are used to overcome various business problems and optimize many business processes (Pal et al. 2021). For example, Knirsch et al. (2018) present a blockchain protocol for protecting privacy while charging electric vehicles.

Blockchain technology also offers advantages in financial auditing operations. It reduces transaction audit (Salah et al. 2019) and intermediation costs (Michelman 2017). It can also contribute in minimizing costs by eliminating intermediaries in tourism, food, agriculture and similar sectors (Hughes et al. 2019). Each node (user) has its own copy of the ledger, and all records are updated at the same time once a new transaction occurs (Pimentel and Boulianne 2020).

10.2.4 Consensus Algorithms

Blockchain is a distributed network system in which consensus protocols have emerged to ensure that transactions are flawless and accurate where trust is established so that potential conflicts are prevented and users who do not know each other can work together (Li et al. 2018). Agreeing on a block of transactions in a given directory of a blockchain to ensure that the transactions of a distributed system are flawless or correct is called consensus. The first consensus protocol was Proof of Work (PoW) used in Bitcoin, and then Proof of Stake (PoS) and Byzantine Fault Tolerance (BFT) were developed and different versions continue to be developed based on these basic protocols. Due to its association with Bitcoin, PoW is the most well-known consensus algorithm (Hasselgren et al. 2020). PoW is often used in public blockchains as it allows anyone to participate in the process of validating blocks for the blockchain. Blockchain nodes running PoW on the network must solve a cryptographic "puzzle" when validating a new block for the blockchain to which the answer is random and expensive to guess by computing which makes it very difficult for a malicious miner to guess the correct answer as it creates millions of possibilities through an intensive and expensive process (Zghaibeh et al. 2020). In fact, for a malicious miner to be successful, he/she must possess more than 50% of the network's computing power, which is known as a 51% attack (Hardin and Kotz 2019). However, PoW is a quite costly process in terms of energy consumption and computing time. Blockchain networks based on PoW have slow transaction verification times.

PoS is an energy-saving alternative to the PoW protocol, and in PoS, the creator of the next block is chosen based on his/her holdings of the related cryptocurrency. PoS provides a way to reach consensus without forcing record keepers to solve difficult computational problems and avoids the waste of resources required by the PoW system (Abadi and Brunnermeier 2018). Ethereum is currently planning to switch from PoW to PoS (Hasselgren et al. 2020).

Inspired by the Byzantine Generals Problem, BFT is based on the strategy of the Byzantine emperor using more than one messenger to send news to his generals. In the adaptation of this strategy for blockchain technology, the nodes in the blockchain represent the generals. Nodes that do not allow attacks on the system represent loyal generals, while nodes that plan attacks to disrupt the blockchain and share false fraudulent information represent treacherous generals (Kardaş 2019). The very first application of this strategy in a blockchain was the Practical Byzantine Fault Tolerance (PBFT) algorithm used on Hyperledger Fabric platform which is a permissioned blockchain, and then different versions such as Istanbul Byzantine Fault Tolerance (IBFT) were developed (Hardin and Kotz 2019). PBFT is used more effectively in a consortium blockchain network where there is a small group of organizations willing to reach consensus (Dhuddu and Mahankalı 2022).

10.2.5 Sectors Where the Blockchain Technology Is Utilized

Literature on the use of blockchain technology demonstrates that it is being used and has the potential to be used in fields including business administration, real estate and land registry, public administration, digital identity, education (Zeadally and Abdo 2019), health, logistics and supply chain (Fachrunnisa and Hussain 2020), manufacturing, health, energy (Demirkan et al. 2020), agriculture, tourism (Önder and Treiblmaier 2018), robotics (Al-Jaroodi and Mohamed 2019), entertainment, finance and banking, accounting and auditing (Özyürek and Etlioğlu 2021). Research indicates that blockchain technology is used to make secure online transactions (Garg et al. 2021), create digital identities (Dogru et al. 2018), increase the efficiency of supply chains (Fachrunnisa and Hussain 2020), keep patient records in the health sector (Hardin and Kotz 2019) and monitor gun owners by crime control agencies (Kumar 2021).

Blockchain can be used in production planning, production control and quality controls as part of operations management which offers the benefits of visibility, validity and automation (Pal et al. 2021). Blockchain is evolving rapidly to help improve the traceability of cargo trade flow across borders (Pournader et al. 2020). For cross-border trade flow management, high-value goods are often subject to close monitoring. Blockchain can play an important role in such examples of high-value goods transport. According to Mishra and Venkatesan (2021), timestamps in blockchain can be used to perform time-sensitive tasks such as just-in-time (JIT) production. Research also shows that blockchain-based recruitment management system and blockchain-based HRM system algorithms have advantages compared to existing recruitment systems (Onik et al. 2018).

10.3 Digital Twins

10.3.1 Definition and Background of Digital Twins

A DT is defined as a real-time digital profile of the past and present behaviour of a physical object or process that helps optimize its operational performance and is based on real-time, real-world, big and cumulative data measurements. It is based on the idea that a digital information structure about a physical system can be created as a stand-alone entity. This digital information creates the "twin" of the information embedded in the physical system and is linked to this physical system throughout its entire life cycle (Kumaş and Erol 2021).

The DT technology, which emerged as a result of NASA's mapping technology, was first introduced in 2002 by Dr. Michael Grieves at University of Michigan in his presentation to the industry for the establishment of the Product Lifecycle Management (PLM) centre (Grieves and Vickers 2017), and then its conceptual model was presented in 2006 (Grieves 2006).

Two identical spacecrafts were built to reflect the conditions of the spacecraft during NASA's Apollo mission and the vehicle that remained on earth was called a twin. During the flight mission, the twin on earth was used to ensure that the data reflected the flight conditions as precisely as possible in order to assist orbiting astronauts in critical situations, while also being used to simulate alternatives. Therefore, any prototype that simulates real-time behaviour and reflects the real work environment can be considered a twin. Grieves' conceptual model was named the "Information Reflection Model" (Grieves 2006), then it was referred to as the "Digital Twin" in the NASA Technology Roadmap in 2010 (Piascik et al. 2010) and thereby renamed as such. The technology that made the DT affordable and accessible for businesses was IoT technology.

Although the idea of creating a digital copy of a product or process is not novel, the work for creating a digital copy that is closest to the real product or process has been going on since the first use of computers by businesses. Computer-aided manufacturing and design technologies (CAD/CAM) have enabled the product to be physically modelled, while it is still in the idea stage, while simulation programmes made it possible to test the effects of the factors affecting the product, VR delivered more realistic results and remote controls were made possible via augmented reality (AR) technologies. However, simultaneous monitoring and control of digital copies and the use of real-time and instant environmental data demonstrate that a DT is not only a digital copy but also an important building block on the way to the smart factories as a self-managing product which enables understanding, learning and reasoning.

DT refers to a real-time virtual copy of a physical product, but since the term was first coined in 2002, many definitions have emerged regarding the areas of use and purposes due to both technological advances and the widespread use of DT technology. Negri et al. (2017) demonstrate that these definitions were initially used for the aviation sector between 2010 and 2013, the first studies for the manufacturing sector were published in 2013 and the studies that included a general product concept for the DT have emerged in 2015 (Negri et al. 2017).

Table 10.1 shows DT definitions with diverse objectives from different sectors and fields.

Abovementioned definitions demonstrate that a DT is used to model a complex physical product or process as advanced interfaces by simplifying it, to increase its performance, to monitor it in real time, to perform understanding-learning and reasoning activities and to optimize security and maintenance activities, and it is defined in different ways according to its purpose.

10.3.2 Concept and Maturity Model

The first conceptual model was put forward by Grieves. It emerged from the need for advanced analysis by bringing together all the data and information of a physical asset. The model includes the flow and processing of information between the real environment, the virtual environment and both:

- Real Space,
- Virtual Space,
- Connection from the real space to the virtual space for data flow,
- Connection from virtual space to the real space for information flow and virtual sub-spaces.

Grieves' definition of a DT is a three-dimensional structure that describes the physical asset, the real model, and the data connection between them. Tao et al. (2022) have added the concept of services to this structure (Tao et al. 2022). The DT data represent the asset that collects information from both the physical asset and the virtual asset. The structure that processes these data is called a service.

Arup (2019) published a study called "Digital Twin Towards a Meaningful Framework" which involved Dr. Michael Grieves and classified DTs based on their levels. According to this study, DTs are classified into five levels under four dimensions which are:

- *Autonomy:* Ability to act without human intervention.
- *Intelligence:* It can copy human cognitive processes and perform the tasks assigned to it.

TABLE 10.1

Definitions of Digital Twins

Source	Definition	Year	Title of the Study
NASA	The Digital Twin provides highly accurate simulation of product lifecycle and maintenance times, using all current and historical data to ensure a high level of safety and reliability.	2009	Apollo 13
General Electric	The Digital Twin, as a living model, drives the outcome of things.	2016	Meet the Digital Twin
Microsoft	A Digital Twin is a real-time virtual model of a process, product, production or service, powered by IoT, machine learning, and data analytics.	2017	The Promise of a Digital Twin Strategy
Deloitte	A Digital Twin is a real-time digital image of a physical product or process and enables to optimize business performance.	2017	Industry 4.0 and the Digital Twin
Grieves and Vickers	A Digital Twin is a one-to-one creation of a physical product in a virtual environment, and this virtual image has all the information that the physical product contains under optimum conditions.	2017	Digital Twin: Mitigating Unpredictable, Undesirable Emergent Behaviour in Complex Systems
Accenture	A Digital Twin is a digital model of a physical object, system or process that helps businesses monitor the condition of equipment at remote facilities, optimize product designs for specific needs and even model the business itself.	2018	What Is a Digital Twin?
Centre for Digital Built Britain	A Digital Twin is a literal virtual representation of a physical object.	2019	National Digital Twin Programme
IBM	A Digital Twin is a simulated virtual representation of a product or system throughout its entire lifecycle, using real-time data to aid decision-making.	2019	What Are Digital Twins?
SIEMENS	A Digital Twin is a simulated virtual representation of a physical product or process to understand, predict and optimize performance characteristics throughout the product lifecycle.	2019	Siemens Glossary: Digital Twin
MIT Sloan	A Digital Twin is a virtual copy of a physical object or system that can be continuously updated with real data.	2020	How Digital Twins Are Reinventing Innovation

TABLE 10.2

Levels of Digital Twins

Level	Description of Level	Dimension	Description of Dimension
Level 1	Connects to a real-world system and collects data, but lacks autonomy, learning and decision-making abilities	Autonomy	There is no complete autonomy.
		Intelligence	The twin has no intelligence
		Learning	The twin has no learning component.
		Fidelity	It can be accepted as a conceptual model, its accuracy is low.
Level 2	At this level, Digital Twins are usually limited to modelling small-scale systems, with some capacity for feedback and control.	Autonomy	Autonomy is limited but can provide feedback on the effectiveness of the system.
		Intelligence	It responds to stimuli but cannot draw meaningful conclusions from previous data.
		Learning	It can be programmed using a long list of commands.
		Fidelity	It can measure in the low to medium accuracy range.
Level 3	At this level, Digital Twins can analyse the data they collect and provide predictive maintenance and insights, in addition to the characteristics of level 2 Digital Twins.	Autonomy	It has the ability to stimulate and control the system in certain ways and is partially autonomous.
		Intelligence	It can learn from historical data to improve and make decisions.
		Learning	It can predict and give feedback.
		Fidelity	It has moderate accuracy, capable of representing the physical world.
Level 4	At this level, Digital Twins learn from the data they collect from the environment, draw conclusions, optimize the process and give smart advice to users.	Autonomy	It can monitor conditions and perform critical tasks with little or no human intervention.
		Intelligence	It can understand the needs of systems.
		Learning	It can make sense of the environment by itself.
		Fidelity	It can provide precise measurements.
Level 5	This level of Digital Twins reason autonomously thanks to artificial intelligence and users can make decisions instead.	Autonomy	It can operate completely safely without human intervention.
		Intelligence	It can have self-awareness similar to human intelligence.
		Learning	It learns from historical data and experience to find the optimum result that will improve performance.
		Fidelity	It has a high degree of accuracy, which can be used in vitally important situations such as life safety and critical operational decisions.

Source: Adapted from Arup (2019).

- *Learning:* It can learn from data automatically to improve performance. It can classify the properties of systems through machine learning.
- *Fidelity:* It can reach results close to actual value of measurements, calculations or specifications of a system.

10.4 Literature Review

In the literature, studies have been conducted on the use of blockchain-based DTs for different purposes and fields. In Table 10.3, the studies are presented by grouping them according to their aims.

10.5 Integration of Digital Twin and Blockchain Structures and Model Proposal

This study explains the process of creating a DT and its integration with the blockchain ecosystem. It presents a general framework that can be customized to meet the specific needs and functions of businesses requiring a digital model that can synchronize with the real world to achieve specific goals. Multi-disciplinary teams collaborate to create a DT, which is a virtual model of a physical entity. Reliable management and monitoring tools are essential in this process to ensure the tamper-proof preservation of historical information. That's why blockchain, a decentralized technological protocol for data exchange among multiple stakeholders, is used to meet stringent security requirements.

The study explains the integration between the blockchain and the DT with a conceptual model and technological framework consisting of five layers. Figure 10.1 shows the realization of the integration of blockchain with the key technologies of Industry 4.0. It emphasizes how ecosystems and sub-ecosystems are built on the mutually used blockchain platform. The model offers various information technology components, management components, data about physical objects, potentially into real-time business models. The outermost layer contains data from the physical world, decision mechanisms, designers and planners. Data enter the DT system by using connection points through IoT devices, RFID, and deep learning applications. The dynamic nature of the DT is realized within the design and computing layer. The goal is to eliminate intermediaries between the real world and the DT and reduce storage space. To do this, it leverages blockchain ledgers and

TABLE 10.3

Literature Review

Author(s)	Year	Aim	Category
Huang, Wang, Yan and Fang	2020	In the study, a data management method based on blockchain technology has been proposed to solve the problems that may occur during the acquisition, storage, sharing, authenticity and protection of data.	*Data Management*
Li, Li, Huang and Qu	2021	In the study, a blockchain-based Digital Twin sharing platform was created to simplify the integration of decentralized and heterogeneous environments, and protect software copyright.	*Data Security*
Tao, Zhang, Cheng, Ren, Wang, Qi and Li	2022	The purpose of the article is to propose a collaboration mechanism developed with DT-BC for the Industrial Internet platform for data security–related issues.	
Putz, Dietz, Empl and Pernul	2021	In the study, in order to ensure data integrity, availability and confidentiality, a Digital Twin model based on blockchain is presented.	
Mandolla, Petruzzelli, Percoco and Urbinati	2019	The purpose of this article is to explain how you can use blockchain to build a secure and interconnected additive manufacturing infrastructure.	
Shen, Hu, Zhang and Ma	2021	This paper proposes a blockchain-based approach to securely share large Digital Twin data, aiming to address the challenges posed by the exponential growth and timely sharing of such data.	
Zhang, Zhou, Li and Cao	2020	The aim of the study is combining IoT with a blockchain, and it introduces a new architecture of Manufacturing Blockchain of Things (MBCoT) to configure a secure, traceable and decentralized IMS.	
Lee, Lee, Masoud, Krishnan and Li	2021	The aim of the study is to realize transparent information sharing about the project between the stakeholders with the integrated Digital Twin and blockchain model.	*Monitoring*
Guo, Ling, Li, Ao, Zhang, Rong and Huang	2020	The purpose of this article is to provide guidance and reference point for personalized products.	*Personalization*
Wang, Cai and Li	2022	The purpose of the article is to eliminate latency while guaranteeing sustainable energy and sustainable information by optimizing measurements that ensure data accuracy.	*Sustainability*

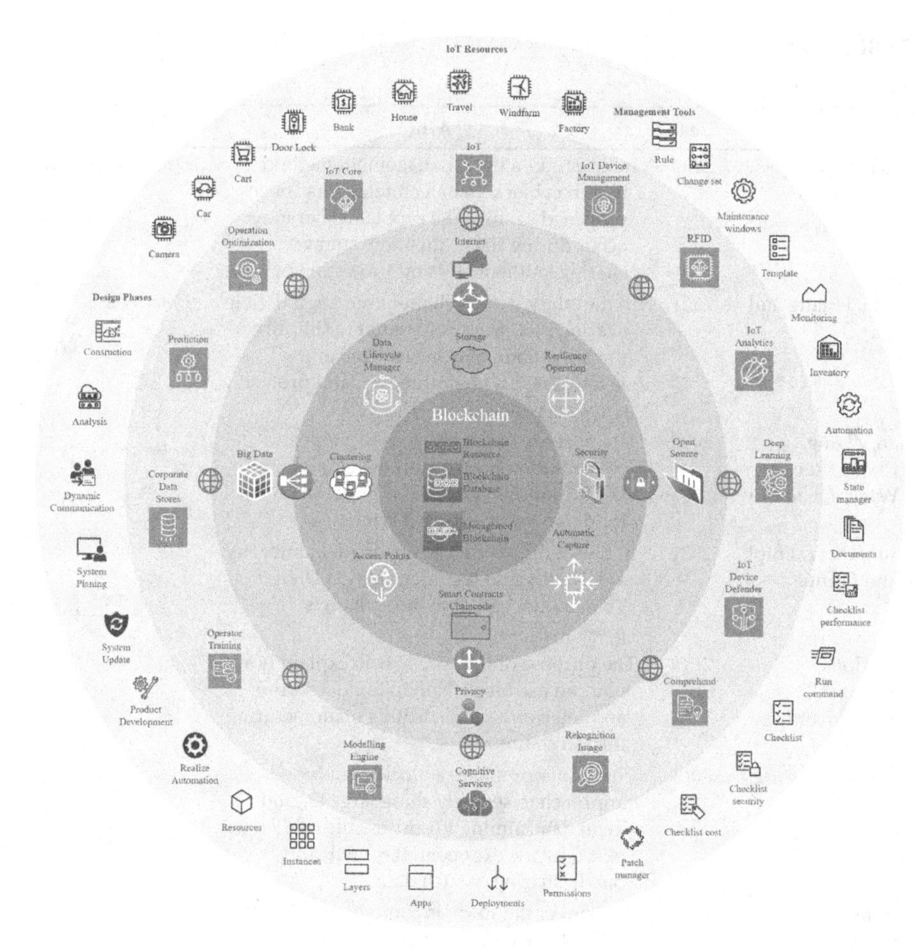

FIGURE 10.1
Digital Twin and blockchain integration.

decentralized applications that interact with each other using open source distributed and encrypted identities and direct P2P communication and do not need intermediaries. The data management layer provides secure connectivity and access to participants interacting with the DT system. It also works with unclassified big data from the real world. The collected big data are standardized and classified before they enter the blockchain through connectivity software. At this layer, data from participants IoT sensors and various sources are validated before entering the blockchain layer and leverages decentralized protocols. These protocols provide transparency between participants and increase security on a global scale. Based on the automated verification process, technology components can trigger payments or participant requests immediately if the agreed contractual arrangements are met, as specified in smart contracts. Transactions are enabled by end-to-end encryption within a network of distributed ledgers that must be approved by the majority of the network. The approval process prevents manipulation and fraudulent behaviour by a single entity. Manipulation becomes increasingly difficult when stakeholders are included in the network. When standardized data enter the blockchain, it becomes secure, preserving historical integrity. The blockchain is an important element of the data layer in terms of integrating smart contracts, security protocols and increasing visibility. Smart contracts are programmed to compute the main processes within the DT and are used by decentralized key factors, such as the DT's stakeholders or device identities, to improve access control mechanisms. A decentralized platform allows a large number of network participants to interact directly with each other without the need for a third party. With the information provided by DTs, data regarding trade, production, consumption, storage, etc. are transmitted directly to the targeted party, improving transparency and communication and preventing inefficiencies. For example, by adding physical network components – an important element for predictive maintenance models – to the blockchain, objects can be tracked and guaranteed that all their data are verifiable. As another example, personal data are represented in the real world by an ID card. ID cards can be stored as DTs, providing transparency about who has access to a particular area, room, building, documents, and machines. At the core of the model, the blockchain ensures that the historical data flow between the DT and the physical world is securely recorded and the integrity of the data is preserved. The model as a whole works by taking privacy and encryption mechanisms into account.

Among the data that can be audited, public records such as environmental records and political records can serve. Public blockchain transaction times and costs will be low for any consumer or user. Transactions that require high confidentiality can be processed on private networks. Only data of public interest can be used in a fixed way on public blockchains. With combinations of both private and public blockchains, the DT system creates a model that can meet the requirements. The model aims to record historical data obtained at each stage of the project on the blockchain. For example, updates of models created in real time with the real world during design can be recorded on the blockchain. In the maintenance planning phase, IoT sensor data and maintenance data can be pinned to the blockchain. The data management layer of the DT also includes key processes that can be managed automatically with smart contracts. The exchange of data and value using DTs can be determined by global protocols or by widely accepted rules and agreements between stakeholders. Projects or products need to determine which data should be stored on-chain or stored in data silos without being included in the chain. Decentralized storage systems should be in place for data stored off-chain. This layer of the ecosystem also provides the necessary bandwidth for data transmission, taking into account stakeholder demands.

Another objective of the model is to optimize the analysis of data from IoT sensors, enabled by blockchain technology, encouraging the effective use of both financial and management tools by all stakeholders involved. The model aims to centralize data, management tools, modelling, design, planning processes with blockchain-based protocols. The connections between the layers facilitate the use of blockchain-based protocols and the decentralization of the analysis of data from IoT sensors, resulting in a reduction in the number of points of failure. Finally, the model avoids complex loops by combining unclassified big data into a single source with key data securely anchored in the blockchain.

10.6 Conclusion and Suggestions

This study develops a framework for the integration of DT and blockchain in the context of technological innovations and developments. The aim is to propose a model that enables the adoption of blockchain within the DT technology and minimizes the gaps. The ecosystem of the model includes traceability of information, management of networks, classification of data, security risks, standardization of management, transfer of the physical world to the virtual world, centralization of the system, and integration of smart contracts. The integration of the DT with the blockchain ensures the security and integrity of the data accumulation generated by the IoT from the stakeholders involved in the system by verifying transactions in blockchain ledgers. The major advantages of blockchain over DT technology are security and the reduction of data stores. The transaction volume of a blockchain is sufficient and there are no or very low transaction fees. Stakeholders coming together under the DT project will be able to monitor and transact data together through a blockchain that is interoperable with the companies' own networks. The transaction speed will vary depending on the volume and type of data. Stakeholders will be able to use smart contracts to increase efficiency, reduce costs and save time. Experimental studies are recommended for the implementation of the developed model. It may reveal that energy savings can be achieved with selected blockchain protocols for DTs. Moreover, the DT and blockchain mechanisms considered in this study should be tested with high data volume, velocity and variety simulating the real conditions of the Industry 4.0 ecosystem. Therefore, future work should test whether the DT can perform adequately in terms of its integration with key technologies such as blockchain, IoT, simulation, deep learning, etc. Finally, future work should also identify categories that require security, transparency and immutability for each process of the project or product lifecycle.

References

Abadi, J., Brunnermeier, M. (2018). *Blockchain economics*. National Bureau of Economic Research.

Al-Jaroodi, J., Mohamed, N. (2019). Blockchain in industries: A survey. *IEEE Access*, 7, 36500–36515.

Alammary, A., Alhazmi, S., Almasri, M., ve Gillani, S. (2019). Blockchain-based applications in education: A systematic review. *Applied Sciences*, 9 (12), 2400.

Arup. (2019). *Digital Twin: Towards a Meaningful Framework. Technical Report*, Arup, London.

Ashton, K. (2009). That 'internet of things' thing. *RFID Journal*. https://www.rfidjournal.com/that-internet-of-things-thing. RFID journal, 2009, 22(7): 97–114.

Bahga, A., Madisetti, V. K. (2016). Blockchain platform for industrial internet of things. *Journal of Software Engineering and Applications*, 9(10), 533–546.

Brynjolfsson, E., & McAfee, A. (2014). *The second machine age: Work, progress, and prosperity in a time of brilliant technologies*. WW Norton & Company.

Chang, V., Baudier, P., Zhang, H., Xu, Q., Zhang, J., Arami, M. (2020). How Blockchain can impact financial services-the overview, challenges and recommendations from expert interviewees. *Technological Forecasting and Social Change*, 158, 120166.

Cong, L. W., He, Z. (2019). Blockchain disruption and smart contracts. *The Review of Financial Studies*, 32(5), 1754–1797.

Demirkan, S., Demirkan, I., McKee, A. (2020). Blockchain technology in the future of business cyber security and accounting. *Journal of Management Analytics*, 7(2), 189–208.

Dhuddu, R., Mahankalı, S. (2022). *Blockchain A to Z Explained, Become A Blockchain Pro With 400+ Terms*, BPB Publications, India.

Dogru, T., Mody, M., Leonardi, C. (2018). *Blockchain Technology & Its Implications for the Hospitality Industry*. Boston University, Boston.

Fachrunnisa, O., Hussain, F. K. (2020). Blockchain-based human resource management practices for mitigating skills and competencies gap in workforce. *International Journal of Engineering Business Management*, 12, 1847979020966400.

Ferguson, T. (2002). Have your objects call my object. *Harvard Business Review*, https://hbr.org/2002/06/have-your-objects-call-my-objects. Accessed August 2022.

Garg, P., Gupta, B., Chauhan, A. K., Sivarajah, U., Gupta, S., Modgil, S. (2021). Measuring the perceived benefits of implementing blockchain technology in the banking sector. *Technological Forecasting and Social Change*, 163, 120407.

Graglia, J. M., Mellon, C. (2018). Blockchain and property in 2018: At the end of the beginning. *Innovations: Technology, Governance, Globalization*, 12(1–2), 90–116.

Grieves M.. (2006). Product Lifecycle Management: Driving the Next Generation of Lean Thinking. McGraw-Hill, NewYork.

Grieves, M., Vickers, J. (2017). Digital twin: Mitigating unpredictable, undesirable emergent behavior in complex systems. In *Transdisciplinary Perspectives on Complex Systems* (pp. 85–113). Springer, Cham.

Gupta, A., Patel, J., Gupta, M., Gupta, H. (2017). Issues and effectiveness of blockchain technology on digital voting. *International Journal of Engineering and Manufacturing Science*, 7(1), 20–21.

Hardin, T., & Kotz, D. (2019). Blockchain in health data systems: A survey. In *Sixth International Conference on Internet of Things: Systems, Management And Security (IOTSMS)*, pp. 490-497, IEEE.

Hasselgren, J., Munkberg, J., Salvi, M., Patney, A., Lefohn, A. (2020). Neural temporal adaptive sampling and denoising. *Computer Graphics Forum*, 39(2), 147–155.

Huang, S., Wang, G., Yan, Y., & Fang, X. (2020). Blockchain-based data management for digital twin of product. *Journal of Manufacturing Systems*, *54*, 361-371.

Hughes, L., Dwivedi, Y., Misra, S., Rana, N., Raghavan, V. & Akella, V. (2019). Blockchain research, practice and policy: Applications, benefits, limitations, emerging research themes and research agenda. International Journal of Information Management, 49, 114–129

Kamilaris, A., Fonts, A., & Prenafeta-Boldύ, F. X. (2019). The rise of blockchain technology in agriculture and food supply chains. Trends in Food Science & Technology, 91, 640–652.

Kardaş, S. (2019). Blokzincir teknolojisi: Uzlaşma protokolleri. *Dicle Üniversitesi Mühendislik Fakültesi Mühendislik Dergisi*, 10(2), 481–496.

Knirsch, F., Unterweger, A., ve Engel, D. (2018). Privacy-preserving blockchain-based electric vehicle charging with dynamic tariff decisions. *Computer Science-Research and Development*, 33(1), 71–79.

Kumar, R. (2021). Advance concepts of blockchain. In S. S. Tyagi & S. Bhatia (Eds.) *Blockchain for Business: How It Works and Creates Value*,361–372, Wiley.

Kumaş, E., & Erol, S. (2021). Digital Twins as Key Technology in Industry 4.0. *Journal of Polytechnic*, 24(2).

Kurzweil, R. (2016). İnsanlık 2.0: Tekilliğe doğru biyolojisini aşan insan (çev. M. Şengel). Türkiye: Alfa, (Original work was published in 2004), 23–24.

Li, Y., Bienvenue, T.M., Wang, X., Pare, G. (2018). Blockchain technology in business organizations: A scoping review. In *Proceedings of the 51th Hawaii International Conference on System Sciences* (pp. 4474–4483).

Lin, Y. P., Petway, J. R., Anthony, J., Mukhtar, H., Liao, S. W., Chou, C. F., Ho, Y. F. (2017). Blockchain: The evolutionary next step for ICT e-agriculture. *Environments*, 4(3), 50.

Mandolla, C., Petruzzelli, A. M., Percoco, G., Urbinati, A. (2019). Building a digital twin for additive manufacturing through the exploitation of blockchain: A case analysis of the aircraft industry. *Computers in Industry*, 109, 134–152.

Michelman, P. (2017). Seeing beyond the blockchain hype. *MIT Sloan Management Review*, 58(4), 17–19.

Miraz, M.H., Ali, M. (2018). Applications of blockchain technology beyond cryptocurrency. *Annals of Emerging Technologies in Computing (AETiC)*, 2(1), 1–6.

Mishra, H., Venkatesan, M. (2021). Blockchain in human resource management of organizations: An empirical assessment to gauge HR and non-HR perspective. *Journal of Organizational Change Management*, 525–424. https://www.emerald.com/insight/0953-4814.htm.

Nakamoto, S. (2008). *Bitcoin: A Peer-to-Peer Electronic Cash System*. https://assets.pubpub.org/d8wct41f/31611263538139.pdf. Accessed December, 2023

Nam, K., Dutt, C. S., Chathoth, P., and Khan, M. S. (2021). Blockchain technology for smart city and smart tourism: latest trends and challenges. *Asia Pacific Journal of Tourism Research*, 26(4), 454–468.

Negri, E., Fumagalli, L., Macchi, M. (2017). A review of the roles of digital twin in CPS based production systems. *Procedia Manufacturing*, 11, 939–948.

Nikolakis, W., John, L., Krishnan, H. (2018). How blockchain can shape sustainable global value chains: An evidence, verifiability, and enforceability (EVE) framework. *Sustainability*, 10(11), 3926.

Önder, I., Treiblmaier, H. (2018). Blockchain and tourism: Three research propositions. *Annals of Tourism Research*, 72(C), 180–182. DOI: 10.1016/j.annals.2018.03.005.

Onik, M. M. H., Miraz, M. H., Kim, C. S. (2018). A recruitment and human resource management technique using blockchain technology for industry 4.0. Smart Cities Symposium 2018. IET, 2018: 1–6.

Özyürek, H., Etlioğlu, M. (2021). *İşletmeler İçin Blockchain*, Efe Akademi Yayınevi, İstanbul.

Pal, A., Tiwari, C.K. and Behl, A. (2021). Blockchain technology in financial services: a comprehensive review of the literature. *Journal of Global Operations and Strategic Sourcing*, 14(1), 61–80.

Piascik R., Vickers J., Lowry D., Scotti S., Stewart J., Calomino A. (2010). Technology Area 12: Materials, Structures, Mechanical Systems, and Manufacturing Road Map. National Aeronautics and Space Administration (NASA). USA.

Pimentel, E., Boulianne, E. (2020). Blockchain in accounting research and practice: Current trends and future opportunities. *Accounting Perspectives*, 19(4), 325–361.

Pournader, M., Shi, Y., Seuring, S, ve Koh, S. L. (2020). Blockchain applications in supply chains, transport and logistics: a systematic review of the literature. *International Journal of Production Research*, 58(7), 2063–2081

Putz, B., Dietz, M., Empl, P., & Pernul, G. (2021). Ethertwin: Blockchain-based secure digital twin information management. Information Processing & Management, 58(1), 102425.

Radochia, S. (2019). *How Blockchain Is Improving Compliance and Traceability in Pharmaceutical Supply Chains*. https://samantharadocchia.com/blog/how-blockchain-is-improving-compliance-and-traceability-in-pharmaceutical-supply-chains. Accessed October 23, 2021.

Salah, K., Rehman, M. H. U., Nizamuddin, N., Al-Fuqaha, A. (2019). Blockchain for AI: Review and open research challenges. *IEEE Access*, 7, 10127–10149.

Shen, W., Hu, T., Zhang, C., Ma, S. (2021). Secure sharing of big digital twin data for smart manufacturing based on blockchain. Journal of Manufacturing Systems, 61, 338–350.

Suri, M. (2021). The scope for blockchain ecosystem. In *Blockchain for Business: How It Works and Creates Value.* Wiley Online Library. 29–58.

Tao, F., Zhang, Y., Cheng, Y., Ren, J., Wang, D., Qi, Q., Li, P. (2022). Digital twin and blockchain enhanced smart manufacturing service collaboration and management. *Journal of Manufacturing Systems, 62*, 903-914.

Tonta, Y. ve Küçük, M. E. (2005). Main dynamics of the transition from industrial society to information society. Proceedings of the Third International Symposium on "Society, Governance, Management and Leadership Approaches in the Light of the Technological Developments and the Information Age". Ankara: The Turkish General Staff Directorate of Military History, Strategic Studies and Inspection Publications.

Veuger, J. (2020). Dutch blockchain, real estate and land registration. *Journal of Property, Planning and Environmental Law*, 12(2), 93–108. https://www.emerald.com/insight/content/doi/10.1108/JPPEL-11-2019-0053/full/html. Accessed August, 2022.

Wang, C., Cai, Z., Li, Y. (2022). Sustainable blockchain-based digital twin management architecture for IoT devices. *IEEE Internet of Things Journal. 10(8),* 6535–6548.

Yli-Huumo, J., Ko, D., Choi, S., Park, S., Smolander, K. (2016). Where is current research on blockchain technology? - A systematic review. *PLoS One,* 11(10), e0163477. DOI: 10.1371/journal.pone.0163477.

Zeadally, S., Abdo, J. B. (2019). Blockchain: Trends and future opportunities. *Internet Technology Letters,* 2(6), e130. DOI: 10.1002/itl2.130.

Zghaibeh, M., Farooq, U., Hasan, N. U., Baig, I. (2020). Shealth: A blockchain-based health system with smart contracts capabilities. *IEEE Access,* 8, 70030–70043.

Zhang, C., Zhou, G., Li, H., Cao, Y. (2020). Manufacturing blockchain of things for the configuration of a data-and knowledge-driven digital twin manufacturing cell. *IEEE Internet of Things Journal,* 7(12), 11884–11894.

11

Physics-Based Digital Twins Leveraging Competitive Edge in Novel Markets

Emil Kurvinen
University of Oulu

Antero Kutvonen and Päivi Aaltonen
LUT University

Jussi Salakka
University of Oulu

Behnam Ghalamchi
California Polytechnic State University

11.1 Introduction

This chapter focuses on the utilization of technology of physics-based simulation beyond product development and utilizes the digital twin (DT)-enabled approaches. We further draw attention to the product, process, and innovation creation potential of DTs and discuss the managerial implications of such technologies. The concept of the Digital Twin dates back to a University of Michigan presentation to the industry in 2002 for the formation of a Product Lifecycle Management (PLM) center [1], and has since been increasingly adopted by technologically advanced manufacturers globally. During the past decade, the development of physics-based simulation, especially real-time capable dynamic calculations and the methodologies to conduct the analyses have been developed rapidly. Currently, the emergence of virtual machines and real-world machines are actively focused and demonstrations are reported (see, e.g., [2,3]). Subsequently, the literature focused on understanding the operational benefits of DTs from the managerial perspective, such as increasing transparency and aiding in decision-making [4,5], has increased over the decades discussing both industrial applications benefits (further, see, e.g., [6]) and the strategic integration of advanced technologies [7]. Previously, many of the economic theories discussed the potential of

DOI: 10.1201/9781003425724-14

augmentation, whereas currently an increasing number of scholars highlight the significance of automation and encourage management to take part in research conducted specifically focusing on intelligent technologies utilization in firms [8]. However, the most common benefit of using advanced technologies in firms is increased product and process innovations as well as 'world-first' innovations [9]. Both product and technology development, and utilization of the systematically structured information in the business scenarios can be studied, yet the maturity level of DT defines the achievable objectives along firm-specific factors, such as R&D expenditure, market, and strategy (see, e.g., [9]).

To introduce the practical advances of DT technology in creating a competitive edge, this book chapter is written based on research work conducted with non-road mobile machinery (NRMM) and high-speed electric machine (HSEM) applications [10], where dynamics modeling and simulation are applied, thereby giving a solid basis to build the DTs on top of the extensive R&D knowledge. In these applications, the final product can be assessed in the virtual domain at high technology readiness level (TRL); for example, in high-speed applications, the controls can be model-based (see, e.g., [11]), i.e., the dynamics model that is used to analyze the machine performance is also used as the basis for the control of the final product. Dynamic modeling is enabling to combine information from design data and verifying the correctness of data as it is fundamentally based on the mass properties of the studied product and its structural stiffness. Experimentally validating the dynamics can be done with modal analysis [12], where the inertia measurement unit, accelerometer, microphone sensors, or laser scanning vibrometer (LSV) [13] are required to capture the vibration in time-domain and transform it to frequency domain (Fast Fourier transformation). From the frequency response, the peaks, i.e., natural frequencies, can be evaluated and compared to the simulation ones. The simulation of the dynamics is a straightforward procedure when using finite element method (FEM) calculation [14], as the geometry and its mass and stiffness properties are needed. Based on these application experiences, the strong physics-based DT is relevant in the applications where the dynamics are relevant in the application.

In the chapter, the simulation-based virtual product is the basis of development, i.e., machine manufacturer R&D knowledge tied to a virtual product. The main applications which the earlier work covers are related to HSEM and NRMM, where the computationally efficient dynamic models are critical from the perspective of the final product and its behavior. The chapter introduces the approaches of different level DTs and exemplary concepts of how they can be leveraged in business to gain a competitive edge in novel markets. The focal point of discussion is linking technology with business thinking and emphasizing how the DT's maturity affects its ability to capture business value. Therefore, the chapter takes an important position in bridging the technology- and business-oriented issues and experts through building mutual understanding and addressing not only the development but also the business application of DTs.

The chapter is structured as follows. Section 11.2 explains the context and background to the viewpoint and theories; Section 11.3 describes physics-based DT and competitive edge and value creation; and Section 11.4 discusses the benefits, opportunities, and challenges of the physics-based DT in novel markets. Section 11.5 is the concluding section.

11.2 Context and Background

Digital twin as a terminology has been expanding rapidly during the last decade [15]. Digital twins can be prototypes (DTP), instances (DTI), or aggregates (DTA) that operate in the digital twin environment (DTE) [1], where the uses for the DTs can be various, e.g., reporting, predicting, analyzing, and autonomous decisions [1]. Depending on the increasing coordination, synchronization, and pooled synergies between various uses within a firm, these can be described with the term 'digital twin maturity', and understood as industrial applications equivalent to firm-level artificial intelligence (AI) maturity (see further, e.g., [6,7]). In the last few years, the maturity model and viewpoint on DTs have been developed to discuss the content from the right viewpoint between different experts.

In this book chapter, the DT is viewed from the simulation perspective, where the modeling, computing capacity, and implementation have been enhanced significantly during the last 5 years. For the DT, real-time behavior is desired, but also faster-than-real-time models can give additional insight to the machine, e.g., when utilizing reinforcement learning algorithms [16].

Also, one distinguishing feature when utilizing simulation-based DTs comes from the products which are tailor-made. For mass products, the available data are easily available in large quantities, which is fruitful for AI and neural networks to identify patterns between different measured quantities. The building of a DT from the physics-based models is a bit different viewpoint than the majority of DT literature; however, it has great potential to assess the system behavior beyond the limited scenarios, e.g., in bearing fault classification [17]. In addition, the physics-based simulation serves also for the company process and planning perspective, e.g., human tasks or specific application process optimization.

The cost of failure of an individual machine in a typical product development process becomes higher the longer the machine has been designed, procured and assembled, commissioned, and produced. The design part can be done to a great extent in the virtual world.

The engineering is mainly done in the virtual world, as there the iteration loops are rapid to operate, and also while having the physical real-life conditions accounted the manufacturing process can be pre-planned to a high level, i.e., the product can be assembled completely in the virtual world and thereby avoid costly mistakes in the actual assembly process [18]. The physics-based simulation can be used to support the transition phase from design to actual product, where the costs are already quite high in newly commissioned products. After the commissioning phase, the data collected from the individual machine become a valuable source of information. Currently, data-based techniques are where most of the research is focused. However, the combination of both techniques is where from a reasonable amount of data the relevant information can be collected, see, for example, Ref. [19], where the neural networks were taught solely with physics-based simulation domain, and then real-world data were used to assess the parameter which is difficult to measure directly from the system.

The growing trend with DTs currently is that the maturity level thinking has found its place, as the DT term itself covers practically everything related to the product (further see, e.g., [1]). Thereby the importance of understanding the depth in which the product is discussed is important to have a mutual understanding regarding it. There are similarities when considering, e.g., the TRL, where the low TRL is 0 and the highest is 9. Similarly, firms' AI maturity is described with five to seven levels [7], yet very few established managerial theoretical frames still exist. However, it has been suggested that the focal point should be on the higher maturity level technologies' implications, in order to create a competitive advantage [9]. Thus, by using a maturity level, the discussion and steps to enhance the complexity become easier to communicate beyond different disciplines and focus on the relevant matter. The DT maturity models are currently being defined and depending on the source – for example, what is understood as the DT concept may vary a lot across industries [20]. In general, between four and six maturity levels can be identified, similar to AI maturity [7]. Further, common to all definitions is that lower levels are unstructured and the highest level represents the most advanced and accurate model, used for decision-making, integration of processes, and higher level autonomy and predictive operations (see further, e.g., [1,6]).

As physics-based simulations can artificially create experience-based information, and the dynamics-based models can correlate data between different parameters to the behavior of the modeled system, the utilization of simulation methods with industrial AI has great potential for building DT applications that can provide a competitive edge. In explaining how organizations can exploit AI solutions, recent research in the areas of management [21] and information systems [22] notes that AI offers qualitatively different opportunities for organizing, scaling business processes, analyzing data, and automating decisions, effectively eliminating the growth constraints imposed by previous-generation systems. Based on the cumulative research, it has been shown that AI involves several features that make it distinct from other technologies, creating specific implications for organizational issues such as trust [4] and decision-making [5]. To leverage business data and become data-driven, firms need to complement their technical analytical resources with multiple different organizational resources and involve several managerial levels [23,24]. As a crucial part of it, firms need to establish a data-driven culture and decision-making practices to tap the potential of their business data [25,26]. Nonetheless, technologies such as DTs are associated with higher levels of product and process innovation, and ones with a longer history with the techniques yield higher output figures [9]. However, theories are currently unclear on the push-and-pull impact of technological maturity and managerial capabilities. For instance, key aspects of organizational "AI readiness" include a firm's innovativeness, collaborative work, and change management [27]. Innovativeness sets the tempo for rapid AI adoption and helps large firms to change the status quo, collaborative work teams up the business, AI, and IT experts to identify the potential use cases for AI [27,28]. On the other hand, despite increasing the competitive edge, only 5.8% of all firms in Germany were actively using AI [9].

From a manufacturer's perspective, the historical data enables the rapid design and testing of future products, both for novel solutions but also for incremental changes in current lines, such as added elements, suitability for custom conditions, and connected novel product sales. Further, the training potential of DTs is significant and can be further leveraged to additional firm personnel, such as sales, to deal with added sales. Finally, this enables the communication and co-creation of products (also similarly [29]) jointly with the customer from early on, as well as maintenance and service support. Customers can test the product in different conditions in the digital space and get much more information about the product before any physical prototype has been manufactured. For the customer, the DT creates similar leveraging potential, also in terms of cost-efficiency optimization, sustainability evaluation, maintenance cost reduction, and custom solution creation. The end-user training can be especially beneficial, as well as the co-creation of solutions jointly with the manufacturer. The challenge regarding industrial AI comes from the need for systematic and well-defined data. Figure 11.1 depicts DTs and different maturity levels connected to exploitation in different business scenarios under the assumption that currently most sales are coming from selling physical products, and DT1–DT6 represents the lowest maturity level to the highest.

The physics-based DTs have been used in dynamic applications [30] as there the importance is to understand the loading influence on the machine behavior, e.g., to avoid resonances. While these models are developed, especially in cases for example where a human is operating the machine, the real-time dynamic model is required. Accuracy and its definition play an important role in the model and refer to what physical phenomena are taken into account in the model and what can be ignored. On the other hand, model partitioning also plays an important role. Overall, when a model is being built, the model should be subjected to a sensitivity analysis, i.e., which

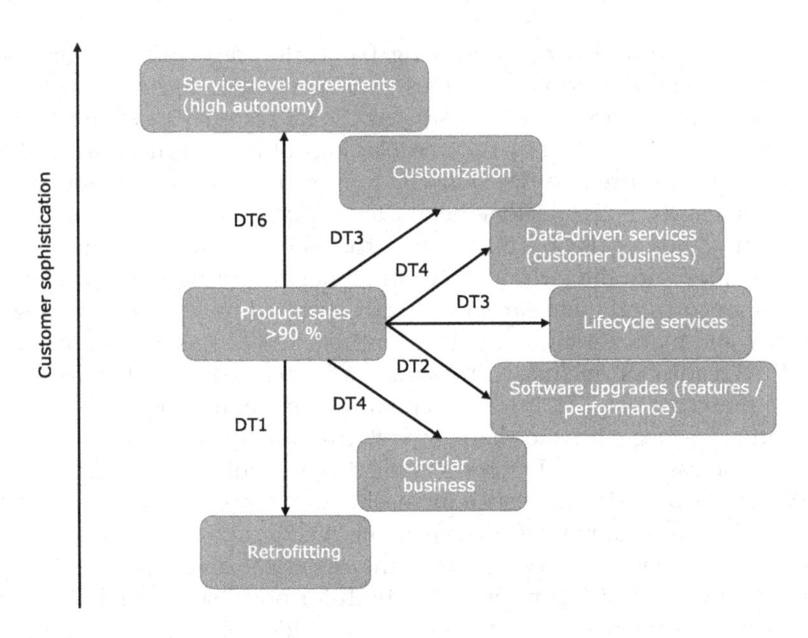

FIGURE 11.1
Digital twin maturity levels and exploitation for different business cases.

factors have the most significant effect on model accuracy [31]. In addition to this, the model designer must evaluate what margin of error they are willing to accept.

One part of the model is constructing its operating environment. This brings a new element to physics-based simulation, where, for example, the reactions of machine automation to information coming from the environment can be simulated using a virtual twin. If the machines delivered to customers can be monitored comprehensively enough, this collected history data can be used to create an image of the operating environment. With these data, it is possible to simulate what-if type scenarios for the virtual twin and create a better understanding of the machine's behavior.

Another option is to study the use of the machine in exceptional environments so that the operator uses the machine in a simulator and learns how to respond to challenging conditions [32]. Again, it makes sense to collect real data from the operating environment so that the simulator environment can be built to best reflect reality.

Reliability analysis, fault detection, and health monitoring are essential in machinery operation in modern industry. Traditional maintenance was limited to periodical checkups, evaluation of measurable output variables, and replacing critical components according to the predetermined schedule. This process is unable to provide insight and in-depth information regarding the operating conditions which is needed for the health monitoring process. To overcome these issues, fault detection and condition monitoring has been introduced. A fault can be defined as a deviation of one or more features or parameters from normal behavior [33]. In this approach, measured signals are compared with the information that can be executed directly from the mathematical modeling or the dynamic model of the system. Another advantage of model-based fault detection is to expand the data to the other states of the system where the measurement is not available using the state estimations such as the Kalman filter. And again, this is possible if we have access to an accurate model of the system. An example of model-based fault detection has been discussed in Ref. [34].

PLM has been a well-known concept for over two decades and is widely used in product development in technology companies [35]. In a wider perspective, it is a business strategy that connects product data, tools, processes, and resources [35,36]. The goal for the company is to be able to manage all the elements of product development in one system. PLM systems were not envisioned to manage large quantities of dynamic data or active models running within the system, but rather they were designed for managing static technical documents, document lifecycle, business processes, and human resources. Ideas of a closed-loop PLM system have been proposed but not yet put into active use in the industry [36].

The implementation of physics-based DTs needs more features than what PLM systems may be able to offer today, but PLM is perhaps the most suitable of such platforms where DTs could be introduced in the future. It may also be possible that DT-specific platforms are taken into use, and integrations between systems (e.g., PLM, ERP) are tailored according to needs. The main elements of a DT platform are a data management system that supports IoT technology and operates in a cloud based manner [37]. The second main element is an MBSE-supporting model that can be operated together with the actual data. In this way, for example, what-if analyses can be performed by an expert [38].

11.3 Physics-Based Digital Twin Competitive Edge and Value Creation

This central role in supporting, coordinating, and planning exploitation activities with DTs affords an excellent opportunity to exploit it, for example, in business and management in relation to Industry 4.0 solutions, business model studies and technology, and innovation management. In addition, it enables actively identifying opportunities to multiply and transfer technologies that will have a real-world impact. A unique result of attaining advanced process simulation capability is the ability to de-couple current process and machine-based constraints from the development of products and associated processes. This is depicted in Table 11.1.

TABLE 11.1

De-Coupling Process and Machine-Based Constraints from the Development of Products and Associated Processes

	Real-World Process	Digital Process
Real-world machine	Continual improvement process • Only small incremental co-evolution/improvement possible • Ease of implementation is critical	Traditional process re-engineering • Constrained by machine design • Process changes costly to validate in real life
Digital machine	Digital twin and machine simulation • Constrained by process design • Co-evolutional designs unattainable incremental innovation	Fully digital simulated co-evolution of machine and process • Enables radical co-evolutional design • Experimentation without strategic constraints • Simulated validation of process changes

So far process innovation has been constrained by re-engineering processes with regard to constraints of the utilized machines and the requirement to keep processes online thus reducing room for experimentation. On the other hand, machine design even with the aid of DTs is constrained by conservative process design. With the ability to develop, design, and test both machines and processes in the digital realm, these constraints can be lifted and simulated experimentation with radical co-evolved designs becomes possible. The physics-based DT is based on software. The software enables new business opportunities. Especially when real physics is involved, the inertia of building prototypes and conducting tests can be avoided, and systems can be studied in real time. This enables rapid prototyping and benchmarking of different solutions in a virtual environment.

The commercial impact of DT exploitation can be described at three levels. First, on the product level, DT delivers new service products and improves on current offerings by enabling cost-effective, process aware machine and service design. Improving the quality and range of service products (for example, AR/VR training) works together with the global trends of servitization and digitalization of industrial products to significantly enhance export potentials for companies. At the company level, organizational renewal is driven by both new service product provision and simplified product process planning and experimentation. Finally, through working in the ecosystem with complementary partners, competitiveness in the global markets is achieved by sharing competencies and co-creating better services. Noticeably as DTs enable the creation of new products and services that facilitate process innovation at the customer end, it will produce significant downstream competitiveness gains. This strengthens the ecosystem and reinforces the relationships between active participants.

11.4 Benefits, Opportunities, and Challenges of Physics-Based Digital Twin in Novel Markets

The benefits, opportunities, and challenges of the DTs are clearly divided between two groups: customers and manufacturers. In practice, the introduction of DTs is on the shoulders of manufacturers, and therefore the benefits for them must be foreseen. For manufacturers, the use of DTs can produce many benefits. Above all, manufacturers benefit from the fact that they can accumulate their product knowledge and that DTs can offer them a competitive advantage over customers who prefer personalized services and user experiences. In some cases, public authorities may also be interested in exploiting data generated by DTs, but this poses a challenge to the protection of customers' privacy. The opportunities related to DTs are endless; however, the approach to a given industry and product varies significantly, and quantifying the business value requires a case-specific approach. Table 11.2 gives a summary of DT-created competitive edge and customer benefits.

For customers, the benefits of DTs are focused on individual customization and product maintenance. Individual customization is meant herein that the DT can propose a user-specific interface and upgrades to individualize operating modes. In this way, the user can get a better user experience or savings with the fine-tuned operating modes. Product maintenance means that, based on the historical data collected, the DT can suggest a suitable

TABLE 11.2

Summary of Digital Twin Created a Competitive Edge to Company and Customer Benefits

Company	Customer
• Improving future design based on historical data	• Machine tuning and testing in special conditions
• Creation of novel custom products	• Machine resilience estimations
• Improving design for custom conditions	• Machine cost-efficiency and sustainability evaluation
• Remote updates	• Historical machine data collection
• Customer communication platform	• Cost-efficiency optimization
• Added sales potential through platform	• Reducing insurance and maintenance cost
• Internal training and educational tool, also for added sales potential visualization	• Support and assistance platform
• Innovation co-creation	• Participation in building custom solution
• Estimation of manufacturing cost-efficiency	• End-user training easiness

maintenance frequency for the user, allowing for even less maintenance in certain cases compared to the common annual maintenance procedure [38]. On the other hand, in more demanding use, it is possible to extend the life of the machine by maintaining it in a preventive maintenance manner.

For manufacturers, the benefits of introducing DTs and DT platforms will be focused on the comprehensive collection of historical data, the improved consumer experience, and the better service opportunities it provides. Historical data allow manufacturers to simulate the behavior of the DT under different conditions by different customers. This enables the development of customized and personalized automation systems as well as robust design for new machines. The historical data collected in the DT instances can serve as a data bank for simulation models with which AI models can be trained efficiently.

The challenge with regard to the utilization of data from DTs will be ownership and use of data [38]. In practice, the ownership of data is still undefined in most cases. By default, the machine owner is the strongest candidate for ownership, but on the other hand all the services for data are provided by the manufacturers, possibly in collaboration with third parties. Many owners may experience data collection as a violation of their privacy and refuse to disclose the information. This could be alleviated by informing the owners so that they clearly understand the purpose of the data and that the collection of data will also be beneficial them.

Theories on creating competitive edge via AI technologies, described in this chapter with the concept of DTs, emphasize the development of AI, yet there are significant other determinants impacting firm performance. Further, as illustrated also in Table 11.1, potential may be dependent on degree of novelty or degree of cost savings [9]. Figure 11.2 illustrates the factors for managerial and manufacturers competitive edge creation.

Novel market entry and firm capabilities depend on multiple levels of operations (see, e.g., [21–24]), yet may enable overcoming of constrains of previous-generation technologies (see Tables 11.1 and 11.2). However, advantages may be dependent on industry and target market. For example, the machinery manufacturing industry is classified as a low- or medium-technology industry, i.e., the ratio between production and R&D funding is typically less than 5% (see [39], Table 6.2, p. 157). High-technology industries, such as pharmaceutics, computing machinery, and spacecraft-related manufacturing, have typically diverted more funds to R&D. In the pharmaceutical industry, novel drugs are discovered every day (see [40]). Because it dramatically reduces development costs, virtual simulation makes it possible for the machine manufacturing industries to move toward the higher-technology spectrum, introducing a further example of novel potential. The adoption of a physics-based simulation environment enables a significant increase in the number of innovations, potentially translating to increased novel market discovery, particularly through leveraging service and software offerings.

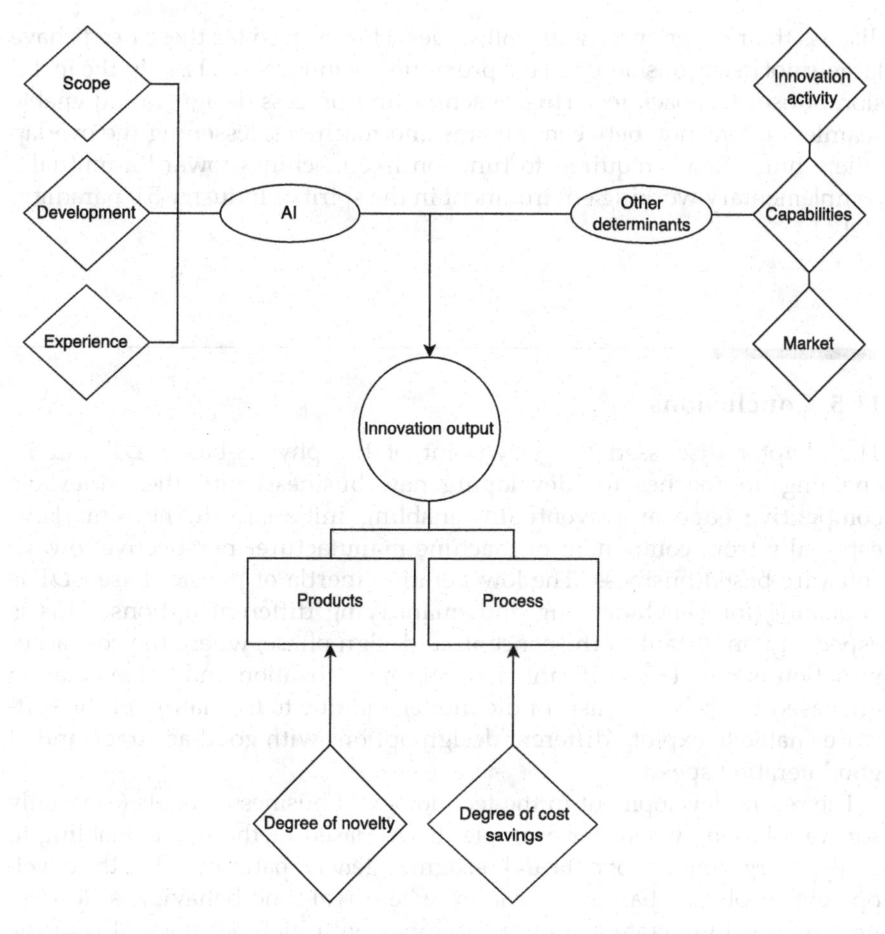

FIGURE 11.2
Factors impacting manufacturer competitive edge. (Source: Based on [9].)

Therefore, the adoption of fully virtual machine and process design leverages the number of innovations beyond typical industry figures. Because they remove boundaries, virtual innovative environments are shown to encourage engagement and innovative performance (see, e.g., [29]), and may be especially beneficial in reacting to disruptive or turbulent environment, i.e., political challenges and emerging economies operations (see more, e.g., [41]). Moreover, they provide a fruitful base for co-evolutionary development between machine manufacturing and the manufacturing process. Including the end-user in the development of machines and processes and their testing enables the rapid realization of innovative solutions targeted to fit the true needs of the user. For example, it becomes much easier for machine operators to get workflow up to speed if they are provided with audiobooks and can discuss their experiences with colleagues. However, so far these needs have largely not been considered in the promotion of innovation. Finally, the inclusion of user-feedback to virtual machine and process design would enable seamless interaction between humans and machines, lessening the overlap where humans are required to function like machines toward a mutually complementary working environment in the spirit of Industry 5.0 paradigm [42].

11.5 Conclusions

The chapter discussed the viewpoint of the physics-based DT and its enabling approaches for developing new business with the achievable competitive edge and eventually enabling initiations in new markets, especially from component or machine manufacturer perspective toward software-based business. The low iteration inertia of physics-based DT is appealing for benchmarking and quantifying different options. This is especially important in the conceptual design phase, where the cost accumulation is not yet significant. The commercialization and value creation are based on the solid basis of the model and due to the nature of the software enable to explore different design options with good accuracy and at good iteration speeds.

The recent development in the technology of business models (especially software based), where for example the AI-based methods are enabling to analyze large amount of data and recognize general patterns. Also, the development of physics-based simulation, where real-time behavior is desired, has enabled to operate complete machines with high accuracy also in the virtual world similar to how real machines operate. This gives a solid base for data production for different types of analyses, e.g., business, R&D, or innovations [43].

The DT has currently found its shape; however, the strategies and venues for exploiting it in the companies are still under research. Technologically, the possibilities are rigorous, but the implementation in the company organization requires further research. For example, transforming from machine manufacturer to software type of organization is a profound change that can involve re-engineering the skills base, capabilities, business models, and innovation processes of the company to various extents. Given how many manufacturing companies will soon be at the brink of this transition, it needs to be better understood and managed, both theoretically and in practice, possibly through exploiting examples of software companies and manufacturing pioneers that have been among the first to opt for this transition.

Bibliography

[1] Michael Grieves and John Vickers. Digital twin: Mitigating unpredictable, undesirable emergent behavior in complex systems. *Transdisciplinary Perspectives on Complex Systems*, pages 85–113. Springer, 2017.

[2] Emil Kurvinen, Antero Kutvonen, Juhani Ukko, Qasim Khadim, Yashar Shabbouei Hagh, Suraj Jaiswal, Neda Neisi, Victor Zhidchenko, Juha Kortelainen, Mira Timperi, Kirsi Kokkonen, Juha Virtanen, Akhtar Zeb, Ville Lamsa, Vesa Nieminen, Jukka Junttila, Mikko Savolainen, Tuija Rantala, Tiina Valjakka, Ilkka Donoghue, Kalle Elfvengren, Mina Nasiri, Tero Rantala, Ilya Kurinov, Eerik Sikanen, Lauri Pyrhonen, Lea Hannola, Heikki Handroos, Hannu Rantanen, Minna Saunila, Jussi Sopanen, and Aki Mikkola. Physics-based digital twins merging with machines: Cases of mobile log crane and rotating machine. *IEEE Access*, 10:45962–45978, 2022.

[3] Qasim Khadim, Yashar Shabbouei Hagh, Lauri Pyrhonen, Suraj Jaiswal, Victor Zhidchenko, Emil Kurvinen, Jussi Sopanen, Aki Mikkola, and Heikki Handroos. State estimation in a hydraulically actuated log crane using unscented kalman filter. *IEEE Access*, 10:62863–62878, 2022.

[4] Ella Glikson and Anita Williams Woolley. Human trust in artificial intelligence: Review of empirical research. *Academy of Management Annals*, 14(2):627–660, 2020.

[5] Erik Brynjolfsson and Kristina McElheran. The rapid adoption of data-driven decision-making. *American Economic Review*, 106(5):133–139, 2016.

[6] Jay Lee. *Industrial AI: Applications with Sustainable Performance* by Jay Lee, Springer Singapore, 2020, 1–162 pp., ISBN 978-981-15-2143-0 (Hardcover) ISBN 978-981-15-2144-7 (eBook) https://doi. org/10.1007/978-981-15-2144-7, Copyright Holder: Shanghai Jiao Tong University Press." (2023): 404–405.

[7] Ulrich Lichtenthaler. Agile innovation: The complementarity of design thinking and lean startup. *International Journal of Service Science, Management, Engineering, and Technology (IJSSMET)*, 11(1):157–167, 2020.

[8] Sebastian Raisch and Sebastian Krakowski. Artificial intelligence and management: The automation-augmentation paradox. *Academy of Management Review*, 46(1):192–210, 2021.

[9] Christian Rammer, Gaston P Fernandez, and Dirk Czarnitzki. Artificial intelligence and industrial innovation: Evidence from german firm-level data. *Research Policy*, 51(7):104555, 2022.

[10] Emil Kurvinen, Amin Mahmoudzadeh Andwari, and Juho Konno. Physics-based dynamic simulation opportunities with digital twins. *Future Technology*, 1(3):03–05, 2022.

[11] Rafal P Jastrzebski, Atte Putkonen, Emil Kurvinen, and Olli Pyrhonen. Design and modeling of 2 mw amb rotor with three radial bearing-sensor planes. *IEEE Transactions on Industry Applications*, 57(6):6892–6902, 2021.

[12] D.J. Ewins. *Modal Testing: Theory, Practice and Application*. John Wiley & Sons, 2009.

[13] Emil Kurvinen, Miia John, and Aki Mikkola. Measurement and evaluation of natural frequencies of bulk ice plate using scanning laser Doppler vibrometer. *Measurement*, 150:107091, 2020.

[14] Robert D Cook et al. *Concepts and Applications of Finite Element Analysis.* John Wiley & Sons, 2007.
[15] Yuqian Lu, Chao Liu, Kevin I-Kai Wang, Huiyue Huang, and Xun Xu. Digital twin-driven smart manufacturing: Connotation, reference model, applications and research issues. *Robotics and Computer-Integrated Manufacturing,* 61:101837, 2020.
[16] Ilya Kurinov, Grzegorz Orzechowski, Perttu Hamalainen, and Aki Mikkola. Automated excavator based on reinforcement learning and multibody system dynamics. *IEEE Access,* 8:213998–214006, 2020.
[17] Cameron Sobie, Carina Freitas, and Mike Nicolai. Simulation-driven machine learning: Bearing fault classification. *Mechanical Systems and Signal Processing,* 99:403–419, 2018.
[18] Felix Gaisbauer, Philipp Agethen, Michael Otto, Thomas Bar, Julia Sues, and Enrico Rukzio. Presenting a modular framework for a holistic simulation of manual assembly tasks. *Procedia CIRP,* 72:768–773, 2018. *51st CIRP Conference on Manufacturing Systems.*
[19] Denis Bobylev, Tuhin Choudhury, Jesse O. Miettinen, Risto Viitala, Emil Kurvinen, and Jussi Sopanen. Simulation-based transfer learning for support stiffness identification. *IEEE Access,* 9:120652–120664, 2021.
[20] Tuija Rantala, Kirsi Kokkonen, and Lea Hannola. Selling digital twins in business-to-business markets. *Real-Time Simulation for Sustainable Production,* pages 51–62. Routledge, 2021.
[21] Michael Haenlein, Andreas Kaplan, Chee-Wee Tan, and Pengzhu Zhang. Artificial intelligence (AI) and management analytics. *Journal of Management Analytics,* 6(4):341–343, 2019.
[22] Nicholas Berente, Bin Gu, Jan Recker, and Radhika Santhanam. Managing artificial intelligence. *MIS Quarterly,* 45(3):1433–1450, 2021.
[23] Richard Vidgen, Sarah Shaw, and David B Grant. Management challenges in creating value from business analytics. *European Journal of Operational Research,* 261(2):626–639, 2017.
[24] Eivind Kristoffersen, Patrick Mikalef, Fenna Blomsma, and Jingyue Li. The effects of business analytics capability on circular economy implementation, resource orchestration capability, and firm performance. *International Journal of Production Economics,* 239:108205, 2021.
[25] Eivind Kristoffersen, Fenna Blomsma, Patrick Mikalef, and Jingyue Li. The smart circular economy: A digital-enabled circular strategies framework for manufacturing companies. *Journal of Business Research,* 120:241–261, 2020.
[26] Philipp Korherr, Dominik K Kanbach, Sascha Kraus, and Paul Jones. The role of management in fostering analytics: The shift from intuition to analytics-based decision-making. Journal of Decision Systems, 2022: 1–17.
[27] Jan Johnk, Malte Weißert, and Katrin Wyrtki. Ready or not, AI comes-an interview study of organizational AI readiness factors. *Business & Information Systems Engineering,* 63(1):5–20, 2021.
[28] Tim Fountaine, Brian McCarthy, and Tamim Saleh. Building the AI-powered organization. *Harvard Business Review,* 97(4):62–73, 2019.
[29] Angelos Kostis and Paavo Ritala. Digital artifacts in industrial cocreation: How to use VR technology to bridge the provider-customer boundary. *California Management Review,* 62(4):125–147, 2020.

[30] D. J. Wagg, K. Worden, R. J. Barthorpe, and P. Gardner. Digital twins: State-of-the-art and future directions for modeling and simulation in engineering dynamics applications. *ASCE-ASME Journal of Risk and Uncertainty in Engineering Systems, Part B: Mechanical Engineering*, 6(3):030901, 2020.

[31] David Ullman. *The Mechanical Design Process*. McGraw Hill, 2009. - ds.amu.edu.et

[32] Pedro Tavares, Joao Andre Silva, Pedro Costa, Germano Veiga, and Antonio Paulo Moreira. Flexible work cell simulator using digital twin methodology for highly complex systems in industry 4.0. ROBOT 2017: Third Iberian Robotics Conference: Volume 1. Springer International Publishing, 2018: 541–552.

[33] Rolf Isermann. Model-based fault-detection and diagnosis - status and applications. *Annual Reviews in Control*, 29(1):71–85, 2005.

[34] Behnam Ghalamchi, Zheng Jia, and Mark Wilfried Mueller. Real-time vibration-based propeller fault diagnosis for multicopters. *IEEE/ASME Transactions on Mechatronics*, 25(1):395–405, 2020.

[35] Fernando Mas, Rebeca Arista, Manuel Oliva, Bruce Hiebert, Ian Gilkerson, and Jose Rıos. A review of plm impact on us and eu aerospace industry. *Procedia Engineering*, 132:1053–1060, 2015.

[36] Hong-Bae Jun, Dimitris Kiritsis, and Paul Xirouchakis. Research issues on closed-loop plm. *Computers in Industry*, 58(8):855–868, 2007.

[37] Victor Zhidchenko, Egor Startcev, and Heikki Handroos. Reference architecture for running computationally intensive physics-based digital twins of heavy equipment in a heterogeneous execution environment. *IEEE Access*, 10:54164–54184, 2022.

[38] Azad M Madni, Carla C Madni, and Scott D Lucero. Leveraging digital twin technology in model-based systems engineering. Systems, 2019, 7(1): 7.

[39] Jan Fagerberg, David C Mowery, Richard R Nelson, et al. *The Oxford Handbook of Innovation*. Oxford University Press, 2005. books.google.com

[40] Pierpaolo Andriani, Ayfer Ali, and Mariano Mastrogiorgio. Measuring exaptation and its impact on innovation, search, and problem solving. *Organization Science*, 28(2):320–338, 2017.

[41] Paivi Aaltonen, Lasse Torkkeli, and Maija Worek. The effect of emerging economies operations on knowledge utilization: The behavior of international companies as exaptation and adaptation. International Business and Emerging Economy Firms: Volume I: Universal Issues and the Chinese Perspective, 2020: 49–87. Springer, 2020.

[42] Francesco Longo, Antonio Padovano, and Steven Umbrello. Value-oriented and ethical technology engineering in industry 5.0: A human-centric perspective for the design of the factory of the future. Applied Sciences, 2020, 10(12): 4182.

[43] Mira Holopainen, Minna Saunila, Tero Rantala, and Juhani Ukko. Digital twins' implications for innovation. Technology Analysis & Strategic Management, 2022: 1–13.

Part 4

Digital Twins Design and Standard

12

Digital Twin Model Formal Specification and Software Design

Yevgeniya Sulema
National Technical University of Ukraine

Andreas Pester
The British University in Egypt

Ivan Dychka and Olga Sulema
National Technical University of Ukraine

12.1 Introduction

The technology of digital twins implements an approach [1–7] that enables digital representation of a physical object or process, specifically its past, current, and future state, behavior, characteristics, appearance, etc. According to the definition proposed by the authors of the digital twin concept, Michael Grieves and John Vickers, in [1], a *digital twin* is a collection of virtual information constructs that fully describes a physical object: from the micro level (level of an individual element) to the macro level (general appearance, geometric representation, general properties of the object as a whole). In other words, a digital twin is a model of a physical object or process (*physical twin*) that most fully reflects its characteristics in dynamics during a specific time. The concept of a digital twin involves the presentation, processing, and manipulation of all data that characterize a physical twin, as well as obtaining specific components of these data as the need for them arises when solving problems of analyzing and forecasting the state of a physical twin, optimizing procedures for physical twin control, etc.

The analysis of the digital twin data makes it possible to detect anomalies in the behavior of the components of the physical twin before an emergency occurs in the physical twin and, thus, prevent the emergency in time. For example, a digital twin of an aircraft [8–11] integrates data from sensors of an embedded system located on board a physical twin and contains data on the history of maintenance of the research object, the history of its

DOI: 10.1201/9781003425724-16

emergencies, etc. Thus, a digital twin is an *individualized* realistic virtual model of a particular researched object – a physical twin.

From the point of view of practical implementation, a digital twin is a complex software system that includes data storage and software modules that provide data acquisition, processing, and visualization.

Let us consider the advancement of the digital twin concept by assuming that a digital twin consists of *visual* and *behavioral* models of the research object. These models are implemented based on appropriate mathematical and data presentation models and ensure synchronization between the virtual and real systems at the level of data from sensors installed for constant monitoring of the research object. Further development of the concept of a digital twin is possible in several directions, in particular, in the direction of using artificial intelligence tools and the transition from graphic representation to multimedia and mulsemedia [12,13] representation of the research object.

12.2 Related Works

Most studies, which have been published on digital twins, are devoted to general questions about the technology of digital twins, the field of its application, the advantages provided by this technology, and organizational issues regarding its implementation. Such focus in publications can be explained by the fact that this field of research has emerged relatively recently, so scientific works at this stage are still focused on solving more global issues related to the concept of digital twins.

Nonetheless, some researchers offer solutions to more specific, practical-oriented scientific problems. For instance, works [14–16] present the basic architecture of a software system for the data processing of a digital twin. The architecture proposed in [14] assumes the presence of the following main components:

1. A software component that implements an information model providing an abstract specification of the technical characteristics of a physical object.
2. A software component that implements the communication mechanism between digital and physical twins.
3. A software component that implements procedures for processing, analyzing, and searching for multimodal data to obtain up-to-date information about a physical twin.
4. A key-value storage designed to store digital twin data.

The proposed architecture does not include a component for rendering a digital twin. The result of the operation of such a software system is only analytical data.

The paper [15] presents the basic architecture of a digital twin for cloud-based cyber-physical systems. One of the components of the proposed architecture is an intelligent interaction controller that allows dynamic reconfiguration of the cyber-physical system.

The authors of [16] propose a generalized software and hardware architecture of a six-level digital twin. The first and lowest level is the level of the physical object (physical twin), on which its characteristics are recorded using a set of sensors, and interaction with the physical twin takes place through actuators. The second layer is the layer of local controllers that provide the specific functionality of the digital twin. The third level is the local database level, while the fourth is communication, which ensures the interaction between local and cloud data storage. The fifth level is cloud data storage. The sixth level, the last, is where a physical twin's emulation (modeling of current behavior) and simulation (modeling of future or possible behavior) occur.

In the work in [17], a basic digital twin model is proposed; it is intended to analyze and model production processes. It consists of four levels: the level of the physical space of the manufacturing process, the level of the communication system, the level of the digital twin, and the level of the user space. The software system for implementing a digital twin includes a control module, a simulation module, an anomaly prediction, and detection module, and a data storage module (cloud storage). The disadvantage of the proposed model is the impossibility of using it for other fields of application of digital twins.

Storing of digital twin data can be done in the AutomationML format [18], which is proposed in [19]. This format is based on the XML format and allows storing data of industrial object models. The object is presented as a hierarchy of its components. The obtained model can be used for data exchange between software platform components.

Another format that can be used to save data of industrial object models is B2MML [20]. This format is also based on the XML format and the IEC/ISO 62264 standard [21]. B2MML consists of schemas written using the XSD language [22].

The studies presented in [23,24] are devoted to creating visual models of digital twins. The authors demonstrate the solution to this problem by creating a digital twin of a micro-manufacturing unit [23] and a digital twin of a vertical milling machine [24]. To obtain a visual model, it is suggested to use the SolidWorks software platform [25]. The created model [24] can be saved in the STL format [26], which allows its visualization not only on the monitor screen but also in the form of a plastic model made by 3D printing. Software visualization is performed using the OpenGL library [27]. The physical twin's characteristics are monitored using a set of sensors, the data from which are sent for storing in the local data storage – the PostgreSQL database [28]. Remote access to data occurs via the Internet. A multimodal data set for a digital twin of a vertical milling machine includes information on vibration, acceleration, and electrical characteristics (amperage, energy consumption). The processing of temporal multimodal data in this study aims to predict the quality of mechanical processing using a machine tool. A corresponding mathematical model is proposed in [24] to achieve this goal.

The analysis of these and other literary sources allows us to assert the absence of established theoretical foundations for developing digital twins. The existing approaches are not universal but have an applied nature and solve the problem of creating a digital twin for a specific application.

In contrast, in our research, we propose an overall approach that can be employed for the development of a digital twin in a wide range of applications.

12.3 A Multi-Image Concept for a Formal Specification of Digital Twin Models

Let us consider the object S which reveals its essence through a set of properties $F_1, F_2, \ldots, F_N$ that can be measured (recorded). These properties are semantically interconnected as they characterize the same object. We can present the properties $F_1, F_2, \ldots, F_N$ as an aggregate A_S according to the Algebraic System of Aggregates (ASA) rules [29,30]; the components of this aggregate are tuples of values $f_i^{j}{}_{i=1}^{n_j}$ of features F_j $(j = 1 \ldots N)$:

$$A_S = \left[\!\left[M_1, M_2, \ldots, M_N \mid \left\langle f_i^1 \right\rangle_{i=1}^{n_1}, \left\langle f_i^2 \right\rangle_{i=1}^{n_2}, \ldots, \left\langle f_i^N \right\rangle_{i=1}^{n_N} \right]\!\right], \tag{12.1}$$

If the measurement of the values of different properties is simultaneous, then all tuples of the aggregate A_S have the same length. In this case, the aggregate A_S can be considered as a sequence of multicomponent values. This sequence of multicomponent values can also be considered a function of many variables: $y = x(t, F_1, F_2, \ldots, F_N)$. Such a view of aggregated data allows to apply to them the apparatus of calculation of many variables in tasks in which it is appropriate to present and process data using operations and relations defined in ASA.

If the data about the object are obtained considering the time of observation or measurement, then the aggregate includes a tuple of corresponding temporal values that correspond to the moments of time when the values of the multimodal data tuples were obtained. Such an aggregate is called a *multi-image* of a digital twin [31]. Let us define the multi-image formally.

Definition 12.1

Multi-image is a non-empty aggregate such as:

$$I = \left[\!\left[T, M_1, \ldots, M_N \mid \langle t_1, \ldots, t_\tau \rangle, \langle a_1^1, \ldots, a_{n_1}^1 \rangle, \ldots, \langle a_1^N, \ldots, a_{n_N}^N \rangle \right]\!\right], \tag{12.2}$$

where T is a set of time values; $\tau \geq n_j, j \in [1, \ldots, N]$.

So, in the mathematical sense, a multi-image is an aggregate whose first tuple is a non-empty tuple of time values. These values can be natural numbers or any other values that provide clarity and unambiguous presentation of information about the moments of time at which the elements of other tuples of the multi-image were obtained.

12.4 Digital Twin Model Formal Specification

The digital twin is provided in the software system through the data obtained in the process of observing its physical twin – the object under study. These data can be considered at different levels of abstraction, based on the purpose of the study of the physical twin. Depending on the context of the specific problem being solved, the digital twin can represent the physical twin as:

- A solid object, then the object can be considered indivisible, but at the same time, it can be heterogeneous in its structure, material, physical properties, etc.
- An object defined by a set of states dynamically changing over time.
- An object that is a composition of its components, each of which, in turn, can be considered as a separate object, forming a hierarchy of objects.

These levels of abstraction can be combined to produce a more informative way of describing a physical twin.

Let us consider a model representing the physical object as a digital twin in terms of time. Such a model, called temporal linked model (TLM) [32], presents the object through a set of its discrete states, determined at unique moments in time by a set of characteristics about objects, which are multimodal data. At each moment in time, the object has one and only one state, which is determined by its multi-image. At the same time, the time tuple of each multi-image contains only one value, which determines the moment in time when the object had a certain specific state. This approach enables using the time value as a key that uniquely determines this specific state in the ordered sequence of states of the object under study (Figure 12.1).

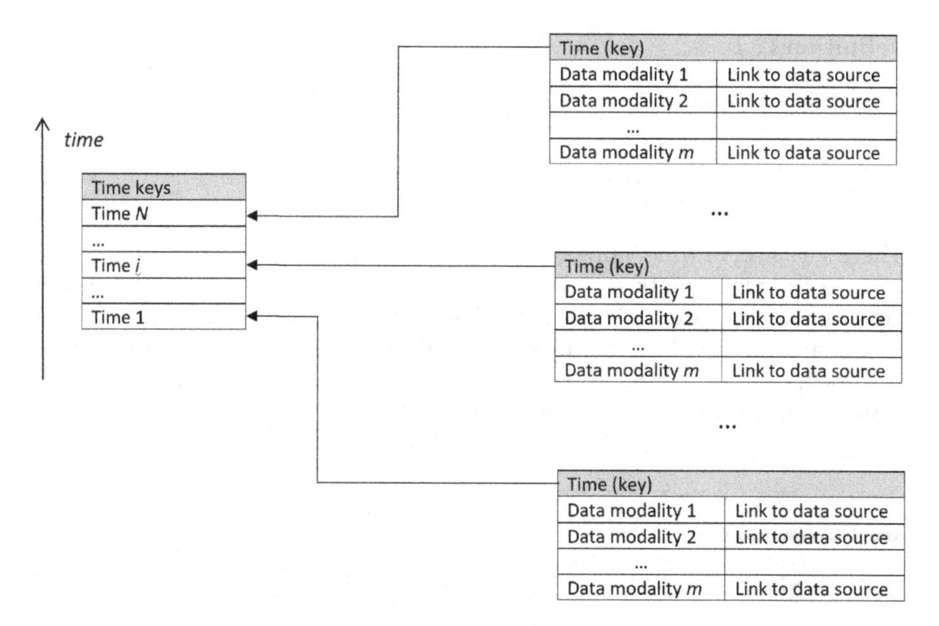

FIGURE 12.1
Temporal linked model.

To create the object's TLM, it is necessary to determine the structure of the object's multi-image. At the same time, a multi-image for a TLM differs in that its time tuple contains only one value of time, which is associated with data from other models that determine the object.

To define the object's state at a certain point in time, it is necessary to set the data sources of each modality that determine this state. This can be done by assigning unique identifiers that represent links. If the data of a certain modality cannot be obtained or is not needed at the current time, then the attribute *unavailable* is assigned to the data set instead of the link identifier.

The states of the object are discrete, which means that on the time scale of the object's description, separate discrete moments of time (*time keys*) are defined, with which specific states of the object are associated. When each discrete state is determined, the data that arrived or began to arrive at a certain discrete moment in time, which is the key to this current time, is synchronized and aggregated.

Thus, we receive data from each source and form a partial multi-image of the following type:

$$I_j = \left[\!\left[T, M_j \mid \langle t \rangle^j, \left\langle d_i^j \right\rangle_{i=1}^{n_j} \right]\!\right], \tag{12.3}$$

where t^j is a time key of j-state of the object, $t^j \in T$; d_i^j is the data of j-modality associated with the key t^j, $d_i^j \in M_j$; T is a set of key values; and M_j is a set of data of j-modality, $j = [1 \ldots N]$.

Partial multi-images are combined into a single multi-image that determines the current state of the object, while during the synchronization and aggregation of data of partial multi-images, one-time key is selected from among all-time keys, which will determine this current state of the object, and the synchronization procedure is fulfilled according to the rules of fuzzy synchronization. As a result, to determine the state of the object, we get a multi-image, the time tuple of which contains one single value:

$$I = \bigcup_{j=1}^{N} I_j = \left[\!\left[T, M_1, \ldots, M_N \mid \langle t \rangle, \langle d_i^1 \rangle_{i=1}^{n_1}, \ldots, \langle d_i^N \rangle_{i=1}^{n_N} \right]\!\right] \tag{12.4}$$

where t is defined according to Eqs. (12.5)–(12.7), or semantically.

$$\langle t \rangle = \max_{j=1\ldots N} t^j \tag{12.5}$$

$$\langle t \rangle = \min_{j=1\ldots N} t^j, \tag{12.6}$$

$$\langle t \rangle = \frac{\sum_{j=1}^{N} t^j}{N}, \tag{12.7}$$

Equation (12.5) means that the moment of time, when the entire set of data is received, is taken as the time key of the state. Equation (12.6) means that the moment of time, when the data just started to arrive, is taken as the key. Equation (12.7) determines the average time of data arrival.

The data source can be a device that generates streaming data or a storage (local, cloud) that stores data as files.

Let us consider the case when the data sequence $d_{i\,i=1}^{j\,n_j}$ is saved as a file of a certain format. In some file formats (e.g., medical imaging formats NIFTI [33] and DICOM [34]), the acquisition times of the components of this data sequence are stored in the metadata or directly with the underlying data.

However, in many cases, the temporal information related to the data sequence is limited to the time of creation (modification) of the file, and information about the time of receipt of individual components of the data sequence is stored separately. Therefore, the task of synchronizing data of different modalities, which must be performed when creating a TLM, can be complicated by the lack of a single method of storing time values.

To simplify the processing of temporal multimodal data when creating a TLM, it is advisable to modify the files that store the data of each modality used in the model with a wrapper file. The structure of this file (Figure 12.2) includes three sections: a metadata section, a time values section, and the main section where the wrapped file resides.

The metadata section contains the "TIME" token, the file type identifier (e.g., "nii" for the NIFTI file format, "dcm" for the DICOM file format, etc.),

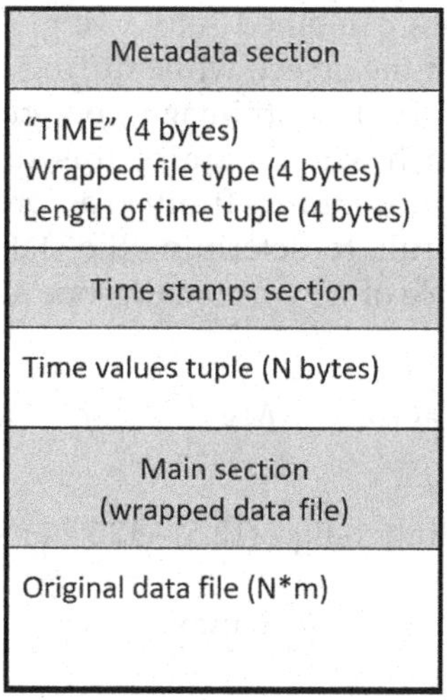

Metadata section

"TIME" (4 bytes)
Wrapped file type (4 bytes)
Length of time tuple (4 bytes)

Time stamps section

Time values tuple (N bytes)

Main section
(wrapped data file)

Original data file (N*m)

FIGURE 12.2
A wrapper file structure.

and the size of the tuple of time values. The time values section contains the time values in ascending order.

The main section contains the wrapped file; this file contains ready-to-use data. This procedure for modifying the original file enables unifying the procedure for synchronizing sequences of data received from sources of different types and, at the same time, saving the data in their original format.

It is advisable to use the TLM of the digital twin in cases where the study of the physical twin takes place through the determination of its characteristics at discrete moments of time, resulting in the formation of a sequence of discrete states of the object, each of which is characterized by a set of temporal multimodal data associated with a specific moment in time that can serve as a time key of a certain state of the object.

12.5 Software Architecture Design for Digital Twin Applications

The software implementing the digital twin technology must cover all aspects of the presentation and processing of temporal multimodal digital twin data. Therefore, the software system for processing temporal multimodal data of digital twins should include components that provide:

- Implementation of rules and templates for synchronization of temporal data.

- Synchronization of temporal multimodal data.
- Aggregation of temporal multimodal data.
- Obtaining data about the object (physical twin) from sensors and other data sources.
- Conversion of data formats.
- Implementation of the behavioral model of the digital twin.
- Implementation of a visual model of a digital twin.
- Formation of a multi-image of the object.
- Digital twin data processing.
- Data analysis of the behavioral model of the digital twin.
- Reproduction of the data of the visual model of the digital twin.
- Visualization of the results of physical twin research.
- Ensuring the confidentiality of data submission of the digital twin.
- Digital twin data storage.

Then, the generalized architecture of the software system for processing temporal multimodal data of digital twins is presented in Figure 12.3.

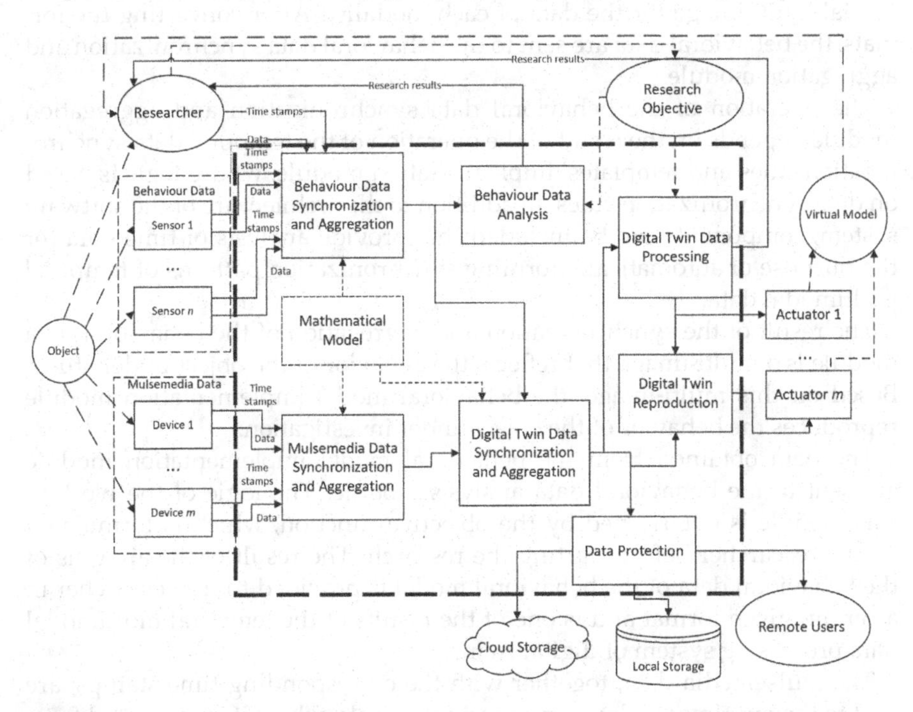

FIGURE 12.3
Generalized architecture of the software system for processing temporal multimodal data of digital twins.

This architecture implements the following logic for working with a digital twin. The researched object manifests itself through several physical processes, phenomena, and sensory effects that characterize it and can be recorded in the form of objective parameters of the researched object using a set of sensors and information recording devices, as well as subjective parameters that are evaluated by the researcher. Objective characteristics are data divided into behavioral and multimedia.

Behavioral data in the further process of researching the object are considered and processed as some technical indicators of the researched object. Mulsemedia data are used for virtual visualization of the researched object and are intended for human sensory perception.

All data obtained because of the registration of the characteristics of the object, regardless of their type, are accompanied by a sequence of time stamps that determine the moments of time when these data were obtained. The procedure for obtaining labels depends on the modality of the data, the way these data are registered, and the format in which it enters the software system for processing temporal multimodal data of digital twins.

Behavioral data, along with the corresponding time stamps, are sent to the behavioral data processing module, where they are converted from a specific format to an internal representation format, which involves the creation of a partial multi-image for the data of each modality. After converting the formats, the behavioral data are sent to the behavioral data synchronization and aggregation module.

The operation of the behavioral data synchronization and aggregation module depends on the result of the operation of the temporal data synchronization rules and templates' implementation module, whose logic is based on data synchronization rules. In addition to the architecture of the software system, components can be included that provide analysis of time data for the purpose of automatically forming synchronization patterns of temporal multimodal data.

The result of the synchronization and aggregation of the behavioral data module is a multi-image that reflects the behavior of the object under study. Based on this multi-image, the behavioral model implementation module reproduces the behavior of the object under investigation.

The data obtained from the behavioral model implementation module are sent to the behavioral data analysis module. The logic of the work of this module is determined by the objective function, which is formulated by the researcher before starting the research. The result of the analysis of the behavioral data of the behavioral model is provided to the researcher in a certain given format and is one of the results of the temporal multimodal data processing system of digital twins.

The mulsemedia data, together with the corresponding time stamps, are sent to the multimedia data reprocessing module, where it is converted from a certain format to the format of the internal presentation, which involves the creation of a partial multi-image for the data of each modality. After converting the formats, the mulsemedia data are sent to the mulsemedia data synchronization and aggregation module.

The operation of the mulsemedia data synchronization and aggregation module depends on the result of the operation of the temporal data synchronization rules and templates' implementation module. The result of the mulsemedia data synchronization and aggregation module is a multi-image that reflects the characteristics of the external manifestations of the object, including the characteristics of the appearance.

Based on this multi-image, the visual model implementation module reproduces the appearance and other external manifestations of the object, including creating a virtual model of this object. This virtual model is one of the results of the software system for processing temporal multimodal data of digital twins.

The virtual model is transmitted for reproduction to the reproduction module of the digital twin. In addition to the architecture of the software system, components can be included that ensure the improvement of the quality of visualization and the application of other types of data processing of the visual model, in particular, perform the processing of graphic data.

The multi-images obtained because of the behavioral data synchronization and aggregation module and the mulsemedia data synchronization and aggregation module are sent to the digital twin's data synchronization and aggregation module, the operation of which depends on the result of the temporal data synchronization rules and patterns' implementation module.

The result of the operation of the digital twin data synchronization and aggregation module is the consolidated data of the digital twin of the object. These data are transferred to the data processing module of the digital twin, the logic of which is determined by the given target function of the study. The result of processing is provided to the researcher in some certain defined form (format) and is one of the results of the system.

In addition, the consolidated data of the digital twin are sent to the reproduction module of the digital twin, in which the digital twin is corrected according to the results of its data processing in the data processing module of the digital twin and in accordance with the new data coming from the sensors that register the multimodal parameters of the object.

Dynamically changing data of the visual model of the digital twin received by the reproduction module of the digital twin is fed to a set of actuators, each of which allows to reproduce the data of a certain modality in the form of a certain aspect of the image of the digital twin, that is, in the form of a virtual model of the object, which is shown to the researcher, which is another result of the software system for processing temporal multimodal data of digital twins.

The data of the digital twin of the research object are sent to the data protection module of the digital twin for encryption and/or steganographic concealment of the confidential data of the digital twin.

After processing the confidential data, the digital twin data are sent for storage to local and/or cloud storage and can also be sent to remote users of the digital twin data of the object.

Let us consider the application of this generalized architecture for the development of software systems for processing temporal multimodal data of digital twins in the field of health care.

12.6 Case Study: Software for Medical Diagnostic Systems Based on the Technology of Digital Twins

The implementation of information technology of digital twins in the field of health care creates conditions for the development of support systems for medical decision-making based on synchronized and consolidated temporal multimodal medical data on the patient's health, which will provide the doctor with information on the patient's health at a qualitatively new level. A medical decision support system based on the technology of digital twins will allow the doctor:

- To analyze the information provided in terms of time.
- To understand better the dependencies between treatment events and treatment results (including side effects for other organs and systems).
- To look for similar cases that happened to the patient in the past.
- To model the possible impact of new treatment regimens, etc.

The technology of digital twins involves the accumulation of data that is registered because of long-term monitoring of a real object. This corresponds to the case of long-term monitoring of the patient's health status, especially for patients with chronic diseases. Thus, the idea behind the use of digital twins in healthcare is to create and dynamically update a patient's individual data set. This personalized data set contains synchronized and aggregated multimodal data obtained from various medical studies and presented according to a personalized semantic model of the patient, organs, and organ systems of his body. This personalized semantic model should reflect the patient's chronic diseases and individual body conditions. The human body as a whole, as well as its individual organs and organ systems, can be characterized by many parameters measured using a wide range of medical equipment. The data obtained from various medical studies and tests are multimodal because they describe a specific aspect of the functioning of the human body. This multimodal data must be synchronized and aggregated to create a digital twin of the physical twin.

Multimodal data to create a digital twin of a patient can be obtained using a wide variety of devices, instruments, and tools that generate multimodal data in specialized, sometimes unique formats. To adapt the generalized architecture of the software system for processing temporal multimodal data of digital twins to the specifics of the medical field, it is necessary to analyze the types of medical diagnostic equipment and the types of data that can be obtained with its help.

All medical devices and instruments can be divided into several groups, including diagnostic equipment, medical laboratory equipment, medical monitors, therapeutic equipment, life support equipment, etc. Diagnostic devices and instruments can be classified as measuring medical instruments (thermometers, sphygmomanometers, glucometers, etc.), laboratory equipment (biochemical analyzers, hematology analyzers, etc.), medical imaging equipment (ultrasound machines, Magnetic Resonance Imaging (MRI) machines, Positron Emission Tomography (PET) and Computed Tomography (CT) scanners, X-ray machines, etc.), and functional diagnostic equipment (electrocardiographs, electroencephalographs, electromyographs, rheographic devices, spirometers, etc.). These devices and tools can be digital or non-digital. Measurements taken with non-digital instruments (e.g., mercury thermometer, hand-held sphygmomanometer) are usually recorded manually either as paper or electronic records. Measurements obtained using digital equipment are provided in the form of files of certain formats depending on the type of medical equipment used. In some diagnostic procedures, such as endoscopy, in particular, colonoscopy, where the result of the examination is a stream of video data, general-purpose file formats are used. The most common medical data file formats [33–41] are presented in Table 12.1.

Analysis of Table 12.1 allows us to conclude that most medical data formats provide a metadata field for storing timestamps; however, the file structure is significantly different. Therefore, it is advisable to include in the software architecture of the medical diagnostic system based on the technology of digital twins a conversion module that is to convert each type of file that is supposed to be used in this system into the wrapper file format proposed above.

Taking into consideration the specificity of healthcare applications, the software architecture of the medical diagnostic system based on the technology of digital twins is proposed as shown in Figure 12.4.

To simplify the visual modeling, the use of template models of the human body, its parts, and individual organs is assumed. Based on measurements and medical studies performed for a specific patient, the template model is customized by changing the dimensions, proportions, and other indicators that affect the visual model. In addition to the architecture of the medical software system, components can be included that ensure the application of procedures for improving the quality of medical graphic data and other types of visual model data processing.

TABLE 12.1

Medical Data File Formats

Data File Type	Extension	The Main Type of Data Stored	Type of Research	Availability of Time Stamps
FEF	.FEF	Biological signal	Electrocardiogram	There is a field for a sequence of time values
UFF	.BDF	Multichannel biological signal	Electroencephalogram	There are four fields for time values: • "Start date of recording", • "Start time of recording", • "Number of data records", • "Duration of a data record, in seconds"
LabPas HL7	.HL7	The result of a blood test	Blood test	There is the field "Date/Time of the Observation"
UFF	.UFF	Ultrasound data	Ultrasound	There is the field "local_time"
NIFTI	.NII	MRI data	Magnetic resonance imaging	There is a reserved field for time values
DICOM	.DC3, .DCM, .DIC	Medical images and metadata	Computed tomography, radiography, etc.	There is the field "Study date" та "Study time"
CSV	.CSV	Integers and real numbers (ASCII)	Spirometry, thermometry, sphygmomanometry, etc.	No

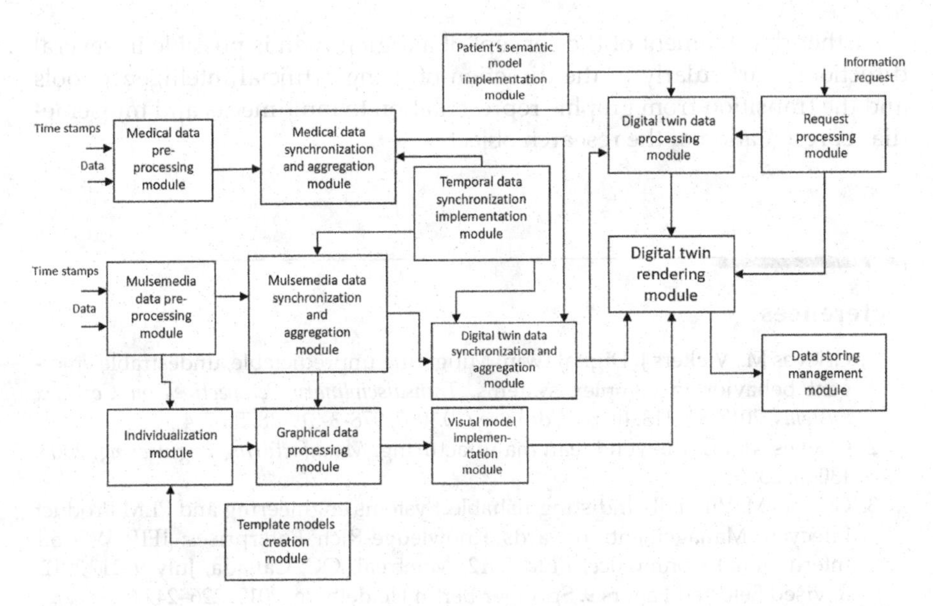

FIGURE 12.4
Software architecture of a medical diagnostic system based on a digital twin.

The proposed architecture of a medical software system based on the technology of digital twins can be applied to the development of expert medical systems with advanced functionality, e-Health [42] and m-Health system software [43], Hospital-at-Home software [44], etc.

12.7 Conclusion

This chapter provides the classification and analysis of digital twin's types based on recent research in this scientific area.

The advancement of the digital twins' concept supposes that a digital twin is defined by both behavioral and visual models of the object. These models can be implemented based on appropriate mathematical and data representation models.

One of the mathematical approaches assumes using an ASA. This enables the formal specification of digital twin models based on the object's multi-image, an ordered set of synchronized data.

In terms of data representation, the object can be defined using a TLM that serves to present the object in the dynamics of changes in its state and is based on the use of time keys to form an ordered sequence of object states and determine links with data sources for dynamic registration of information about the object.

Further development of the concept of a digital twin is possible in several directions, particularly in the direction of using artificial intelligence tools and the transition from graphic representation to multimedia and mulsemedia representation of the research object.

References

1. Grieves M., Vickers J. Digital twin: Mitigating unpredictable, undesirable emergent behavior in complex systems. *Transdisciplinary Perspectives on Complex Systems*, 2017, 85–113. https://doi.org/10.1007/978-3-319-38756-7_4
2. Grieves M. PLM-Beyond lean manufacturing. *Manufacturing Engineering*, 2003, 130(3), 23.
3. Grieves M. Virtually indistinguishable: Systems engineering and PLM.Product Lifecycle Management. Towards Knowledge-Rich Enterprises: IFIP WG 5.1 International Conference, PLM 2012, Montreal, QC, Canada, July 9–11, 2012, Revised Selected Papers 9. Springer Berlin Heidelberg, 2012: 226–242.
4. Grieves M. *Virtually Perfect: Driving Innovative and Lean Products Through Product Lifecycle Management*. Vol. 11. Cocoa Beach: Space Coast Press, 2011.
5. Grieve M. Product lifecycle management: The new paradigm for enterprises. *International Journal Product Development*, 2005, (2), 71–84.
6. Grieves M. *Product Lifecycle Management: Driving the Next Generation of Lean Thinking*. McGraw-Hill, 2006. 319 p. New York.
7. Grieves M. *Virtually Intelligent Product Systems: Digital and Physical Twins. Complex Systems Engineering: Theory and Practice*. American Institute of Aeronautics and Astronautics, 2019, pp. 175–200.
8. Mike S, Mike C, Rich D, et al. *Modeling, Simulation, Information. Technology & Processing Roadmap. Technology Area 11*. National Aeronautics and Space Administration, 2010. 32 p.
9. NASA Technology Roadmaps. *Technology Area 12: Materials, Structures, Mechanical Systems, and Manufacturing*. National Aeronautics and Space Administration, 2015. 138 p.
10. Glaessgen E. H., Stargel D. The digital twin paradigm for future NASA and US Air Force vehicles. 53rd AIAA/ASME/ASCE/AHS/ASC structures, structural dynamics and materials conference 20th AIAA/ASME/AHS adaptive structures conference 14th AIAA. 2012: 1818.
11. Tuegel E. J., Ingraffea A. R., Eason T. G., Spottswood S. M. Reengineering aircraft structural life prediction using a digital twin. *International Journal of Aerospace Engineering*, vol. 2011, Article ID 154798, 14 pages, 2011. https://doi.org/10.1155/2011/154798.
12. Ghinea G., Andres F., Gulliver S. Multiple sensorial media advances and applications: New developments in mulsemedia. *Information Science Reference*, 2012, 320 p.
13. Ghinea G., Timmerer C., Lin W., Gulliver S.R. Mulsemedia: State of the art, perspectives, and challenges. *ACM Transactions on Multimedia Computing, Communications, and Applications*, 2014, 11, 17:1–17:23.

14. Lu Y., Liu C., Wang I-K., Huang H., Xu X. Digital twin-driven smart manufacturing: Connotation, reference model, applications and research issues. *Robotics and Computer Integrated Manufacturing*, 2020, 61, 1–14.
15. Alam, K. M., El Saddik, A. C2PS: A digital twin architecture reference model for the cloud-based cyber-physical systems. *IEEE Access*, 2017, 5, 2050–2062.
16. Redelinghuys A. J. H., Basson A. H., Kruger K. A six-layer digital twin architecture for a manufacturing cell. *Studies in Computational Intelligence*, 2018, 803, 412–423.
17. Bevilacqua M. et al. Digital twin reference model development to prevent operators' risk in process plants. *Sustainability*, 2020, Paper 1088 (12), 17 p.
18. Ye X., Hong S. An automationML/OPC UA-based industry 4.0 solution for a manufacturing system. 2018 IEEE 23rd International Conference on Emerging Technologies and Factory Automation (ETFA). Vol. 1. IEEE, 2018.
19. Schroeder G. N., Steinmetz C., Pereira C. E., Espindola D. B. Digital twin data modeling with automationML and a communication methodology for data exchange. *IFAC-PapersOnLine*, 2016, 49(30), 12–17.
20. Business to Manufacturing Markup Language (B2MML). https://isa-95.com/b2mml/
21. IEC 62264-1:2013. Enterprise-control system integration - Part 1: Models and terminology. https://www.iso.org/standard/57308.html.
22. XML Schema Tutorial. https://www.w3schools.com/xml/schema_intro.asp.
23. Lohtander M., Ahonen N., Lanz M., Ratava J., Kaakkunen J. Micro manufacturing unit and the corresponding 3D-model for the digital twin. *Procedia Manufacturing*, 2018, 25, 55–61.
24. Cai Y., Starly B., Cohen P., Lee Y.-S. Sensor data and information fusion to construct digital-twins virtual machine tools for cyber-physical manufacturing. *Procedia Manufacturing*, 2017, 10, 1031–1042.
25. SolidWorks. https://www.solidworks.com/.
26. The STL Format. https://www.fabbers.com/tech/STL_Format.
27. Open GL. https://www.opengl.org/.
28. Postgre SQL: The World's Most Advanced Open Source Relational Database. https://www.postgresql.org/.
29. Dychka I.A., Sulema Ye S. Ordering operations in algebraic system of aggregates for multi-image data processing. *KPI Science News*, 2019, (1), 15–23.
30. Dychka I.A., Sulema Ye S. Logical operations in algebraic system of aggregates for multimodal data representation and processing. *KPI Science News*, 2018, (6), 44–52.
31. Pester A., Sulema Y. Multimodal data representation based on multi-image concept for immersive environments and online labs development. Cross Reality and Data Science in Engineering: Proceedings of the 17th International Conference on Remote Engineering and Virtual Instrumentation 17. Springer International Publishing, 2021: 205–222.
32. Sulema Y., Dychka I., Sulema O. Multimodal data representation models for virtual, remote, and mixed laboratories development. *Lecture Notes in Networks and Systems*, 2018, 47, 559–569.
33. NIf TI: Neuroimaging Informatics Technology Initiative. https://nifti.nimh.nih.gov/.
34. DICOM. https://www.dicomstandard.org/.

35. Värri A. File exchange format for vital signs and its use in digital ECG archiving. *Proceedings of 2nd Open ECG Workshop "Integration of the ECG into the EHR & Interoperability of ECG Device Systems"*, 2004. 2 p.
36. Which File Format Does BioSemi Use? https://www.biosemi.com/faq/file_format.htm.
37. *Specification for the HL7 Lab Data Interface, Oracle(r) Health Sciences LabPas Release 3.1*. Oracle Health Sciences LabPas Release 3.1. Part Number: E48677-01, 2013. 39 p.
38. Olivier B. et al. The ultrasound file format (UFF). 2018 IEEE International Ultrasonics Symposium (IUS). IEEE, 2018: 1–4.
39. Robert W Cox, Robert W Cox. et al. *The NIFTI-1 Data Format*. NIFTI. Data Format Working Group, 2004. 30 p.
40. Digital Imaging and Communications in Medicine. https://www.dicomstandard.org/.
41. CSV Files. https://people.sc.fsu.edu/~jburkardt/data/csv/csv.html.
42. eHealth. https://www.who.int/ehealth/en/.
43. The Rise of mHealth Apps: A Market Snapshot. https://liquid-state.com/mhealth-apps-market-snapshot/.
44. Hospital at Home. https://www.johnshopkinssolutions.com/solution/hospital-at-home/.

13

Layering Abstractions for Design-Integrated Engineering of Cyber-Physical Systems

Thomas Ernst Jost, Richard Heininger, and Christian Stary
Johannes Kepler University Linz

13.1 Introduction

Many organizations experience a variety of challenges when digitally transforming to cyber-physical systems (CPS) and Internet of Things (IoT). These challenges range from location factors to interoperability due to the heterogeneity of components and also encompass managerial problems [31–33]. CPS are socio-technical systems that link physical devices through sensors or actuators with digital information processing systems in the course of operation. The IoT is a network of physical devices and other objects embedded with electronic items, software, sensors, actuators, and network connectivity that enable them to collect and exchange data.

Digital twins (DTs) have increasingly turned out as development support instruments, as the variety of application domains reveals. They concern societally significant areas like agriculture [18], medicine [29], logistics, and digital production [9]. Originally, DTs have been designed as virtual model to mirror a physical object. It can be equipped with functionality beyond the physical object's performance, such as predicting intervention needs based on Big Data and Machine Learning applications. Moreover, DTs enable behavior modifications and simulations to study various processes before applying them back to the original physical object [1,12,13].

The Digital Twin Consortium also puts the emphasis on objects and processes that are represented in digital form for simulation of 'predicted futures':

> A digital twin is a virtual representation of real-world entities and processes, synchronized at a specified frequency and fidelity: Digital twin systems transform business by accelerating holistic understanding, optimal decision-making, and effective action; Digital twins use real-time

DOI: 10.1201/9781003425724-17

> and historical data to represent the past and present and simulate predicted futures; Digital twins are motivated by outcomes, tailored to use cases, powered by integration, built on data, guided by domain knowledge, and implemented in IT/OT systems [3].

Process twins also refer to the knowledge that is needed to synchronize components for future system operation [19]. They concern the 'macro level of magnification, reveal how systems work together to create an entire production facility. Are those systems all synchronized to operate at peak efficiency, or will delays in one system affect others? Process twins can help determine the precise timing schemes that ultimately influence overall effectiveness' [13]. Once the macro level has been addressed by process twins, for model execution and simulation, refinements to the individual component behavior are required, in the sense of 'abstractions (formalisms) and architectures to enable control, communication and computing integration for the rapid design and implementation of CPS' [23, p. 467].

Recently, a study on how DTs are developed across various sectors has been published [11]. Although there seems to exist some unifying understanding on twin generation at a high level of abstraction, there are a variety of refinement strategies involving different layers of abstractions to develop operational DTs. Consequently, our goal is to offer systems developers a framework from which to consider design and engineering in light of integrated process twins. This chapter tackles two basic questions:

1. Is there a way to express the interplay of process design and system interactions of a CPS?
2. Are there refinement strategies and methods that combine high-level process abstractions with behavior details for DT operation in a consistent way?

Such a framework needs to 'allow the integration and interoperability of heterogeneous systems that composed the CPSs in a modular, efficient and robust manner' [23, p. 467]. We propose a behavior-centered approach to layer process twins based on various application experiences with subject-oriented modeling and execution [5,6,16].

The literature review pertaining to the study is presented in Section 13.2. Section 13.3 shows the proposed framework and generic development questions due to the pragmatic approach taken in this study. The case study performed demonstrates the practicality of the approach. Practical and methodological implications are discussed in Section 13.4. Section 13.5 concludes the chapter by summarizing the objectives, achievements, limitations, and future research directions.

13.2 Abstraction and Refinement in CPS DTs

The sophistication of DT models in product development and production engineering has led to enriching these models with production and operation data in cyber-physical settings [21]. These enrichments 'allow the efficient prediction of the effects of product and process development as well as operating and servicing decisions on the product behaviour without the need for costly and time-expensive physical mock-up. Particularly in design, such realistic product models are essential to allow the early and efficient assessment of the consequences of design decisions on the quality and function of mechanical products' [21, p. 141]. In order to provide a conceptual foundation for the implementation of DTs supporting design and production engineering, Schleich et al. [21] proposed a DT development reference model focusing on essential model qualities, including interoperability and adaptation of the DT model along the product life cycle (comprising design, manufacturing, assembly, and inspection).

Its subject of concern is not a 1:1 relationship between a physical asset and some digital representation, but 'a set of linked operation data artefacts and simulation models, which are of suitable granularity for their intended purpose and evolve throughout the product life-cycle' [21, p. 142]. The framework has been tested in a geometrical variations management use case. The different views on the product along the life cycle were supported through the operations composition, decomposition, conversion, and evaluation. In this way, the DT enabled the operation of the physical product in an integrated way along design and production engineering. Hence, the DT model did not only capture product details and variants in a digital representation but was part of a development process, including design and engineering activities.

However, with respect to the chosen level of abstraction in DT model creation and development, it seems to be a case-to-case decision, as VanDerHorn et al. [28] concluded in the summary of their analysis. However, both the goal that needs to be achieved using a DT representation and the scope of applying DTs seem to be decisive for abstraction and operation support. For modeling 'all relevant interacting and interrelated entities that form a unified whole' should be considered. Regarding behavior, all 'functions being performed by the physical system that result in the system undergoing changes in state' should be captured.

The analysis provided by Jones et al. [14] reveals that many DT model developments follow a generic DT transition scheme. According to their findings, DTs are generated after some brainstorming and ideation phases in product development. A prototype is tested in its context (environment), followed by instantiation and aggregation processes to achieve system-level behavior. In this way, the entire life cycle of a product can be captured.

From the perspective of heterogeneous system components, as for CPS, such a development design allows the successive integration of system components and adaptation until a system operates as desired or imagined. In any case, the process DT is defined by a product life cycle and its main stages from creation to retirement.

With respect to operational DT development support, composition seems to be essential in Cyber-Physical Production Systems. As the study by Park et al. [17] shows, it can be used on an abstract level – DT-based service composition based on a common information model is of benefit for personalized production processes. The information model is mainly based on function trees decomposing functional elements required for service and production engineering.

In light of highly volatile settings in manufacturing, Seok et al. [22] have addressed unexpected real-time problems by DT discrete-event models for wafer fabrication. Their hierarchical aggregation/disaggregation scheme enables substituting complex event-driven operations with two-layered abstracted models. Its operation on a single-group and multiple-group abstraction layer leads to faster response to production problems. The key component, the abstraction-level converter, mainly allocates and de-allocates machines to groups and extends or splits groups. It also controls the event flow of each group's input lot.

In addition to data-driven abstraction, communication abstraction has been proposed recently by Bellavista et al. [2]. The scheme aims at simplifying communication management between system components. The proposed intent-based messaging system refers to so-called high-level 'intents'. They represent goals in system operation. These goals need to be achieved by the respective intent receiver. The receiver is expected to deliver a specific outcome denoted by the goal. Since the intent does not contain information on how the goal is achieved, it subsumes task accomplishments in a declarative way, abstracting from operational data, including users and system components, and the required functions to produce a certain deliverable.

Aside from showing one way to make CPS systems better understandable for humans [30], aligned layered abstraction allows handling complex interactions and architectures of heterogeneous systems [27]. The latter has recently been tackled by Kreutz et al. [15] by addressing DevOps. The findings indicate research needs to couple higher-level processes, such as business operations, to design and engineering processes. Hence, for process DT models, the findings by Sobrahjan et al. [23] still seem to be valid with respect to the need for abstraction in CPS design and engineering. They advise a high-level language that compiles down to the detailed behavior of individual CPS components. Such a language needs to go beyond existing applications of process modeling to represent business models for CPS [26], or IoT process modeling [10] on a single layer of abstraction. Development support requires capabilities for virtual validation and verification due to the transformation needs triggered by technological advances [8].

Different stakeholder groups, users, domain experts, designers, engineers, and programmers will be in need of proper specification and development support. Hence, several levels of details, ranging from low-level communication to high-level resource management and business optimization, need to be addressed in a networked way.

13.3 A Layered Architecture for Digital Process Twins

In this section, we introduce a layered architecture for process DT based on the findings on communication-based abstraction. We use behavior encapsulations which are specified in a subject-oriented notation as proposed by Fleischmann et al. [6], as it allows for automated execution of models.

We first introduce the modeling approach before we detail the layered architecture and discuss pragmatic design issues.

13.3.1 Subject-Oriented CPS Modeling

Subject orientation has been introduced in the field of Business Process Management to capture business operations in such a way that people can easily understand how tasks are accomplished and organized in an organizational context, in order to meet business objectives [6]. The interaction- and behavior-centered encapsulation technique has been successfully applied in service and production industries [5,16] and is of benefit for the design and engineering of Internet-of-Behavior applications [24] and CPS development [25].

A behavior encapsulation of an active system component is termed subject. It includes activities that result in communication and functional behavior. They concern objects of manipulation. The latter are exchanged between subjects to complete operational processes. The pattern of interactions and the internal functions of subjects constitute a system's behavior. Due to the choreographic nature of interaction, subjects operate in parallel and are synchronized through exchanging messages.

Subject-oriented behavior models contain CPS components as subjects that can either be digitally or physically implemented. The interaction represents the connection between physical and/or digital components via message exchanges. Messages can contain data in the form of business objects and trigger further actions in the addressed subject. Hence, a CPS is modeled as a set of autonomous, concurrent behaviors of distributed components.

Subjects are modeled using two types of diagrams and a set of five symbols along integrated system design and engineering, namely Subject Interaction Diagrams (SIDs) and Subject Behavior Diagrams (SBDs). SIDs provide an integrated view of a CPS, comprising the CPS components involved and the messages they exchange. SBDs provide a local view of the behavior from the perspective of individual subjects. They include refinements of the SID for each subject in terms of sequences of internal actions representing system states.

As validated behavior specifications, SBDs can be executed without further model transformation, either by using vendor-specific solutions, such as the Compunity Tool-Suite (https://compunity.eu), or academic tools like SiSi [4]. For each subject or CPS component, the behavior can be checked interactively and adapted dynamically by replacing individual subject behavior specifications during run-time.

Subject-oriented CPS development consists of several phases: setup, refinement, validation, implementation assignments, and run-time, according to [25]:

Setup: The CPS developer generates a digital behavior model, i.e., digital process twin, structured by a SID: Each concerned service and CPS component device is represented by a subject.

Refinement: Each subject, i.e., each node of the SID, is further detailed in terms of its task- and communication-specific behavior, specifying an SBD.

Validation: The system behavior is checked along the message paths determining the interaction between individual subjects, when the digital process twin model (i.e., a network of SBDs) is executed.

Implementation details: After successful validation, organizational and technical details need to be specified by assigning a human or CPS technology component as role or task carrier to each subject for CPS operation.

Run-time: During run-time (based on the executable behavior models), additional CPS components and services can be added, or replace existing ones, to adapt to changed requirements or to changing environmental conditions.

In this way, design and engineering tasks operate both on the digital process twin, enabling seamless CPS life-cycle management. Since the digital representation of a CPS exists independently of its counterparts in the physical world, human role carriers and technical CPS components can be assigned dynamically to execute subject behavior. In this way, existing work organizations can be transformed successively and systematically, starting with a DT prototype of a CPS, as proposed by Jones et al. [14].

13.3.2 The Layered Architecture

In this section, we introduce the layered subject-oriented architecture to allow for structured and contextual refinement of digital CPS process twins.

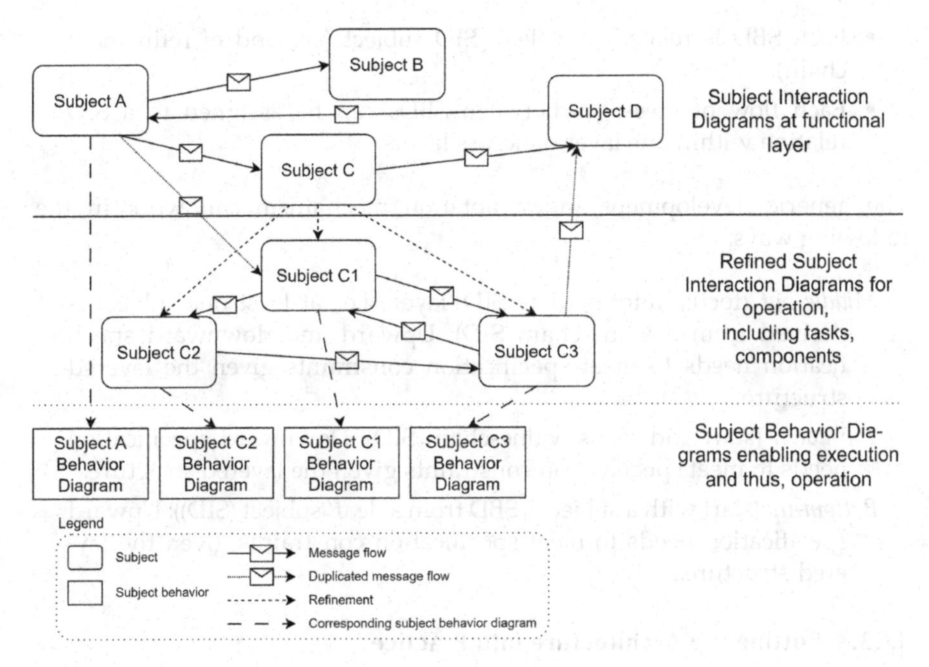

FIGURE 13.1
Subject-oriented abstraction layer architecture.

Although it is described in a top-down way, it can be applied through a middle-out or bottom-up approach, e.g., in case a specific CPS component, such as a robotic device for production, is the starting point of CPS development. Figure 13.1 shows the fundamental idea and mechanism of subject-oriented layering of digital CPS twins.

The architecture is based on specifying a SID:

- At least one layer (SID on a single layer of abstraction) needs to be given, providing a SID representing value-generating subjects and their interaction.
- Each subject of a SID on a specific layer of abstraction can be refined under the given constraints of interactions (passing of messages) on a lower layer.
- The last subject refined ('leaf' subject) needs to be detailed by a SBD. Layering means the following for SBDs:
- At least one layer needs to be given, …
  …providing a set of SBDs derived from the SID,…
  …with each SBD derived from a SID subject,…
  with each SBD being part of at least one interaction with another SBD.

- Each SBD is related to a 'leaf' SID subject (i.e., end of refinement chain).
- Each flow of messages between SBDs can be assigned to a SID relation within one layer or across layers.

The generic development and adaptation mechanism can work in the following ways:

Middle-out (focus: intermediate SID-'layer', i.e., at least one subject is derived from a value-chain SID): Upward and downward specification needs to meet specification constraints given the layered structure.

Top-down (start and focus: value-chain SID): Downward specification needs to meet specification constraints given the layered structure.

Bottom-up (start with a subject's SBD from a 'leaf' subject (SID)): Upward specification needs to meet specification constraints given the layered structure.

13.3.3 Putting the Architecture into Practice

In this section, we refine the generic architecture for operational DT practice and demonstrate its use through applying the layering to a dispatching use case. We start with DT components and their interaction representing the value-adding chain and successively refine it to operate the process twin.

13.3.3.1 Outline of the Use Case

The demonstration case concerns a dispatching unit as part of a smart logistics process including a CPS [11]. It is designed to handle the transportation of critical goods, such as vaccines to be distributed from some storage location to a vaccine center. The intermediate dispatching unit is based on a CPS to implement packaging according to the transport requirements based on a specific vaccine order.

The most important part for DT design and engineering is the vaccine being packaged involving CPS components. It starts with delivering the vaccine to the dispatching station with (order) information on what needs to be taken into account for its transportation, such as positioning in a transportation box and cooling.

The dispatching unit has the task to package the vaccine in a transportation box fitting to the order and transportation requirements, so it can be picked up by a delivery service. Once the vaccine is received by the dispatching station, a robot system carries it to a dispatching bench monitored by sensors to ensure that proper temperature conditions are met. There, the vaccine is packaged using storage containers that can be equipped with IoT devices to ensure the transportation requirements are satisfied.

The robot system equips the transportation container according to the transportation requirements and puts the vaccine into the container, before carrying it to a pick-up place when the dispatching process has been completed, and the subsequent transport process can start.

13.3.3.2 Exemplification of Modeling According to the Top-Down Approach

According to the presented concept in the previous section, the proposed DT layering architecture consists of one or multiple SID layers and one or multiple SBD layers. The most abstract SID layer is termed the functional value creation layer and depicts a crucial process of the value chain of a company. This entails that the proposed architecture can be created for each value-chain process of an organization. Each SID layer consists of at least two subjects, as implied by the structure of a value chain. Subjects are connected through messages and there can be no 'loose' subjects on a SID layer, i.e., subjects that are not connected to any other subject through a message (since this would make subjects 'unreachable' during process model execution). Furthermore, aside from the subject starting the process, each other subject must receive at least one message, since this is needed to kick off their process-specific behavior.

Considering the outlined use case, the dispatching/distribution of vaccines constitutes a value-chain process. This process could be depicted at the highest possible abstraction layer, that of functional value creation. Without further refinement, each of the constituting subjects of the resulting SID would be a leaf subject and require an associated SBD. In case of multiple SID layers, each subject of a lower layer must refine a subject of a higher layer and each subject of a layer can be refined by one or multiple subjects of one or multiple lower layers. Subjects on the lowest layer cannot be refined further. This way, for each subject of a higher layer, a tree structure of refining subjects is formed. In this tree hierarchy, the leaf subjects (i.e., subjects that are not further refined on lower SID layers) need to be detailed by a SBD.

These SBDs are part of the SBD layers, with each SBD pertaining to one leaf subject part of one of the SID layers and the SBD layer hierarchy corresponding to the SID layer hierarchy. This principle means the following: if a leaf subject is part of SID layer two, its corresponding SBD is part of SBD layer two. Following the rule of no loose subjects, each SBD needs to be part of at least one interaction with another SBD. Each flow between messages between SBDs can thus be assigned to a message connecting two subjects on the same SID layer or different SID layers.

The refinement of a subject on a higher layer through subjects of a lower layer indicates the 'replacing' of the abstract higher layer subject with more concrete subjects, bringing the value-chain functionality of the highest layer subject one step closer to operation. It also entails that the (incoming and outgoing) message flow needs to be diverted from the higher layer subject to the more concrete refining lower layer subjects.

The following occurs if the subjects from which the higher layer subject receives messages and to which it sends messages have not been refined themselves: for each incoming message of the higher layer subject from a nonrefined subject, one refining subject on the corresponding lower layer needs to be selected, which also receives the exact same message (the sender remains the same), creating the incoming link between SID layers. Similarly, for each outgoing message of a higher layer subject to a nonrefined subject, a refining subject on the corresponding lower layer needs to be chosen, which also sends the exact same message (the receiver remains the same), creating the outgoing link between SID layers. This means that the role of the refined subject in the process flow can now be entirely replaced by the refining subjects of the lower layer, while still adhering to the original message flows.

In case that the subject from which a higher-level subject receives a message (both on the same SID layer) has already been refined itself, the message in question can be received from the corresponding refining subject of the lower-level subject. The same can also be done if a receiving subject on the same layer as the higher-level subject that should be refined has already been refined previously. The application of the presented refinement rules is illustrated in Figures 13.2 and 13.3 for different constellations of subjects across layers.

Since the process-based approach is based on executing behavior diagrams (SBDs) refined from interaction diagrams (SIDs), once all subjects of a higher SID layer have been refined, there are no message links left between the higher layer and the lower layer. If all refined subjects of the lower layer are detailed in terms of SBDs, then the lower-level model can be put into operation and the higher-level model is not needed for execution. However, the refinement hierarchy allows stakeholders to understand the design, both, in terms of the behavior specified on the refinement layers for engineering (i.e., DT operation), and their relation to value-chain processes – all subjects that are not leaf subjects can be represented as placeholders or intermediate subjects for later refinement.

According to the integrated design and engineering approach, SBDs of the leaf subjects on the higher layer need to send messages to the refining leaf subjects of the lower layer. This relation ensures behavior is executed correctly. It is denoted by additional (engineering) message links between layers, such as the ones depicted in Figures 13.2 and 13.3. It should be noted that with regard to refinement situations not all possible constellations between subjects of different layers were depicted, but the rules for the remaining options can be inferred from the given examples.

For the demonstration case, a model version including the SBD layers was created, to demonstrate the behavior refinement of leaf subjects. It is depicted in Figure 13.4. As can be observed from the modeled internal behavior, further possibilities for behavior refinement exist. Do-states can be split into more fine-grained actions, which could again be modeled explicitly

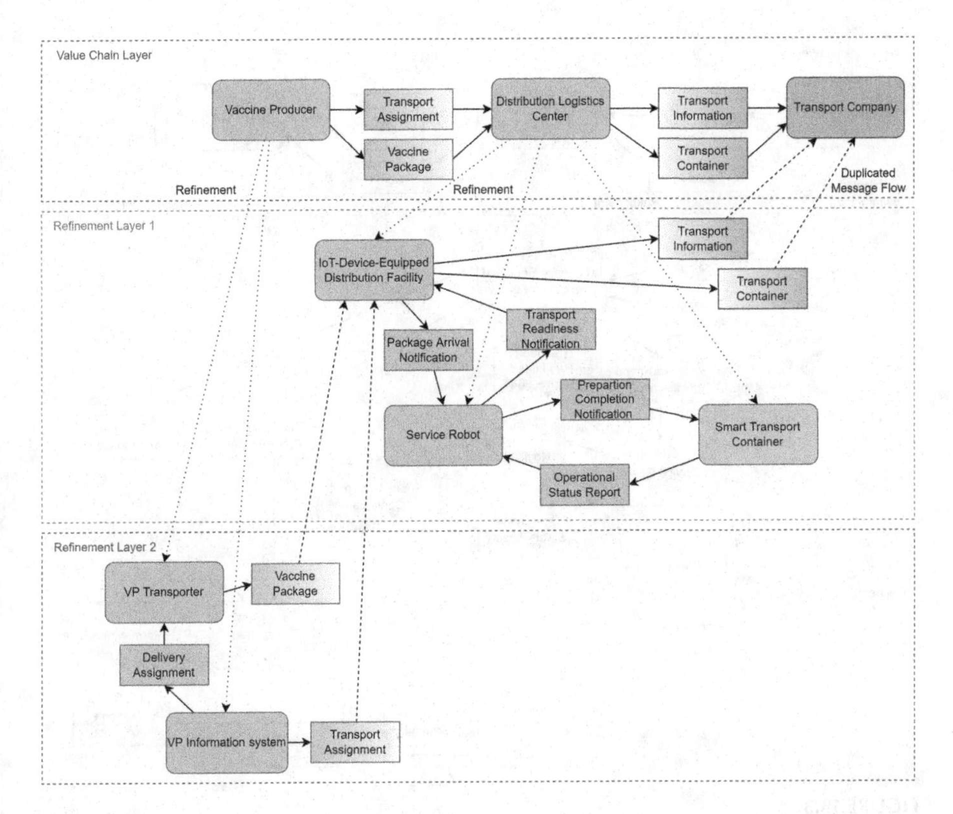

FIGURE 13.2
Abstraction layer architecture – diversion of message flow to refining subjects. Subject that is being refined (distribution logistics center) receives messages from refined subjects on a different layer than the one that it should be refined at (VP Transporter, VP Informationsystem) and sends messages to unrefined subject (transport company).

as states. Currently, only the 'happy path' of behavior was modeled, i.e., the intended and optimal occurrence of states. In case that, e.g., the operational status report sent by the transport robot states that some of the installed sensors are not working correctly, additional behavior would be necessary. In subject-oriented languages, such as the Parallel Activity Specification Scheme, exceptional cases could be modeled through the optional guard behavior of a subject [7].

The created models can furthermore be seen as 'containers' for other models and data, or as an embodiment of a possible structure indicating the relationships between them. Other models and data can be assigned to each of the elements in the abstract layer structure, i.e., to a state, to an SBD, to a message, to a subject, or to a layer. This way, the abstract layer architecture can be used as an integration concept for DT data. The service robot subject can, e.g., have an associated 3D model of the robot, a kinematic model, and sensor data produced by the robot.

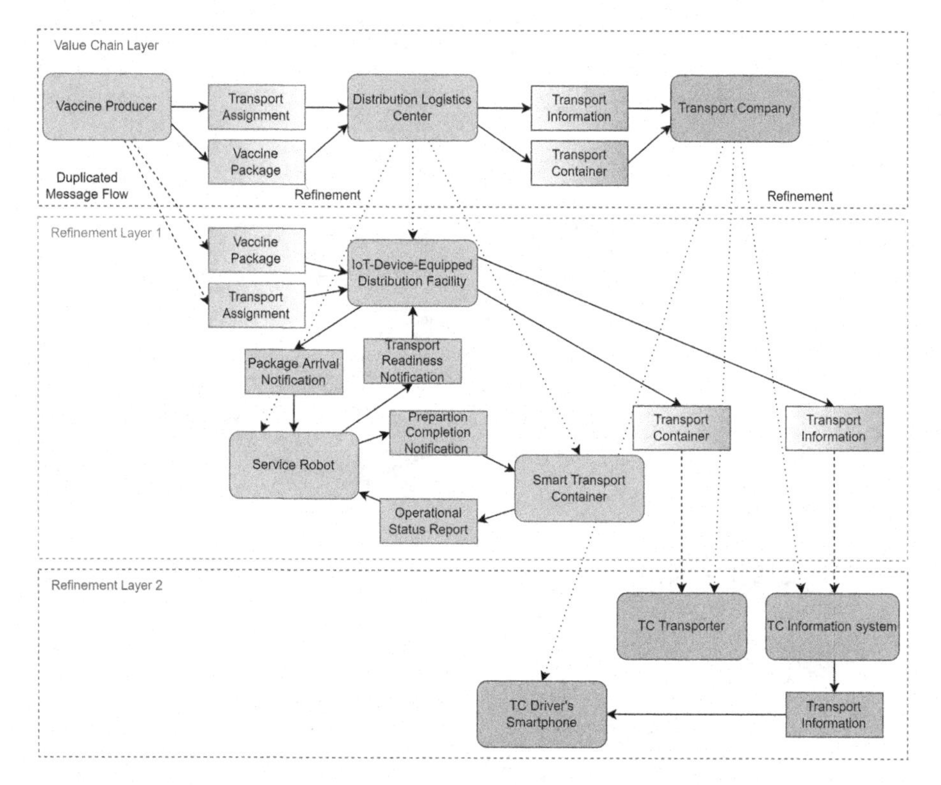

FIGURE 13.3
Abstraction layer architecture – diversion of message flow to refining subjects. subject that is being refined (distribution logistics center) sends messages to refined subjects on a different layer than the one that it should be refined at (TC Transporter, TC Informationsystem) and receives messages from unrefined subject (vaccine producer).

13.4 Discussion

In this section, we reflect on the architecture and its application presented in the previous section. We clarify design and engineering challenges that popped up in the course of application and may need further research.

Since the proposed architecture consists of one or multiple SID layers and one or multiple SBD layers, an organization dealing with various value-added processes may need further structures to organize the process diagrams. When the proposed architecture can be created for each value-chain process of an organization, communication across processes is likely to occur and some SIDs and SBDs might be needed to show the involvement of particular roles or tasks in various value chains.

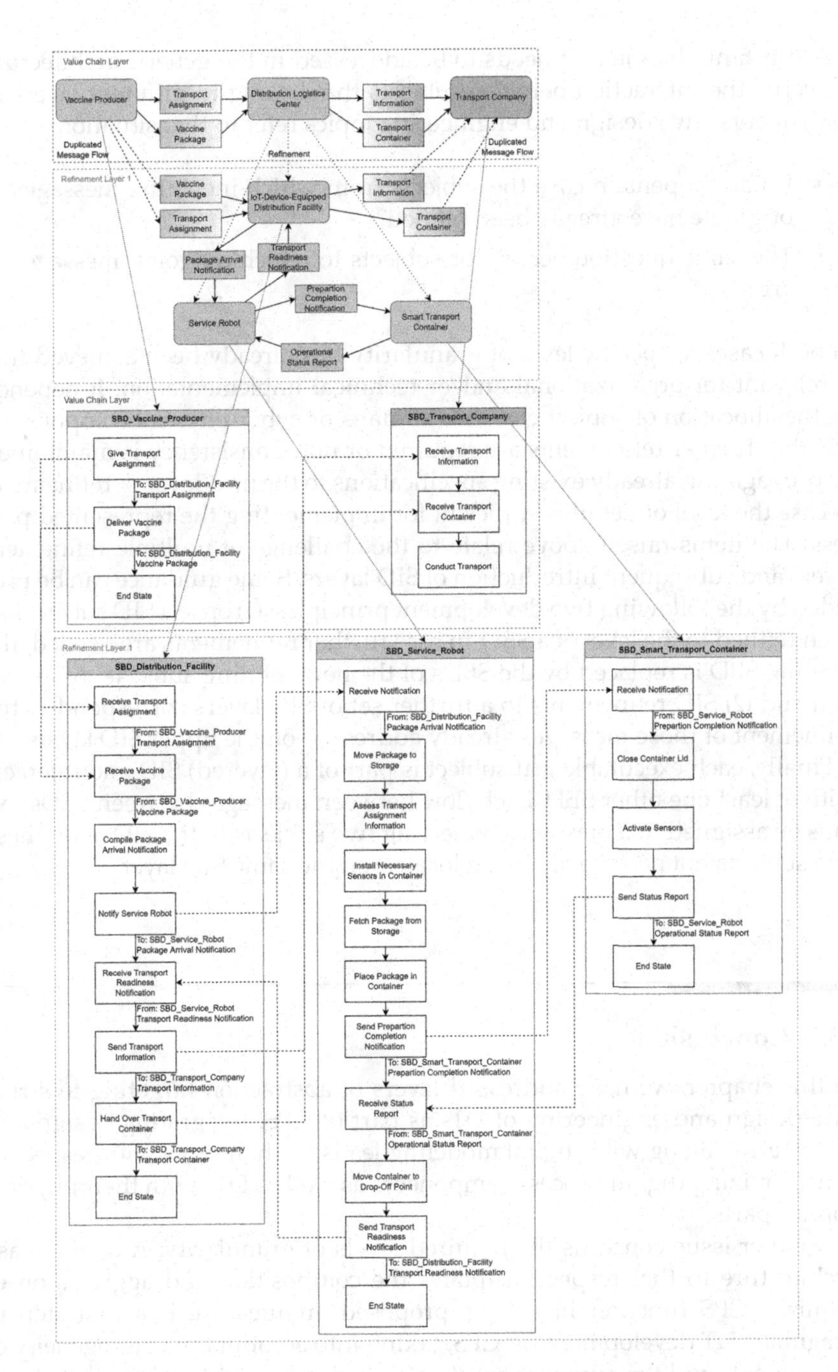

FIGURE 13.4
Abstraction layer architecture – SID layers and SBD layers containing the internal behavior of leaf subjects.

A substantial issue that needs to be addressed in the generic architecture concerns the interaction between subjects that belong to different layers of refinements. Two design and engineering topics refer to that situation:

- What happens in case the subjects from which incoming messages originate have already been refined?
- The same question occurs for subjects to which outgoing messages are sent.

In both cases, a specific level of granularity has already been achieved that is relevant for organizational and/or technical implementation. It depends on the allocation of subject carriers (humans or cyber-physical components) whether further refinements are required or not. Consistency is maintained by propagating already existing specifications to the next layer of refinement, in case the level of detail is sufficient for implementing the represented process. The items raised above relate to the challenge of multiple refinement layers and subsequent introduction of SID layers. Some guidance can be provided by the following two development principles: (1) once a SID subject has been refined to the level of a SBD, in case further refinements are needed, the existing SBD is replaced by the SBDs of the new refining subjects for execution and (2) SID refinements to a further set of SID layers correspond to the refinement of those messages already addressed on the upper SID layers.

Finally, each executable leaf subject is part of a (layered) SID and interacts with at least one other SBD. Each flow between messages between SBDs can thus be assigned to a message connecting two subjects of the SID, with these two subjects not necessarily being located on the same SID layer.

13.5 Conclusion

In this chapter, we have addressed layers of abstraction targeting to structure design and engineering of DTs as part of CPS. Integrating design and engineering along with digital modeling leads to challenges with respect to synchronizing digital process components as well as DTs with their physical counterparts.

Another issue concerns the required levels of granularity. A case-to-case architecture to that respect facilitates the composition and aggregation of required CPS functionalities. The proposed architecture is a first step to pragmatic DT development for CPS, taking into account the heterogeneity of components and the dynamic nature of CPS design and operation.

The presented architecture is based on two intertwined layers of abstraction, designed to tune for engineering a declarative interaction network to functional behavior components. The architecture can be used in several way, including top-down design to engineering and middle-out development based on given CPS components. The practical application on a transportation case demonstrated the effectiveness of the approach, however, revealed several challenges to find the proper balance between precision in specification and dynamic control of CPS development and operation. The results will trigger further research on modeling principles and usefulness of modeling constructs and principles for abstract DT representation.

Bibliography

[1] Michael Batty. Digital twins. *Environment and Planning B: Urban Analytics and City Science*, 45(5):817–820, September 2018.

[2] Paolo Bellavista, Carlo Giannelli, Marco Mamei, Matteo Mendula, and Marco Picone. Digital twin oriented architecture for secure and QoS aware intelligent communications in industrial environments. *Pervasive and Mobile Computing*, 85:101646, September 2022.

[3] The Digital Twin Consortium. Website - The Definition of a Digital Twin. https://www.digitaltwinconsortium.org/initiatives/the-definition-of-a-digital-twin/, retrieved 18.08.2022.

[4] Matthes Elstermann and Jivka Ovtcharova. Sisi in the ALPS: A simple simulation and verification approach for PASS. Proceedings of the 10th International Conference on Subject-Oriented Business Process Management. 2018: 1–9.

[5] Albert Fleischmann, Werner Schmidt, and Christian Stary, editors. *S-BPM in the Wild*. Springer International Publishing, Cham, 2015.

[6] Albert Fleischmann, Werner Schmidt, Christian Stary, Stefan Obermeier, and Egon Borger. *Subject-Oriented Business Process Management*. Springer, Berlin, Heidelberg, 2012.

[7] Albert Fleischmann, Werner Schmidt, Christian Stary, and Florian Strecker. Nondeterministic events in business processes. In Marcello La Rosa and Pnina Soffer, editors, *BPM 2012: Business Process Management Workshops*, pages 364–377. Springer, Berlin, Heidelberg, 2013.

[8] Didem Gurdur Broo, Ulf Boman, and Martin Torngren. Cyber-physical systems research and education in 2030: Scenarios and strategies. *Journal of Industrial Information Integration*, 21:100192, March 2021.

[9] Dirk Hartmann and Herman Van der Auweraer. Digital twins. In Manuel Cruz, Carlos Pares, and Peregrina Quintela, editors, *Progress in Industrial Mathematics: Success Stories*, pages 3–17. Springer International Publishing, Cham, 2021.

[10] Faruk Hasic, Estefanıa Serral, and Monique Snoeck. Comparing BPMN to BPMN + DMN for IoT process modelling: A case-based inquiry. Proceedings of the 35th Annual ACM Symposium on Applied Computing. 2020: 53–60.

[11] Wolfgang Heindl and Christian Stary. Structured development of digital twins-a cross-domain analysis towards a unified approach. *Processes*, 10(8):1490, July 2022.

[12] Christoph Herwig, Ralf Portner, and Johannes Moller. *Digital Twins*. Springer, 2021.

[13] IBM. Website - What is a Digital Twin? https://www.ibm.com/topics/what-is-a-digital-twin/, retrieved 18.08.2022.

[14] David Jones, Chris Snider, Aydin Nassehi, Jason Yon, and Ben Hicks. Characterising the Digital Twin: A systematic literature review. *CIRP Journal of Manufacturing Science and Technology*, 29:36–52, May 2020.

[15] Andreas Kreutz, Gereon Weiss, Johannes Rothe, and Moritz Tenorth. *DevOps for Developing Cyber-Physical Systems*. Fraunhofer Institute for Cognitive Systems, Munich, 2021.

[16] Matthias Neubauer and Christian Stary, editors. *S-BPM in the Production Industry*. Springer International Publishing, Cham, 2017.

[17] Kyu Tae Park, Jehun Lee, Hyun-Jung Kim, and Sang Do Noh. Digital twin-based cyber physical production system architectural framework for personalized production. *The International Journal of Advanced Manufacturing Technology*, 106(5–6):1787–1810, January 2020.

[18] Christos Pylianidis, Sjoukje Osinga, and Ioannis N. Athanasiadis. Introducing digital twins to agriculture. *Computers and Electronics in Agriculture*, 184:105942, May 2021.

[19] Roland Rosen, Georg von Wichert, George Lo, and Kurt D. Bettenhausen. About the importance of autonomy and digital twins for the future of manufacturing. *IFAC-PapersOnLine*, 48(3):567–572, 2015.

[20] Andrey Rudskoy, Igor Ilin, and Andrey Prokhorov. Digital twins in the intelligent transport systems. *Transportation Research Procedia*, 54:927–935, 2021.

[21] Benjamin Schleich, Nabil Anwer, Luc Mathieu, and Sandro Wartzack. Shaping the digital twin for design and production engineering. *CIRP Annals*, 66(1):141–144, 2017.

[22] Moon Gi Seok, Wentong Cai, and Daejin Park. Hierarchical aggregation/disaggregation for adaptive abstraction-level conversion in digital twin-based smart semiconductor manufacturing. *IEEE Access*, 9:71145–71158, 2021.

[23] Pooja Sobhrajan and Swati Y Nikam. Comparative study of abstraction in cyber physical system. *International Journal of Computer Science and Information Technologies*, 5(1):466–469, 2014.

[24] Christian Stary. Digital twin generation: Re-conceptualizing agent systems for behavior-centered cyber-physical system development. *Sensors*, 21(4):1096, February 2021.

[25] Christian Stary, Matthes Elstermann, Albert Fleischmann, and Werner Schmidt. Behavior-centered digital-twin design for dynamic cyber-physical system development. *Complex Systems Informatics and Modeling Quarterly*, 30:31–52, 2022.

[26] Nicola Terrenghi, Johannes Schwarz, and Christine Legner. Towards design elements to represent business models for cyber physical systems. *Proceedings of the 26th European Conference on* Information *System, Portsmouth*(ECIS 2018), *UK*, pages 1–16, 2018.

[27] Martin Torngren and Paul Grogan. How to deal with the complexity of future cyber-physical systems? *Designs*, 2(4):40, October 2018.

[28] Eric VanDerHorn and Sankaran Mahadevan. Digital Twin: Generalization, characterization and implementation. *Decision Support Systems*, 145:113524, June 2021.

[29] Isabel Voigt, Hernan Inojosa, Anja Dillenseger, Rocco Haase, Katja Akgun, and Tjalf Ziemssen. Digital twins for multiple sclerosis. *Frontiers in Immunology*, 12:669811, May 2021.

[30] Baicun Wang, Huiying Zhou, Geng Yang, Xingyu Li, and Huayong Yang. Human digital twin (HDT) driven human-cyber-physical systems: Key technologies and applications. *Chinese Journal of Mechanical Engineering*, 35(1):11, December 2022.

[31] Vishal Ashok Wankhede and S. Vinodh. Analysis of barriers of cyber-physical system adoption in small and medium enterprises using interpretive ranking process. *International Journal of Quality & Reliability Management*, 39(10):2323–2353, October 2021.

[32] Vishal Ashok Wankhede and S. Vinodh. Analysis of Industry 4.0 challenges using best worst method: A case study. *Computers & Industrial Engineering*, 159:107487, September 2021.

[33] Shun Yang, Nikolay Boev, Benjamin Haefner, and Gisela Lanza. Method for developing an implementation strategy of cyber-physical production systems for small and medium-sized enterprises in China. *Procedia CIRP*, 76:48–52, 2018.

14

Issues in Human-Centric HMI Design for Digital Twins

Vivek Kant and Jayasurya Salem Sudakaran
Indian Institute of Technology Bombay

14.1 Introduction: Why Is Human-Centric Design Required for Digital Twins?

The current chapter presents the issues and challenges of human–machine interaction (HMI) design with a special emphasis on the human-centric interaction design of digital twins. Digital twins are digital models mimicking the functioning of technologies and machines (Hsu, Chiu, & Liu, 2019; Liu, Fang, Dong, & Xu, 2021; Qi et al., 2021; Tao et al., 2021). These models can be used for a variety of purposes ranging from understanding the failure of parts in certain conditions to predicting the asset performance of the technology through the life cycle. While, in most cases, digital twins (and even digital shadows) are designed to be machine-centric, they often need to present some information to the users/operators/managers/other stakeholders who are using the digital twin technologies.

HMI design is not a straightforward task of representing information in the form of graphs and charts; rather, the emphasis is to present meaningful information to particular groups for effective decision-making (Bennett & Flach, 2011; Hollifield, Habibi, & Oliver, 2013; Rasmussen, 1979, 1986; Rasmussen, Pejtersen, & Goodstein, 1994). Decision-making in scenarios related to digital twins involves how information is displayed and under what circumstances. The effective manner in which information is displayed and interaction designed falls under the realm of HMI. In our current chapter, we revisit a case study of the HMI considerations of a digital twin (see Kant & Sudakaran, 2021, Section 5) with a special emphasis on interactional features. While the case study has been developed elsewhere in greater detail, its interactional challenges have not been described and emphasized in the manner it is developed here in terms of challenges of HMI design for digital twins (Kant & Sudakaran, 2021 for a discussion of the design process). Thus, this chapter draws extensively from the disciplines of interaction

 DOI: 10.1201/9781003425724-18

design to highlight key issues and challenges in the human-centered design of HMIs.

The rest of the chapter highlights the key issue of human-centeredness. In this direction, it briefly presents a case study of the design of an HMI for a digital twin (Section 14.2). This case study is then used to explore four main issues related to human-centricity in Section 14.3. This chapter concludes with a discussion of challenges related to human-centric design (Section 14.4).

14.2 Case Study: HMI for Digital Twin and Its Interactional Elements

A digital twin was created for a Hardinge 600 Vertical Machining Centre (VMC) II by a university-based engineering research group working in the manufacturing sector. The authors of this current chapter were involved in the creation of the HMI for this digital twin (Figure 14.1). Here, we briefly recount this case study to further highlight four generic challenges for HMI design for digital twins. The first issue is designing for the human users/operators of the digital twins. The second involves the representation of the systems complexity of digital twins in human-centered terms. The third and fourth challenges refer to the issues related to interaction design and information design pertaining to the HMI of digital twins.

In order to design the HMI, the focus was on understanding the machine in human-centered terms. This was accomplished using design frameworks to create a space juxtaposing functional-abstraction and structural-decomposition elements. The abstraction-decomposition space (discussed in Section 14.2) allows for addressing technology in human-centered terms and enables the progress of the design. Based on these insights and using a design process, the final interface was designed.

TABLE 14.1

Multiple Stakeholders and Their Activities

Operator	Supervisor	Maintainer	Administration
Operates one machine. Produces a target number of products within the deadline.	Overlooks a cluster of machines. Divides work among operators.	Repairs a set of machines.	Overlooks production efficiency and finance aspects of the industry.
Expert in operating machines.	Expert in planning and management.	Expert in maintaining machine health.	Expert in inputs and outputs of industry.
Concerned about finishing work quickly.	Concerned about finishing work efficiently.	Concerned about fixing issues quickly.	Concerned about the well-being of the industry.

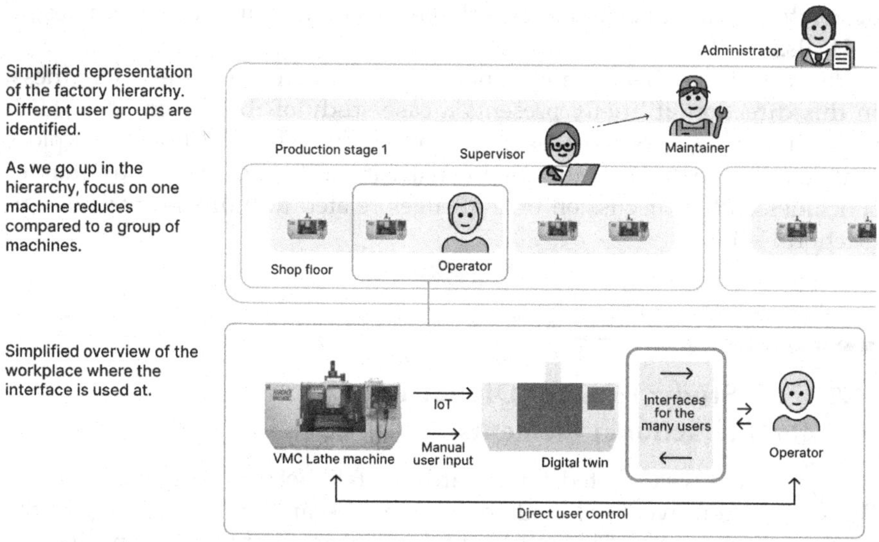

FIGURE 14.1
Multiple stakeholders and the HMI for a digital twin.

The focus of the designed HMI was the following:

1. To help operators get insights into their existing machines without needing newer machines with inbuilt digital twin systems and
2. To help operators in keeping track of the short- and long-term health of the machines.

With a list of data points obtainable from the digital twin, the key challenge was to best represent the data such that it helps the operator in the most optimal manner possible. The interface was planned to reflect the three major areas of interest for the operators. These include:

1. Overview of their workflow,
2. The efficiency of the operation, and
3. Better visibility during operation.

As a result, the final interface (Kant & Jayasurya, 2021, Figures 14.1 and 14.4) is comprised of two sections, one reflecting the efficiency of the system at all times and the other allowing the user to switch between an overview of the whole operation (for most of the time) and a detailed visualization mode (for specific use cases). The important variables are made more salient using a visual hierarchy. Further, relatively less prominent variables are made less salient in terms of size but placed spatially close to other related data so that it is easier for the operator to find. We will use this interface (Kant & Jayasurya, 2021, Figure 14.4) from the case study briefly to highlight certain challenges for the HMI design of digital twins from the perspective of interaction design as a discipline.

14.3 Issues in Human-Centric HMI Design for Digital Twins

The above case study briefly recounted the design process of the HMI for the digital twin of the VMC. A few key aspects that became prominent in the design process can be comprehended more generically for the design of digital twins in general. These four aspects also show us the possible challenges faced while designing HMIs for digital twins. These include designing for human users and other stakeholders (Section 14.3.1), representation of systemic complexity in human-centered terms (Section 14.3.2), along with issues related to interaction design (Section 14.3.3) and information design (Section 14.3.4).

14.3.1 Issue 1: Designing for Human Users (Trained Operators and Other Stakeholders) of Digital Twins

A digital twin is able to provide detailed insights about the machine itself and enables data collection, analysis, and even prediction in various ways that can leverage the many developments in the field of data science. However, the data that is obtained from the digital twin is in its raw form and is not information; i.e., it can be processed in multiple ways. *Not all data is useful to all the users in all situations.* In other words, this data has to be "digested" and provided in such a way that it actually is of help to the user. Each "user group" or "operational role" has a different need and a different end goal; therefore, it is crucial that we have a deep understanding of each user's work, problems, and requirements before deciding on how the digital twin can help them. The interface between the user and the digital twin has to be custom-made to aid the user in their specific tasks and reduce the chances of a mistake.

Let's envision these varieties of "users" in the form of spheres. The users at the core of the system have absolute control over the operation of the system, and as we move outwards, the users require fewer amounts of control and more refined experiences. For example, a user booking a train ticket (the vast majority of users, the general users) are at the outer surface of the sphere, they need an easy and convenient interface to ensure they get a ride from points A to B. A train ticket booking service representative (hypothetically) is somewhere in the middle of the sphere, they require a bit more visibility over the details of the train operations. Their interface should be easy and convenient as well, but should be a bit more detailed—they should be able to observe all the available seats in all the trains that go from A to B and allot one for the general user. In addition, their need is to help the general user in the best way, even though the goals are different in nature.

The third kind of user, the one at the sphere's core, is the train operator. They are the ones that get the general user from A to B and are in full control of the train itself—their interface can be easy and convenient, but more importantly, it should be pragmatic. It should help them to obtain efficient control of each function of the train and help them monitor all the relevant information. This kind of expert user is well versed with the inner workings of the system and has acquired this expertise over a period of time. We can observe that as close the user is to the core of the system, the more control and visibility the user needs, which also increases the learning curve.

In industrial systems, there are multiple kinds of human users working together—they interact directly with the machines in the system at various capacities. An operator of a single machine directly works with that machine during operation, whereas the floor manager interacts with and keeps track of all the machines on their floor; furthermore, the industry administrator might be monitoring the system as a whole. Each of the users has different goals and needs that must be catered specifically to them—as the "distance" from the core operation increases, the users need an increasingly wider view, but the users closest to the operation need a very deep view. The operator needs to keep track of all the changes in the system pertaining to their machine during operation at the moment of the operation itself. A floor manager requires a wider view of the live state of all machines on the floor, not necessarily the intricate details of each machine. Similarly, the administrator might require a bird's-eye view of the overall operation of the system and aggregated insights on the operations, and more details on the inputs and outputs throughout the day.

In addition to multiple stakeholder groups, within each group, the HMI design needs to cater to both novices and experts. A multitude of studies on the topic of expertise has shown that experts perceive and think differently. They may be looking at the same scene, but they will be parsing it differently and attending to various elements that novices would not. As a result, during both normal and emergency operations, experts and novices have to be supported by the same HMI. In addition, the HMI design should support both in such a manner that the novices can become experts over a period of time.

In general, trained operators in technology-intensive sectors need information support to enhance their *in situ* operations (Romero, Bernus, Noran, Stahre, & Fast-Berglund, 2016). The focus is on improving their perception of the system and helping simplify their decision-making. The critical information needs to be available at once without increasing the cognitive load of the operators. This is accomplished by making proper interface navigation decisions. In order to support the multi-modal information requirements of the operator, this information can be organized in the form of an "information heterarchy", where the user can understand the bulk of information relevant to their mental model at once. This will also help display the interconnectedness of different parts of the system. The "information heterarchy" can be derived from the various work analysis methods, as well as the understanding of user objectives and their mental models. Designing interfaces based on these ideas has been slowly gaining currency as the HMI requirements in the industry are being developed to manage complexity through meaningful representation.

14.3.2 Issue 2: Representation of Systems Complexity of Digital Twins in Human-Centered Terms

Industrial systems for which digital twins are built are, oftentimes, complex in nature. There is a lot of emergent behavior simply because of the sheer scale and intricacy of each part of the system. In many cases, operations in these systems are essentially pre-planned, where human input is required in course-correcting any deviations from the plan. How each deviation is controlled depends on the human operator; they are trained to understand the system and react appropriately to the changes. These changes can be (1) in the output itself and (2) in the inner workings of the system. While the information is directly available in the first case, the operator has to base their decisions solely on the data exposed by the system via an interface in the second case. Ineffective representation or misrepresentation of this data can lead to a loss of situational awareness. In addition, the operator uses the available data to make an informed decision based on their understanding of the system and its emergent behavior, i.e., mental models (Norman, 2013; Norman & Stappers, 2015). The cognitive load on the operator is high, and the decision-making skills of the human user make them an inseparable part of the system.

In order to successfully design for humans in these complex systems, design processes have been developed by Human Factors engineering researchers as well as HMI experts (Burns & Hajdukiewicz, 2004; Hollifield et al., 2013; Vicente, 2002). One such process is the Integrated Ecological Interface Design (IEID) process, which derives from notable approaches of both design and engineering disciplines. A crucial part of this approach is the step of converting the technical aspect of the system into human-centered terms. In other words, how can the machine be represented in human-centered terms to represent its complexity as well as to support the operator's mental models? In order to accomplish this issue, in the IEID design process, the first step is to understand the machine from the viewpoint of the operator's activities. Next, the insights from these interviews, as well as the technical background literature, are used to chart the system in terms of functional abstraction as well as structural decomposition.

In terms of functional abstraction, the system can be conceptualized in terms of various abstract layers ranging from the purpose of the machine to its physical form. The system can be broken down into five functional layers of abstraction, as the emphasis is on "what it does" (Vicente, 1999):

1. Functional purpose—directly relates to the objectives of how the system is created.
2. Abstract function—gives a theoretical understanding of the system in terms of its base levels, laws, and principles governing it.
3. Generalized function—gives a general area of the system in terms of various processes.
4. Physical function—identifies the various sub-components and sub-assemblies.
5. Physical form—identifies the attributes of the physical parts and sub-assemblies.

Next, the technical system is also charted in the form of its structural constitution. This structural composition gives an insight into how the system is related to its subsystems and parts. With this deconstructed view of the technical system and the mental models obtained from the user research, it is possible to identify the exact set of relevant factors and relationships in the overall functioning of the system (Kant & Sudakaran, 2021).

From the two descriptions of the system, the key idea is to understand (1) what the expected purpose of each part is and (2) how these parts interact with each other in various ways. The system can also be simply broken down structurally, from the bigger parts of the machinery to single components; a cross matrix between the structural and abstract layers effectively charts out the system for users to then map out the interactive connections both functionally and structurally. These aspects have been derived from studies of operator experience and are currently considered a standard manner of conceptualizing technical systems in human-centered terms.

In our case of the VM, the systems were conceptualized into different levels of functional abstraction. This resulted in a complicated network of interconnections where some parts were connected to others but not all. In addition, the parts above and below were connected by a why–what–how relationship. Any selected level tells you what exists in that level. The level above tells you why and the level below tells you how. This why–what–how relationship is called means-ends reasoning. It provides some information for the designer in terms of how the functional nodes at various levels can be traced to the overall purpose of the system.

The functional-abstraction hierarchy was then mapped onto a hierarchy consisting of the structural make-up of the system. This resulted in the functional-abstraction–structural-decomposition space (Figure 14.2).

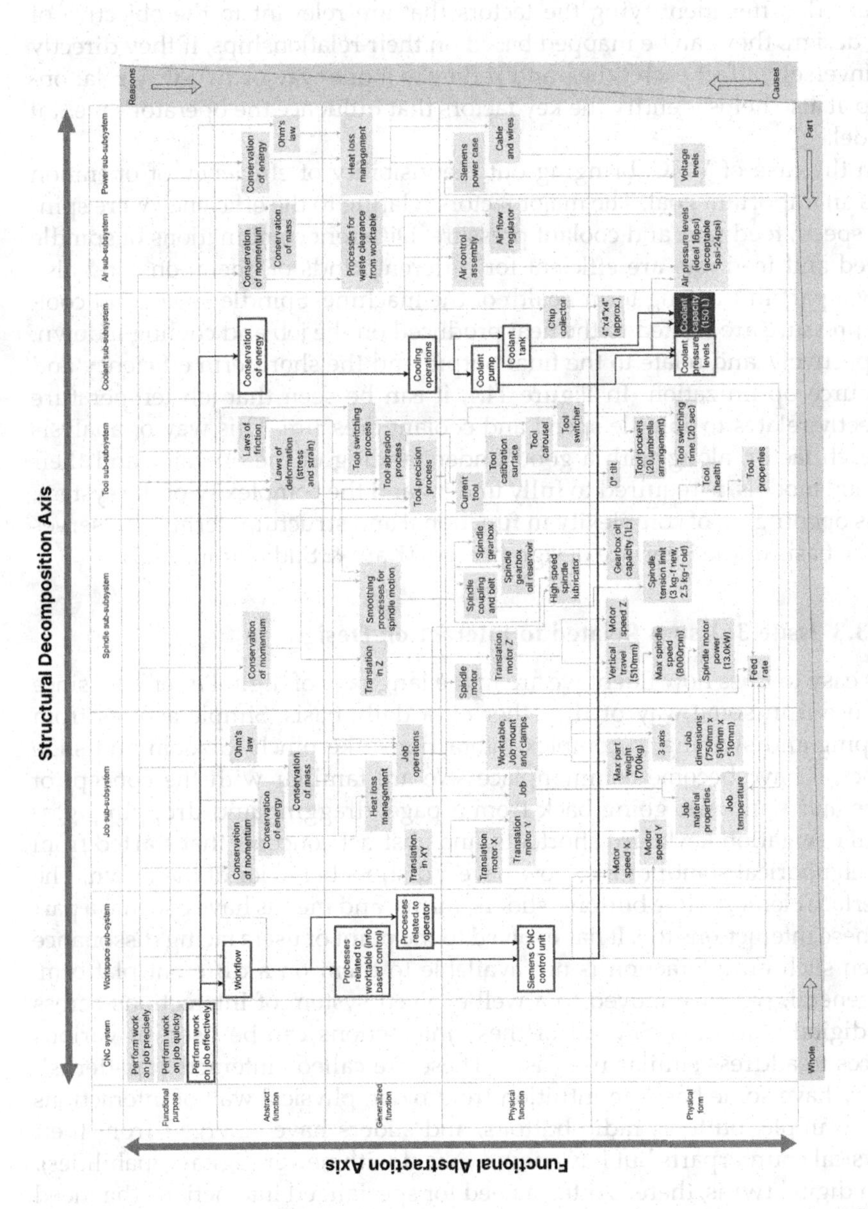

FIGURE 14.2
Abstraction-decomposition space for charting out inter-relationships (adapted from Kant & Sudakaran, 2021).

The result is a deconstructed map of all the interconnections within the system, from which relevant connections were taken to be represented in the interface and the level of abstraction at which it was required to be represented. After identifying the factors that are relevant to the objective of the design, they can be mapped based on their relationships, if they directly or inversely affect each other, and if they're a one-way or two-way relationship. It also helps identify the key factors that influence the operator's mental model.

In the case of VMC, bringing out the visibility of efficiency of operation was an important goal. The major factors relating to the efficiency were spindle speed, feed rate, and coolant pressure. Different combinations of spindle speed and feed rate are efficient for different kinds of operations and also play a part in the long-term health of the machine. Spindle speed and coolant pressure are related to the heat produced on the job and cooling it down, respectively, and relate to the final output and the short-term efficiency and resource optimization. In Figure 14.4, it can be seen that job temperature directly relates to spindle speed and coolant pressure. This way of analysis of each factor, along with a good understanding of the operator and their mental models, is required to fully understand the complexity of the system. This opening up of complexity in functional and structural terms also serves as the basis of the interface design process (Kant & Sudakaran, 2021).

14.3.3 Issue 3: Issues Related to Interaction Design

It is easy to miss how fluent we are in the language of digital interfaces since we use it in some way or the other on a daily basis. Simple actions from tapping and swiping to complex interactions like pinch to zoom and slide to scroll have become commonplace. We are familiar with the concept of apps and windows, going back from a page, dragging and dropping, gestural navigation keyboard shortcuts, and abstract concepts that started from a metaphorical standpoint to now have a distinct behavior of their own. The interface elements like buttons, sliders, forms, and menus have become a part of these interactions too. It has evolved to the point of users facing dissonance when such an interaction is not available to them on a different platform. In general, we have moved to a well-evolved system of interactions across all digital platforms. Several of these interactions can be used in various places to address similar use cases. These are called "interaction patterns". Many have some basis on intuition from older, physical way of interactions (for example, buttons, radio buttons, and sliders have emerged from their physical counterparts but have now evolved with newer digital capabilities).

In digital twins, there is often a need for specialized interactions that need to reflect the complex relationships in the system. In these cases, it is important that the general interactions are considered standard and the foundational basis upon which newer interactions are designed. More importantly, the designer should always ensure if the new interactions they design are going against the intuitions of a similar general interaction, such as in a challenging and stressful operation, then this issue might cause cognitive dissonance and even force the operators to commit ungainful action, being labeled as "errors".

The HMI design should consider simplicity and clarity with equal measure and ensure that neither affects the understanding of the mental model and situation awareness of the user negatively. This is depicted through the design of effective interactions. In the case of VMC HMI, almost every piece of information was organized into different groups and subgroups and was shown on one screen. This provided an outlined view of the work progress, the status of the subsystems of the machine, and an efficiency section. At any point during the operation, the user can access all the relevant information in one place—the overview screen. The only major navigation in the interface was to and from a detailed 3D view—where the user can view and interact with the 3D visualization of the output in detail, in real time. A simpler 3D view is still available in the overview screen, but if the user wants to monitor in detail any deviations and further complications that can occur because of it, they can switch to the detailed 3D view. The most relevant information is still kept on the other screen as well; only a part of the screen changes. This is also in line with the switch in the intent of the user—the overview will be used for the most part of the operation, and when the user wants to inspect the output, other information is subdued to focus on the output.

One of the most important considerations while designing for such situations is to ensure the user is well aware of the output they receive. Regardless of whether the change is in the form of a user action or a notification of a change in the system, it is crucial to be brought to the attention of the busy operator. These are achieved by several micro-interactions that provide "feedback" to a "trigger" from the operator or the machinery itself (Saffer, 2014).

These micro-interactions can happen in many modes—visual animations, audio, and haptics, among others. When the micro-interactions are a response to a user action, micro-interactions must happen within 400 ms, as per Doherty's threshold (Saffer, 2014), from the time of trigger to ensure the user notices it. In the case of VMC HMI, the interface always has a specific colored "glow" around it, which can help the operator understand the overall status of the operation at any point just by looking at the screen quickly (Figure 14.3). The border around the interface and the background of the notification area are colored solid green when the operation is going okay, yellow when there's a suggestion, and red if there's a warning. This color change, along with a sound alert as feedback, would keep the operator alerted about any major change in the system status, which is a trigger.

In addition to the micro-interactions-based feedback, spatial thinking is an important part of mental models, especially when the system occupies physical space and the operations are at least partly hands-on. In this case, it is very important to accommodate the spatial understanding of the operator into the design. For example, an operator handling heavy machinery makes a

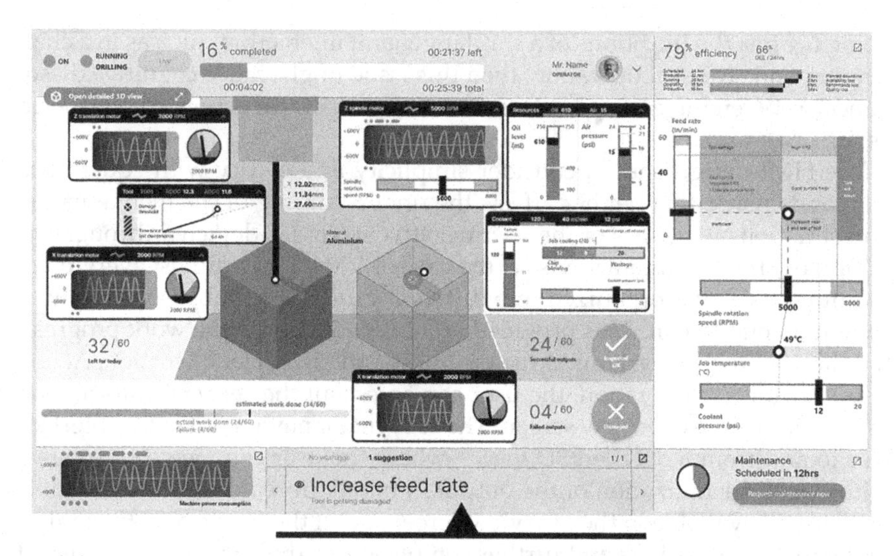

(a) Increase feed rate
(depicted as suggestion in yellow)

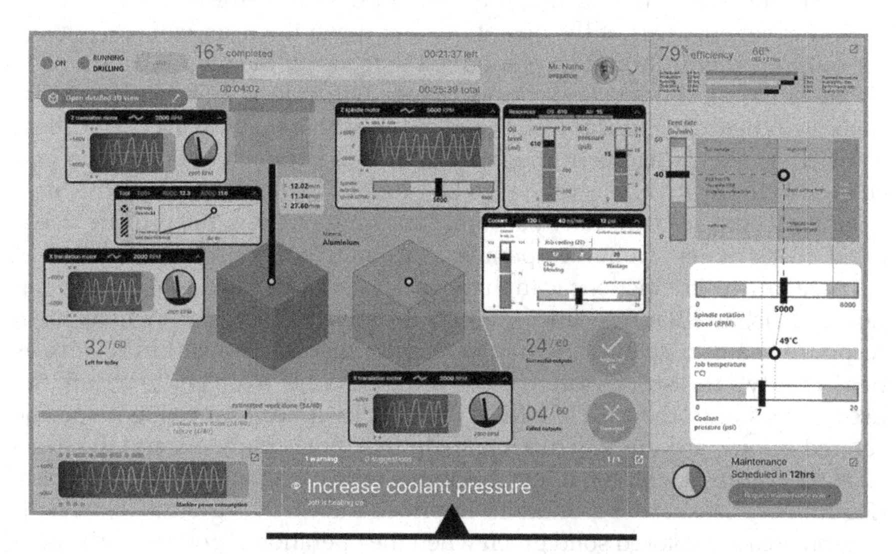

(b) Increase coolant pressure
(depicted as warning in red)

FIGURE 14.3
Supporting situational awareness of operators for understanding change.

lot of decisions based on the physicality of the machine; a lot of the information can simply be made understandable by visualizing it spatially; i.e., over a digital replica of the machine or virtually above the machine itself.

Some mental models can also be spatially represented, although they're not physically relevant. For example, the user workflow in the case of VMC HMI can be metaphorically compared to an assembly line—where the user takes up a raw job, runs it through a series of operations in the machine, obtains a completed job, inspects and approves it. Representing this process like an assembly line is well encapsulating the workflow of the operator without having to explain it indirectly in a written form. This flows well in providing a visual representation of the routine work the operator performs and relevant information in context, like how many raw materials are still left and how many outputs were good vs. damaged. In addition to these issues of Interaction, digital twin HMI also requires a discussion of information design.

14.3.4 Issue 4: Issues Related to Information Design

Given specific data, information design addresses how to represent it best for different users (Goodwin, 2011). In the case of digital twins, there is a huge set of raw data that needs to be constantly addressed and displayed for human consumption for insight discovery and analytics. However, only some of the data is useful in its raw form, whereas others can be used to extract useful insights for the user. Once this step is completed, the information must be structured and visualized well for the user to perceive and understand. The issues of perception can be supported using graphic design principles regarding color and typography and the gestalt laws, whereas issues of understandability can be solved with effective data visualization.

In a complex system, relationships between factors cannot be simplified beyond a level; this is because doing so might lead to a loss of representation of emergent behavior. Here, we can make the distinction between essential complexity, which is necessary for the functioning of the system, and accidental complexity, which is a result of the manner in which the system is *perceived* to be functioning. Depending on the objectives of the user, accidental complexities can be avoided.

When an irrelevant relationship between two factors is represented with the same importance as a relevant relationship, it might lead the user to interpret a false connection between the factors and lead to unwanted cognitive load while making a decision. While the relationship might still be important to understand the machine's workings, it might be irrelevant to the user and causes accidental complexity in the representation of a relationship. On the other hand, a cluster of factors might have a complex relationship that needs to be fully represented to be correctly interpreted. The essential and accidental depends on the tasks and objectives of the user.

In essence, the complexities in the relationships of the factors can be dealt with in two ways: (1) ignore the complexity by design and simplify representation or (2) present complexity for the user to comprehend. Further, depending on the user's needs, option 1 might be useless if the simplicity fails to communicate the relationship, and option 2 might cause unwanted confusion when it is used on an irrelevant relationship.

This issue can be explored further from the viewpoint of the digital twin. In the case of the HMI for the digital twin of the VMC (Figure 14.4), the efficient operation of the machine depends on various factors—the operation type depends on the feed rate and the spindle speed, and various combinations lead to efficiency in different operation types. The job temperature depends on the spindle speed and the coolant level, whereas the coolant level always has to compensate for the heating caused by spindle speed. Therefore, feed rate and coolant levels both put restrictions on spindle speed; operation type efficiency and coolant use both affect the operation efficiency of the system, whereas the job temperature affects the output as well as the health of the tooling used. This relationship is an essential complexity—while the operator can manipulate the three factors in multiple ways to achieve similar outputs, the visualization communicates what exactly this exact combination is doing to the system and insights on which factors to manipulate instead to obtain the same output while keeping the efficiency high. This is an essential complexity that has to be expressed to let the operator understand the role of these factors in efficiency.

The feed rate and spindle rotation (Figure 14.4, Section A) have a complicated relationship across the CNC efficiency matrix, and the spindle rotation speed and coolant pressure have another complicated relationship. Excess coolant pressure leads to wastage of coolant liquid, which affects the efficiency of the workflow, but excess spindle rotation speed overheats the job and causes the work to fail. The coolant pressure defines the pale green region, which is the acceptable area for the job temperature to fall within; the pale red area is the temperature the coolant cannot handle. The spindle rotation speed directly affects the job temperature in the absence of coolant, and therefore the coolant sets an upper limit on how much the spindle rotation speed can affect the efficiency matrix. When the temperature is in the acceptable range, the meter is green, and when it exceeds that, the meter turns red—the ideal operation is when there is just enough coolant to counter the heat caused by the spindle rotation speed. The various shades of green and red show the intensity of the impact that region of the component has; the grays show that the region is not optimal as the white region.

The Spindle rotation speed slider directly impacts the job temperature—hence the line connecting them. The coolant level slider sets an upper limit of temperature it can cool off—hence the line defining the pale green region in the job temperature component. The relation extends further, too, in the form of dotted lines—the job temperature marks the lowest point the coolant level can reach; the line from the coolant level slider reaches up to mark the

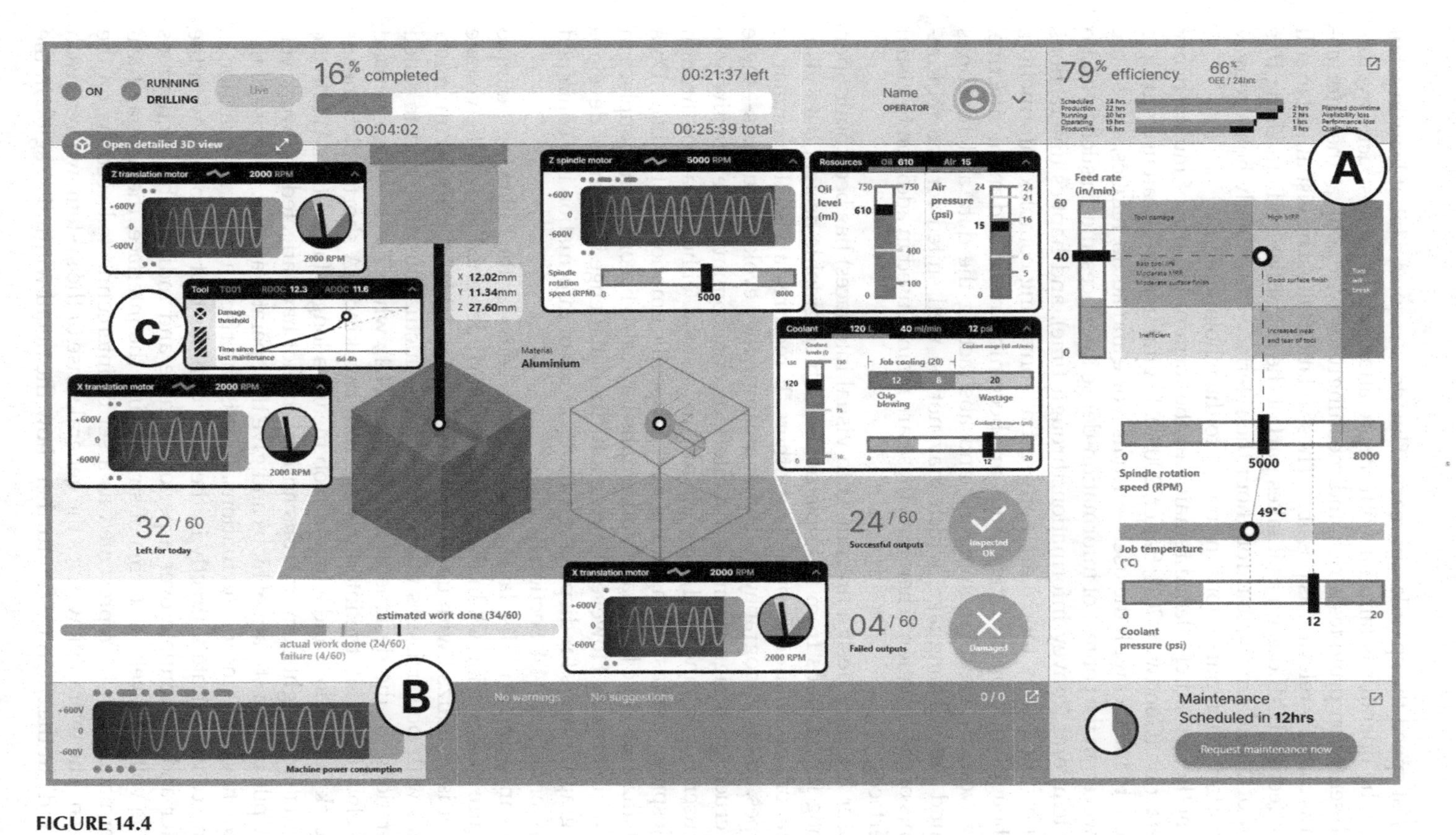

FIGURE 14.4
Visual logic for representing inter-relationships A, B, and C (adapted from Kant & Sudakaran, 2021).

highest speed the spindle can reach. This does require a bit of learning, but since the operator is familiar with this relationship in an abstract way, this representation provides a rich and meaningful way to look at the information as directly relevant as possible. These relations must be visualized in full for effective HMI design as it takes part of the cognitive load off. Using this kind of visual language throughout the interface consistently helps convey important interconnections in context to the whole screen.

Another broad challenge remains in the form of the use of multisensory interaction (Goodwin, 2011; Saffer, 2009a). Digital interfaces can use multiple media to interact with the user—visuals, voice and audio, haptics, and so on. However, in the context of information-dense interfaces, media such as audio are temporal (convey information through time), and haptics (interactions that can be felt by touch) are not great at conveying complex information. Although voice-based interactions are becoming more and more effective, they work by greatly automating and abstracting the data and functions behind natural language, which may not be well suited for the requirements of the situation. Therefore, these are mainly used in addition to visual interfaces.

User Interface design (in context to visual interfaces) has been built upon strong foundations from multiple fields such as graphic design, data visualization, ergonomics, optics, and others. It requires a visual grammar of shapes, colors, typography, and what meaning they evoke in humans—these are crucial in designing impactful interfaces that communicate effectively to the user. For instance, red is commonly understood (either innate or through widespread use) as a danger and evokes specific emotions in the user. Data visualizations like bar charts and graphs have also become commonplace in digital interfaces. However, there is a further challenge in terms of developing graphic forms for displaying various kinds of information in a conjoined manner (Figure 14.4, Section B).

As an example of displaying information, consider the following example (Figure 14.4, Section C). In the case of the HMI designed above, the voltage meter is visualized with upper and lower limits. VMC machines run in factories with power issues often face voltage fluctuations that lead to failed operations. In the designed HMI (Figure 14.4), when the levels go beyond or below the limit, the interface logs it in two kinds. A shorter blob means a quick fluctuation, and a longer blob means a longer fluctuation. While the fluctuations might be numerous and might not have affected the work yet, they pile up as a stream of blobs and give the operator a quick glimpse of the frequency and kind of fluctuations. In this interface, a lot of information is being communicated directly in the form of blobs that are floating with the meter and accumulate over a period of time and indicate the fluctuations of the voltage levels. In terms of structuring the interface, the heterarchical information can be represented as a set of interconnected sections on the screen. The overall view gives the general sense of the system, whereas each "island" in the interface can further show the detail within its region in the interface. This will allow the user to metaphorically "drill down" or "zoom out" of the heterarchy as required (Saffer, 2009a, b, 2014; also see Cooper et al., 2014; Goodwin, 2011; Tidwell, Brewer, & Valencia, 2020).

In the case of VMC HMI, if there were critical errors, the entire screen was tinted red, except for the relevant information unintended. This helped tone down irrelevant information to directly draw attention to the relevant ones, thereby (1) drawing user attention by displaying an overall redness on display and (2) making it easier to locate the relevant information. It was important not to block access to other information, and therefore, only a red tint was used, and not opaque red, as well as multiple ways to easily dismiss the tinting—mouse shake, click, touch, key press, or any other interactions that the users may try.

In the case of the VMC HMI, change blindness was addressed in two ways:

1. Whenever there was a critical change, that information was notified and logged for the user to address. This also enabled the user to ignore frequent irrelevant changes without worrying about critical changes. For example, whenever the voltage fluctuated above or below the safe limit, a red dot/line was logged, and the user was notified.
2. Animation and micro-interactions were used to draw attention to continuous changes which were relevant to the user at all times.

14.4 Conclusion

While addressing the sheer scale and complexity of digital twins, it is quite possible to forego human-centric design as it may be categorized as a secondary goal. However, it is important to note that in the end, all the additional data and capabilities from digital twins help us in two major ways: (1) provide new valuable insights into the system, which provides better visibility and control to the human user and (2) offload manual work and provide additional means to automate it, making it easier for human users to focus on more important tasks. In both cases, human users are the intended beneficiaries—it is important that they are given adequate consideration at each step of the digital twin design process. Good human-centered design reduces cognitive load and "human errors". As a result, good HMIs can have an enormous impact on the success of digital twins. The key ideas discussed throughout this chapter can be considered as a glimpse into the broad aspects of human-centric design—each topic is an enormous field of its own, with applications throughout our daily lives. A human-centric approach to designing digital twins and their interfaces would be crucial in ensuring that the technology is used effectively and provides the highest possible level of benefits to its human users.

References

Bennett, K. B., & Flach, J. M. (2011). *Display and Interface Design Subtle Science, Exact Art*. Boca Raton, FL: CRC.

Burns, C. M., & Hajdukiewicz, J. R. (2004). *Ecological Interface Design*. Boca Raton, FL: CRC Press.

Cooper, A., Reimann, R., Cronin, D., Noessel, C., Csizmadi, J., & LeMoine, D. (2014). *About Face: The Essentials of Interaction Design*. Indianapolis, IN: Wiley.

Goodwin, K. (2011). *Designing for the Digital Age How to Create Human-Centered Products and Services*. Hoboken, NJ: Wiley.

Grudin, J. (2017). From tool to partner: The evolution of human-computer interaction. *Synthesis Lectures on Human-Centered Informatics*, 10(1), 1–183. https://doi.org/10.2200/S00745ED1V01Y201612HCI035.

Hollifield, B., Habibi, E., & Oliver, D. (2013). *The High Performance HMI Handbook*. Plant Automation Services, Inc.

Hsu, Y., Chiu, J., & Liu, J. S. (2019). Digital twins for industry 4.0 and beyond (pp. 526–530). 2019 IEEE International Conference on Industrial Engineering and Engineering Management (IEEM). IEEE, 2019: 526–530.

Kant, V., & Sudakaran, J. S. (2021). Extending the ecological interface design process-integrated EID. Human Factors and Ergonomics in Manufacturing & Service Industries, 2022, 32(1): 102–124.

Liu, M., Fang, S., Dong, H., & Xu, C. (2021). Review of digital twin about concepts, technologies, and industrial applications. *Journal of Manufacturing Systems*, 58, 346–361.

Norman, D. (2013). *The Design of Everyday Things*. Cambridge, MA: Basic Books.

Norman, D, & Stappers, P. J. (2015). DesignX: Complex sociotechnical systems. She Ji: The Journal of Design, Economics, and Innovation, 2015, 1(2): 83–106.

Qi, Q., Tao, F., Hu, T., Anwer, N., Liu, A., Wei, Y., et al. (2021). Enabling technologies and tools for digital twin. *Journal of Manufacturing Systems*, 58, 3–21.

Rasmussen, J. (1979). *On the Structure of Knowledge - A Morphology of Metal Models in a Man-Machine System Context*. Risø National Laboratory, 1979.

Rasmussen, J. (1986). *Information Processing and Human-Machine Interaction: An Approach to Cognitive Engineering*. New York: North-Holland.

Rasmussen, J., Pejtersen, A. M., Goodstein, L. P. (1994). *Cognitive Systems Engineering*. New York: John Wiley & Sons.

Romero, D., Bernus, P., Noran, O., Stahre, J., & Fast-Berglund, Å. (2016). The Operator 4.0: Human cyber-physical systems & adaptive automation towards human-automation symbiosis work systems. In I. Nääs, O. Vendrametto, J. Mendes Reis, R. F. Gonçalves, M. T. Silva, G. von Cieminski, & D. Kiritsis (Eds.), *Advances in Production Management Systems. Initiatives for a Sustainable World* (pp. 677–686). Cham: Springer International Publishing.

Saffer, D. (2009a). *Designing for Interaction: Creating Smart Applications and Clever Devices*. Indianapolis, IN; London: New Riders.

Saffer, D. (2009b). *Designing Gestural Interfaces*. Sebastopol, CA: O'Reilly.

Saffer, D. (2014). *Microinteractions: Designing with Details*. Sebastopol, CA: O'Reilly.

Tao, F., Anwer, N., Liu, A., Wang, L., Nee, A. Y. C., Li, L., & Zhang, M. (2021). Digital twin towards smart manufacturing and industry 4.0. *Journal of Manufacturing Systems*, 58, 1–2.

Tidwell, J., Brewer, C., & Valencia, A. (2020). *Designing Interfaces (3rd ed.)*. Sebastopol, CA: O'Reilly.
Vicente, K. J. (1999). *Cognitive Work Analysis: Toward Safe, Productive, and Healthy Computer-Based Work*. Boca Raton, FL: CRC Press.
Vicente, K. J. (2002). Ecological interface design: Progress and challenges. *Hum Factors*, 44(1), 62–78. https://doi.org/10.1518/0018720024494829.

15

Toward a New Generation of Design Tools for the Digital Multiverse

Chiara Cimino, Gianni Ferretti, and Alberto Leva
Politecnico di Milano

15.1 Introduction and Motivation

Digital technologies are nowadays of fundamental importance as enablers of Cyber-Physical Systems (CPSs) within the Industry 4.0 context (Oztemel & Gursev, 2020). The cyber part of those intelligent systems is enabled by several digital twins (DTs) for decision-making that can be used during the whole life cycle phases of a production asset: from design (Lo et al., 2021; Tao et al., 2018a), engineering, commissioning and control (Leng et al., 2021) to maintenance and management (Tao et al., 2018b; Dinter et al., 2022). A DT can enhance both asset monitoring and analysing throughout its lifecycle (Leng et al., 2021). Nevertheless, the existent DT applications are strictly related to their application purposes, they generally offer a limited set of services usually not related to their physical counterpart (Davila Delgado & Oyedele, 2021; Tao et al., 2019).

Considering the design phase of an asset, the DT – being a twin – should be born together with the real system to simulate it, to take decisions in its current state and grow with it, shaping a history of DTs that correspond to the counterpart with the applied modifications through time. This concept becomes more complex when many professional figures work and take decisions on the same system since a DT can be based on different modelling approaches: i.e. a Building Information Model database, a system of differential and algebraic equations, a neural network. And the same must hold for the level of detail, the model can refer to a single component, machine, process, or whole asset. When a professional takes some decisions through a DT – using various tools possibly at different levels of detail – he/she cannot know whether he/she is taking decisions in conflict with those made previously and/or in parallel by other professionals. Hence, the lack of consistency in terms of the relationships that must be established among different DTs can cause a lack of consistency in the decision-making during the design stage, especially when DT applications are so heterogeneous, i.e. of a

DOI: 10.1201/9781003425724-19

dynamic model and a set of recorded data plus their statistical analysis tools, that even just imagining such a bidirectional exchange appears problematic.

This chapter proposes the specification of a tool that can help ensure consistency among such a heterogeneous set of DTs, making consistent the set of models and data that is processed during the design phase to create a knowledge base of each system from its design phase. The mentioned tool is based on the digital multiverse (DM) paradigm, structured by Cimino et al. (2021). Starting from recalling the basic concepts of the paradigm in the related work of Section 15.2, we identify the key actors at the basis of the tool creation in Section 15.3. We continue our treatise illustrating a UML specification to build a standard Model and Data Base (MDB) in Section 15.4, and we propose a specification to delineate the access profiles (Section 15.5), the MDB management tool and the version control specification (Section 15.6). Finally, we provide some concluding remarks and future works.

15.2 Related Work

To provide coherent tool specifications that include all the dynamics envisioned by the DM paradigm, we resume in this section its main pillars and the technologies chosen for the paradigm realisation. Finally, we point out some key concepts concerning the novelty of our proposal and the specification of the tool.

15.2.1 The Digital Multiverse Paradigm and the Enabling Technologies

The paradigm structured in Cimino et al. (2021) is based on the *twin* concept mentioned in Section 15.1. The term *twin* suggests that the physical and virtual systems of a CPS should be born together and live somehow parallel lives, but this does not correspond to how the DT applications are currently constructed. The most important characteristic of the digital world – where different virtual systems cohabit – is its multiverse nature. After this nature, we name our paradigm, and on its basis, we introduce an entity – that we call a Digital Meta Twin (DMT) – that has the property of presenting – when observed from a certain viewpoint – a DT in the sense corresponding to that viewpoint. A DMT must contain both models and data: herein, we concentrate on the model-based DTs as the equation-based models as opposed to the data-based DTs that require a database (historical and/or real-time acquisitions). To relate all the DTs among each other, we introduce the idea of data and model relations – as an extension to that of relations among data as per the database theory and jargon – as a necessity for ensuring consistency in an MDB, which is the basic entity that implements a DMT in the digital world. Summing up, the paradigm relies on creating and managing MDBs as repositories for DMT in the DM.

The DM paradigm research (Cimino et al., 2021) enhances the consistency among the models by combining different existing technologies:

- The Model Description Language of the paradigm is object-oriented modelling (OOM), abstracting the entities needed to instate the relationships (Scaglioni & Ferretti, 2018). OOM tools are fundamental for their capability of checking a model (i.e. Open Modelica or Dymola) for mathematical correctness (i.e. that it has as many equations as variables, contains no structural singularity and so forth).
- The Data Description Language is the Relational Data Model (RDM) – i.e. an example in a related field as integrated design (Feng et al., 2009). This requires some structure to be properly coordinated with the OOM, but this choice relieves the RDM from storing all the mathematical relations confined in OOM.
- The Operational Language is Structured Query Language in coordination with a Version Control System (VCS) to keep track of the MDB history and multiple user modifications.

15.2.2 The Digital Multiverse Tool Specification

The paradigm unification as the one here proposed – which is much more than just transferring data among domain-specific tools, as will be shown – is therefore of undoubted value, and highly desirable. Each MDB collects consistency functions and scripts, that are used to instate relations between models that can eventually be checked (i.e. through special simulation campaigns) whenever required by a user. Hence, consistencies correspond not only to data exchanges but to specific actions that the tool must perform to check that all the MDB's models stay consistent with each other. It is important to specify the matter since those mechanisms provide additional benefits over current approaches to connecting different DTs, such as the Asset Administration Shell – considered an enabler for the digital transition – or the other existing architecture listed in (Boyes & Watson, 2022).

The rest of this chapter is focused on how the overall management of an MDB leads to the specification of a DM tool using the UML language, considered the industry standard for object-oriented system development (Bashir et al., 2016).

15.3 The Actors for a DM Tool Specification

As introduced, the DM tool will be used to consistently link several types of models, data and consequently analysis. To this aim, each professional user should be able to include data and models – that must be consistent and subjected to a specific version control ordering – into an MDB, depending on his/her profile. When it comes to describing the dynamics of such complex procedures, we need to add more details, specifying firstly the actors involved in the MDB management.

Assuming the existence of a set of MDBs used by, i.e. teamwork, as depicted in Figure 15.1a, each professional can use his/her workspace to access all the models and data present in the considered MBD. Each professional should be able to open, simulate and/or analyse a system through the online/offline DTs of its speciality, comparing also different universes (hence MBDs) in a DMT as prescribed by the DM approach.

For example, access to, i.e. the MDB_1 of Figure 15.1a from a certain DMT seems to be a relatively easier procedure: the user can copy on the local workspace the version(s) of the MDB(s) he/she is interested in – in compliance to the VCS versioning – and start his/her own analyses on it (them). Instead, when a user wants to add the analyses performed in an existing version of MDB and save it in the storage: (1) a consistency check tool, between each workspace and the MDBs storage must check again the involved consistency relationships, mainly for safety reasons, using some set of simulations, as prescribed by some consistency script and functions, and (2) the new/updated MDB must be saved and contextually catalogued by the versioning control tool before reaching the storage.

The example identifies the main actors that must be considered to further expand the DM tool specification: a simulation tool; a workspace handling system; and an MDB management tool. The realisation of a proper tool that reflects the DM paradigm requires the specification of such a complex structure. Problems that can be cast into "more classical" ones such as locking and allowing multiple user access will not be discussed to not stray from the main presentation path.

15.4 The MDB Structure

The objective of this paragraph is to specify the structure of an MDB, the core element of the whole approach, in compliance with the OOM language. Recalling that we want to focus on implementing a tool that deals with the creation of OOM model-based DTs (as per the definition given herein for the models), those must be modular and must allow the replaceability of the same model: for example, a component model in an aggregate one, as could be a machine model, must keep the same interface to enable, i.e. increasing the level of detail in terms of physical phenomena described. This concept is one of the key points of the research to which we refer, this MDB organisation allows us to instate more easily the consistencies among different models. Herein, we limit to recall it to propose a proper tool specification.

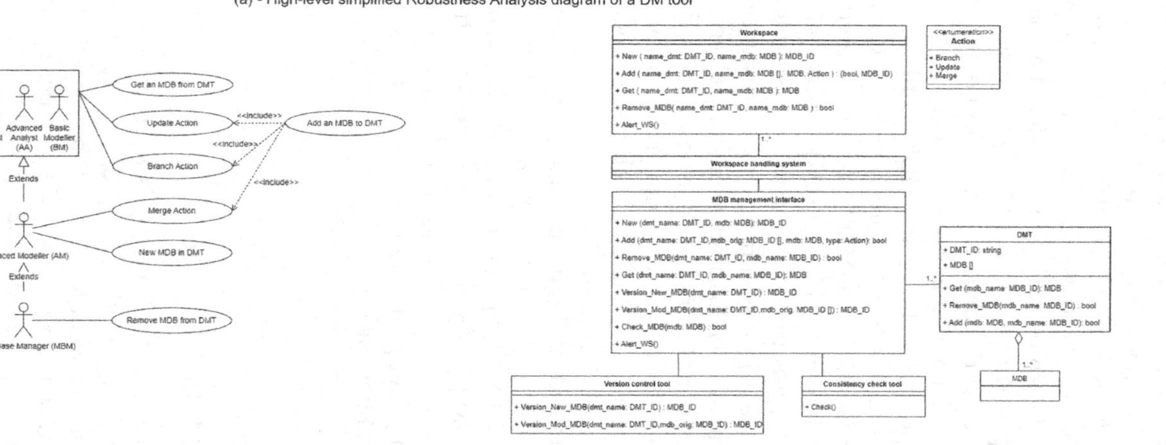

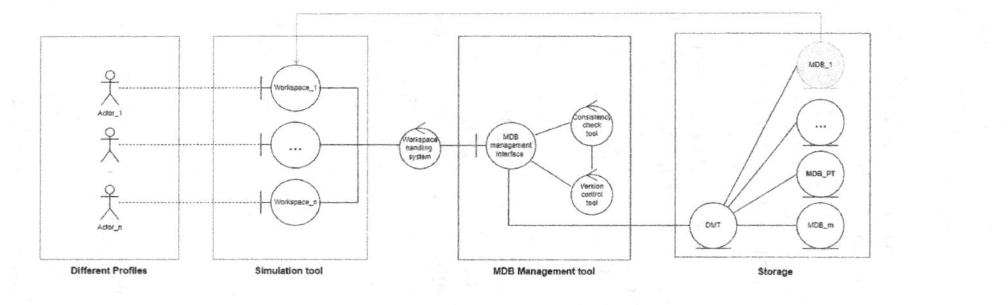

FIGURE 15.1
(a) High-level simplified Robustness Analysis diagram of a DM tool. (b) UML Specification of the Access Profiles interaction with the MDB management tool – Use Cases. (c) UML Specification of an MDB Management tool – Class Diagram.

The specification of the MDB structure is illustrated in detail in Figure 15.2, it is the base to implement a tool that can comprehend the creation and storage of OOM model-based DTs through a standard procedure (starting also from existing OOM libraries, such as Modelica). We focus on the OOM-based DT structure inside an MDB, to which we will refer from now on as the *Model root* of an MDB. All the model-based DTs contained in the MDB must have this structure to be stored, exported locally and imported to the storage. The UML classes included in the model root of an MDB (Figure 15.2) are the ones prescribed by (Cimino et al., 2021), the long-term research to which this paper belongs. Each model root is related to an MDB indicating the MDB_ID code and includes all the major classes detailed in the following.

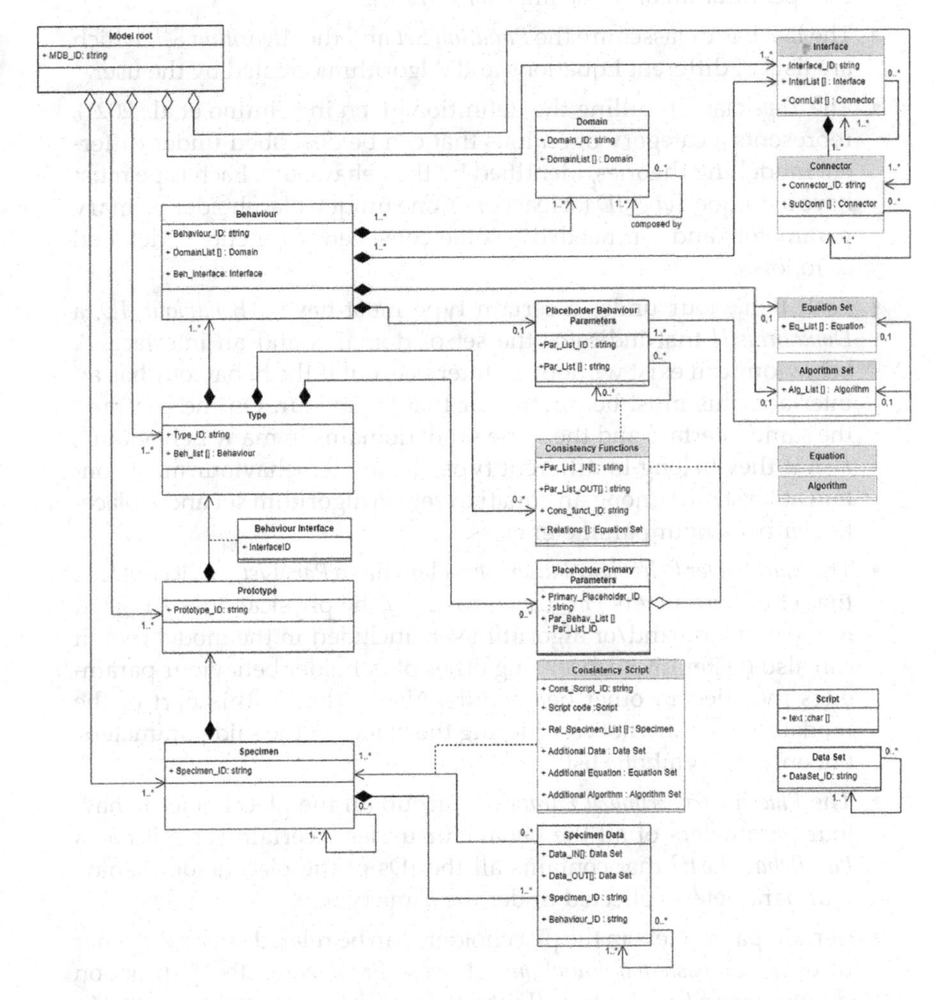

FIGURE 15.2
Structure of the Model root of an MDB – UML Class Diagram

- The most basic class contained by a model root is a *Connector*. It represents the physical terminal of a model, in terms of the variables accessible from the outside. Each connector in the model root must have its *Connector_ID*, which can be composed of a list of connectors *SubConn[]*, i.e. when using a hierarchic connector.
- The model root allows the grouping of the connectors into an *Interface*. Each interface must have an *Interface_ID* and the string of the included connectors *ConnList[]* (at least one) or the string of the iteratively included interfaces *InterList[]* (at least one).
- Another basic class is the *Domain*. Each domain has of course a *Domain_ID*. A domain can be made by a group of domains through the specification of the string *DomainList[]*.
- The last basic classes are the *Equation Set* and the *Algorithm Set*, which are lists of different Equations and Algorithms created by the user.
- The *Type* class, recalling the definition given in (Cimino et al., 2021), represents a category of entities that can be described under different modelling theories, identified by the behaviours. Each type must have a unique *Type_ID* (at least one), one unique placeholder primary parameters and – if needed – some consistency functions, detailed as follows.
- Each Behaviour under a certain type must have a *Behaviour_ID*, a *DomainList[]* that indicates the set of domains and an interface. A behaviour can exist without an interface, but if the behaviour has an interface, this must be unique for that behaviour. On the contrary, the same interface and the same set of domains in many behaviours, also if they belong to different type classes. A behaviour must contain at least one among an equation set, an algorithm set and a placeholder behaviour parameter class.
- The *Placeholder Behaviour Parameters* class has a *Par_List_ID*, it contains the set of parameters *Par_List[]* – usually the physical ones – related to the equation and/or algorithm sets included in the model root. It can also be iterative concerning other placeholder behaviour parameters included in other behaviours. Notice that in this part of the model root, we are not considering the values of the said parameters, but only the symbolic list.
- The *Placeholder Primary Parameters* group all the placeholder behaviour parameters of all the behaviour under a certain type, it has a *Par_Behav_List[]* that contains all the IDs of the placeholder behaviour parameters collected under the same type.
- Certain parameters in the placeholders can be related one to the other through *Consistency Functions* classes. Each consistency function should have a *Const_funct_ID*, the *Relation[]*, the set of the equation(s) that establishes the connection between the parameters, and the

Par_List_IN[] and *Par_List_OUT[]* that specify which set of parameters in input gives the ones in output through the relations stated. Those types of consistencies can be described in UML as dependencies with a dashed line.

- The model root also contains the *Prototype* class; each prototype has its own *Prototype_ID* and collects all the behaviours – at least one for the existence of the prototype – of the same type that share the same interface.
- Finally, the model root includes the *Specimen* class. As said above, all the parameters considered for each type and behaviour in the placeholders are symbolic to keep the models generic and applicable in different cases. The specimen is considered the first parameterised object: it is a prototype with an assigned set of parameters. The specimen is the parametrised object that can be instantiated in a simulation tool, and the same holds when more specimens grouped and connected compose a specimen iteratively.
- The data associated with a specific specimen with certain behaviour, through both the *Specimen_ID* and the *Behaviour_ID*, are structured in the *Specimen Data* class. The data associated comprehend the Data_IN, the list of parameters in input to a specimen – the one that enables the simulation, which can also be exported from an external database – and the list of Data_OUT[] resulting from the specimen simulation. Moreover, as a specimen can be composed iteratively by connecting many specimens, the specimen data can be composed by a list of specimen data of an aggregate specimen.
- Each specimen data (i.e. Data_OUT[]) in the model root is associated with one or more specimens (i.e. Data_IN[]) depending on the relations stated in the *Consistency Script*. Hence, each script has its *Cons_ScriptID*, a script code – needed, i.e. to establish a simulation campaign – and the *Rel_Specimen_List[]* the list of all the specimens that use the simulation output data as input. Relating the data to the specimen that generates them is of major importance to establish all the possible relationships among models and data. The same must be ensured also if the data should be processed with the use of additional data, equations and/or algorithms.

Overall, as illustrated in Figure 15.2, it is possible to divide all the classes into three groups:

- The objects in bold are the ones that already exist in OOM systems, i.e. Modelica language, and can be created inside the OOM environment, i.e. it is possible to create new interfaces using different connectors;

- All white objects help to group the models of the MDB into the classes identified by the paradigm: the main classes (Type, Prototype, Specimen), the sub-classes (Behaviour, Domain), the classes for symbolic parameter collection (Placeholder Behaviour Parameters and Placeholder Primary Parameters), the data collection for the parametrisation of models (Specimen Data and Data Set) and script collection to instate the consistencies (Script);
- The shaded objects of the model root represent the main novelty introduced by the paradigm: the consistency functions relate equation- and/or algorithm-based models, keeping consistencies among the parameters; the consistency scripts enable simulation campaigns when changes and checks take place, i.e. changing the level of detail.

Within the given specification the proposed tool will document *how* different models are connected, and *how* to instate those relations for checking purposes. Additionally, we must stress that the tool herein specified is not made to create the consistencies; rather, the MDB is designed to allow the user to enforce all the consistencies, while the tool must trigger the checks to control that the consistencies are observed when simulating.

15.5 The Access Profiles

This section describes how each profile interacts from the workspace with the MDB, according to the permissions granted in Figure 15.3.

15.5.1 Interactions with the Storage

Using and maintaining an MDB requires assigning and managing access privileges properly. As described in Section 15.1, a user can create from scratch an MDB or he/she can add models and data to an existing MDB from its workspace. We report in this section a synthetic proposal of user access profiles within their assigned privileges, accordingly to the organisation of the MDB outlined so far.

- *Basic analyst (BA)*: can only run simulations instantiating one existing specimen (clearly, that of a "complete" runnable model). This is the least privileged profile, it could be attributed to somebody who conversely has high privileges on the data that are generated by simulations.
- *Advanced analyst (AA)*: in addition to the attributions of a BA, this person can create new specimens, however only by changing

FIGURE 15.3
UML Specification of the Access Profiles – Use Cases

parameters. For example, an AA could take a specimen, taking a certain prototype with specific behaviour, and store it in the MDB as a new specimen, setting the primary parameters as per the datasheet; if the information contained therein is not complete, an error will be generated at the check level to signal and request back to the user the missing information.

- *Basic modeller (BM)*: in addition to the attributions of an AA, a BM can create new types (thus prototypes and specimens as well) but only by aggregating instances of specimens that already exist, or of new specimens created with power limited to that of the advanced analyst (i.e. by just changing parameters). A BM can also create new interfaces, most typically for new behaviours, only with existing connectors. The limitations for the BM profile prevent the creation of new behaviours by acting on equations and/or on algorithms, hence admitting no interaction with consistency functions or generation/consistency scripts either. For example, a BM can assemble and store as type of circuit (say an RC filter) with existing components (elementary such as resistors and capacitors or also aggregate), create and store the interface for its pinout using standard pin connectors, create a resistor specimen with say a new proper inductance but not create a resistor type with new behaviour in terms of equations (if not connection ones) or algorithms. Also, a BM shall not be allowed to create types spanning more than one domain, as it avoids a BM interacting with consistency functions: since components are validated individually, the consistency functions for an aggregate with one domain only are the union of the consistency functions of the components. For example, a BM can create a first-principle model of a machine but not its DEVS companion as the latter would belong to a second domain, and BM cannot create the proper consistency functions between the two hypothetical behaviours of the same type having different domains.

- *Advanced modeller (AM)*: in addition to the attributions of a BM, this profile can also create new behaviours by interacting directly with equations (other than connection ones, clearly) and algorithms, within new or existing types, domain(s) and interfaces, and having responsibility on all the required consistency functions and scripts.
- *Model base manager (MBM)*: this person has full authority, having only to abide (wherever this is applicable) by the rules of the database system hosting the MDB; ultimately, in synthesis, the MBM is the person who can even bypass – temporarily and knowledgeably – the MDB interface to which all the others confined, typically for recovery in the case of an MDB corruption.

To provide a picture of how a DM tool should be specified while maintaining the high-level overview, we present the UML use cases sketch to translate the access profiles presented so far in a completely abstract manner in Figure 15.3.

15.5.2 Interactions with the MDB Management Tool

In compliance with the actions allowed by its access profile, a user can create, import and export an MDB. The set of high-level UML in Figure 15.1c allows interactions of the users with the storage through the following methods:

- Include a new MDB – created from scratch – with the *New()* method into a DMT, identified by the *DMT_ID*; this method returns the *MDB_ID* elaborated by the version control tool.

- Export an MDB from the storage with the *Get()* method, providing both the *DMT_ID* and the *MDB_ID* and returning the exact MDB.
- Add an up-to-date MDB version of a locally modified MDB in a DMT through the *Add()* method; this functionality requires the *DMT_ID*, the *MDB_ID* of the original MDB, and, specifying the action performed, returns a successful indicator and the MDB_ID chosen by the version control tool. The *Action* is related to what the user needs to do when uploading a new version of MDB in a DMT and can be of different types:
  - *Branch*: if the new version of the MDB represents a parallel analysis, i.e. an alternative to the original one;
  - *Update:* if the new version of the MDB is an evolution of the original one; and
  - *Merge:* if the new MDB version is the result of a merge of multiple MDBs (for this reason the MDB_ID of the original MDB in the Add method can also be a list of MDBs).
- Remove an MDB version from a DMT through the *Remove()* method providing both the *DMT_ID* and the *MDB_ID* and returning a successful indicator.
- We do not enter the details of the *Workspace handling system*, and we just specify that its role is to manage the communications, advising the user when the MDB management tool is busy.

Figure 15.1b illustrates the specification regarding the users' interactions with the MDB management tool, regulated following their access profile:

- The BA, AA and BM profiles can only perform the Get and Add methods, but only using the Update and Branch action; since those profiles can only act on existing models, they can only use the specimens already present in other existent MDBs to create new Specimen entities; also, if in some cases the merge action could not involve any consistency, when merging different works performed by multiple users, consistencies could appear. For this reason, those profiles need the supervision of an AM user to perform any merge.
- The AM can also create new MDBs from scratch using the New method and can perform the Add method with the Merge action.
- The MBM user has full authority; he/she can also use the Remove method, to remove an MDB from a DMT.

Overall, the specifications illustrated to show how to manage the MDB tool in working groups. Multiple users can work on different MDBs and also on different analyses, subject to the limits of their roles. The high-level users (BA, AA and BM in Figures 15.1b and 15.3) can work in parallel on different analyses, but when all the analyses serve to take a decision, an – or a group of – AM will merge all of their work, inserting new specimens with new behaviours and/or identifying the consistencies if needed. As a consequence, the AMs that work to take different decisions on the same DMT would identify together if it is the case to add some consistency functions inside the constructed MDBs under the advice of an MBM.

It is evident that the whole unification works to replicate the PT cannot be a completely automatic process, it requires the collaboration of different professionals with more or less expertise in their field that must communicate and collaborate among themselves. The tool will support their work by checking if all the inserted models are consistent.

15.5.3 Complete Specification

Overall, we can say that each user has access to his/her workspace from which he/she can use methods to import, create, edit and save back to a DMT in the storage his/her work, i.e. new models in the form of MDB structure and simulation results. Figure 15.4 depicts the UML class diagram of the different workspaces in which the different profiles interact depending on their privileges.

All the profiles inherit the possibility to login(), and then, each profile is allowed to visualise a specific workspace with the privileges required by the class, as listed in the following (coherently with all the specifications described in this chapter).

- The BA has the Workspace_BA class: he/she can import an MDB with the correspondent function *Get()* knowing the DMT_ID and the MDB_ID; he/she can open a specimen selecting the MDB_ID and simulate the said specimen respectively with the *Open()* and with the *Simulate()* functions. The user can also add a new specimen as a list of other specimens with the function *NewSpecimen()*, or add a specimen at a time to an existing one with *AddSpecimen()*. Finally, he/she can add the modified MDB to the storage through the use of the *Add()* function, but only using the actions *branch* or *update*.
- The AA can add from its Workspace_AA a new set of parameters for a certain dataset with the function *AddParData()* indicating the DataSet_ID to which the parameters refer. Then, he/she can create a new specimen with the function *AddSpecimen()*, selecting an already existent specimen through the Specimen_ID and adding the new DataSet_ID that must be coherent with the said behaviour of the specimen.
- The BM uses the Workspace_BM, which allows the BM profile to create new specimens as aggregates of other specimens, eventually adding a new interface and new behaviour placeholders, but always

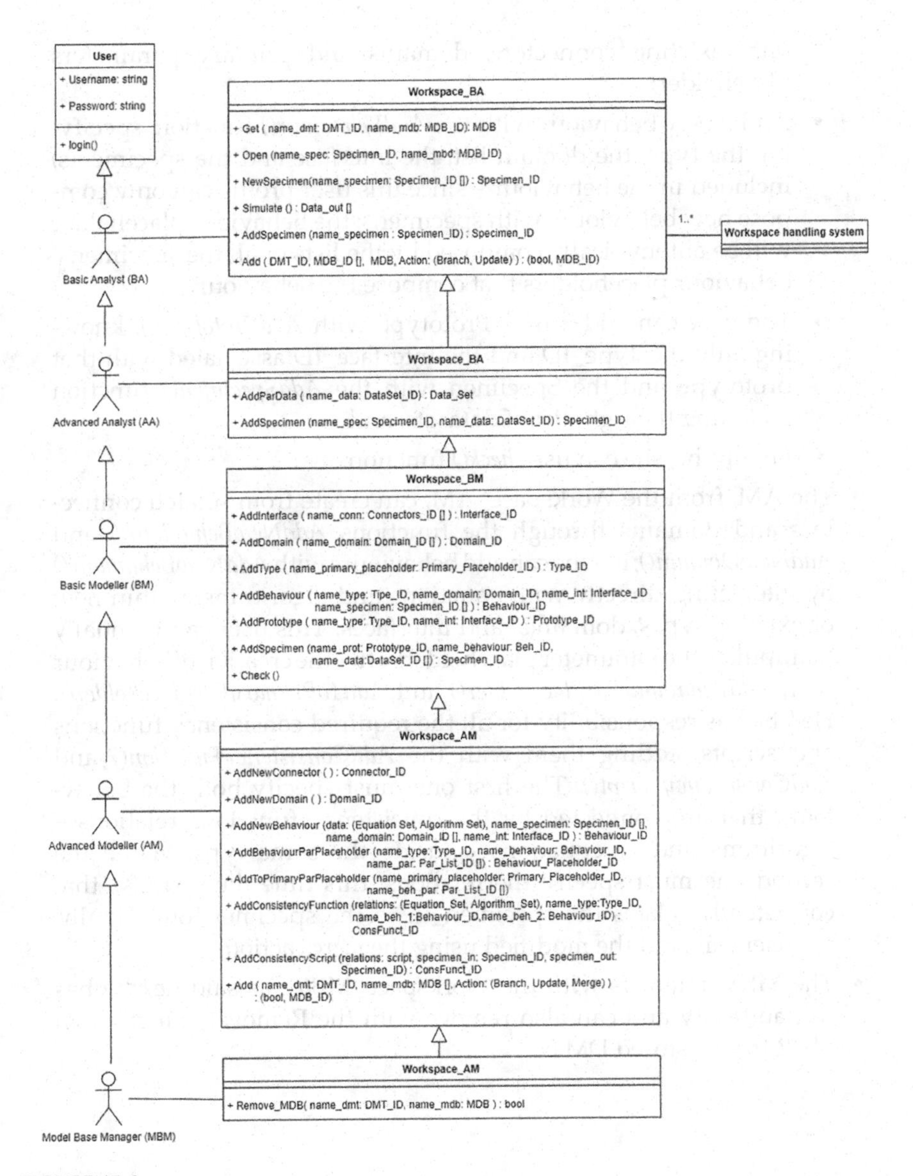

FIGURE 15.4
UML Specification of the workspaces' interaction with the MDB Management tool depending on the profile – Class Diagram

considering a single domain, also as a list of other domains avoiding the use of consistency functions and scripts. The possible interactions include:

- Add a new interface specifying the list of connectors used with *AddInterface()*, *AddDomain()* and *AddType()*, respectively, using only existing connectors, domains and primary parameters placeholder;

- Add a new behaviour with the *AddBehaviour()* function, specifying the type, the domain list, the interface and the specimen(s) included in the behaviour – since this user profile can only compose new behaviours with specimens; the behaviour placeholder will be automatically composed by the lists of all the specimens' behaviour placeholders that compose the behaviour;
- The user can add a new Prototype with *AddPrototype()*, knowing only the Type_ID and the Interface_ID associated with that prototype and the Specimen with the *AddSpecimen()* function including the right list of datasets; and
- Finally, he/she can use *check()* function.

- The AM, from the Workspace_AM, can create from scratch connectors and domains through the functions *AddNewBehaviour()* and *AddNewDomain()*; he/she can add behaviours with *AddNewBehaviour()* by interacting directly with equations and algorithms, within new or existing types, domain(s) and interfaces. This user can manually manipulate the parameter placeholders after the creation of behaviour with *AddBehaviourParPlaceholder()* and *AddToPrimaryParPlaceholder()*. He/she has responsibility for all the required consistency functions and scripts, adding them with the *AddConsistencyFunction()* and *AddConsistencyScript()*. The first one must specify both the behaviours that are connected by the consistency functions relations – algorithms and/or equations – inside the same type. While the second one must specify the relations, this time the script(s), that consistently relates the specimen_in to the specimen_out. Finally, the user can Add the modified using the *merge* action.
- The MBM interacts with the Workspace_MBM, as said he/she has full authority and can also remove with the Remove() function an MDB from a stored DMT.

15.6 MDB Management Tool Specification

Each MDB composes the knowledge base for the creation of a digital universe where all the analyses must keep consistent; therefore, a management tool is needed to check those consistencies when required by the user in a workspace, or before the affected MDB is stored in the relative DMT. Each workspace is interfaced – through a handling system – with the MDB management tool (Figure 15.1c) and can use different methods offered by the workspace depending on its access profile in Figure 15.1b. This section illustrates more in detail the UML specification for an MDB management tool compliant with the approach proposed in Figure 15.1c. At large, when the MDB management interface receives a request from a workspace, it can only order an MDB export or import from/to a specific DMT, dealing with those requests as explained in the following.

- An export request can be of two types, *Get()* and *Remove()* are used respectively to use a specific existent MDB in a local workspace and eliminate an MDB indicating both the MDB_ID and the DMT_ID. These two functionalities are inherited by the DMT profile class connected to the interface.
- An import request is made when a user wants to include an MDB in a specific DMT. The user can import an MDB created from scratch or he/she can upload a new version of an MDB existent in the same DMT. In both cases, such a request is more complex and requires the use of a sequence of operations that involve the other tools as follows:
    - When the user wants to import a new MDB the method *New()* is called inserting the DMT_ID of the destination and the created MDB; the MDB management tool calls first the versioning tool that returns the MDB_ID through the Version_New_MDB method, then the Check() method is called for safety reasons; finally, the MDB is added in the DMT through the Add method.
    - In case the MDB is a modified version of an existing one, the previous Add() method allows obtaining a new MDB version. The versioning tool will take care of the list of original MDBs in the method Version_Mod_MDB to confer an MDB_ID; then, the MDB is added in a DMT.

In both import cases, when the Check function checks all the consistency as prescribed (Figure 15.1c) and generates the new MDB, the interface generates an alert – through the Alert_WS() function – to all the workspaces that work on an older version of that MDB.

Without entering the details of the UML specification for the version control tool, we specify again that the DM tool is not oriented – at least at this development stage – to manage efficiently the work of the different users. The groups that work on the same MDB should work synchronously on it or use parallel MDBs to be merged at a later stage by higher user profiles. The final objective of the DM approach is to build a consistent and updated knowledge base, common to all the various professionals that work on the same asset.

15.7 Concluding Remarks and Future Work

This chapter illustrates a possible UML specification of the DM tool to prove the feasibility of its realisation. Although the specification is still in its early stages, the UML sketches provide the basis to understand the roles, functionality and potentiality of the tool. Future work will be devoted to the specification of each profile use case into specific Sequence Diagrams to evaluate the internal dynamics and the communications among all the classes.

Regarding tool usage, it is possible to develop models – virtual twins of a real system – that change through time. Each physical system can be accompanied by its MDB – e.g. machine or component directly from the manufacturer – which can be used in different multiverses of a larger system to collect the whole history in that system, useful as end-of-life information to be used in other similar systems.

References

Bashir, R. S., Lee, S. P., Khan, S. U. R., Chang, V., & Farid, S. (2016). UML models consistency management: Guidelines for software quality manager. *International Journal of Information Management, 36*(6), 883–899. https://doi.org/10.1016/j.ijinfomgt.2016.05.024.

Boyes, H., & Watson, T. (2022). Computers in industry digital twins: An analysis framework and open issues. *Computers in Industry, 143*(February), 103763. https://doi.org/10.1016/j.compind.2022.103763.

Cimino, C., Ferretti, G., & Leva, A. (2021). Harmonising and integrating the Digital Twins multiverse: A paradigm and a toolset proposal. *Computers in Industry, 132*, 103501. https://doi.org/10.1016/j.compind.2021.103501.

Davila Delgado, J. M., & Oyedele, L. (2021). Digital Twins for the built environment: learning from conceptual and process models in manufacturing. *Advanced Engineering Informatics, 49*(May), 101332. https://doi.org/10.1016/j.aei.2021.101332.

Dinter, R., Tekinerdogan, B., & Catal, C. (2022). Predictive maintenance using digital twins : A systematic literature review. *Information and Software Technology, 151*(February), 107008. https://doi.org/10.1016/j.infsof.2022.107008.

Feng, G., Cui, D., Wang, C., & Yu, J. (2009). Integrated data management in complex product collaborative design. *Computers in Industry, 60*(1), 48–63. https://doi.org/10.1016/j.compind.2008.09.006.

Leng, J., Wang, D., Shen, W., Li, X., Liu, Q., & Chen, X. (2021). Digital twins-based smart manufacturing system design in Industry 4.0: A review. *Journal of Manufacturing Systems, 60*(May), 119–137. https://doi.org/10.1016/j.jmsy.2021.05.011.

Lo, C. K., Chen, C. H., & Zhong, R. Y. (2021). A review of digital twin in product design and development. *Advanced Engineering Informatics, 48*(July 2020): 101297. https://doi.org/10.1016/j.aei.2021.101297.

Oztemel, E., & Gursev, S. (2020). Literature review of Industry 4.0 and related technologies. *Journal of Intelligent Manufacturing*, *31*(1), 127–182. https://doi.org/10.1007/s10845-018-1433-8.

Scaglioni, B., & Ferretti, G. (2018). Towards digital twins through object-oriented modelling: a machine tool case study. *IFAC-PapersOnLine*, *51*(2), 613–618. https://doi.org/10.1016/j.ifacol.2018.03.104.

Tao, F., Cheng, J., Qi, Q., Zhang, M., Zhang, H., & Sui, F. (2018a). Digital twin-driven product design, manufacturing and service with big data. *International Journal of Advanced Manufacturing Technology*, *94*(9–12), 3563–3576. https://doi.org/10.1007/s00170-017-0233-1.

Tao, F., Zhang, M., Liu, Y., & Nee, A. Y. C. (2018b). Digital twin driven prognostics and health management for complex equipment. *CIRP Annals*, *67*(1), 169–172. https://doi.org/10.1016/j.cirp.2018.04.055.

Tao, F., Zhang, H., Liu, A., & Nee, A. Y. C. (2019). Digital Twin in Industry: State-of-the-Art. *IEEE Transactions on Industrial Informatics*, *15*(4), 2405–2415. https://doi.org/10.1109/TII.2018.2873186.

16

A Service Design and Systems Thinking Approach to Enabling New Value Propositions in Digital Twins with AI Technologies

Shaun West
Lucerne University of Applied Sciences and Arts

Cecilia Lee
Royal College of Art

Utpal Mangla
IBM Cloud Platform

Atul Gupta
Merative

16.1 Introduction

Our intention is to guide readers through widely accepted and used service design approaches to help build a better understanding of how smart service providers could integrate the human side into the development of digital twins.

With the advancement of emerging technologies, such as artificial intelligence, digital twins have also attracted much interest from industry and academia. The term 'digital twin' has been around since it was introduced by Grieves (2014). The term is still loosely used, resulting in varying definitions and interpretations, especially within academic literature (Tao et al., 2019). A digital twin is a virtual representation of a physical system that articulates the physical processes, which mirrors the system's static and dynamic attributes (Batty, 2018; Löcklin et al., 2021). As the purpose of the digital twin is to mirror the physical system to identify areas within a system that may require improvements, most studies have examined the technical assets of the system. As emerging technologies, such as AI, the Internet of Things (IoT), and

DOI: 10.1201/9781003425724-20

big data, penetrate at a phenomenal rate, the adoption of digital twins has rapidly expanded to sectors such as energy, healthcare, and agriculture.

A service system represents multiple actors, such as humans, technologies, organizations, and information designed to deliver value for the actors within the system (Medina-Borja, 2015; Maglio & Spohrer, 2008). The emergence of AI, the IoT, and big data means that technology actors in a service system can now autonomously interact with other technology actors and human actors in a service system to co-create value. The actors' ability to learn and adapt instills smartness and creates a smart service system.

A smart service system shows the complex nature of the interactions between actors that are enabled by emerging technologies. Recent studies address the importance of integrating soft elements, such as human actors' needs, desires, cultural beliefs, and behavior, to build a deeper understanding of interaction dynamics for the system's sustainability (Medina-Borja, 2015). The physical processes defined in a smart service system involve human actors who interact with technology actors to co-create value and realize value-in-use. However, most studies of digital twins have focused on the technical side of the processes and system, undermining the human side. A digital twin that simulates a smart service system should represent both the technical and human sides of the system.

Although Shostack first introduced service design in the 1980s, it has only recently gained traction with industry practitioners and academics. In current literature, service design is described as a human-centered approach to operationalize highly abstract conceptual thinking (Vargo & Lusch, 2019, cited by Meierhofer et al., 2020). Although previous studies that have used service design to explore the dynamics of service systems in a service-dominant context do not explicitly introduce it as an enabler for the operationalization of SDL, it is evident that service design in these studies was used as a bridge between a high-order concept, SDL, and an empirical observation that takes place at micro-level (e.g., Yu and Sangiorgi, 2018; Sudbury-Riley et al., 2020).

16.2 Smart Service System

Previous studies describe a service system as the configurations of people, technologies, organizations, and information designed to deliver services that create value for the beneficiaries (Medina-Borja, 2015; Maglio & Spohrer, 2008). Technology actors are given material agency (Leonardi, 2011) enabled by AI in a service system. This allows them to autonomously interact with humans and other technology actors and deliver smart services via intelligent objects (Wunderlich et al., 2015). The technology actors with material agency have instilled smartness into a service system, as technology actors' newly acquired skills – self-learning and self-adaptability – enable a service system to evolve continuously and respond to a constantly changing environment. The concept of a smart service system is still relatively new to industry and academia. Previous service science SDL research introduced many competing definitions for a smart service system. Although there is yet to be a definition widely accepted by both industry and academia, most previous studies agree that a smart service system is capable of self-learning, self-adapting, self-replicating, and decision-making (Beverungen et al., 2017; Medina-Borja, 2015). West, Gaiardelli, and Rapaccini (2018) used the lens of SDL to create a framework to assess digitally enabled solutions for 'Smartness' by considering the service ecosystem, service platform, and the opportunity offered for value co-creation (Table 16.1). This framework helps operationalize the SDL, as shown in Table 16.1, and assesses smartness broader than the smart, connected products approach described by Porter and Heppelmann (2014). In that, the emphasis was on the technology and products rather than on value co-creation that could be made possible (Frost, Cheng, & Lyons, 2019).

In a smart service system, technology is no longer considered only an operand resource, acted upon for the value co-creation process in a system. Instead, technology actors have now taken a new additional role as an operant resource that human actors use to co-create value (e.g., knowledge and skills). As technology has become an actor that can directly interact with users for the value co-creation process in smart service systems, the system can facilitate the communications between human to human, human

TABLE 16.1

SD-Logic Framework to Assess Smartness (West et al., 2018)

Major Dimensions	Key Characteristics	Score	
		1 (worst)	**5 (best)**
Service ecosystem	Flexibility & integrity	No flexibility, no additional integration	Open system built on a flexible and integrated architecture
	Shared view	Limited understanding today, no future view	Clear shared view, today and in the future
	Actor roles	Not defined	Multi-roles
	Architecture	Closed	Open and secure
Service platform	Modular structure	No	Highly modular, with third-party integration
	Rules of exchange	Poorly defined	Clearly defined
Value co-creation	Value creation between actors	One-way, single-actor	Two-way, multi-actor
	Interactions between diverse actors	Two actors	Multi-actor
	Accommodation of roles	Two roles only	Multi-roles/multi-actor
	Resource integration	Single resource	Integration of many resources

to machine, and machine to machine. From an ecosystem perspective, the viability of a smart service system depends on its ability to seamlessly facilitate the value co-creation process among the actors for their mutual benefit.

Although technologies with the material agency have become smarter and can autonomously interact with users based on the contextual data they can read, their intelligence is still far away from enabling them to fully serve the needs and expectations of today's demanding users. So, it has become more critical for smart service providers to gain deeper insights into the enablers of, and deterrents to, value co-creation between actors in a smart service system (Wieland et al., 2012). Previous studies in SDL that offer a service ecosystem perspective (e.g., Akaka, Vargo & Lusch, 2013; Storbacka et al., 2016) emphasize the importance of understanding the interaction between actors at a micro-, meso-, and macro-level and their interrelationships. However, this is not always easy. A previous study by Storbacka et al. (2016) first explored the interaction between actors at a micro-level to see how this interaction in a service system relates to meso- and macro-level interactions to enable value co-creation.

A service system can remain viable only when the actors in it can seamlessly facilitate value co-creation for mutual benefit. The users of smart services who benefit determine its value-in-use. This again emphasizes the importance of putting a user or human actor at the center of a (smart) service system operation (Maglio, 2015; Medina-Borja, 2015). This further encourages the integration of soft elements, such as human actors' behavior, needs, and desires, because technology alone cannot explain the emerging phenomena in a smart service system. A human-centered smart service system that articulates both the human actors' complex needs and desires and the involvement of the technology actors can offer more accurate and reliable insights into the dynamics of the interaction that takes place across all three system levels.

To realize a human-centered smart service system that generates mutual benefits for system actors, smart service providers may need to consider adopting a human-centered approach in building digital twins with AI technologies to enable new value propositions in a smart service system. Human-centered design will allow smart service providers to look closely at the micro-level interaction between human and technology actors and how it translates into the value co-creation process.

Recent studies in smart service systems have taken a human-centered approach to understanding how to design an adaptive manufacturing system (Peruzzini & Pellicciari, 2017) and how to design smart housing (Agee et al., 2021). Growing complexity often becomes a challenge to designing a human-centered smart service system. Some organizations have adopted service design to mitigate this challenge, as it offers an innovative approach to untangle the complexity of the smart service system and find new ways to enable the creation of value propositions.

16.2.1 Systems Thinking and Service Design

As the world economy has shifted from goods-dominant logic toward service-dominant logic (SDL), the perspective on innovation has also shifted from an output-based view toward value co-creation (Sudbury-Riley et al., 2020). The emergence of an SDL economy has also contributed to the rapid traction of service design by both design practitioners in industry and design researchers and academics.

A heightened emphasis on the outcome of value co-creation as an emergent approach to innovating has further elevated the role of design in industry. With the development of SDL economy, service design has emerged as a human-centered approach exploring opportunities to recon Figure 16.1 resources and enhance innovation through value co-creation (Sun & Runcie, 2017; Andreassen et al., 2016). Most recent studies in service design describe it as a problem-solving process that can explore complex problems in a messy world, allowing actors in a service system to collaboratively identify potential solutions (e.g., Salgado et al., 2017). The service design process is iterative and incorporates a reflection-in-action approach (Schön, 1983). This iterative and adaptive process offers a robust approach to examining emerging challenges in an SDL economy in which value-in-use continuously evolves with the context. Also, more importantly, service design helps smart service providers to understand problems from a user's perspective rather than from a service provider's perspective, reinforcing the notion of value-in-use.

The first step toward leveraging a systems thinking-oriented service design process introduces a service ecosystem framework that can help smart service providers identify relevant actors at each system level and build a holistic view of how their interactions will affect one another.

The SDL research community and service design community widely use the Service Ecosystem Framework in Figure 16.1. It emphasizes inter- and intra-linkage of the value co-creation process across each level of service ecosystem. It represents a high-level view of how value co-creation occurs in a service system and contributes to its viability. The logic of system operation is also directly applied to the context of a smart service system operation. For example, service designers from the smart healthcare services sector can use the framework to build a shared understanding of the interrelationship between different service system layers. This exercise will give designers a holistic view of the enablers and deterrents to value co-creation, to develop relevant design interventions where necessary.

A holistic understanding of the interactions between actors within a system would give smart service providers the foundational knowledge of the key actors and the dynamics of the interactions. Based on this, smart service providers can use service design tools to generate deeper insights into human actors' motivations, needs, and desires that shape their interaction patterns with others. This process can be guided by the Double Diamond

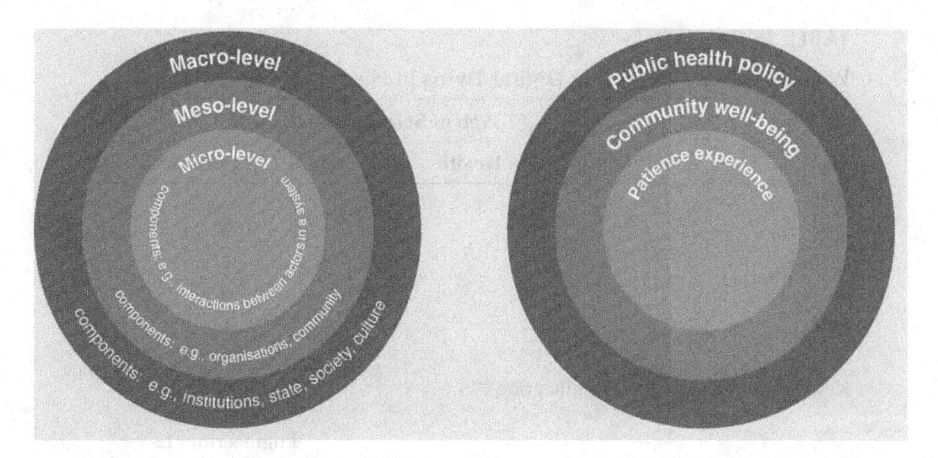

FIGURE 16.1
Service ecosystem framework. (*Source:* Artwork by Stefanie Fink, Adapted from Akaka, Vargo, & Lusch, 2013.)

Design Process introduced by the British Design Council in the early 2000s (Ball, 2019).

The Double Diamond Design Process is iterative and deeply embedded in Schon's reflection-in-action approach. During the discovery stage in the first diamond, smart service providers may use research methods such as interviews and shadowing to gather contextual data about human actors. They then broadly describe how human actors' desires, needs, and expectations shaped their interaction with other actors within a system for value co-creation. This process will help narrow the problem space and show where a digital twin can offer benefits. Once the problem has been defined, smart service providers can leverage insights on human-to-machine interaction at the micro-level of the system to develop human-centered digital twin prototypes for simulation testing. At this stage, most smart service providers are expected to go through a steep learning curve to iterate continuously, to ensure their solutions target the problems at hand.

Although human-centered digital twins help smart service providers expand their view on smart service systems, integrating the human side into digital twins requires carefully assessing the ethical implications.

16.3 The Human and Its Digital Twin – *The Digital Human Body*

When we consider the digital human body, we quickly move from the technology toward the value proposition and then to ethical issues. These issues are much more profound and fundamental than data ownership.

TABLE 16.2
Web of Science's Interest in Digital Twins in Health

	Web of Science Search Code	
Year	**Digital Twin Health**	**Digital Twin Ethics Health**
2022	114	3
2021	111	3
2020	69	1
2019	42	0
2018	12	1
2017	3	0
Key disciplines	Engineering /IS	Medical Social sciences Engineering/IS

Table 16.2 confirms that the emergence of digital twins within health has been growing, based on published paper counts, since 2017. The three searches confirm that the application of digital twins has been based on technology and the consideration of ethics within the health environment needs to catch up.

16.3.1 Digital Human Body and Possible Value Propositions Enabled by Digital Twins

In effect, the human body digital twin is an intelligent avatar of a digital human body (Figure 16.2) capable of sending and receiving data to and from the human body. In the service context, once this relationship between the physical and digital human body is established it can, for example, evolve a new dimension to preventative care (i.e., a new value proposition).

Preventative care in the digital twin world can provide guidance ahead of symptoms and body reflexes. The collected data will work with SDOH (Social Determinants of Health) and evolving AI/ML algorithms to deliver the digital twin-enabled value proposition.

For instance, an intelligent avatar (digital twin) can predict a 'severe headache for the physical human body' in the next few hours based on historical and real-time human data analysis and IoT devices capable of measuring vital signs. The digital twin can guide the human to either go and take a half-hour power nap, drink water, or walk outdoors to relieve stress, minimizing the need for traditional pain medication. Minimizing medication is not only more sustainable for the environment but also will have positive impacts on human health.

A digital twin can gather intelligence via the environment, human behavior, IoT, medical equipment, and healthcare facilities and bring all the data feeds into a unified environment to enable AI services and downstream capabilities. This will benefit patients who receive guidance, advice, and

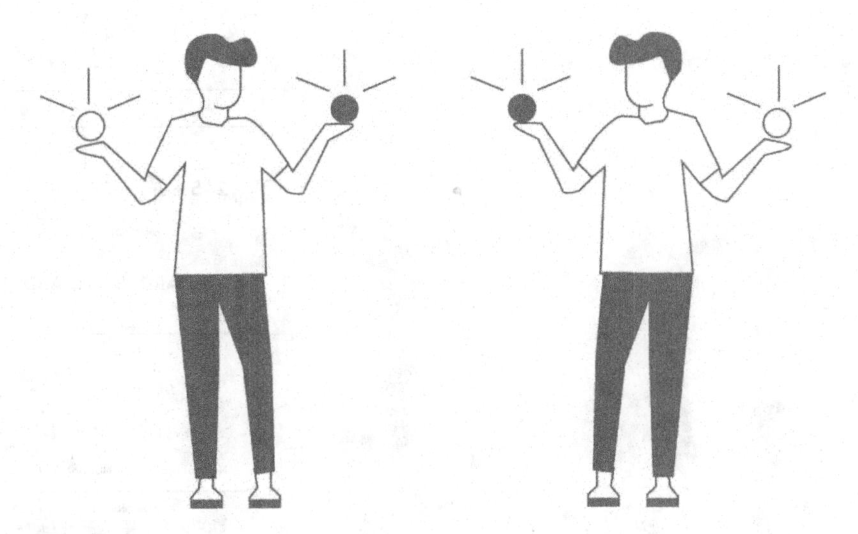

FIGURE 16.2
Intelligent avatar of the human body. (*Source:* Artwork by Stefanie Fink.)

future considerations for their physical and mental health issues. The areas of preventive care, precision care, and personalized patient treatment plans can all be enabled using digital twins.

Moving outside the purely medical environment, we can consider the digital twin within healthcare insurance. Already, aggregated data are being used for actuarial calculations that support setting insurance premiums and allocating resources based on actual and forecast demands. With the technology enabled by a digital twin, we can suddenly imagine new value propositions that help us stay healthy and reduce our insurance costs. Conversely, we can imagine another world where our behavior is controlled by 'Big Brother,' and we are penalized for 'bad' behaviors.

For additional insights, we have considered the application of digital twins to improve athletic performance. Sports provide a complex system of systems that allows for the integration of many digital twins as we have the participant, their equipment, the team, the environment where the team is playing, and the opposition, not forgetting seasonal aspects. Suddenly, we have a system of systems; some aspects are measured directly, whereas others are indirectly observed. Sport offers an extreme case where participants freely participate, providing an environment where more can be willingly tried and tested.

16.3.2 The Context of Sports

Sports are closely aligned to healthcare, as an athlete's performance is a function of personal health. Some firms are using scanning systems to build a digital human with the exact measurements of the body (Figure 16.3). In this case, such a system allows the exact matching of feet with shoes. Apple has gone one step further with its Fitness and Health apps in the

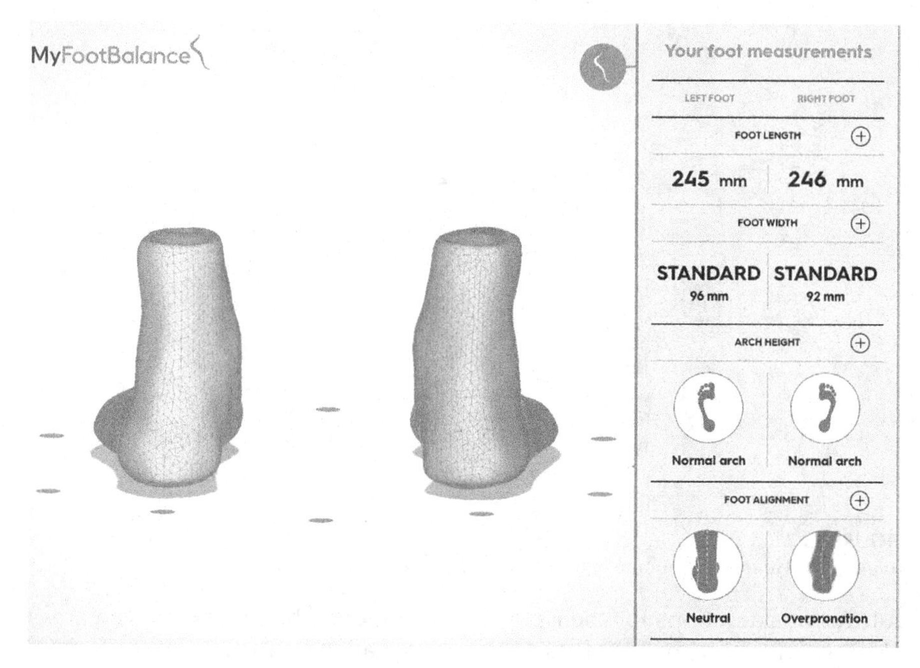

FIGURE 16.3
One of the authors' digital feet.

consumer environment. Moving into the professional athlete's world, monitoring is critical to performance in competition. Consider Formula 1 without telematics systems that take data from both the car and the driver to support tactical and operational decision-making. The decision-making is supported by an underlying model, which is, in effect, a set of digital twins that capture the current status and forecast possible future options. What is clear is that the market is undergoing rapid rates of innovation as sensors and AI get cheaper. Examples include the Carv app and ski boot sensors to 'coach' skiers (getcarv.com).

The intelligent avatar is being built from the data collected from individual athletes. Currently, much of this is fragmented, with islands of specialisms that are weakly connected (e.g., CAD-like feet or full-body scans can be made, or weight collection). Nevertheless, these can be built upon, and slowly digital twin technology with its AI will be better able to provide insight into actual and predicted performance. This can move into closely associated areas such as nutrition and health.

16.3.2.1 Perspective 1: Individual Athlete to Many Athletes

Today, your watch tells you to 'move'; in the near future, it can tell you what to eat and drink or how to move to achieve athletic performance. This can be very helpful for individual athletes to improve their personal performance. Again, the digital twin can create institutional structures for value co-creation.

With a small jump, health information gathering will be possible from consumer-grade products with the mass collection. As the data set grows and the data capture duration lengthens, the AI models underpinning them will improve to a point where individual segmentation will become possible, raising major ethical questions in addition to the value co-created with the athlete.

The value co-creation here is mainly individual but can be multi-actor at the collective level as new insights can be captured with AI and fed back into the digital twin, providing more/better advice. Ethically, this is simpler as it is based on comparative performance. Nevertheless, some of the information can end up with 'dual uses,' and it is important to allow control of the use of the insights.

16.3.2.2 Perspective 2: Human and Machine System

The most basic question is, 'will the shoes fit?' Until now, we have considered the individual athlete. However, the digital twin allows us to replicate the system. By expanding the system to include the equipment, we link our digital twin with the equipment digital twin. Once linked, the athlete's performance using the equipment can be compared, and improvements recommended. This also allows the equipment designers to understand the long-term human/machine performance better.

The actors here can be limited to those in perspective 1 and additionally the designer. They should all, at some point, be beneficiaries of the co-creation web that is created. Ethically, it remains like perspective 1.

16.3.2.3 Perspective 3: The Coach's Perspective

The coach wants to know how the individuals and the team will perform; therefore, a 'family' of digital twins must be created to support the team, based on each individual's performance data (a health proxy). To be valid, there must be both a sufficiently detailed underlying model and data that is up to date to support decision-making. Integrating third-party data, such as weather, into the individual digital twins further improves the team selections (i.e., decision-making), as some athletes are better in cooler or wet weather. In contrast, others are better in hot, dry conditions. Further expansion of data integration comes from performance data on the other athletes. Integration of publicly available, albeit with poorer granularity and reliability, nevertheless supports the decision-making that the digital twin can provide.

Integrating the coach, the team, and the individual athlete creates different actors, yet here the coach and team are the direct beneficiaries. This is where zooming in/out is helpful. In many sports, the team is key and the individual secondary, so the predicted optimal team performance may be detrimental to individuals because they may not be selected for a particular event.

16.3.3 The Context of Healthcare

The intersection of digital twins and the healthcare industry is well-timed with the recent rise of data usage within organizations and with external partners. Typically, most healthcare data are very siloed, making it challenging for AI and ML algorithms to use it collaboratively. Further, data privacy, trust, and accountability challenges enforce essential blocks to the use of collected data besides vetting the algorithms, tools, and techniques. Capabilities in the healthcare business have matured and are developing quickly in many areas, often solving complex enterprise, business, regulatory, and compliance requirements.

The value proposition for digital twins in healthcare is not only to generate a digital equivalent of a human body at the cellular and organ levels but also to collaborate with incoming real-time feeds from medical and surgical equipment. A step forward in this would be to implement IoT devices to monitor and feed data into a digital twin equivalent of a human body (Figure 16.4). Typically, the human body is not only generating reflexes, observation, and experience data. The reflexes are because the human body is in a particular environment, physical location, temperature, humidity, etc. IoTs can fill this gap and bring in this wealth of data. A digital twin with real-time bi-directional data feeds can provide AI predictions and guidance to the human body ahead of actual events. The challenges for the digital twin ecosystem will be gathering the data feeds, as the data is spread across many organizations and healthcare providers.

A digital twin for healthcare will need the 'service context' more than ever. To put this in perspective, health imaging data is very costly and time intensive. Running AI analysis and deriving inferences at a large scale with big-data initiatives are mostly not in time for end-user consumption. A single static image or photograph has an average size of 5 MB; a genetic profile for a single cancer patient can amount to about 1 PB. Collecting this huge dataset and running an analysis on it are not sustainable, with high costs and delayed results. Instead, in the *service context*, a digital twin can feed real-time data to AI services and inform healthcare professionals only about the inferences that are necessary. Additionally, continuous data feeds to AI services using a digital twin and sending inferences to actors will bring new dimensions to healthcare. With responsible AI and respectful ethics, privacy, and trust standards, healthcare professionals can use digital twins in real time for diagnosis, preventative care, and research activities.

Privacy concerns may prohibit all the necessary information from being fed to the digital twin ecosystem. For example, location data give the patient's coordinates to feed social data to the digital twin, but privacy may not allow

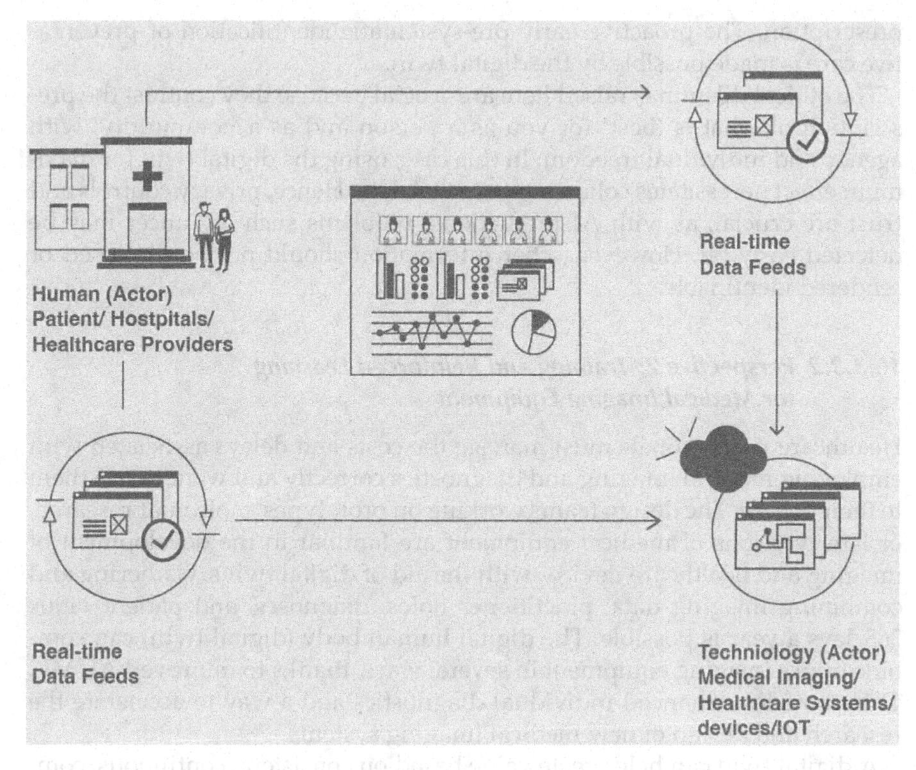

FIGURE 16.4
Building the digital human in healthcare. (*Source:* Artwork by Stefanie Fink.)

this. In the service context, this can be made possible in a federated ecosystem to let the data stay where it is and only use inferences for AI services. Trust from healthcare professionals can be developed over time as the consumption of AI services and inferences increases and matures. Digital twins for healthcare will need an overarching governance layer to ensure trust and build people's confidence.

16.3.3.1 Perspective 1: Preventative Care and Minimizing Medicine Use

Traditional medical care and treatment are prescriptive and begin with a patient conversation, fact-finding, and an examination of the symptoms to establish the diagnosis. The digital twin of the human body can give a steady stream of the patient's pertinent bodily data, input/output data, social characteristics, and linked data to medical professionals. The patient should never enter the symptomatic stage for preventable conditions if everything is set up correctly, monitored, and predicted based on historical and existing vital signs. By continuously monitoring the subject, the digital twin might also be able to target testing (such as NMR, endoscopy, and X-rays) or drug prescription. The proactive early pre-systematic identification of preventative care is made possible by the digital twin.

The ethical dilemmas raised here are crucial because they contrast the prescription of what is 'best' for you as a person and as a 'community' with agency and individual freedom. In this case, using the digital twin for maximum effect necessitates collecting personal data. Hence, privacy controls and trust are crucial, as with AI predictions, problems such as cancer may be detected early on. However, other information should not be disclosed or rendered identifiable.

16.3.3.2 Perspective 2: Training and Reinforced Learning for Medical Imaging Equipment

Healthcare professionals must manage the costs and delays associated with employing medical imaging and diagnostics correctly and want to use them to their benefit. The design teams working on prototypes, molecular research, or later versions of medical equipment are familiar to the development of imaging and healthcare devices with the aid of digital twins. Gathering and combining imaging data, practitioner notes, diagnoses, and patient vitals 365 days a year is possible. The digital human body (digital twin) can combine or use imaging equipment in several ways, thanks to improved AI/ML. This provides enhanced individual diagnostics and a way to accelerate the research and design of new medical imaging systems.

A digital twin can help create value based on consistent, continuous, complete, accurate, timely, and reliable data from various sources. It is clear that better imaging technology is required, and integration with additional data sources could lead to the development of novel, minimally invasive health treatments. In terms of ethics, we need to understand how this may pose hazards to individual data. Transparency regarding data collection, handling, and storage techniques is essential for maintaining the AI's ethics, which will effectively manage public expectations and trust.

16.3.3.3 Perspective 3: The Context of Healthcare Insurance

Given its importance to the general public, healthcare insurance has always been pivotal to the healthcare industry. Financial projections, estimates, and margins often drive insurance and the human factor needs to be noticed, even though insurance firms have access to essential health data. Fraud and falsifications of claims have also abused traditional healthcare insurance. Human body digital twins can feed anonymized data to insurance providers, enabling them to see behind the claims and trends and run broad-scope AI-enabled analyses. To add to this data, SDOH parameters can provide another dimension to insurance analysis and give healthcare providers and patients feedback. Ethically, there are huge issues in providing insurance firms with access to medical data.

Insurance looks through a financial lens and needs to pay attention to the inconsistent, difficult-to-interpret, expensive, and time-consuming practitioner and patient data. Imagine a digital twin where insurance companies, medical equipment manufacturers, and healthcare providers collaborate to overlay the digital twin service environment. In such instances, the data produced can feed AI/ML algorithms, advising insurance companies on costs and determining the best course of action for specific patients. The expense of post-operative care and insurance for knee and hip surgery are two examples of a digital twin being used in the insurance setting.

This example shows how a smart healthcare service system's patients (or human players) can use a human-centered digital body to enable new value propositions and reciprocal advantages for all actors. Healthcare insurance service providers become more patient-centric and help to sustain smart healthcare service systems as they use the digital twin data of each of their customers to verify claims and offer more customized insurance plans to their clients.

The patient can lose agency in this situation, which raises ethical issues. The intricate mix of participants and benefits makes the problem more difficult. How can we stop patients from being instructed by the insurance company to receive a specific treatment that will only cut costs for the insurer, rather than being advised on which options are best for their long-term health by their doctor?

16.4 Closing

Service Design provides an approach to understanding the digital twin and how it can contribute to value co-creation. When we integrate the ecosystem and value co-creation aspects of SD Logic, a complete understanding of the digital twin can be created; we have provided two use cases that help to demonstrate this. Digital twins enable new value propositions. The underpinning technology is complex and demanding, yet the value co-creation process can be described from different perspectives. Within the value constellation, value destruction should be considered. Here, our proxy for value destruction was the ethical aspect. AI ethics can be leveraged to manage the ethical aspects of digital twins. The data collection, processing, and digital twins' tools and techniques will need to embed ethical governance. Digital twins enable us to design a 'smart service system' that represents the integration of human and technology actors and organizations that co-create value to deliver smart services. Using the lens of SD Logic, we have a framework that can help us to focus on the value co-creation aspects of smart services.

Recent studies in smart service systems emphasize the importance of integrating soft elements, such as human behavior, cultural beliefs, and human actors' needs and desires, better to understand the system's interaction dynamics and viability. However, in most digital twin studies, the role of human actors is ignored. Smart service system literature suggests it is important to articulate human–technology interaction in digital twins because the digital twin's physical processes often involve interaction between human actors and machines (and processes) for value co-creation. This observation suggests that when building digital twins, the knowledge of human behavior in human–technology interactions will help identify the processes that may work more effectively for human actors.

In a nutshell, the human-centered smart service system is a growing research area, and digital twins should be able to articulate the human side of the physical processes within a system. We need to consider system thinking and how the service design process can be used to develop a human-centered digital twin that explores both the technical and human sides of the system.

Acknowledgments

The authors would like to thank their organizations for supporting this work and Stefanie Fink (stefanie.fink@hslu.ch) for the fantastic artwork.

References

Agee, P., Gao, X., Paige, F., McCoy, A., and Kleiner, B. (2021). A human-centred approach to smart housing. *Building Research & Information*, 49(1), 84–99.

Akaka, M.A., Vargo, S.L., and Lusch, R.F. (2013). The complexity of context: A service ecosystems approach for international marketing. *Journal of International Marketing*, 21(4), 1–20.

Andreassen, T.W., Kristensson, P., Lervik-Olsen, L., Edvardsson, B, and Colurcio, M. (2016). Linking service design to value creation and service research. *Journal of Service Management*, 27(1), 21–29.

Ball, J. (2019). The Double Diamond: A universally accepted depiction of the design process. Retrieved from https://www.designcouncil.org.uk/our-work/news-opinion/double-diamond-universally-accepted-depiction-design-process/.

Batty, M. (2018). Digital twins. *Environment and Planning B: Urban Analytics an City Science*, 45(5), 817–820.

Beverungen, D., Muller, O., Matzner, M., Mendling, J., and vom Brocke, J. (2017). Conceptualising smart service systems. *Electronic Markets*, 29, 7–18. https://doi.org/10.1007/s12525-017-0270-5.

Frost, R. B., Cheng, M., and Lyons, K. (2019). A multilayer framework for service system analysis. Handbook of Service Science, Volume II, 2019: 285–306

Grieves, M. (2014). *Digital Twin: Manufacturing Excellence through Virtual Factory Replication. Technical Report*. White paper, 2014." *Online: 03–01*.

Leonardi, P.M. (2011). When flexible routines meet flexible technologies: Affordance, constraint, and the imbrication of human and material agencies. *MIS Quarterly*, 35(1), 147–167.

Löcklin, A., Jung, T., Jazdi, N., Ruppert, T., and Weyrich, M. (2021). Architecture of a human-digital twin as common interface for operator 4.0 applications. *Procedia CIRP*, 104, 458–463.

Maglio, P. (2015). Editorial: Smart service systems, human-centred service systems, and the mission of service science. *Service Science*, 7(2), 2–3, https://doi.org/10.1287/serv.2015.0100.

Maglio, P.P. and Spohrer, J. (2008). Fundamentals of service science. *Journal of the Academy of Marketing Science*, 36(1), 18–20.

Medina-Borja, A. (2015). Editorial: Smart things as service providers: A call for convergence of disciplines to build a research agenda for the service systems of the future. *Service Science*, 7(1), 1–5.

Meierhofer, J., West, S., Rapaccini, M., and Barbieri, C. (2020). The digital twin as a service enabler: From the service ecosystem to the simulation model. In: Nóvoa, H., Drăgoicea, M., Kühl, N. (eds) *Exploring service science. IESS* 2020. *Lecture Notes in Business Information Processing* (p. 377). Springer, Cham. https://doi.org/10.1007/978-3-030-38724-2_25.

Peruzzini, M. and Pellicciari, M. (2017). A framework to design a human-centred adaptive manufacturing system for aging workers. *Advanced Engineering Informatics*, 33(2017), 330–349.

Porter, M. E. and Heppelmann, J. E. (2014). How smart, connected products are transforming competition. *Harvard Business Review*, 92(11), 64–88.

Salgado, M., Wendland, M., Rodriguez, D., Bohren, M.A., Oladapo, O.T., Ojelade, O.A., Olalere, A.A., Luwangula, R., Mugerwa, K., and Fawole, B. (2017). Using a service design model to develop the "Passport to Safer Birth" in Nigeria and Uganda. *International Journal of Gynecology and Obstetrics*, 139(1), 56–66.

Schön, D. A. (1983). *The Reflective Practitioner: How Professionals Think in Action*. New York: Basic Books, 1983.

Storbacka, K., Brodie, R.J., Bohmann, T., Maglio, P.P., and Nenonen, Suvi (2016). Actor engagement as a microfoundation for value co-creation. *Journal of Business Research*, 69(2016), 3008–3017.

Sudbury-Riley, L., Hunter-Jones, P., Al-Abdin, A., Lewin, D., and Naraine, M.V. (2020). The trajectory touchpoint technique: A deep dive methodology for service innovation. *Journal of Service Research*, 23(2), 229–251.

Sun, Q. and Runcie, C. (2017). Is service design in demand? *Design Management Journal*, 11(1), 67–78.

Tao, F., Sui, F., Liu, A., Qi, Q., Zhang, M, Song, B., Guo, Z., Lu, S., and Nee, A.Y.C. (2019). Digital twin-driven product design framework. International Journal of Production Research, 2019, 57(12): 3935–3953.

Vargo, S.L. and Lusch, R.F. (2019). *The SAGE Handbook of Service-Dominant Logic*. SAGE, Los Angeles, CA.

West, S., Gaiardelli, P., and Rapaccini, M. (2018). Exploring technology-driven service innovation in manufacturing firms through the lens of Service Dominant logic. *Ifac-Papersonline*, 51(11), 1317–1322.

Wieland, H., Polese, F., Vargo, S.L., and Lusch, R.F. (2012). Toward a service (eco) systems perspective on value creation. *International Journal of Service Science, Management, Engineering, and Technology*, 3(3), 12–25.

Wunderlich, N., Heinonen, K., Ostrom, A.L., Patricio, L., Sousa, R., Voss, C. and Lemmink, J.G.A.M. (2015). "Futurising" smart service: Implications for service researchers and managers. *Journal of Services Marketing*, 29(6/7), 442–447.

Yu, E. and Sangiorgi, D. (2018). Service design as an approach to implement the value cocreation perspective in new service development. *Journal of Service Research*, 21(1), 40–58.

17

Tokenized Digital Twins for Society 5.0

Abdeljalil Beniiche and Martin Maier
Institut national de la recherche scientifique

17.1 Introduction

The Corona crisis has been acting like a mirror that makes us see the vulnerabilities and shortcomings of ourselves as individuals and our societies. Throughout human history, crises have created bifurcations where civilizations either regress (e.g., tribalism) or, alternatively, progress by raising their level of complexity through the integration of internal contradictions. In doing so, dichotomies are transcended and societies are rewired, giving rise to "the new normal."

In his critically acclaimed book "Social Physics" [1], a term originally coined by Auguste Comte, the founder of modern sociology, Alex Pentland argues that social interactions (e.g., social learning and social pressure) are the primary forces driving the evolution of *collective intelligence (CI)*. According to Pentland, CI emerges through shared learning of surrounding peers and harnessing the power of exposure to cause desirable behavior change and build communities. Further, he observes that most digital media are better at spreading information than spreading new habits due to the fact that they don't convey social signals, i.e., they are socially blind. However, electronic reminders are quite effective in reinforcing social norms learned through face-to-face interactions. He concludes that humans have more in common with bees than we like to admit and that future techno-social systems should scale up ancient decision-making processes we see in bees.

This conclusion is echoed by Max Borders through his concept of the *social singularity* that defines the point beyond which humanity will operate much like a hive mind (i.e., collective consciousness) [2]. Currently, two separate processes are racing forward in time: (1) the technological singularity: machines are getting smarter (e.g., machine learning and AI) and (2) the social singularity: humans are getting smarter. In fact, he argues that these two separate processes are two aspects of the same underlying process waiting to be woven together toward creating new human-centric industries,

DOI: 10.1201/9781003425724-21

where human labor will migrate into more deeply human spheres using the surpluses of the material abundance economy and the assistance from CI.

It's interesting to note that CI will also play an important role in the vision of future 6G mobile networks. In contrast to previous generations, 6G will be transformative and will revolutionize the wireless evolution from "connected things" to "connected intelligence" [3]. In fact, according to [4], 6G will play a significant role in advancing Nikola Tesla's prophecy that "when wireless is perfectly applied, the whole Earth will be converted into a huge brain." Toward this end, the authors of [4] argue that 6G will provide an ICT infrastructure that enables human users to perceive themselves as being surrounded by a huge artificial brain offering virtually zero-latency services, unlimited storage, and immense cognitive capabilities. Further, according to the world's first 6G white paper published by the 6Genesis Flagship Program (6GFP) in September 2019, 6G will become more human-centered than 5G, which primarily focused on industry verticals. The authors of [5] observed that the ongoing deployment of 5G cellular systems is exposing their inherent limitations compared to the original premise of 5G as an enabler for the Internet of Everything (IoE). They argue that 6G should not only explore more spectrum at high-frequency bands but, more importantly, converge driving technological trends, thereby ushering in the 6G post-smartphone era. Specifically, they claim that there will be novel driving applications behind 6G, including wireless brain–computer interaction and blockchain and distributed ledger technologies. Note that blockchains such as Ethereum are commonly referred to as the *word computer,* which may naturally lend itself to help realize Nikola Tesla's aforementioned prophecy of converting the whole Earth into a huge brain.

Recently, in [6], we introduced the *Internet of No Things* as an important stepping stone toward ushering in the 6G post-smartphone era, in which smartphones may not be needed anymore. We argued that while 5G was supposed to be about the IoE, to be transformative 6G might be just about the opposite of Everything, that is, Nothing or, more technically, No Things. The Internet of No Things offers all kinds of human-intended services without owning or carrying any type of computing or storage devices. It envisions Internet services to appear from the surrounding environment when needed and disappear when no needed. The transition from the current gadgets-based Internet to the Internet of No Things is divided into three phases: (1) bearables (e.g., smartphone), (2) wearables (e.g., Google and Levi's smart jacket, virtual reality head-mounted devices), and then finally, (3) nearables. Nearables denote nearby computing/storage technologies and service provisioning mechanisms that are intelligent enough to learn and react according to user context and history in order to provide user-intended services.

In this chapter, we inquire how this world computer may be put in practice in emerging *cyber-physical-social systems (CPSS)* as a prime example of future techno-social systems. We put a particular focus on advanced blockchain technologies such as tokenized digital twins and on-chaining oracles, which leverage also on human intelligence. Toward this end, we first introduce the evolution of mobile networks as well as the Internet toward the 6G vison. Next, we elaborate on the various blockchain technologies in technically greater detail. Further, we borrow ideas from the biological superorganism with brain-like cognitive abilities observed in colonies of social insects for realizing internal communications via feedback loops, whose integrity is essential to the welfare of Society 5.0, the next evolutionary step of Industry 4.0. In addition, we elaborate on the role of *tokenomics* in Society 5.0. We outline our proposed path to a human-centered society and explain purpose-driven tokens and token engineering in technically greater detail. Further, we introduce our CPSS-based bottom-up multilayer token engineering framework for Society 5.0. Specifically, we explain how our proposed framework can be applied to advancing human CI in the 6G era. Finally, we describe its experimental implementation and highlight some illustrative results, before concluding the chapter.

17.2 Blockchain and Distributed Ledger Technologies

The radical potential of blockchain technology has long spread outside the world of crypto into the hand of the general public. We've all heard through one way or another that it is most likely the most revolutionary technology that is presently available in any known market and that includes the real world as well as the digital space. Blockchain technology is principally behind the emergence of Bitcoin [7] and many other cryptocurrencies that are too numerous to mention [8]. A blockchain is essentially a distributed database of records (or public ledger) of all transactions or digital events that have been executed and shared among participating parties [9]. Each transaction in the public ledger is verified by consensus between the majority of the participants in the system. Once entered, information can never be erased. The blockchain contains a certain and verifiable record of every single transaction ever made. At the point when the block reaches a certain size, it is timestamped and linked to the previous block through a cryptographic hash, thereby forming a chain of timestamped blocks (hence the name blockchain).

Blockchain technology is being successfully applied in both financial and non-financial applications. It has the potential to reduce the role of one of the most important economic and regulatory actors in our society, the middleman [10,11]. Blockchain technology was initially linked to the decentralized cryptocurrency Bitcoin, as it is the main and first application of the network (known as Blockchain 1.0 [12]). However, there exist many other use cases and several hundred different applications besides Bitcoin that use blockchain technology as a platform such as Ethereum. The rise of Ethereum and smart contracts heralded Blockchain 2.0 [12].

As the hype of blockchain technology advanced, Blockchain 3.0 aims to popularize blockchain-based solutions by expanding the traditional sectors (finance, goods, transactions, etc.) to government, IoT, decentralized AI, supply chain management, smart energy, health, data management, and education [13,14]. Therefore, the applications of blockchain have evolved to much wider scopes. However, these new applications introduce new features to next-generation platforms, including key aspects such as platform interconnection or more advanced smart contracts that provide higher levels of transparency, while reducing bureaucracy with self-enforcing code. These new technologies, therefore, promise more decentralized and spontaneous coordination over the Internet between users who do not know or trust each other, often referred to as decentralized autonomous organizations (DAOs). A DAO exists as open-source, distributed software for executing smart contracts built within the Ethereum project. A DAO is like a decentralized organization, except that autonomous software agents (i.e., smart contracts) make the decisions, not humans. In a more decentralized setup, the governance rules automatically steer behavior with tokenized incentives and disincentives [15]. In such cases, programmable assets called tokens, managed by a special smart contract, act as governance rules to incentivize and steer a network of actors without centralized intermediaries [15]. Further, tokens issued by the DAO enable their respective holders to vote on matters about the development of the organization and decision-making. As a result, the decision-making process is automated and a consensus is reached among the participants.

The blockchain in fact is an enclosed system, where interactions are limited to the data available on it. Hence, it is a practical problem commonly referred to as the "oracle problem" that defines how real-world data can be transferred into/from the blockchain [16]. Toward this end, oracles (also known as data feeds) act as trusted third-party services that send and verify the external information and submit it to smart contracts to trigger state changes in the blockchain [17]. Oracles may not only relay information to the smart contracts but also send it to external resources. They are simply contracts on the blockchain for serving data requests by other contracts. Without oracles, smart contracts would have limited connectivity. Hence, they are vital for the blockchain ecosystem by broadening the operational scope of smart contracts.

17.3 From Industry 4.0 toward Society 5.0

17.3.1 Cyber-Physical-Social Systems (CPSS)

Smart factories under Industry 4.0 have several benefits such as optimal resource handling, but also imply minimum human intervention in manufacturing. When human beings are functionally integrated into a CPS at the social, cognitive, and physical levels, it becomes a so-called CPSS, whose members may engage in cyber-physical-social behaviors that eventually enable metahuman beings with diverse types of superhuman capabilities. CPSS belongs to the family of future techno-social systems that by design still require heavy involvement from humans at the network edge instead of automating them away. For a comprehensive survey of the state of the art of CPSS, we refer the interested reader to [18].

17.3.2 Industry 5.0

One of the most important paradigmatic transitions characterizing Industry 5.0 is the shift of focus from technology-driven progress to a thoroughly human-centric approach. An important prerequisite for Industry 5.0 is that technology serves people, rather than the other way around, by expanding the capabilities of workers (up-skilling and re-skilling) with innovative technological means such as VR/AR tools, mobile robots, and exoskeletons.

Currently, two visions emerge for Industry 5.0. The first one is human–robot co-working, where humans will focus on tasks requiring creativity and robots will do the rest. The second vision for Industry 5.0 is bioeconomy, i.e., a holistic approach toward the smart use of biological resources for industrial processes [19]. The bioeconomy has established itself worldwide as a mainstay for achieving a sustainable economy. Its success is based on our understanding of biological processes and principles that help revolutionize our economy dominated by fossil resources and create a suitable framework so that economy, ecology, and society are perceived as necessary single entities and not as rivals. More specifically, *biologization* will be the guiding principle of the bioeconomy. Biologization takes advantage of nature's efficiency for economic purposes— whether they be plants, animals, residues, or natural organisms. Almost every discipline shares promising interfaces with biology. In the long term, biologization will be just as significant as a cross-cutting approach as digitalization already is today. Biologization will pave the way for Industry 5.0 in the same way as digitalization triggered Industry 4.0. It is also obvious that the two trends—biologization and digitalization—will be mutually beneficial [20].

17.3.3 Society 5.0

The Industrial Revolution reduced the agricultural population from more than 90% to less than 5%. Similarly, the IT revolution reduced the manufacturing population from more than 70% to approximately 15%. The Intelligence Revolution of the 6G era will reduce the entire service population to less than 10%. Upon the question of where will people go and what will they do then, the author of [21] gives the following answer: Gaming! Not leisure, but scientific gaming in cyberspace. Artificial societies, computational experiments, and parallel execution—the so-called ACP approach—may form the scientific foundation, while CPSS platforms may be the enabling infrastructure for the emergence of intelligent industries. In the ACP approach, intelligent industries will build all kinds of artificial societies, organizations, and systems in order to perform different types of computational experiments and conduct numerous scientific games for analyzing, evaluating, and optimizing decision-making processes, as well as mastering skills and resources required for the completion of tasks in the shortest time with the least energy and cost through the parallel execution of and interaction between real and artificial dual entities, who we can play, work, and live with. Everything will have its parallel avatar or digital twin in cyberspace, such that we can conduct numerous scientific games before any major decision or operation. This new, yet unknown CPSS-enabled connected lifestyle and working environment will eventually lead to high satisfaction as well as enhanced capacity and efficiency. Further, the author of [21] foresees that the *Multiverse* or parallel universes based on Hugh Everett's many-worlds interpretation of quantum physics will become a reality in the age of complex space infrastructures with the emergence of intelligent industries, which calls for a new profession of scientific game engineers. However, he warns that the capability of CPSS to collect tremendous energy from the masses through crowdsourcing in cyberspace and then release it into the physical space can bring us both favorable and unfavorable consequences. Therefore, one of the critical research challenges is the human-centric construction of complex spaces based on CPSS.

Similar to Industry 4.0/5.0, Society 5.0 merges the physical space and cyberspace by leveraging ICT to its fullest and applying not only social robots and embodied AI but also emerging technologies such as ambient intelligence, VR/AR, and advanced human–computer interfaces. However, Society 5.0 counterbalances the commercial emphasis of Industry 4.0. If the Industry 4.0 paradigm is understood as focusing on the creation of the smart factory, Society 5.0 is geared toward creating the world's first *supersmart society*. More interestingly, according to [22], Society 5.0 also envisions a paradigm shift from conventional monetary to future *non-monetary economies* based on technologies that can measure activities toward human co-becoming that have no monetary value (to be explained in more detail shortly).

17.4 Token Economy: From Cognitive Psychology to Web3

The shift from conventional monetary economics to non-monetary tokenomics and the central role tokens play in blockchain-based ecosystems were analyzed recently in [23]. Decentralized blockchain technologies have been applied in a non-monetary context by exploiting a process known as tokenization in different value-based scenarios. The tokenization of an existing asset refers to the process of creating a tokenized digital twin for any physical object or financial asset. The resultant tokens are tradeable units that encapsulate value digitally. They can be used as incentives to coordinate actors in a given regulated ecosystem in order to achieve a desired outcome. According to [23], tokens have a disruptive potential to expand the concept of value beyond the economic realm by using them for reputation purposes or voting rights. Through tokenization, different types of digitized value can be exploited in an ecosystem of incentives by sharing the rewards and benefits among its stakeholders.

Tokens might be the killer application of blockchain networks and are recognized as one of the main driving forces behind the next-generation Internet referred to as the Web3 [15]. While the Web1 (read-only web) and Web2 (read-and-write web) enabled the knowledge economy and today's platform economy, respectively, the Web3 will enable the token economy where anyone's contribution is compensated with a token. The token economy enables completely new use cases, business models, and types of assets and access rights in a digital way that were economically not feasible before, thus enabling completely new use cases and value creation models. Note that the term token economy is far from novel. In cognitive psychology, it has been widely studied as a medium of exchange, and arguably more importantly, as a positive reinforcement method for establishing desirable human behavior, which in itself may be viewed as one kind of value creation. Unlike coins, however, which have been typically used only as a payment medium, tokens may serve a wide range of different non-monetary purposes. Such purpose-driven tokens are instrumental in incentivizing an autonomous group of individuals to collaborate and contribute to a common goal. According to [15], the exploration of tokens, in particular different types and roles, is still in the very early stages.

The token economy plays a central role in realizing the emerging DAO, which has become a hot topic spawned by the rapid development of blockchain technology in recent years [24]. The DAO may be viewed as a social system composed of intelligent agents coevolving into human–machine integration based on real-world and artificial blockchain systems. In the DAO, all the operational rules are recorded on the blockchain in the form of smart contracts. Token economy incentives together with distributed consensus protocols are utilized to realize the self-operation, self-governance, and self-evolution of the DAO. In fact, according to the authors of [24], the use of tokens as incentives is the main motivator for the DAO, whereby the so-called incentive mechanism layer of their presented multilayer DAO reference model will be key for the token design (to be discussed in more detail shortly).

17.5 The Path (DAO) to a Human-Centered Society

17.5.1 Purpose-Driven Tokens and Token Engineering

We are still in the very early stages of exploring different roles and types of tokens. For instance, so-called purpose-driven tokens incentivize individual behavior to contribute to a certain purpose or idea of a collective goal. This collective goal might be a public good or the reduction of negative externalities to a common good, e.g., reduction of CO_2 emissions. Purpose-driven tokens introduce a new form of public goods creation without requiring traditional intermediaries, e.g., governments. Blockchain networks such as Ethereum took the idea of collective value creation to the next level by providing a public infrastructure for creating an application token with only a few lines of smart contract code, whereby in principle any purpose can be incentivized. However, given that operational use cases are still limited, this new phenomenon of tech-driven public goods creation needs much more research and development [15].

Proof of work (PoW) is an essential mechanism for the maintenance of public goods. Even though the collective production of public goods can result in positive externalities, it does not necessarily exclude other negative externalities, e.g., energy-intense mining process of blockchains. When designing purpose-driven tokens as a means to provide public goods, behavioral economics methods, e.g., the well-known nudging technique and behavioral game theory, provide important tools to steer individuals toward certain actions. For further information about our recent work on playing a blockchain-enabled version of the widely studied trust game of behavioral economics with on-chaining oracles and persuasive social robots, we refer the interested reader to [25].

In the following, we focus on the important problem of token engineering, which is an emerging term defined as the theory, practice, and tools to analyze, design, and verify tokenized ecosystems [15]. It involves the design of a bottom-up token engineering framework along with the design of adequate mechanisms for addressing the issues of purpose-driven tokens. Note that mechanism design is a subfield of economics that deals with the question of how to incentivize everyone to contribute to a collective goal. It is also referred to as "reverse game theory" since it starts at the end of the game (i.e., its desirable output) and then goes backward when designing the (incentive) mechanism.

17.5.2 Proposed Token Engineering DAO Framework for Society 5.0

Recall from Section 17.4.2 that for the token design, a multilayer DAO reference model was proposed in [24], though it was intentionally kept generic without any specific relation to Society 5.0. The bottom-up architecture of the DAO reference model comprises the following five layers: (1) basic technology,

(2) governance operation, (3) incentive mechanism, (4) organization form, and (5) manifestation. Due to space constraints, we refer the interested reader to [24] for further information on the generic DAO reference model and a more detailed description of each layer. In the following, we adapt the generic DAO reference model to the specific requirements of Society 5.0 and highlight the modifications made in our CPSS-based bottom-up token engineering DAO framework.

Figure 17.1 depicts our proposed multilayer token engineering DAO framework for Society 5.0 that builds on top of state-of-the-art CPSS. While the Internet of Things as prime CPS example has ushered in Industry 4.0, advanced CPSS such as the future Internet of No Things, briefly mentioned

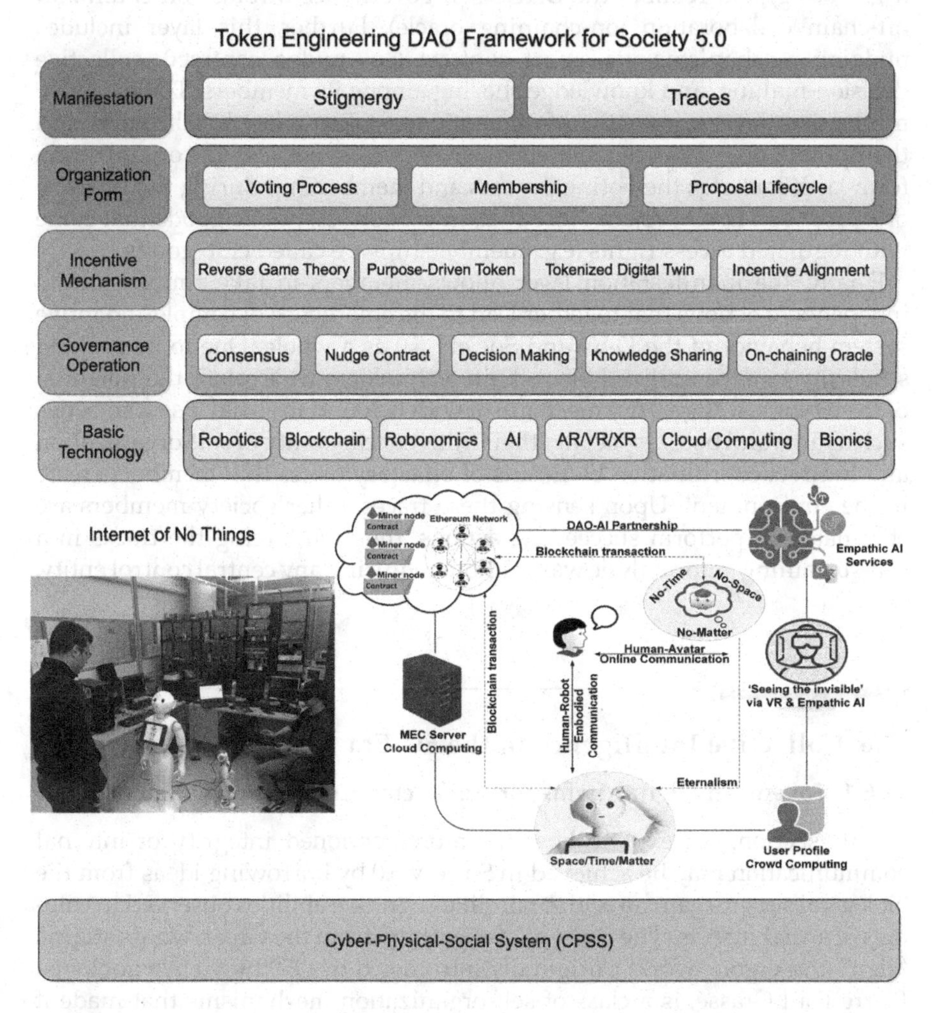

FIGURE 17.1
CPSS-based bottom-up token engineering DAO framework for Society 5.0.

in the introduction, will be instrumental in ushering in Society 5.0. As explained in more detail in [6], the Internet of No Things creates a converged service platform for the fusion of digital and real worlds that offers all kinds of human-intended services without owning or carrying any type of computing or storage devices. The basic technology layer at the bottom of Figure 17.1 illustrates the key enabling technologies (e.g., blockchain) underlying the Internet of No Things. In addition, this layer contains future technologies, most notably bionics, that are anticipated to play an increasingly important role in a future supersmart Society 5.0 (see also Figure 17.1).

Above the basic technology layer, there exists the governance operation layer. Generally speaking, this layer encodes consensus via smart contracts (e.g., voting) and realizes the DAO's self-governance through on-chain and off-chain collaboration (on-chaining oracle). Further, this layer includes nudging mechanisms via smart contract (i.e., nudge contract), collective decision-making, and knowledge sharing among its members. The incentive mechanism layer covers the aforementioned token-related techniques and their proper alignment to facilitate token engineering. Next, the organization form layer includes the voting process and membership during the lifecycle of a proposed DAO project. Note that, in economics, public goods that come with regulated access rights (e.g., membership) are called club goods.

Finally, the manifestation layer allows members to take simple, locally independent actions that together lead to the emergence of complex adaptive system behavior of the DAO and Society 5.0 as a whole. Due to its striking similarity to decentralized blockchain technology, we explore the potential of the biological stigmergy mechanism widely found in social insect societies such as ants and bees, especially their inherent capability of self-organization and indirect coordination by means of olfactory traces that members create in the environment. Upon sensing these traces, other society members are stimulated to perform succeeding actions, thus reinforcing the traces in a self-sustaining autocatalytic way without requiring any central control entity.

17.6 Collective Intelligence in the 6G Era

17.6.1 Tokenized Digital Twins for Advancing Collective Intelligence

In this section, we explore how the aforementioned integrity of internal communication may be achieved in Society 5.0 by borrowing ideas from the biological superorganism with brain-like cognitive abilities observed in colonies of social insects. The concept of *stigmergy* (from the Greek words stigma "sign" and ergon "work"), originally introduced in 1959 by French zoologist Pierre-Paul Grassé, is a class of self-organization mechanisms that made it possible to provide an elegant explanation to his paradoxical observations that in a social insect colony, individuals work as if they were alone while their collective activities appear to be coordinated. In stigmergy, traces are left by individuals in their environment that may feed back on them and thus incite their subsequent actions. The colony records its activity in the environment using various forms of storage and uses this record to organize and constrain collective behavior through a feedback loop, thereby giving rise to the concept of *indirect communication*. As a result, stigmergy maintains social cohesion by the coupling of environmental and social organization. Note that with respect to the evolution of social life, the route from solitary to social life might not be as complex as one may think. In fact, in the AI subfield of swarm intelligence, e.g., swarm robotics, stigmergy is widely recognized as one of the key concepts.

17.6.2 Illustrative Case Study: The Internet of No Things

In the following, we describe five bottom-up design steps of suitable purpose-driven tokens and mechanisms that are instrumental in converting our CPSS of choice, the Internet of No Things, into a stigmergy-enhanced Society 5.0, by using ESPN's online environment based on advanced Ethereum blockchain technologies and involving different types of ESPN's offline agents (social robot, embodied AI, human):

- *Step1: specify purpose*: The design of any tokenized ecosystem starts with a desirable output, i.e., its purpose. As discussed before, the goal of Society 5.0 is to provide the techno-social environment for CPSS members that (1) extends human capabilities and (2) measures activities toward human co-becoming supersmart. Toward this end, we advance AI to CI among swarms of connected human beings and things, as widely anticipated in the 6G era.
- *Step 2: select CPSS of choice*: We choose our recently proposed Internet of No Things as state-of-the-art CPSS, since its final transition phase involves nearables that help create intelligent environments for providing human-centered and user-intended services. In [6], we introduced our ESPN, which integrates ubiquitous and persuasive computing in nearables (e.g., social robot, virtual avatar) to change the behavior of human users through social influence. In this chapter, we focus on blockchain and robonomics as the two basic technologies to expand ESPN's online environment and offline agents, respectively.
- *Step 3: define PoW*: PoW is an essential mechanism for the maintenance of tech-driven public goods. Specifically, we are interested in creating club goods, whose regulated access rights avoid the well-known "tragedy of the commons." To regulate access, we

exploit the advanced blockchain technology of on-chaining oracles. On-chaining oracles are instrumental in bringing external off-chain information onto the blockchain in a trustworthy manner. The on-chained information may originate from human users. Hence, on-chaining oracles help tap into human intelligence [25]. As PoW, we define the oracles' contributions to the governance operation of the CPSS via decision-making and knowledge sharing, which are both instrumental in achieving the specific purpose of CI.

- *Step 4: design tokens with proper incentive alignment*: Most tokens lack proper incentive mechanism design. The use of tokens as incentives lie at the heart of the DAO and their investigation has started only recently. Importantly, recall that the tokenization process creates tokenized digital twins to coordinate actors and regulate an ecosystem for the pursuit of a desired outcome by including voting rights. The creation of a tokenized digital twin is done via a token contract that incentivizes our defined PoW, involving the following two steps: (1) create digital twin that represents a given asset in the physical or digital world and (2) create one or more tokens that assign access rights/permissions of the given physical/digital asset to the blockchain address of the token holder:
- *Step 5: facilitate indirect communication among DAO members via stigmergy and traces:* Finally, let the members participating in a given DAO project (1) record their purpose-driven token incentivized activities in ESPN's blockchain-enabled online environment and (2) use these blockchain transactions (e.g., deposits) as traces to steer the collective behavior toward higher levels of CI in a stigmergy-enhanced Society 5.0.

Figure 17.2 illustrates the functionality of each of these five steps in more detail, including their operational interactions.

17.6.3 Experimental Setup and Results

A general definition of human intelligence is the success rate of accomplishing tasks. In our implementation, human intelligence tasks (HIT) are realized by leveraging the image database ImageNet (www.image-net.org) widely used in deep learning research and tokenizing it. Specifically, humans are supposed to discover a hidden reward map consisting of purpose-driven tokens by means of image tagging, which is done by relying on the crowd intelligence of Amazon Mechanical Turk (MTurk) workers and the validation of their answers via a voting-based decision-making blockchain oracle. We measure CI as the ratio of discovered/rewarded number and the total number of purpose-driven tokens.

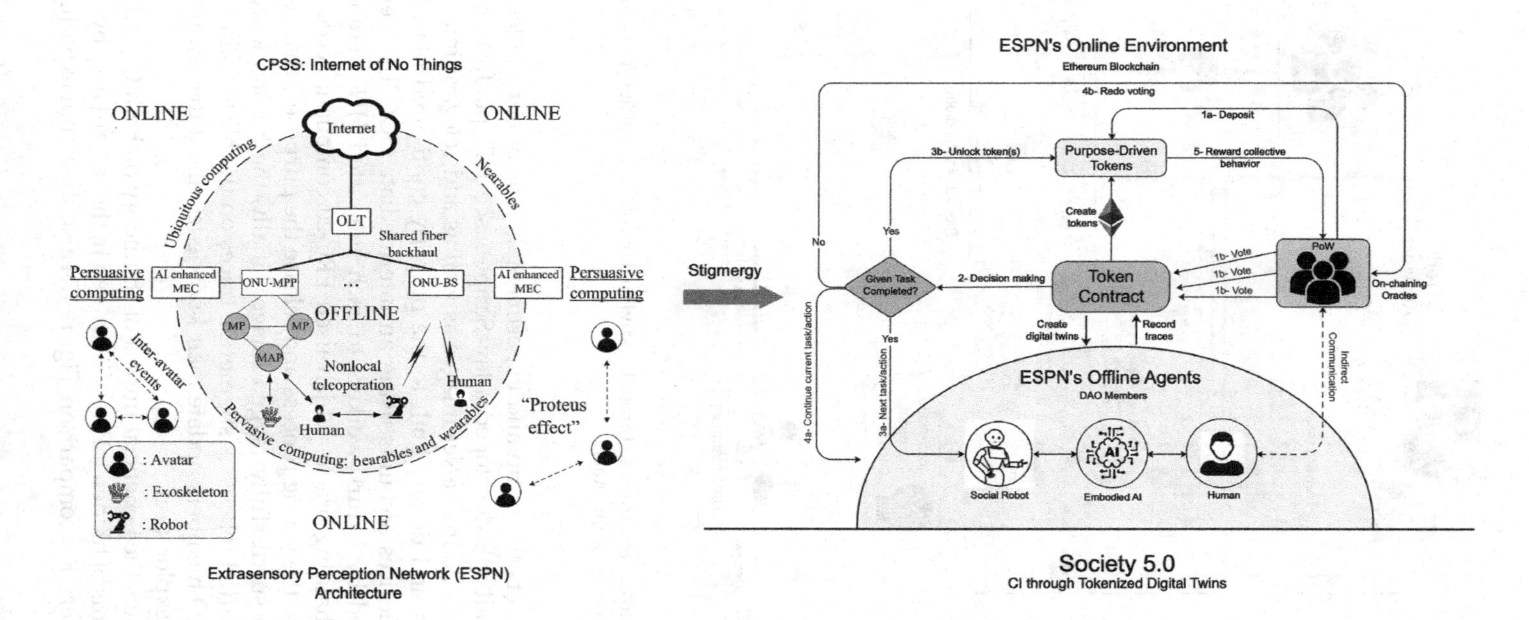

FIGURE 17.2
Stigmergy-enhanced Society 5.0 using tokenized digital twins for advancing collective intelligence in our CPSS of choice, the Internet of No Things, and underlying ESPN architecture.

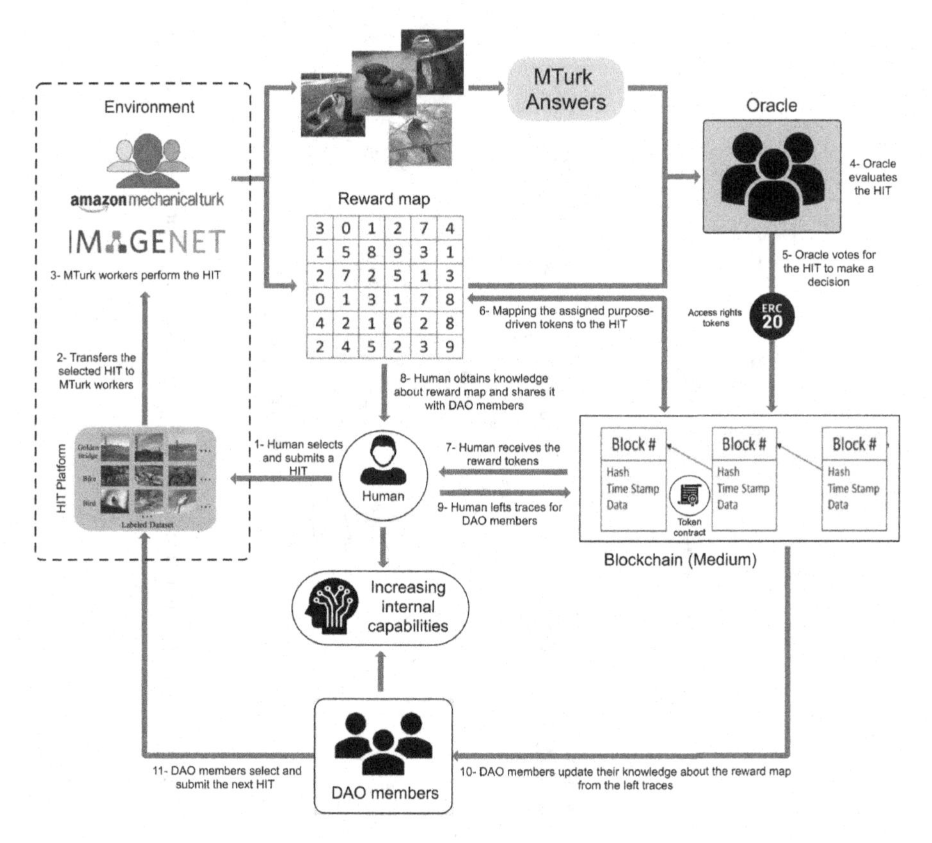

FIGURE 17.3
Discovery of hidden token reward map through individual or collective ImageNet tagging via Amazon MTurk and on-chaining oracle.

Figure 17.3 depicts the setup and experimental steps of our implementation in more detail. We developed a JavaScript-based HIT platform to let a human select from 20 ImageNet images as well as add relevant image tagging information and deliver both to the properly configured MTurk and Amazon Web Services accounts using an intermediate OOCSI server. The answers provided by MTurk workers to each submitted HIT were evaluated by an on-chaining oracle, which used ERC-20 compliant access right tokens to regulate the voting process and release the purpose-driven tokens assigned to each successfully tagged image. Finally, the human leaves the discovered/rewarded tokens as stigmergic traces on the blockchain to help participating DAO members update their knowledge about the reward map and continue its exploration.

Figure 17.4 shows the beneficial impact of stigmergy on both CI and internal reward in terms of hidden tokens discovered in the reward map by a DAO with eight members. For comparison, Figure 17.4 also shows our experimental

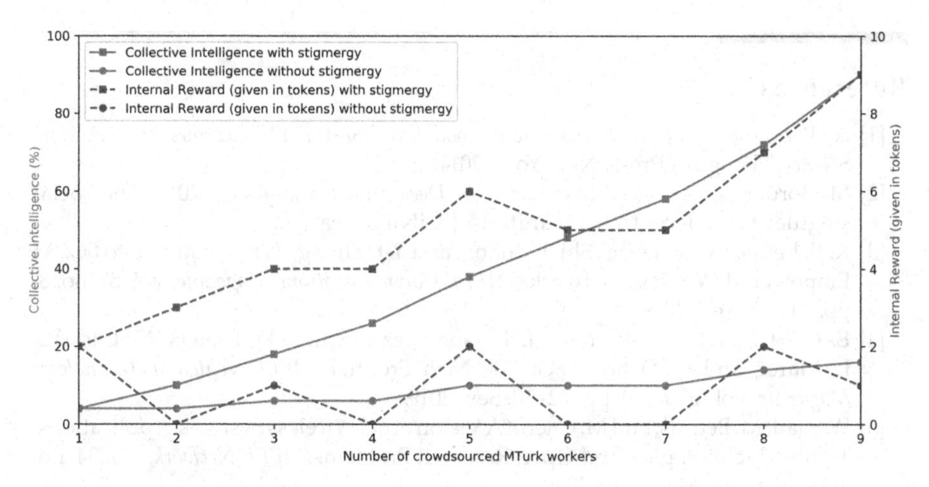

FIGURE 17.4
Collective intelligence (given in percent) and internal reward (given in tokens) with and without stigmergy vs. number of crowdsourced Amazon MTurk workers.

results without stigmergy, where the DAO members don't benefit from sharing knowledge about the unfolding reward map discovery process.

17.7 Conclusion

In this chapter, we have shed some light on the potential of advanced blockchain technologies—most notably, tokenized digital twins, purpose-driven tokens, and on-chaining oracles—to rewire Society 5.0 and advance its CI via state-of-the-art CPSS such as our considered Internet of No Things, whose nearables give rise to intelligent environments. Different CPSS members may use these environments to create tech-driven public or club goods, where human-centered and user-intended services materialize in stigmergic techno-social systems such as Society 5.0. Ironically, despite all the technological prowess, old or even ancient ideas seem to resurface in the 6G era, including but not limited to (1) ancient decision-making processes in social insect colonies, (2) Nikola Tesla's prophecy of converting the whole Earth into a huge brain, (3) Society 5.0's return to the unpredictability, wildness, and continual encounters with the other thanks to the prevalence of diverse non-human agency (social robots, AI agents), and (4) non-monetary economies such as the token economy. Among others, key open research challenges include the exploration of different roles and types of tokens as well as the delivery of advanced XR experiences for the extension of human capabilities, given that the external world consists of conscious agents who favor interactions that drive the evolution of CI while enjoying and acting on experiences.

References

[1] A. Pentland, *"Social Physics: How Good Ideas Spread-The Lessons from A New Science,"* Penguin Press, New York, 2014.
[2] M. Borders, *"The Social Singularity: A Decentralist Manifesto,"* 2018. The social singularity : a decentralist manifesto | CiNii Research.
[3] K. B. Letaief, W. Chen, Y. Shi, J. Zhang, and Y-J. Zhang, "The Roadmap to 6G: AI Empowered Wireless Networks," *IEEE Communications Magazine*, vol. 57, no. 8, pp. 84–90, Aug. 2019.
[4] E. C. Strinati, S. Barbarossa, J. L. Gonzalez-Jimenez, D. Ktenas, N. Cassiau, L. Maret, and C. Dehos, "6G: The Next Frontier," *IEEE Vehicular Technology Magazine*, vol. 14, no. 3, pp. 42–50, Sep. 2019.
[5] W. Saad, M. Bennis, and M. Chen, "A Vision of 6G Wireless Systems: Applications, Trends, Technologies, and Open Research Problems," *IEEE Network*, vol. 34, no. 3, pp. 134–142, May/June 2020.
[6] M. Maier, A. Ebrahimzadeh, S. Rostami, and A. Beniiche, "The Internet of No Things: Making the Internet Disappear and "See the Invisible"," *IEEE Communications Magazine*, vol. 58, no. 11, pp. 76–82, Nov. 2020.
[7] S. Nakamoto, "Bitcoin: A Peer-to-Peer Electronic Cash System," *Decentralized Business Review*, pp. 1–9, Oct. 2008.
[8] Y. Hu, H. G. A. Valera, and L. Oxley, "Market Efficiency of the Top Market-Cap Cryptocurrencies: Further Evidence from a Panel Framework," *Finance Research Letters*, vol. 31, pp. 138–145, Dec. 2019.
[9] F. Tschorsch and B. Scheuermann, "Bitcoin and Beyond: A Technical Survey on Decentralized Digital Currencies," *IEEE Communications Surveys & Tutorials*, vol. 18, no. 3, pp. 2084–2123, 2016.
[10] R. Beck, "Beyond Bitcoin: The Rise of Blockchain World," *IEEE Computer*, vol. 51, no. 2, pp. 54–58, Feb. 2018.
[11] D. Tapscott and A. Tapscott, *"Blockchain Revolution: How the Technology Behind Bitcoin Is Changing Money, Business, and the World,"* Portfolio, Toronto, May 2016.
[12] W. Yang, S. Garg, A. Raza, D. Herbert, and B. Kang, "Blockchain: Trends and Future," Knowledge Management and Acquisition for Intelligent Systems: 15th Pacific Rim Knowledge Acquisition Workshop, PKAW 2018, Nanjing, China, August 28-29, 2018, Proceedings 15. Springer International Publishing, 2018: 201–210.
[13] J. Suzuki and Y. Kawahara, "Blockchain 3.0: Internet of Value - Human Technology for the Realization of a Society Where the Existence of Exceptional Value is Allowed," Human Interaction, Emerging Technologies and Future Applications IV: Proceedings of the 4th International Conference on Human Interaction and Emerging Technologies: Future Applications (IHIET–AI 2021), April 28–30, 2021, Strasbourg, France 4. Springer International Publishing, 2021: 569–577.
[14] L. Seok-Won, I. Singh, and M. Mohammadian, (eds), *"Blockchain Technology for IoT Applications,"* Springer Nature, 2021. https://doi.org/10.1007/978-981-33-4122-7.
[15] S. Voshmgir, *"Token Economy: How the Web3 reinvents the Internet (Second Edition),"* BlockchainHub, Berlin, June 2020.

[16] G. Caldarelli, "Real-World Blockchain Applications under the Lense of the Oracle Problem. A Systematic Literature Review," *In Proceedings of IEEE International Conference on Technology Management, Operations and Decisions (ICTMOD)*, pp. 1–6, Nov. 2020.

[17] H. Al-Breiki, M. H. U. Rehman, K. Salah, and D. Svetinovic, "Trustworthy Blockchain Oracles: Review, Comparison, and Open Research Challenges," *IEEE Access*, vol. 8, pp. 856750–85685, May 2020.

[18] Y. Zhou, F. R. Yu, J. Chen, and Y. Kuo, "Cyber-Physical-Social Systems: A State-of-the-Art Survey, Challenges and Opportunities," *IEEE Communications Surveys & Tutorials*, vol. 22, no. 1, pp. 389–425, Firstquarter 2020.

[19] K. A. Demir, G. Döven, and B. Sezen, "Industry 5.0 and Human-Robot o-Working," *Procedia Computer Science*, vol. 158, pp. 688–695, 2019.

[20] G. Schütte, "What Kind of Innovation Policy Does the Bioeconomy Need?" *New Biotechnology*, vol. 40, Part A, pp. 82–86, Jan. 2018.

[21] F.-Y. Wang, "The Emergence of Intelligent Enterprises: From CPS to CPSS," *IEEE Intelligent Systems*, vol. 25, no. 4, pp. 85–88, July/Aug. 2010.

[22] Hitachi-UTokyo Laboratory (H-UTokyo Lab), *"Society 5.0: A People-Centric Super-Smart Society,"* Springer Open, Singapore, 2020.

[23] P. Freni, E. Ferro, and R. Moncada, "Tokenization and Blockchain Tokens Classification: A Morphological Framework," In Proceedings of IEEE *Symposium on Computers and Communications (ISCC), Rennes, France*, pp. 1–6, July 2020.

[24] S. Wang, W. Ding, J. Li, Y. Yuan, L. Ouyang, and F. -Y. Wang, "Decentralized Autonomous Organizations: Concept, Model, and Applications," *IEEE Transactions on Computational Social Systems*, vol. 6, no. 5, pp. 870–878, Oct. 2019.

[25] A. Beniiche, S. Rostami, and M. Maier, "Robonomics in the 6G Era: Playing the Trust Game with On-Chaining Oracles and Persuasive Robots," *IEEE Access*, vol. 9, pp. 46949–46959, March 2021.

18

Urban Digital Twin as a Socio-Technical Construct

Timo Ruohomäki, Heli Ponto, Ville Santala, and Juho-Pekka Virtanen
Forum Virium Helsinki Oy

18.1 Introduction

The development of digital twins has been active during the past years. In urban planning, the digital twin approaches have been seen as promising, due to their ability to digitize physical environments. While digital twins, with background in industry, can be powerful tools for design purposes, they generally seem to undermine the role of human perspectives and social interaction. The transition from technology-oriented digital twins to socio-technological transition is not, however, an 'automation' (Nochta et al., 2021) but more a result of political decision-making.

The definition of digital twin in the industrial setting is somewhat mature and use cases can be found across several industries. In this chapter, the concept of the urban digital twin is presented to discuss the emergence of virtual representations in cities. We define urban digital twins as a system of systems, i.e., a collection of data and tools that aim to virtually represent the city. Most focus in city-wide digital twins has been on developing realistic models; however, this industry-based approach has faced critique on only providing means for visualization and not providing sufficient tools for urban planning. Cities are more than just the physical infrastructure (Yossef Ravid & Aharon-Gutman, 2022), and thus, in order to build representative urban digital twin, the need to understand the citizen perceptions, processes that involve users, and social relationships better is motivating the cities to invest in data capabilities in order to get better insight and make data-driven decisions and to assess and understand their consequences.

Recently, more focus has been put on how to visualize and model citizen perceptions, to engage citizens via urban digital twins, and to simulate and model population. This positions citizens simultaneously as active subjects in the creation of virtual representation of the city and objects and inputs for models and simulations for decision-making. The aim of this chapter is

 DOI: 10.1201/9781003425724-22

to be a discussion starter to the socio-technical viewpoints that are revolving around urban digital twin development. We explore urban digital twin as a socio-technical construct and give examples on how social aspects and people are considered in current development directions regarding urban digital twins. With this paper, we want to extend the current understanding of urban digital twins beyond the industrial definition and potentially contribute to the discussion on their role in data-driven decision-making and participatory urban governance.

This chapter is divided into three main sections. The next section gives a brief introduction to the concept of urban digital twins. Then, we provide a conceptual overview of the urban digital twin from a socio-technical perspective. The digital twin is seen here as a specific type of information system, and the key concepts presented here provide input for the conceptual design of the digital twin as a socio-technical system. The link between the concepts and application areas of socio-technical design is presented with some case examples, namely (1) participation and engagement, (2) simulations and population modeling, and (3) personal digital twin, which have recently drawn attention in the development of urban digital twin. Finally, in the last section, we present the key findings. In the context of the urban digital twin, the socio-technical approach can be seen in two dimensions: (1) the design of the digital twin as an information system supporting participation and engagement actions with citizens as subjects and (2) the social dimension of the information acquired from a person as a population unit with citizens as objects.

18.2 Urban Digital Twins

Several information systems are typically used in the management and planning of urban environments. These include the conventional two-dimensional GIS/CAD tools, the tools applied for 3D city modeling, but also the various information systems used in different asset management tasks in cities. In geospatial sciences, the term spatial data infrastructure has been used for describing the combination of data, tools, and background systems applied in the geographic information system domain.

The concept of digital twins has also been brought to the urban domain. Originally, the concept of digital twin emerged after a University of Michigan presentation in 2002 as part of the formation seminar of the university's Product Lifecycle Center (Grieves & Vickers, 2017). The presentation did not name the concept as a digital twin but in a white paper published the following year, the digital twin concept as a *virtual representation of what has been produced* (Barachini & Stary, 2022) was presented. An essential part of the concept is the word Twin. It was expected that the set of virtual information constructs would fully describe the potential or actual physical manufactured product from the micro-atomic level to the macro-geometrical level (Grieves & Vickers, 2017). Ideally, any information that could be obtained from inspecting the physically manufactured product was obtained from its digital twin.

The concept further evolved toward representations of more complex cyber-physical systems. A digital twin is then a virtual representation of a process, service, or product with parts of it in the real world. The concept has no predefined restrictions and when representing a process, the target object can have context relevant to the core functionality: a plane is related to all kinds of operational characteristics and urban infrastructures and biological processes can copy the human organism (Barachini & Stary, 2022).

As no widely adopted, precise definition exists for a digital twin, different implementations with varying focus have emerged (Van der Valk et al., 2020). In the urban context, the concept of *an urban digital twin* or *a digital twin of a city* has emerged, once again, with varying terminology and implementation (Ketzler et al., 2020). When defining the digital twin as a multi-physical, multi-scale, and probabilistic simulation model of a complex product or system (Durão et al., 2018), similarities can be seen between the applications of urban systems and the expectations on how different future scenarios could be tested with the digital twins in the urban context.

However, this complexity of the urban systems has not been properly reflected in the information systems used to manage them. In many cases, innovation and design of information systems have been driven by engineering activities and technocratic ideals, with the social aspects becoming somewhat neglected. This has been criticized by socio-technical researchers (Ghaffarian, 2011). Furthermore, the human-centered approach cannot be reduced to perceived usability, customer journey, or user experience but should also support fundamental system design decisions such as defining information entities, concepts, and models.

18.3 Urban Digital Twin from a Socio-Technical Perspective

The socio-technical approach was further refined in the 1970s as part of organization design frameworks. All the organizations are socio-technical systems, but how that reflects on the design of information systems has been questionable. Cherns (1976) defined nine principles of socio-technical design that still can be considered relevant to the development of socio-technical systems. In addition to more technical principles, in this chapter, we focus on expanding the principles that have a link to participatory and iterative aspects. According to Cherns (1976), the organizational objectives are best met not by the optimization of the technical system and then adapting the social system with it but as a joint, iterative effort to optimize the technical and social aspects, thus exploiting the adaptability and innovativeness of people in attaining goals instead of over-determining technically how the goals should be attained. The link between the organizational theories above and urban development's social structures and objectives is clear.

The ETHICS (Effective Technical & Human Implementation of Computer-Based Systems) methodology by Enid Mumford provided a framework for humanizing work with the help of technology (Stahl, 2007). Mumford specifically stated that an information system only designed to meet technical requirements is likely to have unpredictable human consequences (Mumford & Weir, 1979). While the socio-technical approach provides guiding questions for the practitioners, it does not provide a specific blueprint for action (Cherns, 1976). The framework Mumford developed was to provide a more specific formal structure. The ETHICS method has three principal objectives: (1) to enable future users to play a major role in system design, (2) to ensure that the new systems are acceptable to users, and (3) to assist users in developing skills for managing organizational change (Singh et al., 2007).

According to Ghaffarian (2011), a critical principle that was a byproduct of the socio-technical approach in the field of information systems was user participation. When the socio-technical principles were adopted as design guidelines, all the intended users were expected to get involved with all the system development tasks and stages. In reality, the involvement seldom reached such a level but was reduced to users being consulted on the technical systems to support them on their tasks (Ghaffarian, 2011). As Mumford explained, the outcome fell short of the original idea of human values and individuals' rights: *'The most important thing that socio-technical design can contribute is its value system. This tells us that although technology and organizational structures may change, the rights and needs of the employee must be given as high a priority as those of the non-human parts of the system'* (Mumford, 2006).

The potential of design science research in the development of urban digital twins is promising for those looking for more specific tools. Design science provides constructive methods, including conceptual development that, according to Järvinen (2005), refers to the development of various models and frameworks which do not describe any existing reality but rather help to create a new, synthetic one. Conceptual development shall describe the desired state of the new information system, while technical development is an implementation of that objective. In the case of the digital twin concept, a major part of the conceptual work focuses on defining key entities, their relationships, and the overall dimensions of reality included in the model produced as a digital twin. In the industrial digital twin, the scope of work can often be reduced in the simulation of a single function and its input and output.

In the 1990s, new research interests in the social dimension of information systems emerged. One specific new requirement that encouraged this transition was the need to understand better the relationship between information systems development, their usage, and the resultant social and organizational changes (Ghaffarian, 2011). While the role of technology in the socio-technical approach was somewhat reduced to being just a tool, the recent position has also emphasized that ICT should not be seen as a ready-made tool that can be readily applied for specific organizational objectives. In that sense, the position of Cherns on the socio-technical approach being more of a list of principles than a specific list of requirements remains valid. The context of urban systems, however, sets some specific scenarios where it is useful to look at socio-technical principles as part of the user story and requirements formation. Examples of such scenarios are presented in the following sections.

18.3.1 Participation and Engagement

The inclusion of 'social' and citizens' perspectives has been a target for urban planning from the 1960s onwards, when modernist urban planning was seen to be excluding human action in urban space (Jacobs 1993, see Hall 2002). Later, for example, the United Nations' Agenda 21 (United Nations, 1992) has forwarded citizen participation, sustainability, and social perspectives. More recently, technical development has provided novel opportunities to better understand social space in urban planning, starting from the 2000s. As urban digital twins are increasingly suggested as platforms facilitating urban planning activities, the 'social dimension' should also be adequately present in them.

Geographic information systems provide a common framework to organize, visualize, and analyze spatial and aspatial information in research. Due to the nature of cities as public organizations, the geographic information system tools quickly became core information systems of the cities. The geographic information is the basis of urban planning, zoning, permits, and service network planning, e.g., in the scope of school districts. The interoperability of geographic information was also one of the first major tasks taken by the European Union as part of the digitalization initiatives when the INSPIRE spatial data infrastructure directive came into force in May 2007. One of the main objectives of the INSPIRE directive was to support environmental policies, which naturally have a strong spatial dimension.

Researchers criticized geographic information system methods in the 1990s and 2000s for being overly technology-led, asking them to be more 'democratic' and socially inclusive (Dunn, 2007; Fagerholm et al., 2021). Approaches, such as participatory geographic information system and public participation geographic information system (PPGIS) (Dunn, 2007), and later also volunteered geographic information (Sui & Cinnamon, 2016) and virtual geographic environments (Lin et al., 2013, 2022), were seen as key to emphasizing citizens' involvement in the production and use of geographic information system methods. They allowed deeper insights into human life and were giving interactive elements and opportunities to participate.

Development in computing and communication technologies has enabled the creation of new techniques to engage citizens and potentially empower citizens to get involved in various stages of urban governance. New participatory tools and processes were to mobilize citizens, attract new stakeholders to make participation more inclusive, and build community consensus (Afzalan et al., 2017). Furthermore, new digital tools allow citizens to feed in information and for cities to crowdsource information from the citizens. Alongside top-down governed, facilitated engagement activities, technology provides opportunities for bottom-up and grassroots innovations such as citizen science that can potentially shape urban governance models.

These approaches have provided novel insights into urban planning, and digital participation tools have advanced significantly during the past years. It has been, nevertheless, argued that participatory-based geographic information system methods tend to lack an organized framework, and wider connection to scientific research and practice (Fagerholm et al., 2021), despite the plethora of PPGIS methods available. Many PPGIS projects have not contributed to citizen participation and are based on a limited set of variables and processes (Charitonidou, 2022). There are also calls to increase the diversity of participating citizens, use geographic information system-based participatory tools more effectively, and keep the quality and usability of data high (Kahila-Tani & Kyttä, 2019).

Similar needs have been identified in the context of urban digital twins. Despite the targets of urban digital twins to increase citizen participation in the decision-making processes, these processes have remained unclear (Charitonidou, 2022). Similarly, urban digital twins should include citizens' perspectives in technical development and clarify its connections to research more widely.

18.3.2 Simulations and Population Modeling

Starting from the 1980s, there is a well-established tradition of using modeling for urban and transport planning. Traditionally, traffic simulations, as part of urban planning, have targeted to advance (car)-traffic flow, often at the expense of sustainability and human perspectives. More recently also population modeling has raised interest in urban governance to create better insights for, e.g., service network planning. Various terms have been used to describe simulation approaches; however, work evolving around urban digital twins is connected with agent-based modeling. Urban digital twins can visualize, simulate, and manage urban traffic and city environments (Nochta, 2021). Similar to urban digital twins, agent-based simulations aim to simulate 'real' life and help traffic and urban planners to understand the impact of different measures better.

Agent-based simulation models refer to models simulating actions and interactions of individuals or groups of people and are seen as flexible and detailed in depicting human interactions. Agent-based modeling is a useful approach to observing phenomena at the population level. It has been widely used in epidemiology and ecological research. According to Xu, they often investigate the processes, mechanisms, and behaviors of many complex social systems due to their ability to capture the nonlinear dynamics of social interactions (Xu et al., 2017).

In the case of integrated population models, the main assumptions made in each model are classified into two categories: parameters of independence and common demography (Schaub & Kéry, 2022). The assessment of the model's fitness currently mainly relies on predictive checks with small sample sizes. However, developing a more comprehensive population model would open new possibilities for the goodness of fit testing.

A synthetic population can be defined as a simplified representation of the actual population. The synthetic population matches various statistical distributions of the actual population and thus is close enough to the true population to be used in modeling. The main steps of the synthetic population methodology involve generalizing from some of these observations of phenomena to produce a theory, based on which an artificial system can be constructed and then used to test predictions deriving from the theory (Jones et al., 2013). In human-centric systems, the observed phenomena often are qualitative, and the methods of creating repeatable and interoperable indicators for perception and expectations require special attention. Finally, to some extent, it has to be accepted that there is a need to simplify the principal social concepts involved. Simplification, or aggregation, maybe also due to the concerns on privacy and trust the citizens have raised. As detailed population data about the actual population might raise political questions, synthetic data could help to mitigate such issues and make it possible to model and do analyses that consider equality, ethics, and privacy issues.

The synthetic population is used for analyzing action in urban space. Furthermore, since population data are usually mostly static and based on census or other statistical functions, synthetic population enables modeling commuting and transportation in an urban context. The synthetic population has been applied in transportation planning for estimating transportation patterns and demand (Hörl & Balac, 2021).

In a similar way to other technical approaches discussed in this chapter, there is a need to clarify connections to citizen participation, social perspectives, and inclusion in the context of agent-based simulations and population modeling. With in-depth participation, societal and environmental phenomena, such as social segregation, climate change, pollution, and noise reduction, could be more directly named as topics of agent-based simulations (Nochta et al., 2021). Like the contexts of urban digital twin and PPGIS, 'human' perspectives might help to dispel 'technology hype' revolving around simulation work (Nochta et al., 2021) and help to build a deeper and more profound theoretical background (Fagerholm et al., 2021) for agent-based simulation work.

18.3.3 Personal Digital Twin

A personal digital twin has been defined as a virtual version of an individual that is built from her digital footprint (Nativi et al., 2022). Slightly different flavors of personal digital twin can be distinguished from different disciplines: in healthcare, the personal digital twin is steered toward a merger of all health data, supported by live sensor devices (Shengli, 2021). The resulting human digital twin can be applied to visualization, simulation, and prediction (Shengli, 2021). The approach in retail, digital marketing, and e-commerce is similar: by accumulating data from consumer behavior, a digital twin of the customer can be built and applied to produce better-targeted recommendations (Vijayakumar, 2020).

Cities also gather data about individual citizens. Various registries record their basic information, such as place of residence, use of city services, and possible involvement in participatory processes. Following the logic of Nativi et al. (2022), this data hold the potential to form a personal digital twin of a citizen. Corresponding with the logic from medical sciences or commerce, having a citizen's personal digital twin seems to offer several benefits both for the city and the citizen. For the city acting as a service provider, a better understanding of the use and users of services supports their development. When appropriate, combining multiple records to form a bigger picture of an individual's situation might support targeting actions and better guiding people toward the available services. Understandably, the personal digital twin closely relates to managing personal data and the MyData principles (Ministry of Transport and Communications, 2015). Here, applying the personal digital twin concept might support the individual's rights to manage their data by supporting a better curation of data across different systems. The principles established in the MyData work also help form a balance between utilitarian consumption of all private information for maximal efficiency and minimizing one's digital footprint for privacy reasons (Peltz, 2018) by offering a controlled way of sharing and using data.

In addition, combining personal digital twins that share common properties (e.g., all users of a specific service in a given area) would enable defining populations. From the perspective of city data, this would form a significantly more flexible source of information than current statistical grids. For example, a population density for a given area could be obtained by merging all citizen personal digital twins from it. This also builds a conceptual bridge between the personal digital twins and the urban digital twins – if the role of the personal digital twins is to form a population that is applicable to either directly forming or indirectly by calibrating a synthetic population used for an agent-based simulation, the personal digital twins will begin to hold a role in the urban digital twin.

There are, understandably, limitations to how far the concept of a citizen's personal digital twin and derived population data can be taken. Personal data require adhering to privacy and data protection regulations and managing

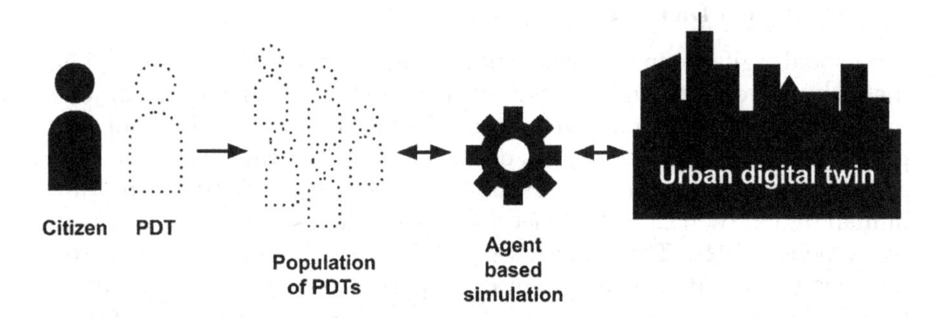

FIGURE 18.1
The conceptual link between personal digital twins (PDT) and urban digital twins (UDT), formed by populations and agent-based simulation.

user consent. Records portray the citizens as static entities and only include the people present in registers (excluding, for example, tourists). Thus, the population properties, such as density in a particular area, do not describe the actual occupancy of an urban area at a given moment in time.

18.4 Conclusions

Integration of cities' geospatial data and social artifacts is well established and present in many existing urban information systems. Most commonly geographic information systems features depict real-world artifacts containing attributes related to various processes. A street polygon in geographic information systems can carry information on stakeholders responsible for its maintenance process. Social processes can also produce their own geospatial feature sets that do not adhere strictly to the objects present in the physical world. For example, a point of interest identified in a PPGIS study does not necessarily correspond to the physical features of the urban environment. The tighter integration of geospatial features with modeled processes and other data is a topical development direction in industrial facilities management (Love & Matthews, 2019).

The current vision of urban digital twin has four main domains: (1) geospatial information, (2) dynamic observations, (3) asset management, and (4) population model. The asset management domain is about modeling of events and processes with related stakeholders and can support the action and behavioral model of the population (Gartner, 2021). The socio-technical perspective adds people, their actions, processes, and behavior as requirements that could be considered extensions to the current ideal. The socio-technical aspects explained in previous sections could provide a starting point for the conceptual modeling of a digital twin as an information system.

From the perspective of urban information systems, the macro social structures and macro dynamics fall into the scope of the urban digital twin, while micro processes likely can be implemented as part of the *personal digital twin*, the digital twin of individuals (Nativi et al., 2022). The elemental support for microanalysis, such as agent-based simulation, can be part of the attributes of population units in the population model of the urban digital twin. An emerging pattern can be seen here: microanalysis of an individual as part of a larger structure, the population, as a dynamically defined population unit. The key finding is that we cannot fully understand the related patterns on the macro level by only sampling interactions and micro-level situations. As a potential next step, realizing the integration of citizens' personal digital twins with an urban digital twin would expand the concept of bringing social processes, stakeholder information, and digital geometric models together and scale it, not only to include specific predefined stakeholders but also the general public. However, this requires further development of the information model applied in the urban digital twin to maintain the key artifacts of social structures such as populations. The design of the population domain in urban digital twin is a highly iterative process (Cherns, 1976), and thus the exact entities and properties will naturally depend on the objectives. Finally, considering urban digital twins from a socio-technical perspective sheds light on how data on social aspects and people are interconnected to urban information systems. Recently, with initiatives such as the Built4People Partnership and the New European Bauhaus; since digital twins are one major development direction of cities' digitalization agenda, they provide a fruitful lens to explore the potential shift on a city and European level. Urban digital twins could provide valuable tools and insights to urban planning and decision-making and support the shift from technology-centric to human-centric design of smart cities. However, integration of various sources of information to extend urban digital twins with participatory data

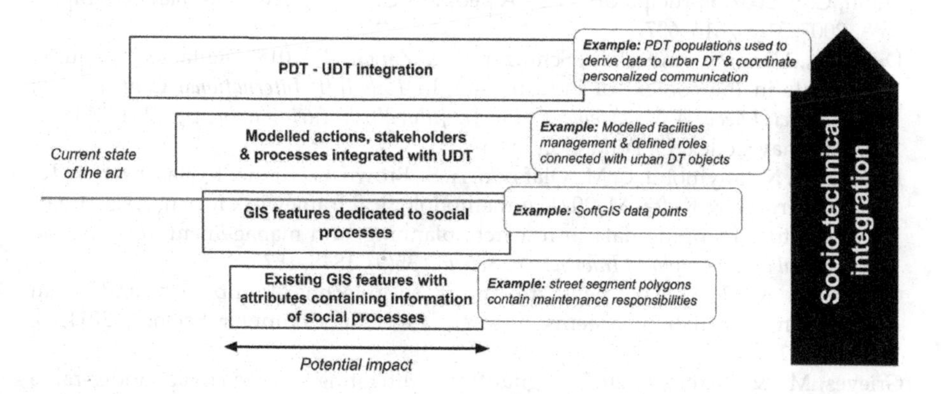

FIGURE 18.2
Levels of socio-technical integration in urban digital twins. The highest level is expressing the integration of personal digital twins (PDT) and urban digital twins (UDT).

sources or more detailed data on populations leads to ethical and political questions that need to be carefully dealt with to ensure citizens' equality and build simulations and models that comply with EU General Data Protection Regulation.

Acknowledgments

The research is connected to the following projects by the European Union Horizon 2020 program: mySMARTLife (731297), Urbanite (870338), and Urbanage (101004590). The city of Helsinki's digitalization program also supported the work.

References

Afzalan, N., Sanchez, T. W. & Evans-Cowley, J. 2017. Creating smarter cities: Considerations for selecting online participatory tools, *Cities*, 67, 21–30.

Barachini, F., & Stary, C. 2022. *From Digital Twins to Digital Selves and Beyond*. Cham: Springer.

Charitonidou, M. 2022. Urban scale digital twins in data-driven society: Challenging digital universalism in urban planning decision-making. In Ed. Brown, A., *International Journal of Architectural Computing*, vol 20, Issue 2, pp. 283–253. Thousand Oaks, CA: SAGE Publishing.

Cherns, A. 1976. The principles of sociotechnical design. Human relations, 1976, 29(8): 783–792.

Dunn, C. E. 2007. Participatory GIS - A people's GIS? Progress in human geography, 2007, 31(5): 616–637.

Durão, L., Haag, S., Anderl, R., Schützer, K. & Zancul, E. 2018. Digital twin requirements in the context of industry 4.0. In *15th IFIP International Conference on Product Lifecycle Management (PLM), July 2018, Turin, Italy*, pp. 204–214. Le Chesnay Cedex: La Fondation Inria.

Fagerholm, N., Raymond, C. M., Olafsson, A. S., Brown, G., Rinne, T., Hasanzadeh, K., Broberg, A. & Kyttä, M. 2021. A methodological framework for analysis of participatory mapping data in research, planning, and management. *International Journal of Geographical Information Science*, 35(9), 1848–1875.

Ghaffarian, V. 2011. The new stream of socio-technical approach and main stream information systems research. Procedia Computer Science, 2011, 3: 1499–1511..

Grieves, M., & Vickers, J., 2017. Digital twin: Mitigating unpredictable, undesirable emergent behavior in complex systems. Transdisciplinary perspectives on complex systems: New findings and approaches, 2017: 85–113.

Hall, P., 2002. *Urban and Regional Planning*. 4th edition. London: Routledge. 248 p.

Hörl, S. & Balac, M. 2021. Synthetic population and travel demand for Paris and île-de-france based on open and publicly available data. In ed. N. Geroliminis. *Transportation Research Part C 130*, Amsterdam: Elsevier B.V.

Jacobs, J. (1993). *The Death and Life of Great American Cities*. Vintage Books. Jstor: 1597–1602.

Jones, A., Artikis, A., & Pitt, J. 2013. The design of intelligent socio-technical systems. In eds. G. Jezic, S. Ossowski, F. Toni, & G. Vouros. *Artificial Intelligence Review*, vol 39, pp. 5–20. New York: Springer Nature.

Järvinen, P. 2005. Action research as an approach in design science. In *Design, Collaboration and Relevance in Management Research, The EURAM Conference Proceedings*. DEPARTMENT OF COMPUTER SCIENCES UNIVERSITY OF TAMPERE TAMPERE 2005.

Kahila-Tani, M., & Kyttä, M. 2019. Does mapping improve public participation? Exploring the pros and cons of using public participation GIS in urban planning practises. Landscape and urban planning, 2019, 186: 45–55.

Ketzler, B., Naserentin, V., Latino, F., Zangelidis, C., Thuvander, L., & Logg, A. 2020. Digital Twins for cities: A state of the art review.Built Environment, 2020, 46(4): 547–573.

Lin, H., Chen, M., Lu, G., Zhu, Q., Gong, J., You, X., Wen, Y., Xu, B., & Hu, M. 2013. Virtual geographic environments (VGEs): A new generation of geographic analysis tool. Earth-Science Reviews, 2013, 126: 74–84.

Lin, H., Xu, B., Chen, Y., Li, W., & You, L. 2022. VGEs as a new platform for urban modeling and simulation. In ed. G. D. Bathrellos, *Sustainability*. Basel: MDPI.

Love, P., & Matthews, J. 2019. The 'how' of benefits management for digital technology: From engineering to asset management. Automation in Construction, 2019, 107: 102930.

Ministry of Transport and Communications, 2015. *MyData - A Nordic Model for Human-Centered Personal Data Management and Processing*. http://urn.fi/URN:ISBN:978-952-243-455-5.

Mumford, E. 2006. The story of socio-technical design: Reflections on its success, failures and potential. In *Information Systems Journal*, vol 16 Issue 4, pp. 317–342.

Mumford, E. & Weir, M.. *Computer Systems in Work Design – The ETHICS Method*. Associated Business Press, 1976.

Nativi, S. et al. 2022. *MyDigitalTwin: Exploratory Research Report*. Luxembourg: Publications Office of the European Union.

Nochta, T. et al. 2021. A socio-technical perspective on urban analytics: The case of city-scale digital twins. *Journal of Urban Technology*, 28(1–2), 263–287.

Peltz, J. 2018. *The Algorithm of You: Your Profile of Preference or an Agent for Evil?* Newport, RI: Naval War College.

Schaub, M. & Kéry, M. 2022. Assessment of integrated population models. *Integrated Population Models*, pp. 271–306.

Shengli, W. 2021. Is human digital twin possible? In *Computer Methods and Programs in Biomedicine Update*, vol 1. Amsterdam: Elsevier B.V.

Singh, R., Wood, B., & Wood-Harper, T. 2007. Socio-technical design of the 21st century. In eds. McMaster, T., Wastell, D., Ferneley, E., and DeGross, J. *IFIP International Federation for Information Processing, Volume 235, Organizational Dynamics of Technology-Based Innovation: Diversifying the Research Agenda*, pp. 503–506. Boston, MA: Springer.

Stahl, B.C. 2007. Ethics, morality and critique: An essay on Enid Mumford's socio-technical approach. In ed. K. Lyytinen. *Journal of the Association for Information Systems*, vol 8, Issue 9. 470–489. Atlanta, GA: Association for Information Systems.

Sui, D., & Cinnamon, J. 2016. Volunteered geographic information. *nternational Encyclopedia of Geography: People, the Earth, Environment and Technology: People, the Earth, Environment and Technology*, pp. 1–13.

United Nations. 1992. *Results of the World Conference on Environment and Development: Agenda 21. UNCED United Nations Conference on Environment and Development, Rio de Janeiro*, United Nations, New York.

van der Valk, H, Haße, H., Möller, F., Arbter, M.,Henning, J-L. & Otto, B. 2020. A Taxonomy of digital twins. AMCIS 2020 Proceedings, vol 4. https://aisel.aisnet.org/amcis2020/org_transformation_is/org_transformation_is/4.

Vijayakumar, S. 2020. Chapter 11 – Digital twin in consumer choice modeling. *Advances in Computers*, vol 117, Issue 1, 265–284. .

Xu, Z., Glass, K., Lau, C. L., Geard, N., Graves, P. & Clements, A. 2017. A synthetic population for modelling the dynamics of infectious disease transmission in American Samoa. In *Scientific Reports*, 2017.

Yossef Ravid, B., & Aharon-Gutman, M. (2022). The social digital twin: The social turn in the field of smart cities. Environment and Planning B: Urban Analytics and City Science, 2023, 50(6): 1455–1470.

Niederman, Fred. "Project management: openings for disruption from AI and advanced analytics." *Information Technology & People* 34.6 (2021): 1570–1599.

19

Design and Operationalization of Digital Twins in Robotized Applications: Architecture and Opportunities

Tobias Osterloh
RWTH Aachen University

Eric Guiffo Kaigom
Frankfurt University of Applied Sciences

Jürgen Roßmann
RWTH Aachen University

19.1 Introduction

Modern robotic systems rapidly evolve and increasingly adapt to varying application demands. Advances in digital engineering, high-performance computing, and data intelligence significantly contribute to this success. Nevertheless, the resulting complexity is not solely restricted to the robot performance, but also inherent to the robotized application as a whole. Applications examples include the prospective transfer and usage of intelligent robots in emerging production markets and robotized servicing in the space environment. To manage the ever-increasing complexity, modern robotized applications rely on digital twins (DTs). Even though there is no consensus on and a consistent definition of the term DT yet, various applications emphasize the added value of DTs for robotic systems [16]. To develop and employ DTs, current approaches fuse data sensed from the physical system in a comprehensive digital replica. This digital surrogate is then used to, for instance, monitor and even adapt the behavior of the paired system as it deviates from its desired state. Therefore, DTs reflect operational data of robotic systems and provide a static (i.e., recorded) digital model of the system itself. DTs also reason about data to control the physical system.

Previously mentioned approaches to develop and use DTs for the support of robotized applications face some limitations. First, sensed data are available

DOI: 10.1201/9781003425724-23

and relevant only for a given scenario with hardly reproducible constraints (e.g., disturbances, uncertain system parameters) and specific robots. Second, how the robot and its operational environment influence each other about an operating point that differs from the current production-relevant one is rarely recorded. This is especially the case when the environment or robot is not easily accessible and downtime is critical for economic reasons. Finally, although sophisticated and mature technologies are available in intelligent robotics, understanding their usefulness and emerging properties through distinctly customized transfers to and on-site adaptations in prospective environments is not supported. Consequently, using a DT-based prospection to contextually enhance the employment of physical intelligent robots as part of the life-cycle management of robotized applications is hindered.

This chapter describes the development of a physics-based environment in which dynamic DTs fed with both real and simulation data capture relevant properties of ongoing and planned robotized applications as a whole. Since the simulation of naturally interacting and semantically cooperating DTs is loosely coupled with real applications, obtained DTs enrich distinct robotized applications throughout their life cycle. Information harnessed to this end stems from the prospection skills of DTs, such as exploration, exploitation, and optimization. These skills build upon our previous work, including projective virtual reality [7], experimentable [18], and value-driven [10] DTs for robotized applications. This chapter shares details about the physics-engine framework which has propelled these endeavors so far and presents a holistic DT-driven support for robotized applications.

19.2 State of the Art

19.2.1 Standardized Foundation for Digital Twins

Originally, the DT concept emerged from product life-cycle management (PLM). It defines a digital representation of an asset based on gathered operational data [9]. The NASA independently introduced the term DT as a highly accurate model that integrates multiple subsystems to yield a holistic simulation [8]. Nowadays, DTs are regarded as a fusion of a digital shadow (operational data) and a digital model (collection of models describing the assets, e.g., simulation models), which differently behave depending on applications. Despite their immense potential, standardizing DTs remains challenging. A reason is the variety of definitions, interpretations, and concepts for their realization.

In industry 4.0 applications, DTs are closely related to the asset administration shell. They operate at the edge or cloud level and are at the heart of cyber-physical systems run in the Industrial Internet importance of simulation approaches to unleash the full potential of DTs. To emphasize this view-point, the concept of *Experimentable Digital Twins* has been introduced [19]. The concept inherently interlaces DTs with modern simulation technology. Recent advances in DT technology agree upon practical opportunities of DTs [10,16,18]. Nevertheless, the foundation of DTs is not yet sufficiently standardized.

This gap links the contributions of this chapter to the current state of the art. We provide details about a proven methodology to formally describe DTs of robotic systems and their operational environments. We describe the underlying architecture and outline relevant components for the simulation-based operationalization of DTs to support real applications.

19.2.2 Modularization of Simulation Back-Ends

Simulation technology is a key enabler to bring DTs to life and take advantage of their service-based skills to support real applications. Therefore, a simulation back-end needs to interpret information provided by a DT. Also, a simulation back-end must be capable of automatically deriving the executable simulation model that suits the respective simulation engine being supported. Taking a detailed look at robot dynamics leads to standardization approaches like URDF [12] and the Physics Abstraction Layer (PAL) [4] for the description of robotic systems. These standards are supported by multiple simulation engines like ODE, Bullet Physics, NVIDIA PhysX, or MuJoCo, which are worth considering for the realization of the simulation back-end.

Nonetheless, comparing the capabilities of these well-known simulation engines with the multitude of robotic applications (e.g., industrial robotics, mating tasks, locomotion) and the requirements of DTs (e.g., semantically related and extensible submodels, real-time capabilities, physical accuracy, interaction) reveals potentials for improvement. Aiming at an evolving simulation platform for robotic DTs, adaptable and configurable capabilities of the simulation back-end are crucial since allowing to support the cross-domain application by using DTs. In fact, specific requirements of distinct use cases can be thereby fulfilled. Nowadays, however, most simulation engines hardly provide configuration options for single simulation components (e.g., inverse dynamics formulation and solver, integration of forward dynamics, contact models for physical interactions, friction models, compliant elements) [6]. In conclusion, the required modularization of simulators for the comprehensive realization of DTs is not yet available. We contribute to filling this modularization gap in this chapter.

19.3 Simulation Architecture for Digital Twins

The proposed DT architecture is based upon a central and important design decision: a semantic decoupling of the DT modeling from simulation functionalities. This enables flexible and toolagnostic developments of DTs.

Also, it allows for supporting unique requirements from applications, such as additional robot skills and environment features, with specialized and application-specific simulation algorithms.

19.3.1 Ontology for Digital Twins

Being similar to well-known standards like URDF and the PAL, the developed ontology builds upon commonly used model elements and supplements them with properties that can be customized. The competitive advantage of the ontology is its simplicity and generic nature. Arbitrary kinematics can be described without any restriction to robotized systems. In other words, the description of operational environments of robots is supported as well. For that, the ontology is based on a flat design pattern, i.e., not limited to serial kinematics, and thus extends the range of applications.

19.3.1.1 Model Elements

Atomic building blocks of the ontology are the model elements M_i describing, e.g., single bodies of a robot. Properties, such as inertia tensors $\mathbf{\Theta}_i$ function of the mass m_i and center of gravity $\vec{r}_{cog,i}$, are given with respect to a reference frame $\mathbf{R}_i$. The *Geometry* (provided as CAD-file) and *Material* are optional references used to integrate robots in operational environments, i.e., to model their spatial and physical interactions with environments.

$$M_i = \left\{\mathbf{R}_i, m_i, \Theta_i, \vec{r}_{\mathrm{cog},i}, (\text{Geometry, Material})\right\} \tag{19.1}$$

19.3.1.2 Coupling Elements

Coupling elements C_i reflect the kinematic design of robotic systems and beyond. The frame $\mathbf{R}_i$ identifies the pose of coupling element and the *Type* classifies the joint type (e.g., 43, hinge or prismatic joint). Furthermore, references to connected model elements M_i and M_j are given. Optional properties f_i encode, e.g., friction parameters. On this basis, a model transformation to either minimal (i.e., joint) or maximal (i.e., Cartesian) coordinates representation can be performed, enabling to realize well-suited simulation back-ends for each individual DTs.

$$C_i = \left\{\mathbf{R}_i, M_i, M_j, \text{Type}, (fi)\right\} \tag{19.2}$$

19.3.1.3 Actors

Driving systems is achieved by using active components A_i. The goal is to simplify the quick replication of real-world behaviors of robotic systems. An actor is assigned to a respective axis $\vec{n}$ of the coupling element C_i.

Additionally, the *Type* identifies the motor model and its specific parameters a_i (e.g., drive train and compliance [14]).

$$A_i = \{C_i, \vec{n}_i, \text{Type}, a_i\} \tag{19.3}$$

19.3.1.4 Sensors

To measure and feedback simulated data to a DT (or physical robot) and thus close the loop from the simulation back-end to the DT (or physical robot), sensors S_i are used. Sensors capture the current states of simulated DTs and their environments. Sensors can also be fed with measured data from physical systems. Thus, sensors inject multi-modal data into simulations, measure simulation states (e.g., joint positions of DTs or simulated contact forces) in which case they are called *virtual sensors*, and return simulated data to DTs or real robot controllers [10]. Hence, *virtual sensors* enrich DTs with not yet available data, fostering novel robotic skills based on simulation-driven artificial intelligence/machine learning. A sensor is fixed on the frame $\mathbf{R}_i$ and linked to an element E_i, which can either be a model element M_i (e.g., IMU), an actor A_i (e.g., force torque sensor), or a coupling element C_i (e.g., phase-angle sensor). Optional static information on the sensor is s_i.

$$S_i = \{\mathbf{R}_i, E_i, (s_i)\} \tag{19.4}$$

19.3.2 Modular Simulation Back-End

The modular simulation back-end is the core component to realize DTs. It allows for experimenting DTs described by means of the previous ontology in simulation environments. The basic idea behind the modular architecture is to disassemble simulation algorithms in interoperable distinct components, which are then combined for each application individually. Thus, the configuration of the simulation back-end is specific and highly suited for cross-domain applications. An overview of the overall architecture is in Figure 19.1.

19.3.2.1 Model and Data Transformation

Model and data transformations are the connective links between the DT and the dynamics simulator and vice versa. The ontology presented in Section 19.3.1 is vital for the realization of required interfaces. It ensures the validity of the mapping from the formal description to an executable simulation model. All parameters of the digital model are resolved and mirrored to parameterized models (e.g., contact or actor models). Finally, the model transformation initializes the state vector (replicating the current state of the robotic system) and arranges entities for their simulation. Data transformation converts

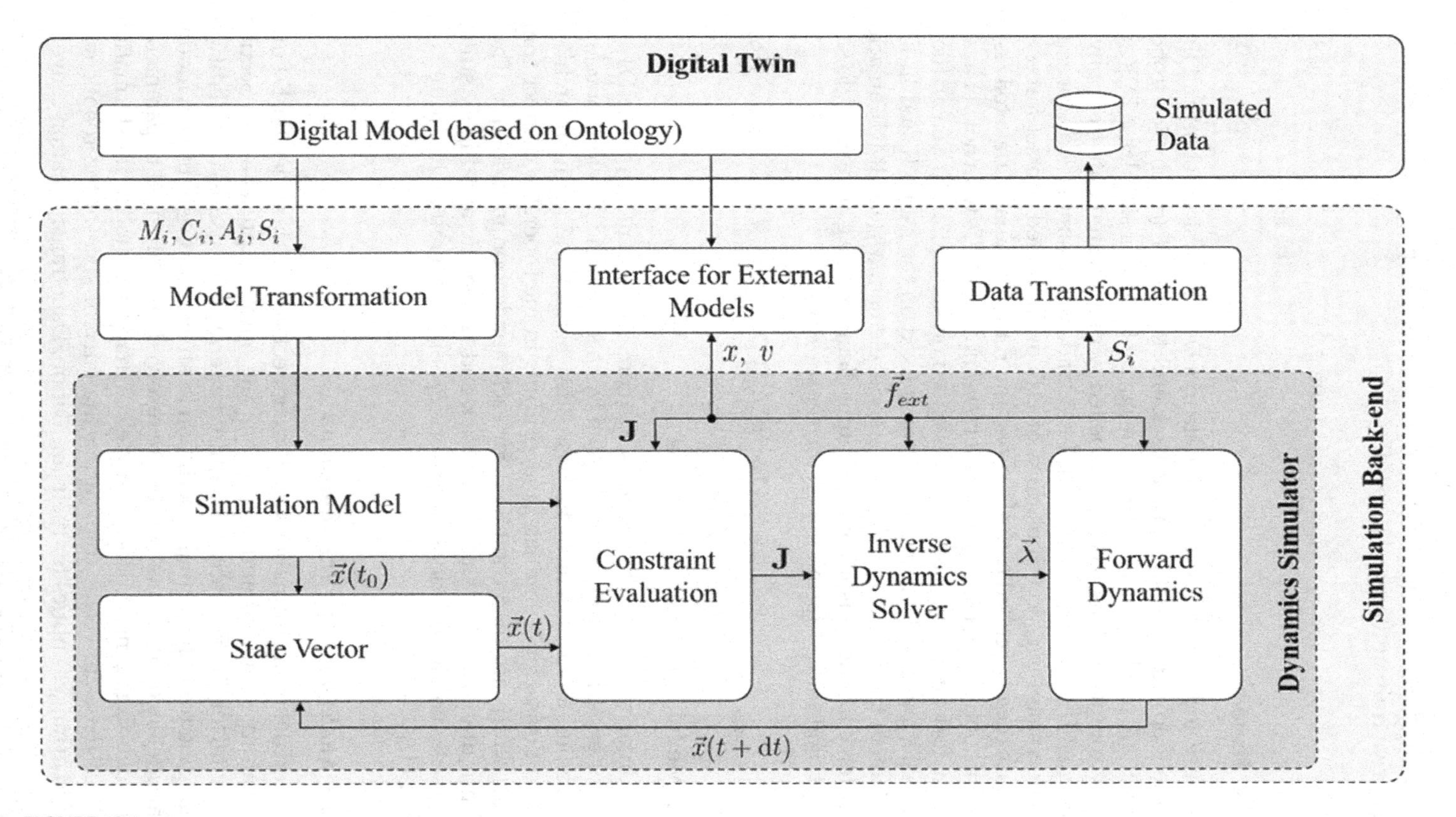

FIGURE 19.1
Architecture of the modular simulation back-end.

simulated data to the data model provided by the DT. This way, operational data and simulated data are easily processed, matched, and analyzed for further applications.

19.3.2.2 Dynamics Simulator

To realize a versatile and sustainable approach to DTs, the simulator support composability and expandability requirements from robotized applications. We developed a modular architecture for a dynamics simulator that offers a variety of configuration capabilities. These cover a wide range of robotized applications. Since we aim at high-performance analysis options (e.g., analysis of dynamic bearing loads) and to provide a generic as well as deep insight into the multi-body dynamics of robots and their environments, we harnessed an impulse-based simulation scheme in maximal coordinates [20]. Relevant building blocks for the dynamics simulator design are highlighted in Figure 19.1.

Constraint evaluation: Evaluating a current DT state, explicit couplings described by the ontology (e.g., existing kinematic constraints) are systematically transformed into $\mathbf{J} \cdot \vec{v} = \vec{b}$ with constraint Jacobian $\mathbf{J}$, searched velocity $\vec{v}$, and known variable $\vec{b}$ implicit interactions between DTs (e.g., a robot with its environment) are detected automatically and modeled as unilateral constraints $\mathbf{J} \cdot \vec{v} \geq \vec{0}$. Friction is modeled using a linearized friction cone [21] and compliance is described according to the constraint force mixing principle.

Inverse dynamics solver: Using constraints, the inverse dynamics is formulated according to the $\mathbf{JM}^{-1}\mathbf{J}^T$ approach [2]. Contact dynamics and friction are covered as inequality constraints, resulting in either LCP-based [22] or QP-based formulations [23] of the inverse dynamics.

$$\mathbf{JM}^{-1}\mathbf{J}^T \cdot \mathrm{d}t \cdot \vec{\lambda}\left(\vec{v} + M^{-1} \cdot \mathrm{d}t \cdot \vec{f}_{\text{ext}}\right) = \vec{b} \tag{19.5}$$

Especially QP-based formulations are of great interest for robotic systems physically interacting with their environments, since they minimize kinetic energy and thereby realize the principle of maximal dissipation [23], while simultaneously avoiding the penetration of interacting bodies. In order to cope with redundant constraints, which result in singular system matrices, different measures have been developed and used within the frame of the proposed architecture [13].

Forward dynamics: Based on the Lagrangian multipliers $\vec{\lambda}$, the forward dynamics $\vec{v} = \mathbf{M}^{-1}. \left(\vec{f}_{\text{ext}} + \mathbf{J}^T \cdot \vec{\lambda}\right)$ is obtained. For stability reasons, most approaches taking advantages from maximal ·coordinates implement a semi-implicit Euler integration scheme [1]. Nevertheless, applying higher-order integration schemes is an advantageous alternative [3].

Interface for external models: External models extend the simulation scope by including additional components (e.g., a drive train) to enhance both the relevance and significance of simulation results. They are thus important building blocks for high-fidelity DTs. Developed bidirectional interfaces inject external forces (i.e., as inputs) into and return, e.g., motion values (i.e., as outputs) from the simulation.

19.4 Applications and Opportunities

The architecture described so far has led to a comprehensive simulation paradigm for the realization and usage of DTs of robots in operational environments. Configuration capabilities of the dynamics simulator span various robotic applications, supporting the entire PLM of robotic systems (i.e., DTs evolve over time). The following applications provide results from two large-scale projects in which the framework was extensively and successfully deployed and evaluated.

19.4.1 Continuous Calibration of Digital Twins

Taking a detailed look at the calibration of robotic DTs, the added value of the defined ontology becomes obvious, as it comprises relevant properties of robotic systems and thus provides an interface for their calibration. Over time, mechanical wear affects the dynamic properties of robotic systems (e.g., friction parameters). Therefore, a calibration process has to be implemented to continuously tune and sharpen system parameters (as defined by the ontology) and ensure that the DT accurately mirrors its physical sibling at any given time. The initial parameterization of the DT is based on dedicated reference experiments, taking into account the operational data. The continuous adjustment of time-dependent parameters is performed periodically.

Figure 19.2 methodologically summarizes experiments conducted to align simulated resp. sensed operational joint torques of a robotic DT (bottom) and its physical sibling (top). The alignment identifies dynamic parameters (e.g., mass, inertia, viscous, and static friction) of the physical robot to parameterize the DT. For this purpose, the error $\epsilon = \left\|\vec{\tau}_{\text{ref}} - \vec{\tau}_{\text{sim}}\right|$ is minimized by carrying out a particle swarm optimization (PSO) [11]. ϵ is the cost function. Particle positions are vectors of robot parameters being identified. Dedicated services of the DT apply PSO to continuously explore and exploit high-dimensional particle positions contributing to minimize ϵ. Particles are initially distributed as a Gaussian model during pre-calibration. Valid parameter intervals are specified to speed up convergence. Despite non-trivial and large joint positions, velocities, and accelerations, Figure 19.3 shows that the convergence is accurately met during 55[s].

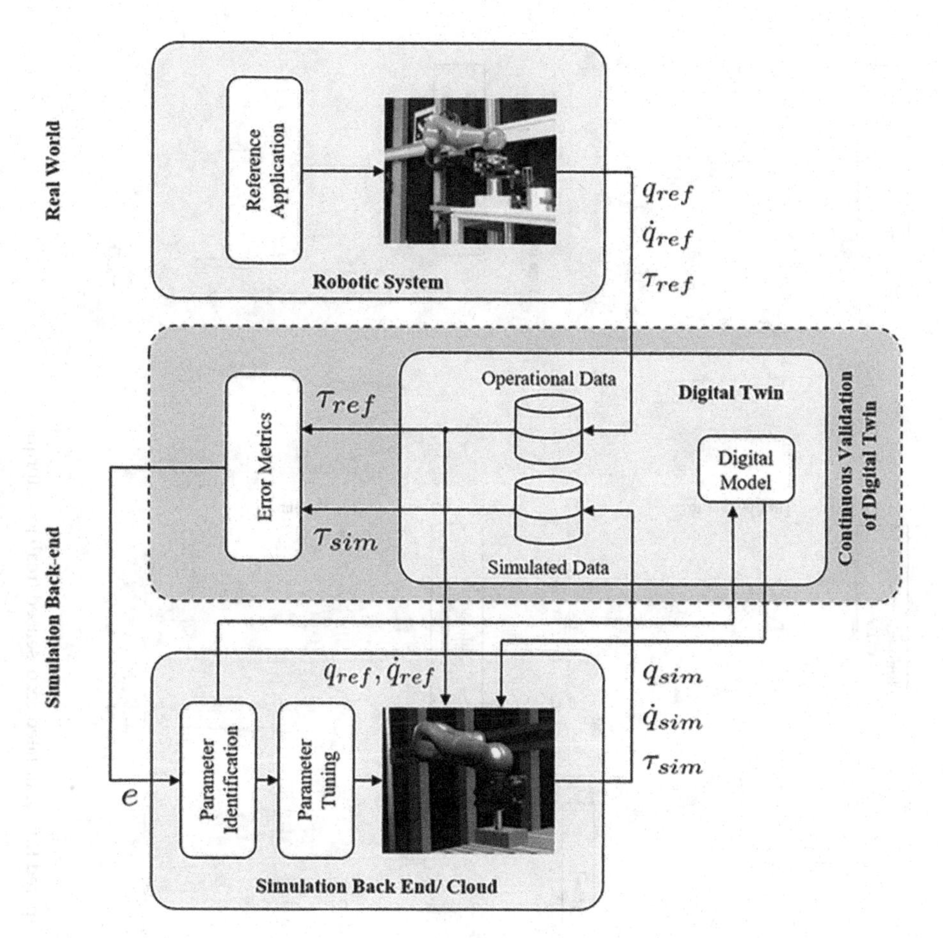

FIGURE 19.2
Continuous DT calibration. q_{ref}, q_{sim}, τ_{ref}, and τ_{sim} are sensed and simulated joint positions q and torques τ.

19.4.2 Bringing Robotized Manipulation On-Orbit at Low Costs

The potential and importance of results obtained in Section 19.4.1 become more apparent when it comes to understand and enhance the feasibility of robotized maneuvers for on-orbit servicing. This industrial market is estimated to have a volume of approximately 4.4 billion $ by 2030 [17]. The framework presented so far was used within the scope of the project iBOSS-III. The goal was to assess the robotized reconfiguration of satellites to prolong their operating life.

Whereas a generic modeling of robots physically interacting with operational environments including satellites was realized in maximal (i.e., Cartesian) coordinates, minimal (i.e., joint) coordinates on the other side

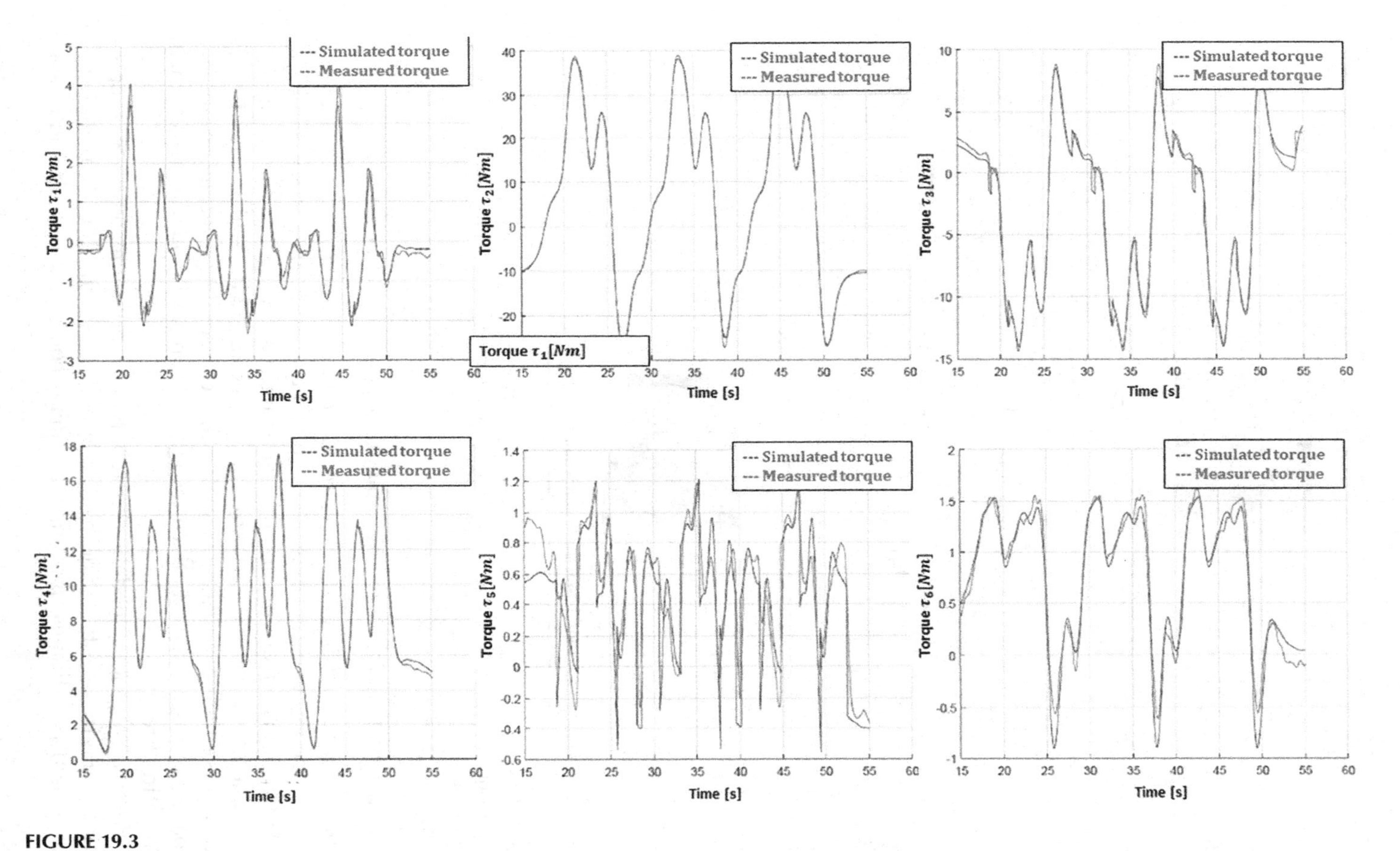

FIGURE 19.3
Comparing torque profiles after a DT calibration. Blue resp. red line: simulated resp. sensed (real) joint torques.

captured the Newton-Euler dynamics of robots only. For instance, mass and Coriolis matrices as well as gravity and frictional forces in joint and operational spaces were constructed and made available on demand at each simulation step. This was done for *any* existing or prospective robot modeled and parameterized on the maximal coordinates side.

One advantage of this coupling is insights gained from the automatic model-based abstraction. For instance, the effective inertia felt at an end-effector along or about a desired direction was optimized by exploring and exploiting rewarded joint configurations to mitigate contact forces during docking maneuvers (see Figure 19.4). Also, single contributions to the energy consumption and regeneration of robots were optimized through trajectory scaling to yield sustainable motions. The second advantage of coupling coordinates is that the robot modeling and dynamics model reconstruction were *not* restricted to any specific robot unlike, e.g., the Fast Robot Interface tailored to the KUKA LWR IIWA robot. The third advantage is the usage of the available robot model to develop and optimize motion control algorithms (e.g., torque-based impedance and hybrid force/position control).

These algorithms are decoupled from the underlying robot model via suitable interfaces. Hence, the control library was shared by distinct robots. When an existing manipulator was dexterity-limited, its control capability was applied to another structurally enhanced DT arm prospectively extended with a micro- or macro-arm to further assess servicing feasibility.

FIGURE 19.4
Cyber-physical assessment and optimization of robotized servicing on-orbit. This is achieved by using digital twins (left upper side) of physical robots (r.h.s.) designed and operated with the framework presented in Section 19.3.

19.4.3 Decision Support for the Operation of Robotic Systems

Finally, the simulation back-end offered analytic and predictive capabilities, allowing to examine the current state of the robotic system in detail and to safely evaluate different actions of a robotic system in its virtual operational environment. Within the scope of the project KImaDiZ, the DT served as a mediator between the robotic system and its real-world counterpart [5] and empowered operators to control a robot via its DT. Especially in the teleoperation of robotic systems (e.g., inaccessible environments with large time delays), this approach offers novel opportunities by providing not-yet-accessible information to the operator and thereby creating a deep insight into the robotic system (often outperforming capabilities of integrated sensor systems).

By creating various instances of a DT, it was possible to initiate multiple test actions in parallel to support fast decision-making before applying obtained efficient motions to physical robots. This way, performance indexes were evaluated w.r.t. specific objectives. Best-possible motion alternatives were then selected to be executed by the robot controlled by its DT in real time.

19.5 Conclusions

Integrating DTs with modern simulation technology offers a major benefit for the development and operation of robotic systems in challenging environments. In particular, it allows to create novel applications in industry and research. The proposed DT architecture conflates the physical robotic system with simulation technology in a virtual physics-based environment, forming a virtual test bench. To this end, DTs are formally described by using an ontology that decouples the representation of the DT from the specific realization of the simulation components. This, in turn, enables flexibility that manifests itself in a versatile applicability. In the future, integrating big data to the concept will offer new possibilities for predictive maintenance and further matching simulated data to available operational data. In addition, extending the architecture with further artificial intelligence skills will allow DTs to autonomously validate, predict, and extrapolate their actions in simulations, thus fostering intelligence in complex robotic systems at substantially reduced efforts and costs.

Acknowledgment

This work is part of projects "KImaDiZ" and "IBOSS-III", supported by the German Aerospace Center (DLR) with funds from the German

Federal Ministry of Economics and Technology (BMWi), support codes 50RA1934/50RA1203.

Bibliography

[1] Mihai Anitescu and Florian A Potra. Formulating dynamic multi-rigid-body contact problems with friction as solvable linear complementarity problems. *Nonlinear Dynamics*, 14(3):231–247, 1997.

[2] David Baraff. Linear-time dynamics using Lagrange multipliers. Proceedings of the 23rd annual conference on Computer graphics and interactive techniques. 1996: 137–146.

[3] Jan Bender and Alfred Schmitt. Constraint-based collision and contact handling using impulses. Proceedings of the 19th international conference on computer animation and social agents. 2006: 3–11.

[4] Adrian Boeing and Thomas Braunl. Evaluation of real-time physics simulation systems. Proceedings of the 5th international conference on Computer graphics and interactive techniques in Australia and Southeast Asia. 2007: 281–288.

[5] Torben Cichon, Heinz-Jurgen Roßmann, and Jochen Deuse. *Der digitale zwilling als mediator zwischen mensch und maschine. Technical report,* Lehrstuhl und Institut fur Mensch-Maschine-Interaktion, 2020.

[6] Tom Erez, Yuval Tassa, and Emanuel Todorov. Simulation tools for model-based robotics: Comparison of bullet, havok, mujoco, ode and physx. 2015 IEEE international conference on robotics and automation (ICRA). IEEE, 2015: 4397–4404.

[7] Eckhard Freund and Juergen Rossmann. Projective virtual reality: Bridging the gap between virtual reality and robotics. *IEEE Transactions on Robotics and Automation*, 15(3):411–422, 1999.

[8] Edward Glaessgen and David Stargel. The digital twin paradigm for future NASA and US Air Force vehicles. Structural Dynamics, and Materials Conference: Special Session on the Digital Twin. 2012: 1–14.

[9] Michael W Grieves. Product lifecycle management: the new paradigm for enterprises. *International Journal of Product Development*, 2(1–2):71–84, 2005.

[10] Eric Guiffo Kaigom and Jurgen Rossmann. Value-driven robotic digital twins in cyber-physical applications. IEEE Transactions on Industrial Informatics, 2020, 17(5): 3609–3619.

[11] James Kennedy and Russell Eberhart. Particle swarm optimization. In Proceedings of ICNN'95-International Conference on Neural Networks. IEEE, volume 4, pages 1942–1948, 1995.

[12] Open Source Robotics Foundation. Xml robot description format (urdf). Online: https://wiki.ros.org/urdf/XML/model, 2012. Accessed: 18.02.2021.

[13] Tobias Osterloh and Jurgen Roßmann. Versatile inverse dynamics framework for the cross application simulation of rigid body systems. Modelling and simulation 2020 : the 2020 European Simulation and Modelling Conference : ESM '2020 : October 21–23, 2020, Toulouse, France, Seiten/Artikel-Nr: 245–252,.

[14] Hae-Won Park, Koushil Sreenath, Jonathan W Hurst, and Jessy W Grizzle. Identification of a bipedal robot with a compliant drivetrain. *IEEE Control Systems Magazine*, 31(2):63–88, 2011.

[15] Plattform Industrie 4.0. Details of the asset administration shell, 2022. Accessed: 25.02.2021.

[16] Qinglin Qi, Fei Tao, Tianliang Hu, Nabil Anwer, Ang Liu, Yongli Wei, Lihui Wang, and AYC Nee. Enabling technologies and tools for digital twin. *Journal of Manufacturing Systems*, 58:3–21, 2021.

[17] Satellite Applications Catapult. UK opportunity for in-orbit services, 2021. Accessed: 24.12.2021.

[18] Michael Schluse, Marc Priggemeyer, Linus Atorf, and Juergen Rossmann. Experimentable digital twins-streamlining simulation-based systems engineering for industry 4.0. *IEEE Transactions on Industrial Informatics*, 14(4):1722–1731, 2018.

[19] Michael Schluse and Juergen Rossmann. From simulation to experimentable digital twins: Simulation-based development and operation of complex technical systems. In 2016 IEEE International Symposium on Systems Engineering (ISSE). IEEE, pages 1–6, 2016.

[20] David Stewart and Jeffrey C Trinkle. An implicit time-stepping scheme for rigid body dynamics with coulomb friction. In *Proceedings 2000 ICRA. Millennium Conference. IEEE International Conference on Robotics and Automation. Symposia Proceedings (Cat. No. 00CH37065)*. IEEE, volume 1, pages 162–169, 2000.

[21] David E Stewart. Rigid-body dynamics with friction and impact. *SIAM Review*, 42(1):3–39, 2000.

[22] Jeffrey C Trinkle, J-S Pang, Sandra Sudarsky, and Grace Lo. On dynamic multi-rigid-body contact problems with coulomb friction. *ZAMM-Journal of Applied Mathematics and Mechanics/Zeitschrift fur Angewandte Mathematik und Mechanik*, 77(4):267–279, 1997.

[23] Samuel Zapolsky and Evan Drumwright. Quadratic programming-based inverse dynamics control for legged robots with sticking and slipping frictional contacts. 2014 IEEE/RSJ International Conference on Intelligent Robots and Systems. IEEE, 2014: 3266–3271.

Part 5

Digital Twins in Management

20

Management of Digital Twins Complex System Based on Interaction

Vladimir Shvedenko
Russian Institute of Scientific and Technical Information of the Russian Academy of Sciences (VINITI RAS)
Federal Agency for Technical Regulation and Metrology ROSSTANDART

Valeria Shvedenko
T-INNOVATIC Ltd

Oleg Schekochikhin
Kostroma State University

Andrey Mozokhin
Department of Automated Systems of Process Control of SMGMA Group

20.1 Introduction

In accordance with the concept of Industry 4.0, many organizational and technical systems will interact with each other based on the system of systems (SoS) concept. At the same time, some of them will be represented as cyber-physical systems interconnected by a global computer network Internet and focused on the use of cloud technologies, quantum computing and artificial intelligence. The trend of such development is obvious and requires theoretical and engineering study of data exchange and storage, reasonable synchronization of the targets of interacting systems and efficient allocation of shared resources.

It is obvious that these systems are complex heterogeneous structures that have specifics of a particular subject area, operate with their own established concepts, parameters and indicators, and are independent or weakly dependent on each other.

The process of interaction of these systems can be ambiguous, both leading to the formation of positive synergistic effects, and deteriorating the characteristics of one or both systems.

DOI: 10.1201/9781003425724-25

To reduce the risks of adverse results, it is advisable to be able to conduct virtual testing of interaction processes in order to identify their most effective parameters, determine the "bottlenecks" of cooperation and assess the degree of emerging risks and their consequences. Such a tool of interacting cyber-physical systems can become a digital twin as a virtual environment for conducting experiments and debugging of real-world tasks based on modeling of predictive scenarios of achieving alternatives.

The main complexity of the interaction of independent cyber-physical systems is associated with their different goal setting, as well as heterogeneity of physical, technological, informational, organizational and managerial environments. This requires the creation of conditions for interoperability of interacting systems to ensure the coherence of information flows and to determine the rules and conditions of data exchange.

Part of these functions can be assumed by digital twins. However, if you create a digital twin of the whole system, it will have high complexity, monolithic and inertia to changes, which ultimately leads to inefficient management. The most promising way from our point of view is to create digital twins of individual objects of the system and their association.

Issues of integration of the information space of interacting systems belong to the knowledge area of SoS design and management.

20.2 Methodological Foundations of the Polystructural Representation of SoS

Two or more independent complex systems can be combined into a polystructural system (PS), which corresponds in its classification features to a directed, component, or recognized SoS. A polystructured system is inherently heterogeneous. Its constituent elements are independent systems that differ in the characteristics of controlled objects and processes, the sources used, methods and mechanisms for obtaining, processing and interpreting data, have different structures, relationships, targets, resource and other constraints. PS elements can simultaneously be part of several SoS and take part in solving several classes of problems.

PS has the following set of properties:

1. It is open and dynamically changeable. New elements can be added to the system, existing elements can be excluded or existing elements can be made latent.
2. Allows achieving synergistic effects or implementing new functions, generating new regularities, generalizing experience of

different subject areas and revealing new connections and forms of influence on subordinate or otherwise connected elements of the system.

It is characterized by the presence of a set of rules and a language of interaction of PS elements through built-in mechanisms of cleaning, aggregation, routing and synchronization of input and output data flows.

3. Represents a system of objects, processes, functions and links between them, the depth of decomposition of which is determined by the type of tasks being solved and the details of which can be expanded.
4. The vector of target development of each PS element is represented in the form of a tree of goals and a set of indicators assessing the state of its objects and processes.
5. The structure and complexity of each element of PS, as well as its relationships with other elements of PS and systems that are not part of it, can be transformed under the influence of external and internal factors, which requires the presence in PS of a system for monitoring and forecasting of emerging deviations, the mechanism for their adjustment.
6. The object state indicators of each PS element, as well as their transformed values form the data bank of PS indicators, and through the integrating PS element can be available to each other.
7. The life cycle of the PS is defined by the objectives facing it.
8. The presence of the integrating element PS, which we will hereinafter refer to as the body of the polystructure (hereinafter—BPS). BPS performs the functions of developing, adjusting and programming the orchestral score of interaction of PS elements, coordinating their target benchmarks and jointly used resources to obtain a positive result and level out the emerging risks. At the same time, BPS supports multi-level PS management within the information space of the PS and organizes the interaction procedures of PS elements without affecting the independence of their systems of indicators, measurements and thesauri.

Each PS element (EPS) has its own organizational form and uses process or functional management, as well as their combination (Shvedenko, 2019).

A distinctive feature of the proposed PS model is the way it organizes the integrating element—BPS, whose functions are focused on synchronizing or reducing the divergence of the target settings of PS elements to achieve a common goal.

20.3 Organization of Information Flows in a Polystructured System

The BPS information space is subdivided into segments of the outer and inner layers. Each PS element corresponds to a segment of the outer layer. The internal integrating layer of BPS consists of the domain ontology, the goal tree and the information model.

The structural components of PS, which form its basic functions, are the following aggregates:

- polymetric system of the polystructure body (PMS_BPS),
- metric system of a polystructure element (MS_EPS),
- target tree of the polystructure body (GT_BPS),
- target tree of the polystructural system element (GT_EPS),
- validator of filtering, grouping and data transformation ($V_{PMS\_BPS-MS\_EPS}$).

A schematic diagram of the relationships of these units is shown in Figure 20.1.

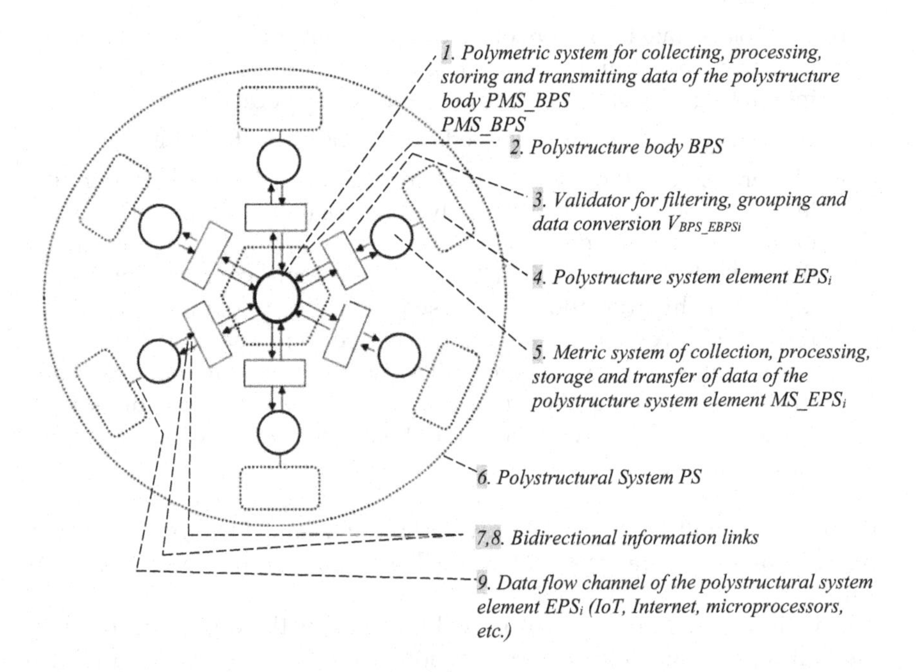

FIGURE 20.1
An enlarged scheme of organization of information flows between the elements of a polystructured system and the body of a polystructure.

PMS_BPS acts as a tool for forming and coordinating the information data flows both in BPS and between BPS and PS elements and performs the following functions:

- assigning conditions for "cleaning" and completeness of a set of input data streams based on the specified business logic;
- defining the rules for translating input data into a single measurement system and their further processing (conversion, aggregation) to obtain metrics—key indicators used in decision-making at the operational, tactical and strategic levels of PS management;
- monitoring of deviations for specified time intervals of actual values of metrics from their target values, which are normative values of the objectives tree PS;
- calculation of the size of deviations of actual metric values to determine the type of management action;
- establishment of rules for transmitting signals on the emergence of an abnormal situation;
- routing of data on existing or predicted deviations of metrics values to relevant decision-making centers.

PMS_BPS includes the following:

- a set of metrics—aggregated and transformed metrics reflecting the state of PS objects and processes, which are managed in accordance with the BPS goal tree;
- a set of methods of aggregation and transformation of MS_EPS input metrics into PMS_BPS metrics;
- a set of methods to compare actual and predicted values of PMS_BPS metrics and their normative values;
- conditions and rules for routing PMS_BPS metrics deviation values to decision-making centers (points of active influence on changes in the properties of the managed system objects) to effectively achieve the specified PS goals.

Metrics are formed from the number of indicators that characterize the state of a property or group of properties of the system objects.

Each metric is paired with a goal tree sheet to evaluate the deviation from the normative indicators of the system.

Depending on the deviation, a signal is formed to be sent to the appropriate decision-making center.

A distinction is made between the magnitude of deviations for sending a signal to the areas of operational, tactical and strategic centers of formation and decision-making. The ranges of these critical deviations are set for each control object.

The connection between BPS and the metric systems of its elements MS_EPS is established through the validators of filtering, grouping and data conversion $V_{PMS\_BPS - MS\_EPS}$.

Depending on the types of tasks to be solved, BPS can interact with two or more PS elements. Any interaction implies the adjustment of management parameters, which implies the adjustment of

- the structure of the goal tree;
- the system for measuring changes in the state of the objects under study, as well as deviations of their values from the established standards;
- values of normative indicators, which correspond to the values of the leaves of the objectives tree.

The construction of GT_BPS is carried out in accordance with the goal facing PS, and may change over time, depending on environmental conditions, the parameters of resource availability of one or more of its elements, as well as a number of other factors.

The hierarchical structure of GT_EPS contains the branches of the goal tree corresponding to the functions of the EPS objects, and its leaves are the normative values of indicators characterizing the state of individual properties of objects, when performing one or more processes. Each object of the managed system can correspond to several leaves in the goal tree structure.

Like GT_BPS, GT_EPS can be periodically restructured, which is associated with the change of PS targets or PS elements. Similarly, their alignment methods may change. The parameters of the BPS goal tree are set in relation to the goal trees of PS elements. Each GT_BPS sheet corresponds to a subset of the BPS indicators.

The schematic diagram of the information interaction of GT_BPS, GT_EPS, PMS_BPS, and MS_EPS is shown in Figure 20.2.

PS control depends on the accuracy of its state prediction for a given time lag.

The most promising technologies for the realization of this function have become digital twins.

Therefore, the digital twin is an important element of PS.

Figure 20.3 shows a scheme of interaction of digital twins with other PS components.

The GT_BPS configuration component sets the hierarchy of PS goals and subgoals, as well as the links between them, and sets the normative values of the GT_BPS leaf indicators. It consists of several modules:

- a module for constructing the goal tree and defining the rules and conditions for its decomposition;

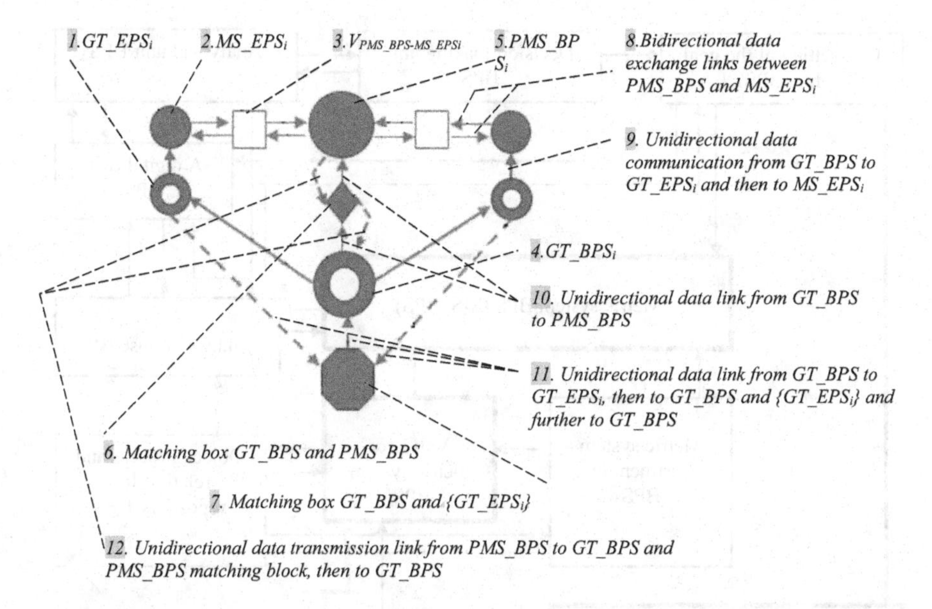

FIGURE 20.2
The basic scheme of information interaction of GT_BPS and GT_EPS, as well as PMS_BPS and MS_EPS.

- module for filling the leaves of the objectives tree with normative values;
- a module for the coordination of normative values of the leaves of the objectives tree, which provides the selection of the best option for their combination;
- a library of formed tree of objectives and their versions;
- a library of methods to set the normative values of GT_BPS leaves;
- module for the establishment of rules and conditions for loading a particular GT_BPS.

GT_BPS leaves are mapped to the leaves of each GT_EPS.

Based on the BPS target tree, the structure of the polymetric BPS system is formed.

Through the PMS_BPS configuration block and the VPMS_BPS-MS_EPS validator, the two-way data transmission between PMS_BPS and MS_EPS is configured.

By performing a variation analysis of the ratio of GT_BPS and GT_EPS leaf values and considering the directional vectors of the BPS and EPS targets, the digital twin simulates the values of the predicted outcome and offers decision-makers several options. There are possible situations of automatic decision-making according to the given parameters of the choice of the best alternative.

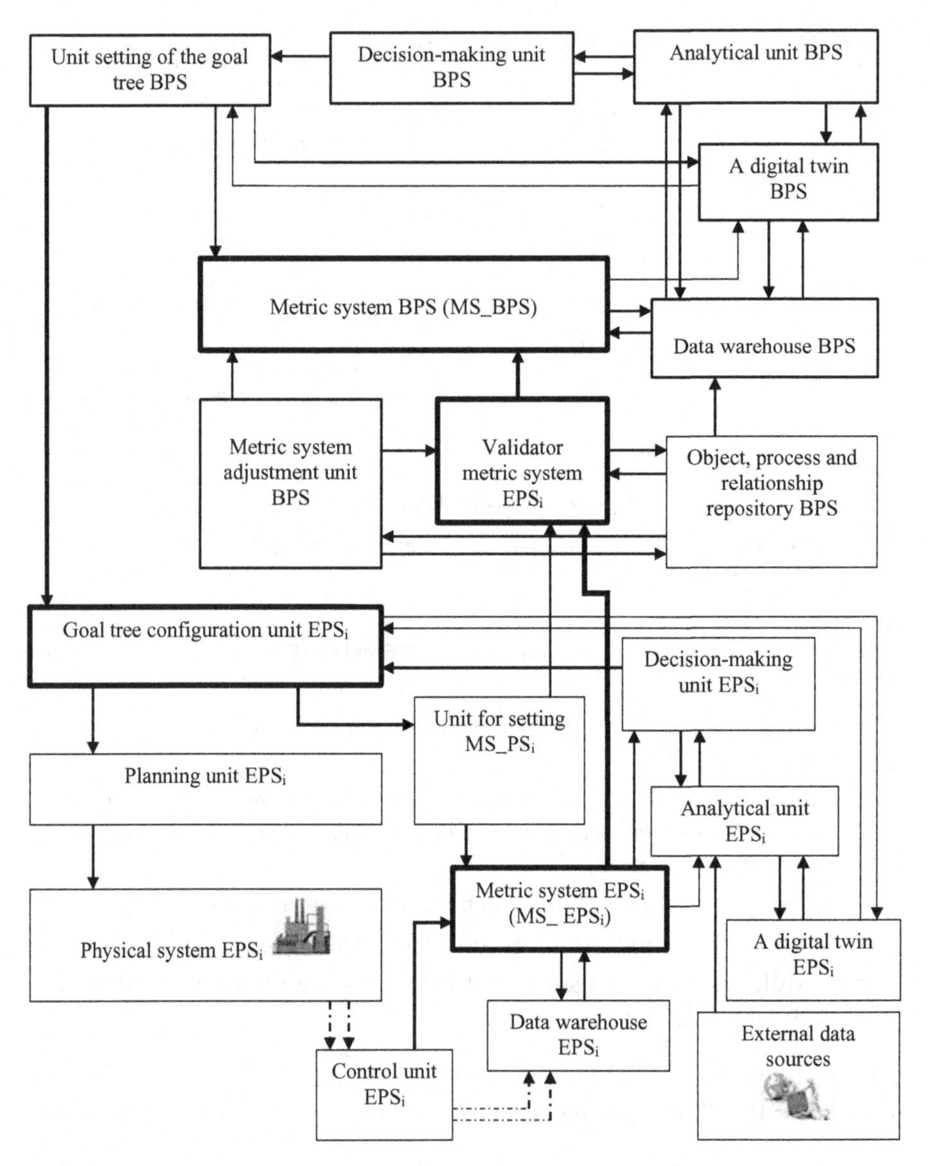

FIGURE 20.3
Architectural solution of interaction between BPS and EPSi.

In a cyber-physical system, the digital twin now becomes an element of system control and is a multiphysical, multiscale probabilistic model of a physical object or system, for predicting its state. This model uses the best available physical models, certified or standardized methods of engineering calculations, including industry standards and guidance documents focused on the design and production of specific products or assemblies, statistical data on similar objects, etc.

A digital twin will be effective when highly accurate simulation is provided and linked to the physical facility condition management system, maintenance history, and all available historical and similar facility data to reflect the condition of the controlled facility and provide a given level of safety and reliability.

Since a production system has a large variety of constituent elements, it can contain many digital twins, which will be in interaction with each other.

20.4 Metrology of the Digital Twin

In the digital twin laid the mechanism of comparison of predicted and actual values, as well as their synchronization in accordance with a given regulation by adjusting the models of objects and processes. The digital twin assumes the obligatory presence of the virtual environment which describes a part of the phenomena and the processes proceeding in the physical environment and the mechanism which provides the interaction of physical and virtual environments. One of the main elements of this mechanism is the metrology of the digital twin. Metrology distinguishes between physical and virtual processes. Physical processes accompany the processes of functioning of the object whose digital twin is created. Parameters of physical processes are measured by, for example, IIoT and IoT sensors.

The basic tasks of metrology are as follows:

1. measurement by sensors of the current state of the controlled system to correct the exact model of the digital twin. Here the key function should be the periodicity of measurement and transfer of the digital twin in a new transaction.
2. evaluation of a mathematical model of PS state prediction depending on external environment conditions and internal processes occurring in EPS.
3. changing regulated parameters to minimize the probability of an emergency situation.

It is proposed to consider the option when digital twin metrology becomes a tool for creating and using self-developing systems.

In most cases, high-precision modeling can be provided by qualitative physical models, possibility of correct assignment of boundary conditions, as well as necessary volume of statistical data. This will allow taking into account both deterministic and random processes observed in production systems.

Probabilistic processes, nevertheless, require their mathematical calculations for determination of operability of design and technical product as a whole, as well as to make certain decisions to change the modes of its operation.

The effect of using a digital twin is the ability to predict the response of the system to perturbing influences that lead to events critical to the efficiency of management and safety, and identify previously unknown problems before they occur in a real object and become probable, as well as to prevent unwarranted overuse of material, financial, labor and other resources.

The digital twin is considered from a position of constant improvement of model, its accuracy and prediction accuracy taking into account the probabilistic nature of properties of physical objects, which are collected in a digital twin.

Metrology of digital twin allows us to receive a forecast of a state of system at the decision of a wide range of problems from carrying out engineering calculations to an estimation of economic efficiency of various processes in cyber-physical system, including an estimation of economic feasibility of application of digital twins.

Qualitative metrology of digital twin gives the opportunity to predict the state of the system for a given time lag. The more exact the model, the more effective will be management decisions. High accuracy of digital twin measurements allows us to reach three models of behavior.

The algorithmic behavior model allows maintaining the stationary state of the controlled system, according to predetermined algorithms.

The trigger behavior model allows the system to move from one steady state to another steady state.

The search behavior model is used to find a solution in conditions not previously encountered or not present in the historical database of the digital twin.

Construction of a digital twin of a cyber-physical system is performed in five stages:

1. Construction of the goal tree of the controlled system.
2. Creating a metamodel of objects.
3. Creating a metamodel of processes.
4. Formation of mathematical models, linking indicators of objects and technological processes with controlled parameters and properties of the controlled system.
5. Creation of the object data warehouse.

To describe the metamodels of objects and processes, an object-process data model is developed, which is described in detail in Shchekochikhin (2018).

To describe the structures of objects and processes, the set-theoretic representation and the mathematical apparatus of graph theory are used, and to model semantics the language of predicate calculus is used.

The basic entities of the model are represented by sets, and the tree of goals and the structures of business processes are described with the use of graphs. Using the predicates of the first order allowed to develop the basic structures of the language of representation and manipulation of data in the object-process model (Shchekochikhin, 2017).

A reference model of the information system based on the object-process data model is proposed. The basic idea is that the construction of the information system (IS) is top-down—from the goal tree to the stages of business processes. And the movement of data during the operation of the information system is bottom-up—from IIoT data providers and controllers. The data is processed and aggregated during the execution of business processes, taking into account the business functions inherent in each stage of the business process. The aggregated data is further transferred to the leaves of the goal tree as the actual values of the cyber-physical system state indicators.

This way of organizing information flows involves the use of digital twins in the aggregation of data and the formation of predictive values.

The way the digital twin is implemented in this case is not decisive. The main factor will be the prediction accuracy and the time period for which the aggregated values will coincide with the predicted values with a given accuracy.

This is the main function of the digital twin metrology. The digital twin metrology provides a highly accurate measurement of the characteristics of a technical object and physical environment and supplies this data to the information system database.

To perform typical calculations, we will use a hierarchy of classes of objects of the subject area. As well as three types of polymorphism: structural, parametric and interclass, as detailed in Shvedenko (2019a, b).

One manifestation of polymorphism can be considered methods for calculating the probability of failure of objects of cyber-physical system.

One of the functions of the digital twin metrology should be the adjustment of the digital model depending on the primary data obtained, as detailed in Shvedenko and Shchekochikhin (2022).

The time to update the digital model must be justified by the costs of development, storage, measurement, transmission, adequacy assessment and other overheads.

Two options for implementing digital twin model transactions are proposed. The first approach is based on a constant time cycle of updating the digital model. The second approach is based on the idea of changing the digital model with periodicity, which depends on the value of the probability of occurrence of one or another event, for example, the probability of failure of one of the system objects.

20.5 An Example of Creating a Polystructural System with Control of Interacting Digital Twins

Human living space can be considered a PS. The following elements of polystructure are considered heating system, air conditioning system, etc.

Each of the cyber-physical systems of a "smart house" acts as a source of information or as a recipient of information, and often as both.

However, depending on the types of work performed, the properties of controlled objects, the methods used to assess the adequacy of the results achieved, the devices used, sensors and so on, each cyber-physical system corresponds to its own set of indicators, the measure and frequency of measuring their values, a set of ontologies and target benchmarks for the results achieved.

The problem is overconsumption of energy resources by components of smart house systems, connected with the inconsistency of modes of operation of devices, in the course of their functioning. So the joint work of heating, ventilation and air conditioning systems leads to the growth of power consumption at their uncoordinated work on achievement of the purpose of maintenance of a microclimate of inhabited space.

The main reason that reduces the efficiency of joint functioning of climate control devices in the premises of a "smart house" is the presence of hierarchical control systems, aimed at performing private tasks. This leads to inconsistency in the joint operation of air conditioning, ventilation and air heating devices and also creates conditions for conflicts between dispersed control systems in the allocation and consumption of resources at the stages of technological processes in the living space of a smart house.

20.6 Creating a Digital Model of a Smart Home

We will consider technology of the smart home in relation to places of human living, as well as objects of social and cultural life. The physical environment of a smart home includes elements of the building structure, supply and exhaust ventilation, temperature and humidity conditions of the room air environment, as well as the human factor that affects the indicators of the physical environment.

In the polystructural smart home system (SHP), the data received from the objects of the physical environment are digitized and transmitted through communication channels into the digital environment. The digital SHP environment is a virtual analog of the physical environment, in which data on the state of real objects and the human environment are used for analysis, synthesis and processing, in order to model real processes and develop controlling effects on the objects of the physical environment. The digital environment uses useful data to create digital models of smart home components and the polystructure as a whole.

When creating digital SHP models, the following system decomposition methods are used:

- physical process decomposition, functional decomposition;
- decomposition based on the life cycle of smart home components.

A SHP is created on the basis of digital models of components connected to each other by means of energy and information flows.

The main components of SHP are as follows:

1. the management and monitoring subsystem;
2. communication subsystem, intended for formation of infrastructure of interconnection of components of the smart home by means of the integration data bus;
3. subsystem of the premises microclimate;
4. lighting subsystem;
5. the power supply subsystem that provides uninterrupted power to all other subsystems;
6. subsystem of security.

A distinctive feature of the method of designing a digital model of a smart home is the organization of connections between its components by determining the jointly used energy and information flows. In this case, it is appropriate to use collinear connections (Shvedenko 2019a, b).

The goal of SHP is to provide a comfortable physical environment for the people who are there while ensuring a minimum expenditure of electrical energy.

20.7 The Digital Twin in a Smart Home Polystructure Management System

The feature of application of the digital twin in the management system of the polystructure of the smart home (DT SHP MS) is its openness for the inclusion of new components, during the life cycle of the smart home.

The functional purpose of the DT SHP MS is as follows:

- customizable validation of information data streams from SHP components;
- consolidation of DT SHP projections obtained by different methods of its components decomposition;
- provide information interaction of digital models of smart home components through the physical environment by means of collinear connections (aggregation of model versions);
- data synthesis to form SHP digital shadows;
- DT SHP data aggregation;
- simulation modeling of operation modes of SHP components during operational and strategic planning;
- analysis of energy and information connections of smart home subsystems;
- generation of control actions on the components of SHP.

Thus, we can select several functional blocks included in DT SHP MS: block of customizable validation of input data, block of data consolidation, block of provision of information interconnection, block of simulation, block of analysis and management block.

The structural diagram of DT SHP MS interaction with smart home components and the physical environment is shown in Figure 20.4.

Data collection from different digital devices of smart home components is implemented by means of an integration bus, which realizes the principle of information consolidation for different protocols of devices included in the SHP. Before the primary data from the physical to the digital environment comes in, it is validated. As a result, noises, incorrect or false data are filtered out. Further processing and analysis of the input data takes place in the digital environment.

The main difference of management using DT SHP consists in economy of resources of the components themselves, due to modeling of the management process in the space of digital models SHP. The control action is sent to the component actuator only after evaluating its interaction with other components and their influence on the parameters of the physical environment. The influence is evaluated by recursive processing of information from IoT sensors of components and the physical environment. Modeling of the state of the physical environment is implemented in a digital environment by means of artificial neural networks (Mozohin, 2021).

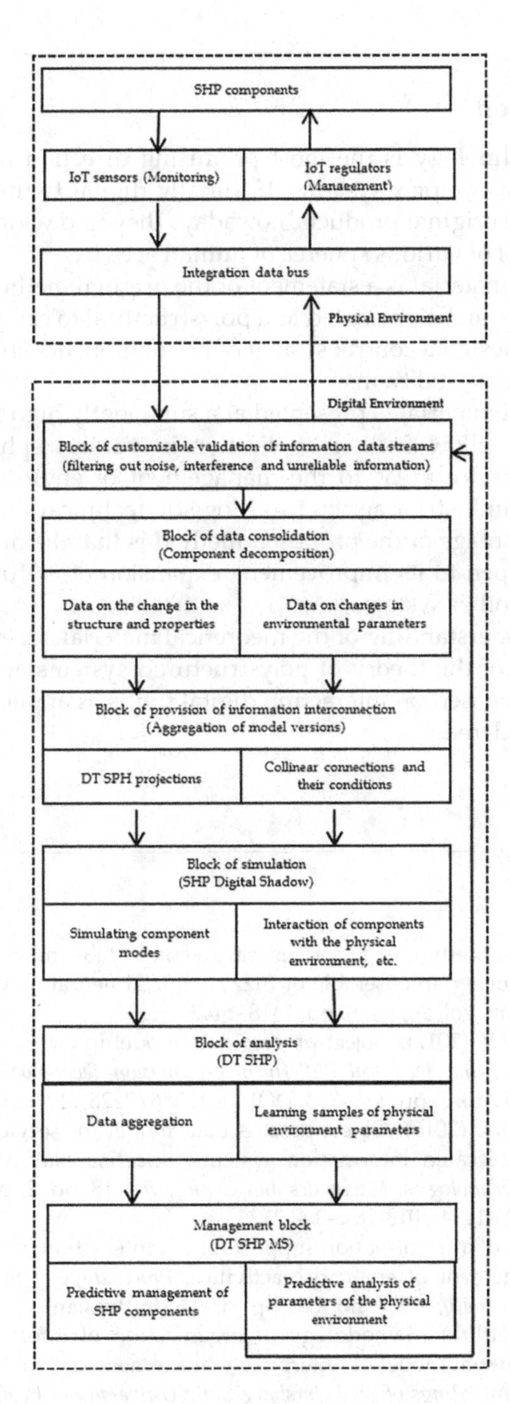

FIGURE 20.4
Structure diagram of DT SHP MS interaction with smart home components and the physical environment.

20.8 Conclusion

Digital twin technology is the most promising direction in the organization of control of complex systems. If initially digital twins were used for science-intensive original products, nowadays they find wide application for automatic control of various spheres of human activity.

The proposed material is a statement of the original method by which the managed system can be presented as a polystructural to carry out its decomposition and to design a control system of its components on the basis of the use of interacting digital twins.

The theoretical material is presented at a sufficiently high level of abstraction, which will allow further application in various spheres of human activity: from service areas to the management of environmental, social, socio-economic, industrial, agricultural, logistic, technical and other systems.

The main advantage of the proposed method is that the management system is built as open to its improvement, expansion of its functionality and interaction with other systems.

For a better understanding of the theoretical material, an example of practical application of the theory of polystructured systems and management of their elements based on interacting digital twins as applied to the energy industry is considered.

References

Mozohin, A. (2021). Methodology for ensuring a comfortable microclimate state in a smart home using an ensemble of fuzzy artificial neural networks. *Informatics and Automation*, vol. 20, no. 6, pp. 1418–1447.

Shchekochikhin, O.V. (2017). Object-process data model in control information systems. *Scientific and Technical Bulletin of Information Technologies, Mechanics and Optics*, vol. 17, no. 2, pp. 318–323. DOI: 10.17586/2226-1494-2017-17-2-318-323.

Shchekochikhin, O.V. (2018). Object-process data model for service-oriented architecture of integrated information systems. *Scientific and Technical Bulletin of Information Technologies, Mechanics and Optics*, vol. 18, no. 2, pp. 307–312. DOI: 10.17586/2226-1494-2018-18-2-307-312.

Shvedenko, V.V. (2019a) Information support for the interaction of process and functional management of enterprise activities. *Proceedings of St. Petersburg State Economic University*, vol. 6, no. 120, pp. 90–94. (In Russian).

Shvedenko, V.V. (2019b) Methodology of organization of information flows and process-functional model of enterprise management and tools for their implementation. *Proceedings of St. Petersburg State University of Economics*, vol. 5, no. 119(Part 1), pp. 128–132. (In Russian).

Shvedenko, V. and O. Shchekochikhin (2022) A new view on the metrology of digital twin objects of production processes on the example of assessing the serviceability of metal cutting tools. URL: https://www.researchsquare.com/article/rs-2587376/v1.

Shvedenko, V.N., V.V. Shvedenko, and O.V. Shchekochikhin (2019) Using structural and parametric polymorphism in the creation of digital twins. *Automatic Documentation and Mathematical Linguistics*, vol. 53, no. 2, pp. 81–84.

21

Artificial Intelligence Enhanced Cognitive Digital Twins for Dynamic Building Knowledge Management

Gozde Basak Ozturk and Busra Ozen
Aydin Adnan Menderes Universitya

21.1 Introduction

Nonaka and Takeuchi (1995) discussed knowledge management (KM) from the systematical point of view, since then it has been recognized as an important component to achieve organizational goals and an organizational asset for sustainable competitiveness (Farooq, 2018; Birzniece, 2011). Throughout the industrial revolutions, developments in technology transformed the knowledge-intensive industries. This transformation has attracted great attention in architecture, engineering, construction, operation, and facility management (AECO/FM).

The improvements in Information and Communication Technologies (ICTs) have brought benefits to the AECO-FM industry, such as the emergence of the transition to construction 4.0 (Kozlovska et al., 2021). The fact that the AECO-FM industry plays an important role in the economy of countries has revealed the necessity of digitalization in the industry. However, the AECO-FM industry is conservative in adopting new technologies and the presence of many stakeholders slows this transition. Although the application of technologies in construction project processes has been a challenge, digitalization is triggered in the AECO-FM industry because of the difficulty in communication among stakeholders and of the fragmented structure in the construction industry. Delivery time can be reduced, and efficient processes can be achieved by providing integration between stakeholders via digital technologies (García de Soto et al., 2022).

The AECO-FM is a highly knowledge-intensive industry with its fragmented nature, organizations' divergent structure, uniqueness of projects, ad-hoc project teams, and other industry-related challenges that have negative effects on KM throughout the project lifecycle (Kim, 2014).

DOI: 10.1201/9781003425724-26

Storing, sharing, and evolving knowledge is a challenge. In the computing era, these technologies have generated enormous amounts of data flow traffic, which triggers the creation of big data. The organizations in the AECO-FM industry have to deal with a vast amount of data continuously streaming among many stakeholders, processes, divergent professional areas, and information platforms. A construction project's chaotic knowledge transaction environment needs to be digitized for effective and efficient management of the construction project-related knowledge throughout the project lifecycle (Lin, 2014). Like the other industries in the digital age, the AECO-FM organizations also prioritize data collection, segregation, and supervision. KM refers to the process that includes different activities such as information capture, sharing, storage, use, and reuse to achieve the goals and objectives of an organization (Ozturk and Yitmen, 2019; Wang and Meng 2019; Farooq, 2018; Majumder and Dey, 2022; Rundquist, 2014). KM throughout the project lifecycle requires a computational perspective to minimize error and maximize efficiency in KM (Lin, 2014). Therefore, in the era of digital transformation, digital resources can be enriched and collaboration can be achieved with the integration of KM and Industry 4.0 technologies. Artificial intelligence (AI) technologies enhance the data-to-knowledge transformation. AI is an increasingly focused subject in the AECO-FM industry-based research studies associated with KM. Researchers suggest that semantic web and ontology studies boosted the ICT use in KM. Therefore, Big Data Technologies and ICT will be unitedly used for the future applications of KM (Yu and Yang, 2018).

AI-based systems contribute to decision-making processes and the protection of data integrity (Birzniece, 2011). In addition to the input and output units, information storage environments are also considered the main elements of KM. The input, which is the subject of information processes, is the data at the bottom of the data-information-knowledge-wisdom (DIKW) model and represents the characteristics of events and their environments. Knowledge is objective, transferable, transformable, and measurable. It may be formatted, processed, accessible, produced, transmitted, stored, sent, distributed, searched, used, compressed, and copied. In information storage processes, "information" which has an object can be stored and shared. However, it provides "knowledge" to transform the information obtained from the data into instructions for deciding which information to keep in the storage process. After the conversion to knowledge, the information becomes ready for the decision-making process. The right knowledge must be conveyed to the right place at the right time to make decisions. A decision support system, which is used to make fast and robust decisions, integrates human intelligence and information technology to solve complex problems. Since AI helps to analyze very large and complex datasets, it can increase the effectiveness of decision support systems or can be designed for complete human replacement systems.

AI endows machines with the capability of human-like reasoning, learning, communicating, perceiving, planning, thinking, and behaving

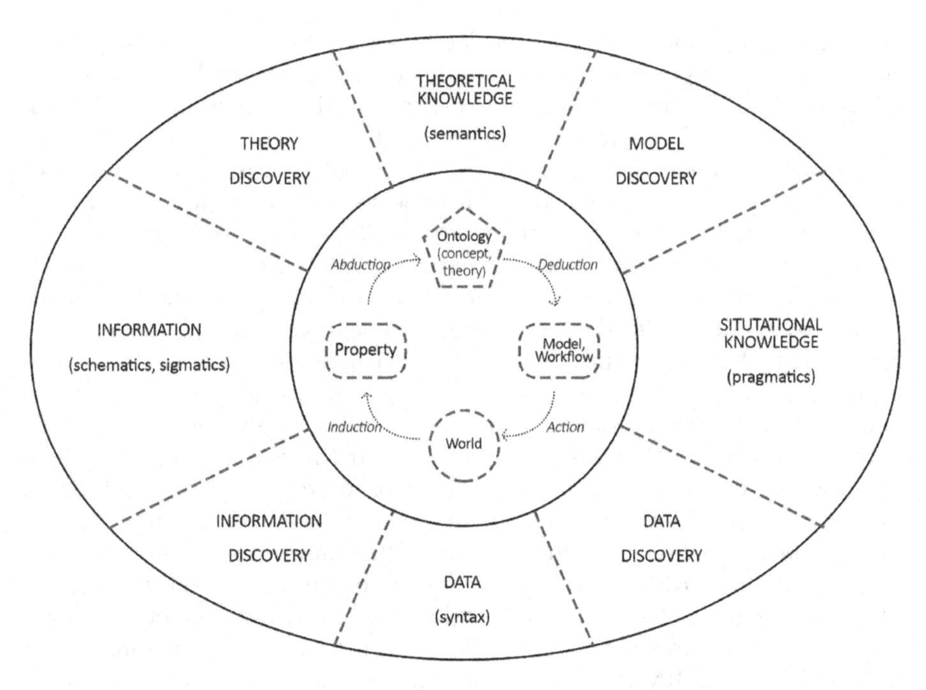

FIGURE 21.1
Knowledge transformation cycle in information systems (Brodaric, 2005).

(Liao et al., 2022). Learning from big data can be performed, and large volumes of data can be analyzed quickly and accurately via AI. The active connection between physical and virtual worlds for KM is ensured with AI technologies (Figure 21.1).

21.2 Artificial Intelligence in the AECO-FM Industry

Adoption of technologies in the AECO-FM industry is slower than in other industries because productivity is not as stable as in other industries and investment in technology is somewhat less (Akinosho et al., 2020). Another reason is that the construction professionals are mostly on the move. One of the challenges faced in the AECO-FM industry is that most industry contributors accept a culture of resistance to digital technologies. This has caused the construction industry to be one of the least digitized industries. Lack of digitization has caused problems such as low productivity, cost overrun, schedule overrun, scope creep, and poor quality (Abioye et al., 2021), leading to a poor decision-making process (Delgado and Oyedele, 2021). In the AECO-FM industry, the subfields of AI such as Natural Language Processing

(NLP), Machine Learning (ML), Deep Learning (DL), robotics, computer vision, automated planning & scheduling, and optimization were enhanced to increase the efficiency and effectiveness of industry applications and operations. However, the AI applications are data-driven, which is a challenge in the industry. Building information modeling (BIM), as an information digitalization platform for construction projects, has been an effective technology to overcome these challenges, indeed ensuring the integration of AI-based applications (Ozturk and Tunca, 2020; Ozturk, 2021a,b,c). AI methods utilize trained algorithms to recognize and infer predefined target sets of concepts. Thus, semantic enrichment of BIM models is ensured (Sacks et al., 2020). AI helps the AECO-FM industry catch up with the rapid pace of automation and digitization by contributing to issues such as forecasting, optimization, and decision-making (Figure 21.2).

ML is related to learning computer programs from historical data for modeling, controlling, and predicting without programming. The subfields of machine learning are supervised, unsupervised, reinforcement ML, and DL. Supervised ML is designed through the learning of datasets for decision-making via classification and regression solutions for chaotic problems (Baskarada and Koronios, 2013). Unsupervised ML is applied to enhance machines to learn from unlabeled datasets to derive the essential structure via clustering and dimension reduction techniques (Gentleman and Carey, 2008). Reinforcement learning aims to maximize the reinforcement signal via mapping information from the interactions with the environment. DL is an accurate predictive one among other ML techniques while working with big data (Oyedele et al., 2021). It is learning from examples of automation in predictive analytics and modeling. Learning rate decay, transfer learning, training from scratch, and dropout are among the techniques to create deep learning models. ML Computer Vision is the artificial mimic method of the human visual perception system. Computer vision concerns processing digital and multidimensional images using algorithms to achieve high-performance decision-making. Automated Planning and Scheduling are used mainly for efficient management of the existing scarce resources in terms of selecting, sequencing, and allocating to achieve desired goals in a complex environment that requires timely and the best solutions. Robotics is a science field that builds on to mimic physical human actions to realize highly specialized tasks by interacting with the environment through sensors and actuators (Delgado and Oyedele, 2021). Knowledge-based systems are an AI technique for machine decision-making using existing domain expert knowledge, past lessons learned, and experiences. The system has a knowledge base, an inference engine, and a user interface that are interacting together to get heuristic, flexible, and transparent conclusions to increase productivity and efficiency in accessing large domain knowledge. Knowledge-based systems can be categorized as expert systems and case-based reasoning systems, intelligent tutoring systems, database management systems, and linked systems. NLP is applied in machine translation, text processing, summarization,

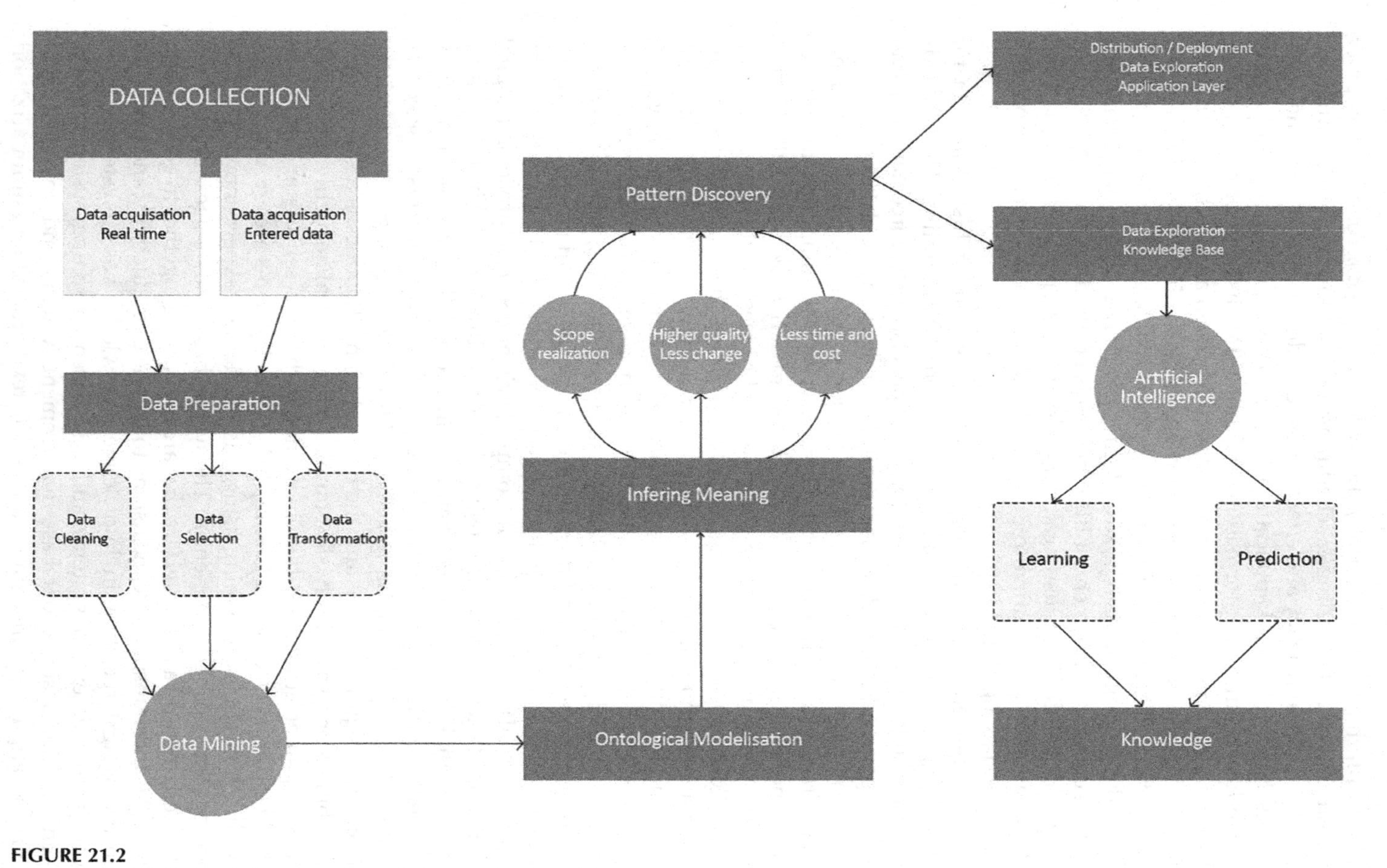

FIGURE 21.2
AI-enabled DIKW (data-information-knowledge-wisdom) model.

cross-language information transfer, expert systems, and speech recognition to mimic the linguistic capabilities of humans. Optimization is used to achieve the best decision to be made in a specific situation with a given set of constraints.

The application of AI techniques in the AECO-FM industry has many benefits mentioned above with the limitations sourcing from the nature of the industry-specific restraints, namely, incomplete data, data stream, high-dimensional data, distributed computing, scalability of models, implementation cost in the field, integration difficulties, high initial cost, unstructured work environment, intellectual property security, data privacy, and computing power (Abioye et al., 2021). Despite the industry-specific and general computing difficulties, there are promising research studies of AI-driven application opportunities in the AECO-FM industry related to holistic waste management, DL for cost and time estimation, BIM-based construction site analytics, chatbot for site information, systems trainers and testers, automation tools, BIM for IoT, building management systems, Augmented Reality-based virtual site exploration system, as-planned vs as-built, blockchain applications for supply chain management, chatbot for supply chain, holistic supply chain management, DL for predictive health and safety management, contract management, etc. (Abioye et al., 2021). In the future process of discovery and diagnosis (Kouhestani and Nik-Bakht 2020), pattern recognition (Pan et al., 2020; Pan and Zhang, 2020), social network analysis (Pan et al., 2020), and time-series analysis (Pan and Zhang, 2020) may be in the focus of the AECO-FM researchers to utilize in smart robotics, cloud systems, immersive systems, IoT, digital twin (DT), 4D printing, and blockchain applications (Pan and Zhang, 2020).

21.2.1 Building Information Modeling

BIM is a technology that can create, share, and use nD information throughout the project and building lifecycle. Condensable project information throughout its lifecycle flows from individual professionals, teams, and organizations to the BIM model, which is a digitalization mediator for the built environment. A BIM model is a digital representation of a building. It can be integrated with multidimensional information as a means of creating and sharing knowledge (Ozturk, 2020). The promise of BIM is to increase efficiency in construction projects by reducing time and cost through its potential such as clear and fast analysis, multipurpose service visualizations, and increased cooperation among all stakeholders. Nowadays, BIM is widely used for performance improvement, reliable predictions, design improvement, and the ability to integrate endless amounts of data (Chen and Tang, 2019).

Data transfer is possible with BIM, enabling interoperability, streamlining, and automation of workflow. The quality and size within a BIM model are expressed as BIM Maturity. BIM Maturity, in other words, is an application improvement level. Progression from low to higher levels provides control, competence, consistency, and remarkable applicability with execution goals (Ozturk, 2021a). Many researchers consider BIM as the main enabler of the construction of the digital twin of a building (Sacks et al., 2020). However, the information provided is out of date because of the absence of a data feedback loop from the real-world building, which prevents the data information and knowledge transformation. The data feedback from the existing building to the BIM model can be acquired via the sensors and actuators. Then, the data can be processed by the AI methods to further be generated, derived, stored, shared, used, and managed by the intended people, platforms, and systems.

21.2.2 Digital Twin as KM Platform

Cyber-physical systems (CPSs) are systems that can sense, manage, and control the physical world and provide a continuous automatic connection between intelligent digital components. CPS forms the basis of Industry 4.0. CPSs create a two-way link between the physical environment and the cyber environment to manage methods and information. They work with sensors and processors that provide feedback with physical means (Klinc and Turk, 2019).

CPS has a five-level architecture. CPS consists of connectivity that provides real-time data retrieval from the physical world and information feedback from the cyber environment, intelligent data management, and analytical computing capability that make up the cyber environment. The first stage is the collection of precise and reliable information from different components by sensors. The second stage, "covering", consists of forecasts, statistics, and data stored in the first stage to support decision-making. The "Cyber" stage is the DT, which is the digital representation of the object that exists or will exist in the physical world. Transformation is provided from data stored at the level of smart links to information. The "Cognition" phase uses AI technologies to do learning and diagnostics. The final phase, "Configuration", provides intelligent responsive, self-learning, and automatic configuration (Lee et al., 2015).

In the AECO-FM industry, CPS offers the potential to solve recurrent problems. Connected autonomous systems can be created to improve the communication, operation, safety, and performance of construction projects in a very short time and to provide a two-way interaction between physical building and digital building models as connected components of a system (Tao et al., 2018). The analyzed data is processed and then transmitted to the necessary information. Built assets are considered components of the physical counterpart. Real-time interaction is a challenge because of the interoperability between BIM and CPS. However, it is used in smart building systems (Linares-Matás et al., 2019).

The DT concept, as a CPS, first emerged in Industry 4.0 for product lifecycle and management. Although the AECO-FM industry has slowly adapted to the digitalization that comes with Industry 4.0, there have been improvements in automated decision-making, processes, and use of information. Match physical assets and virtual assets with DT; a system that matches both virtual and physical assets with their virtual counterparts is envisaged. The concept of DT as a virtual model of a physical entity can help overcome some of the complex issues in AECO-FM operations. The duplication of physical assets in a digital environment for remote viewing, monitoring, and control will affect routine processes and procedures in all organizations. The integration of DT and other cognitive technologies could redefine the future vision of the AECO-FM industry. Today, there is a huge amount of information generated from the building and its subsystems throughout the building lifecycle. Integration with embedded systems can provide an opportunity to capture, store, and share critical building information (Ozturk, 2021a, b, c).

The use of DTs in the AECO-FM industry is a highly desirable need for lean processes and effective decision-making throughout the project lifecycle and building lifecycle. The DT subject is newly presented to the AECO-FM industry (Ozturk, 2021a). The gaps and trends in the DT subject were proposed. The main research topics in the DT subject are stated as building lifecycle management, virtual-based information utilization, virtual–physical building integration, information-based predictive management, and information-integrated production. DT research mainly encompasses subjects such as (Ozturk, 2021a,b,c)

- Building lifecycle data use for project management
- Design, construction, and risk assessment, code compliance check, walkthrough experiences for monitoring, controlling, and real estate marketing, and user interaction through virtual data
- Twinning the digital model with real-time data through BIM model integration
- Stakeholder collaboration to improve processes for sustainable approaches
- Information management via enhancing AI applications
- Real-time data use for dynamic monitoring
- Digital transformation
- Interoperability

While the below-listed ones are in the scope of the near future research potential:

- IoT for data collection
- Data analytics for predictive management
- ML applications to related data to enhance the data-information-knowledge cycle

- Improved and integrated BIM model
- Smart DT applications
- Smart city applications
- Interactive DT-enabled buildings
- City-level DT subjects

The Defense Acquisition University defines a DT as "an integrated multi-physics, multiscale, probabilistic simulation of an as-built system, enabled by Digital Thread, that uses the best available models, sensor information, and input data to mirror and predict activities/performance over the life of its corresponding physical twin". DT can be generated by a high level of integrated and equipped BIM model with the real-world building interaction. BIM model data is the static data that is integrated into the model as an information representation platform and can be extracted manually from the information repository model in case of need. However, a DT is integrated with the real-time building to collect data from sensors and actuators. The dynamic data collection enhances continuous big data acquisition from the real-world environment about the building. Dynamic data/transactional data flow input can be processed by AI techniques generating output of information or knowledge for use and reuse.

The gap between the digital model and its physical counterpart is decreasing through the development of integration technologies and tools. Innovations in AI in the last decade have provided a rich source for the problem of coping with the collected enormous data repository. As the data collection capabilities increase with sensors and actuators, the embedded processors improve and become less costly. Thereby, more realistic representations of the physical world can be created and stored more efficiently in cloud systems. Sensors enable the collection of real-world data, ensuring the accuracy of the digital model into which real-world data can be integrated. DT consists of three main parts: A physical building in the real environment, a digital model in the virtual environment, and data connection units between physical and virtual objects. Digital models contain the designers' mental expectations and abbreviations of the physical world. DT enables the refinement of design patterns and the iterative optimization of design schemes in product design. Moreover, virtual validation, product function, structure, and fast and easy manufacturability can be predicted. Thus, the defect of the design in the virtual world can be found correctly, the design can be improved and rapid changes can be made. The entire process, from raw material input to finished building output, can be managed via the DT. Digital models and their production strategies can be simulated and evaluated. Real-time monitoring is provided with the repetitive digital–physical interaction. Digital models can update themselves based on data in the physical world (Qi and Tao, 2018).

Modern industrial systems are becoming more and more complex under the influence of technological developments. With DT, it is important to comprehensively network and integrate the entire lifecycle of a system. However, complex systems may contain multiple subsystems and components that have their own DT models. These systems can be created by various stakeholders based on different standards whose data structures are heterogeneous. This may make it difficult to integrate DT models. To overcome these difficulties, the idea of combining the DT with cognitive abilities and advanced semantic modeling technologies was proposed. The concept defined as the cognitive digital twin (CDT) is seen as a promising trend of the DT (Zheng et al., 2021).

Semantic technologies have been used as key enabling tools in smart systems to enable the interoperability of heterogeneous data and information. With semantic models, system information can be captured intuitively and a short and unified description of information can be provided (Rivas et al., 2018). Information is defined in standardized ontology language to determine direct interrelationships between various systems and models. As one of the advanced semantic technologies, the infographic makes it possible to create new knowledge. It enables the modeling of model information in the form of entities and relationships. Thus, by semantic modeling and infographic modeling, heterogeneous DT models in a complex system can be integrated between different domains and lifecycle stages (Psarommatis, 2021).

Research on CDT started with research on giving cognitive abilities to DTs using semantic technologies. First, a developed DT system based on semantic modeling and ontologies was proposed. With this DT proposal, the characteristics and state of the system are captured while interacting with other components in a complex system. Semantic digital twin based on IoT and knowledge graph allows for collecting data that is spread across complex systems and making inferences dynamically in task-oriented frameworks. Afterward, DTs' cognitive abilities were tried to be developed with semantic technologies. CDT has been defined as the digital representation, amplification, and intelligent companion of the physical entity as a whole, including its subsystems, throughout all lifecycles and evolutionary stages. CDT includes cognitive features that enable the perception of complex and unpredictable behaviors and optimization strategies for an ever-evolving system (Zheng et al., 2021) (Figure 21.3).

21.3 AI-Enhanced Digital Twin: A Conceptual Framework

The AECO-FM industry is an inclusive industry with diverse stakeholders from different disciplines and many stakeholders such as the design team, owners, contractors, subcontractors, users, sponsors, and local authorities. The stakeholders aim to produce structures that perform well. The multidisciplinary and fragmented nature of the AECO-FM industry poses challenges

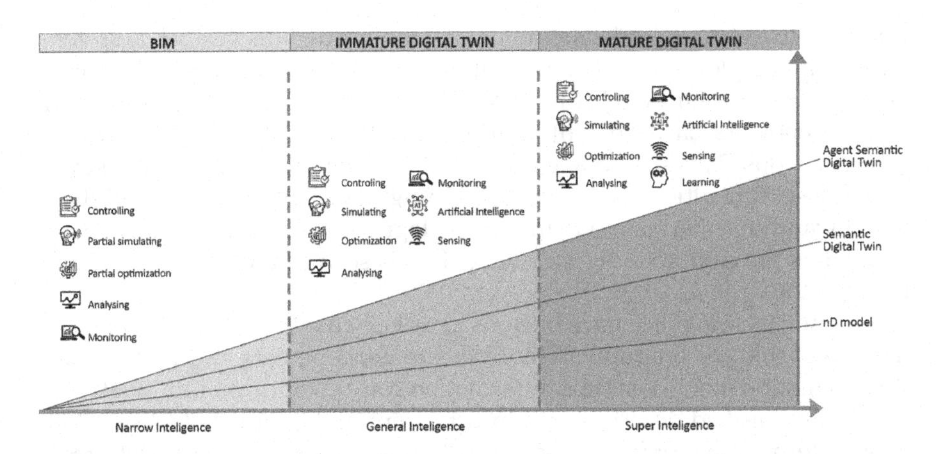

FIGURE 21.3
BIM model to digital twin transition.

in all aspects of the project lifecycle. Therefore, the AECO-FM industry has a complex structure. This fragmentation in the construction industry poses a challenge in communicating information among project stakeholders throughout the project lifecycle. Ensuring cooperation between stakeholders is a requirement to overcome these challenges. This causes difficulties in KM in the AECO-FM industry. Collaboration requires knowledge sharing, coordination, and communication to achieve a common goal. Collaboration can be achieved with new technologies and methods. Even the cumbersome environment of the AECO-FM industry cannot resist the latest opportunities that BIM offers for competitiveness such as less time and cost requirements, higher quality and scope availability, greater collaboration, and less change in construction projects. BIM contains the information of a project and can share this information among stakeholders. Along with Industry 4.0, developments in ICT affected BIM technologies and improved KM (Ozturk and Yitmen, 2019; Ozturk et al., 2016).

The ontology-based semantic enriched and AI-integrated BIM model may provide a better adaptation to the information age. Ontology and semantic web technology achieve representation and reuse of domain knowledge. The integration of BIM technology and AI methods triggers the dynamic DIKW transformation and rapid use of information. While the information is reused, the model size is enriched with related data without excessive growth. The amount of information produced in buildings is increased (Ozturk, 2019). Advances in ICT offer opportunities for online interaction and updating of the information produced. Methods such as IoT and big data analysis accelerate smart systems integration into the BIM model. The AI-enhanced information model provides intelligent and real-time decision-making. With this model, information about the building can be obtained, created, shared, used, and reused throughout the project lifecycle. Models can be linked to each other to enhance learning from each other. Continuing learning provides the ability for continuous self-improvement in the AECO-FM industry. Developments in AI, BIM, and KM will increase the use of learning methods together with techniques such as ML and DL (Ozturk and Yitmen, 2019). The DIKW model in the decision-making processes of AI systems can be reconciled with the processes of the AECO-FM industry. The BIM model was created by integrating data into the design process. An information model was created with nD information integrated into the BIM. Semantic BIM was obtained with enrichment methods to ensure the promotion of data integration, merge heterogeneous information from various sources, formal representation of standardized knowledge to explicit information, search and discovery options, the taxonomic hierarchy for broad exploration, linking related content for serendipitous discovery, hooks to allow connection to external sources and to exchange information across applications, connecting users and content, etc. There is much research in the AECO-FM industry about semantic web technology integration into the BIM model (Niknam and Karshenas, 2017). The machine-readable data is stored in the knowledge databases in IFC format linking concepts to ontologies with a unique resource identifier. The semantic web uses these identifiers to enhance effective and efficient information retrieval. This has transformed into a model that includes knowledge. Afterward, the model is supported throughout the building lifecycle via AI methods generating new information and knowledge for improving the lifecycle management process of the building. Evolution is provided from the data input contained in the model to the transformed information and knowledge. The BIM model evolves into a cognitive model with integrated AI technologies and methods (Figure 21.4). Communication can be established between the physical building and the digital model. As the feedback loop is closed by the two-way communication between physical and digital entities, data can be transferred from the physical asset to the digital model, which is then transformed into information to feed-forward from the self-deciding digital model to the physical building.

21.4 Conclusion

Emerging digital technologies have triggered better KM processes throughout the building lifecycle in the AECO-FM industry. Integrating new technologies such as BIM, DT, and AI together can improve knowledge generation, storage, retrieval, sharing, use, and reuse. With the contribution of digital technologies, smart processes can be achieved throughout the building lifecycle. This chapter presented AI integration to BIM to create a DT for improved KM processes in the AECO-FM industry. Advances in ICT and AI technologies have enhanced the capacity of BIM models

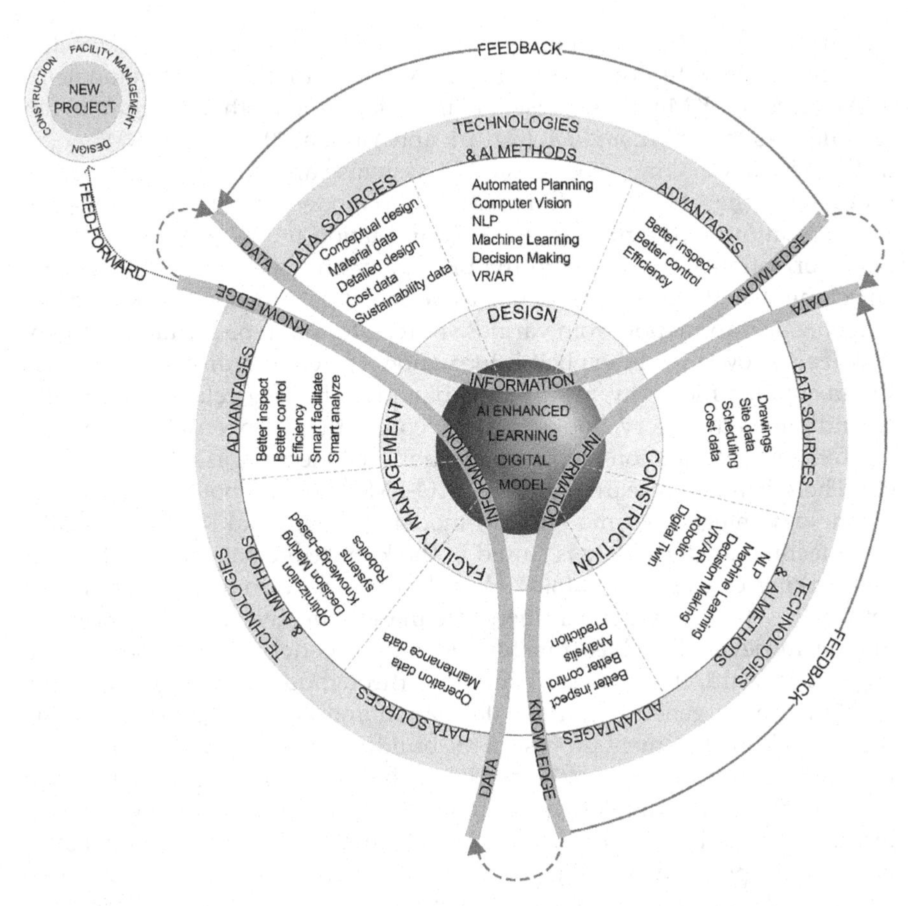

FIGURE 21.4
A conceptual framework of AI-enabled digital twin for AECO-FM activities.

transforming static BIM models into dynamic DTs. Therefore, dynamic information and knowledge flow from the digital model to the physical building and vice versa. Features such as reuse, feedback, feed-forward, updating, communication, and networking for the information management process can be provided by these technologies. However, hardware and software update related to the continuous improvement in digital technologies is needed to establish these systems. Apart from this, there are difficulties such as adapting to new technologies, designing new systems, and accessing very unique and novel technologies. Future work will focus more on the integration of BIM and AI technologies to turn BIM into a CDT of the existing physical building. Thus, communication can be improved between the physical building and the digital model. The digital model can update itself with the presented dynamic physical building data. Thus, the DIKW cycle works in a continuous improvement manner for more efficient KM processes throughout the building lifecycle.

References

Abioye, S. O., Oyedele, L. O., Akanbi, L., Ajayi, A., Delgado, J. M. D., Bilal, M., … Ahmed, A. (2021). Artificial intelligence in the construction industry: A review of present status, opportunities and future challenges. *Journal of Building Engineering*, 44, 103299. doi:10.1016/j.jobe.2021.103299.

Akinosho, T. D., Oyedele, L. O., Bilal, M., Ajayi, A. O., Delgado, M. D., Akinade, O. O., & Ahmed, A. A. (2020). Deep learning in the construction industry: A review of present status and future innovations. *Journal of Building Engineering*, 32, 101827. doi:10.1016/j.jobe.2020.101827.

Baskarada, S., & Koronios, A. (2013). Data, information, knowledge, wisdom (DIKW): A semiotic theoretical and empirical exploration of the hierarchy and its quality dimension. Australasian Journal of Information Systems, 2013, 18(1).

Birzniece, I. (2011). Artificial intelligence in knowledge management: Overview and trends. Computer Science (1407-7493), 2011, 46.

Brodaric, B. (2005). "Representing Geo-Pragmatics", PhD Thesis, The Pennsylvania State University, The Graduate School, College of Earth and Mineral Sciences.

Chen, C., & Tang, L. (2019). BIM-based integrated management workflow design for schedule and cost planning of building fabric maintenance. *Automation in Construction*, 107, 102944. doi:10.1016/j.autcon.2019.102944.

Delgado, J. M. D., & Oyedele, L. (2021). Deep learning with small datasets: using autoencoders to address limited datasets in construction management. *Applied Soft Computing*, 112, 107836. doi:10.1016/j.asoc.2021.107836

Farooq, R. (2018). A conceptual model of knowledge sharing. International Journal of Innovation Science, 2018, 10(2): 238–260.

García de Soto, B., Agustí-Juan, I., Joss, S., & Hunhevicz, J. (2022). Implications of construction 4.0 to the workforce and organizational structures. *International Journal of Construction Management*, 22(2), 205–217. doi:10.1080/15623599.2019.1616414.

Gentleman, R., & Carey, V. J. (2008). Unsupervised machine learning. *ioconductor Case Studies* (pp. 137–157). doi:10.1007/978-0-387-77240-0.

Kim, S. B. (2014). Quantitative evaluation on organizational knowledge implementation in the construction industry. *KSCE Journal of Civil Engineering*, 18(1), 37–46. doi:10.1007/s12205-014-0190-2.

Klinc, R., & Turk, Ž. (2019). Construction 4.0-digital transformation of one of the oldest industries. *Economic and Business Review*, 21(3), 4. doi:10.15458/ebr.92.

Kouhestani, S., & Nik-Bakht, M. (2020). IFC-based process mining for design authoring. *Automation in Construction*, 112, 103069. doi:10.1016/j.autcon.2019.103069.

Kozlovska, M., Klosova, D., & Strukova, Z. (2021). Impact of industry 4.0 platform on the formation of construction 4.0 concept: A literature review. *Sustainability*, 13(5), 2683. doi:10.3390/su13052683.

Lee, J., Bagheri, B., & Kao, H. A. (2015). A cyber-physical systems architecture for industry 4.0-based manufacturing systems. *Manufacturing Letters*, 3, 18–23. doi:10.1016/j.mfglet.2014.12.001.

Liao, L., Quan, L., Yang, C., & Li, L. (2022). Knowledge synthesis of intelligent decision techniques applications in the AECO industry. *Automation in Construction*, 140, 104304. doi:10.1016/j.autcon.2022.104304.

Lin, Y. C. (2014). Construction 3D BIM-based knowledge management system: A case study. *Journal of Civil Engineering and Management*, 20(2), 186–200. doi:10.3846/13923730.2013.801887.

Linares-Matás, G. J., Yravedra, J., Maté-González, M. Á., Courtenay, L. A., Aramendi, J., Cuartero, F., & González-Aguilera, D. (2019). A geometric-morphometric assessment of three-dimensional models of experimental cut-marks using flint and quartzite flakes and handaxes. *Quaternary International*, 517, 45–54. doi:10.1016/j.quaint.2019.05.010.

Majumder, S., & Dey, N. (2022). Artificial intelligence and knowledge management. *AI-Empowered Knowledge Management* (pp. 85–100). doi:10.1007/978-981-19-0316-8_5.

Niknam, M., & Karshenas, S. (2017). A shared ontology approach to semantic representation of BIM data. *Automation in Construction*, 80, 22–36. doi:10.1016/j.autcon.2017.03.013.

Nonaka, I., & Takeuchi, H. (1995). *The Knowledge Creating Company*. Oxford: Oxford University Press.

Oyedele, A. O., Ajayi, A. O., & Oyedele, L. O. (2021). Machine learning predictions for lost time injuries in power transmission and distribution projects. *Machine Learning with Applications*, 6, 100158. doi:10.1016/j.mlwa.2021.100158.

Ozturk, G. B. (2019). The relationship between BIM implementation and individual level collaboration in construction projects. *OP Conference Series: Materials Science and Engineering* (Vol. 471, No. 2, p. 022042). doi:10.1088/1757-899X/471/2/022042.

Ozturk, G. B. (2020). Interoperability in building information modeling for AECO/FM industry. *Automation in Construction*, 113, 103122. doi:10.1016/j.autcon.2020.103122.

Ozturk, G. B. (2021a). Digital twin research in the AECO-FM industry. *Journal of Building Engineering*, 40, 102730. doi:10.1016/j.jobe.2021.102730.

Ozturk, G. B. (2021b). The evolution of building information model: Cognitive technologies integration for digital twin procreation. *BIM-Enabled Cognitive Computing for Smart Built Environment* (pp. 69–94).

Ozturk, G. B. (2021c). The integration of building information modeling (BIM) and immersive technologies (ImTech) for digital twin implementation in the AECO/FM industry. *BIM-Enabled Cognitive Computing for Smart Built Environment* (pp. 95–129).

Ozturk, G. B., Arditi, D., Yitmen, I., & Yalcinkaya, M. (2016). The factors affecting collaborative building design. *Procedia Engineering*, 161, 797–803. doi:10.1016/j.proeng.2016.08.712.

Ozturk, G. B., & Tunca, M. (2020). Artificial intelligence in building information modeling research: Country and document-based citation and bibliographic coupling analysis. *Celal Bayar University Journal of Science*, 16(3), 269–279. doi:10.18466/cbayarfbe.770565.

Ozturk, G. B., & Yitmen, I. (2019). Conceptual model of building information modelling usage for knowledge management in construction projects. In IOP Conference Series: Materials Science and Engineering (Vol. 471, No. 2, p. 022043). doi:10.1088/1757-899X/471/2/022043.

Pan, Y., & Zhang, L. (2020). BIM log mining: Exploring design productivity characteristics. *Automation in Construction*, 109, 102997. doi:10.1016/j.autcon.2019.102997.

Pan, Y., Zhang, L., & Skibniewski, M. J. (2020). Clustering of designers based on building information modeling event logs. *Computer-Aided Civil and Infrastructure Engineering*, 35(7), 701–718. doi:10.1111/mice.12551.

Psarommatis, F. (2021). A generic methodology and a digital twin for zero defect manufacturing (ZDM) performance mapping towards design for ZDM. *Journal of Manufacturing Systems*, 59, 507–521. doi:10.1016/j.jmsy.2021.03.021.

Qi, Q., & Tao, F. (2018). Digital twin and big data towards smart manufacturing and industry 4.0: 360 degree comparison. *IEEE Access*, 6, 3585–3593. doi:10.1109/ACCESS.2018.2793265.

Rivas, A., Martín, L., Sittón, I., Chamoso, P., Martín-Limorti, J. J., Prieto, J., & González-Briones, A. (2018). Semantic analysis system for industry 4.0. In *International Conference on Knowledge Management in Organizations* (pp. 537–548). doi: 10.1007/978-3-319-95204-8_45.

Rundquist, J. (2014). Knowledge integration in distributed product development. International Journal of Innovation Science, 2014, 6(1): 19–28.

Sacks, R., Girolami, M., & Brilakis, I. (2020). Building information modelling, artificial intelligence and construction tech. *Developments in the Built Environment*, 4, 100011. doi:10.1016/j.dibe.2020.100011.

Tao, F., Cheng, J., Qi, Q., Zhang, M., Zhang, H., & Sui, F. (2018). Digital twin-driven product design, manufacturing and service with big data. *The International Journal of Advanced Manufacturing Technology*, 94(9), 3563–3576. doi:10.1007/s00170-017-0233-1.

Wang, H., & Meng, X. (2019). Transformation from IT-based knowledge management into BIM-supported knowledge management: A literature review. *Expert Systems with Applications*, 121, 170–187. doi:10.1016/j.eswa.2018.12.017.

Yu, D., & Yang, J., (2018). Knowledge management research in the construction industry: A review. *Journal of the Knowledge Economy*, 9(3): 782–803. doi:10.1007/s13132-016-0375-7.

Zheng, X., Lu, J., & Kiritsis, D. (2021). The emergence of cognitive digital twin: Vision, challenges and opportunities. International Journal of Production Research, 2022, 60(24): 7610–7632.

22

On the Design of a Digital Twin for Maintenance Planning

Frits van Rooij
University of Salford

Philip Scarf
Cardiff University

22.1 Context and Motivation

A digital twin (DT) is a virtual representation of a system. A DT is virtual in the sense that it exists as an executable code, with a user interface (Autiosalo et al., 2019; Wang, 2020; Huang et al., 2021). The system for our purposes is an engineered object (EO) that is typically a subsystem of a technical system. The DT is a "twin" in the sense that it is a close representation of some aspect of the EO. The DT provides a platform, or environment, for experimentation, understanding, and prediction. One (or more) of these is the purpose of the DT (Grieves & Vickers, 2017). The DT should provide a representation of the EO that behaves like the real EO so that the behaviour of the DT is close enough to the real EO to be deemed good for its purpose. Thus, the DT might be used during the design and development of an EO (experimentation), whereby the performance of digital prototypes encoded as DTs is studied (Tuegel et al., 2011), and/or used to study how the EO behaves or interacts with its environment (understanding) (Haag & Anderl, 2018), and/or used to represent future behaviour of the EO under different conditions given its current state (prediction) (Weyer et al., 2016). Predictions can be useful for studying different scenarios. Different scenarios might arise as variations in environment, operations, and maintenance interventions, some controllable, some not. The state of the EO is the quantification of age, wear (degradation), operating parameters, and operating environment.

Strategically, DT development is a core component of Industry 4.0 (Uhlemann et al., 2017), because Industry 4.0 deals with digital innovation (Zhou et al., 2015), connectivity of EOs, data analytics, and automation (Zonta et al., 2020). Tactically (operationally), a DT is useful because it can provide

 DOI: 10.1201/9781003425724-27

a convenient platform on which to study a system. Thus, one can study an EO in the lab before it is built (digital prototyping) (Grieves & Vickers, 2017), or study a system that is out of reach (e.g. the interior of a star) (Yadav et al., 2015), or study interventions that are impractical to carry out on an EO itself; our case study has this purpose. Real interventions may be deemed impractical for reasons of time, money, or risk (to life, property, or society).

In this chapter, we suppose the purpose of the DT is prediction, and the purpose of prediction is to study controllable scenarios. Thus, one may be interested to know, given the current state of an EO, when and how to maintain the EO. This is the essence of maintenance planning (MP) (Dwight & Gordon, 2022).

"When" and "how" are matters of decision-making. Thus, we suppose the DT provides a platform for decision support. In aside, we note that decision support may be just that (to support a human decision-maker) or it may be more than support, wherein the DT would make decisions and plan interventions. Related to this is the scope of a DT. If one considers the other chapters (and projects) in this book, some regard a decision-support platform as a DT, while others view a DT as an element of a decision-support platform. Materially, this is a point of semantics rather than design, although consistent terminology is important for understanding. Regardless of terminology, we suppose that a decision-support platform has a user interface and (at least) three modules: (1) data collection and processing; (2) a representation (model) of the EO of interest; and (3) a scenario generator. Module 1 collects data from the real EO in real time, thereby quantifying the current state of the EO. Module 3 specifies a future scenario by setting the operating conditions and time and nature of interventions, say, and randomises environmental conditions. The state and scenario are then input to Module 2, which outputs the response. This module encodes the behaviour of the DT, and the response quantifies the behaviour of the DT, thus estimating the behaviour of the EO. Then, the platform can in theory optimise operation and maintenance (OM) parameters or in practice at least compare competing OM parameters. Finally, on the point about terminology, for the purpose of this chapter, we shall call the decision-platform the DT, although under a more traditional definition, Module 2 is the DT.

Returning to the matter of usefulness, the justification for using DTs for MP is many-fold. Firstly, decision support must be timely. For example, the comparison of the effectiveness of replacement of traction motors in railway vehicles on a five-year cycle with the same on a ten-year cycle would require many years for data for evaluation, and by this time the vehicles may be obsolete (Mayisela & Dorrell, 2019). Such a comparison is only practical in a modelling environment. Secondly, a DT necessitates simplification so that interventions are specified consistently, whereas real interventions tend to be messy, not least because EOs and the management priorities for them are continuously evolving (Haag & Anderl, 2018). Thirdly, tacit human knowledge about the behaviour of EOs and the likely effect of interventions is receding (Dwight et al., 2012; VanHorenbeek & Pintelon, 2013). Modern technical EOs are complicated, and it is increasingly important to encode knowledge about EOs digitally (Isaksson et al., 2018). Finally, MP interacts with many other managerial functions, e.g. operations (Silvestri et al., 2020), spare parts inventory (Scarf et al., 2023), and digital control rooms are being developed to manage these interactions (Topan et al., 2020).

So far, we have dealt with what is a DT and why a DT might exist generally and specifically for MP. The remainder of this chapter is concerned with how to design a DT for MP. We shall present the principles for its design. And we shall provide an example of how a DT looks in the context of MP in reverse-osmosis (RO) desalination, with a particular focus on the restoration of membranes. In the description of the case, we shall link design aspects of the DT to the principles. Before that we shall review the literature on DT design for MP and in so doing show that the ideas in this chapter are original (novel) and important.

22.2 Literature Review

A critical part of Industry 4.0 is the digitalisation of maintenance, technologically and managerially (Silvestri et al., 2020). As maintenance strategy has evolved from time-based and condition-based maintenance to predictive and prescriptive maintenance, the importance of maintenance modelling has increased. Predictive and prescriptive maintenance use field data but are underpinned by mathematical modelling (Zonta et al., 2020). This tendency is reflected in the discussion of DTs in maintenance. Indeed, after production planning and control, the primary utilisation of DTs is maintenance (Kritzinger et al., 2018; Errandonea et al., 2020).

In MP, decision-making support is vital (Vatn, 2018). While MP theory is well developed, viz *maintenance concept* in Gits (1992), Pintelon and Parodi-Herz (2008), and Ben-Daya et al. (2016), *maintenance requirements analysis* in Liyanage et al. (2009) and Dwight et al. (2012), and *maintenance policy* in Burhanuddin et al. (2011), decision-making support is often neglected. Labib (2008), in an early work, evaluated the implementation of computerised maintenance management systems and found little on the development of decision-support systems. More recently, the literature on DTs in maintenance has been dominated by anomaly detection, failure prediction, and diagnosing the causes of failures and the symptoms (Errandonea et al., 2020; Silvestri et al., 2020: Zonta et al., 2020). In these cases, the DT informs when to undertake action. In some cases, the DT supports the decision-making of what to undertake. Thus, Liu et al. (2020) present a DT of a drilling riser that estimates riser fatigue and makes recommendations for inspections and for rearranging degraded riser joints (swapping). Other DTs provide support for MP for windfarms (Fox et al.; 2022), manufacturing (Neto et al., 2021; Frantzén et al., 2022), urban facilities (Bujari et al., 2021), tunnel operations and maintenance (Yu et al., 2021), a cutting tool monitoring and replacement (Xie et al., 2021), and aviation (Lorente et al., 2022).

Works published so far are application-specific. A current gap in the literature is that there does not exist a general framework for the design of maintenance decision support driven by a DT. We will fill this gap and demonstrate the framework we propose with a practical example in industry.

22.3 Principles for the Design of a DT Maintenance Planning

To establish principles for the design of a DT for MP, we must first know the principles for maintenance and MP itself. Then, DT design can be a subprinciple in such a set of principles. Therefore, this section proposes a framework of maintenance principles in which low-level, detailed principles are derived from high-level, fundamental principles (Figure 22.1).

22.3.1 Level 1. General Principles of Maintenance

We propose the first layer as the fundamental maintenance principles. Gits (1992), who pioneered maintenance theory, and Ben-Daya et al. (2016) are clear in their discussions of principles that an unmaintained EO will eventually fail (Principle 1.1). Safe failure may be acceptable. If not, the cost of unsafe failure must be bearable by the operator and society; otherwise, the EO must cease operation and be made safe or scrapped (Principle 1.2) (Waeyenbergh & Pintelon, 2002; Riane et al., 2009). Maintenance itself bears partial responsibility for safe operation because there may exist uncontrollable risks (Hansson, 2013; Verhulst, 2014). Thus, marine transportation is risky no matter how well ships are maintained (Principle 1.3). Principle 1.4 is self-explanatory.

22.3.2 Level 2. Principles of Planned Maintenance

That better MP derives from a better understanding of the EO and its operational environment (Principle 2.1) is uncontroversial (Riane et al., 2009; Liyanage et al., 2009; Dwight et al., 2012; Ben-Daya et al., 2016). Many agents may be stakeholders (e.g. the operator, the maintainer if different from the operator, the original equipment manufacturer, the asset owner, clients and customers, warranty providers, consultants, regulators). They may act according to their own priorities (Söderholm et al., 2007), and MP needs to take account of multiple priorities (Principle 2.2). However, these many priorities may be competing (Principle 2.3). Finally, it is rational that MP should use knowledge of both degradation and the effect of restoration (Ben-Daya & Duffuaa, 2000) (Principle 2.4).

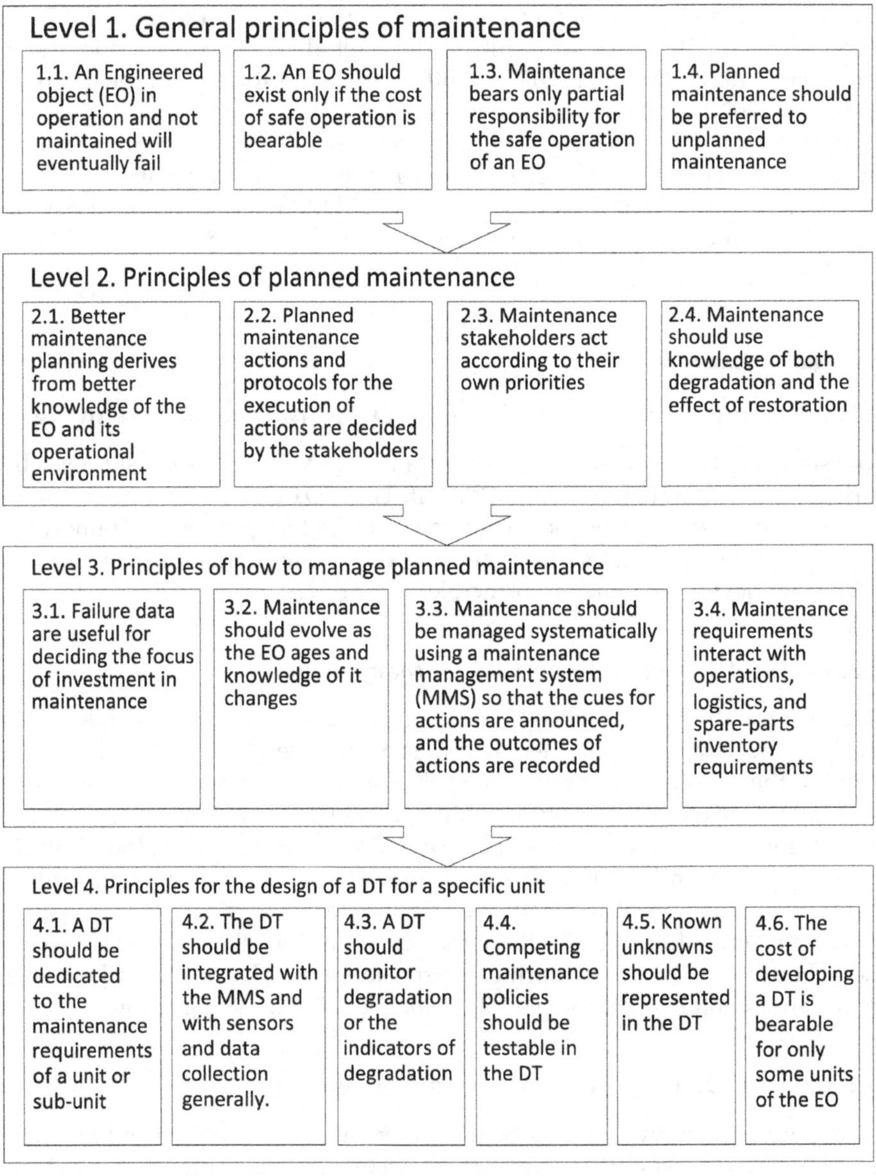

Level 1. General principles of maintenance

- 1.1. An Engineered object (EO) in operation and not maintained will eventually fail
- 1.2. An EO should exist only if the cost of safe operation is bearable
- 1.3. Maintenance bears only partial responsibility for the safe operation of an EO
- 1.4. Planned maintenance should be preferred to unplanned maintenance

Level 2. Principles of planned maintenance

- 2.1. Better maintenance planning derives from better knowledge of the EO and its operational environment
- 2.2. Planned maintenance actions and protocols for the execution of actions are decided by the stakeholders
- 2.3. Maintenance stakeholders act according to their own priorities
- 2.4. Maintenance should use knowledge of both degradation and the effect of restoration

Level 3. Principles of how to manage planned maintenance

- 3.1. Failure data are useful for deciding the focus of investment in maintenance
- 3.2. Maintenance should evolve as the EO ages and knowledge of it changes
- 3.3. Maintenance should be managed systematically using a maintenance management system (MMS) so that the cues for actions are announced, and the outcomes of actions are recorded
- 3.4. Maintenance requirements interact with operations, logistics, and spare-parts inventory requirements

Level 4. Principles for the design of a DT for a specific unit

- 4.1. A DT should be dedicated to the maintenance requirements of a unit or sub-unit
- 4.2. The DT should be integrated with the MMS and with sensors and data collection generally.
- 4.3. A DT should monitor degradation or the indicators of degradation
- 4.4. Competing maintenance policies should be testable in the DT
- 4.5. Known unknowns should be represented in the DT
- 4.6. The cost of developing a DT is bearable for only some units of the EO

FIGURE 22.1
Maintenance principles: a framework.

22.3.3 Level 3. Principles of How to Manage Planned Maintenance

Operational failures (failures of components during operation) are typically rare, an individual circumstance is often unique, and analyses of failure data are obsolete because, meanwhile, an EO has evolved. These can explain the large growth in research of subjective methods in reliability analysis

(Aslansefat et al., 2020). Nonetheless, analysis of past operational failures can be useful for deciding the focus of investment in maintenance (Principle 3.1) (Dwight et al., 2012). Further, a maintenance plan should evolve as the EO evolves (Principle 3.2). Thus, for example, maintenance priorities for an established plant will differ for a retiring plant (Ruiz et al., 2014). Maintenance should be systematically managed so that cues for action are announced, and the outcomes of actions are recorded (Principle 3.3). Such control implies the use of a maintenance management system (Labib, 2008; Bakri & Januddi, 2020; Catt, 2020). Finally, maintenance requirements interact with operations, logistics, and spare parts inventory (Principle 3.4). Therefore, maintenance activities must be coordinated with these functions so that maintenance pitstops do not unnecessarily interfere with the production output (van Rooij & Scarf, 2019). Thus, unavailability of spare parts leads to inefficient maintenance (excessive downtime), and excessive spare parts inventory is costly (Carnero Moya, 2004; Topan et al., 2020).

22.3.4 Level 4. Principles for the Design of a DT for Maintenance Planning

We assume the existence of a maintenance concept (the set of rules that recommend what maintenance is required and when) for an EO and that the maintenance concept is operationalised digitally on a Computerised Maintenance Management System (CMMS). The argument for Principle 4.1 is that the particulars of an EO under specific environmental and operational conditions are diverse and current DT architecture is inflexible (Catt, 2020). An EO will have many units so that Principle 4.1 implies an architecture like Figure 22.2, in which DTs are integrated with but external to the CMMS (Principle 4.2).

Principle 4.3 follows from Principle 2.4 because it is not untypical that large volumes of sensor data provide little by way of information about condition (Assaf et al., 2018) and that processing and modelling of sensor data is necessary for decision-making.

Maintainers are faced with many choices because EOs are complicated and different interventions have different effects, which must be learned. Testing maintenance policies on a physical unit is costly and potentially will sacrifice the unit. Therefore, testing requires a simulator, the DT, that can test competing maintenance policies (Principle 4.4) (Nakagawa, 2000; Ben-Daya et al., 2016).

Degradation processes are stochastic to a greater or lesser extent (Dohi et al., 2000). The effects of maintenance interventions are likewise. These factors are the known unknowns in maintenance planning, and they must be modelled. Thus, the state of a unit and the effect of a repair are not completely known to the decision-maker, and its future state will be less well known (Duffuaa & Raouf 2015). Thus, the DT should encode these uncertainties (Principle 4.5). An implication is that DTs are limited because they cannot encode unknown unknowns.

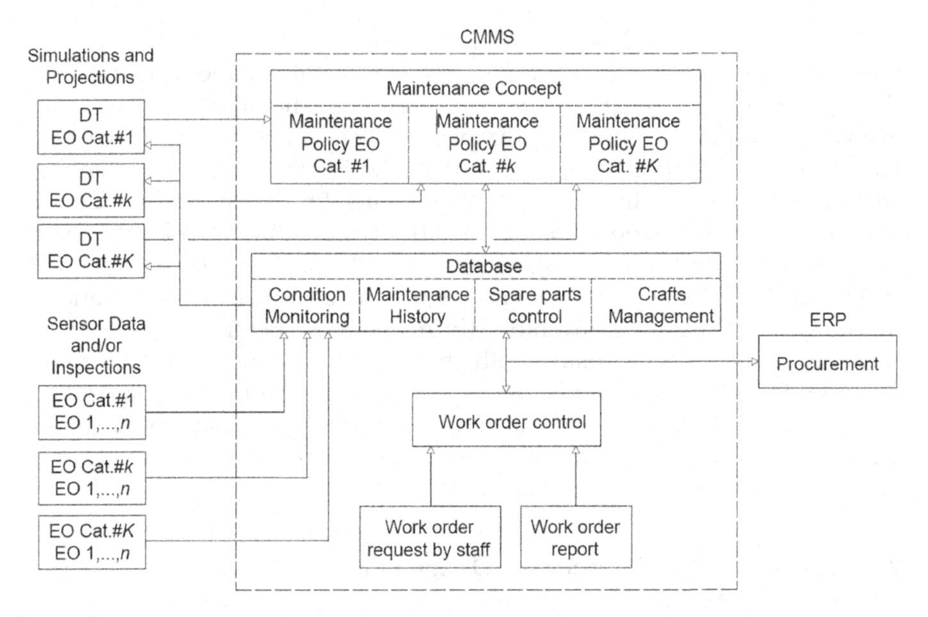

FIGURE 22.2
Positioning of a DT in the maintenance management system.

Finally, time, knowledge, skills, and high implementation costs are often significant barriers to the development of useful solutions (Noor et al., 2021). A production facility has numerous EOs each with many subunits. The plant in our case study, for example, has approximately 23,500 assets registered in the CMMS, including items such as tools. Developing a DT for even a fraction of these units will be too costly. Therefore, the cost of developing a DT is bearable for only some units of the EO (Principle 4.6).

22.4 Case Study

Now we show how the principles in our framework guide the design of a DT for MP. The case considers long-term MP of RO membranes in seawater desalination. A significant factor in the degradation of these membranes is biofouling due to seasonal algae blooms, and biofouling is estimated to cost the desalination industry an estimated $15 billion annually (Matin et al., 2021). The DT quantifies the hidden wear-state of membrane elements over time (Principle 4.3). The effect of restorations (swapping elements or inserting a new element at a particular position or both) is uncertain and modelled in the DT (Principle 4.5). The DT is used to evaluate different, competing maintenance strategies (Principle 4.4) that it would not be practical to evaluate directly on the EO. A preliminary analysis of maintenance performance at the plant (van Rooij & Scarf 2019) justified the investment necessary to develop the DT (Principle 4.6). The DT itself models an idealised pressure vessel (Principle 4.1), and maintenance policies tested on the DT are regarded as providing a good indication of the effectiveness of maintenance of all the seawater vessels in the plant.

22.4.1 System Description

There are 1,792 pressure vessels in the seawater RO system in the Carlsbad Desalination Plant, the biggest and most sophisticated desalination plant in the Western Hemisphere. The vessels are divided into 14 stacks, with 128 vessels in parallel in each stack. Each stack is referred to as an RO train. Each vessel is loaded with eight identical (when new) elements arranged in series. Clarified seawater enters the feed and the concentrated rejected brine is discharged at the tail of the vessel. The front permeate goes directly to the post-treatment for re-mineralisation. The rear permeate goes first to a brackish water RO system for further desalination. The membrane elements are spirally wounded with a fine feedwater carrier to allow the seawater to be pushed across the membrane surface of all the elements from the feed of the lead element to the tail of the last element. Resistance in the feedwater carrier results in a small hydraulic pressure drop across each element. This is the pressure differential (PD) of a new element.

Each train is operated and monitored independently of every other, and demand for permeate (drinking water) is typically met when 13 of the 14 trains are operating. More precisely, the mean peak demand for water from the plant is 204 Km^3/day and recommended allowable maximum supply per train is 631 m^3/hour.

22.4.2 Modelling Degeneration and Restoration of Membranes

Biofouling increases the PD due to the build-up of the biofilm between the membrane surface and the feedwater carrier. Since PD varies without deterioration of the membranes due to changes in feedwater temperature, salinity, and flows, the environmental effects must be filtered out. This correction gives a normalised PD (NPD), which is used as the wear indicator. Further temporal processing provides each train's daily mean NPD measurement ("P observed" in Figure 22.3).

The degradation of membranes is heterogeneous. This is because algae adhere more to more worn elements, and the effect of algal contamination is indirect. The lead seawater side degrades faster than the tail product side. From a maintenance modelling standpoint, an RO vessel is a multi-component system with non-identical components (Nicolai & Dekker, 2008), with stochastic dependence between components (Iung et al., 2016), in which the rate of wear of one component (membrane element) depends on both its

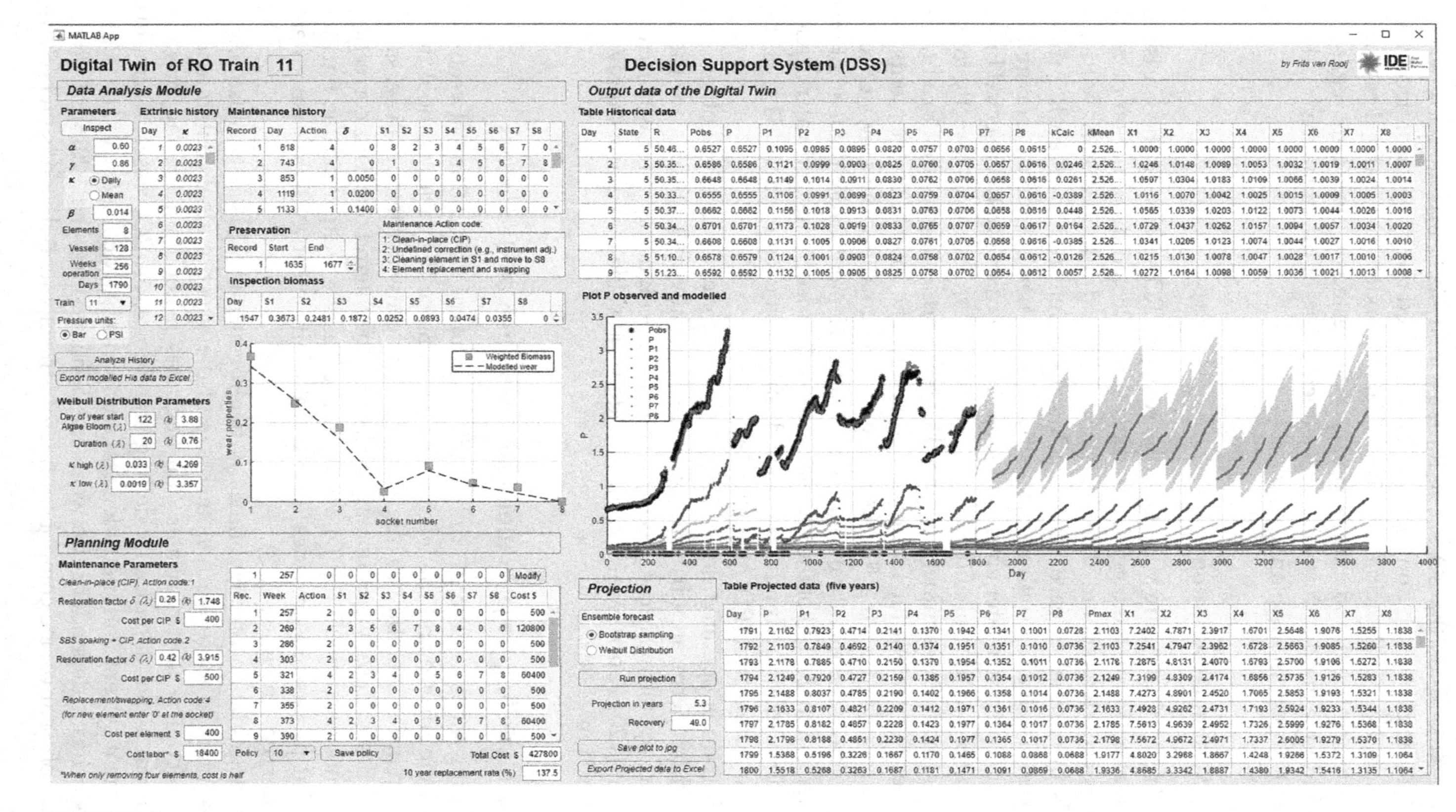

FIGURE 22.3
The user interface.

position and the state of the others (Bian & Gebraeel, 2014; Assaf et al., 2018; Do et al., 2019). This complexity necessitates a bespoke DT (Principle 4.1).

Restoration is multifaceted. Biomass can be partially removed by cleaning membranes in-situ, and there are, broadly, two cleaning methods: in C1, low pH rinsing follows high pH rinsing; in C2, the same process is preceded by soaking vessels in sodium bisulphate. Vessels can be restored by either partial or complete replacement of elements. The latter is ruled out on the basis of cost. In the former, heterogeneous wear makes cascading elements attractive. In general, *r* elements can be replaced, and 8-*r* cascaded. Another component-reallocation intervention that partially restores a vessel is swapping, whereby elements in leading sockets are systematically swapped with elements in trailing sockets, and no new elements are used.

With replacement, cascading, and swapping, elements must be inserted at the lead end (for technical reasons that we omit here), so that when an element in *Sn* is removed, elements in S1 to *Sn-1* must be removed. This is the so-called structural dependence in maintenance theory.

While wear of elements is heterogeneous, wear of vessels tends to be homogeneous because worn vessels degrade more slowly than newer vessels (because the load on worn vessels is lower). Therefore, all the vessels in a train are similar. Production and maintenance are managed at a train level so that, over time, trains are heterogeneous. This justifies modelling a single, idealised vessel in a train in the DT.

The NPD (monitored condition indicator) is the PD across the entire vessel. Nonetheless, we can model the wear (state) of each of the eight elements using a model of saline flow in an RO vessel (van Rooij et al., 2021). Parameters are estimated using particle filtering (Doucet et al., 2001). Parameter estimation is aided by selective opening and weighing individual elements, and by measuring PD across elements using a test rig.

Algae blooms are forecast in a separate model so that element states assuming no maintenance can be forecast. Essentially, the backcast of seasonal, daily feedwater quality measures was bootstrapped (Efron & Tibshirani, 1994) to provide different projected scenarios so that the DT quantifies another known unknown (Principle 4.5), the condition of the feedwater in the near and medium terms.

Next, restoration is modelled. Replacement, cascading and swapping are deterministic because the states of elements are known prior to a replacement or swap or cascade. The effect of cleaning is measured at a vessel level, but we assume the effect of cleaning on an element is proportional to the wear of the element. The mean restoration was 23% (sd 13%) for C1 and 35% (sd 11%) for C2.

Theoretically, the mathematical model permits the comparison of different restoration policies (Principle 4.4). Practically, a platform is needed that implements the model. This platform is the DT, the decision-support system, so that by implication the DT will estimate and project degradation of a vessel (Principle 4.3).

22.4.3 Digital Twin of a RO Train

The DT has three modules: a data analysis module, a planning module, and a simulator. The data analysis module imports data (observed NPD, product flow) and maintenance history from the CMMS (Principle 4.2) and estimates parameters. At the user interface (Figure 22.3), simulated trajectories can be compared with a real trajectory. In the planning module, restoration policies are compared, using simulated NPD trajectories and the costs of interventions. The user interface displays the modelled NPD (red pen, Figure 22.3), calculated using the simulator that runs in the background, the observed NPD (black pen), and the NPD implied by the modelled wear of each element in each socket (other coloured pens). Then, a restoration policy is selected, and the DT simulates an ensemble forecast over the period of projection (100 simulations, grey ribbon). Policies can be compared on the basis of cost and risk (Figure 22.4). Van Rooij et al. (2021) give the details of the policy schedules.

Replacement, cascade, and swapping policies are numerous, and we omit the details. Our purpose is to demonstrate that the DT is consistent with the principles of the framework. A policy is deemed admissible if the projected NPD is always below the critical threshold of 3.5 bar, above which elements can fail irrecoverably. Policies 10, 11, and 12 have an acceptable risk (median <15% on the conservative risk measure), and Policy 12 stands out in terms of risk and cost. Policy 10 replaces 83% of the membranes over 5 years, Policy 12 62.5%, and Policy 11 50%. The latter two policies have three cleans per train per year rather than two. The DT also facilitates sensitivity analysis. Thus, the risk measure decreases with the length of the smoothing window of the feed quality, while there is no change in the risk ranking of the policies. Choosing this window is a step in the analysis, with a longer smoothing window giving clearer differentiation of periods of algae blooms.

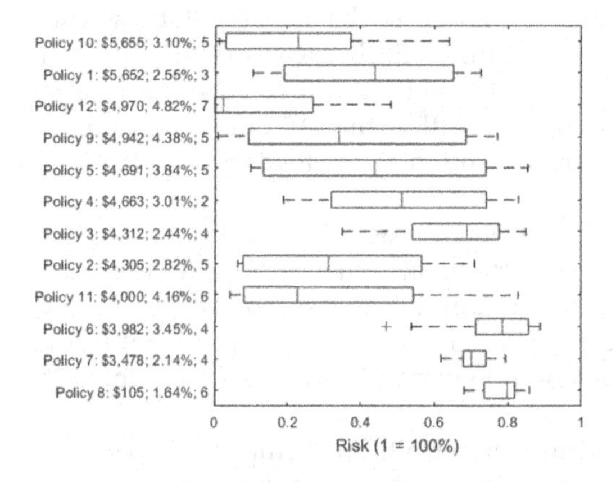

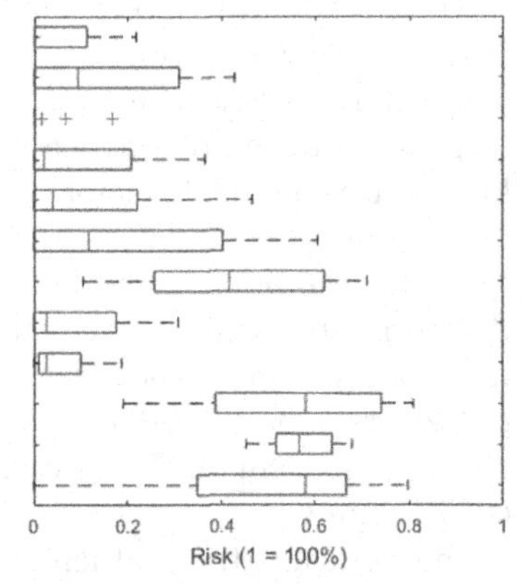

FIGURE 22.4
Evaluation of 12 policies (over 2020–2025). Policies ranked by total cost ($000s) and showing downtime per train (%), number of stops per train per year, and boxplots of risk measure (left, 3 bar; right, 3.5 bar) across 14 trains.

22.5 Discussion and Conclusion

We propose a set of principles for the design of a DT for MP and demonstrate the development of a DT according to these principles. The basis of the DT is a mathematical model of wear and restoration of the membrane elements of a pressure vessel in an RO train. The wear of the elements is principally due to biofouling. The DT allows the operator to prepare good long-term cost projections for different, competing restoration plans. The principles we propose are logically structured so that low-level principles (about DT development) follow from high-level principles (about maintenance generally).

Specifically, the DT described in the case study compares policies for risk, cost, and downtime, and projections indicate that the replacement rate can be reduced by 50%–80% without compromising the integrity of a train. The operator is exploring the use of the DT at other plants. The DT might be further developed in two ways: from decision-support to decision-making; and from a static to a dynamic decision support. From a research standpoint, both of these would be interesting. However, practically, their utility is questionable. For the former, trains are "big-ticket" items, and the operator would be unlikely to give up its oversight. For the latter, specifying a finite planning horizon is difficult. Nonetheless, decisions about the timing of cleaning that are dynamic (based on the current state of a train) rather than static (according to a time schedule) may provide further cost-savings.

Generally, there are lessons to be learned from the case study that impact the principles. Firstly, scaling is a challenge. Principle 4.1 (bespoke DT development) and Principle 4.6 (DTs are costly) are barriers to a more widespread use of DTs for decision-support. The challenge is to develop more generic decision-support systems that can be adapted cheaply to specify EOs. Secondly, if a DT is to be effective its mathematical model of degradation must be well-specified and estimated. A good model requires significant investment in both data collection and model building. Data collection can be automated, but there exists something of a gap between output from sensors and data that is useful for model building.

References

Aslansefat, K., S. Kabir, Y. Gheraibia, and Y. Papadopoulos. 2020. "Dynamic fault tree analysis: State-of-the-art in modeling, analysis, and tools." *Reliability Management and Engineering*, edited by Harish Garg and Mangey Ram, pp. 73–112.

Assaf, R., P. Do, S. Nefti-Meziani, and P. Scarf. 2018. "Wear rate-state interactions within a multi-component system: A study of a gearbox-accelerated life testing platform." *Journal of Risk and Reliability*, 232(4): 425–434.

Autiosalo, J., J. Vepsäläinen, R. Viitala, and K. Tammi. 2019. "A feature-based framework for structuring industrial digital twins." IEEE access, 2019, 8: 1193–1208.

Ben-Daya, M., and S.O. Duffuaa. 2000. "Overview of maintenance modeling areas." Maintenance, Modeling and Optimization, 2000: 3–35.

Ben-Daya, M., U. Kumar, and D.N.P. Murthy. 2016. *Introduction to Maintenance Engineering: Modelling, Optimization and Management*. Dhahran, Lulea, Brisbane: Wiley.

Bian, L., and N. Gebraeel. 2014. "Stochastic framework for partially degradation systems with continuous component degradation-rate-interactions." *Naval Research Logistics*, 61: 286–303.

Bujari, A., A. Calvio, L. Foschini, A. Sabbioni, and A. Corradi. 2021. "A digital twin decision support system for the urban facility management process." *Sensors*, 21(24): 8460.

Burhanuddin, M.A., S.M. Halawani, and A.R. Ahmad. 2011. "An efficient failure-based maintenance decision support system for small and medium industries." In *Efficient Decision Support Systems: Practice and Challenges from Current to Future*, edited by C. Jao, pp. 195–210. BoD-Books on Demand.

Moya, M.C.C. 2004. "The control of the setting up of a predictive maintenance programme using a system of indicators." *Omega, The International Journal of Management Science*, 32: 57–75.

Catt, P.J. 2022. "A tailorable framework of practices for maintenance delivery." *Journal of Quality in Maintenance Engineering* 28.1 (2022): 233–251.

Do, P., R. Assaf, P. Scarf, and B. Iung. 2019. "Modelling and application of condition-based maintenance for a two-component system with stochastic and economic dependencies." *Reliability Engineering & System Safety*, 182: 86–97.

Dohi, T., N. Kaio, and S. Osaki. 2000. "Basix preventive maintenance policies and their variations." In *Maintenance, Modeling and Optimization*, edited by M. Ben-Daya, S. O. Duffuaa and A. Raouf, pp. 155–183. Kluwer Academic Publisher.

Doucet, A., N. de Freitas, and N. Gordon. 2001. "An introduction to sequential Monte Carlo methods." In *Sequential Monte Carlo Methods in Practice. Statistics for Engineering and Information Science*, edited by A. Doucet, N. de Freitas and N. Gordon, pp. 3–14. New York, NY: Springer.

Duffuaa, S.O., and A. Raouf. 2015. *Planning and Control of Maintenance Systems. Modelling and Analysis*. Second ed. Dhahran, Lahore: Springer.

Dwight, R., and P. Gordon. 2022. Maintenance requirements analysis and whole-life costing. In *Multicriteria and Optimization Models for Risk, Reliability, and Maintenance Decision Analysis. International Series in Operations Research & Management Science*, edited by A.T. de Almeida, L. Ekenberg, P. Scarf, E. Zio and M. J. Zuo, Vol. 321. Cham: Springer. doi:10.1007/978-3-030-89647-8_16.

Dwight, R., P.A. Scarf, and P. Gordon. 2012. "Dynamic maintenance requirements analysis in asset management." In *Advances in Safety, Reliability, and Risk Management*, edited by Christophe Berenguer, Antoine Grall and Carlos Guedes Soares, pp. 847–852. London: Taylor and Francis.

Efron, B., and R.J. Tibshirani. 1994. *An Introduction to the Bootstrap*. CRC Press.

Errandonea, I., S. Beltrán, and S. Arrizabalaga. 2020. "Digital Twin for maintenance: A literature review." *Computers in Industry* 123 (2020): 103316.

Fox, H., A.C. Pillai, D. Friedrich, M. Collu, T. Dawood, and L. Johanning. 2022. "A review of predictive and prescriptive offshore wind farm operation and maintenance." *Energies*, 15(504): 1–28.

Frantzén, M., S. Bandaru, and A.H. Ng. 2022. "Digital-twin-based decision support of dynamic maintenance task prioritization using simulation-based optimization and genetic programming." *Decision Analytics Journal*, 3: 100039.

Gits, C.W. 1992. "Design of maintenance concepts." *International Journal of Production Economics*, 24(3): 217–226.

Grieves, M., and J. Vickers. 2017. "Digital twin: Mitigating unpredictable, undesirable emergent behavior in complex systems." In *Transdisciplinary Perspectives on Complex Systems*, edited by F.-J. Kahlen, S. Flumerfelt and A. Alves, pp. 85–113. Springer International Publishing.

Haag, S., and R. Anderl. 2018. "Digital twin - Proof of concept." *Manufacturing Letters*, 15: 64–66.

Huang, Z., Y. Shen, J. Li, M. Fey, and C. Brecher. 2021. "A survey on AI-driven . digital twins in Industry 4.0: Smart manufacturing and adva+nced robotics." *Sensors* 21.19 (2021): 6340.

Isaksson, A.J., I. Harjunkoski, and G. Sand. 2018. "The impact of digitalization on the future of control and operations." *Computers & Chemical Engineering*, 114: 122–129.

Iung, B., P. Do, E. Levrat, and A. Voisin. 2016. "Opportunistic maintenance based on multi-dependent components of manufacturing system." *CIRP Annals - Manufacturing Technology*, 65: 401–404.

Kritzinger, W., M. Karner, G. Traar, J. Henjes, and W. Sihn. 2018. "Digital Twin in manufacturing: A categorical literature review and classification." Ifac-PapersOnline, 2018, 51(11): 1016–1022.

Labib, A. 2008. "Computerised maintenance management systems." In *Complex System Maintenance Handbook*, edited by Khairy A. H. Kobbacy and D. N. Prabhakar Murthy, pp. 416–435. Springer Series in Reliability Engineering.

Liu, S., J. Guzzo, L. Zhang, U. Kumar, and G.J. Myers. 2020. "Ultra-deepwater drilling riser lifecycle management system." *Procedia Manufacturing*, 49: 211–216.

Liyanage, J.P., J. Lee, C. Emmanouilidis, and J. Ni. 2009. "Integrated e-maintenance and intelligent maintenance systems." In *Handbook of Maintenance Management and Engineering*, edited by Mohamed Ben-Daya, Salih O. Duffuaa, Abdul Raouf, Jezdimir Knezevic and Daoud Ait-Kadi, pp. 499–544. Springer.

Lorente, Q., E. Villeneuve, C. Merlo, G.A. Boy, and F. Thermy. 2022. "Development of a digital twin for collaborative decision-making, based on a multi-agent system: Application to prescriptive maintenance." *INCOSE International Symposium*, 32: 109–117.

Matin, A., T. Laoui, W. Falath, and M. Farooque. 2021. "Fouling control in reverse osmosis for water desalination & reuse: Current practices & emerging environment-friendly technologies." *Science of the Total Environment*, 2021, 765: 142721.

Mayisela, M., and D.G. Dorrell. 2019. "Application of reliability-centred maintenance for dc traction motors-a review." In *2019 Southern African Universities Power Engineering Conference/Robotics and Mechatronics/Pattern Recognition Association of South Africa (SAUPEC/RobMech/PRASA)*, pp. 450–455. IEEE.

Nakagawa, T. 2000. "Imperfect preventive maintenance models." In *Maintenance, Modeling and Optimization*, edited by M. Ben-Daya, S. O. Duffuaa and A. Raouf, pp. 201–214. Kluwer Academic Publisher.

Nicolai, R.P., and R. Dekker. 2008. "Optimal maintenance of multi-component systems: A review." In *Complex System Maintenance Handbook*, edited by Khairy A. H. Kobbacy and D. N. Prabhakar Murthy, pp. 263–286. London: Springer Series in Reliability Engineering.

Noor, H.M., S.A. Mazlan, and A. Amrin. 2021. "Computerized maintenance management system in IR4.0 adaptation - A state of implementation review and perspective." *IOP Conference Series: Materials Science and Engineering*, 1051, 012019.

Pintelon, L., and A. Parodi-Herz. 2008. "Maintenance: An evolutionary perspective." In *Complex System Maintenance Handbook*, edited by K. A. H Kobbacy and D. N. Prabhakar Murthy, pp. 21–48. London: Springer.

Riane, F., O. Roux, O. Basile, and P. Dehombreux. 2009. "Simulation based approaches for maintenance strategies optimization." In *Handbook of Maintenance Management and Engineering*, edited by Mohamed Ben-Daya, Salih O. Duffuaa, Abdul Raouf, Jezdimir Knezevic and Daoud Ait-Kadi, pp. 133–153. Springer.

Ruiz, P.P., B.K. Foguem, and B. Grabot. 2014. "Generating knowledge in maintenance from experience feedback." *Knowledge-Based Systems*, 68: 4–20.

Scarf, P., A. Syntetos, and R. Teunter. 2023. "Joint maintenance and spare-parts inventory models: A review and discussion of practical stock-keeping rules." IMA Journal of Management Mathematics, 2023: dpad020.

Silvestri, L., A. Forcina, V. Introna, A. Santolamazza, and V. Cesarotti. 2020. "Maintenance transformation through Industry 4.0 technologies: Asystematic literature review." Computers in Industry, 2020, 123: 103335.

Söderholm, P., M. Holmgren, and B. Klefsjö. 2007. "A process view of maintenance and its stakeholders." Journal of quality in maintenance engineering, 2007, 13(1): 19–32.

Topan, E., A.S. Eruguz, W. Ma, M.C. van der Heijden, and R. Dekker. 2020. "A review of operational spare parts service logistics in service control towers." *European Journal of Operational Research*, 282: 401–414.

Tuegel, E.J., A.R. Ingraffea, T.G. Eason, and S.M. Spottswood. 2011. "Reengineering aircraft structural life prediction using a digital twin." *International Journal of Aerospace Engineering*, 2011.

Uhlemann, T.H.J., C. Lehmann, and R. Steinhilper. 2017. "The digital twin: Realizing the cyber-physical production system for industry 4.0." *Procedia Cirp*, 61: 335–340.

van Rooij, F., and P. Scarf. 2019. "Towards a maintenance requirements analysis for maximizing production." *Proceedings of the 29th European Safety and Reliability Conference (ESREL), 2019.*

van Rooij, F., P. Scarf, and P. Do. 2021. "Planning the restoration of membranes in RO desalination using a digital twin." Desalination, 2021, 519: 115214.

VanHorenbeek, A., and L. Pintelon. 2013. "Development of a maintenance performance measurement framework-using the analytic network process (ANP) for maintenance performance indicator selection." *Omega*, 42: 33–46.

Vatn, J. 2018. "Industry 4.0 and real-time synchronization of operation and maintenance." In *Safety and Reliability-Safe Societies in a Changing World*, edited by Stein Haugen, Anne Barros, Coen van Gulijk, Trond Kongsvik and Jan Erik Vinnem, pp. 681–686. Trondheim: CRC Press.

Verhulst, E. 2014. "Applying systems and safety engineering principles for antifragility." *Procedia Computer Science*, 32: 842–849.

Waeyenbergh, G., and L. Pintelon. 2002. "A framework for maintenance concept development." *International Journal of Production Economics*, 77: 299–313.

Wang, Y. 2020. "A rigorous cognitive theory for autonomous decision making." 2020 IEEE International Conference on Systems, Man, and Cybernetics (SMC). IEEE, 2020: 1021–1026.

Weyer, S., T. Meyer, M. Ohmer, D. Gorecky, and D. Zühlke. 2016. "Future modeling and simulation of CPS-based factories: An example from the automotive industry." *Ifac-Papers Online*, 49(31): 97–102.

Xie, Y., K. Lian, Q. Liu, C. Zhang, and H. Liu. 2021. "Digital twin for cutting tool: Modeling, application and service strategy." *Journal of Manufacturing Systems*, 58: 305–312.

Yadav, R.K., U.R. Christensen, J. Morin, T. Gastine, A. Reiners, K. Poppenhaeger, and S.J. Wolk. 2015. "Explaining the coexistence of large-scale and small-scale magnetic fields in fully convective stars." *The Astrophysical Journal Letters*, 813(L31): 1–6.

Yu, G., Y. Wang, Z. Mao, M. Hu, V. Sugumaran, and Y.K. Wang. 2021. "A digital twin-based decision analysis framework for operation and maintenance of tunnels." *Tunnelling and Underground Space Technology* 116: 104125.

Zhou, Keliang, Taigang Liu, and Lifeng Zhou. "Industry 4.0: Towards future industrial opportunities and challenges." *2015 12th International conference on fuzzy systems and knowledge discovery (FSKD)*. IEEE, 2015.

Zonta, T., C.A. da Costa, R. da Rosa Righi, M.J. de Lima, E.S. da Trindade, and G.P. Li. 2020. "Predictive maintenance in the Industry 4.0: A systematic literature review." Computers & Industrial Engineering, 2020, 150: 106889.

23

Organizational Barriers and Enablers in Reaching Maturity in Digital Twin Technology

Päivi Aaltonen and Laavanya Ramaul
LUT University

Emil Kurvinen
University of Oulu

Antero Kutvonen
LUT University

Andre Nemeh
Rennes School of Business

23.1 Introduction

Digital twin (DT) paradigm has been actively researched during the past decade and currently the maturity of DTs and their definition are developing. Yet, the DT maturity is not a systematic and standardized concept. What constitutes a mature DT may differ between industries and firms (e.g., [1]). DTs are an example of the category of technologies known as artificial intelligence (AI), also referred to as Intelligent Industry (II) or Industrial AI (IAI) (see further [2]). Data-driven modelling, i.e., DTs, is a new paradigm in science and engineering for dealing with design problems in a systematic and defined structure. In the past, collection of data was problematic, and it was mainly done to document experiments and verify research hypotheses. Today, we are facing an information explosion, as data has become affordable and simple to gather, store and distribute. This became possible due to the availability and prevalence of sensors, high-volume storages, cloud services, Internet-of-Things (IoT) solutions and high wire and wireless data transfer technology. These technologies drive the productivity, procedural improvement and innovation across diverse industries and companies [3–5], moving

 DOI: 10.1201/9781003425724-28

from augmentation of human decisions to autonomy [6]. However, several industries suffer from functional barriers along this path [7], causing hindrance for worker safety and sustainability [8–10], as operational functions of large industrial firms often lack agility to transform rapidly. While abundant data accessibility causes serious challenges to modern engineering, there is enormous potential in learning how to take advantage of them in design processes and thus reach higher levels of maturity. Reaching higher levels of maturity and the utilization of data-driven models (DDMs) has many important applications – complete models can be created based on gathered information without knowing the underlying physics. Existing, working models can be improved, verified, and extended cost-efficiently and rapidly. Furthermore, for many important applications where strict simulation times are crucial, like in control and human in the loop systems, DDM can be an instant solution by providing highly efficient models.

This chapter will firstly illustrate different perspectives on AI and DT maturity, such as technologies used and strategic integration. Second, we will discuss shortly the macro-economic common barriers and enablers in reaching maturity, from global and national perspectives, e.g., government influence in market openness [11] and global expansion strategies based on technical advances [12]. Finally, we compare the differences between high- and low- level maturity in terms of economic, technical, organizational and personnel parameters, and discuss the specific limitations and potential this introduces in development towards higher maturity in DTs.

23.2 Context and Background

The terminology and definitions around the subject of this chapter are diverse and lack a unified conceptualization. In general, AI tends to refer to either macro-economic development, e.g., labour market [13], 'world-first' innovations [14], i.e., Uber and Netflix [15] or deep conceptualizations [16,17], whereas DTs refer to a firm-specific application along with Cyber-Physical-Systems (CPS), IoT [2] and Internet of Industrial Things (IIoT). IoT focuses on managing home appliances, e.g., managing electric usage. IIoT focuses on industrial machines including, e.g., data analytics. In such II terms, several overlapping categories for technologies exist, such as Data Technology (DT), Analytic Technology (AT), Platform Technology (PT), Operations Technology (OT) and Human–Machine Technology (HMT) [2]. The explanation is simple. In the AI literature, empirical material and popular media, two different timelines exist: the development and the application of AI technology.

In the 1950s, two simultaneous events occurred. Alan Turing developed the Turing test, defining "AI", while elsewhere George Devol and Joseph

Engelberger invented the world's first industrial robot [2]. The Turing test inspired the more 'popular' branch of the usage AI terminology, connected to events such as IBM's computer defeating chess world champion in the late 1990s and finally a chat robot passing the Turing test in 2014 (see [2], Figure 3.1, p. 34). Subsequently, this perspective has since remained the focal point of economic researchers [18]. However, industries have been developing and using AI and robotics that has little to do with the Turings test, nor do they need to. The latter event is connected to system management theories such as Standard Process Control in the 1950s and the Six Sigma in the 1980s, further continuing as Product Lifecycle Management [2,19] – leading to the development of the concept of 'digital twins' [19]. In short, the former has focused more on the top-of-the-line advances in understanding the concept and limitations of AI, while the latter has focused on industrial applications utilizing AI. However, the concept of 'maturity' brings these in current day discussion relatively close to one another.

Evaluation of a company's AI maturity is somewhat subjective [20]. One example is an AI framework of various levels from lowest (0 or 1) to highest [21]. Level 1 or lower maturity levels indicate the company has taken initial steps towards experimentation with selected technologies, but the implementation is limited. Next levels indicate increasingly ongoing initiatives, but such that are typically efficiency orientated and in selected units (i.e., exploitation [22]). Further up, multiple solutions are exploited and there is a distinct coordination between organizational units. These levels describe how AI is utilized beyond efficiency and the highest levels indicate truly novel solutions (i.e., exploration [22]). However, depending on the perspective of the source, the frameworks for evaluation rarely take into consideration both the AI technologies, (e.g., Data, see [7]; further, e.g., [12]), and the AI applications. Figure 23.1 illustrates AI maturity based on AI technology and strategic integration, excluding the individual AI applications in wide use. This framework highlights the differences in maturity levels. For example, moving from level 0 to level 1 requires initiating experimentation with AI technologies. Level 2 focuses on exhibiting different AI activities across various organizational units and establishing improved efficiency through advanced automation. With respect to level 3, 'AI-first' firms focus on the successful implementation of AI activities with organizational coordination. Advancing AI initiatives to ensure innovation beyond advanced automation is the purpose of level 4. Level 5 integrates human and machine intelligence for new innovations and to enable a sustainable competitive advantage for the business [23].

One of the industrial applications of AI is the DT. Recent managerial literature has focused on AI technology development [17,18], the role of AI applications in industries, occasionally called II, Industry 4.0 or Industrial AI [2,24]. Industry 4.0 refers to a Germany-led strategic initiative in the manufacturing industry, describing the currently used technologies in the manufacturing industry – such as CPS, IoT, IIoT and cloud computing.

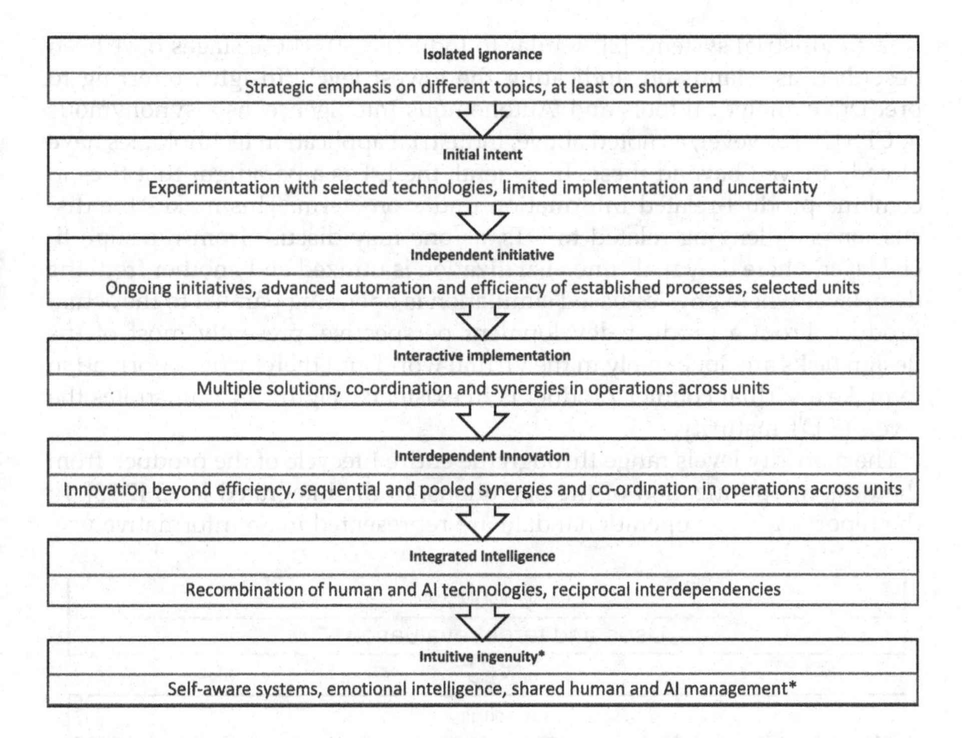

FIGURE 23.1
Artificial Intelligence maturity framework (based on [21]). *Beyond individual experiments not possible with current technologies.

However, current industrial applications are far more advanced. For example, the first data centre locations were established in 1994, and multinational corporations' high AI maturity levels are currently common [21]. Due to these advantages, The European Commission launched the term Industry 5.0 to credit the impact of technology development on society as a whole, increasing resilience, prosperity and increasing employee well-being – the society of human–machine work symbiosis (e.g., [24]). In particular, technologies from virtual simulation, DTs and autonomous AI actions are examples connected to Industry 5.0. [24]. Definitions of DTs may depend upon individual firms or industries [1]. It has been suggested, that there are two types of DTs, the prototype (DTP) and the instance (DTI) (e.g., [19]). The former refers to the duplication of the physical product, and the latter is a product life time link between the digital duplicate and its physical counterpart. Moving forward from this, the next step is the aggregate (DTA), an aggregation of all the DTs, that is not always an independent data structure such as DTI. All the versions operate in a digital environment (DTE) where they can be used for a variety of purposes (see further, e.g., [19]). These purposes then translate to levels of DT maturity, ranging from predictive to autonomous [2]. These levels, understood in this chapter as maturity, have also been described as stages towards smart industrial systems [2], similar to Industry 5.0. These stages have been described as 'Hands-on', indicating the lowest level, 'Insight', referring to predictive analytical tools and 'Autonomous Intelligence' as a synonymous to CPS [2]. However, as noted above, industrial application technologies have already moved beyond these. In general, the DT is a paradigm that tries to combine product-related information under one term. This makes the discussions challenging related to DTs, as one may discuss from the superficial layer where only real-time visualization is utilized and another from the deep layer where physics-based simulation is operating parallel to the actual product. From a product development perspective, presently most of the design tasks are done solely in the virtual world and thereby the information to make a virtual counter product is in existence. Figure 23.2 illustrates the levels of DT maturity.

The maturity levels range through the entire lifecycle of the product: from the design, optimization of the operations, to the integrated level [2,23]. In the reporting stage, operational data are represented in an informative way

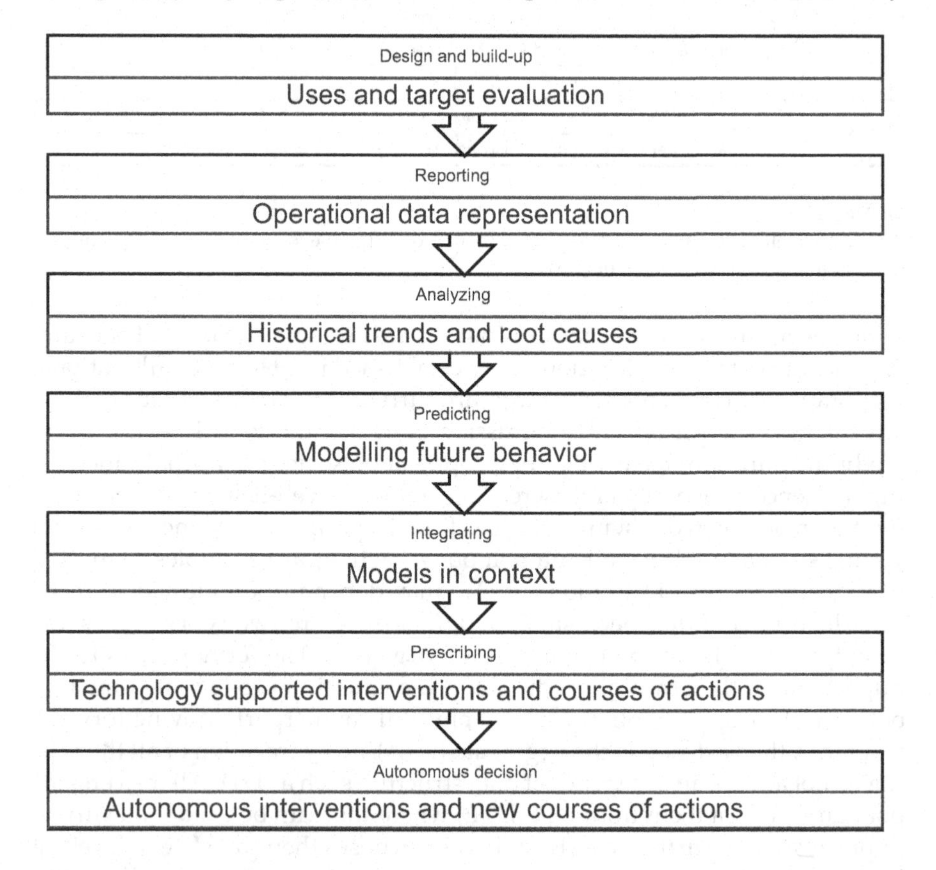

FIGURE 23.2
Elements in Digital Twin maturity (based on [2]).

to understand how the model performs. The second stage includes analysis of historical data which helps determine root causes for potential failures. Utilizing enterprise data, predictive models can be created based on operational variables to schedule repairs and reduce downtime, for example. However, integrating the operational data with external data would enhance the existing predictive models for strategic use such as creating cost models for optimum management, predicting future labour and procurement needs [2]. Moving towards advanced analytical capabilities, a firm at this stage of DT maturity may explore technology support interventions in business operations such as supply chain, quality control or resource management. Such multi-dimensional analytics would aid in decision-making and defining further course of actions. At more advanced stages of DT maturity, integrated and data-driven businesses will enable autonomous interventions and new courses of actions, for example, a possible visualization of the entire business process, from raw material to sales contracts in a new enterprise DT [19].

Despite AI and DTs remaining far from synonyms, much of the elements related to their maturity (low–high) are highly similar, as illustrated in this section and Figures 23.1 and 23.2. However, as a limited amount of theoretical paradigms touch on the concept of AI and DT maturity, our understanding barriers and enablers in reaching that maturity remains scarce. Further, the question remains, what are the differences between such barriers between AI and DTs, and how can we leverage DT technologies towards increasing maturity. While previously direct comparison between barriers and enablers and maturity levels has not been made, availability of resources, right strategy and successful management of operations are some indicators.

23.3 External Limitations and Enablers

Organizational barriers and enablers may refer to macro-economic or firm-specific factors. Institutional theories are an example of theories discussing various implications in the socio-technical landscape (see, e.g., [25–27]), and the implications of AI have been an increasing subject in the field [28,29].

The practical implications of such factors from a production perspective can be, for example, regulatory actions, technology and knowledge resources (see, e.g., materials, tools, research and educated personnel). Governmental actions may indirectly impact the initial experimentation's with novel technologies. Manufacturing industries with high degree of AI maturity, such as automobile manufacturing and heavy machinery, make up most of the export market in the European Union (mainly in Germany and Sweden), Russia and the United States. Machinery producers such as John Deere (USA), Hitachi (Japan), Liebherr (Germany), Volvo (Sweden), Caterpillar (USA), and CNH

Industrial (Netherlands) are global companies exporting products around the globe. However, global expansion, entrepreneurship opportunities and competition vary between each country due to differences in governance. Governmental elements impacting AI maturity can range from law to government size to regulatory efficiency and market openness. The latter two in particular limit and enable firm and individual possibilities such as the possibility to open a business and the degree of government interference in trade and investment. For example, Finland and Australia are ranked among the top 15 amongst all countries in terms of individual and firm-level global economic freedom (e.g., [11]). While this translates to increased individual freedom (i.e., health, labour, rights), Australia is more beneficial for entrepreneurs due to investment freedom, and faces less governmental involvement in foreign trade and stock exchange, allowing the freedom of competition [11].

Especially, for small to medium-sized firms, such country-specific attributescan have a significant role in enabling or constraining growth. For larger global manufacturers, changing regulatory demands may be complex and costly. For example, the non-road mobile machinery (NRMM) producers are facing heavy international competition and must challenge it with R&D and new inventions. Due to the increasing move towards carbon-neutral and cost-effective solutions, the global demand for new electric and hybrid designs is at an all-time high. Tier V regulations came into effect in the USA in 2020 and will come in EU stage V in 2025 which will place a new tight level of emission limits for CO_2 and other air pollutants of diesel-operated work machines. Answering to these tightening regulations requires rapid measures from the machine manufacturers to strengthen their competitive positions in the world market. Sustaining this competitive strategy in the face of major green and digital disruption requires both high-level competence and innovative development of new sustainable technologies to create and capture additional green value on the export markets. As regulation and competition focus on distinctly different categories, e.g., immaterial rights and carbon dioxide emissions, manufacturing industries need to simultaneously play two fields in digital transformation. Furthermore, the technological resources between countries vary. This can mean supporting infrastructure, such as reliable electricity, political stability or general technology development level in a given country – industrial firms with capabilities for fully autonomous operations also require a certain degree of maturity from their supply chain and customer base. Based on the technological 'status' of a country, firms may, for example, balance between one of the four strategic expansion choices (see, e.g., [12,30]). For example, a country with 'high' technology, such as many western industrialized nations, may opt to expand operations towards ones with lower technology in order to improve efficiency and cut cost. Two firms located in similar technology readiness locations may look for increasing knowledge and radical innovative solutions [12].

23.4 Internal Barriers and Enablers

Digital transformation requires a certain process of planning, integrating and analysing. In integrating AI and DT applications to the organization strategy or business models, the process has been described as sequential and incremental. In the strategy formulation phase, external analysis of micro- and macro-environment of a firm are determined, i.e., customers and competitors and political, economic and social trends, respectively [31]. Next, the factors influencing the macro- and micro-environment are prioritized according to their future potential and subsequent scenarios are developed [32]. The internal analysis of the firm includes defining the technological maturity level of the company and thus formulating a digital strategy most suited to the corporate strategy and mission statement. Strategy formulation forms the basis of developing new business models. In the business model transformation phase, the ambition and fit of the existing business model is evaluated against the customer requirements and business objectives [33]. The implementation phase includes identifying resources and capabilities of the firm and defining the customer experience and value-creation network. Initiatives are defined and categorized according to technology, organization, skills and culture which are evaluated in terms of their impact [31]. Finally, these initiatives are implemented utilizing agile methods and are measured to provide digital and operational excellence.

Adopting any technological innovation such as DT as part a of the existing IT structure of a firm would require incurring costs for installation, repair and maintenance, in addition to computational expense for training data models. The size of the firm defines the availability of financial resources that determine the investments in AI and DT; hence, smaller firms have limited investment capabilities and deployment potential [34]. Competitive firms seek to maximize adoption of DT beyond the domestic market and improve the proximity to the customer with developed digital capabilities to increase customer satisfaction and loyalty [35]. Technological challenges include matching the existing IT infrastructure to realize the full potential of AI and DT projects, low-volume or low-quality data and lack of data governance to define the protection, accessibility and lifecycle of the existing data [35,36]. Manufacturing complex products requires a multi-stage production process and research shows that the organization of manufacturing complex products is more challenging than manufacturing simple products in large batches [37]. DT technologies could be implemented in small scale production of complex products to make process such as workload planning more efficient, thus ensuring the economic efficiency of a large batch production for a smaller more complex production [35].

Different competitive strategies reflect the strategic vision and business potential of DT relevant to a particular firm. For example, firms which view product design or customization as a competitive advantage are more willing to adopt AI and DT faster than manufacturing firms aiming at mass production with reliable and zero defect production [35,38]. Organizational culture and a favourable attitude towards adoption is an enabler to the adoption of DT and AI technologies. Firms focused on research and development activities, prefer integrating new technologies to enhance existing routines and production processes [39,40]. To ensure successful DT adoption, employees must be equipped with specific digital skills [41]; hence, top management must support the development of technical skills to remain competitive. Sufficient digital skills ensure effective use of AI and DT technologies, in addition to managing large sets of data and drawing appropriate knowledge using such high-level technologies. In addition, insufficient supply of talent such as data engineers and data scientists acts as a barrier to the successful implementation of AI [42,43]. Providing on-the-job training and specialized recruitment processes could encourage the development of these skills [43].

23.5 Balancing between Barriers and Enablers

Based on the previous sections in this chapter, the barriers and enablers can be categorized as either economic, technology, organizational or people related. The former two refer to external factors, as discussed in Section 1.3, and the latter two to internal, based on Section 1.4. Here, 'Technology' can be understood as the general level of technology in the firm surrounding, such as supply chain, resources and knowledge, based on Figure 23.1. 'Organizational' aspects illustrate the internal processes, strategies and management practices, similar to Figure 23.2. 'People' refer to personnel, the required HR practices (i.e., training, job description) and top management mentality in designing personnel-related issues, also as described in Figure 23.3. In Figure 23.3, these are connected to different levels of DT maturity, with practical details illustrated related to each end of the maturity scale.

For low-maturity technologies, the expectation is that market changes are rapid and one must adapt to such changes. Thus, new pilots and products are pushed forward, simultaneously accepting the failures of launches as a part of the game. In mature technology levels this orientation changes. The focal point is on research and long-term orientation, also paving way to increasing number of custom solutions and expertise, i.e., product diversification. From technical perspective, in low-maturity situations, there is a general assumption of all parties having a basic knowledge of a subject, for example, the internet. This further translates to regulatory and legislative institution understanding of the necessary support for said technologies, enforcing basic infrastructure and initiatives, as well as schools and educational programme levels (e.g., long distance learning). On the other end

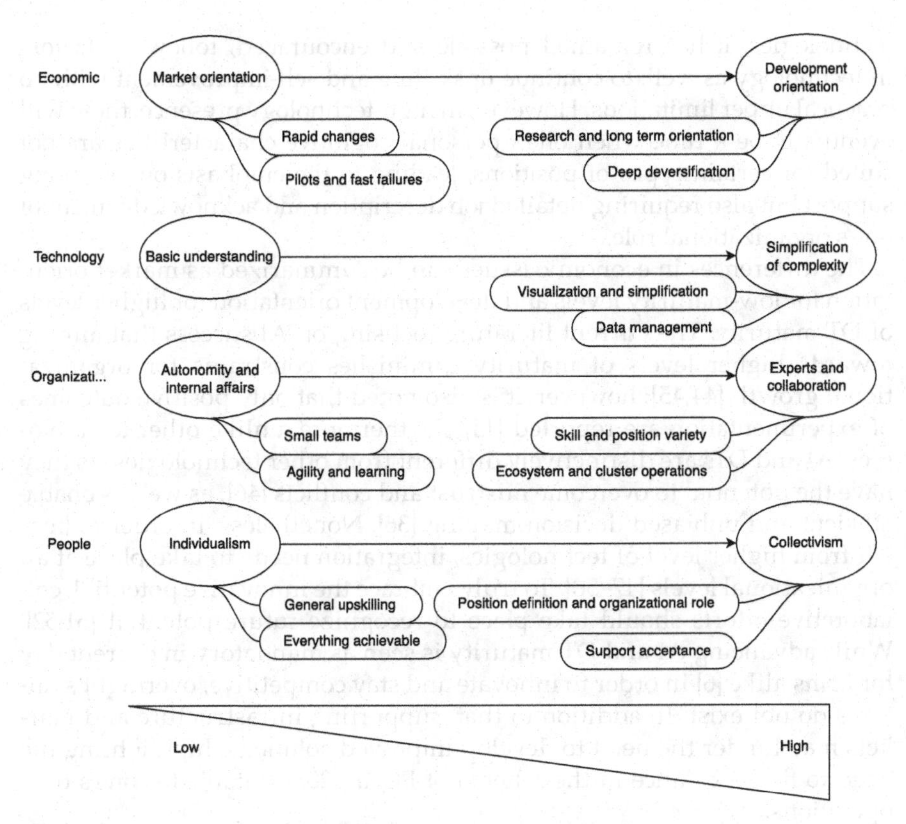

FIGURE 23.3
Differences between high- and low-level Artificial Intelligence maturities.

of the spectrum, it is likely that the complexity of the technologies falls beyond the cognitive capabilities of all, and communication of complex elements to large audiences becomes more significant. This also leads to data collection, management and ownership aspects. From an organizational, e.g, firm internal processes perspective, in low-maturity technologies business units and teams can remain independent, and organizational borders are defined. Firm strategic perspectives, following the market logic, emphasize agility and constant learning – the expectation is that a firm should manage on its own and create competitive edge with internal organization. In higher levels of maturity, echoing the technology complexity, the need for experts and ad-hoc interfirm collaboration becomes mandatory, as do RI&D cluster networks. For internal processes this further means the design of positions for very specific skill-sets and the lack of repetition of highly similar positions. Instead of planning to employ four to do a similar job, one should plan for four complementary positions. From a personnel perspective, this means a mentality change, as well as HR and management support in creating positions, following the organizational perspective. In low-maturity technologies, it has remained possible and encouraged, following factors in technology as well, to continue upskilling and self-improvement with no practical upper limitations. However, in high-technology presence, there will eventually be a time when one's personal cognitive characteristics are not suited for certain types of positions, leading to the emphasis on accepting support but also requiring detailed job description and acknowledgement of one's organizational role.

The differences in economic issues can be summarized as market orientation for low-maturity levels and development orientation for higher levels of DT maturity. The current literature focusing on AI success that aiming towards higher levels of maturity diminishes constrains for organizational growth [44,45]; however, it is also noted that only positive outcomes of experimentation are reported [14]. Furthermore, unlike other technologies, AI and DTs are distinctively different from other technologies, as they have the potential to overcome mistrust and conflicts [46], as well as enable efficient and unbiased decision-making [36]. Nonetheless, in order to benefit from higher level of technologies, integration needs to take place at all organizational levels [47–50]. To truly embrace the innovative potential, collaborative efforts should take place to recognize future potential [51,52]. While advancing AI and DT maturity is seen as mandatory in current day for firms alike [6] in order to innovate and stay competitive, overnight solutions do not exist. In addition to that, supporting infrastructure and markets may hinder the need to develop improved solutions, highlighting the need to find a balance in the schema of Figure 23.3 suitable for one's own operations.

23.6 Conclusions

In this chapter, we have introduced theories and perspectives for understanding DT maturity and impacting external and internal factors in reaching a milestone in maturity. In Section 23.2 of this chapter, we have discussed the concept of AI and DT maturity and their relation to each other. While AI and DTs as technologies have clear differences, their application and integration to business processes and organizations share the same non-technical barriers and enablers that can be captured through the discussion of maturity levels in the adoption of these technologies. In Section 23.3, we focused on some aspects of global and national challenges, such as EU or national regulatory challenges, supporting infrastructure and market openness. These follow the lines of AI and DT maturity scholarly research, and highlight the complex field of operations in organizations. Furthermore, the impact of constrains may depend on firm size and complexity – while small firms may be agile and adapt to changes, decades old global players face a variety of demands internally and externally that are challenging to change. In Section 24.4, we focus on firm-specific barriers and enablers in reaching DT maturity, such as top management orientation, talent acquisition and product complexity. Section 24.5 covers in more detail the differences in managerial orientation when advancing from less mature technologies to higher ones, DDMs, DTs or other AI technologies share a transformative property in how they fundamentally alter the way that organizations acquire, process, analyse, exploit and share information. Their true potential is initially restricted by organizational designs, managerial practices and business models created to function in conditions of severely limited information and data. Therefore, to unlock the true potential, their adoption needs to be accompanied by truly transformative business change that promotes inter-organizational and crossfunctional collaboration, employee autonomy and the creation of new deeply digital products, services and systems. Eventually, the organization's willingness and capacity to embrace this change determines the digital maturity [53] who define digital maturity as "the status of a company's digital transformation". As we have detailed in this chapter, each organization faces this journey of transformation from a different starting point dictated by their idiosyncratic mix of context, external factors and internal barriers and enablers.

Bibliography

[1] Tuija Rantala, Kirsi Kokkonen, and Lea Hannola. Selling digital twins in business-to-business markets. In *Real-time Simulation for Sustainable Production*, pages 51–62. Routledge, 2021.

[2] Jay Lee. *Industrial AI: Applications with Sustainable Performance*. Springer, 2020.

[3] Erik Brynjolfsson and Andrew Mcafee. Artificial intelligence, for real. *Harvard Business Review*, 1:1–31, 2017.

[4] Jun Liu, Huihong Chang, Jeffrey Yi-Lin Forrest, and Baohua Yang. Influence of artificial intelligence on technological innovation: Evidence from the panel data of china's manufacturing sectors. *Technological Forecasting and Social Change*, 158:120142, 2020.

[5] Serge-Lopez Wamba-Taguimdje, Samuel Fosso Wamba, Jean Robert Kala Kamdjoug, and Chris Emmanuel Tchatchouang Wanko. Influence of ar tificial intelligence (ai) on firm performance: the business value of ai-based transformation projects. *Business Process Management Journal*, 26(7):1893–1924, 2020.

[6] Sebastian Raisch and Sebastian Krakowski. Artificial intelligence and management: The automation-augmentation paradox. *Academy of Management Review*, 46(1):192–210, 2021.

[7] Sulaiman Alsheiabni, Yen Cheung, and Chris Messom. Towards an artificial intelligence maturity model: From science fiction to business facts. *PACIS 2019 Proceedings*, 2019: 46.

[8] Kymalainen Heli, Laitila Juha, Vaatainen Kari, and Jukka Malinen. Workability and well-being at work among cut-to-length forest machine operators. *Croatian Journal of Forest Engineering: Journal for Theory and Application of Forestry Engineering*, 42(3):405–417, 2021.
[9] Lonnie J Love, Eric Lanke, and Pete Alles. *Estimating the Impact (Energy, Emissions and Economics) of the US Fluid Power Industry*. Oak Ridge National Laboratory, Oak Ridge, TN, 2012.
[10] Jacek Paraszczak, Erik Svedlund, Kostas Fytas, and Marcel Laflamme. Electrification of loaders and trucks-a step towards more sustainable underground mining. *Renewable Energy and Power Quality Journal*, 12(12):81–86, 2014.
[11] T Miller, AB Kim, JM Roberts, et al. *2023 Index of Economic Freedom*. The Heritage Foundation, 2022.
[12] Christian Le Bas and Christophe Sierra. Location versus home country advantages' in R&D activities: Some further results on multinationals' locational strategies. *Research Policy*, 31(4):589–609, 2002.
[13] Nestor Duch-Brown, Estrella Gomez-Herrera, Frank Mueller-Langer, and Songul Tolan. Market power and artificial intelligence work on online labour markets. *Research Policy*, 51(3):104446, 2022.
[14] Christian Rammer, Gaston P Fernandez, and Dirk Czarnitzki. Artificial intelligence and industrial innovation: Evidence from german firm-level data. *Research Policy*, 51(7):104555, 2022.
[15] Marco Iansiti and Karim R Lakhani. *Competing in the Age of AI: Strategy and Leadership When Algorithms and Networks Run the World*. Harvard Business Press, 2020.
[16] Dirk Lindebaum, Mikko Vesa, and Frank Den Hond. Insights from "the machine stops" to better understand rational assumptions in algorithmic decision making and its implications for organizations. *Academy of Management Review*, 45(1):247–263, 2020.
[17] Vern L Glaser, Neil Pollock, and Luciana D'Adderio. The biography of an algorithm: Performing algorithmic technologies in organizations. *Organization Theory*, 2(2):26317877211004609, 2021.
[18] Katherine C Kellogg, Melissa A Valentine, and Angele Christin. Algorithms at work: The new contested terrain of control. *Academy of Management Annals*, 14(1):366–410, 2020.
[19] Michael Grieves and John Vickers. Digital twin: Mitigating unpredictable, undesirable emergent behavior in complex systems. Transdisciplinary perspectives on complex systems: New findings and approaches, 2017: 85–113.
[20] Raghad Baker Sadiq, Nurhizam Safie, Abdul Hadi Abd Rahman, and Shidrokh Goudarzi. Artificial intelligence maturity model: A systematic literature review. *PeerJ Computer Science*, 7:e661, 2021.
[21] Ulrich Lichtenthaler. Five maturity levels of managing ai: From isolated ignorance to integrated intelligence. *Journal of Innovation Management*, 8(1):39–50, 2020.
[22] James G March. Exploration and exploitation in organizational learning. *Organization Science*, 2(1):71–87, 1991.
[23] Ulrich Lichtenthaler. Agile innovation: The complementarity of design thinking and lean startup. *International Journal of Service Science, Management, Engineering, and Technology (IJSSMET)*, 11(1):157–167, 2020.

[24] Xun Xu, Yuqian Lu, Birgit Vogel-Heuser, and Lihui Wang. Industry 4.0 and industry 5.0-inception, conception and perception. *Journal of Manufacturing Systems*, 61:530–535, 2021.
[25] Paul J DiMaggio and Walter W Powell. The iron cage revisited: Institutional isomorphism and collective rationality in organizational fields. *American Sociological Review*, 147–160, 1983.
[26] Frank W Geels. *Technological Transitions and System Innovations: A Coevolutionary and Socio-Technical Analysis*. Edward Elgar Publishing, 2005.
[27] R Edward Freeman. Divergent stakeholder theory. *Academy of Management Review*, 24(2):233–236, 1999.
[28] Bengt-Ake Lundvall and Cecilia Rikap. China's catching-up in artificial intelligence seen as a co-evolution of corporate and national innovation systems. *Research Policy*, 51(1):104395, 2022.
[29] Robin Mansell. Adjusting to the digital: Societal outcomes and consequences. *Research Policy*, 50(9):104296, 2021.
[30] Pari Patel and Modesto Vega. Patterns of internationalisation of corporate technology: Location vs. home country advantages. *Research Policy*, 28(2–3):145–155, 1999.
[31] Daniel R. A. Schallmo and Christopher A. Williams. An integrated approach to digital implementation: Tosc-model and dpsec-circle. Digitalization: Approaches, Case Studies, and Tools for Strategy, Transformation and Implementation. Cham: Springer International Publishing, 2021: 371–380.
[32] Carlos Cordon, Pau Garcia-Mil'a, Teresa Ferreiro Vilarino, and Pablo Caballero. From *Digital Strategy to Strategy Is Digital*, pages 9–45, 2016.
[33] Daniel R. A. Schallmo and Christopher A. Williams. *Roadmap for the Digital Transformation of Business Models*, pages 41–68. Springer International Publishing, Cham, 2018.
[34] Dora Horvath and Roland Zs. Szabo. Driving forces and barriers of industry 4.0: Do multinational and small and medium-sized companies have equal opportunities? *Technological Forecasting and Social Change*, 146:119–132, 2019.
[35] Steffen Kinkel, Marco Baumgartner, and Enrica Cherubini. Prerequisites for the adoption of ai technologies in manufacturing-evidence from a worldwide sample of manufacturing companies. *Technovation*, 110:102375, 2022.
[36] Erik Brynjolfsson and Kristina McElheran. The rapid adoption of datadriven decision-making. *American Economic Review*, 106(5):133–139, 2016.
[37] Eva Kirner, Steffen Kinkel, and Angela Jaeger. Innovation paths and the innovation performance of low-technology firms-an empirical analysis of german industry. *Research Policy*, 38(3):447–458, 2009. Special Issue: Innovation in Low-and Meduim-Technology Industries.
[38] Fatima Gillani, Kamran Ali Chatha, Muhammad Shakeel Sadiq Jajja, and Sami Farooq. Implementation of digital manufacturing technologies: Antecedents and consequences. *International Journal of Production Economics*, 229:107748, 2020.
[39] Jaime Gomez and Pilar Vargas. Intangible resources and technology adoption in manufacturing firms. *Research Policy*, 41(9):1607–1619, 2012.
[40] Lara Agostini and Anna Nosella. The adoption of industry 4.0 technologies in smes: Results of an international study. *Management Decision*, 58(4):625–643, 2019.

[41] Patrick Mikalef, John Krogstie, Ilias O Pappas, and Paul Pavlou. Exploring the relationship between big data analytics capability and competitive performance: The mediating roles of dynamic and operational capabilities. *Information & Management*, 57(2):103169, 2020.
[42] Bernd W Wirtz, Jan C Weyerer, and Carolin Geyer. Artificial intelligence and the public sector-applications and challenges. *International Journal of Public Administration*, 42(7):596–615, 2019.
[43] Sam Ransbotham, David Kiron, Philipp Gerbert, and Martin Reeves. Reshaping business with artificial intelligence: Closing the gap between ambition and action. MIT Sloan Management Review, 2017, 59(1).
[44] Nicholas Berente, Bin Gu, Jan Recker, and Radhika Santhanam. Managing artificial intelligence. *MIS Quarterly*, 45(3):1433–1450, 2021.
[45] Michael Haenlein, Andreas Kaplan, Chee-Wee Tan, and Pengzhu Zhang. Artificial intelligence (ai) and management analytics. *Journal of Management Analytics*, 6(4):341–343, 2019.
[46] Ella Glikson and Anita Williams Woolley. Human trust in artificial intelligence: Review of empirical research. *Academy of Management Annals*, 14(2):627–660, 2020.
[47] Richard Vidgen, Sarah Shaw, and David B Grant. Management challenges in creating value from business analytics. *European Journal of Operational Research*, 261(2):626–639, 2017.
[48] Eivind Kristoffersen, Patrick Mikalef, Fenna Blomsma, and Jingyue Li. The effects of business analytics capability on circular economy implementation, resource orchestration capability, and firm performance. *International Journal of Production Economics*, 239:108205, 2021.
[49] Eivind Kristoffersen, Fenna Blomsma, Patrick Mikalef, and Jingyue Li. The smart circular economy: A digital-enabled circular strategies framework for manufacturing companies. *Journal of Business Research*, 120:241–261, 2020.
[50] Philipp Korherr, Dominik K Kanbach, Sascha Kraus, and Paul Jones. The role of management in fostering analytics: The shift from intuition to analytics-based decision-making. Journal of Decision Systems, 2022: 1–17.
[51] Tim Fountaine, Brian McCarthy, and Tamim Saleh. Building the ai-powered organization. *Harvard Business Review*, 97(4):62–73, 2019.
[52] Jan Johnk, Malte Weißert, and Katrin Wyrtki. Ready or not, ai comes-an interview study of organizational ai readiness factors. *Business & Information Systems Engineering*, 63(1):5–20, 2021.
[53] Simon Chanias and Thomas Hess. How digital are we? Maturity models for the assessment of a company's status in the digital transformation. *Management Report/Institut fur Wirtschaftsinformatik und Neue Medien*, (2):1–14, 2016.

24

Digital Twin Development – Understanding Tacit Assets

Petra Müller-Csernetzky
Lucerne School of Engineering and Architecture

Shaun West
Lucerne University of Applied Sciences and Arts

Oliver Stoll
Lucerne School of Engineering and Architecture

24.1 Introduction

Digital servitization describes manufacturers' transformation from being product vendors to becoming suppliers of digitally enabled product-service systems (PSSs). It comes with the promise of creating new revenue streams during the lifecycle of the products sold. The change in organizations due to digitization projects usually puts existing assets and process management to the test. Can the existing asset creation still be mapped or must all assets such as equipment, services, people, and knowledge be created from zero?

Manufacturers who sell products with a long lifespan have many opportunities for offering Smart Services bundled with their products, or to provide them separately. Due to changes in technologies, market expectations, or regulations (West et al., 2018), they also have the chance to adapt, upgrade, or modify the system to better achieve the outcomes required. With digital technologies, value creation can be improved through Smart Services that enhance collaboration between different actors. With the appropriate use of technology, businesses can achieve new goals. This requires a cross-functional system understanding. The successful operation of assets depends on complex decisions by many actors; many decisions will have unforeseen (or unwanted) consequences.

The development of the digital twin (DT) makes it possible to show the consequences of decisions through simulations. However, to understand causality in its entirety, the simulations connect to other systems through

DOI: 10.1201/9781003425724-29

knowledge carriers and expose information or knowledge that is not fully captured in the DT. The combination between the DT, which is not 100% correct, becomes effective through the link to other systems via people and creates a real competitive advantage (Neely & Barrows, 2011).

24.1.1 Lifecycle

One of the key drivers of industrial services is the lifecycle of equipment. Equipment is purchased, commissioned, and operated; it wears and ages, needing maintenance to sustain its performance. Unplanned failures can require repairs (West et al., 2021). During its lifespan, technologies, regulations, or market conditions may change, necessitating upgrades or modifications. For the owner or operator of an industrial plant, the equipment must be available with an optimum or known performance when needed. Some plant items may be operated every day, whereas other subsystems may be operated intermittently.

Regarding performance, the supplier can provide support enabled by digital technologies to help operators achieve the outcomes they require. This may vary from routine or operational maintenance support to production scheduling, integrated operations/maintenance planning, consumable materials planning, or general operations support and training. Obsolescence planning is part of lifecycle asset management that is demanded by technology changes, regulations, or market conditions. Here again, the supplier can identify changes to operation or technology replacements (West et al., 2020a, b).

From these steps, we see that service strategy planning is related to risk management: planned inspections are risk mitigation actions that avoid unplanned shutdowns and consequential costs. At the same time, over-maintenance will result in higher costs while reducing availability. Operational support and asset management can enhance the business outcomes by improving system productivity, creating more for the same input. Over the equipment's lifecycle, many tasks can be planned while other tasks may require different actors and have different value propositions associated with them. Around the equipment, there are changes to the people, the markets, and technologies (Frost, Cheng, & Lyons, 2019). All these changes within the system have consequences – everything is connected, and nothing happens in isolation.

24.1.2 Knowledge and People

Interconnectivity is a particular fascination that we have cultivated since we realized that information only becomes valuable for systems and people in a certain context, and that knowledge is a very special entity (Bratianu, 2015). The DIKW model (Data-Information-Knowledge-Wisdom) (Ackoff, 1988) defines information as a unit, based on data. Information can only be regarded as such if it is placed in a frame that can be used, recognized, decoded, and evaluated. Knowledge can result from perceptions based on experience using the right information, and this is often based on interactions between two or more people (Nonaka & Toyama, 2003). On this basis, value has been co-created between two actors, with the information usually provided digitally.

In developing a DT, the appearance of information relies on physical and psychological aspects. Shannon represents an engineering perspective toward information: *semantic aspects of communication are irrelevant to the engineering problem* (Shannon, 1948). He refers to the meaning of data as a technical signal and contrasts his statements with the relevance of meaning. The content or the reason for information and communication only gains relevance if it can be used or transformed into knowledge. A genuine theory of information would be about the content of our messages and not about the form in which it is embedded (Bratianu, 2015).

Knowledge management, in contrast, understands that knowledge building is a creative, cognitive process, which relates concepts that can be far apart in an individual's memory and are accessible through metaphors (Nonaka & Takeuchi, The Knowledge-Creating Company, 1995). Tacit knowledge can be transformed into communicative knowledge by recognizing contradictions through metaphor and resolving them through analogy (Nonaka, Toyama, & Konno, 2001). For this reason, knowledge is agile and changing. Making use of the right information at the right time enables its development.

Today, the success of companies lies more in recognizing, promoting, and utilizing the intellectual and systemic capabilities of its employees than in placing its material assets at the forefront of an image. However, it is this deeply embedded knowledge that is not openly shared or cannot be shared easily, often because of how individuals perceive and communicate with each other (Quinn, Anderson, & Finkelstein, 2009). A decisive change takes place in recognizing the transformation of information into knowledge as an active exchange between people, or between systems and people.

24.1.3 Value Co-Creation

Adding value in the context of servitization describes the process of adding services to existing products. Smith et al. (1977) explained in their book *An Inquiry into the Nature and Causes of the Wealth of Nations*:

> The word VALUE, it is to be observed, has two different meanings, and sometimes expresses the utility of some particular object, and sometimes the power of purchasing other goods which the possession of that object conveys. The one may be called 'value in use'; the other, 'value in exchange.'
>
> *(Smith, Cannan, & Stigler, 1977, p. 42)*

The idea of value creation goes beyond selling goods (e.g., value in exchange) and toward providing outcomes that support the customer's business. Value creation and co-creation are essential aspects of servitization; many fundamental characteristics must be considered for use in PSS-related frameworks. Value creation depends on the extent to which participants engage in co-creative activities in a problem-solving framework (Aarikka-Stenroos & Jaakkola, 2012).

Service-Dominant Logic (SDL) (https://en.wikipedia.org/wiki/Service-dominant_logic) is based on value in use (or context) and is seen as essential to service science and the study of value creation in service systems (Lightfoot et al., 2013). SDL is a concept that describes a fundamental shift from goods-centred logic that focuses on value in exchange, to a service-centred logic focusing on value co-creation with different actors. Because of this, PSSs often lead to more complex offerings based on multiple value propositions.

Like the creation of knowledge, value propositions also seem to be dynamic. Smith, Maull, and Ng (2014) suggest four nested value propositions based on the analysis of product-service attributes:

- Asset value proposition
- Recovery value proposition
- Availability value proposition
- Outcome value proposition.

In designing operations to deliver these value propositions, the importance of context increases while moving from asset value toward outcome value. SDL is a concept that helps to identify, create, and deliver value in offerings based on PSSs. Within Service Innovation, SDL highlights the importance of ecosystems (networks), processes, actors, and value co-creation (Frost, Cheng, & Lyons, 2019).

24.1.4 Technology

Data Science is a concept that combines data analysis, its statistics, machine learning (ML) and related methods to understand real-world causal phenomena (Hayashi, 1998). It uses techniques and theories from many fields. 365 Data Science (2018) uses the term "advanced analytics" to summarize Business Analytics, Data Science, Data Analytics, ML, and AI (Figure 24.1). In addition, they attach a time component, providing an insight into the maturity of the different approaches. The advanced analytics perspective describes all concepts from business to data analytics arching across to different kinds of ML and AI.

Business analytics consist of business case studies or qualitative analysis based on the intuition and knowledge of individuals without the need for

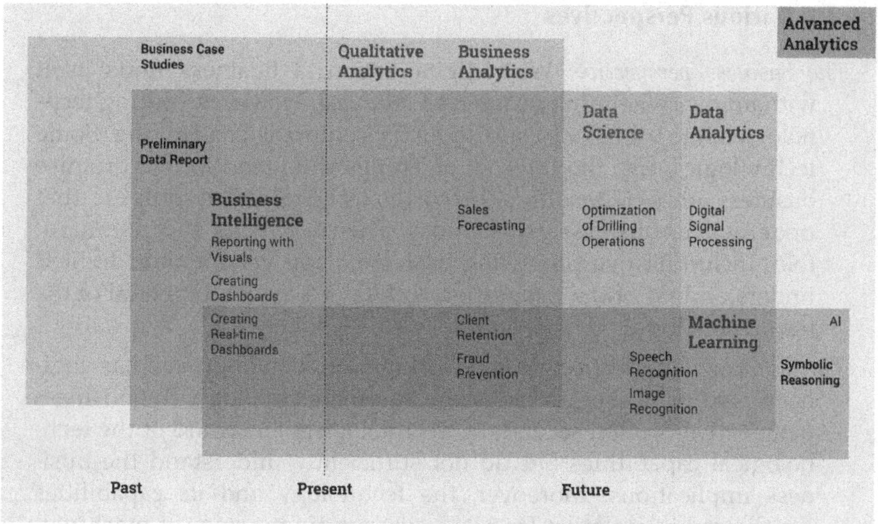

FIGURE 24.1
Data Science vs. Machine Learning vs. Data Analytics vs. Business Analytics.

(Source: Retrieved from 365 DataScience, 2018; own illustration based on DataScience.)

data. Data Science in turn is a discipline reliant on data availability, whereas business analytics do not completely rely on data as they contain more contextual factors. Finally, advanced analytics such as ML is a subset of AI.

ML describes the ability of machines to identify outcomes without being explicitly programmed to do so. The algorithms of ML let the machine collect, analyse, and interpret data. Based on this, the machine can give recommendations for actions. AI is a greatly hyped topic; however, it is an enabling technology for the fourth stage of the "capabilities of smart connected products" (Porter & Heppelmann, 2015). Based on this, the "Digital Worker" becomes a co-worker that helps and facilitates the job of the human.

24.2 Digital Twin Development

The development of a DT relies on the understanding of complex ecosystems in which people, processes, and things exist. The development therefore aims to tackle the difficulties in the existing decision-making due to limited understanding of the underlying causes and effects. This is because PSSs are complex and multidimensional, with many actors with different perspectives (and conflicting roles), all of which impact different timeframes, assets, and values themselves.

24.2.1 Various Perspectives

The business perspective: With digitalization, a business finds itself with an overwhelming number of changes. However, putting technologies into a business and industry context is challenging. Some technologies, e.g., the Internet of Things (IoT) tend to overpromise business impact, resulting in costly technology-led projects that underdeliver business expectations. The reasons for this are manifold, including overpromising by technology vendors and limited understanding of the integration and value creation potential of the new technology.

The technology perspective: People working in technology, such as engineers and computer specialists, value the technology; the business perspective is often secondary. Technologists are aware of the technological capabilities but do not sufficiently understand the business implications. Moreover, the technology and its capabilities are viewed in isolation from the surrounding system of machines. Limited awareness of the technological system of interconnected machines, the people, and the processes, often hinder technology's full potential value from being achieved.

The operations perspective: People working in operations management are concerned with operations excellence. This is achieved when actions from people and machines are deployed to achieve best business outcomes. This area of expertise requires a high degree of system thinking.

However, operations managers rely narrowly on business goals and technology. Setting business aims related to digital technologies is challenging and complex, over or under engineering affects operational effectiveness. In digital solutions, operations managers can play a vital role in creating bridges between the business and technology perspectives (Figure 24.2).

24.2.2 Tackling Complexity

These different roles imply a large set of activity and performance indicators to represent what is relevant for the various actors. For the best outcomes from PSSs that links technical input with business questions, it is essential to understand that decisions are taken on operational, tactical, and strategic grounds.

The complexity is created by the different interactions between the different layers, actors, and organizations. This makes taking the right decision at the right time difficult due to conflicting demands that must be balanced out. Here, the use of different data sources (e.g., ERP, CRM) and the system understanding embedded within the DT can help to deliver the *right information,*

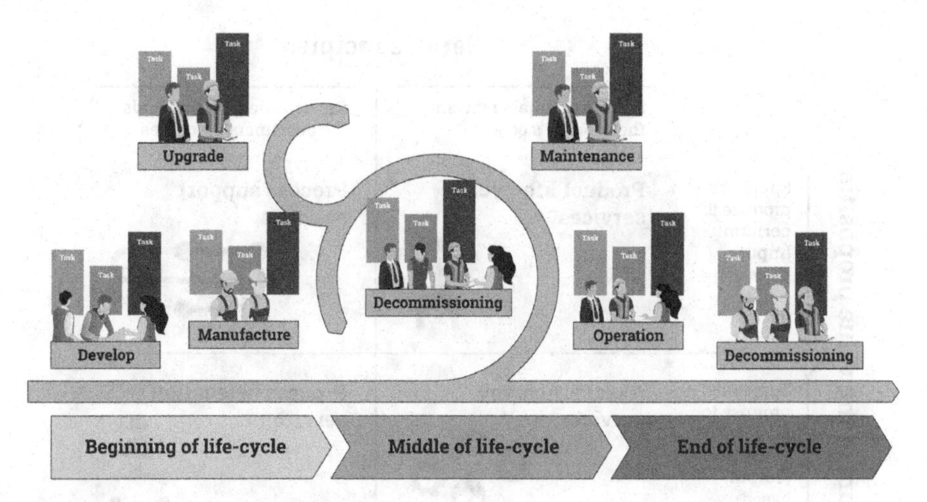

FIGURE 24.2
Lifecycle model and roles (West et al., 2020a, b).

at the right time, in the right form, to the right person or team, to support taking the right action (Frost, Cheng, & Lyons, 2019).

It is the opportunity, and the promise, of Smart Twin development to acquire the right data, evaluating different strategical, tactical, and operational options, while considering the system's feedback dynamics. In this, the DT is like a brain running in parallel with the real physical system. In contrast to a simulation, DTs are coupled to their physical counterparts using datalinks in both directions: DTs gather data from the system (close to real time) while providing the system with both data and actionable information.

This helps the physical system to operate more efficiently. To enable this, DTs implement data models and system models. Data models provide the necessary structure to foster additional information from the mix of data obtained from the physical system. Simulation models can be used to compare the performance of the physical system with expected performance. Insights from the DT can provide actors with advice on how to improve performance and operations.

24.2.3 Digitally Enabled Solutions

DTs can also be given delegated autonomy, allowing them to automate tasks and report back when necessary – in effect, becoming co-workers. Together, the data and simulation models provide a virtual representation of the physical system, creating the cyber-physical environment. The DT offers an excellent opportunity to help solve different problems in the planning and operation of PSSs. By incorporating the system dynamics and different timeframes, different actors enable well-founded decisions and actions toward a

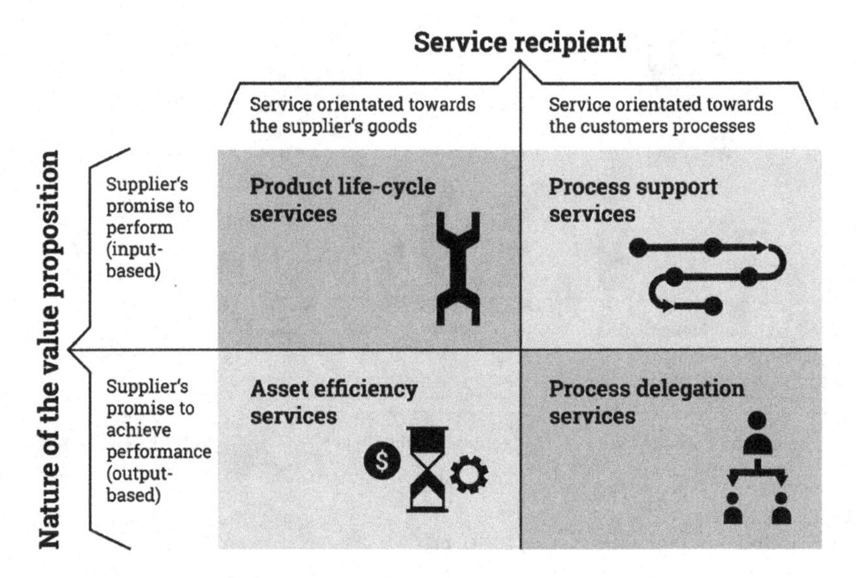

FIGURE 24.3
Service–value relationship. (Source: Own illustration based on Kowalkowski & Witell, 2020.)

common goal. In summary, DTs provide an excellent opportunity for digitally enabled solutions (Figure 24.3):

Verification and monitoring of the system performance: Using a DT, a system's behaviour can be forecast based on different scenarios or assumptions. This enables verification of performance and identification of potentially hazardous developments, or an understanding of potential consequences.

Assisting operational decision-making: Many operational decisions take place on subsystems while the general goal is to optimize the surrounding system's performance. By combining relevant data from the subsystems with a model of the overall system, decisions at the subsystems level can be made that consider the whole system.

Training of personnel: DTs can be used to train new operators (e.g., a flight simulator). This can be done if the physical system is not available yet, or if training on the physical system is too risky or costly.

24.3 Five Case Studies

This chapter is based on 5 of more than 20 cases examined. In each case, we follow the different perspectives of the given challenges of the company and the impact of a DT. Each selected case represents an industry sector.

This approach allows an overview of the benefits and challenges of digitally enabled servitization, from the different system environments.

24.3.1 Ship Manufacturing

The firm designs, manufactures, and maintains a fleet of midsized ships that carries passengers and tourists throughout the year on lakes in Switzerland. The vessels they make are important assets with a very long lifecycle that their customers operate for many years. The oldest ships in service today are over 110 years old.

From a business perspective, the company needs to sell new ships. Selling, designing, engineering, and building new ships require many skilled people. The services these people provide are also sold separately. The lifecycle of the vessel requires regular servicing, some of which are mandated by regulation. This offers the firm "one-time" product revenue as well as "recurring" service-based revenues.

Ships require many disciplines such as mechanical engineering, electrical engineering, designers, automation engineers, shipbuilders, and HVAC experts, due to the technical complexity of vessels. The inter-dependency of systems makes it hard to balance the right amount of technological functionalities and capabilities.

24.3.1.1 Impact of Digital Twin

The firm's vessels operate on Swiss lakes for both industrial and commercial activities. The decisions made in the design, construction, or operation of the ship can lead to unintentional consequences. These can be in terms of performance costs, or revenue of the other different areas, which may not show up until later. Also, linking the consequences of the decisions taken to the analysis layer is essential to enable continuous learning and improvement.

Requirements: The business goals such as selling new ships, consulting, and lifecycle management services are dependent on the efficient use of resources, including people, machines, and materials. Operations management is an important role as service demands require customer management and the integration of customer operation goals.

Business challenges: The challenge is to meet the requirements ahead of new product development.

1. How to best operate the vessel based on the schedule (or could the schedule be adapted based on performance)?
2. What are the actual and anticipated costs (e.g., people, machines)?
3. What is the actual and anticipated fuel consumption based on operations (and how does this compare with design)?

24.3.1.2 Business Intelligence

- Optimized deployment of resources (people, machines, and materials).
- Cost-efficient resource usage.
- Vessel fuel efficiency based on actual conditions and routings.
- Repair, replace, and upgrade decisions based on actual subsystem component wear and tear.

24.3.1.3 Consequences

- Increased efficiency of resources.
- Reduced energy and equipment costs.
- Increased planning accuracy for maintenance.
- More efficient upgradability of vessels.

Findings: Business and data analytics enable ML on new product development and support ships that are operational today as well as helping with the design of new vessels.

24.3.2 Facility Management

A small firm providing facility management services offers end-to-end services to building owners and occupiers, from the design and planning to the construction and operation of a complex office building. With third-party suppliers, they bundle products and services together into complex offerings. Value is created from the services provided, especially the stability and availability of server rooms.

The value proposition was based on the server room's availability. This is managed through climate control and field service staff making the necessary changes when a problem is detected. The management of different machines and people within this environment to achieve the expected outcome is difficult. Building Information Management (or BIM) was used within the building, with links to procurement for replacement parts. Spares holding, call-out time, and mean time to remedy failures, all impact the overall cost and availability of the servers.

The facility management business is focused on the design, building, and operation of a server room with tight limits of temperature and humidity. The server room climate control is provided to customers, as problems in the server room are an issue that is mission-critical to their business. The customer's operational staff work 24/7 with remote support from the facility management firm to maintain climate control.

24.3.2.1 Impact of Digital Twin

The DT supported decision-making and provided a degree of automation to control the server room. This automation simplified the operations, informing the operators of the decisions it was making and the likely impact on the facility's operation. Procedures were preplanned based on the DT, allowing standard operating procedures to be developed and tested. This sped up the mobilization of the overall service and modelling of the site's availability based on operational assumptions. Callouts for service technicians were simplified as the error states could be more quickly identified or narrowed down.

Requirements: It was important for the provider to ensure their performance commitment. The level of spares in stock must be carefully planned to meet needs, to avoid failure and obsolescence.

Business challenges: The challenge is to handle fallout times in an intelligent way to support decision-making in time-sensitive situations.

1. Mean time to failure (MTTF) of the machines?
2. How long does it take to repair the broken machine?
3. Manage redundancy and contingency plans for server rooms.

24.3.2.2 Business Intelligence

- Real-time monitoring and statistical time to failure calculations.
- Process control and testing.
- Complex PSSs control.

24.3.2.3 Consequences

- Increased availability of operational conditions for server rooms.
- Minimized Mean Time to Repair (MTTR).
- Continuous evolution of the systems.

Findings: The technology had to link several different aspects together as well as the supplier and the customer. Depending on ambient conditions, the failure of a subsystem in the server room may not cause any immediate issues. Using the technology to estimate the time to exceed allowable conditions, the firm could offer and price a performance-based contract. They also understood more about the maintenance requirements for the system.

24.3.3 Rail Transport

The largest railway transportation company in Switzerland operates public transport and cargo trains. This includes operating and maintaining the infrastructure, such as stations, tracks, and tunnels, which the company develops and manages. The government partially owns the company.

The rail operator had a capacity problem at a station and decided to expand it to provide more platforms. Redesigning a station means considering many different stakeholders and risks of unintentional consequences. The real estate industry requires high customer footfall to maximize retail shopping opportunities. The railway company wants a simple and regular timetable, passengers with transfer connections want easy transfers between different trains and some want to go shopping. The station manager must ensure a safe environment with minimal risk of accidents. Safety is prioritized through both train planning and platform allocation.

24.3.3.1 Impact of Digital Twin

The DT supports safe planning and provides a simulation for different scenarios, to allow the firefighters, police, ambulance, etc., to test different approaches to deal with incidents while the station is being expanded. This short-term period can be highly complex to plan and may need other solutions depending on the schedule. The DT can support changes to train schedules, helping ensure optimal arrangements of trains for travellers. The shops (and rents) can be sold based on expected footfall, with the DT checking simulated footfall against actual video recordings.

Requirements: Existing floor plans and path use were crucial inputs for the DT. The BIM data supported building the twin as it provided detailed plans for the station layout. The existing station plans provided the counterpart for the new layouts. Also important were the initial train schedules and platform arrangements, along with passenger forecasts.

Business challenges: The challenge is to establish an excellent traveller experience and, at the same time, focus on key metrics such as convenience, cost, and safety.

1. How to redevelop stations at lowest cost to support future footfalls?
2. How to best integrate the schedule into the extended station?
3. How to optimize flow with shopping?

24.3.3.2 Business Intelligence

- Optimized routing of passengers
- Optimized arrivals
- Optimized use of the infrastructure (elevator, escalators, shops).

24.3.3.3 Consequences

- Increased capacity within the allowable budget
- Improved customer and traveller experience
- Improved station safety.

Findings: The technology can help make integrated design and planning visible while considering multiple perspectives. In costly capital projects, this is particularly important as the station will remain in operation for many years. Once the station is built, costs for even small changes are very high.

24.3.4 Joinery and Carpentry

The firm produces carpentry products on a made-to-order basis. The firm's production lines consist of modern automated machines within a semi-automated production environment. The orders are highly customized (single-piece flow) with a production batch size of one. This makes procurement, manufacturing, and assembly planning very difficult, creating challenges for sales, cost estimation, and lead times.

The firm's advanced production line consists of many machines with different levels of automatization. The interplay of people, machines, and raw materials is complex and comes with the potential of optimization. The firm manufactures architectural interior furnishings, where the products are integrated into a building. Customer satisfaction while offering wood-based products for a competitive price is the main challenge from a business perspective.

There are many production challenges. The production line is complex, with many machines, people, and raw materials. The variables make aligning production with the business goals challenging. Here, the Smart Twin offers production managers the opportunity to test different scenarios, which helps with planning and manufacturing processes, cost production, and cost estimations.

24.3.4.1 Impact of Digital Twin

The use of the DT supports joint decision-making within the operations team by providing limited options and the consequences of each option. The ability to rapidly reschedule production based on the illness of a machine operator, a breakdown of a machine, or planned maintenance allows the company to understand the impacts on the production order and minimize the effects. The aim here is to have an adaptive plan rather than a fully optimized one. Similarly, discussing scheduling with the sales teams helps all parties understand the consequences and results of pricing. The DT also shows the loadings on people and machines, allowing resource allocations to be more effective. The visual adaptive plan allowed the firm to negotiate the tactical and operational schedule and routing, putting them into a more competitive position and improving customer satisfaction.

Requirement: The capabilities and the resources of the production environment are necessary. This includes all machines and staff, remembering that not everyone can work on every machine. The inputs are based on the customer's orders and the delivery scheduled. The scheduler aggregates the individual orders based on delivery constraints. The constraints of the production environment are also significant (e.g., planned maintenance requirements). The DT needs to have these as inputs; otherwise, the resource allocation will be incorrect.

Business challenges: The challenge is to produce according to the individual customer orders.

1. How to schedule production in the most efficient and feasible way?
2. How to allocate resources in the production setup?

24.3.4.2 Business Intelligence

- Customer orders can be classified, and the best option produced.
- Have an efficient and feasible production schedule.
- Optimized resource allocation (people, machines, and materials).

24.3.4.3 Consequences

- Improved quality of customer service.
- Reliable schedule.
- Cost reduction due to resource utilization.

Findings: The main output of the DT's adaptive scheduling was a visual production plan. This enabled the operations team and sales managers to assess options for new orders and discuss the impact of sickness or loss of production. Assigned scheduling enabled improvements in production utilization. The ability to "see" the production gaps allowed everyone involved to identify new opportunities and reduced the time and stress for planning.

24.3.5 Transport Solutions

The firm was the supplier of the water pumping system for a rail tunnel and also provided operational support. The removal of water from a tunnel is a critical safety aspect for its operation. In winter, trains carry substantial amounts of water as snow, which then melts and must be removed. Rainwater also percolates through the mountain into the tunnel, to be pumped out.

For the rail company, the significant risk occurs when a pump fails, coupled with major water ingress in the tunnel and storage sumps start to fill. The tunnel operator needs to know how long they can safely operate the tunnel before they must close it. The supplier needs to follow the operation of the tunnels and how this affects the planning to support unplanned repairs. The tunnel operator needs to understand how they can control and limit water ingress and how operational decisions may buy them time until either the repair is made or the weather changes and water ingress is substantially reduced.

24.3.5.1 Impact of Digital Twin

The DT provided three cases. The first was to support the operations team with planning and dealing with challenges based on water ingress and lack of pumping capacity, to understand how long before the tunnel would become unsafe. The second case was to develop a simulator to help the operations team pre-plan for once-in-a-lifetime events and draft procedures to deal with them. The third case supported the suppliers by identifying call-out field service teams and integrating road with location data to forecast the average call-out time and confirm the system's safety margin.

Requirements: The BIM and equipment performance data provided critical inputs to the model. To capture the relationship between weather and water ingress, train schedules were used to model the ingress and then support call-out times for field service technicians.

Business challenges: The challenge is to develop a training simulator for the operators to deal with once in a life incident planning.

1. What is the effective safety margin for operation?
2. Where should the OEM's field service team be stationed for effective call-out?

24.3.5.2 Business Intelligence

- Mean time to failure/repair
- Maintenance schedules
- Train schedules.

24.3.5.3 Consequences

- Increased and reliable tunnel availability
- Minimized MTTR
- Safe operation.

Findings: The system developed provided multiple uses, initially with the build and configuration of the system and the ongoing activities and operational support provided by the OEM.

24.4 Lessons Learned

24.4.1 Comparison of All Five Cases

The cases provide different perspectives on developing DTs in various industrial settings. They all describe how digital technologies can help connect different decision-making perspectives (Figure 24.4), and how DTs can support this. Linking cause and result in a structured way means decision-making is backed by data, and supports the integration of intangible know-how into the broader business.

The level of computational power needed to support the DT application was not related to the value that the beneficiaries identified in the individual cases. Transparency and visibility provide excellent support to decision-making in multi-stakeholder environments. Common views enabled actions to be taken based on solid understanding.

Creating a DT helps to integrate tacit know-how and leads to the creation of knowledge and wisdom. The simulations can then be integrated into the

Type of Service	Ship Manufacturing	Facility Management	Rail Transport	Joinery & Carpentry	Transport Solutions
What is happening?	Link of technical and business implications for single events.	New build, so not appliable.	The flows of people around the station could be visualized.	The flow of wood around the factory was tracked.	New build, so not appliable.
Why is it happening?	Insights applicable to operational ships and new ships.	A model provided the basis for cause and effect within the facility.	The model developed integrated the station with the trains schedules allowing insights to be visualized and compared with reality.	The tracking allowed detailed understanding to the machining times per step.	The model provided the foundation to the water flows and system limitations.
What will happen?	Advisory models to support ship operation.	An Advisory model developed that integrated routine, preventive maintenance and breakdown maintenance with standard operational procedures.	The model allowed footfalls around the station to be forecast based on changes to train arrivals.	Based on the understanding a job scheduler was developed to understand what could happen, based on machine breakdowns, staff illnesses or additional sales.	Based on the boundary conditions the future was modeled. This was the foundation to the advisory service offer for three different groups of stakeholders.
Delegated autonomy?	Not tested.	Within defined boundaries autonomy was assigned to the system.	Not tested.	The scheduler could support autonomy, in particular automation of materials.	The autonomy was limited due to safety management issues.

FIGURE 24.4
Cross case analysis of all five cases.

system to support "what if…" analysis and delegated automation. It supports and enables people to make better decisions and act more swiftly.

24.5 Conclusions

DTs offer us the opportunity to create or codify tacit knowledge from existing physical systems. In doing so, they enable improved decision-making. The object boundaries of the DTs we develop are important, as they limit the causes/effects and consequences of the system. They do not and cannot provide insights outside their boundaries, so we cannot remove the manager – the ultimate arbitrator. This does not mean decision-making capabilities cannot be delegated to the DT, rather, that once the system boundaries have reached, the DT must ask for them. Ultimate agency must rest with people (mangers) and not with machines.

When designing DTs, it is important to be able to zoom in and out with scale and on a temporal dimension. The speed of decision-making required must be reflected in the data flow to and from the DT. Where the decision-making horizon is in days, then data flows of hours or minutes may be sufficient to support decision-making, but in emergencies it must be exponentially faster.

Finally, the DT is a concept based on many technologies that enable us to deliver and build new value propositions. It provides the framework of institutional arrangements that integrate different process, machines, and people. Loosely coupled families of DTs may coexist to orchestrate operation, and support decision-making though the codification of tacit system knowledge. DTs allow people to collaborate, to integrate new information, and create new knowledge leading to new tacit assets at every level of a firm, improving efficiency and allowing new value propositions to be delivered.

References

365 DataScience. (2018). Von data science vs machine learning vs data analytics vs business analytics. https://365datascience.com/data-science-vs-ml-vs-data-analytics/ abgerufen.

Aarikka-Stenroos, L., & Jaakkola, E. (2012). Value co-creation in knowledge intensive business services: A dyadic perspective on the joint problem solving process. *Industrial Marketing Management*, 41, 12–26.

Ackoff, R. (1988). From data to wisdom. *Journal of Applied Systems Analysis*, 16, 3–9.

Bratianu, C. (2015). Knowledge creation. In C. Bratianu, *Organizational Knowledge Dynamics: Managing Knowledge Creation, Acquisition, Sharing, and Transformation*. Hershey: IGI Global.

Dalkir, K. (2005). *Knowledge Management in Theory and Practice*. Burlington and Oxford: Elsevier.

Davenport, T., & Prusak, L. (1998). *Working Knowledge: How Organizations Manage What They Know*. Boston, MA: Harvard Business School Press.

Dawson, R. (2005). *Developing Knowledge-Based Client Relationships*. Butterworth-Heinemann.

Frost, R. B., Cheng, M., & Lyons, K. (2019). A multilayer framework for service system analysis. In P. Maglio, C. Kieliszewski, J. Spohrer, K. Lyons, L. Patrício, & Y. Sawatani, Handbook of Service Science, Volume II, 2019: 285–306.

Hayashi, C. (1998). What is data science? Fundamental concepts and a heuristic example. *Data Science, Classification, and Related Methods*, 40–51.

ISS. (2017). SS1.3: What is data science and how will it impact rehabilitation science? Von. *Handbook of Service Science*, Volume II, 2019: 285–306. https://www.iss.pitt.edu/ISS2017/ISS2017Pro/ISS2017ProIC/ISS2017ProICD1/ISS2017ProICD1_SS01_3/icSS1_3.htmabgerufen.

Kowalkowski, C., & Witell, L. (2020). Typologies and Frameworks in Service Innovation. In The Routledge Handbook of Service Research Insights and Ideas (pp. 109–130). Routledge.

Mitchell-Guthrie, P. (2014). Looking backwards, looking forwards: SAS, data mining, and machine learning. Von. https://blogs.sas.com/content/subconsciousmusings/2014/08/22/looking-backwards-looking-forwards-sas-data-mining-and-machine-learning/ abgerufen.

Neely, A., & Barrows, E. (2011). *Managing Performance in Turbulent Times: Analystics and Insight*. Wiley Online Library.

Nonaka, I., & Takeuchi, H. (1995). *The Knowledge-Creating Company*. New York, Oxford: Oxford University Press Inc.

Nonaka, I., & Toyama, R. (2003). The knowledge-creating theory revisited: Knowledge creation as a synthesizing process. *Knowledge Management Research & Practice*, 1, 2–10.

Nonaka, Ikujiro, Ryoko Toyama, and Noboru Konno. "SECI, Ba and leadership: a unified model of dynamic knowledge creation." *Long range planning* 33.1 (2000): 5–34.

Polanyi, M. (1966). *The Tacit Dimension*. New York: Doubleday & Company, Inc.

Porter, M., & Heppelmann, J. (2015). How smart, connected products are transforming companies. *Harvard business review* 93.10 (2015): 96–114.

Quinn, James Brian, Philip Anderson, and Sydney Finkelstein. "Managing professional intellect: making the most of the best." *The strategic Management of Intellectual capital*. Routledge, 2009. 87–98.

Reason, P., & Bradbury, H. (2008). *The SAGE Handbook of Action Research*. Los Angeles, London, New Delhi, Singapore: Sage.

Shannon, C. E. (1948). A mathematical theory of communication. The Bell system technical journal, 1948, 27(3): 379–423.

Shotter, J. (1993). *Cultural Politics of Everyday Life, Social Constructionism, Rhetoric and Knowing of the Third Kind*. Buckingham: Open University Press.

Smith, A., Cannan, E., & Stigler, G. J. (1977). *An Inquiry into the Nature and Causes of the Wealth of Nations*. Chicago: University of Chicago Press.

Smith, L., Maull, R., & Ng, I. C. (2014). Servitization and operations management: A service dominant-logic approach. *International Journal of Operations & Production Management*, 34(2), 242–269.

West, S., Gaiardelli, P., Resta, B., & Kujawski, D. (2018). Co-creation of value in Product-Service Systems through transforming data into knowledge. *IFAC-PapersOnLine*, 51(11), 1323–1328.

West, S., Stoll, O., Meierhofer, J., & Züst, S. (2021). Digital twin providing new opportunities for value co-creation through supporting decision-making. *Applied Sciences*, 11(9), 3750.

West, S., Stoll, O., & Müller-Csernetzky, P. (2020a). Avatar journey mapping' for manufacturing firms to reveal smart-service opportunities over the product life-cycle. *International Journal of Business Environment*, 2020, 11(3): 298–320.

West, S., Stoll, O., Østerlund, M., Müller-Csernetzky, P., Keiderling, F., & Kowalkowski, C. (2020b). Adjusting customer journey mapping for application in industrial product-service systems. *International Journal of Business Environment*, 11(3), 275–297.

25

Digital Twins for Lifecycle Management: The Digital Thread from Design to Operation in the AECO Sector

Sofia Agostinelli
Sapienza University of Rome

25.1 Introduction and Background

Recently, a relevant increase in the use and availability of digital technologies has occurred in the architecture, engineering, construction and operation (AECO) sector improving efficiency and productivity at different levels (Sacks et al., 2020).

With the rising adoption of building information modelling (BIM) methods and tools (ISO, 19650-1, 2018) within industries and organizations in the AECO sector, digital tools started enhancing process and information management towards a full digitization of process management (Kim et al., 2018).

Some of the main factors driving the transformation's paradigm in lifecycle management of the built environment are the advanced capabilities in digital modelling and representation of assets (Stojanovic et al., 2018) together with data collection and management across different lifecycle phases, and business process management (Tang et al., 2020).

The digital twin (DT) concept is defined differently depending on the domain context such as manufacturing, automotive, aerospace, etc., and it has been recently explored and introduced in the built environment sector (McKinsey Global Institute, 2017). As the DT concept rapidly gained attention and widespread interest, it became a research area for the AECO sector and contributions in literature increased over time after 2016 towards an effective conceptualization. In this regard, in the United Kingdom, the National Infrastructure Commission (NIC) issued the *Data for the Public Good* report in 2017, and the Centre for Digital Built Britain (CDBB) introduced *The Gemini Principles* report in 2018, in order to define policy standards and frameworks for the implementation of digital technologies as an interconnection between society and the built environment (Batty, 2018).

 DOI: 10.1201/9781003425724-30

Moreover, the DT concept is also closely related to the domain of smart cities and buildings as it enables predictive insights into a city management approach based on digitalization (Mohammadi and Taylor, 2017), providing big data management capabilities in urban spaces (Oliver et al., 2018).

However, the DT term was firstly introduced in 2003 by Grieves in product lifecycle management (PLM), and later in 2012 it was further evolved in the aerospace industry related to modelling and simulation as it was defined by Glaessgen and Stargel as

> an integrated multiphysics, multiscale, probabilistic simulation of a vehicle or system that uses the best available physical models, sensor updates and fleet history, among others, to mirror the life of its flying twin. The digital twin is ultrarealistic and may consider one or more important and interdependent vehicle systems.
>
> *(Shafto et al., 2012; Tuegel et al., 2011)*

25.2 Definitions and Principles

Many DT definitions have been developed over time and several research studies introduced the concept of cyber-physical connection (Haag and Anderl, 2018; Tomko and Winter, 2019) and real-world mirroring at different levels from products, assets, buildings (Buckman et al. 2014), districts to national ecosystems (Bolton et al., 2018).

Control and monitoring capabilities combined with intelligence skills are some of the main goals for DTs in the built environment throughout its lifecycle.

While looking at specific definitions of the concept related to process management in the AECO sector, many conceptualizations from different points of view can be observed.

The DT Consortium defines it as "a virtual representation of real-world entities and processes, synchronized at a specified frequency and fidelity" introducing the synchronization concept as a fundamental element.

The Gemini Principles report defines the DT as a dynamic model of the physical asset, controlled and operated through real-time sensor data measuring performance. As a recently born concept, DT is often resulting in confusion about its intended uses, performances and outcomes, so identifying the main levels of DT in the built environment could be valuable in obtaining clearer definitions.

Digital twinning is generally conceived as virtually simulating the relevant behaviour of physical objects in real-world environments (Hochhalter et al., 2014) as the main components of a typical DT are the physical entities in the physical world, the digital models in the virtual world and the data that tie the two worlds together (Grieves, 2014).

Moreover, many definitions emphasize the concept of synchronization between the digital and virtual world, such as Garetti et al. (2012) who define DTs as "the virtual representation of a production system that is able to run on different simulation disciplines that is characterized by the synchronization between the virtual and real system".

Moreover, a selection of 31 definitions from research literature are included in Table 25.1 and grouped according to six different key layers, which specifically identify the main key points according to the field of interest.

According to Table 25.1, it emerges that the DT concept is constantly evolving as the main definitions have been provided in the USA and Europe between 2017 and 2019, and the K.5 conceptualization is predominant.

Moreover, cyber-physical systems (CPS), the internet of things (IoT), machine learning (ML) and artificial intelligence (AI) are also included in the domain of Industry 4.0 and the relationships between DTs and such concepts is quite relevant in order to identify their specific role in the evolution of the AECO sector.

In this regard, some other definitions are related to levels of complexities, in order to identify general and common languages in the development of DTs in different contexts.

In this regard, the CDBB–DT Hub provided the DT Toolkit Report in 2022 defining some of the main capabilities and key objectives of different DT configurations in an outcome-oriented approach as summarized below:

1. *Descriptive* DTs are able to collect and visualize data through dynamic data aggregation, with the main objective of responding to questions such as *What happened?*;
2. *Informative* DTs are able to generate insights through data aggregation and analysis using advanced digital techniques and providing a response to questions such as *Why did it happen?*;
3. *Predictive* DTs provide real-time monitoring and prediction capabilities through data science and ML techniques, with the main objective of responding to *What will happen?*;
4. *Prescriptive* DTs are able to propose interventions based on prescriptive analytics, simulations and AI-based learning answering to *What should I do?*;
5. *Cognitive* DTs are able to autonomously take actions and interventions, focusing on decision-making automation based upon prediction through ML and deep learning.

TABLE 25.1

DT Definitions and Key Layers

Key Layers	Definition	Field	References
Integration [K.1]	[D.1] Integrated multiphysics, multiscale, probabilistic simulation composed of physical product, virtual product, data, services and connections between them.	Aerospace	Glaessgen and Stargel (2012)
	[D.2] Ultrarealistic integrated multiphysics, multiscale, probabilistic simulation of a system.	Complex equipment	Tao and Qi (2017)
	[D.3] Comprehensive physical and functional description of a component, product or system, together with all available operation data.	Product lifecycle	Boschert and Rosen (2016)
	[D.4] A means to link digital models and simulations with real-world data, creates new possibilities for improved creativity, competitive advantage and human-centred design.	Built environment	Arup Group Limited (2019)
	[D.5] Big collection of digital artefacts that has a structure, all elements are connected; there exists metainformation as well as semantics.	Manufacturing	Rosen et al. (2015)
Connection [K.2]	[D.6] New mechanisms to manage IoT devices and IoT systems-of-systems.	Industrial IoT	Canedo (2016)
	[D.7] One where the virtual object exchanges data flows with the physical one in both directions.	Manufacturing	Kritzinger et al. (2018)
Information [K.3]	[D.8] The notion where the data of each stage of a product lifecycle is transformed into information.	Product lifecycle	Abramovici et al. (2017)
	[D.9] Comprehensive physical and functional description of a component, product or systems.	Smart manufacturing	Shao et al. (2019)
	[D.10] Digital information construct about a physical system.	Aerospace	Tuegel et al. (2011)
Simulation model [K.4]	[D.11] Simulation based on expert knowledge and real data collected from existing systems.	Machine engineering	Gabor et al. (2016)
	[D.12] Reengineering computational model of structural life prediction and management.	Product lifecycle	Grieves and Vickers (2017)
	[D.13] Virtual models for physical objects to simulate their behaviours.	Smart manufacturing	Qi and Tao (2018)

(*Continued*)

TABLE 25.1 (*Continued*)

DT Definitions and Key Layers

Key Layers	Definition	Field	References
Virtual replica [K.5]	[D.14] Computerized clones of physical assets.	Industrial production	Banerjee et al. (2017)
	[D.15] Virtual and computerized counterpart of a physical system.	Production systems	Negri et al. (2017)
	[D.16] Functional system formed by the cooperation of physical production lines with a digital copy.	Industrial production	Vachálek et al. (2017)
	[D.17] Cyber copy of a physical system.	System of systems	Alam and Saddik (2017)
	[D.18] Digital model that dynamically reflects the status of an artefact.	Healthcare	Bruynseels et al. (2018)
	[D.19] Digital replica of physical entity with two-way dynamic mapping.	Manufacturing	El Saddik (2018)
	[D.20] Virtual representation of production system that is able to run on different simulation disciplines.	Product lifecycle	Garetti et al. (2012)
	[D.21] Digital mirror of the physical world.	Smart manufacturing	Guo et al. (2019)
	[D.22] Virtual model of the physical object.	Smart manufacturing	Mabkhot et al. (2018)
	[D.23] Dynamic digital representation of a physical system.	System engineering	Madni et al. (2019)
	[D.24] Virtual representations of physical manufacturing elements, such as personnel, products, assets and process definitions.	Manufacturing	Shao and Helu (2020)
	[D.25] Virtual representation of real products.	Industrial component	Schroeder et al. (2016)
	[D.26] Virtual model of physical assets.	Manufacturing	Talkhestani et al. (2018)
	[D.27] Digital copy of a physical system.	Industrial production	Wärmefjord et al. (2017)
	[D.28] A realistic digital representation of assets, processes or systems in the built or natural environment.	Built environment	Bolton et al. (2018)
	[D.29] Dynamic digital replica of physical assets, processes and systems, involving internet of things (IoT) devices and information feedback from citizens.	Built environment	Lu et al. (2020)
Capabilities [K.6]	[D.30] Dynamic virtual representation of a physical object or system across its lifecycle, using real-time data to enable understanding, learning and reasoning.	Built environment	NIC (2017)
	[D.31] Systematic approach consisting of sensing, storage, synchronization, synthesis and service.	Manufacturing	Lee et al. (2013)

25.3 Cyber-Physical Systems and Digital Twins

CPS are very closely related to DTs as their configuration is based one to each other (Liu et al., 2017; Kleissl and Agarwal, 2010). The concept of CPS is defined as "systems of collaborating computational entities which are in intensive connection with the surrounding physical world and its ongoing processes, providing and using, at the same time, data accessing and data processing service available on the internet" (Monostori, 2015). The key objective of CPSs is to enhance computation, communication and control systems in the interactions between physical and virtual worlds, in a variety of different industries from aerospace, manufacturing, healthcare, energy, automotive and civil engineering (Khaitan and McCalley, 2014) through the use of technologies such as autonomous vehicles, robotics, control systems, etc. (Saracco, 2019).

The use of CPSs is widely implemented in the manufacturing sector as it provides flexible automation pyramid-based systems (Iarovyi et al., 2016) and IoT enables pervasive connectivity towards real-time synchronized simulations.

In this regard, DTs can be considered as complementary to CPSs in terms of digital representation. As such, CPSs consists of two elements: the computational and simulation part, which is composed of (1) a synchronizing part between physical and virtual and (2) a simulation part which is demanded for digital replication (Cimino et al., 2019; Rajkumar et al., 2010).

The synchronization between the simulation part and the physical part of CPSs enables DTs' requirements; in fact, while CPSs allow to collect data from the physical world, DTs replicate the physical counterpart (Negri et al., 2017) working as a high-fidelity representation of the functional operation of the physical (or potential) counterpart.

25.4 Digital Twin–Based Lifecycle Management

Information and communication technology (ICT) allows asset modelling (Amadi-Echendu et al., 2010) and provides the opportunity to gain insights from data (Silva et al., 2018) at different levels and sub-levels of assets (building, infrastructures, systems, products, etc.) as an embedded dynamic replicas.

City DTs can be divided into different levels of linked DTs (city, building, asset, etc.) interacting one to each other in a bidirectional way and providing cooperative services without limiting other sub-DTs' autonomy (Lu et al., 2020).

This type of framework enables the realization of DTs of the built environment leveraging on the value of heterogeneous data coming from different sources and basing on existing multi-tier architectures that can be specifically applied to building, cities and built environments. In this regard, Industry

Foundation Classes (IFC) is used to integrate building information models (BIM) as 3D geometric and georeferenced entities among other data sources through distributed systems. Hence, cloud computing and IoT-based services such as real-time sensor data acquisition and distribution provide interaction and communication with end-users, bridging connections between humans and urban environments.

25.4.1 Information Management and Digital Thread

Currently, the BIM model plays a relevant role as a data source (Boje et al., 2020) and geometrical information model for building, infrastructure and asset management (Lu et al., 2019). However, much of the information and data generated across the lifespan goes unused or missed, and only a limited amount of data is transferred to the next stage. That's why data sources need to be more connected and integrated in order to properly address the information management challenges in the built environment (Boje et al., 2020).

Digital thread and simulation are concepts relatively close to, because it integrates real-time data and synchronization, simulation, process monitoring and control, diagnostic and predictive capabilities moving forward to real-time data analysis (Schleich et al., 2017).

As such, the digital thread concept is related to the system development lifecycle (SDLC), which involves the entire lifecycle of the asset from planning to design, construction, operation and maintenance (Singh and Willcox, 2018). Hence, digital threads can generate insights from the entire lifecycle process and enhance predictivity, decision-making and traceability.

Hence, the digital thread of data allows DTs to collect, manage and use data from the pre-project stage to the end-of-life of the physical object, as a tool to use connected information safely and effectively from design to operation and maintenance. Then data related to every component and sub-component of the real-world object needs to be trusted, accurate and reliable as well as structured correctly and consistently, in order to maintain it across the entire lifecycle and raise the value of DTs (Dawood et al., 2020).

In this regard, Grieves evidences fallacies in the concept of DTs only existing until physical counterparts are created. Duality and strong similarity are the only requirements for a DT as an existing model related to some point of the asset's lifecycle. As such, it can exist prior or even after the physical counterpart as the actual existence of a physical object is not a strict requirement, but its intended realization is the key differentiator of DTs from digital models (Grieves, 2022).

As such, a DT is composed of a repository which stores and links data, keeping an audit trail and maintaining its current record. This feature is related to the concept of golden thread of data and information. The concept was derived from the business world as a metaphor for organizational strategies and the UK Government introduced it in the construction sector in 2016 within the BIM mandate and with a more recent report by the UK Ministry of Housing, Communities and Local Government. The concept of transferring critical information from one stage to another is essential and it ensures a full understanding of the project by the stakeholders. In addition, it allows avoiding costly re-works in discovering information which had been already generated during the previous stage.

A huge amount of data and information are generated, combined, managed and used across the lifecycle and its complexity requires a robust information management process. In fact, the main inefficiencies in the management process of the built environment are related to fragmentations.

An information management framework is essential in order to properly apply DT strategies in the construction sector, as it has been defined by the CDBB as a "formal mechanism to ensure that the right information can be made available at the right time, to the right people and that the quality of the information is known and understood, is required".

As defined by the ISO 19650 series related to BIM implementation (Tang et al., 2020; Eastman et al., 2011), smart buildings and information management in construction, the need of defining structured processes is essential in order to embed and integrate data over the object lifecycle as a virtual replica. In this regard, Table 25.2 shows how the information management process and digital thread are structured in the construction sector through agreed data standards handling real-time, as-built and performance data (sensor data, IoT, point clouds, etc.). Actually, the amount of "dark data" is around 95% of data produced over the lifecycle of an asset which goes unused.

As mentioned above, digital models can replicate different stages or processes of the assets like historical, current or forward-looking states, while integrating data along temporal dimensions is the main objective of more mature DTs. Digital Thread is a tool for integrating data and information across multiple dimensions based on solid and substantial real-world identifier, ensuring the effectiveness of configuration management capabilities, recording historical, intended, or real-time configurations enabling a sort of "configuration history" dimension. Hence, Digital Threads are composed of identifiers and the software component which allows cohesive state views of information about product, systems or processes.

Identifying the key capabilities for each stage of the virtual replica is essential as they may differ in terms of technology and intended outcomes throughout the lifecycle. Capability is defined as "the ability to perform certain actions or achieve certain outcomes" and in the DT domain some capabilities may change across the lifespan, becoming obsolete or being renewed (e.g., capabilities related to the construction stage may not be similar to the operation and maintenance stage of a HVAC plant). The clear identification of capabilities enhances the development of the digital thread throughout different lifecycle stages.

In Table 25.2, the main components of digital thread and information management processes in the built environment are shown.

TABLE 25.2

Digital Thread and Information Management in Building Lifecycle

Phase	Outcome	Digital Thread Lifecycle Data		Model	
Plan	Simulation and what-if scenarios	As planned	Customer data		Systems
	Optimized configurations		Demand data		
	Procurement		Design specifications		
Design	Simulation and what-if scenarios	As designed	Customer data	Time series data and models	
	Asset/district/city modelling		Demand data		
	Maintenance simulations and improved operational efficiency		Design specifications		
Construct	Site-based process management	As constructed	Process data	Meta or master data	Services
	Prefabrication and industrialized construction		Planning data		
	Production management		Real-time status data		
	Supply-chain management		Capacity data		
	Data and information management		Class library		
Commission	Compliance	As commissioned	Certifications	Business systems data	
	Design verification		Class libraries		
Operate	Optimization	As operated	Operational data	Physics models	Sensors
	Simulation and what-if scenarios		Energy data		
	Safety		Status data		
	Real-time monitoring		Behaviour data		
Maintain	Prescriptive maintenance	As maintained	Historical fault data	Analytic and visual models	
	Predictive maintenance		Time series data		
	Condition-based maintenance		Spare parts data		
Retire	Decommissioning	As retired	Operational data		

25.4.2 Digital Twin Architectures

DT architecture is the main core of DTs (Aheleroff et al., 2021) as a reference model which deconstructs complex relations and hierarchies into simple clusters, layers or blocks (Kritzinger et al., 2018). Multi-tier architectures have been proposed in literature in order to support DT environments making data accessible (Wärmefjord et al., 2017; Zhuang et al., 2017), integrated and usable through CPSs, IoT and big data platforms (Silva et al., 2020). A multi-tier architecture is based on dividing the main components into a set of independent layers (Terrazas et al., 2019) sorted according to logical and functional aggregations, communicating with the adjacent layers and deriving functionality only from the lower layer.

As a result, layers act as a whole and independent component in a modular system architecture and components could be used as services within other layers, resulting in simple and traceable data flows.

In order to acquire relevant information from the physical world, connection layers are implemented as a direct thread to physical assets or sensors.

Bluetooth, near-field communication (NFC) and RFID are some of the network layers' components which are implemented to transfer data from connection layers to processing layers. Upper layers are defined according to different application uses.

CPSs architectures are based on the interaction between machines and human beings integrating ICT, sensing and control capabilities, while IoT architectures (Oliver et al., 2018) are based on big data and cloud computing technologies.

In summary, a hierarchical and flexible architecture needs to be developed for data collection and acquisition across the lifecycle of assets, buildings and cities. Moreover, computational capabilities and information management are essential in order to manage a great amount of multidomain data.

Also, the human communication and interaction layer plays a relevant role as data or condition status helps users in monitoring as-is feedbacks.

25.4.2.1 Data Acquisition and Ingestion Layer

The data acquisition layer is essential in the DT architecture for the built environment considering the huge number of heterogeneous sources and volume of datasets. In this scenario, defining clear information requirements is necessary to properly understand the needs of DT users during the entire asset lifecycle according to *The Gemini Principles*, as well as collecting data and maximizing their cost-effective value.

Hence, the adoption of distributed sensor techniques at an urban scale is gradually increasing especially for infrastructure, power grids, telecommunication networks, etc.; IoT and micro-electromechanical systems (MEMS) are used at building scales.

According to the Federation Principle, DTs are a federated amount of shared data from interconnected DTs in a common architecture where data are accessible, integrable and federated.

25.4.2.2 Transmission and Network Layer

The transmission layer is essential for the security and functionality of the DT as it is necessary to transfer data to the model and analysis layers through short-range technologies such as Zigbee, Bluethoot and WiFi, and wider-range such as 5G (Ge et al., 2016; Huang et al., 2012; Ohmura et al., 2013).

The main metrics to evaluate communication performances in indoor and outdoor conditions are frequency bands, bandwidth, range of communication, energy consumption and security protocols.

25.4.2.3 Digital Model Layer

The digital model layer contains technical, physical and functional multi-scale information (Borrmann et al., 2015) such as BIM and city information modelling (CIM) (Gil et al., 2011) related to the asset, acquiring data both at the building (e.g. building management systems) and urban level (e.g. geographic information system), supporting interoperability and data exchange towards upper layers (Gil et al., 2011). As DTs could be multi-purpose, data are collected through defined structures as information need to be interchanged and interoperated across sub-DTs. As such, model-centred ontology-based information standards enable interoperability and shared asset information tools as well as semantic integrations of federated infrastructure, building or city DTs (Lu et al., 2018).

Ontologies are defined as "explicit specifications of a conceptualisations", where conceptualization consists of relationships between objects and entities in some specific domain (Gruber, 1993). Model-driven data interchange and semantic-based interoperability (Abanda et al., 2013) allows ontologies defining knowledge-based meaning enabling collaboration across heterogeneous DTs on a wider application range.

25.4.2.4 Data-Model Integration Layer

The data-model integration layer is the core element of the entire DT architecture as it integrates data coming from heterogeneous sources according to a specific schema. This is the data layer enabling analysis, processing and AI-driven learning and simulation capabilities (Glaessgen and Stargel, 2012). As the heterogeneous nature of collected data, a hierarchy in model and data integration, model visualization and cloud storage (Lin et al., 2013) are required in order to allow DTs' effectiveness (e.g., in updating as-is conditions in energy management or traffic flows as well as in monitoring building maintenance condition status).

In this regard, knowledge engines (KEs) describe dynamic behaviours of physical objects by tracking them and acquiring data continuously, resulting in delivering better-informed services through data integration.

25.4.2.5 Service Layer

The service layer is the implementation layer providing interaction with people and users according to KEs enabling services, analysing metrics related to performances and improvements in terms of social, environmental and urban sustainability, and providing decision-making support to technical professionals in FM and end-users.

Figures 25.1 and 25.2 show the proposed DT multi-tier architecture for data thread and information management application in the built environment.

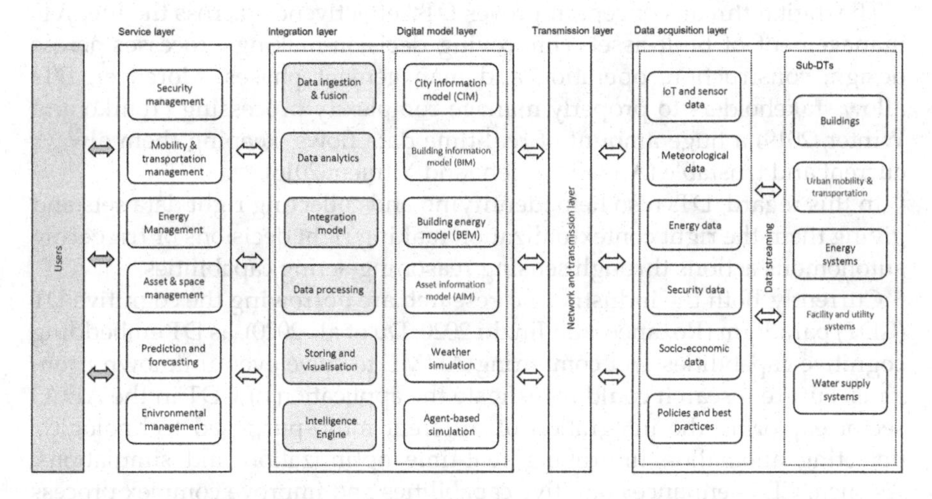

FIGURE 25.1
Digital twin architecture for the built environment.

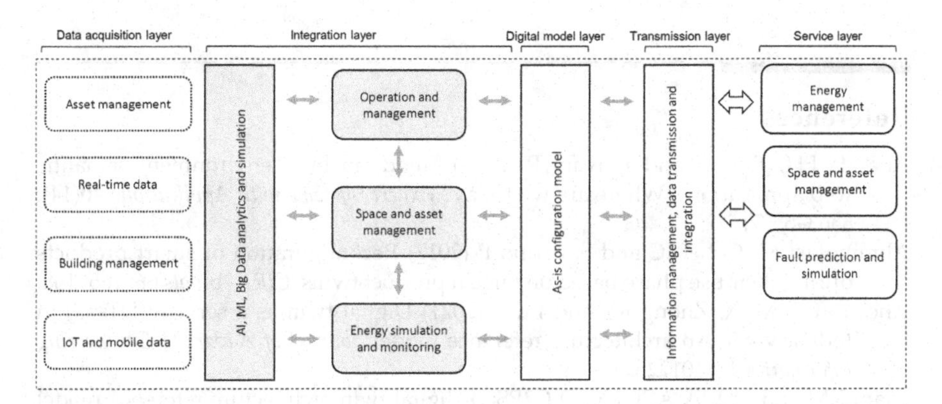

FIGURE 25.2
Building sub-digital twin architecture.

25.5 Conclusions

DT implementation in the AECO sector is still behind the manufacturing, automotive and aerospace field. One of the main concepts to be addressed in the AECO sector is the federation of DTs in order to deploy city/region/nation ecosystems including a network of interconnected dynamic DTs. In this scenario, intelligent engines such as AI, ML and big data analytics (Rathore et al., 2016; Woodall, 2017) enable self-learning and predictive capabilities providing feedback loops.

The digital thread concept improves DTs' effectiveness across the lifecycle management of built assets, improving decision-making processes across design, construction, operation and management phases. Moreover, DTs allow stakeholders to properly manage complexity processing (Tomko and Winter, 2019) a huge amount of real-time data flows, keeping them always current and trustable (Alizadehsalehi and Yitmen, 2016).

In this regard, DTs also help identifying and collecting right data sets and giving them the right contextualization making right decisions or triggering autonomous actions through sensing-reasoning-acting capabilities.

Currently, both the industry and research are borrowing the cognitive DT (CDT) paradigm (Rozanec and Jinzhi 2020; Du et al., 2020), as DT embedding cognitive capabilities and combining AI/ML to solve even unknown problems. Future research could investigate the application of CDT in the AECO sector exploring the integration of different multi-purposed technologies, detecting anomalies, improving real-time optimization and simulations. As such, CDTs enhance cognitive capabilities and improve complex process management systems across planning, design, construction, operation and maintenance of built assets integrating ML and analytics.

References

Abanda FH, Tah JH and Keivani R (2013) Trends in built environment semantic web applications: Where are we today? *Expert Systems with Applications* 40(14): 5563–5577.

Abramovici M, Göbel JC and Savarino P (2017) Reconfiguration of smart products during their use phase based on virtual product twins. *CIRP Annals* 66: 165–168.

Aheleroff S, Xu X, Zhong RY and Lu Y (2021) Digital twin as a service (DTaaS) in Industry 4.0: An architecture reference model. *Journal of Advanced Engineering Informatics* 47: 101225.

Alam KM and Saddik AEL (2017) C2PS: A digital twin architecture reference model for the cloud-based cyber-physical systems. *IEEE Access* 5: 2050–2062. DOI: 10.1109/ACCESS.2017.2657006.

Alizadehsalehi S, and Yitmen I (2016) The impact of field data capturing technologies on automated construction project progress monitoring. *Procedia Engineering* 161: 97–103.

Amadi-Echendu JE, Willett R, Brown K et al. (2010) What is engineering asset management? Echendu J., Brown K., Willett R. and Mathew J. (eds). *Definitions, Concepts and Scope of Engineering Asset Management. Engineering Asset Management Review*, Springer, London, vol. 1. pp. 3–16. DOI: 10.1007/978-1-84996-178-3_1.

Banerjee A, Mittal S, Dalal R and Joshi K (2017) Generating digital twin models using knowledge graphs for industrial production lines. 9th International ACM Web Science Conference. Association for Computing Machinery, New York, pp. 425–430.

Batty M (2018) Digital twins. *Environment and Planning B: Urban Analytics and City Science* 45: 817–820. DOI: 10.1177/2399808318796416.

Boje C, Guerriero A, Kubicki S and Rezgui Y (2020) Towards a semantic construction digital twin: Directions for future research. *Automation in Construction* 114: 103179.

Bolton A, Butler L, Dabson I et al. (2018) *The Gemini Principles*. Centre for Digital Built Britain, Cambridge. DOI: 10.17863/CAM.32260.

Borrmann A, Kolbe TH, Donaubauer A et al. (2015) Multi-scale geometric-semantic modeling of shield tunnels for GIS and BIM applications. *Computer-Aided Civil and Infrastructure Engineering* 30(4): 263–281.DOI: 10.1111/mice.12090.

Boschert S and Rosen R (2016) Digital twin - the simulation aspect. Hehenberger P and Bradley D (eds). *Mechatronic Futures*, Springer, Cham, pp. 59–74. DOI: 10.1007/978-3-319-32156-1_5.

Bruynseels K, Sio FSD and Hoven JVD (2018) Digital twins in health care: Ethical implications of an emerging engineering paradigm. *Frontiers in genetics* 9 (2018): 31. DOI: 10.3389/fgene.2018.00031.

Buckman AH, Mayfield M and Beck SBM (2014) What is a smart building? *Smart and Sustainable Built Environment* 3(2): 92–109. DOI: 10.1108/SASBE-01-2014-0003.

Canedo A (2016) Industrial IoT lifecycle via digital twins. Proceedings of the Eleventh IEEE/ACM/IFIP International Conference on Hardware/Software Codesign and System Synthesis. 2016: 1–1. DOI: 10.1145/2968456.2974007.

Dawood N, Rahimian F, Seyedzadeh S, and Sheikhkhoshkar M (2020) Enabling the development and implementation of digital twins. *Proceedings of the 20th International Conference on Construction Applications of Virtual Reality*. Tesside University Press, Middlesbrough, 2020.

Du J, Zhu Q, Shi Y, Wang Q, Lin Y, and Zhao D (2020) Cognition digital twins for personalized information systems of smart cities: Proof of concept. *Journal of Management in Engineering*, 36, 04019052.

Eastman CM, Teicholz P, Sacks R and Liston K (2011) *BIM Handbook: A Guide to Building Information Modeling for Owners, Managers, Designers, Engineers, and Contractors*. Wiley, Hoboken, NJ.

Gabor T, Kiermeier M, Beck MT and Neitz A (2016) A simulation-based architecture for smart cyber-physical systems. *IEEE International Conference on Autonomic Computing (ICAC)*. IEEE, Piscataway, NJ, pp. 374–379. DOI: 10.1109/ICAC.2016.29.

Garetti M, Rosa P and Terzi S (2012) Life cycle simulation for the design of product-service systems. *Computers in Industry* 63: 361–369.

Ge X, Tu S, Mao G, Wang CX and Han T (2016) 5G ultra-dense cellular networks. *IEEE Wireless Communications* 23(1): 72–79.

Gil, Jorge, Júlio Almeida, and José Pinto Duarte. "The backbone of a city information model (CIM)." *Respecting fragile places: Education in computer aided architectural design in Europe* (2011): 143–151.

Glaessgen E and Stargel D (2012) The digital twin paradigm for future NASA and US Air Force vehicles. *53rd AIAA/ASME/ASCE/AHS/ASC Structures, Structural Dynamics and Materials Conference, Honolulu, Hawaii*, p. 1818.

Grieves M (2014) Digital twin: manufacturing excellence through virtual factory replication. https://www.researchgate.net/publication/275211047_Digital_Twin_Manufacturing_Excellence_through_Virtual_Factory_Replication (accessed 06/07/2021).

Grieves M (2022) Intelligent digital twins and the development and management of complex systems. *Digital Twin*, 2: 8. DOI: 10.12688/digitaltwin.17574.1.

Grieves, M, and Vickers, J (2017). Digital twin: Mitigating unpredictable, undesirable emergent behavior in complex systems. Kahlen, J., Flumerfelt, S., Alves, A. (eds) *Transdisciplinary Perspectives on Complex Systems*. Springer, Cham. DOI: 10.1007/978-3-319-38756-7_4.

Gruber TR (1993) A translation approach to portable ontology specifications. *Knowledge Acquisition* 5(2): 199–220.

Guo J, Zhao N, Sun L and Zhang S (2019) Modular based flexible digital twin for factory design. *Journal of Ambient Intelligence and Humanized Computing* 10: 1189–1200. DOI: 10.1007/s12652-018-0953-6.

Haag S and Anderl R (2018) Digital twin - proof of concept. *Manufacturing Letters* 15: 64–66. DOI: 10.1016/j.mfglet.2018.02.006.

Hochhalter JD, Leser WP, Newman JA et al. (2014) *Coupling Damage-Sensing Particles to the Digital Twin Concept*. National Aeronautics and Space Administration, Langley Research Center, Hampton, VA.

Huang J, Qian F, Gerber A et al. (2012) A close examination of performance and power characteristics of 4G LTE networks. *Proceedings of the 10th International Conference on Mobile Systems, Applications, and Services*. Association for Computing Machinery, New York. pp. 225–238.

Iarovyi S, Mohammed WM, Lobov A, Ferrer BR and Lastra JLM (2016) Cyber-physical systems for openknowledge- driven manufacturing execution systems. *Proceedings of the IEEE* 104: 1142–1154. DOI: 10.1109/JPROC.2015.2509498.

ISO19650-1 (2018) Organization and digitization of information about buildings and civil engineering works, including building information modelling (BIM) - Information management using building information modelling - Part 1: Concepts and principles.

Khaitan SK and McCalley JD (2014) Design techniques and applications of cyberphysical systems: A survey. *IEEE Systems Journal* 9: 350–365.

Kim K, Cho YK and Kim K (2018) BIM-based decision-making framework for scaffolding planning. *Journal of Management in Engineering* 34(6): 04018046.

Kleissl J and Agarwal Y (2010) *Cyber-Physical Energy Systems: Focus on Smart Buildings. Design Automation Conference*. IEEE, Piscataway, NJ, pp. 749–754.

Lee J, Lapira E, Bagheri B and Kao H (2013) Recent advances and trends in predictive manufacturing systems in big data environment. *Manufacturing Letters* 1: 38–41. DOI: 10.1016/j.mfglet.2013.09.005.

Lin J, Zha L and Xu Z (2013) Consolidated cluster systems for data centers in the cloud age: A survey and analysis. *Frontiers of Computer Science* 7(1): 1–19.

Liu XF, Shahriar MR, Al Sunny SN, Leu MC and Hu L (2017) Cyber-physical manufacturing cloud: Architecture, virtualization, communication, and testbed. *Journal of Manufacturing Systems* 43: 352–364.

Lu Q, Lee S and Chen L (2018) Image-driven fuzzy-based system to construct as-is IFC BIM objects. *Automation in Construction* 92: 68–87.

Lu Q, Parlikad AK, Woodall P et al. (2020) Developing a digital twin at building and city levels: Case study of West Cambridge campus. *Journal of Management in Engineering* 36(3): 05020004.

Lu Q, Xie X, Heaton J, Parlikad AK, and Schooling, J (2019) From BIM towards digital twin: Strategy and future development for smart asset management. *International Workshop on Service Orientation in Holonic and Multi-Agent Manufacturing*, Springer: Cham.

Mabkhot MM, Al-Ahmari AM, Salah B and Alkhalefah H (2018) Requirements of the smart factory system: A survey and perspective. *Machines* 6: 23.

Madni AM, Madni CC and Lucero SD (2019) Leveraging digital twin technology in model-based systems engineering. *Systems* 7: 7.

McKinsey Global Institute (2017) *Reinventing Construction: A Route to Higher Productivity*, McKinsey & Company. https://www.mckinsey.com/~/media/mckinsey/business%20functions/operations/our%20insights/reinventing%20construction%20through%20a%20productivity%20revolution/mgi-reinventing-construction-aroute-to-higher-productivity-full-report.pdf (accessed 27/07/2022).

Mohammadi N and Taylor JE (2017) Smart city digital twins. *2017 IEEE Symposium Series on Computational Intelligence (SSCI)*. IEEE, Piscataway, NJ. https://doi.org/10.1109/SSCI.2017.8285439.

Monostori L (2015) Cyber-physical production systems: Roots from manufacturing science and technology. *At-Automatisierungstechnik* 63: 766–776.

Negri E, Fumagalli L and Macchi M (2017) A review of the roles of digital twin in CPS-based production systems. *Procedia Manufacturing* 11: 939–948. DOI: 10.1016/j.promfg.2017.07.198.

NIC (National Infrastructure Commission) (2017) Data for the Public Good. https://nic.org.uk/app/uploads/Data-for-the-Public-Good-NIC-Report.pdf (accessed 31/06/2022).

Ohmura N, Takase E, Ogino S, Okano Y and Arai S (2013) Material property of on-metal magnetic sheet attached on NFC/HF-RFID antenna and research of its proper pattern and size on. 2013 Proceedings of the International Symposium on Antennas & Propagation. IEEE, 2013, 2: 1158–1161.

Oliver D, Adam D and Hudson-Smith AP (2018) Living with a Digital Twin: Operational management and engagement using IoT and Mixed Realities at UCL's Here East Campus on the Queen Elizabeth Olympic Park. Giscience and Remote Sensing. GIS Research UK (GISRUK), 2018.

Qi Q and Tao F (2018) Digital twin and big data towards smart manufacturing and Industry 4.0: 360 degree comparison. *IEEE Access* 6: 3585–3593. DOI: 10.1109/ACCESS.2018.2793265.

Rajkumar R, Lee I, Sha L and Stankovic J (2010) Cyber-physical systems: The next computing revolution. /Proceedings of the 47th design automation conference. 2010: 731–736.

Rathore MM, Ahmad A, Paul A and Rho S (2016) Urban planning and building smart cities based on the internet of things using big data analytics. *Computer Networks* 101: 63–80.

Rosen R, Von Wichert G, Lo G and Bettenhausen KD (2015) About the importance of autonomy and digital twins for the future of manufacturing. *IFAC - PapersOnLine* 48: 567–572.

Rozanec JM, and Jinzhi L (2020) Towards actionable cognitive digital twins for manufacturing. SeDiT@ ESWC, 2020, 2615: 1–12.

Sacks R, Girolami M, and Brilakis I (2020) Building information modelling, artificial intelligence and construction tech. *Developments in the Built Environment* 4: 100011. DOI: 10.1016/j.dibe.2020.100011.

Saracco R (2019) Digital twins: Bridging physical space and cyberspace. *Computer* 52: 58–64.

Schleich B, Anwer N, Mathieu L and Wartzack S (2017) Shaping the digital twin for design and production engineering. *CIRP Annals* 66(1): 141–144. DOI: 10.1016/j.cirp.2017.04.040.

Schroeder GN, Steinmetz C, Pereira CE and Espindola DB (2016) Digital twin data modeling with Automation ML and a communication methodology for data exchange. *IFAC - PapersOnLine* 49: 12–17. DOI: 10.1016/j.ifacol.2016.11.115.

Shafto MM, Conroy M, Doyle R et al. (2012) Modeling, simulation, information technology & processing roadmap. National Aeronautics and Space Administration, 2012, 32: 1–38.

Shao G, Jain S, Laroque C et al. (2019) Digital twin for smart manufacturing: The simulation aspect. 2019 Winter Simulation Conference (WSC). IEEE, 2019: 2085–2098.

Silva BN, Khan M and Han K (2018) Towards sustainable smart cities: A review of trends, architectures, components, and open challenges in smart cities. *Sustainable Cities and Society* 38: 697–713.

Silva BN, Khan M and Han K (2020) Integration of big data analytics embedded smart city architecture with RESTful web of things for efficient service provision and energy management. *Future Generation Computer Systems* 107: 975–987.

Singh V and Willcox KE (2018) Engineering design with digital thread. *AIAA Journal* 56: 4515–4528.

Stojanovic V, Trapp M, Richter R, Hagedorn B and Döllner J (2018) Towards the generation of digital twins for facility management based on 3D point clouds. *ARCOM 2018: 34th Annual Conference, Belfast, UK*, Management, 2018, 270: 279. https://www.arcom.ac.uk/-docs/proceedings/b65e593d342a8de045cf05698677e600.pdf (accessed 28/07/2022).

Talkhestani BA, Jazdi N, Schlögl W and Weyrich M (2018) Consistency check to synchronize the digital twin of manufacturing automation based on anchor points. *Procedia CIRP* 72: 159–164.

Tang S, Shelden DR, Eastman CM, Pishdad-Bozorgi P and Gao X (2020) BIM assisted building automation system information exchange using BACnet and IFC. *Automation in Construction* 110: 103049. DOI: 10.1016/j.autcon.2019.103049.

Tao F and Qi Q (2017) New IT driven service-oriented smart manufacturing: framework and characteristics. *IEEE Transactions on Systems, Man, and Cybernetics: Systems* 49: 81–91.

Terrazas G, Ferry N and Ratchev S (2019) A cloud-based framework for shop floor big data management and elastic computing analytics. *Computers in Industry* 109: 204–214.

Tomko M and Winter S (2019) Beyond digital twins - A commentary. *Environment and Planning B: Urban Analytics and City Science* 46: 395–399. DOI: 10.1177/2399808318816992.

Tuegel EJ, Ingraffea AR, Eason TG and Spottswood SM (2011) Reengineering aircraft structural life prediction using a digital twin. *Journal of Aerospace Engineering* 2011: 154798. DOI: 10.1155/2011/154798.

Vachálek J, Bartalský L and Rovný O (2017) The digital twin of an industrial production line within the Industry 4.0 concept. 2017 21st international conference on process control (PC). IEEE, 2017: 258–262.

Wärmefjord K, Söderberg R, Lindkvist L, Lindau B and Carlson JS (2017) Inspection data to support a digital twin for geometry assurance. ASME international mechanical engineering congress and exposition. American Society of Mechanical Engineers, 2017, 58356: V002T02A101.

Woodall P (2017) The data repurposing challenge: New pressures from data analytics. *Journal of Data and Information Quality (JDIQ)* 8(3–4): 11.

Zhuang CB, Liu JH, Xiong H et al. (2017) Connotation, architecture and trends of product digital twin. *Computer Integrated Manufacturing Systems* 23: 753–768.

Cimino, Chiara, Elisa Negri, and Luca Fumagalli. "Review of digital twin applications in manufacturing." *Computers in industry* 113 (2019): 103130.

Kritzinger, Astrid, et al. "Age-related pathology after adenoviral overexpression of the leucine-rich repeat kinase 2 in the mouse striatum." *Neurobiology of Aging* 66 (2018): 97-111.

Roy, Robin, and James P. Warren. "Card-based design tools: A review and analysis of 155 card decks for designers and designing." *Design Studies* 63 (2019): 125-154.

Padovano, Antonio, et al. "A digital twin based service oriented application for a 4.0 knowledge navigation in the smart factory." *IFAC-PapersOnLine* 51.11 (2018): 631-636.

Shao, Guodong, and Moneer Helu. "Framework for a digital twin in manufacturing: Scope and requirements." *Manufacturing Letters* 24 (2020): 105-107.

Part 6

Digital Twins in Industry

26

Digital Twins for Process Industries

Seppo Sierla
Aalto University

26.1 Introduction

The term process industry is used to refer to a broad range of applications involving the storage and transportation of processing of liquids or gasses. Common components found across such industries include vessels for storage and processing, pipes connecting the vessels and valves and pumps for moving materials between vessels. It is notable that the material that is being processed, e.g. oil, is of a continuous nature, so it is not possible to identify discrete items that are distinct from other items. This is the major difference between the process industry and the manufacturing industry. In manufacturing, a resource such as a robot may perform an operation such as pick-and-place on a workpiece. However, such operations do not occur in an industrial process, as there are no workpieces. Instead, the process acts on the substances being processed through *control loops.* Key elements of a loop are a *measurement, controller, setpoint* and *actuator.* Before discussing digital twins of industrial processes, it is necessary to discuss fundamental concepts such as process *equipment,* models of the physical process, *instrumentation* and *control systems.* Process industry domain terms that are important for the eventual discussion on digital twins are *italicized* when first used.

26.2 Fundamental Concepts of Industrial Processes

26.2.1 An Example Case: A Laboratory Water Process

The simplest kind of process only manages water in its liquid state. Such a laboratory process for heating, pressurizing and circulating water is shown in Figure 26.1. It has been developed at Aalto University for research and educational purposes by Mr. Jukka Peltola (Peltola et al., 2011;

DOI: 10.1201/9781003425724-32

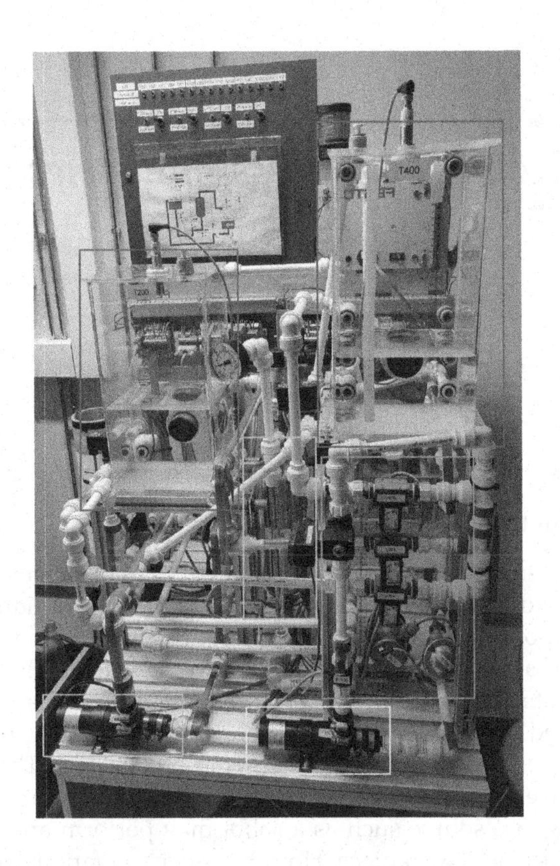

FIGURE 26.1
Laboratory process at Aalto University for heating, pressurizing and circulating water. Process equipment is marked with rectangles for further reference in Figure 26.2.

Vepsäläinen et al., 2010), and it will be used to introduce key concepts and techniques that are encountered in all kinds of industrial processes. This will be the essential background for understanding the subsequent discussion on digital twins for industrial processes.

26.2.2 The Piping & Instrumentation Diagram

Figure 26.2 shows a P&ID (Piping & Instrumentation Diagram) of the process in Figure 26.1. By comparing the figures, it can be seen that the P&ID only includes some of the pipelines, namely the lines that comprise the primary circulation. There are additional pipelines that are needed when executing *sequences* for startup, shutdown, cleaning and maintenance purposes, which have not been modelled in Figure 26.2. As P&IDs are essential source information for building digital twins, this raises an important question: which pipelines of the process should be modelled? The answer depends on the intended usage of the digital twin. However, this illustrates the importance

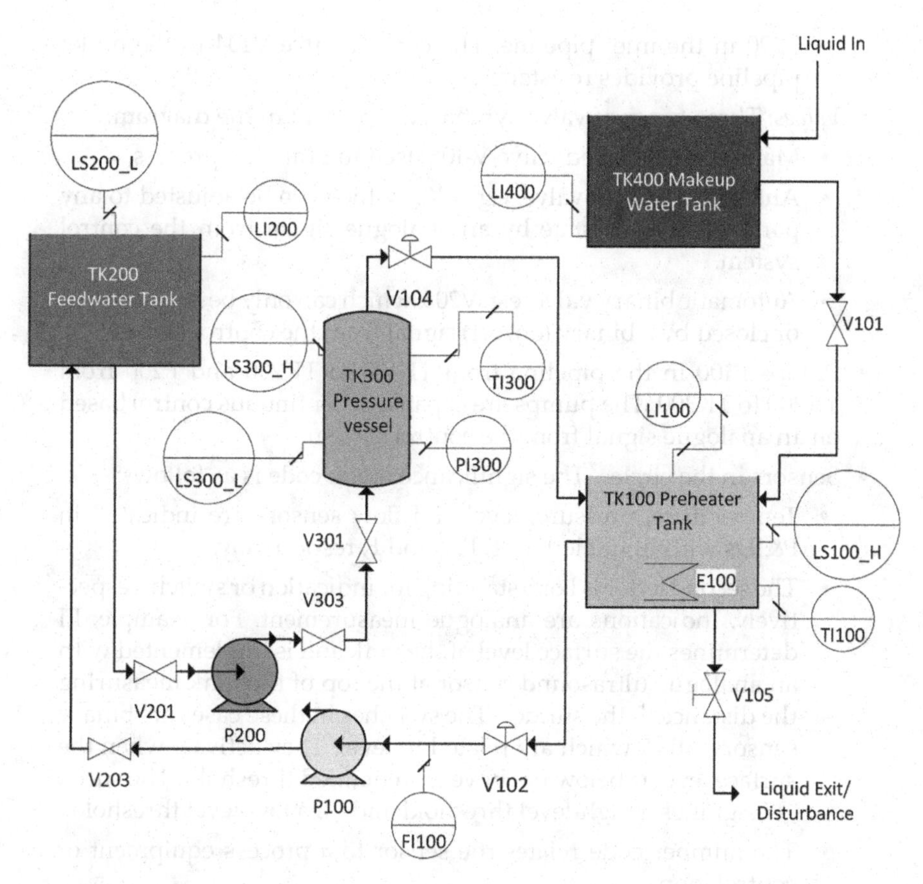

FIGURE 26.2
Process & Instrumentation diagram of the process in Figure 26.1.

of scoping the source information before proceeding with a digital twin project.

The elements in the P&ID in Figure 26.2 are as follows:

- Vessels
  - Atmospheric tanks TK100, TK200 and TK200. The tanks have a lid, but they are not watertight, so overflow must be avoided. It is thus not possible to pressurize the tank above 1 bar.
    - TK100 contains an electrically powered heating element E100
  - Pressure vessel TK300. In Figure 26.2, there is an outlet nozzle at the top of the tank. The layout of the P&ID does not, in general, provide information about pipe routing or nozzle placement, but in this case the nozzle actually is at the top of the tank. In normal operation, the tank is full of water and pressurized by pump

P200 in the inlet pipeline. The control valve V104 in the outlet pipeline provides resistance.

- *Valves:* Three types of valve symbols are present in the diagram:
    - Manually controlled valve V-105 used to drain the process.
    - Automatic control valve, e.g. V102, which can be adjusted to any partially opened state by an analogue signal from the control system.
    - Automatic binary valve, e.g. V203, which can only be fully opened or closed by a binary (on/off) signal from the control system.
- Pumps P100 in the pipeline from TK100 to TK200 and P200 from TK200 to TK300. The pumps are capable of continuous control based on an analogue signal from the control system.
- Sensors in the circles. The significance of the code is as follows:
    - Temperature, pressure, level and flow sensors are indicated in P&IDs with initial letters T, P, L and F, respectively.
    - The second letter is I or S standing for indication or switch, respectively. Indications are analogue measurement. For example, LI determines the surface level of the tank and is implemented with an analogue ultrasound sensor at the top of the tank measuring the distance to the surface. The switches in these cases are binary sensors, all of which are related to level. They activate when the surface level is below or above a predefined threshold. The suffix _H signifies a high-level threshold and _L a low-level threshold.
    - The number code relates the sensor to a process equipment or control loop.

26.2.3 Modelling the Physical Process

Now that the control loops have been discussed, the main functionality of the process is understood, and it is possible to discuss the modelling of the physical process. There are two fundamentally different approaches: *steady state* and *dynamic*. The former does not consider time-dependent phenomena and thus does not model control loops. The latter does consider such phenomena and is useful for investigating how the process reacts to *transients* such as changes to the setpoint of a control loop, operation of binary actuators or various kinds of failures of the process components.

Steady-state modelling can be beneficial, especially for designing new plants and retrofits to reduce operating costs (Pinto-Varela, 2017), energy consumption (Wang et al., 2020), CO_2 emissions (Min et al., 2015), freshwater consumption (Faria & Bagajewicz, 2009) and environmental pollution (Men et al. 2020). It is an open question that what could be considered a steady-state digital twin. Sierla et al. (2020c) presented a roadmap towards generating such a twin, using the process in Figure 26.1 as a case study. In this case, the digital twin would involve an online capability of synchronizing the steady-state model parameters based on recent measurement data from the control system history database.

As digital twins involve a real-time connection to the physical process, the dominant modelling approach in this context is dynamic. The task is simplified in the context of our example process, in the sense that the only substance in the pipes and vessels is water in its liquid phase. The pressure and temperature of the water is of interest in all of the tanks and pipelines. Flow is of interest in the pipelines and surface level is of interest in the vessels. To capture how these quantities vary in time and how they are affected by the actuators requires building a simulation model with a thermo-hydraulic solver (e.g. Hänninen & Ahtinen, 2009). The dimensions of the vessels, the interior diameter of the pipelines and the routing of the pipelines are essential information for building a thermo-hydraulically accurate model, but this information is not available from a P&ID. In particular, the pressure losses within the pipelines are crucial parameters to the model. Calculating them requires detailed information on the elevations of the nozzles at the beginning and end of the pipeline, the interior diameter of the pipeline, the number and angle of bends in the pipeline (known as *elbows*) and junctions where the pipeline branches in two ways (known as *tees*). Martínez et al. (2018a) presented a detailed explanation of the pressure loss calculation and apply it to the process in Figure 26.1.

It is notable that steady-state and dynamic simulation are not mutually exclusive. Some authors use steady-state simulation as a precursor of dynamic simulation analyses (Cui et al., 2019; Yoon et al., 2020; Chisalita & Cormos, 2018; Kender et al., 2021).

In addition to physics-based approaches for modelling the process, it is possible to apply black-box approaches and machine learning models to determine the input–output relationship of industrial processes or subprocesses. As the term digital twin has some marketing value, some such industrial products have been branded as digital twins. However, the twins are only valid when the process is operating in the same conditions as it was when the data was collected for the black-box model. This is counterproductive to the objectives of using digital twins to ensure the desired operation of processes, especially in abnormal situations. Thus, in this section, we adhere to the original NASA definition of digital twins that requires the use of the best possible physical models (Shafto et al., 2010).

26.2.4 Engineering Design Source Information for Modelling the Physical Process

Building an accurate simulation model or digital twin of an industrial process requires relevant engineering design source information beyond a P&ID. At this point, the nature of this source information and its availability for real industrial processes is discussed. It is important to distinguish between *greenfield* plants that are being designed from scratch and *brownfield* plants that are already operational. As process plants frequently have a lifecycle of several decades, the brownfield plant will be an important context for applying digital twins, so the availability of engineering design information in both greenfield and brownfield situation needs to be discussed. There are several ways of obtaining the required engineering design source information, depending on what is available at a specific plant:

- A recent version of a 3D process CAD (computer-aided design) tool is able to export the pipe routing information in an open, standard and machine-readable format. In particular, the PCF (Piping Component File) format for 3D isometrics is supported by major 3D process CAD tool vendors, and Sierla et al. (2020a) developed a software application for parsing such a file for the process in Figure 26.1. The tool was able to extract the elbows, tees and elevations of the nozzles and thus obtained the information required to compute the pressure losses for these pipelines. Unfortunately, at the time of writing, the PCF file can only be exported by recent versions of CAD tools and are thus available only for greenfield plants and very recently commissioned brownfield plants.
- Process plants have a lifecycle of decades, so a plant that has been operational for many years may not have any 3D CAD model, or it may be in an older format, from which it is not possible to export a machine-readable file such as a PCF file (Chen et al., 2018; Arroyo et al., 2016). Constructing a high-fidelity digital twin for such a plant will be laborious. For example, Martínez et al. (2018a) measured the dimensions of the components for the process in Figure 26.1 and manually reverse engineered the 3D CAD model.
- Laser scanning of industrial facilities is a viable approach for brownfield plants. The challenge is to automatically extract the process components from the raw point cloud data from the scan. Xiong et al. (2013) accomplished this for the building, identifying floors, walls, doorways and ceilings. Kawashima et al. (2011) accomplished the same for pipelines, identifying components of interest for our purposes, such as tees and elbows. Even if 3D CAD data is available, one advantage of using point clouds instead of 3D CAD models is that the latter describes the *as-designed* state of the plant whereas the former captures the *as-is* state of the plant. There can be a significant difference between these two configurations at a brownfield plant in which several retrofits have been made over the years. For the purposes of developing a digital twin, the as-is configuration is relevant, as the twin should be synchronized in real time to the physical process rather than to any historical previous embodiment of the process.

26.2.5 Putting It Together: Integrating the Control System to the Physical Process or a Simulation Model Thereof

The control system needs to be integrated into the physical sensors and actuators in the process. For simulation purposes, the integration needs to be done to the sensors and actuators of the simulation software. Understanding these integrations is a starting point for understanding how digital twins can be deployed. The information technologies and architectures for this purpose are beyond the scope of this book; Martínez et al. (2018b) discussed these aspects in the context of the process in Figure 26.1. The capability to interface a process control system to a simulation model of the physical process as well as to the physical process is a prerequisite to building and deploying digital twins. This task is complicated due to the fact that source information for building the control systems comes mainly from the P&ID whereas the source information for the simulation model comes from a 3D CAD or a laser scan. The same components need to be matched in these different sources before the information can be integrated and the correct interfaces can be built. In general, it cannot be assumed that industrial practitioners use consistent naming conventions across different tools such as P&IDs and CADs (Rantala et al., 2019), so automatic matching of tags (e.g. 'V102' or 'E100' in Figure 26.2) is not a viable method for integrating these two sources of information. Tags are also not present in the output of a laser scan. Doing this work manually is very laborious, so the engineering cost of building a digital twin can be very high if even the basic task of correctly integrating the different sources of information poses significant challenges. Recent research towards this end involves the generation of graphs from P&IDs and 3D CAD models, reducing both sources of information to the same abstraction level (Sierla et al., 2020b). Graph matching methods (Wen et al., 2017) can be applied as the next step to identify the same process components from the P&ID and CAD. It is notable that the techniques discussed in this subsection are a field of ongoing research rather than commercially mature technology.

26.3 Types of Industrial Process

The process in Figure 26.1 served the purpose of introducing elements that are generally found in industrial processes across industrial sectors. In this section, some main industrial sectors are discussed. Aspects of these processes that were not present in our example process are discussed.

A *combustion power plant* is in many ways a straightforward extension of the process in Figure 26.1 (Starkloff et al., 2015). Fuel such as coal is burned to generate heat in a vessel containing water at a high temperature and pressure. The water at the top of the steam evaporates despite the high pressure.

The steam is led by a pipeline to a turbine for generating electricity. The low-pressure steam after the turbine needs to be condensed before it can be pumped back to the vessel. A *heat exchanger* is used for this purpose. The exchanger involves a winding pipe going through a tank of cool water. The heat is exchanged through the walls of the pipe, resulting in the steam cooling down and condensing. The high pressures and temperatures justify the investment to a Safety Instrumented System (SIS). The modelling of high-pressure steam is more difficult than the modelling of pressurized and heated liquid water in the example process in Figure 26.1. Thus, a higher fidelity thermo-hydraulic simulator would be required. The accurate modelling of the high-pressure steam is a challenging task, especially for recent efforts to increase the flexibility of coal-fired plants to adapt rapidly to changes in renewable generation to the grid (Zhao et al., 2018; Wei et al., 2021). Dynamic simulation is identified as an appropriate technology for this task (Alobaid et al., 2017). Building on this trend, a recent research direction on dynamic simulation of coal-fired power plants is the solar-aided plant, in which solar generation complements the combustion (Yan et al., 2021).

Another development on the conventional combustion power plant with a steam turbine is a gas turbine driven by high-pressure gas obtained by burning methane or natural gas so that the remaining heat from the gas exiting the turbine is used to generate steam that is fed to another turbine (Henry et al., 2021).

The various types of *chemical processes* involve other substances in addition to water and steam, as well as chemical reactions that occur between the substances. Dynamic simulations have been used for a variety of purposes. Ge et al. (2021) developed and validated environmentally friendly flaring methods reducing the amount of unburned hydrocarbons, NO_x, CO_2 and CO released to the environment. Olivier-Maget et al. (2021) investigated boiler overpressure and flooding scenarios in propylene glycol production process. Khaled et al. (2021) investigated various fault and disturbance scenarios for an offshore gas processing plant. Wanotayaroj et al. (2020) simulated transients in temperature, pressure and tank level to validate and tune controllers for a chemical looping combustion process separating carbon dioxide from flue gas. Yoon et al. (2020) reduced waste and energy consumption in a natural gas liquid recovery process, while investigating potentially hazardous transients. Fluid catalytic cracking is a main process in petroleum refineries converting crude oil into end products such as gasoline. Cui et al. (2019) simulated external disturbances to this highly safety critical process to validate that Safety Integrity Level(SIL) requirements are met. Zhu et al. (2020) simulated potentially hazardous control actions by the human operators of the process. Chisalita and Cormos (2018) used simulation to predict the behaviour of a novel combustion process with carbon capture. Raimondi (2019) modelled the stratification of liquid and gas fractions in an underwater natural gas pipeline.

26.4 Digital Twins in the Process Industry

Based on the introductory material in previous sections and italicized terms, it is possible to analyse recent works on digital twins for process industries. This analysis considers the objectives for the digital twin, the design approach and new capabilities that the digital twin provides above and beyond state-of-the-art industrial process simulations, such as the ones reviewed in the previous section. In the following, recent state-of-the-art works are analysed. At the time of writing, the research on digital twins in process industry is in its early stages. Thus, the goal of this chapter is to demonstrate to the reader how previously introduced concepts of industrial process simulation can be applied to understand and critically assess this literature. A later edition of this book may discuss well-established approaches for developing digital twins for process industries, but at the time of writing, such approaches do not yet exist.

A digital twin for the case process in Figure 26.1 is described in Martínez et al. (2018a, b, c). Sensor measurements are compared to state values of the dynamic simulation of the process that is running in parallel. A PI (proportional-integral) controller computes the error between the state and the measurement and adjusts some parameters of the dynamic simulation model such as the pressure losses in the pipelines. This is the same kind of controller that has been used in the continuous control loops in Table 26.1. The authors reported a very accurate performance at these points in the process. The digital twin can be disconnected from the physical system and run faster

TABLE 26.1

Overview of State-of-the-Art Applications of Digital Twins in Process Industries

Reference	Process	Use case
Martínez et al. (2018c)	Water process in Figure 26.1	Predict future state of the process under specific operating conditions
Wang et al. (2021)	Autoclave	Generate training data for predictive fault maintenance
Kender et al. (2021)	Air separation process	Determine the fastest possible startup and shutdown
Koulouris et al. (2021)	Beverage process	Rescheduling in response to disturbances at plant floor
Aversano et al. (2021)	Furnace operating in flameless combustion conditions	Soft sensor
Yu et al. (2020)	Valve subsystem of steam turbine in a thermal power plant	Online performance monitoring
Maksim et al. (2020)	Producing yellow phosphorus from apatite-nepheline ore waste	Minimize energy consumption

than real time under specified operating conditions, in order to determine the future state of the process if such operator actions would be applied to the physical process. It is unclear how well the dynamic simulation has been synchronized outside of the points at which there was a physical sensor to make an adjustment, so it is not possible to make strong claims about being able to use this digital twin as a soft sensor (i.e. to measure the process state in locations in which there is no physical sensor). However, the procedure for making the adjustment is described in some detail, which is not the case with the majority of the reviewed research discussed next.

Wang et al. (2021) developed a digital twin of an autoclave for fabricating fibre-reinforced plastic composite at a high temperature and pressure. Equipment failures can have significant consequences, motivating investments in predictive fault detection. Data-driven methods such as machine learning could be used, but there is limited data on failure conditions, as the process operator endeavours to keep the autoclave in a healthy and safe state. To obtain such data, the authors use a high-fidelity digital twin to drive a virtual replica of the process to abnormal operating regions. The twin is constructed as a combination of geometric, physical, behavioural and rule-based models. The geometric model corresponds to the 3D process CAD models discussed previously. The physical and behavioural models correspond to a dynamic simulation model. The rule-based model covers the process control system, but also the failure modes of the process equipment. The digital twin is driven to a state that is very close to the current state of the physical process, after which a fitting procedure is applied. The fitting adjusts operating rules, boundary conditions, mesh partitioning, initial conditions and component mesh relationships. The magnitude of the fitting is determined by the deviation between the state of the digital twin and the corresponding measurements from the physical process.

Kender et al. (2021) developed a digital twin of an air separation process to investigate the frequent transients that the process is expected to undergo. The process extracts N_2 and O_2 from the air intake. Whereas many researchers are content to synchronize the twin with the operational process, Kender et al. (2021) considered the evolution of the twin throughout the lifecycle phases of (pre-) sales, equipment and process design, commissioning and operation optimization. Steady-state simulation is used in the first two phases and high-fidelity dynamic simulation in the latter two to accurately capture the nonlinear process behaviour. However, the system is decomposed into subsystem, and it is possible to substitute lower fidelity linear models when such fidelity is acceptable with respect to the objectives of the investigation being carried out with the digital twin. The ability to vary the level of fidelity would make it possible to avoid the use of many separate simulation models throughout the plant lifecycle, which is contrary to the digital twin philosophy. With respect to the method for adjusting the dynamic simulation model to the current state of the physical system, the authors state that it is possible to connect to either historical or live operating data, without elaborating further. Using this digital twin, the authors perform simulations to determine the fastest possible safe startup and shutdown of the process. Other applications of the twin include real-time monitoring and verification of the accuracy of sensor measurements.

Koulouris et al. (2021) considered the special characteristic of the food processing industry, which involves batch processing. Due to the highly seasonal nature of raw ingredients production and rapidly changing market demand, the scheduling of the production is a critical task, involving a combination of make-to-order and make-to-stock. The authors define a digital twin as a synchronization of plant-floor simulations and the scheduling function. The recipe-based representation is identified as a good starting point for the synchronization, which is not yet implemented in practice. It is assumed that a human performs the scheduling function, so the role of the digital twin is to ensure that this work is being carried out against a real-time synchronized model of the situation at the plant floor. The case study shows examples of rescheduling done in response to unexpected occurrences at the plant floor.

Aversano et al. (2021) developed a digital twin on top of physics-based reduced-order models for a furnace operating in flameless combustion conditions. This is an example of a process in which the installation of physical sensors throughout the process is difficult or impossible due to the harsh conditions. The digital twin can be used to predict the state of the process in a three-dimensional space, functioning as a soft sensor. Real-time data from the available sensors is an input to the digital twin, thus synchronizing it with the physical process and enabling it to predict the process state in locations in which there are no physical sensors.

Yu et al. (2020) developed a digital twin for steam turbines in a thermal power plant. As the penetration of renewable generation in the power grid increases, thermal power plants often need to be driven to operating regions in which they were not originally designed to operate, also known as off-design operating modes. In particular, the rapid changes in power generation required from the turbines pose challenges to the operation of the high-pressure control valve systems that govern the steam flow to the turbine, impacting both the turbine and the combustion process. The operation of these control valves is of crucial importance to ensure safety, optimize the energy efficiency and minimize emissions in the off-design mode. The purpose of the digital twin is to enable accurate online performance monitoring. A physics-based steam flow model for the subsystem has been created, corresponding to an as-designed system. Due to factors such as ageing of the valve components, the parameters are adjusted based on operating data to obtain an as-is system.

Maksim et al. (2020) modelled a process for producing yellow phosphorus from apatite-nepheline ore waste. The objective is to reduce energy consumption. A neural network model of the process is run in parallel to the physical system, and its energy consumption output is compared to the energy consumption measurement of the real system. The error between these is used to train the neural network until the error is below an acceptable threshold. After this, the neural network can be used to investigate combinations of process parameters to identify the best combination for the purpose of energy consumption minimization. It is notable that this approach differs from the other approaches that were reviewed previously, in which operating data was used to adjust an as-built simulation, thus achieving an as-is model. In this case, the physics-based model is not adjusted, but the operating parameters are optimized.

The state-of-the-art works that have been reviewed in this section are summarized in Table 26.1, which shows a breadth of different types of processes as well as different use cases for the digital twin, indicating a high potential for this technology in process industries. Further research is required to develop digital twins that can realize several use cases. Further research is also needed to ensure that digital twins are accurate throughout an entire process or subprocess, rather than in specific regions of the process. Although safety has been considered in some works, the existence of safety functions or safety instrumented systems has not been included in the scope of the modelling of digital twins, so the real process may behave very differently than the digital twin in abnormal operating regions, if the safety functions are activated.

Bibliography

F. Alobaid, N. Mertens, R. Starkloff, T. Lanz, C. Heinze, B. Epple (2017) "Progress in dynamic simulation of thermal power plants", *Progress in Energy and Combustion Science*, 59: 79–162, https://doi.org/10.1016/j.pecs.2016.11.001.

E. Arroyo, M. Hoernicke, P. Rodríguez, A. Fay (2016) "Automatic derivation of qualitative plant simulation models from legacy piping and instrumentation diagrams", *Computers in Chemical Engineering*, 92: 112–132, https://doi.org/10.1016/j.compchemeng.2016.04.040.

G. Aversano, M. Ferrarotti, A. Parente (2021) "Digital twin of a combustion furnace operating in flameless conditions: reduced-order model development from CFD simulations", *Proceedings of the Combustion Institute*, 38(4): 5373–5381, https://doi.org/10.1016/j.proci.2020.06.045.

B. Chen, J. Wan, L. Shu, P. Li, M. Mukherjee, B. Yin (2018) "Smart factory of industry 4.0: Key technologies, application case, and challenges", *IEEE Access*, 6: 6505–6519. https://doi.org/10.1109/ACCESS.2017.2783682.

D.-A. Chisalita, A.-M. Cormos (2018) "Dynamic simulation of fluidized bed chemical looping combustion process with iron based oxygen carrier", *Fuel*, 214: 436–445, https://doi.org/10.1016/j.fuel.2017.11.025.

Z. Cui, W. Tian, X. Wang, C. Fan, Q. Guo, H. Xu (2019) "Safety integrity level analysis of fluid catalytic cracking fractionating system based on dynamic simulation", *Journal of the Taiwan Institute of Chemical Engineers*, 104: 16–26, https://doi.org/10.1016/j.jtice.2019.08.008.

D. Faria, M. Bagajewicz (2009) "Profit-based grassroots design and retrofit of water networks in process plants", *Computers & Chemical Engineering*, 33(2): 436–453, https://doi.org/10.1016/j.compchemeng.2008.10.005.

S. Ge, Y. Xu, S. Wang, Q. Xu, T. Ho (2021) "A win-win strategy for simultaneous air-quality benign and profitable emission reduction during chemical plant shutdown operations", *Process Safety and Environmental Protection*, 147: 1185–1192, https://doi.org/10.1016/j.psep.2021.01.044.

M. Hänninen, E. Ahtinen (2009) "Simulation of non-condensable gas flow in two-fluid model of APROS - Description of the model, validation and application", *Annals of Nuclear Energy*, 36(10): 1588–1596, https://doi.org/10.1016/j.anucene.2009.07.018.

S. Henry, J. Baltrusaitis, W.L. Luyben (2021) "Dynamic simulation and control of a combustion turbine process for biogas derived methane", *Computers & Chemical Engineering*, 144: 107121, https://doi.org/10.1016/j.compchemeng.2020.107121.

K. Kawashima, S. Kanai, H. Date (2011) "Automatic recognition of a piping system from large-scale terrestrial laser scan data", *ISPRS - International Archives of the Photogrammetry, Remote Sensing and Spatial Information Sciences*, Volume XXXVIII-5/W12, ISPRS Calgary 2011 Workshop, 29–31 August 2011, Calgary, Canada, pp: 283–288. https://doi.org/10.5194/isprsarchives-XXXVIII-5-W12-283-2011.

R. Kender, F. Kaufmann, F. Rößler, B. Wunderlich, D. Golubev, I. Thomas, A.-M. Ecker, S. Rehfeldt, H. Klein (2021) "Development of a digital twin for a flexible air separation unit using a pressure-driven simulation approach", *Computers & Chemical Engineering*, 151: 107349, https://doi.org/10.1016/j.compchemeng.2021.107349.

M. S. Khaled, S. Imtiaz, S. Ahmed, S. Zendehboudi (2021) "Dynamic simulation of offshore gas processing plant for normal and abnormal operations", *Chemical Engineering Science*, 230: 116159, https://doi.org/10.1016/j.ces.2020.116159.

A. Koulouris, N. Misailidis, D. Petrides (2021) "Applications of process and digital twin models for production simulation and scheduling in the manufacturing of food ingredients and products", *Food and Bioproducts Processing*, 126: 317–333, https://doi.org/10.1016/j.fbp.2021.01.016.

M. Dli, A. Puchkov, V. Meshalkin, I. Abdeev, R. Saitov, R. Abdeev (2020) "Energy and resource efficiency in apatite-nepheline ore waste processing using the digital twin approach", *Energies*, 13(21): 5829. https://doi.org/10.3390/en13215829.

G. S. Martínez, S. A. Sierla, T. A. Karhela, J. Lappalainen, V. Vyatkin (2018a) "Automatic generation of a high-fidelity dynamic thermal-hydraulic process simulation model from a 3D plant model", *IEEE Access*, 6: 45217–45232, https://doi.org/10.1109/ACCESS.2018.2865206.

G. S. Martínez, T. A. Karhela, R. J. Ruusu, S. A. Sierla, V. Vyatkin (2018b) "An integrated implementation methodology of a lifecycle-wide tracking simulation architecture", *IEEE Access*, 6: 15391–15407, https://doi.org/10.1109/ACCESS.2018.2811845.

G. S. Martínez, S. Sierla, T. Karhela, V. Vyatkin (2018c) "Automatic generation of a simulation-based digital twin of an industrial process plant", IECON 2018-44th Annual Conference of the IEEE Industrial Electronics Society. IEEE, 2018: 3084–3089, https://doi.org/10.1109/IECON.2018.8591464.

K.-J. Min, M. Binns, S.-Y. Oh, H.-Y. Cha, J.-K. Kim, Y.-K. Yeo (2015) "Screening of site-wide retrofit options for the minimization of CO2 emissions in process industries", *Applied Thermal Engineering*, 90: 335–344, https://doi.org/10.1016/j.applthermaleng.2015.07.008.

N. Olivier-Maget, F. Berdouzi, C. Murillo, N. Gabas (2021) "Deviation propagation along a propylene glycol process using dynamic simulation: An innovative contribution to the risk evaluation", *Journal of Loss Prevention in the Process Industries*, 70: 104435, https://doi.org/10.1016/j.jlp.2021.104435.

J. Peltola, S. Sierla, T. Vepsäläinen, K. Koskinen (2011) "Challenges in industrial adoption of model-driven technologies in process control application design", 9th IEEE International Conference on Industrial Informatics, pp. 565–572, https://doi.org/10.1109/INDIN.2011.6034941.

T. Pinto-Varela, A. Barbosa-Póvoa, A. Carvalho (2017) "Sustainable batch process retrofit design under uncertainty-An integrated methodology", *Computers & Chemical Engineering* 102: 226–237, https://doi.org/10.1016/j.compchemeng.2016.11.040.

M. Rantala, H. Niemistö, T. Karhela, S. Sierla, V. Vyatkin (2019) "Applying graph matching techniques to enhance reuse of plant design information", *Computers in Industry*, 107: 81–98, https://doi.org/10.1016/j.compind.2019.01.005.

L. Raimondi (2019) "Stratified gas-liquid flow: An analysis of steady state and dynamic simulation for gas-condensate systems", *Petroleum*, 5(2): 128–132, https://doi.org/10.1016/j.petlm.2017.11.002.

M. Shafto, M. Conory, R. Dolye, E. Glaessgen, C. Kemp, J. LeMoigne L. Wang, (2010) DRAFT Modeling, Simulation, Information Technology & Processing Technology Area 11.

S. Sierla, I. Tumer, N. Papakonstantinou, K. Koskinen, D. Jensen (2012) "Early integration of safety to the mechatronic system design process by the functional failure identification and propagation framework", *Mechatronics*, 22(2): 137–151, https://doi.org/10.1016/j.mechatronics.2012.01.003

S. Sierla, M. Azangoo, V. Vyatkin (2020a) "Generating an industrial process graph from 3D pipe routing information", *25th IEEE International Conference on Emerging Technologies and Factory Automation (ETFA)*, pp. 85–92, https://doi.org/10.1109/ETFA46521.2020.9212175.

S. Sierla, M. Azangoo, A. Fay, V. Vyatkin, N. Papakonstantinou (2020b) "Integrating 2D and 3D Digital Plant Information Towards Automatic Generation of Digital Twins", *2020 IEEE 29th International Symposium on Industrial Electronics (ISIE)*, pp. 460–467, doi: 10.1109/ISIE45063.2020.9152371.

S. Sierla, L. Sorsamäki, M. Azangoo, A. Villberg, E. Hytönen, V. Vyatkin (2020c) "Towards semi-automatic generation of a steady state digital twin of a brownfield process plant", *Applied Sciences*, 10: 6959, https://doi.org/10.3390/app10196959.

R. Starkloff, F. Alobaid, K. Karner, B. Epple, M. Schmitz, F. Boehm (2015) "Development and validation of a dynamic simulation model for a large coal-fired power plant", *Applied Thermal Engineering*, 91: 496–506, https://doi.org/10.1016/j.applthermaleng.2015.08.015.

T. Vepsäläinen, S. Sierla, J. Peltola, S. Kuikka (2010) "Assessing the industrial applicability and adoption potential of the AUKOTON model driven control application engineering approach", *8th IEEE International Conference on Industrial Informatics*, pp. 883–889, https://doi.org/10.1109/INDIN.2010.5549626.

B. Wang, J. Klemeš, P. Varbanov, H. Chin, Q.-W. Wang, M. Zeng (2020) "Heat exchanger network retrofit by a shifted retrofit thermodynamic grid diagram-based model and a two-stage approach", *Energy*, 198: 117338, https://doi.org/10.1016/j.energy.2020.117338.

Y. Wang, F. Tao, M. Zhang, L. Wang, Y. Zuo (2021) "Digital twin enhanced fault prediction for the autoclave with insufficient data", *Journal of Manufacturing Systems*, 60: 350–359, https://doi.org/10.1016/j.jmsy.2021.05.015.

T. Wanotayaroj, B. Chalermsinsuwan, P. Piumsomboon (2020) "Dynamic simulation and control system for chemical looping combustion", *Energy Reports*, 6(2): 32–39, https://doi.org/10.1016/j.egyr.2019.11.038.

H. Wei, Y. Lu, Y. Yang, C. Zhang, C. He, Y. Wu, W. Li, D. Zhao (2021) "Research on influence of steam extraction parameters and operation load on operational flexibility of coal-fired power plant", *Applied Thermal Engineering*, 195: 117226, https://doi.org/10.1016/j.applthermaleng.2021.117226.

R. Wen, W. Tang, Z. Su (2017) "Topology based 2D engineering drawing and 3D model matching for process plant," *Graphical Models*, 92: 1–15, https://doi.org/10.1016/j.gmod.2017.06.001.

M. Wen, Q. Wu, G. Li, S. Wang, Z. Li, Y. Tang, L. Xu, T. Liu (2020) "Impact of ultra-low emission technology retrofit on the mercury emissions and cross-media transfer in coal-fired power plants", *Journal of Hazardous Materials*, 396: 122729, https://doi.org/10.1016/j.jhazmat.2020.122729.

X. Xiong, A. Adan, B. Akinci, D. Huber (2013) "Automatic creation of semantically rich 3D building models from laser scanner data", *Automation in Construction*, 31: 325–337, https://doi.org/10.1016/j.autcon.2012.10.006.

H. Yan, M. Liu, D. Chong, C. Wang, J. Yan (2021) "Dynamic performance and control strategy comparison of a solar-aided coal-fired power plant based on energy and exergy analyses", *Energy*, 236: 121515, https://doi.org/10.1016/j.energy.2021.121515.

S. Yoon, J.-S. Oh, J.-K. Kim (2020) "Dynamic simulation and control of natural gas liquids recovery process", *Journal of Cleaner Production*, 257: 120349, https://doi.org/10.1016/j.jclepro.2020.120349.

J. Yu, P. Liu, Z. Li (2020) "Hybrid modelling and digital twin development of a steam turbine control stage for online performance monitoring", *Renewable and Sustainable Energy Reviews*, 133: 110077, https://doi.org/10.1016/j.rser.2020.110077.

Y. Zhao, C. Wang, M. Liu, D. Chong, J. Yan (2018) "Improving operational flexibility by regulating extraction steam of high-pressure heaters on a 660 MW supercritical coal-fired power plant: A dynamic simulation", *Applied Energy*, 212: 1295–1309, https://doi.org/10.1016/j.apenergy.2018.01.017.

C. Zhu, M. Qi, J. Jiang (2020) "Quantifying human error probability in independent protection layers for a batch reactor system using dynamic simulations", *Process Safety and Environmental Protection*, 133: 243–258, https://doi.org/10.1016/j.psep.2019.11.021.

27

Digital Twins in the Manufacturing Industry

Dayalan R. Gunasegaram
CSIRO Manufacturing

27.1 Introduction

The introduction of machines transformed the manufacturing industry, which achieved further gains in productivity through the addition of electrification and automation [1]. It is now well accepted that further increases in manufacturing efficiency can be generated by incorporating artificial intelligence (AI) into machines [2] so that they 'think for themselves,' i.e., learn from historical data to perform a required task in the optimum possible way. Digital twins (DTs) (Figure 27.1) are ideally suited to be the means by which this goal may be achieved [3]. These cyber-physical systems (CPSs) can operate independently and make autonomous decisions, thanks to AI capabilities. DTs are expected to be a part of the Industry 4.0 landscape where high-end products, machines, assembly lines, etc. are embedded with digital sensors and are connected through the Industrial Internet of Things (IIoT). The AI algorithms and digital models help DTs analyse data from sensors and provide control commands to trigger

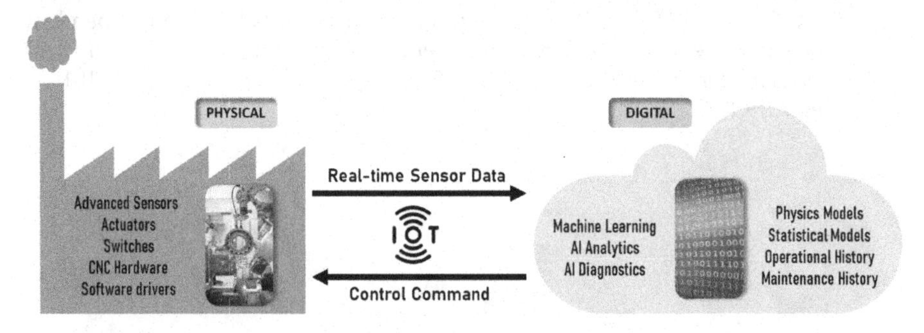

FIGURE 27.1
Components of a digital twin (DT). Note the two-way communication over the Industrial Internet of Things (IIoT) between the physical and virtual counterparts.

DOI: 10.1201/9781003425724-33

corrective or proactive actions. DTs can also optimize the various ancillary operations within factories for improved profitability, sustainability, and safety.

While definitions vary, we consider a DT (Figure 27.1) as a dynamic, self-learning, virtual representation of a physical entity where the digital and physical counterparts maintain two-way communication over the IIoT. A DT digitally models its real-world counterpart's properties, attributes, conditions, and performance [4] and resides in a highly connected ecosystem in cyberspace. The physical entity may be a product, assembly, process, service, asset, system, or even an entire city or country. The DT updates itself using real-time data from connected sensors to accurately mirror its physical twin's current state. It also continuously learns about its physical counterpart from data and updates its embedded digital models and AI algorithms. Thus it gains the ability to predict the potential future states of its physical twin and to recommend optimum process parameters.

27.1.1 Anatomy of a DT and Its Ecosystem

The key components of a DT are shown in Figure 27.1. These include the IIoT, over which data is transferred between the physical and digital counterparts, big data from which machine learning (ML) models are created to provide AI to the DT, digital models that provide process intelligence, AI diagnostics and analytics, sensors, and actuators. The virtual 'living' (i.e., dynamically updated) copy helps mirror the current state of its physical twin. The digital models that contain process intelligence may be based on physics-based computational models or statistical models derived from data gathered from the field, factory, or laboratory experiments. Since physics models can take hours or days to solve (even with parallel processing), their faster-solving surrogate models are usually deployed in DTs for real-time decision-making [1,4]. These surrogate models are reduced dimensional models of their parents; they are simpler and therefore faster but still emulate the main features of the higher order models. ML provides a useful avenue to create surrogate models. Importantly, all models on board a DT learn from real-time data and update themselves continuously to ensure that DT remains a true reflection of its physical counterpart at any given moment and can predict a future state accurately. AI diagnostics and algorithms embedded within the DT provide it with human-like cognitive capabilities that enable it to analyse data in real time and make context-sensitive decisions to achieve the required outcomes. For example, suppose a DT is used in the diagnostic control of a manufacturing process. In that case, it can detect a process excursion (e.g., when a process parameter strays outside its set limits) and take suitable corrective actions through control commands.

27.1.2 DTs Transform the Value of Data: From Automated Factories to Autonomous Factories with Zero Defects and Greener Operations

The introduction of programmable logic controllers in the second half of the 20th century enabled the logging and recording of data from digitally enabled sensors, in addition to allowing the control of robotics and actuators such as motors. This allowed simple corrective measures to be applied in a production environment (e.g., sounding an alarm or stopping a process). However, more sophisticated decisions were made by engineers who analysed the data *offline* (Figure 27.2) for future improvements. However, the presence of a DT capable of making autonomous real-time decisions can guide a process back or even proactively prevent excursions. The DT does this by analysing the sensor data *online* and triggering control commands that would bring the relevant actuators into action for changing the associated process parameters. This ability distinguishes the autonomous factories of the Industry 4.0 era from the merely automated factories of today (Figure 27.2). Its deployment assures quality in the process and, by extension, in the products made by that process, moving factories closer to 'zero defects' operations. Furthermore, DTs are able to suggest optimal process parameters through the 'what-if' type interrogation of its digital models. This knowledge can then be used to optimize the real-world entity's performance to achieve corporate goals.

An example of an Industry 4.0 era smart factory run by a DT at its control centre is depicted in Figure 27.3. The nerve centre may comprised of several interconnected DTs, each mirroring a different operation in the

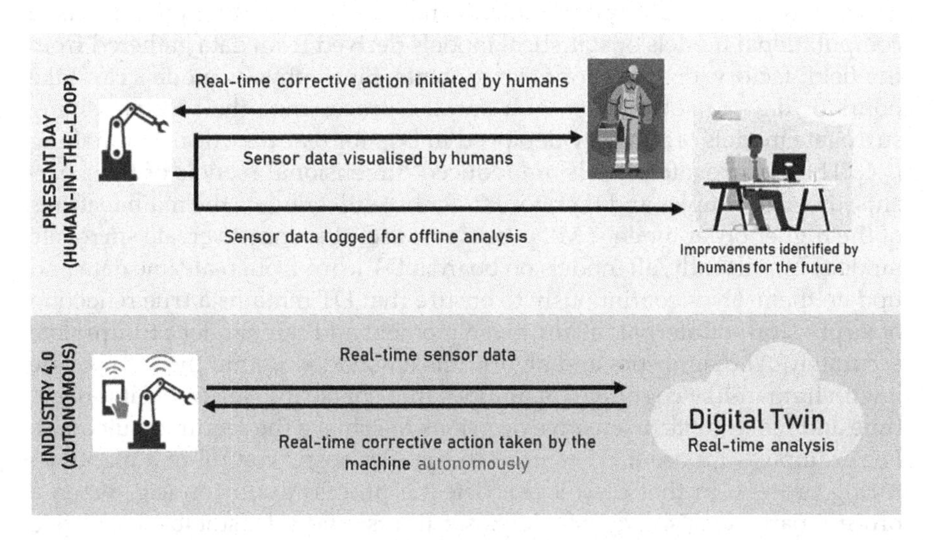

FIGURE 27.2
Traditional automated factories vs the autonomous Industry 4.0 smart factories.

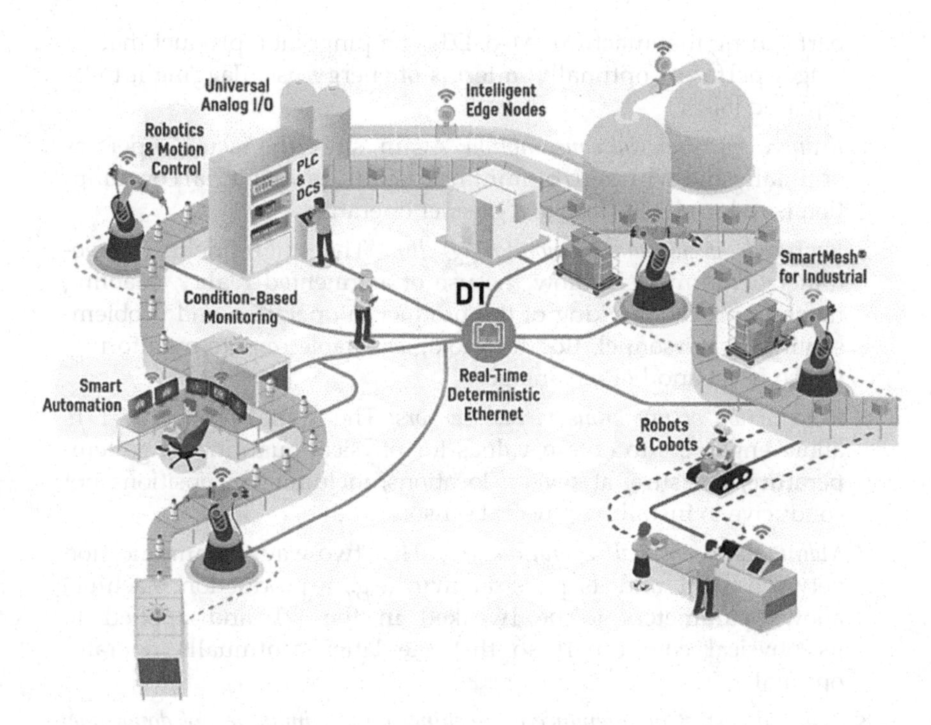

FIGURE 27.3
An example of an Industry 4.0 smart factory where a DT is at the control centre [5].

manufacturing facility. The AI capabilities (e.g., algorithms, process intelligence) embedded in DTs help mimic human cognition, allowing DTs to make real-time decisions autonomously.

The value of a DT to a manufacturing facility may be summarized as follows:

1. *Reduced time to design and to market [6]:* Since the digital models contained within a DT can be used to simulate the performance of a product in service, field trials may be minimized or even eliminated. This is because the digital simulations will be able to, for instance, highlight scenarios where failure is very likely – allowing design engineers to redesign a product even before it is made.
2. *Reduced waste during manufacturing:* Since the optimum parameters to manufacture a product right the first time can be obtained by interrogating a DT, waste in the form of rejects may be avoided. This helps create a greener, advanced manufacturing operation with zero (or minimal) rejects.
3. *Reduced energy consumption:* The DT allows a product to be produced right the first time. This results in a lower energy consumption per

part during manufacture. Also, DTs can pinpoint a product that no longer performs optimally in terms of energy use, flagging it to be replaced [6].

4. *Reduced raw material consumption:* Again, since the factory operates optimally and with zero or minimal defects, raw material consumption is reduced – leading to a greener operation.
5. *Improved performance monitoring [6]:* The high-fidelity three-dimensional models allow the use of augmented reality, enabling much-improved tracking of the product in operation and problem-solving by personnel. IIoT technologies enable remote monitoring, which is yet another advantage.
6. *Introduction of numerous virtual sensors:* The digital models in DTs allow engineers to obtain values for physical quantities (e.g., temperature, pressure) at several locations, including in positions not conducive to installing physical sensors.
7. *Maintaining optimal operation [6]:* The two-way communication between a DT and its physical twin (e.g., a production machine) allows parameters to be tweaked in the DT and applied to its physical counterpart so that the latter continually operates optimally.
8. *Reduced cost of maintenance of machinery and elimination of downtime:* Through its ability to predict future states based on the physical twin's current trajectory (using predictive analytics), a DT can foresee future maintenance issues [6]. This enables the factory to carry out preventative maintenance, eliminating costly shutdowns. Since the DT tunes the assets to operate optimally, the maintenance cost is also reduced.
9. *Improved warehousing/shipping of finished products:* Since a DT that mirrors warehousing and/or shipping can help those services operate optimally, the carbon footprint of the manufacturing facility can be further reduced.
10. *Improved collaboration between teams:* The availability of digital models for the entire operation connected by a common digital thread enables various groups in the factory to collaborate better by exchanging information using a 'single source of truth.' This reduces errors and increases synergies in optimizing the entire manufacturing operation.
11. *Improved safety:* The use of DT-controlled augmented reality [7] can be used to train staff in hazardous trades without exposing them to potential harm. For instance, a DT of the welding process (e.g., [8]) could assist with teaching new students how to use the equipment at the early stages of their education.

27.1.3 DTs Provide Superior Process Controls to Traditional Methods Used in Manufacturing

DTs provide at least two crucial advantages over conventional process controls:

1. *Feedback vs Hybrid Feedback-Feedforward Controls:* The predictive analytics on board a DT can foresee a future state of the process and can work proactively to avoid an undesirable state;
2. *DT controls are programmed using ML:* The programs within a DT are based on ML and thus contain knowledge hidden from human programmers.

These are discussed in turn below.

Feedback vs Hybrid Feedback-Feedforward Controls: Unlike traditional controls that mainly operate as closed-loop (feedback) mechanisms, a DT enables feedforward controls. Such feedforward outputs are based on predictions of the future states of the physical twin as a function of its present trajectory. The DT gains this ability through predictive AI analytics based on up-to-date digital models. These allow the DT to foresee future interruptions to the physical twin and warn the personnel or take steps autonomously to avoid the disruption.

While traditional feedback controls are reactive, feedforward controls are proactive. Generally, hybrid controls that combine feedback with model-assisted feedforward controls are recommended as they comprise the advantages of both. The feedback mechanism in the hybrid strategy will be able to correct any errors inherent in the predictions of the feedforward output and account for unknown/unexpected disturbances [9].

DT controls are programmed using ML: This is another crucial difference between traditional controls used in manufacturing and AI-assisted control commands initiated by DTs. Conventionally, engineers devise and program rigid, hierarchical, 'if-then' type control strategies, which typically provide known solutions for known problems. By contrast, AI systems embedded in the DT can address issues hidden from human expert systems. They can also devise novel responses that humans cannot conceive. This is because the ML models that work on big data can uncover trends and relationships in the manufacturing process that are not visible to factory personnel. Similarly, they can also detect anomalies in the process that may have gone undetected by humans or traditional sensing strategies. Therefore, the programs created by ML models provide superior AI capabilities that transcend the process intelligence contained in expert systems. Thus, the corrective approaches developed by a DT's AI system are likely to be more robust and cover a wider bandwidth of issues in the manufacturing process.

27.1.4 Functions of DTs Deployable in Manufacturing Industries

DTs can be deployed to control several areas of a manufacturing entity that are part of the product lifecycle. These operations are connected by a common digital thread that starts with inputs such as product design and extends to manufacturing processes, products, warehousing/shipping, and performance monitoring (Figure 27.4). This digital thread helps with traceability and learning from experience. For example, if a specific product fails prematurely in the field, its design and manufacture may be modified to address this issue. In this case, the DT for Service/Recycling will need to pass on the necessary data to the Product/Product Design and Manufacturing DT, as shown in Figure 27.4. Note that, in some instances, each instance of a product/assembly/system may have its own DT, e.g., cars. In this scenario, each such item is said to have a unique 'instance' of a DT. It makes sense to have instances for high-value products, which can be remotely monitored and optimized. For example, some jet engines have their exclusive DTs. Similarly, if several machines are available on the factory floor to carry out the same manufacturing operation, each such device will have its unique instance of the manufacturing DT. This would allow each machine to be monitored separately and optimized through measures such as preventative maintenance.

We shall now take a deeper look at each DT shown in Figure 27.4, which collectively covers the entire lifecycle of a product. It is worth noting that these DTs may be developed at different hierarchical levels, which are discussed later in Section 1.5.

27.1.4.1 Product Design DT

The information contained within Product Design DT enables the product design process to be streamlined and the products' field performance to be mirrored. It must also be capable of receiving feedback from the Manufacturing DT and the Service/Recycling DT to improve product design

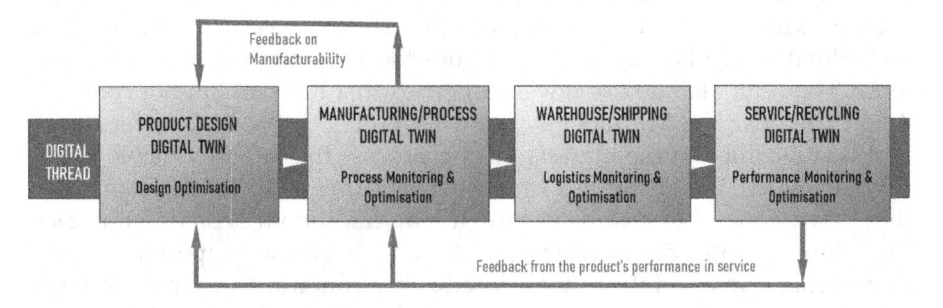

FIGURE 27.4
DTs representing the various stages of a product lifecycle connected by a common digital thread.

based on experience (Figure 27.4). Lo et al. [10] identified four key areas for Product Design DTs: conceptual design, detailed design, design verification, and redesign.

In the conceptual design phase, engineers must consider a vast amount of data related to various influencing factors. These can include customer needs, manufacturability, weight, aesthetics, durability, on-field performance metrics of the previous model, etc. Tao et al. [11] proposed using a DT to analyse the big data containing the required information. Ma et al. [12] suggested using a DT-enhanced augmented reality environment for an improved design experience where the designers can 'see' the product and interact with it as though with a physical object. Illmer et al. [13] noted that the Product Design DT could also be connected with the Manufacturing/Process DT (Figure 27.4) to explore manufacturability and raw material planning.

In the detailed design stage, aspects such as specific product function, dimensions, performance metrics, material properties, and manufacturing process must be considered. At this stage, the designers also typically interact with other internal stakeholders (e.g., raw material purchasers, production engineers, machine operators, and marketing personnel) and external participants (e.g., target customers). Ideally, the Product Design DT must be able to support the following: collation and analysis of big data [14], peer-to-peer exchange of data, e.g., using blockchain technology [15], optimization of material selection [16] based on criteria such as sustainability, and optimization of product dimensions (and mass) based on given constraints. Another critical capability would be the ability to simulate the product's performance in the field [17]. The accuracy of the outcome is likely to be enhanced if this could be done through a connection with the Service/Recycling DT (Figure 27.4).

During design verification, engineers deal with failure analysis and prediction. Product Design DTs must therefore be able to support the prediction of root causes for potential failures [18]. Onboard ML models based on field data and performance simulations would strengthen the prediction outcomes.

In the final redesign phase, designers modify their original design based on failure predictions and feedback from Manufacturing/Process DT and Service/Recyclability DT. Tao et al. [11] have proposed a DT framework where the virtual space would continuously collect, analyse, and accumulate data from the physical space for redesign purposes.

More in-depth studies of the Product Design DTs are available elsewhere (e.g., [10,19–23]).

27.1.4.2 Manufacturing/Process DT

The Manufacturing/Process DTs can be divided into two major categories:

1. DTs that mirror the shopfloor, i.e., the collection of machines, assembly lines, etc. These can help optimize, for example, production planning and scheduling as they contain information on machine

capabilities. They also assist with the preventative maintenance of their physical counterparts.

2. DTs that mirror the manufacturing process. These can carry out diagnostic control of processes to ensure they remain within specified limits. These processes would be quality assured in a way that they produce little or no rejects. By extension, the parts produced by these processes also would be within specifications.

We now take a deeper look into each of the above categories.

DTs that mirror the shopfloor: These can produce optimum production schedules [24] and adjust these schedules in real time based on the current state of the operation [22]. For example, suppose a machine breaks down, creating a backlog in production. In that case, the DT will be able to revise the schedules based on machine capabilities and up-to-the-minute availability to recover at least part of the lost production. The DT can also prioritize production to ensure that urgent orders are fulfilled on time. It can also support flexible manufacturing – where the manufacturer offers product customisation [25] – while keeping the production costs as low as possible. The DT can also help reduce interruptions when introducing new products [26]. This type of DT can also keep the factory running smoothly by predicting the state of each machine in the future and recommending preventative maintenance actions. Significantly, Airbus has developed a DT to optimize its assembly lines [11]. Zhang et al. [27] have proposed a DT-based approach to designing production lines and presented a case study on a glass production line. Further details on this topic may be found elsewhere, e.g., [24,28,29].

DTs that mirror the manufacturing process: DTs in this category are scarce to find in the open domain. This is presumably because of the difficulties associated with digitally representing the complex physics of the processes faithfully. Nevertheless, interest in this topic has steadily improved, given that the production process is at the core of the manufacturing enterprise. DTs operating in a diagnostic control capacity can take real-time (or near real-time) evasive actions to avoid process excursions or failures; consequently, the manufacturing operation may be conducted with minimal rejects. Thus, DTs can accelerate a production facility's movement towards the waste-minimizing 'zero defects manufacturing' paradigm [30] of Industry 4.0. Furthermore, the AI residing within the manufacturing process DTs may be consulted to determine the best processing window for a given product.

The processes that stand to profit the most from DTs have inherent instabilities, which result in highly variable outputs for the same set of input parameters. Powder-based metal additive manufacturing (AM) is a prime example of this [1,4]. It makes sense to avoid rejects in this process through diagnostic control since it is typically expected to produce high-value mission-critical parts (for biomedical, space, aerospace, and defence industries) with strict quality specifications. Additionally, since the process is incremental and requires hours or days to manufacture a complete part, the AM machines' utilization percentage vastly increases when rejects are avoided. DTs in such environments would typically be deployed with hybrid control systems, as discussed earlier.

For the manufacturing process DTs to be effective, the underlying process intelligence must be reliable. Such intelligence is provided by field data or robust multiphysics simulations of the process. However, this knowledge must be converted to rapidly interrogatable surrogate ML models for real-time decision-making by the DT [1].

A detailed account of the steps required to develop manufacturing process DTs of this type for the AM process is available elsewhere [1,31–33]. Other works in the open literature that deal with process DTs include, e.g., DT of laser processes [34], DT of a cutting tool [35], and DT of a welding process [8].

27.1.4.3 Warehouse/Shipping DT

Ancillary services of a manufacturing facility, such as warehouse operations and shipping, can be mirrored using their own DTs [36]. When these are used collaboratively with other DTs of the enterprise, the entire business can be synergistically optimized. For example, shipments can be made according to demand at any given time, and, in turn, production schedules may be amended to suit. Suppose a factory uses external contractors for its shipments. In that case, the DTs of the two businesses (e.g., the third party's Shipping DT and the factory's Warehouse DT and Manufacturing/Process DT) will need to collaborate to achieve maximum gains. The design and development of supply chain DTs have been discussed elsewhere – e.g., [37], where a case study is presented for a pharmaceutical company.

27.1.4.4 Service/Recycling DT

This category of DTs mimics the part's performance in the field and includes the recycling process (where applicable) to complete the product lifecycle. It is by far the most prevalent DT in the manufacturing industry today [38], probably because its benefits are more tangible. For instance, monitoring the health of a high-value or mission-critical product (e.g., jet engines) may be highly advantageous from the point of view of safety, reliability, or the manufacturer's reputation. DT-controlled health monitoring forces a paradigm shift in maintenance strategies through the foreshadowing of potential equipment failures using predictive analytics. Reducing machine downtime using DTs increases the productivity of a manufacturing enterprise. You et al. [39], Nguyen et al. [40], and Hu et al. [41] have covered this topic in good detail. The use of DTs redresses the shortcomings of the conventional strategy, which revolves around designing a product with a sufficiently high safety factor and implementing periodic maintenance. This can result in a waste of resources (e.g., redundant material due to the extra margin of safety built-in) and untimely or too little maintenance [42].

Recyclability is an integral part of sustainable manufacturing. When a product is scrapped, information on the product may be entered into the Service/Recycling DTs to inform future design processes. Service/recycling DTs may collaborate with Product Design DTs and Manufacturing/Process DTs to provide feedback (Figure 27.4). This data flow provides invaluable information from the field to improve product design and manufacturing methods. Such closing of the loop in a product's lifecycle using DTs provides a significant opportunity for sustainable product design and manufacturing.

27.1.5 Hierarchical Levels of DTs Deployable in Manufacturing Industries

A hierarchy in DTs exists because unique DTs may be created for the lowest outputs of an operation (e.g., individual parts) and for higher levels comprising these (e.g., an assembly of parts). For instance, the DT of an aircraft comprises the DT of a rack, the DT of the flight control system, the DT of the propulsion system, etc. [43]. Therefore, a composite DT can be seen as an interconnected network of DTs. The hierarchy may be defined based on several criteria. The hierarchy is well summarized by Singh et al. [44] as follows:

1. *Unit level:* This is the smallest participating unit in manufacturing and can be a part, assembly, piece of equipment, material, or environmental factor. Unit-level DT is based on the unit-level physical twin's geometric, functional, behavioural, and operational model.
2. *System level:* It is an amalgamation of several unit-level DTs in a production system, such as a production line, shop floor, and factory. Interconnectivity and collaboration among multiple unit-level DTs lead to a broader flow of data and better resource allocation. A complex product, e.g., aircraft, can also be considered a system-level DT.
3. *System of Systems (SoS) level:* Several system-level DTs are connected to form SoS-level DT, which helps in collaborating different enterprises or different departments within an enterprise, such as supply chain, design, service, and maintenance. In other words, SoS-level DT integrates different phases of the product throughout its lifecycle.

27.1.6 Examples of DTs Deployed in Manufacturing Industries

Despite the infancy of the DT concept, some early adopters in the manufacturing industry are already enjoying the benefits of the technology. In this section, we catalogue some examples of DTs already operating in the manufacturing industry. It is clear from Table 27.1 that the users are spread across a vast spectrum in terms of the products they make. However, a common attribute is that DTs have been generally deployed in the development of high-value products such as jet engines, aircraft, and cars.

TABLE 27.1

A Non-exhaustive List Containing Examples of DTs Already Deployed in the Manufacturing Industry

Type of DT	Product	Reason for the DT	Company	Source
Product design	Sedan (Ghibli)	To reduce physical testing of prototype	Maserati	[10]
Product design	Aircraft (B777)	To improve the quality of parts	Boeing	[10]
Product design	Rockets (Falcon 1)	For virtual testing of prototypes	SpaceX	[10]
Manufacturing/Process	Aircraft	To monitor production and optimize the operational efficiency of the assembly line	Airbus	[11]
Manufacturing/Process	Aircraft	To reduce downtime and improve the production efficiency of aircraft parts	Airbus	[33]
Warehouse/Shipping	Shipments	To improve efficiency (proposed DT)	DHL	[36]
Service/Recycling	Jet engines (GE60)	Remote health monitoring, performance optimization	General Electric	[10]
Service/Recycling	Jet engines	Remote health monitoring	Rolls Royce	[10]
Service/Recycling	Aircraft (A350 XWB)	Remote health monitoring	Airbus	[10]

This is unsurprising given that, in these circumstances, the relatively high upfront capital outlay required could be more easily justified in terms of returns. The significant investment cost is also why only global giants have been able to implement the advanced solutions offered by DTs so far. However, as an internet search would reveal, consulting companies currently offer assistance with creating DTs in manufacturing for smaller enterprises. Most of these solutions use cloud-based IIoT platforms.

27.1.7 Challenges

Despite the obvious advantages of deploying DTs in smart manufacturing, their widespread adoption is predicated on certain hurdles being addressed:

1. *Lack of big data for training ML models in DTs of manufacturing processes*: ML models play a significant role in DTs as they provide the AI required for the DT to implement autonomous decision-making. Training algorithms depend on large volumes of representative data ('big data') to capture all the possibilities, nuances, and interactions in various trends and relationships. Such data sets do not generally exist in the manufacturing industry. That is because, unlike

consumer industries (e.g., Amazon, Netflix) which are largely homogeneous, manufacturing is heterogeneous, i.e., each manufacturing method is unique and thus has its own data set. It would therefore take a longer time, greater effort, and extra resources to develop large enough data sets for each operation. Other big data-related challenges include difficulty identifying and accessing usable data, transforming data from different sources, poor data quality, and translation loss [45].

2. *Security of data*: Since DTs operate in a highly connected IIoT environment, the security of data in storage and while in transit must be ensured to safeguard intellectual property. In addition, adequate thought must be given to privacy, confidentiality, transparency, and ownership of data [44], especially when collaborating with other businesses.
3. *Lack of standards and regulations and interoperability issues*: For data to be shared, everyone must work to accepted standards. The lack of standards, regulations, and governance around data handling is presently a hindrance to the proliferation of data-centric technologies such as DTs [1,44]. In addition, standards must be developed for interoperability, where data is shared between DTs from different organizations [44]. Also, interoperability issues may arise when DTs at different hierarchical levels (Section 1.5) generate varying types of data and translation is not carried out correctly [46].
4. *Reluctance to share strategic competitive knowledge:* Data is no longer a by-product of business but a strategic asset. Thus, private enterprises may wish to keep their data confidential to secure their competitive advantage [1], reducing the effectiveness of collaboration between the DTs of various organisations.
5. *Observability and controllability issues:* For a control system in a DT to work as intended, the process in question must be adequately observable and controllable [4]. Thus, the sensors must be able to capture the critical quantity (i.e., the process signature) at the speed and resolution necessary. Similarly, the actuators must be capable of taking the corrective action commanded by the DT. Thus, suitable hardware must be available for a DT to be successful.
6. *Hurdles in creating physics models:* The use of physics models in the creation of ML models improves the accuracy of ML models. This is because the physics-informed ML models can be used to identify and eliminate outliers in the training data sets. However, the multiphysics, multiscale nature of physics models makes developing high-fidelity models challenging [1,4].
7. *Lifecycle mismatch:* The lifecycles of products such as aircraft and cars are usually far greater than the useful life of software used for

designing or simulating the DT and storing and analysing data [47]. Software that developers no longer support could thus make the virtual twin obsolete before the physical counterpart does.

8. *Upfront capital outlay:* As seen in Table 27.1, only large international corporations have so far deployed DTs. This is because a significant amount of resources is necessary to create reliable DTs, which only these giants can afford. Thus, unless industry bodies are willing to pool resources together to develop DTs for use by their members, DTs may be out of reach of smaller businesses for several years.
9. *Requirements for additional training:* Conventional engineers need to be trained in new methods involving ML and AI and be convinced of their effectiveness before they will be willing to adopt the use of DTs in the factory.

27.1.8 Outlook

DT offers an ideal method by which operations can be autonomously controlled and optimized in the highly connected smart factories of the Industry 4.0 era. DTs promise significant advantages to the manufacturing industry, including increased process productivity, quality assurance, zero defects, and greener operations. However, the DT technology is still in its infancy and has so far seen limited applications in the manufacturing industry. Only giant corporations are presently enjoying the benefits of the technology, as they possess the necessary resources to develop the DTs. However, going by the number of businesses considering developing DTs, its reach is steadily increasing and is likely to become more widespread in the coming decades. Industry bodies can play a crucial role in pooling the resources of businesses in each manufacturing sector to develop DTs for that sector.

Bibliography

1. Gunasegaram, D.R., et al., Towards developing multiscale-multiphysics models and their surrogates for digital twins of metal additive manufacturing. *Additive Manufacturing*, 2021. **46**: p. 102089.
2. Makridakis, S., The forthcoming Artificial Intelligence (AI) revolution: Its impact on society and firms. *Futures*, 2017. **90**: p. 46–60.
3. Semeraro, C., et al., Digital twin paradigm: A systematic literature review. *Computers in Industry*, 2021. **130**: p. 103469.
4. Gunasegaram, D.R., et al., The case for digital twins in metal additive manufacturing. *Journal of Physics: Materials*, 2021. **4**(4).
5. Mesbahi, M. *How Digital Twins Can Impact the Smart Manufacturing Landscape.* 2020 August 2022; Available from: https://www.wevolver.com/article/how-digital-twins-can-impact-the-smart-manufacturing-landscape.

6. Tao, F., M. Zhang, and A.Y.C. Nee,Background and Concept of Digital Twin, In book *Digital Twin Driven Smart Manufacturing*, F. Tao, M. Zhang, and A. Y. C. Nee, Editors. 2019, Academic Press. p. 3–28. 10.1016/B978-0-12-817630-6.00001-1.
7. Zhu, Z., C. Liu, and X. Xu, Visualisation of the digital twin data in manufacturing by using augmented reality. *Procedia CIRP*, 2019. **81**: p. 898–903.
8. Wang, Q., Toward Intelligent Welding by Building Its Digital Twin, In *Electrical and Computer Engineering*. 2021, Lexington, KY: University of Kentucky. 161. https://uknowledge.uky.edu/ece_etds/161.
9. Juang, J.-N. and K.W. Eure, *Predictive Feedback and Feedforward Control for Systems with Unknown Disturbances*. 1998, Hampton VA: Langley Research Center, NASA. p. 1–39.
10. Lo, C.K., C.H. Chen, and R.Y. Zhong, A review of digital twin in product design and development. *Advanced Engineering Informatics*, 2021. **48**: p. 101297.
11. Tao, F., et al., Digital twin-driven product design framework. *International Journal of Production Research*, 2019. **57**(12): p. 3935–3953.
12. Ma, X., et al., Digital twin enhanced human-machine interaction in product lifecycle. *Procedia CIRP*, 2019. **83**: p. 789–793.
13. Illmer, B. and M. Vielhaber, Synchronizing digital process twins between virtual products and resources - A virtual design method. *Procedia CIRP*, 2019. **84**: p. 532–537.
14. Cheng, J., et al., DT-II:Digital twin enhanced Industrial Internet reference framework towards smart manufacturing. *Robotics and Computer-Integrated Manufacturing*, 2020. **62**: p. 101881.
15. Huang, S., et al., Blockchain-based data management for digital twin of product. *Journal of Manufacturing Systems*, 2020. **54**: p. 361–371.
16. Xiang, F., et al., Digital twin driven green material optimal-selection towards sustainable manufacturing. *Procedia CIRP*, 2019. **81**: p. 1290–1294.
17. Arrichiello, V. and P. Gualeni, Systems engineering and digital twin: A vision for the future of cruise ships design, production and operations. *International Journal on Interactive Design and Manufacturing (IJIDeM)*, 2020. **14**(1): p. 115–122.
18. Detzner, A. and E. Martin. A digital twin for root cause analysis and product quality monitoring. In *Proceedings of the DESIGN 2018-15th International Design Conference*. 2018. Dubrovnik, Croatia: The Design Society.
19. Tao, F., et al., *Digital Twin Driven Smart Design*, ed. F. Tao, et al. 2020: Academic Press.
20. Zheng, P. and K.Y. Hong Lim, Product family design and optimization: a digital twin-enhanced approach. *Procedia CIRP*, 2020. **93**: p. 246–250.
21. Canedo, A., Industrial IoT lifecycle via digital twins, In *Proceedings of the Eleventh IEEE/ACM/IFIP International Conference on Hardware/Software Codesign and System Synthesis*. 2016, Pittsburgh, PA: Association for Computing Machinery. p. Article 29.
22. Schleich, B., et al., Shaping the digital twin for design and production engineering. *CIRP Annals*, 2017. **66**(1): p. 141–144.
23. Zhuang, C., J. Liu, and H. Xiong, Digital twin-based smart production management and control framework for the complex product assembly shop-floor. *The International Journal of Advanced Manufacturing Technology*, 2018. **96**(1): p. 1149–1163.
24. Uhlemann, T.H.J., C. Lehmann, and R. Steinhilper, The digital twin: Realizing the cyber-physical production system for industry 4.0. *Procedia CIRP*, 2017. **61**: p. 335–340.

25. Söderberg, R., et al., Toward a digital twin for real-time geometry assurance in individualized production. *CIRP Annals,* 2017. **66**(1): p. 137–140.
26. Biesinger, F., et al., A digital twin for production planning based on cyber-physical systems: A case study for a cyber-physical system-based creation of a digital twin. *Procedia CIRP,* 2019. **79**: p. 355–360.
27. Zhang, H., et al., A digital twin-based approach for designing and multi-objective optimization of hollow glass production line. *IEEE Access,* 2017. **5**: p. 26901–26911.
28. Rosen, R., et al., About the importance of autonomy and digital twins for the future of manufacturing. *IFAC-PapersOnLine,* 2015. **48**(3): p. 567–572.
29. Beldiceanu, N., et al., ASSISTANT: Learning and robust decision support system for agile manufacturing environments. *IFAC-PapersOnLine,* 2021. **54**(1): p. 641–646.
30. Powell, D., et al., Advancing zero defect manufacturing: A state-of-the-art perspective and future research directions. *Computers in Industry,* 2022. **136**: p. 103596.
31. DebRoy, T., et al., Building digital twins of 3D printing machines. *Scripta Materialia,* 2017. **135**: p. 119–124.
32. Pantelidakis, M., et al., A digital twin ecosystem for additive manufacturing using a real-time development platform. *The International Journal of Advanced Manufacturing Technology,* 2022. **120**(9): p. 6547–6563.
33. Pascual, F.J. and A.G. Aparcero. The Benefits of Building a Digital Twin of Your Factory. 2022 August 2022; Available from: https://capgemini-engineering.com/us/en/insight/the-benefits-of-building-a-digital-twin-of-your-factory/.
34. Papacharalampopoulos, A. and P. Stavropoulos, Towards a digital twin for thermal processes: Control-centric approach. *Procedia CIRP,* 2019. **86**: p. 110–115.
35. Botkina, D., et al., Digital twin of a cutting tool. *Procedia CIRP,* 2018. **72**: p. 215–218.
36. Anon. Digital Twins in Logistics - A DHL perspective 2019 August 2022; Available from: https://www.dhl.com/content/dam/dhl/global/core/documents/pdf/glo-core-digital-twins-in-logistics.pdf.
37. Marmolejo-Saucedo, J.A., Design and development of digital twins: A case study in supply chains. *Mobile Networks and Applications,* 2020. **25**(6): p. 2141–2160.
38. Tao, F., et al., Digital twin in industry: State-of-the-art. *IEEE Transactions on Industrial Informatics,* 2019. **15**(4): p. 2405–2415.
39. You, Y., et al., Advances of digital twins for predictive maintenance. Procedia Computer Science, 2022. **200**: p. 1471–1480.
40. Nguyen, T.N., et al., A digital twin approach to system-level fault detection and diagnosis for improved equipment health monitoring. *Annals of Nuclear Energy,* 2022. **170**: p. 109002.
41. Hu, M., et al., Digital twin model of gas turbine and its application in warning of performance fault. *Chinese Journal of Aeronautics,* 2022. **36**: p. 449–470.
42. Xie, R., et al., Digital twin technologies for turbomachinery in a life cycle perspective: *A Review. sustainability,* 2021. **13**(5): p. 2495.
43. Tuegel, E. The Airframe Digital Twin: Some Challenges to Realization. In 53rd *AIAA/ASME/ASCE/AHS/ASC Structures, Structural Dynamics and Materials Conference 20th AIAA/ASME/AHS Adaptive Structures Conference 14th AIAA.* 2012. Honolulu, HI.
44. Singh, M., et al., Digital twin: Origin to future. *Applied System Innovation,* 2021. **4**(2): p. 36.

45. Tyagi, P. and H. Demirkan. *The Biggest Big Data Challenges*. 2016 August 2022; Available from: https://pubsonline.informs.org/do/10.1287/LYTX.2016.06.05/full/.
46. Mapp, M.-R.G. *Digital Twins, Another Reason to Worry about the IoT and Data Security*. 2020 August 2022; Available from: https://irishtechnews.ie/digital-twins-iot-and-data-security/.
47. Anon. *Digital Twin: 5 Challenges for 7 Benefits*. 2019 August 2022; Available from: https://www.ingenium-magazine.it/en/digital-twin-6-sfide-per-7-benefici/.

28

Cognitive Digital Twins in the Process Industries

Jože Martin Rožanec
Jožef Stefan International Postgraduate School
Jožef Stefan Institute

Pavlos Eirinakis
University of Piraeus

George Arampatzis
Technical University of Crete

Nenad Stojanović
Nissatech Innovation Centre

Kostas Kalaboukas
Gruppo Maggioli—Athens Branch

Jinzhi Lu, Xiaochen Zheng, and Dimitris Kiritsis
Ecole Polytechnique F´ed´erale de Lausanne (EPFL)

28.1 Introduction

Process manufacturing is a branch of manufacturing that converts raw materials through formulas and recipes rather than individual units to process bulk products into other products. We find the food, beverage, pharmaceutical, and petrochemical industries among such industries. The global market share and business performance of the process industry are based on the value of the process sites, people, materials, and intellectual property (product knowledge, process expertise, and physical properties of the materials), with the most operating profit and plant operations.

Optimization of plant operations requires working at two different levels: (1) plant management level (optimizing for performance indicators based on regular plant operation and detected abnormal behaviors) and (2) process

DOI: 10.1201/9781003425724-34

control level (monitoring process variables and adjusting them to expected levels). Such optimization can provide the best value when real-time information that enables tight integration of business systems and plant operations is available, matching information regarding market demand and production processes. Furthermore, it can be realized in the paradigm of Digital Twins (DTs), where information regarding the physical counterpart is fed into a DT Platform. The DT Platform creates a digital shadow whose behavior and relevant insights are modeled and extracted through multiple services supporting operational requirements. Mechanisms must be put in place to gather relevant real-time data and provide feedback from the digital counterparts of a DT back to the physical world. The realization of DTs enables humans to adopt a business-centric approach toward plant management and critical process considerations.

Cognitive Digital Twins (CDTs) aim to enhance DTs, providing a cognition-first approach when combining real-time data, physical dependency models, and intelligence to improve end-to-end processes. To achieve this goal, it is essential to consider what types of knowledge can be realized in the digital domain and the techniques and practices that enable it. Furthermore, it is relevant to understand the limitations of such approaches and the interplay between them to drive value to the end users. In this chapter, we introduce an architecture for CDTs developed for the EU H2020 project FACTLOG (Energy-Aware Factory Analytics for Process Industries), provide a high-level detail of the main components, and describe some of the use cases considered. Throughout the chapter, we share our experiences and insights regarding the CDT development lifecycle and implementation.

28.2 Architecture for Cognitive Digital Twins for Process Industries

To realize such CDTs targeting process industries, a software architecture was devised (see Figure 28.1), linking the data sources to a DTs Platform, analytics, and optimization services, and encoding the data and domain knowledge in a knowledge graph. Communication between the services is realized through application programming interfaces exposing the particular functionalities of each service through multiple protocols. In particular, a messaging bus is used to realize the publish-subscribe model, where service requests are made asynchronously. The services respond by publishing the metadata or result to a particular topic to which interested parties subscribe. Below we describe in detail the services contemplated in the proposed architecture.

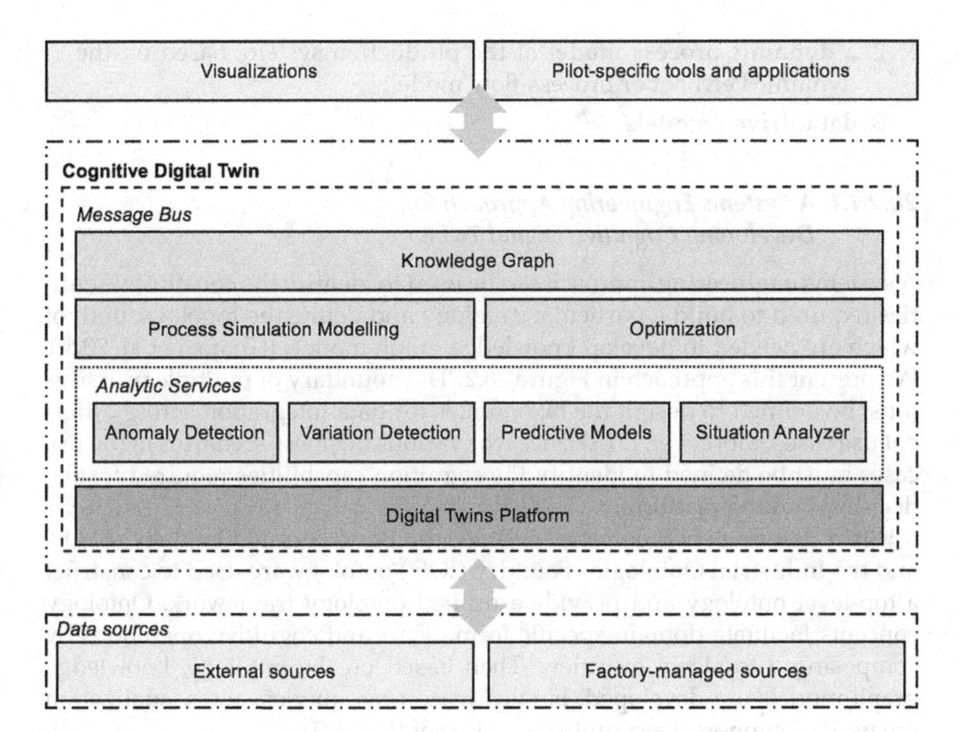

FIGURE 28.1
CDTs high-level architecture, as envisioned in the EU H2020 FACTLOG project.

28.2.1 Knowledge Graphs

The knowledge graph model denotes a unified representation and description of all related pilots and technical solutions based on a top-level ontology. Such ontology is enriched with concepts sourced from domain and application ontologies to provide a formal specification and conceptualization of the cognition needs of each use case. Furthermore, the ontology provides a common interface for importing and exporting knowledge graph models (Lu et al. 2020). The knowledge graph formalizes the related equipment, products, methods, algorithms, and cognition services in a standardized way (Zheng, Lu, and Kiritsis 2021). Furthermore, it describes the platform services and their interrelationships required to interconnect and interoperate with external tools (e.g., tools for optimization and data visualization, among others).

We consider the cognition services to be supported by three components:

1. **knowledge graph models** to describe domain-specific knowledge and data with their interrelationships within the form of subjectives, objectives, and predicates;

2. **a dynamic process model** of the production system, based on the dynamic Petri net or process-flow model;
3. **data-driven models.**

28.2.1.1 A Systems Engineering Approach for Developing Cognitive Digital Twins

A systems engineering approach can be used to identify the constitutive entities required to build a particular ontology and define the topology, both of which are needed to develop knowledge graph models (Gharaei et al. 2020). We present this approach in Figure 28.2. The boundary of multiple use cases must be defined to design the taxonomies for data integration across different use cases. Moreover, DT services for various business scenarios in the use cases must be defined to identify the cognition capabilities required for the decision-making platform.

When defining such ontology entities, the Basic Formal Ontology (BFO[1]), and the Industrial Ontologies Foundry (IOF[2]) ontology are used to construct a top-level ontology and provide a unified ontology framework. Ontology concepts facilitate domain-specific formalisms and cognitive operations by composing a top-level overview. Then based on the ontology, knowledge graph models are developed. Finally, reasoning, queries, and visualization are used to support the cognitive services of the CDT.

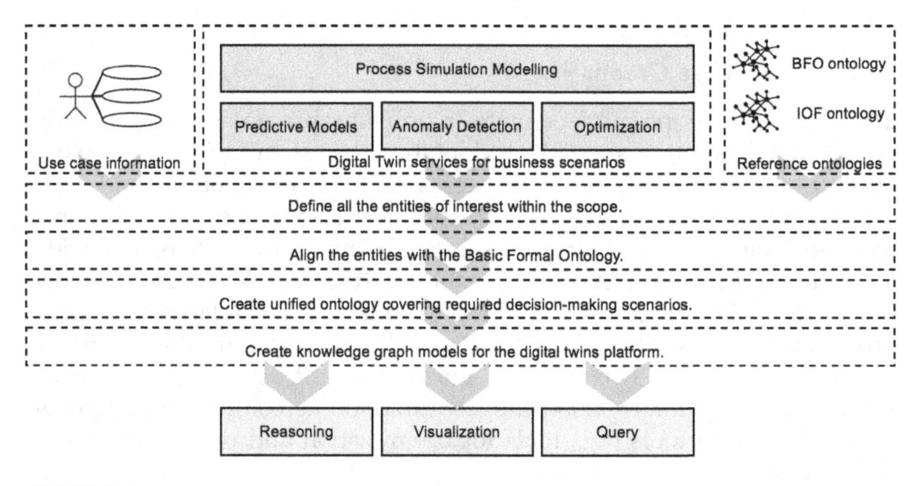

FIGURE 28.2
Systems engineering approach for developing knowledge graph models.

[1] The BFO ontology can be accessed at https://basic-formal-ontology.org/
[2] The IOF ontology can be accessed at https://www.industrialontologies.org/

28.2.1.2 Ontology for Developing Knowledge Graph Models

A BFO-based ontology, considering also concepts defined by IOF (Arp, Smith, and Spear 2015), was developed to support the development of knowledge graphs (see Figure 28.3). It comprises three kinds of concepts: (1) concepts sourced from BFO and IOF, (2) domain concepts, and (3) concepts used to define use cases and business scenarios (not present in Figure 28.3).

Following the BFO-defined taxonomy, we consider all the concepts to be categorized as either continuant or an occurrent. A continuant is defined as an entity that persists or continues to exist through time while maintaining its identity. In contrast, the occurrent is defined as an entity that unfolds itself in time or as a (spatio)temporal region to which entities can relate. We provide the rest of the concept definitions in Table 28.1. Within the definitions, the term *proper parts* refers to parts not equal to the whole.

28.2.1.3 Integration of Knowledge Graph Models and Cognitive Digital Twins

A knowledge graph framework is developed to support the CDT development (Jinzhi et al. 2022). However, given that various formats exist to formalize the ontologies and implement the knowledge graphs, it can be challenging to translate ontologies into knowledge graphs.

For the EU H2020 FACTLOG project, we formalized the ontology models with the Web Ontology Language (OWL). We leveraged Protégé (Musen 2015) to serialize the OWL models into the Resource Description Framework Turtle format (Beckett et al. 2014) and import them into a Neo4j graph database.[3] Then based on the systems engineering approach, we identified different use cases and scenarios for the CDT implementation. Then, we ingested and formalized the use case information and CDT services using knowledge graph models within the Neo4j graph database. Finally, we leveraged the Cypher query language (Francis et al. 2018) to support reasoning within the knowledge graph models. We also leveraged Application Programming Interfaces (APIs) in the Java programming language to support the CDT development. When the stakeholders use this CDT, services capable of parsing Cypher queries are implemented from the CDT platform.

28.2.2 Process Simulation Modeling

28.2.2.1 The Role of Process Modeling

In cognitive manufacturing, the Process Simulation Model denotes a generic model with all related methods integrated into an overall modeling application or platform. In any specialized model (use case), these methods do not change. Only the process model (digital) representation *per se* changes.

[3] Neo4j: https://neo4j.com/

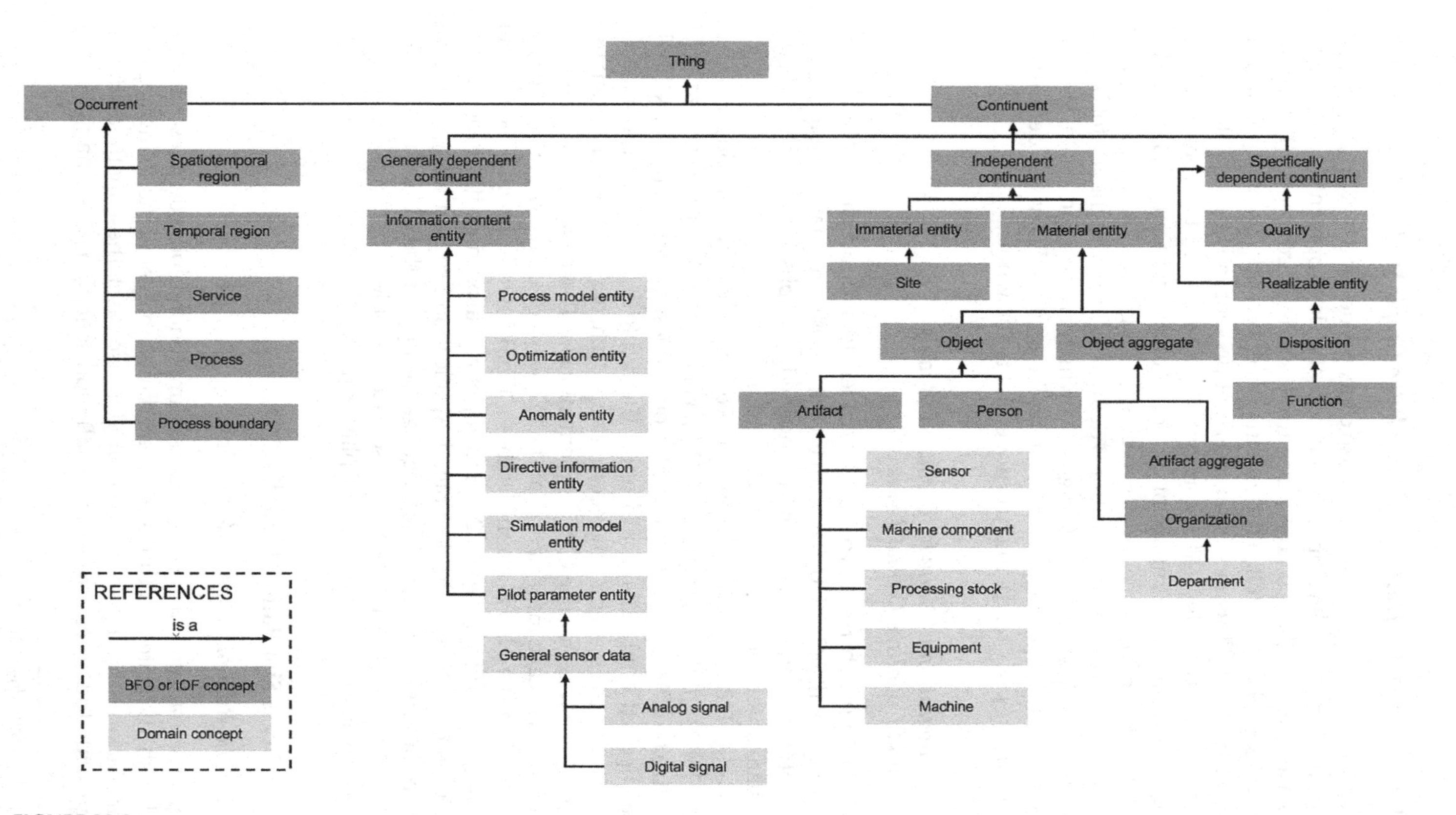

FIGURE 28.3
Ontology for developing knowledge graph models.

TABLE 28.1

Definitions for Ontology Concepts Presented in Figure 28.3

Ontology Concept	Definition
Process	Occurrent that has temporal proper parts and for some time depends on some material entity.
Process boundary	Temporal part of a process with no proper temporal parts.
Service	Implementation of a system function.
Spatiotemporal region	Occurrent entity that is part of space and time.
Temporal region	Occurrent entity that is part of the time as defined relative to some reference frame.
Generically dependent continuant	Continuant that depends on one or more other entities. Can exist in a multiplicity of bearers.
Independent continuant	Bearers or carriers of dependent continuants.
Specifically dependent continuant	Qualities, functions, roles, and dispositions of specific independent continuants.
Information content entity	An entity that can be copied and that is about something in the sense that it bears certain information.
Process model entity	Virtual concept used to define a process model.
Optimization entity	Virtual concept used to define an optimization concept.
Simulation model entity	Virtual concept used to define simulation model concepts.
Pilot parameter concept	Virtual entity to define use case parameters.
General sensor data	Concept used to model sensor data regardless of the use case.
Directive information entity	Specification that describes the inputs and outputs of mathematical functions and the workflow of execution for achieving a predefined objective.
Anomaly entity	Virtual entity to support anomaly detection.
Data analysis entity	Virtual entity used for data analysis.
Quality	A specifically dependent continuant that, in contrast to roles and dispositions, does not require any further process to be realized.
Realizable entity	A specifically dependent continuant that exists essentially or permanently in some independent continuant and is not a spatial region.
Disposition	A realizable entity which is such that, if it ceases to exist, then its bearer is physically changed.
Function	A disposition that exists in virtue of the bearer's physical make-up by coming into being.
Role	A realizable entity exists because the bearer is in some special physical, social, or institutional set of circumstances.
Immaterial entity	Boundaries and sites, bound or demarcated in relation to material entities.
Material entity	Independent continuants that can preserve their identity even while gaining and losing material parts.
Object	Material entity that manifests causal unity of one or many compound units.
Artifact	Object designed by some agent to realize a certain function.

(Continued)

TABLE 28.1 (*Continued*)

Definitions for Ontology Concepts Presented in Figure 28.3

Ontology Concept	Definition
Sensor	Device that produces an output signal to sense a physical phenomenon.
Processing stock	Artifact in an industrial site that corresponds to any material in the process of producing or manufacturing a finished product.
Machine component	Compositions required for constructing machines.
Machine	Physical system using power to apply forces and control movement to perform certain actions.
Equipment	Set of physical resources used in an operation or activity.
Person	Object that is a human being.
Object aggregate	Object where there is a mutually exhaustive and pairwise disjoint partition into other objects.
Artifact aggregate	Collection of artifacts that have been designed or arranged by some agent to realize a certain function.
Organization	Object aggregate that corresponds to social institutions and does something.
Department	Organizational unit for a given use case.

This digital model core can have two components: (1) a knowledge engineering component in the form of a knowledge graph model (see Section 2.1) that represents formal system knowledge and data in the form of rules, relations, associations, and predicates, and (2) a dynamic operational model of the production system. The Process Simulation Model also interconnects and interoperates with other architectural services, including optimization and analytic tools.

Ideally, the role of the central modeling component can be defined concerning the cognition process as follows:

1. **create a digital shadow** of the physical system (Schuh et al. 2019). The digital shadow will provide an accurate image of its operation at any time. It will be updated by getting real-time monitoring data from the analytics platform;
2. **model relevant production processes and entities** along the product's value chain;
3. **assess the performance** of the entire system through Key Performance Indicators (KPIs). Compare KPIs obtained from estimated and real-time data;
4. **provide systemic knowledge** (entities, relations, material flows, process states, performance) and operational data to the knowledge engineering component (e.g., knowledge graph) and the machine

learning tools while using the stored knowledge to make the model more intelligent;
5. **provide support** for root-cause analysis, risk analysis, and hypothesis testing to artificial intelligence models;
6. **provide use case models and data** to support optimization algorithms and tools;
7. **support system adaptability** by building, running, and assessing system adaptation scenarios.

28.2.2.2 Industrial Process Modeling Space

To model any system, its state space needs to be defined. This can be done by specifying the variables that govern the system's behavior concerning the metrics being estimated (Cameron and Hangos 2001). For example, in continuous manufacturing systems in process industries, a flow production method is used to produce (or process) products without interruption by constantly supplying raw materials while the manufacturing process is underway. The finished product cannot be unassembled to its original raw materials. Therefore, the state of such production systems changes continuously with time (Tsinarakis, Sarantinoudis, and Arampatzis 2022).

The main objective of process models is to develop a scientific understanding through the quantitative expression of the knowledge of a system by displaying what we know but may also show up things that we do not know. It is also ideal for testing the effects of changes in a system, making predictions, and helping in decision-making, including both tactical decisions by managers and strategic decisions by planners.

Process models developed in this context can be considered dynamic models. Their dynamism is grounded on the fact that key flows and process control settings (i.e., variable parameters) are continuously updated in near-real time, in connection to a real-time monitoring system, maintaining a digital shadow of the physical system. Flows and settings that are not monitored must be estimable from monitored flows with sufficient accuracy. Based on these flows and settings, the model can identify key system events and estimate the operating state of system processes and the overall status of the modeled system at any time, also enabling the prediction of the next probable state change event.

28.2.2.3 Methodological Basis for Modeling Industrial Production Systems

Typically, the structure of a model represents all the entities involved in its operation and a set of possible states in which the physical system can be found after certain sequences of events appear. Events define the mechanism for the change of state of the system and refer to interactions. The interactions can take place (1) between entities of the system and (2) between the system and the external environment. In addition to the structure of the model, its dynamic state must be defined. This represents the active state at any time and enables the set of possible upcoming events.

A process-flow representational schema, mathematically modeled as a directed Petri net (Toumodge 1995), can be considered the most suitable for modeling production systems. This schema has the advantages of simplicity and strict mathematical definition, enabling easy implementation of simulations, optimizations, and other computational procedures, on top of the Petri net. Production, pre-production, and post-production processes are modeled mathematically as *transitions*, while systemic input or output points, (e.g., warehouses, product outlets) are modeled as *places*. Material, energy, financial, or informational flows are modeled as *arcs*.

Key properties of flows are their unit volumes and their class or type. Processes have technologies and methods that encapsulate the Petri net transitions. These can be modeled simply as transition coefficients, linear relationships between input and output flows, or even complex algorithms (sub-models). The stepwise calculation can be carried out in forward or backward propagation on the directed network to estimate all unknown flows from a key (typically product) flow (or a small number of flows in very complex models). This allows for (1) estimating the resources (and costs) needed for any desired production volume; (2) estimating the production capacity; (3) monitoring a small subset of flows in a dynamic model application and estimating the rest; and (4) using an over-defined subset of monitored flows, to calibrate the model's transitions. This scheme can be applied in static mode, using flows aggregated over time, or in dynamic mode (live (as a shadow model of the physical system with flow monitoring) or in simulations).

28.2.2.4 Simulation of Industrial Production Systems

Simulations are required to develop advanced cognitive systems given their capacity to emulate certain cognitive functions of the human brain. Among such cognitive functions, we find the evaluation of consequences of considered actions or the understanding of causes and evolutionary histories of past events. Process models are used to conduct experiments (scenarios) and understand the system's behavior when certain variables change. The resulting outcomes are helpful for decision-making and evaluating alternative strategies for the operation of the system.

Simulations are performed on either:

- **system change scenarios**, built as part of the system adaptation support mechanism described previously;
- **contingency scenarios (hypothesis tests)**, built in response to feedback from the artificial intelligence tools;
- **optimization models**, either used by optimization algorithms or tools.

28.2.3 Optimization

The Optimization services support decision-makers in responding to problematic events that have risen (or may rise) in their everyday operations in a timely fashion. Considering that different production processes may have utterly different input and output and may require completely different modeling and solution approaches, there is a need for the Optimization services to be designed in a way that meets all varied requirements. The approach followed in the EU H2020 project FACTLOG, and its respective manufacturing use cases, was based on micro-services and structured on predefined templates. The Optimization service is informed of the production process structure and the pilot-specific data and information to enable the relevant Optimization micro-service capable of solving the requested problem.

The architecture enables the Optimization service to provide a set of functionalities through an API that enables interactions as

- *Configuration:* the instantiation is conducted in a pilot-specific manner, and internal updates and monitoring are performed during operation;
- *Management:* manages the interaction with the rest of the services, ensuring access to required data;
- *Feedback:* the output of the Optimization solution is transmitted to subscribed services so that they can consume the outcomes.

The Optimization service exposes an API to manage optimization tasks (e.g., reception, storage, and forwarding). Queues are used to manage the execution of the task, providing status and outcomes to interested parties.

28.2.4 Analytics Services

Analytics services enable a systematic computational analysis of available data. To that end, they leverage statistical, machine learning, and heuristic methods that allow the discovery of meaningful patterns in data to interpret the current state of affairs, identify root causes and expected future scenarios, and perform if-else analysis. The insights enable a deeper understanding of the ongoing processes and better decision-making.

28.2.4.1 Anomaly Detection

Anomaly detection attempts to identify outliers, previously unseen or rare events. Anomaly detection approaches vary based on the characteristics of the data and use case requirements. When considering time series, anomaly detection can be performed through univariate or multivariate anomaly detectors. Such anomaly detectors attempt to determine whether a particular point in time follows a collective pattern from previous and most recent points in the time series. Furthermore, supervised and unsupervised machine learning approaches exist. Most current methods leverage the ability of deep neural networks to work as feature extractors and apply some algorithm on top to determine whether the values in the time series are anomalous. Unsupervised approaches include training autoencoders that learn common patterns and allow finding unusual variations as anomaly candidates. Frequent use cases for anomaly detection on time series in manufacturing are predictive maintenance and proactive alerting regarding whether the processes are executed within usual parameters (e.g., stock or scrap levels or production speed). Anomaly detection for time series can be combined with knowledge graphs to contextualize such anomalies and understand their potential impact along with possible mitigation strategies to reduce adverse effects on the supply chain or production process. Similar anomaly detection approaches were developed for a wide range of types of data (e.g., images, useful for visual quality inspection in the manufacturing context (Kujawin´ska, Vogt, and Hamrol 2016; Schmitt et al. 2020)).

28.2.4.2 Variation Detection

Industrial processes are executed in dynamic environments, leading to variations in different process parameters. Such variations can be reflected in the quality of products. For example, if the pressure of a machine fluctuates above or below certain thresholds, uniform product quality may not be ensured. On the other hand, processes executed in a stable state are predictable and can be improved. Therefore, there is a strong need to detect and understand process variations. A considerable challenge in variation detection is that, in principle, all types of variations cannot be defined in advance. This fact requires a considerable level of autonomy within the variation detection procedure: the detection system should be able to deal with previously unseen or unknown situations.

This section describes an approach for variation detection inspired by a human-like cognition process. This process differentiates two cases:

1. *known changes*: there is no need to understand the changes but only ensure they are recognized. Regarding human psychology, this processing corresponds to *System 1* (*Fast Thinking* (Evans and Stanovich 2013)), which operates automatically and quickly, with little or no effort and no sense of voluntary control.
2. *unknown changes*: requires additional processing to understand causes and impacts. Regarding human psychology, this processing corresponds to *System 2* (*Slow Thinking*). It allocates attention to the effortful mental activities that demand it, including complex computations. Moreover, it enables complex and efficient processing of

complex situations. It considers sensed data to create digital models of observed behavior and support timely and precise decision-making to process such complex situations.

28.2.4.2.1 Fast Thinking

The first case is resolved using statistical methods, finding abnormal observation values (outliers). We applied three types of multivariate control charts: (a) Multivariate Exponentially Weighted Moving Average (MEWMA), (b) Hotelling, and (c) Principal Component Analysis (PCA) control charts.

MEWMA (Jeong and Cho 2012) is a multivariate extension of Exponentially Weighted Moving Average (EWMA) charts. It uses the same logic, but it applies to the multivariate case. The model of a MEWMA chart is given by $Z_i = \lambda \, \Lambda \, X_i + (1 - \Lambda) \, Z_{i-1}$, *for* $i = 1, 2, \ldots, n$, where

- Z_i denotes the ith EWMA vector;
- X_i denotes the ith product vector (values of parameters of the ith product);
- Λ represents the diagonal matrix $diag(\lambda 1, \lambda 2, \ldots, \lambda p)$, where p is the number of parameters; and
- λp is the depth of the pth parameter.

In theory, we can use different values λp for different parameters, but it cannot be done without domain knowledge. Therefore, usually, the same depth is used for all parameters, resulting in $\lambda 1 = \lambda 2 = \cdots = \lambda p = \lambda$.

The MEWMA control chart has advantages over the standard Hotelling T^2 control chart. First, it does not assume that the data comes from a multivariate normal distribution. Second, it can be used on autocorrelated data, although its performance will not be as good as when products are independent. Finally, it can also detect minor shifts in the process.

The Hotelling T^2 control chart (Sullivan 2008) is based on Hotelling T^2 distance. Hotelling T^2 distance is a measure that accounts for the covariance structure of a multivariate normal distribution. It may be considered the multivariate counterpart of the Student's statistic.

The PCA control chart is done by computing Principal Component Analysis on the dataset to reduce the number of features and then using the Hotelling control chart (Lin et al. 2021).

Since each of these three methods focuses on different characteristics of the (anomalous) data, we use the aggregation of the result of each method to improve the accuracy of finding abnormal observation values (outliers). In other words, this method aims to reduce the number of false positive alarms (anomalies). The voting principle decides whether a data point will be considered an anomaly. The highest confidence is achieved when all three methods provide the same result. Figure 28.4 depicts the so-called *process*

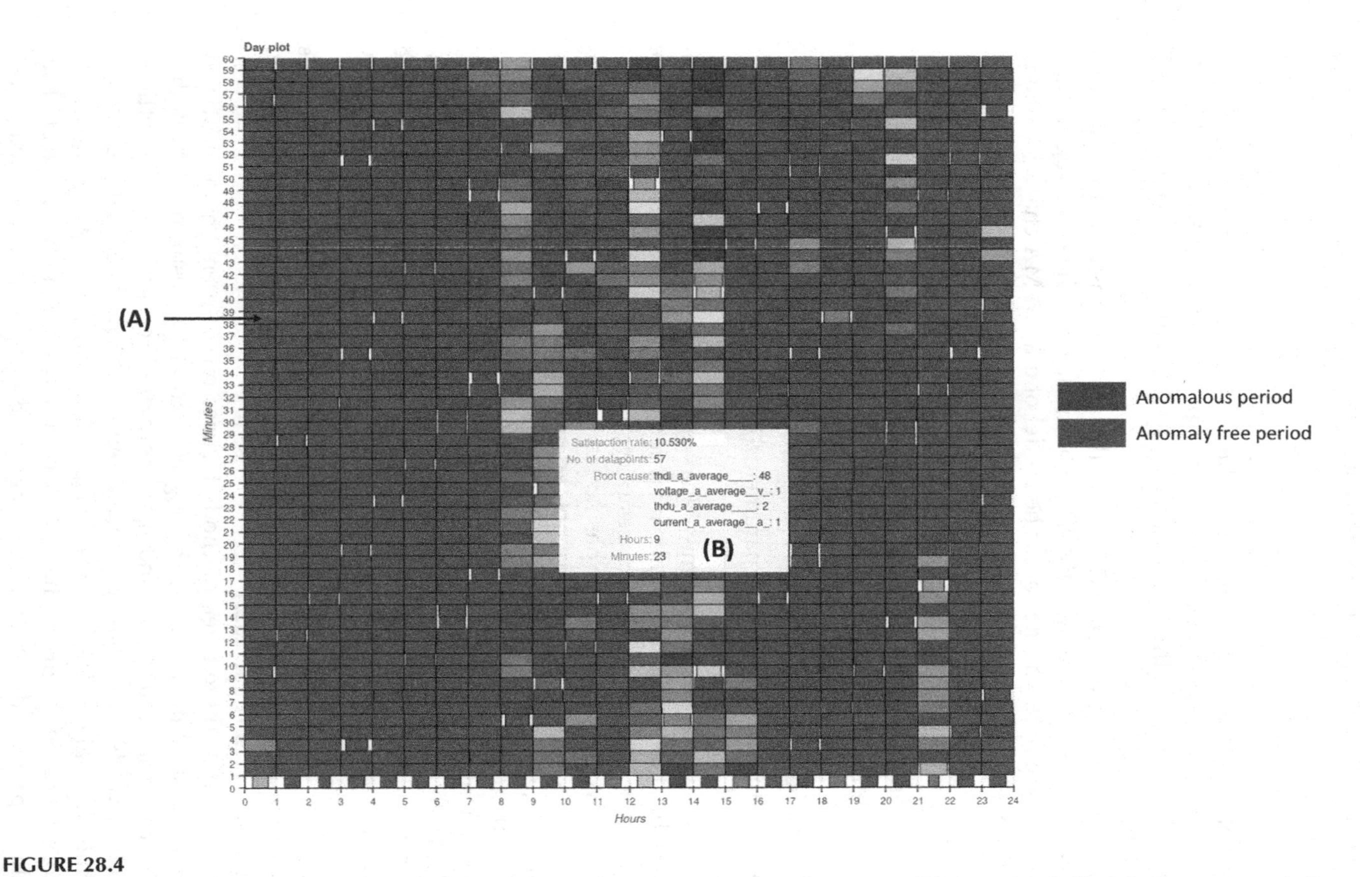

FIGURE 28.4
Process variation map. (A) indicates 1 minute, while (B) provides information regarding anomalies in a period. Black indicates anomaly-free periods, while the other colors signal certain anomaly levels. Strongly anomalous cases are colored in gray.

variation map, which illustrates the distribution of anomalies in a selected period (e.g., one day, as used in the figure).

28.2.4.2.2 Slow Thinking

Slow thinking can be resolved using autoencoder methods to find unexpected changes in one or several parameters. Furthermore, autoencoders enable us to find relationships among multiple variables, which is crucial in describing and understanding the behavior of a system and its cognition capabilities. Therefore, we consider these variations as unusualities (unusual system behavior).

Since autoencoders learn the average representation of a given dataset (Popovic, Fouch´e, and B¨ohm 2019), they fail to reconstruct anomalous data instances properly. Therefore, it will issue a higher reconstruction error for them, providing means to detect them without prior knowledge of the type of anomaly.

The above-stated approach works under the assumption that the process from which the data originated is stable (we can expect a few unusual data points). The stability of a process can be measured with the Stability Index method (Taplin and Hunt 2019; Yurdakul 2018). The abovementioned method should not be applied when the process becomes unstable.

28.2.4.3 Predictive Models

Models in a DT do not attempt to mirror everything about an entire system: they reflect only particular aspects relevant to the stakeholders (Batty 2018). Models can play three roles: (1) descriptive, (2) predictive, and (3) prescriptive (Eramo et al. 2021). Descriptive models represent current and past aspects of an existing system to facilitate understanding and analysis. On the other hand, predictive models predict future outcomes based on given input and provide information on unseen scenarios (Wagg et al. 2020). Such models can include analysis models (e.g., Petri nets), simulation models (e.g., Monte Carlo simulations), and machine learning models (e.g., classification and regression models). The predictive capabilities enabled by such models are considered a distinguishing feature of DTs (Tuegel et al. 2011; Wagg et al. 2020). Finally, prescriptive models are based on a solid theoretical foundation and leverage the outcomes of the descriptive and predictive models to guide decision-making. This way, the three models enable what-if analysis and reactive and proactive decision-making while reducing response times.

Regarding the analysis models, Petri nets are widely used to describe a system and analyze the system's behavior efficiently (Li et al. 2022). In the context of DTs, they have been applied to the shop floor and updated based on real-time data to drive change in the operating state (Zhuang et al. 2021). Friederich et al. (2022) report modeling failures through inhibitor arcs to ensure no transition arcs are fired until the failure is solved. Finally, semantic

Petri nets have been proposed to integrate semantic and temporal models for combined checking of rules concerning domain knowledge (Liu et al. 2022).

Simulation allows projecting the outcomes and performance regarding the operation of complex systems to enable decision support. In DTs, this is ideally done on the fly (Friederich et al. 2022). Data-driven simulation models can capture behaviors that are not easily foreseen and modeled in advance but can only be captured through data. Machine learning models can provide additional insights, leading to a deeper understanding of the system's behavior. Furthermore, in some cases, the machine learning models can leverage simulation data to learn to provide forecasts based on such data while avoiding costly data collection from the real world (Alexopoulos, Nikolakis, and Chryssolouris 2020). Finally, both simulations and machine learning models are expected to provide some quantitative assessment of the level of trust for the outcome produced (Wagg et al. 2020).

28.2.4.4 Situation Analyzer

The *Situation analyzer* service leverages insights obtained from the analytic components along with domain knowledge and heuristics to analyze the given state of the manufacturing plant and provide contextual and relevant information on how unexpected events disrupt the production process. Furthermore, the provided insights can be enriched with information regarding the expected magnitude of the impact of such events and suggested actions that can be taken to mitigate their consequences. Ideally, the service could be evolved toward automating decision-making for everyday, low-risk events. Furthermore, cognitive components could be developed to learn the most frequently taken mitigation actions for a given scenario and decide whether to suggest such actions to the user by providing an adequate explanation or even autonomously taking action in specific circumstances.

28.3 Cognitive Digital Twins – Implementation and Deployment

Common challenges when developing DTs relate to the accurate representation of the reference (physical) object and the contextualization of the DT through replications. Furthermore, given that DTs are based on data, decisions must be made on what type of data to collect, the data collection rate, where to store it, and how to merge disparate kinds of data (Suhail et al. 2021). To answer these challenges, the services mentioned above can be implemented in different contexts following a four-phase methodology

(Kalaboukas et al. 2021): (1) scope definition, (2) system modeling, (3) design and operation, and (4) rollout.

During the scope definition phase, we understand and model the business and functional requirements of the CDTs in the implementation context. More particularly, we create an operational model of the process where the CDTs will function, the underlying processes, and expected functionalities. To do so, we propose the following steps:

1. understand the operational aspects (*as-is process* vs. *to-be process*);
2. identify all involved stakeholders, systems, and assets participating in the new operational model;
3. decide on the target stakeholders consuming the CDT insights and the cognition level required to that end;
4. define the expected functionalities in the desired *to-be situation*. Such functionalities should be in line with the cognition process and stakeholders. It is critical to define how cognition will improve the *to-be process*. Three distinct moments related to the cognition process must be considered: (1) sensing, (2) understanding (through analytics, reasoning, and optimization), and (3) improving behavior;
5. define the cognition level required to address the expected functionalities. The cognition level will determine how each CDT will sense and process the information to provide the required insights. We must (1) understand the inter-relations and needs for information exchange for and among CDTs, (2) provide a detailed account of how analytics, reasoning, and optimization services are expected to meet the cognition requirements of the CDT, and (3) define how the CDT will act upon such insights.

The system modeling phase requires developing an ontology and knowledge graph to model the CDT in the DT Platform. Furthermore, security and privacy policies must be defined at each level of the CDT. The ontology and knowledge graph enable the creation of a *root DT (RDT)* representing the entire manufacturing system (e.g., at a supply chain or factory level). The subsequent CDTs model processes and assets related to the RDT and provide detailed insights into the systemic view enabled by the RDT. Furthermore, they acquire relevant data and properties through their relationship attributes to the RDT.

The basic CDT services are implemented during the design and operation phase. Four distinct moments must be considered:

1. **connect** to the required information sources. The development of such connectivity capabilities can require integrating with third-party software systems and developing *ad-hoc* connectors;

2. **implement** cognition services following the cognition model and needs identified in the Scope Definition phase;
3. **create** other domain-specific functionalities required for a particular context;
4. **integrate** functionalities through an operational storyline concerning the process and operations defined in the *Scope Definition* phase;
5. **iterate** back among the steps described above to enrich the CDT functionality and meet new requirements.

Finally, during the rollout phase, we use standard software engineering practices to deploy and monitor the CDTs' software and its operation. The use case and business requirements, along with the engineering expertise and culture, will determine the deployment process and choices regarding the deployment strategies.

28.4 Industrial Applications

28.4.1 Cognitive Services and Oil Refining Processes

The oil refining industry is considered among the largest energy-consuming process industry sectors worldwide and one of the largest sources of global greenhouse gas emissions (Gu et al. 2015; Lin and Xie 2015; Durrani et al. 2018). Furthermore, growing energy demand and increasingly restrictive environmental quality specifications incentivize the development of new approaches toward greater energy efficiency to (1) maximize the crude oil throughput, (2) reduce the cost of operation, and (3) reduce the overall carbon footprint of the oil refinement operations (Szklo and Schaeffer 2007; Han et al. 2015). Real-time monitoring and optimization are being adopted to achieve the abovementioned goals (Ramli et al. 2014; Subramanian, Moar, and Singh 2021). Furthermore, soft-sensing is being adopted where regular sensors would require high investments, represent maintenance costs, or the aspect of interest cannot be directly measured online (Wang, Shao, and Zhu 2019; Yang et al. 2021; Ro˘zanec et al. 2021, 2022). In this line, DTs can be used to mirror the ongoing oil refining process, provide alerts on current and future states of the distillation process, and perform multi-objective optimization considering (1) hard constraints regarding the expected quality of the product and (2) energy savings that directly impact the greenhouse gas emissions and overall cost of operation of the oil refinery (Teng et al. 2021; Min et al. 2019). Furthermore, the results of such optimization can then be transferred back to the physical process either automatically or with human supervision, based on the maturity of the DT implementation and the criticality of the change introduced into the process.

In terms of Optimization, the problem was formulated as a Mixed Integer Linear Program (MILP) that integrates network flow and blending constraints to identify the most energy-efficient combination of configurations for all process units of the Liquid Petroleum Gas(LPG) purification process to achieve on-specs LPG production (Eirinakis and Koronakos 2022). This approach was based on modeling the different operating conditions of each process unit as corresponding operational scenarios that result in a specific amount of impurity removal, LPG reduction, and energy consumed. The operating conditions of each process unit correspond to values that the process engineers can manipulate.

Hence, the Optimization module calculates the optimal on-specs recovery plan (i.e., the plan that minimizes energy consumption) by appropriately selecting (and returning) the proper operational scenario for each process unit. This operational scenario will correspond to a specific combination of operating conditions for each process unit. Then, the process engineers can manipulate these values (utilizing the corresponding Model Predictive Control of the processing unit) to apply the changes to the process units of the LPG purification process.

The discretization in the aforementioned operational scenarios may lead to many variables affecting solution time. A pre-processing step can be utilized to remove scenarios that are dominated by others. This can be achieved through a two-step procedure by computing first the convex hull of the operational scenarios (e.g., through the *Quickhull algorithm* (Barber, Dobkin, and Huhdanpaa 1996)) to identify the extreme points. Then, a technique based on dominance relationships is applied to derive from those extreme points only the dominant ones. These reduced operational scenarios are then introduced in the MILP model; thus, the optimization service can examine the different combinations of operating conditions in a detailed manner without affecting the overall performance of the overall service.

In this case, Optimization interacts with all other services through the operational scenarios. To obtain the scenarios based on historical data from each process unit, Machine Learning models are created (by the Analytics module) that model the behavior of each process unit (in terms of impurity removal, LPG reduction, and energy consumption). These models are then integrated within the Simulation module, which can be utilized to evaluate the performance of the LPG purification process as a whole or separately for each process unit. For the latter, the Simulation module has included functionality to produce all possible operational scenarios for any process unit of the Tu¨pras LPG purification process with a given step for each corresponding operating condition. These operational scenarios are then transferred via the FACTLOG platform to the Optimization module to be consumed. Finally, the optimization module utilizes them to model the corresponding on-specs recovery problem (i.e., the operational scenarios for each process unit).

28.4.2 Cognitive Services for Metal-Forming Process Analysis

BRC Ltd. was founded in 1908 and is the largest supplier of steel reinforcement and associated products in the United Kingdom, producing up to 2,000 tonnes of fabricated reinforcement for the construction industry per week (Tsinarakis, Sarantinoudis, and Arampatzis 2022). A decision-making platform is developed to provide cognitive services for the metal-forming process planning. When the BRC implements a metal-forming process based on the customers' requirements, the decision-makers need to design the metal-forming process from the perspectives of mission, operation, function, logic, and physical layout definitions. Through such definitions, the decision-makers can provide a solution for the metal-forming process and a solution for analyzing the dynamic performances of the metal-forming process.

This case considers a specific bay of one of the steel-reinforcing installations. This process involves cutting and shaping various diameters of the steel-reinforcing bars using manual or automatic operations. As a result, BRC produces final parts of various diameters in simple straight bars, "U" shaped bars up to complicated shape codes of 3D shapes with different numbers of bents according to customer needs.

To this end, we first used a semantic modeling approach to define the metal-forming solution from requirements to the physical layout using architecture models (She et al. 2021). Then, the models were generated into ontology models developed based on the Graph, Object, Point, Property, Role, and Relationship with Extension (GOPPRRE) ontology (Lu et al. 2021; Ma et al. 2022). Next, a model transformer generated the Petri net models from GOPPRE ontology models for implementing simulations (Meierhofer et al. 2021). Then, ontology models were imported into the decision-making platform to generate knowledge graph models and Petri net models implemented by the Process Simulation Model tool. Once integrated into the CDT, the services enable decision-making for the metal-forming process solution using knowledge graph reasoning and simulation results.

In particular, Optimization in the BRC case handles the flow shop scheduling problems for the coil and bar areas where given the orders in a specific time horizon. The goal is to find the optimal schedule for the predefined KPIs of minimization of makespan and total lateness of jobs, enabling a schedule that facilitates each job to finish as soon as possible while considering potential penalties for late jobs. The general problem can be considered a Flexible multistage flow shop problem with machine-dependent setup times with additional aspects that imitate the actual situation on the shop floor, with a notable example of the incorporation of cranes' movement times. The most relevant previous work on our problem appears in Benda et al. (2019). The authors proposed an elegant methodology for solving large flow shop scheduling instances by proposing a tree-based priority rule in terms of a well-performing decision tree for dispatching jobs. This was solved with a

Mixed Integer Quadratic Programming that lets the optimization consider that the crane movements can affect the job assignments among the machines.

28.4.3 Cognitive Services for Textile Industry

The Lanificio Fratelli Piacenza was founded in 1733 and is an established leading producer of woolen fabrics in their market segment. Suppliers provide them clean wool, which they weave into the fabric on their machine looms. Among the challenges they face are (1) loom setup to support appropriate warping and (2) reducing downtimes due to yarn breakage. As a supplier to a very dynamic and demanding market, Piacenza must continuously adapt to demand changes and therefore adapt the production plans to satisfy such demand changes. The anomaly and variation detection techniques have been applied to data provided by Piacenza. The data corresponds to sensor readings regarding the machines' energy consumption. The detection of anomalies regarding energy consumption allows us to search for the root causes of such anomalies and to propose mitigation measures. The fast and slow thinking presented in Section 2.4.2 is responsible for handling known variations (anomalies) or performing deep analysis of unknown situations, respectively. On the other hand, the optimization service solves the Weaving scheduling problem by solving the Parallel Machine Scheduling problem on unrelated machines with sequence-dependent setup times, job splitting, and resource constraints (Mourtos, Vatikiotis, and Zois 2021). Therefore, in this case, the optimization service provides novel and effective lower bounds and a three-stage heuristic for the makespan minimization problem identified (Avgerinos et al. 2022). Additionally, the optimization service enables the operators to derive informed decisions about creating policies for unexpected events and respond to them.

28.5 Conclusions

This chapter described the architecture and components required to implement CDTs. We presented the role of semantic technologies and the relevance of process simulation modeling, optimization, and analytic services. Next, we shared our experience regarding the methodology we followed to implement and deploy CDTs. Finally, we described the interplay of the above-described services to solve real-world use cases from the oil refining, metal forming, and textile industries. The architecture and services described in this chapter have been validated and tested at a prototype level, with real-world data provided by multiple companies. The CDTs we realized do not include an automated data flow from the digital object to the physical object. Furthermore, multiple components have not been integrated into a single platform at the time of writing. Although the CDTs have not been deployed to production, this paper shares our insights and many lessons learned.

Acknowledgment

The work presented in this chapter has been partially supported by the EU H2020 project FACTLOG (869951) – Energy-Aware Factory Analytics for Process Industries.

References

Alexopoulos, Kosmas, Nikolaos Nikolakis, and George Chryssolouris. 2020. "Digital twin-driven supervised machine learning for the development of artificial intelligence applications in manufacturing." *International Journal of Computer Integrated Manufacturing* 33 (5): 429–439.

Arp, Robert, Barry Smith, and Andrew D Spear. 2015. *Building ontologies with basic formal ontology*. Mit Press.

Avgerinos, Ioannis, Ioannis Mourtos, Stavros Vatikiotis, and Georgios Zois. 2022. "Scheduling unrelated machines with job splitting, setup resources and sequence dependency." *International Journal of Production Research* 61: 1–23.

Barber, C Bradford, David P Dobkin, and Hannu Huhdanpaa. 1996. "The quickhull algorithm for convex hulls." *ACM Transactions on Mathematical Software (TOMS)* 22 (4): 469–483.

Batty, Michael. 2018. "Digital twins." Environment and Planning B: Urban Analytics and City Science, 2018, 45(5): 817–820

Beckett, David, Tim Berners-Lee, Eric Prud'hommeaux, and Gavin Carothers. 2014. "RDF 1.1 Turtle." *World Wide Web Consortium* 18–31.

Benda, Frank, Roland Braune, Karl F Doerner, and Richard F Hartl. 2019. "A machine learning approach for flow shop scheduling problems with alternative resources, sequence-dependent setup times, and blocking." *OR Spectrum* 41 (4): 871–893.

Cameron, Ian T, and Katalin Hangos. 2001. *Process modelling and model analysis*. Elsevier.

Durrani, Muhammad Amin, Iftikhar Ahmad, Manabu Kano, and Shinji Hasebe. 2018. "An artificial intelligence method for energy efficient operation of crude distillation units under uncertain feed composition." *Energies* 11 (11): 2993.

Eirinakis, Pavlos, and Gregory Koronakos. 2022. "A mathematical programming approach for optimizing on-specs production for industrial processes under input uncertainty." *IFAC-PapersOnLine* 55 (10): 2822–2827.

Eramo, Romina, Francis Bordeleau, Benoit Combemale, Mark van Den Brand, Manuel Wimmer, and Andreas Wortmann. 2021. "Conceptualizing digital twins." *IEEE Software* 39 (2): 39–46.

Evans, Jonathan St BT, and Keith E Stanovich. 2013. "Dual-process theories of higher cognition: Advancing the debate." *Perspectives on Psychological Science* 8 (3): 223–241.

Francis, Nadime, Alastair Green, Paolo Guagliardo, Leonid Libkin, Tobias Lindaaker, Victor Marsault, Stefan Plantikow, Mats Rydberg, Petra Selmer, and Andrés Taylor. 2018. "Cypher: An evolving query language for property graphs." Proceedings of the 2018 international conference on management of data. 2018: 1433–1445. Association for Computing Machinery New York, NY, United States. https://doi.org/10.1145/3183713.3190657

Friederich, Jonas, Deena P Francis, Sanja Lazarova-Molnar, and Nader Mohamed. 2022. "A framework for data-driven digital twins for smart manufacturing." *Computers in Industry* 136: 103586.

Gharaei, Ali, Jinzhi Lu, Oliver Stoll, Xiaochen Zheng, Shaun West, and Dimitris Kiritsis. 2020. "Systems engineering approach to identify requirements for digital twins development." In *IFIP International Conference on Advances in Production Management Systems*, 82–90. Springer.

Gu, Wugen, Kan Wang, Yuqing Huang, Bingjian Zhang, Qinglin Chen, and Chi-Wai Hui. 2015. "Energy optimization for a multistage crude oil distillation process." *Chemical Engineering & Technology* 38 (7): 1243–1253.

Han, Jeongwoo, Grant S Forman, Amgad Elgowainy, Hao Cai, Michael Wang, and Vincent B DiVita. 2015. "A comparative assessment of resource efficiency in petroleum refining." *Fuel* 157: 292–298.

Jeong, Jeong-Im, and Gyo-Young Cho. 2012. "Multivariate EWMA control charts for monitoring the variance-covariance matrix." *Journal of the Korean Data and Information Science Society* 23 (4): 807–814.

Jinzhi, Lu, Yang Zhaorui, Zheng Xiaochen, Wang Jian, and Kiritsis Dimitris. 2022. "Exploring the concept of Cognitive Digital Twin from model-based systems engineering perspective." *The International Journal of Advanced Manufacturing Technology* 121 (9): 5835–5854.

Kalaboukas, Kostas, Joze Rožanec, Aljaž Košmerlj, Dimitris Kiritsis, and George Arampatzis. 2021. "Implementation of cognitive digital twins in connected and agile supply networks-an operational model." *Applied Sciences* 11 (9): 4103.

Kujawińska, Agnieszka, Katarzyna Vogt, and Adam Hamrol. 2016. "The role of human motivation in quality inspection of production processes." Advances in Ergonomics of Manufacturing: Managing the Enterprise of the Future: Proceedings of the AHFE 2016 International Conference on Human Aspects of Advanced Manufacturing, July 27-31, 2016, Walt Disney World®, Florida, USA. Springer International Publishing, 2016: 569–579.

Li, Hongcheng, Dan Yang, Huajun Cao, Weiwei Ge, Erheng Chen, Xuanhao Wen, and Chongbo Li. 2022. "Data-driven hybrid petri-net based energy consumption behaviour modelling for digital twin of energy-efficient manufacturing system." *Energy* 239: 122178.

Lin, Boqiang, and Xuan Xie. 2015. "Energy conservation potential in China's petroleum refining industry: Evidence and policy implications." *Energy Conversion and Management* 91: 377–386.

Lin, Shan, Liping Liu, Meiwan Rao, Shu Deng, Jiaxin Wang, Wenfan Zhong, and Li Lun. 2021. "A principal component analysis control chart method for catenary status evaluation and diagnosis." *Advances in Civil Engineering* 2021: Article ID 7703359.

Liu, Han, Xiaoyu Song, Ge Gao, Hehua Zhang, Yu-Shen Liu, and Ming Gu. 2022. "Modeling and Validating Temporal Rules with Semantic Petri-Net for Digital Twins." *arXiv preprint arXiv:2203.04741*.

Lu, Jinzhi, Junda Ma, Xiaochen Zheng, Guoxin Wang, Han Li, and Dimitris Kiritsis. 2021. "Design ontology supporting model-based systems engineering formalisms." *IEEE Systems Journal*.

Lu, Jinzhi, Xiaochen Zheng, Ali Gharaei, Kostas Kalaboukas, and Dimitris Kiritsis. 2020. "Cognitive twins for supporting decision-makings of internet of things systems." In *Proceedings of 5th International Conference on the Industry 4.0 Model for Advanced Manufacturing*, 105–115. Springer.

Ma, Junda, Guoxin Wang, Jinzhi Lu, Shaofan Zhu, Jingjing Chen, and Yan Yan. 2022. "Semantic modeling approach supporting process modeling and analysis in aircraft development." *Applied Sciences* 12 (6): 3067.

Meierhofer, Jürg, Lukas Schweiger, Jinzhi Lu, Simon Züst, Shaun West, Oliver Stoll, and Dimitris Kiritsis. 2021. "Digital twin-enabled decision support services in industrial ecosystems." *Applied Sciences* 11 (23): 11418.

Min, Qingfei, Yangguang Lu, Zhiyong Liu, Chao Su, and Bo Wang. 2019. "Machine learning based digital twin framework for production optimization in petrochemical industry." *International Journal of Information Management* 49: 502–519.

Mourtos, Ioannis, Stavros Vatikiotis, and Georgios Zois. 2021. "Scheduling Jobs on Unrelated Machines with Job Splitting and Setup Resource Constraints for Weaving in Textile Manufacturing." In *IFIP International Conference on Advances in Production Management Systems*, 424–434. Springer.

Musen, Mark A. 2015. "The Protégé Project: A look back and a look forward." *AI Matters* 1 (4): 4–12.

Popovic, Daniel, Edouard Fouché, and Klemens Böhm. 2019. "Unsupervised Artificial Neural Networks for Outlier Detection in High-Dimensional Data." In *European Conference on Advances in Databases and Information Systems*, 3–19. Springer.

Ramli, Nasser Mohamed, Mohamed Azlan Hussain, Badrul Mohamed Jan, and Bawadi Abdullah. 2014. "Composition prediction of a debutanizer column using equation based artificial neural network model." *Neurocomputing* 131: 59–76.

Rožanec, Jože Martin, Elena Trajkova, Jinzhi Lu, Nikolaos Sarantinoudis, George Arampatzis, Pavlos Eirinakis, Ioannis Mourtos, et al. 2021. "Cyber-physical LPG debutanizer distillation columns: Machine-learning-based soft sensors for product quality monitoring." *Applied Sciences* 11 (24): 11790.

Rožanec, Jože Martin, Elena Trajkova, Melike K Onat, Nikolaos Sarantinoudis, George Arampatzis, Blaž Fortuna, and Dunja Mladenić. 2022. "Machine-Learning-Based Soft Sensors for Energy Efficient Operation of Crude Distillation Units." In *2022 International Conference on Electrical, Computer and Energy Technologies (ICECET)*, 1–6. IEEE.

Schmitt, Jacqueline, Jochen Bönig, Thorbjörn Borggräfe, Gunter Beitinger, and Jochen Deuse. 2020. "Predictive model-based quality inspection using machine learning and edge cloud computing." *Advanced Engineering Informatics* 45: 101101.

Schuh, Günther, Christoph Kelzenberg, Jan Wiese, and Tim Ochel. 2019. "Data structure of the digital shadow for systematic knowledge management systems in single and small batch production." *Procedia CIRP* 84: 1094–1100.

She, Shiyan, Jinzhi Lu, Guoxin Wang, Jie Ding, and Zixiang Hu. 2021. "Model-Based Systems Engineering Supporting Integrated Modeling and Optimization of Radar Cabin Layout." In *IFIP International Conference on Advances in Production Management Systems*, 218–227. Springer.

Subramanian, Renganathan, Raghav Rajesh Moar, and Shweta Singh. 2021. "White-box Machine learning approaches to identify governing equations for overall dynamics of manufacturing systems: A case study on distillation column." *Machine Learning with Applications* 3: 100014.

Suhail, Sabah, Rasheed Hussain, Raja Jurdak, Alma Oracevic, Khaled Salah, Choong Seon Hong, and Raimundas Matulevi˘cius. 2021. "Blockchain-based digital twins: Research trends, issues, and future challenges." *ACM Computing Surveys (CSUR)* 54: 1–34.

Sullivan, Joe H. 2008. "Hotelling's T^2 Chart." *Encyclopedia of Statistics in Quality and Reliability* 2. Wiley.

Szklo, Alexandre, and Roberto Schaeffer. 2007. "Fuel specification, energy consumption and CO2 emission in oil refineries." *Energy* 32 (7): 1075–1092.

Taplin, Ross, and Clive Hunt. 2019. "The population accuracy index: A new measure of population stability for model monitoring." *Risks* 7 (2): 53.

Teng, Sin Yong, Michal Touš, Wei Dong Leong, Bing Shen How, Hon Loong Lam, and Vítězslav Máša. 2021. "Recent advances on industrial data-driven energy savings: Digital twins and infrastructures." *Renewable and Sustainable Energy Reviews* 135: 110208.

Toumodge, S. 1995. "Applications of petri nets in manufacturing systems; Modeling, control, and performance analysis [Book review]." *IEEE Control Systems Magazine* 15 (6): 93.

Tsinarakis, George, Nikolaos Sarantinoudis, and George Arampatzis. 2022. "A discrete process modelling and simulation methodology for industrial systems within the concept of digital twins." *Applied Sciences* 12 (2): 870.

Tuegel, Eric J, Anthony R Ingraffea, Thomas G Eason, and S Michael Spottswood. 2011. "Reengineering aircraft structural life prediction using a digital twin." *International Journal of Aerospace Engineering* 2011: Article ID 154798.

Wagg, DJ, Keith Worden, RJ Barthorpe, and Paul Gardner. 2020. "Digital twins: State-of-the-art and future directions for modeling and simulation in engineering dynamics applications." *ASCE-ASME Journal of Risk and Uncertainty in Engineering Systems, Part B: Mechanical Engineering* 6 (3): 030901.

Wang, Zheng, Cheng Shao, and Li Zhu. 2019. "Soft-sensing modeling and intelligent optimal control strategy for distillation yield rate of atmospheric distillation oil refining process." *Chinese Journal of Chemical Engineering* 27 (5): 1113–1124.

Yang, Dan, Xin Peng, Zhencheng Ye, Yusheng Lu, and Weimin Zhong. 2021. "Domain adaptation network with uncertainty modeling and its application to the online energy consumption prediction of ethylene distillation processes." *Applied Energy* 303: 117610.

Yurdakul, Bilal. 2018. *Statistical properties of population stability index*. Western Michigan University.

Zheng, Xiaochen, Jinzhi Lu, and Dimitris Kiritsis. 2021. "The emergence of cognitive digital twin: Vision, challenges and opportunities." *International Journal of Production Research* 60: 7610–7632.

Zhuang, Cunbo, Tian Miao, Jianhua Liu, and Hui Xiong. 2021. "The connotation of digital twin, and the construction and application method of shop-floor digital twin." *Robotics and Computer-Integrated Manufacturing* 68: 102075.

29

Development of the Digital Twin for the Ultraprecision Diamond Turning System and Its Application Perspectives

Ning Gou, Shangkuan Liu, David Christopher, and Kai Cheng
Brunel University London

29.1 Introduction

Ultraprecision machining (UMP) technology is an important indicator of a country's capability in precision engineering, and the associated technological research and development. It is defined as the ultimate capability of a manufacturing process in which removal or addition of materials at the nanometric and/or even lower scale (e.g. atomic scale) is achieved. Within this process, conventional cutting tools do not have the cutting-edge sharpness or the hardness to cope with the highly specific cutting energy and thus tool wear, while UPM uses single crystal diamond (SCD) tools for ultraprecision cutting, grinding or polishing [1].

In the late 1950s, under the demands from aerospace and defence industries, the United States pioneered the development of UPM by using the single-point diamond turning (SPDT) technology for the manufacture of gyroscope components and high-precision spherical and aspheric parts, etc. To date, UPM technology has evolved as a high-precision, machining-related technology developed to meet the manufacturing requirements of high-end cutting-edge and/or high throughput products including nuclear energy producers, ultra-large-scale integrated circuits, freeform optics such as vari-focal lenses, ICT (information and communications technology) hardware, head-up display (HUD) and light detection & ranging (LiDAR) devices for automotive industry, space optics for SpaceX, etc. Specifically, UPM techniques can often be categorised as ultraprecision cutting (diamond turning, diamond milling), ultraprecision grinding, ultraprecision polishing and ultraprecision non-traditional machining (e.g., electron beam figuring and ion beam figuring) [2].

 DOI: 10.1201/9781003425724-35

The development of UPM technology was initially focused on military and aerospace applications and has since evolved into other industry applications in the late 1970s. From the 1990s onwards, the rapid growth of automotive industry, renewable energy, medical devices, optoelectronics, consumer electronics, information and communications industries has led to a dramatic increase in the demand for UPM machines and the manufacturing applications such as aspherical optics, Fresnel lenses, ultraprecision moulds, disk drive reading/writing heads and semiconductor wafer dicing/polishing/processing. During this period, technologies related to ultraprecision diamond turning machines, such as controllers, encoders, sensors, laser interferometers, aerostatic bearing spindles and hydrostatic bearing slideways, direct drives had also become increasingly sophisticated and intrinsically integrated in the context of UPM systems, and UPM machines are commercially available towards ultraprecision production machines and facilities for precision engineering industries. Nowadays, the accuracy of such UPM machines is approaching the nanometric level.

As an industrial application exemplar, the use of an aspherical lens in the Kodak digital camera launched in the early 1980s attracted a great deal of attention from the industry, as it used one aspherical lens to replace at least three spherical lenses. It thus resulted in a miniaturised and lightweight optical imaging system, which could be used in a wide range of applications such as cameras, video recorders, industrial TVs, robot vision, CDs, VCDs, DVDs, projectors and other optoelectronic products. Since then, precision forming of aspherical lenses has become a R&D hotspot and roused great attention in the optics manufacturing industry [3]. Although the research and development emphases or focuses vary among countries as UPM technology is constantly updated and machining accuracy is improved with the times, the factors of promoting the development of ultraprecision machining are essentially the same. These factors can be summarised as the product quests for high quality, miniaturisation, high reliability, high performance, and industrial competitiveness. Nevertheless, UPM technology is not just at the forefront of modern manufacturing technology, it has also constantly been serving as one of key drivers for future manufacturing technologies and paradigms.

With the latest development in power electronics, direct drives, sensors and control technologies, the UPM machines and machining systems aim to continually achieve the higher machining accuracy in an industrial competitive manner. However, the constraints and bottlenecks for the further development of the UPM machines and systems are also becoming obvious. It is essential and much needed to develop an innovative approach to overcoming the nanometric-level hurdle in the context of industrial-scale ultraprecision manufacturing. The approach has to address the multiple factors in multiscale multiphysics, such as cutting mechanics and fundamentals, workpiece materials, machines and tooling design, fixtures, in-process monitoring, measurement and error compensations, in-process and/or real-time diagnosis of the machining conditions (e.g., machining temperature, vibrations, cutting forces, and cleanliness), etc. Furthermore, this often comes at the expense of machining efficiency and increased the technological complexity and costs. At the same time, real-time data monitoring and processing have always been challenging while addressing the need for nanometric-level motion accuracy and higher control resolution at the ultraprecision machining system.

Conventionally, the hardware and software of the UPM machine systems are integrated through encoders as the feedback devices in the systems, while the encoders are feedbacking the positioning data of the slideways and/or spindles but not the direct data of the tool and workpiece engaged, which may impose the limitations in acquisition of in-process real-time machining data and information. These kinds of information asymmetry phenomenon existing at the mechanical and electrical systems of UPM machines, may render the inconsistency between the designed performance and operation performance, reflected in static/dynamic stiffness, thermal stability and motion accuracy of the UPM system, which may also impose the fundamental limitation in achieving higher accuracy beyond the nanometric level. Currently, the next paradigm of the manufacturing industry is coming. The new paradigm is to improve product quality and accuracy through the digital twin (DT) technology of modern UPM technology. DT is widely regarded as the future of modern manufacturing, addresses bottlenecks in areas such as real-time machining process monitoring and diagnosis, and digital and physical data fusion at a higher precision level. Therefore, developing a DT for the UPM machine and/or machining system is an inevitable essential approach for future generation UPM machines and systems. Their design, manufacturing and control, combined with computationally efficient DT design and parallel optimisation algorithms will likely lead to the higher form/dimensional accuracy and finer surface roughness of the ultraprecision components/parts, in a more competitive industrial manner.

This chapter will present the feasibility and innovative application of using a DT for an ultraprecision SPDT machine system. The study and application described in the chapter will demonstrate the significance of developing and applying DT technology to ultraprecision manufacturing of optical components, and in particular those with freeform surfaces. The chapter will include kinematics and dynamics analysis, modelling and analysis of the machine system and the usage of Newton–Raphson method in solving complex equations for multibody dynamics simulations on the SPDT machine system. Automated Dynamics Analysis of Mechanical Systems (ADAMS) will be shown to be a useful tool in the development path of creating a DT within this application, and its ability to run multibody dynamics analysis, and its evaluation and validation. The effect of how contact forces affecting the toolpath will be investigated as well, while it also shows the limitation of using ADAMS only for this application. The chapter will be concluded with a further discussion on the potential and application of a DT-integrated UPM machine system, and how in-process data and information is acquired in real time both virtually and physically, and used to continuously optimise the machine system against the stringent UPM requirements.

29.2 Ultraprecision Diamond Turning System

29.2.1 Background of Ultraprecision Diamond Turning Technology

Diamond turning is defined as a process of ultraprecision mechanical turning of precision elements using natural or synthetic diamond-tipped tools. When turning, the workpiece is rotated and the diamond tool is traversed along X, Z, and/or C axes of motion to produce precise diameters and depths, as shown in Figure 29.1. It is widely used for machining high-quality aspheric optical components from crystals, non-ferrous metals, acrylics and other materials, as well as moulds for plastic optical components. With the assistance of CNC technology, precise control of the machine parts and the processing environment can be achieved, and thus, workpieces can be turned directly with complicated structure and surface quality.

Conventional turning process is limited by technologies such as motion control system and machine tools; it can only be used to machine rotationally symmetrical parts and thus it is impossible to meet the demand for high-precision complex contoured parts, especially those with freeform surfaces. The advantages of diamond turning over conventional turning process are thus obvious: nanoscale smooth freeform surfaces can be obtained under a single, reproducible turning process.

Diamond turning technology is to date relatively mature and has a wide range of applications, for example, the manufacturing of ultra-high-precision aspherical optical components under the combination with ion beam polishing technology, or production of inexpensive precision aspheric mirrors and lenses under the combination with hard carbon plating and epoxy replication, etc.

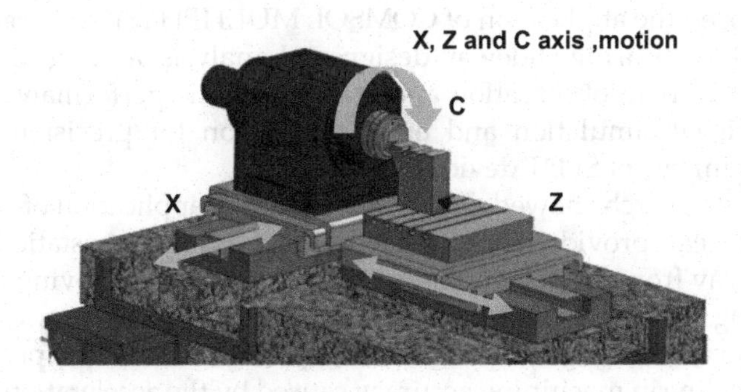

FIGURE 29.1
Motion system of diamond turning machine.

29.2.2 Design and Development of the Ultraprecision Diamond Turning System

The key components of an ultraprecision diamond turning system normally include a high-quality granite base, highly accurate motion system with aero/hydrostatic bearing slideways, linear motor and encoder, air bearing spindle system with electric motor and air chuck, and fast tool servo (FTS) or slow tool servo (STS) system depending on the cutting process. After decades of development, existing design and analysis methods for ultraprecision diamond turning system have matured in terms of computing and simulation capabilities. Technically, there is already a wide range of convenient and powerful commercial computer-aided design (CAD)/finite element method (FEM) software available on the market to support engineering design and analysis, which has directly led to the diversification and economisation of existing devices. Nevertheless, existing design and analysis methods are normally not able to take into account machining errors from multiscale and multiphysics sources; it refers to the transformation of the relative movement of the cutting tool and the workpiece in progress caused by the unbalance of the heat distribution and machining environment of the UPM equipment in the machining state, specifically reflected in the position error, shape error and surface roughness of the workpiece. It generally includes the geometric error, control error, thermal error, force, etc. In order to achieve ultraprecision patterns (2D) and shapes (3D), it is essential that the causes of apparent random errors in processing machines be analysed, upgraded, refined or replaced, i.e., virtually eliminated and that the systematic errors be minimised. Thus, it makes the intervention of the DT technology, with the capability of enabling real-time monitoring and prevention of data from multiple aspects, very relevant and meaningful.

The implementation of DT in design and analysis for UPM is still limited, but many researchers are already exploring active exploratory studies, for instance:

- Regarding motion system development, Gou et al. [4] in their article proposed the application of COMSOL MULTIPHYSICS–based DT in aerostatic bearing slideway design and analysis, aiming to achieve real-time data observation and static/dynamic performance analysis, visual simulation and error prevention for precision motion mechanisms of SPDT devices.

  This research showed a possibility that the application of DT technology can provide many practical solutions for aerostatic bearing slideway from implementation perspectives like improving the processing quality of UPM and optimising end-user experience. For instance, the algorithm preset in DT system can solve the problem of the decline in machining accuracy caused by the acceleration change of the slideway at the end of the running track; by recording and feeding back the usage data and slideway operating parameters, overload and excessive vibration can be interpreted as early warnings; thus, engineers are able to make corresponding optimisation through remote tuning or adjustment of control parameters; the recording data of the full life cycle of aerostatic bearing slideway can help R&D engineers to carry out product iteration and design optimisation.

- In terms of spindle system design, Liu et al. [5] presented a semi-physical simulation-based model for the design of electric spindle drive systems by their research. By designing a digital twin model of the spindle drive system, cost/scrap calculations can be realised in digital world, instead of destructive testing. The designed control system can be easily deployed on a digital signal processing chip to verify its performance in a time- and cost-efficient manner. Thus, the cost/scrap prediction and fault diagnosis of the physical entity are realisable.

Based on the high demands of building accurate multiscale and multiphysics models and breaking through the existing technical bottlenecks, incorporating reconfigurable product structure theory and incorporating with Industry 4.0 concepts especially E-manufacturing, DT and cyber-physical system (CPS) is thus the future research trend and development focus.

29.2.3 Industrial Needs and Enhancement through Digital Twin in the Context of Industry 4.0

Since the beginning of the 21st century, ultraprecision technology has been increasingly used as an enabler for design and manufacturing of high-precision freeform surfaced optics, devices and components particularly driven by the demands and development in precision engineering–related high-tech industries. As ultraprecision machining technology continues to advance, future ultraprecision machining systems are increasingly focused on the machining of micro structured surface features, non-rotationally symmetrical, complex freeform optical parts/devices and are further becoming a vital technology for the machining of high value–added optoelectronic products and optical communication products. Typical industrial applications of UPM include freeform surfaced HUD, vari-focal lenses, medical endoscopy devices and surgeon invasive micro tools, and even the space optics for next generation communications (e.g. SpaceX) and space activities. The multiscale modelling and analysis, and the associated advanced simulations and digital twin integration, have substantial potentials and impacts for extension of simulation from the product design stage to following-up lifecycle stages by in-process or remote tuning, error prediction, higher precision enhancement, continuous improvement of the product, etc. [6].

Nowadays the DT technology is used in a wide range of industrial scenarios, for example, in large production plants and factories to optimise working space layouts and improving safety, effectiveness and ergonomics [7], for production optimisation of the petrochemical industry [8], to improve the vertical and horizontal integration of automotive manufacturing systems[9], etc. However, there are still fewer for DT applications in UPM, as which represents the most advanced machining technology and material processing capability.

29.3 Digital Twin for the Ultraprecision Machining System: Implementation Perspectives

29.3.1 Digital Twin Infrastructure for the Diamond Turning Machine

The overall objective of DT is to identify conditions which are not yet optimal, through looking into every dimension which constitutes the model. This can have significant effects on the physical world in dimensions such as speed and cost, etc. The creation of a DT involves a physical model that can be fitted with an array of sensors, allowing information to be sent on critical points within the physical parts and its surroundings. There are two main types of sensors, which can be categorised as operational and environmental, with the capability to monitor factors that significantly affect the accuracy performance and lifetime of the diamond turning equipment. All sensory information is aimed to provide data that can be communicated with different systems within the manufacturing process, as shown in Figure 29.2:

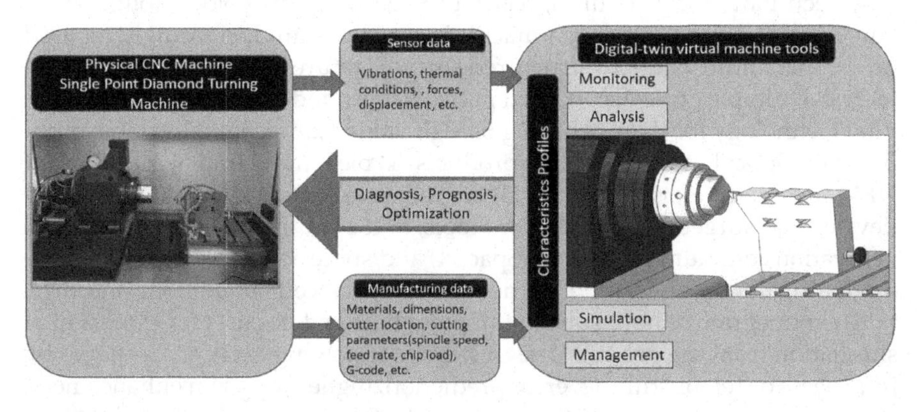

FIGURE 29.2
Digital Twin using sensory data and information integration.

29.3.2 Kinematics Analysis and Integration of Position Level

Through developing a multibody system in ADAMS on the road to DT, it has been seen that error predicting is a major part to the advantages to creating a virtual version of the physical machine. The use of ADAMS is employed for the demonstration of the motion study and to produce the tool path. However, it was validated that it is not just limited to providing high accuracy for the complexity of freeform surfaces, as it can also provide high-quality visuals and an array of outputs such as frictions (dynamic and static), any contact forces, linear and angular accelerations, displacements and velocities. Furthermore, it can benefit the process from being able to provide in-line measurement and insights can be looked at during the entirety of the cutting process.

The main parameters that have a profound effect on the surface finish are the cutting forces and tool wear. For cutting forces, this is down to the interference between the diamond tool and workpiece surface; in this scale, there are numerous influence factors which are involved, namely, feed rate and depth of cut, among others. Understanding the cutting forces can provide a better understanding of its behaviour and its effect. Hence, being able to monitor these contact forces in the physical system is hugely beneficial and can optimise the surface finish. Data retrieval will be done through the metrology systems which are already in use and have had vast amounts of research undertaken into them. The systems are the scanning probing method, optical detection method, microsensors, etc.

29.3.3 'In-Process' Data Retrieval from the Machining System

It is proved by the previous sections that the collection of critical data from the cutting process is essential for realising the communication between physical and digital world. In this aspect, the industrial-scale applications with the aid of advanced in-process monitoring and prediction models, algorithms and digital-enabling technologies for diamond turning system are thus required.

The degradation of surface quality during UPM processes is often attributed to dynamic instabilities. However, the existing methods for predicting instabilities in UPM processes are still in their infancy. In order to applicate the DT from this aspect, the relationship between different parameters and surface characteristics should be investigated using a combination of analytical modelling and real-time monitoring means. Furthermore, even under stable process conditions, the surface roughness of UPM machined workpieces varies considerably for different reasons, due to their complex chip formation processes, thermal effects, motion control accuracy and other uncontrollable factors. Therefore, by combining suitable virtual models and sensing data, it is possible to help select suitable and "stable" process conditions, observe variations in surface roughness while obtaining it at the nanoscale, and predict machining errors that are barely visible, thus reducing post-processing difficulties. In the following sections, some possibilities for in-process data retrieval from the UPM system will be discussed separately.

29.3.3.1 Using the Encoder Outputs within the Machine System

Optical encoders are high-precision positioning sensors based on the interference patterns produced by the relative movement between two gratings. One of the gratings is impressed on a scale, and the other is located in what is known as the scanning head. The encoder is the sensor of choice for the positioning of movable parts in machine tools. Its high accuracy and resolution, combined with its relatively low price, have led to a wide range of applications for position and speed measurement in position or motion control, particularly in UPM equipment.

According to the operating principle, the displacement, velocity and even acceleration for the motion parts of UPM system can be directly obtained during machining in real time based on the approach of encoders. These data provide a very visual representation of the relative movement between cutting tool and workpiece, which directly influences the machining quality. By integrating the collected data into the virtual model of DT, it is possible to observe the state of the motion system in real time throughout the machining process, thus to detect and predict machining defects in advance and to reduce repetitive post-processing works.

29.3.3.2 Using Cutting Forces Data from the Machining Process

Cutting force calculation, modelling and analysis has been always an important process indicator in UPM, which can collectively reflect the various cutting process phenomena and dynamics such as size effect, chip formation, energy consumption and cutting heat partition and the machining instability and chatter. It can also be correlated with the tool cutting performance particularly with the tool wear and tool life. Therefore, cutting force is seen as a key factor to optimise the cutting process variables and tool geometries in micro-cutting processes, and thus make it meaningful to use cutting forces data as an in-process monitoring aspect for the DT.

In ultraprecision and micro-cutting processes, cutting forces are in the range of 0.1–1 N, whereas in conventional machining they are typically hundreds or thousands of times higher. Thus, accurately measuring and analysing their values can be challenging, especially during the machining process. Regarding this, the dynamometer with small temperature error, high sensitivity and natural fraction is nowadays widely used as the dominant cutting force detection device in UPM, which can measure three perpendicular cutting force components simultaneously during machining and the measured numerical values can be stored in computer though the data acquisition system. For instance, by the study of Niu et al. [10], the Kistler dynamometer 9256C2 was used for measuring small force less than 0.002 N during the micro-milling process.

The use of cutting forces as a data resource of DT for in-process monitoring enables real-time detection of the variation such as energy input, environmental disturbances during machining and prediction of tool wear. From another aspect, the integration of pre-established cutting force models in DT enables the observation of unexpected changes of cutting force and thus the prediction of possible defects before post-processing and surface inspection work.

29.3.3.3 Using the Surface 'Signature' Data from the Component Ultraprecision Machined

Compared to the previous in-process monitoring method, which focuses on intermediate parameters, direct inspection of the 3D surface texturing and characteristics of the machined workpiece surface during processing is a relatively more intuitive method. The real-time monitoring and control of surface morphology variations in their incipient stages are vital for assuring nanometric range finish in UPM process.

To date, the method for surface monitoring of UPM are various, for instance, by the analytical approach of non-parametric Bayesian to capture the inherently complex, non-Gaussian and non-stationary sensor signal patterns observed in process [11], or use Acoustic Emission (AE) to collect the signals of the transient elastic waves generated from the rapid release of energy from one or more sources within the material [12], etc. These kinds of in-process monitoring methods have proved to be very beneficial to the UPM DT from application standpoints, for example, during the wafer dicing, which is a typical application scenario of UPM; wafer yield losses will be mitigated to a great extent, if the onset of UPM process drifts can be detected timely and accurately.

29.4 Application Case Studies on Ultraprecision Manufacturing of Freeform Surfaced Devices/Components

29.4.1 Background of Freeform Surface Components

Freeform surfaces, also known as Non-Rotationally Symmetric (NRS), can be defined as surfaces with no axis of rotational invariance (within or beyond the optical part) [13] It requires and leverages three or more independent axes (e.g., the C-axis in diamond machining) to create an optical surface with as-designed asymmetrical features. In practical terms, in the context of design, a freeform surface may be identified by a comatic-shape component or higher-order rotationally variant terms of the orthogonal polynomial pyramids, themselves independent so they can be "dialled-in" at will. These components often come together with an astigmatic surface component. In simple terms, in design, freeform surfaces go beyond spheres, rotationally symmetric aspheric, off-axis conics, and toroid. It is noteworthy that an off-axis conic is not a freeform surface because two axes are sufficient to fabricate and measure it. However, in some cases, making the conic with three axes is more practical. A toroid is a freeform in fabrication and metrology and may serve as a freeform base surface in design.

The freeform technological advances in freeform surfaces have enabled the design and fabrication of optical surfaces with more complex geometrical shapes and configurations, instead of the traditional ones with only the symmetrical revolution, such as conics and other aspherical surfaces. These new shapes and configurations represented in freeform surfaces can reduce aberrations, dimensions and weight of an optical system substantially better than conventional ones [14]. Due to the advantages such as good image quality, simplified instrumentation and cost-effective productivity, freeform optics has become a key component for optoelectronic and communication products.

Today, the emergence of freeform optics has permeated remote sensing and military instruments [15–18], energy research [19], transportation [20], manufacturing [21] and medical and biosensing technologies [22]. Freeform optics has promise in both refractive and all-reflective unobscured systems. They both benefit from high performance and compactness, while all-reflective approaches provide the advantage of lightweight and achromatic solutions.

29.4.1.1 Freeform Surface Machining

Since freeform surfaces lack an axis of rotational invariance, their fabrication requires more than two degrees of freedom typical for conventional methods, with material removal via a sub-aperture mechanism. These two common characteristics of freeform fabrication introduce challenges. Each additional degree of freedom adds error sources and increases the complexity of motion control. Sub-aperture removal increases the fabrication time and introduces mid-spatial frequency (MSF) errors. Thus, error sources occur across a wide range of spatial wavelengths from Figure 29.3 to surface roughness, with processes applied to a wide range of materials, ductile and brittle, reflective and transmissive. The sub-aperture interaction zone affects the mechanisms of material removal. The scientific understanding of the corresponding micro-cutting mechanics in freeform surface machining is essentially important, while constantly achieving the nanometric-level optical surface finishing, multiscale modelling and analysis is likely a useful technique on this combine with advanced algorithms spanned out from NURBS [23,24]. The main freeform fabrication processes are UPM, loose abrasive or bound abrasive finishing, moulding/replication and novel processes [25].

29.4.1.2 Freeform Surface Design

The community has long recognised the disadvantages of a serial design process in optical system development. Yoder quotes Johnson's conviction that "in the design of any optical instrument, optical and mechanical considerations are not separate entities, to be dealt with by different individuals" [26]. Kasunic documents that there can still be a chasm between optical and mechanical design tasks [27]. He states that "the lens designer's deliverable to the optomechanical engineer is a tolerance prescription with alignment and fabrication analyses developed in coordination with the optomechanical engineer; in practice, this is not always done". Johnson's and Kasunic's comments reflect a potential problem area and, indeed, a common challenge in developing state-of-the-art on-axis systems. Experience with off-axis, freeform techniques demonstrates that more than "coordination" between optical and optomechanical design is needed. The entire engineering process must be concurrent.

29.4.2 Ultraprecision Manufacturing of Vari-Focal Lenses

Over the last decade or so, a dramatic application change of freeform surfaces has occurred in design and manufacturing of vari-focal lenses also known as progressive addition lenses (PAL). Vari-focal lenses have been invented and used mainly for the compensation of presbyopia (the age-related loss of accommodative amplitude). The lenses are aspherical and their power could change smoothly from the far to the near prescription. Surveys conducted have shown that, from 2000 to 2015, the global prevalence of presbyopia has increased from 1.4 to 1.8 billion people [28], which means there is a huge potential market for vari-focal lenses. The demands for freeform optics also drive the development of UPM technology. Due to the complexity of freeform surface, the design of vari-focal lens is much more challenging and requires specific expertise compared with that of traditional ophthalmic lens. For lens manufacturers, the machining accuracy, productivity and unit cost in vari-focal lens manufacturing are key challenges. While eye features are unique for an individual customer, it means the prescription of vari-focal lenses for an individual customer is different and personalised. Therefore, mass customisation for personalised lens manufacturing is required, and the quality assurance, delivery time, costs, tracking and tracing of the lens product, and its personalised experience are essential for customers.

The method to generate vari-focal lenses surface is based on nonuniform rational B-splines (NURBS), which is an effective tool for designing a freeform surface, by using a polynomial to represent a curve and the surface allows the three-dimensional model to realise highly precise description of freeform surface, and further simplifies the adjustment of the lens surface according to different customer requirements. After the construct process, the tool path can be easily generated using algorithms or CAM software, and the surface will be further generated by ultraprecision machines with FTS. The whole manufacturing system can be integrated as follows:

Figure 29.3 illustrates the manufacturing system for vari-focal lenses and the role DT plays in it. The information and material flows of the traditional manufacturing system are marked in blue, where customer demands are divided into design requirements and processing requirements and integrated into the UPM process to be realised [29]. In this whole process, information transfer is in a linear/non-crossover system, as there is no reverse feedback between the customer and the "modelling, analysis and design" module, and no real-time data output from the machining process as well, which makes after-sales service and demand adjustment from end-user much more difficult. Based on this situation, the intervention of DT, which is marked in red, makes significant sense. The customer can observe and intervene in the design process through the virtual "twin", which presents the design data in real time, and through the in-process monitoring method mentioned in the above section, e.g., real-time surface observation, to obtain real-time information on the machining progress and quality, thus reducing the production cycle time of the vari-focal lenses, simplifying the quality control work, increasing productivity and improving customer satisfaction.

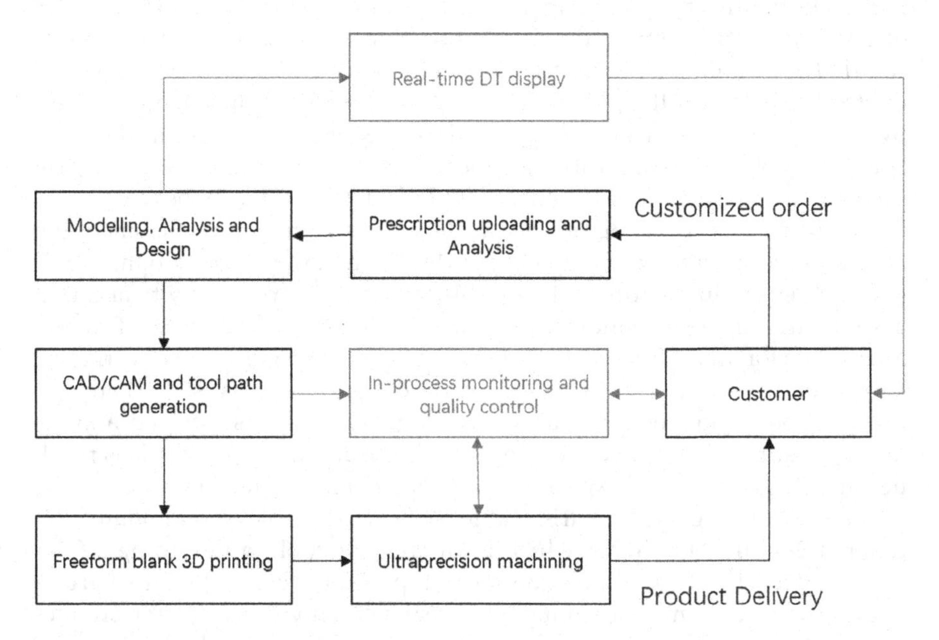

FIGURE 29.3
Vari-focal lenses manufacturing system information processing activities.

29.4.3 Ultraprecision Manufacturing of HUD Devices

HUD is the transparent display that presents data and is designed to enable end-users to read the information they need without taking their eyes off their point of view. The approach of HUD was limited in military and civil aviation applications in the early stage after invention due to the high processing costs. Recently, it is widely used in motor vehicles based on the maturity of the injection moulding technology [30]. As the optical image of a vehicle HUD is formed by reflections on the windshield, and each type of vehicle has its own unique form, it is thus necessary to adapt the image-forming optics uniquely for each variable type, which requires an effective approach in optical design and surface generation.

It is obvious that the HUD manufacturing process has similar limitations like vari-focu lens, although the demand for flexibility in customisation is not as much, the capability and effectivity of mass production are also limited by the rapid design word, data lagging and post-processing. Since the DT has been proved to have great potential in data transparency, in-process machining monitoring and post-processing optimisation, the approach of DT would definitely be the trend of academical research and industrial development for HUD.

29.5 Concluding Remarks

This chapter presents the innovation conception of a DT-based diamond turning machine system and its implementation and application perspectives. It will likely provide new insights into the development of future UPM machines and/or machining systems in the era of Industry 4.0. Although many breakthroughs have been made in the development of ultraprecision diamond turning machine systems, the DT application that is well integrated with ultraprecision machine systems is timely, while it still requires further research and development. General factors affecting the diamond turning machine system include machine tools, machining dynamics and fundamentals, machining parameters/conditions, tool geometry, material properties, chip formation, tool wear, environmental conditions, thermal stability, etc. There are complex intrinsic relationships among these factors, so advanced modelling, design and analysis methods are expected to particularly involve multiscale multiphysics modelling and analysis towards an integrated approach for the machine system design and development. DT has shown the ability to organically map and coordinate the functions, structures, behaviors, control, intelligence and performance of virtual and physical systems.. It further enables the modelling, monitoring and control of manufacturing processes at the machine system, based on the fusion of multidimensional in-process manufacturing data such as geometric deviations, material properties, working conditions, etc. However, there are still systematic gaps in developing multiscale multiphysics design methods for building multidimensional and high-fidelity DT effectively integrated with ultraprecision manufacturing systems. This should be the future research and development direction for applying DT in ultraprecision diamond turning machine systems, albeit work presented in this chapter has made the attempt and efforts along the line.

References

1. Taniguchi, Norio, et al., eds. *Nanotechnology: Integrated Processing Systems for Ultra-Precision and Ultra-Fine Products.* Oxford University Press, 1996.
2. Zhang, Shaojian, et al. "Advances in ultra-precision machining of micro-structured functional surfaces and their typical applications." *International Journal of Machine Tools and Manufacture* 142 (2019): 16–41.
3. Julong, Yuan, et al. "Review of the current situation of ultra-precision machining." *Chinese Journal of Mechanical Engineering* 43, no. 1 (2007): 35–48.
4. Gou, Ning, Kai Cheng, and Dehong Huo. "Multiscale modelling and analysis for design and development of a high-precision aerostatic bearing slideway and its digital twin." *Machines* 9, no. 5 (2021): 85.
5. Liu, Jizhi, Yuhou Wu, Liting Fan, Zhipeng Si, and Zhengwei Jia. "Current hysteresis control design of motorized spindle driven system based on semi-physical simulation model." 2020 Chinese Control And Decision Conference (CCDC). Chinese. IEEE, 2020: 1110-1115.
6. Al-Bender, Farid. *Air Bearings: Theory, Design and Applications.* John Wiley & Sons, 2021. Farid Al-Bender.
7. Kampker, Achim., Stich, Volker, Jussen, Philipp, Moser, Benedikt, and Kuntz, Jan. "Business models for industrial smart services-the example of a digital twin for a product-service-system for potato harvesting." *Procedia Cirp* 83 (2019): 534–540.
8. Min, Qingfei, Yangguang Lu, Zhiyong Liu, Chao Su, and Bo Wang. "Machine learning based digital twin framework for production optimization in petrochemical industry." *International Journal of Information Management* 49 (2019): 502–519.
9. Stark, Rainer, Carina Fresemann, and Kai Lindow. "Development and operation of digital twins for technical systems and services." *CIRP Annals* 68, no. 1 (2019): 129–132.
10. Niu, Zhichao, Feifei Jiao, and Kai Cheng. "Investigation on innovative dynamic cutting force modelling in micro-milling and its experimental validation." *Nanomanufacturing and Metrology* 1, no. 2 (2018): 82–95.

11. Rao, Prahalad, Satish Bukkapatnam, Omer Beyca, Zhenyu James Kong, and Ranga Komanduri. "Real-time identification of incipient surface morphology variations in ultraprecision machining process." *Journal of Manufacturing Science and Engineering* 136, no. 2 (2014): 021008.
12. Chen, Xuemei. *Monitoring and Analysis of Ultra-Precision Machining Processes Using Acoustic Emission*. University of California, Berkeley, 1998.
13. Fang F Z, Zhang X D, Weckenmann A, et al. "Manufacturing and measurement of freeform optics." *CIRP Annals* 62, no. 2 (2013): 823–846.
14. Huerta-Carranza, Oliver, Maximino Avendaño-Alejo, and Rufino Díaz-Uribe. "Null screens to evaluate the shape of freeform surfaces: Progressive addition lenses." *Optics Express* 29, no. 17 (2021): 27921–27937.
15. Chrisp, M., L. Petrilli, M. Echter, and A. Smith. "Freeform surveillance telescope demonstration." MSS Parallel Conference, MSS by BRTRC Federal Solutions under contract (Ed). 2019: 2769-2775.
16. Reimers, Jacob, Aaron Bauer, Kevin P. Thompson, and Jannick P. Rolland. "Freeform spectrometer enabling increased compactness." *Light: Science & Applications* 6, no. 7 (2017): e17026–e17026.
17. Geyl, Roland, Eric Ruch, Remi Bourgois, Renaud Mercier-Ythier, Hervé Leplan, and Francois Riguet. "Freeform optics design, fabrication and testing technologies for Space applications." International Conference on Space Optics—ICSO 2018. SPIE, 2019, 11180: 274-283.
18. Schiesser, Eric M., Aaron Bauer, and Jannick P. Rolland. "Effect of freeform surfaces on the volume and performance of unobscured three mirror imagers in comparison with off-axis rotationally symmetric polynomials." *Optics Express* 27, no. 15 (2019): 21750–21765.
19. Cui, Sifang, Nicholas P. Lyons, Liliana Ruiz Diaz, Remington Ketchum, Kyung-Jo Kim, Hao-Chih Yuan, Mike Frasier, Wei Pan, and Robert A. Norwood. "Silicone optical elements for cost-effective freeform solar concentration." *Optics Express* 27, no. 8 (2019): A572–A580.
20. Wei, Lidong, Yacan Li, Juanjuan Jing, Lei Feng, and Jinsong Zhou. "Design and fabrication of a compact off-axis see-through head-mounted display using a freeform surface." *Optics express* 26, no. 7 (2018): 8550–8565.
21. Liu, Yan, Yanqiu Li, and Zhen Cao. "Design method of off-axis extreme ultraviolet lithographic objective system with a direct tilt process." *Optical Engineering* 54, no. 7 (2015): 075102.
22. Yoon, Changsik, Aaron Bauer, Di Xu, Christophe Dorrer, and Jannick P. Rolland. "Absolute linear-in-k spectrometer designs enabled by freeform optics." *Optics Express* 27, no. 24 (2019): 34593–34602.
23. Sun, Xizhi, and Kai Cheng. "Multiscale simulation of the nanometric cutting process." *International Journal of Advanced Manufacturing Technology*, 47 (2010): 891–901.
24. Huo, Dehong. *Micro-Cutting: Fundamentals and Applications*. John Wiley & Sons, 2013. Editor: Kai Cheng. Brunel University, UK.
25. Hocken, Hocken R, Simpson J A, Borchardt B, et al. "Three dimensional metrology." *Annals of the CIRP* 26, no. 2 (1977): 403–408.
26. Yoder Jr, Paul R. *Opto-Mechanical Systems Design*. Boca Raton. CRC press, 2005. https://doi.org/10.1201/9781420027235
27. Kasunic, Keith J. *Optomechanical Systems Engineering*. John Wiley & Sons, 2015. Founded by Stanley S.ballard, University of Florida.

28. Fricke, Timothy R., Nina Tahhan, Serge Resnikoff, Eric Papas, Anthea Burnett, Suit May Ho, Thomas Naduvilath, and Kovin S. Naidoo. "Global prevalence of presbyopia and vision impairment from uncorrected presbyopia: Systematic review, meta-analysis, and modelling." *Ophthalmology* 125, no. 10 (2018): 1492–1499.
29. Liu, Shangkuan, Kai Cheng, and Liang Zhao. "Development of the framework of an e-portal driven personalized manufacturing system for freeform vari-focal lenses and its precision engineering implementation perspectives." *International Journal of Mechatronics and Manufacturing Systems* 16, no. 1 (2022): 1–21.
30. Ott, Peter. "Optic design of head-up displays with freeform surfaces specified by NURBS." In *Optical Design and Engineering III*, vol. 7100, pp. 339–350. SPIE, 2008.

30

Conceptualization and Design of a Digital Twin for Industrial Logistic Systems: An Application in the Shipbuilding Industry

Giuseppe Aiello, Islam Asem Salah Abusohyon, Salvatore Quaranta, and Giulia Marcon
Department of Engineering, Università degli studi di Palermo

30.1 Introduction

With the objective of improving operational efficiency and worker safety of production organizations, the European roadmap for 2030 (EC, 2019) is a substantial opportunity for achieving the targets of the Fourth industrial revolution and promoting the paradigm of the smart factory as the basis of next generation manufacturing systems. In such a context, the cooperation of intelligent manufacturing resources is a key enabler for an integrated production model featuring real-time production changes in response to the customers' needs and making human operators and logistics systems in a safer, more sustainable and more comfortable working environment. In such a view, logistic operations play a fundamental role in the efficiency of modern supply chains, since they are directly involved in enhancing the flexibility of production systems, while accounting for a relevant share of the production cost. As reported by many authors (Tompkins et al., 2003; Gamberi et al., 2009), a typical manufacturing company dedicates 25% of employees, 55% of all factory space and 87% of the production time to material handling operations which can thus account for up to 75% of an item's total production cost (Sujono & Lashkari, 2007). The impact of logistic operations on the overall efficiency of the production system is particularly relevant in Engineer-To-Order (ETO) supply chains where the high degree of customization and the complexity of the product imply a well-coordinated production system and an accurate management of the flows of information and materials. In particular, the shipbuilding industry is probably the most complex type of ETO manufacturing, because the production of unique (one-of-a-kind) products involves highly dynamic, uncertain and complex

DOI: 10.1201/9781003425724-36

manufacturing and logistics operations; therefore, in order to increase the efficiency, the logistic cost related to the material handling operations must be accurately addressed. In such a context, the Fourth industrial revolution is a substantial opportunity to improve the performance of the shipbuilding supply chain by introducing the model of the "digital Shipyard" or "Shipyard 4.0" (Ramirez-Peña et al., 2019, 2020), to replicate the concept of the smart factory in shipbuilding production. Shipyard 4.0 is an integrated and synchronized production environment where the flows of materials and information are timely coordinated with the manufacturing operations through the implementation of digital technologies. Shipbuilding is also a critical industrial sector in terms of workers' safety and health (Tsoukalas & Fragiadakis, 2016; Efe, 2019; Liu et al., 2022), as many manufacturing and logistics operations are associated with safety risks and hazardous events having the potential to seriously affect the safety and health of the workers. Recent research confirms how the outcomes of shipbuilding manufacturing processes (fumes, spark, asbestos) as well as dust, noise, vibration and volatile organic compounds (VOC) may adversely impact the health of the shipyard workers (Coggon & Palmer, 2016; Cherniack et al., 2004; Gillibrand et al., 2016; Malherbe & Mandin, 2007). Logistic and material handling operations are thus a substantial source of occupational accidents and injuries (struck by a motor vehicle, falling materials from cranes, etc.) (Barlas, 2012), and human factors are a most relevant cause of such events (Barlas and Izci, 2018; Crispim et al., 2020).

In such a context, the Digital Twin (DT) technology, originally introduced in 2002 by prof. M. Grieves at the University of Michigan and recently included in the Industry 4.0 paradigm, is spreading in modern manufacturing systems with the objective of improving their performance, sustainability and safety (Lee et al., 2019; Tao & Zhang, 2017; Parrot & Warshaw, 2017). Roughly speaking, a DT can be considered the virtual duplication of a complex system and includes the relevant data and inherent models designed describing its behaviour (Shao et al., 2019; Tao et al., 2019). In manufacturing systems, the DT refers to a digital model of the production processes and system based on real-time data collected from smart sensors to perform near-real-time analysis and control (Longo et al., 2021). Although frequently considered a modern surrogate of the traditional simulation models (Shao et al., 2019), the DT is a more advanced concept which implies the real-time bidirectional data exchange between the digital system and the physical world, thus offering the opportunity to design, test, and control real-world manufacturing processes. In addition the DT facilitates the seamless integration of collaborative systems enabling humans and robots to work together in shared spaces, and to interact in a more safe, sustainable and productive way.

This research focuses on the conceptualization and development of the DT of a logistic system for the shipbuilding industry, with the objective of improving efficiency and safety of the logistic operations involving the movement of metal plates and sheets, and prefabricated blocks by means of human controlled multi-wheeled hauling vehicles. This research specifically discusses the development of DT for shipyard logistics, based on collaborative transporters, equipped with autonomous guidance systems. Some relevant efforts towards the development of autonomous vehicles have recently been made in industrial logistics leading to the commercialization of automated forklifts (Tamba et al., 2009; Pradnya and Ganesh, 2014; Correa et al., 2010; Pagès et al., 2001) confirms.

In such a context, this research proposes a conceptual model and practical application of the DT of a logistic system for the shipbuilding industry, involving the replacement of the manually operated heavy-duty transporters with new generation AMRs, capable of operating in dynamic production environments. To accomplish such objectives, AMR systems must implement the typical features of cyber-physical systems, enabling them to sense the nearby environment, and to implement decision-making functionalities for navigating the surrounding environment autonomously, selecting the most efficient paths while safely avoiding obstacles and human operators. Within this general framework, the scope of this research embraces the following research objectives:

- Formulating a DT model of a logistics system architecture for the shipyard industry, improving safety and efficiency of operations.
- Establishing a preliminary development and validation approach for the design of a logistic system based upon autonomous transporters operating in an interconnected production environment.

30.2 The DT Framework

This section discusses the conceptualization of a DT framework for the shipbuilding logistic system, and focuses in particular on the material handling processes performed in the shipyard area during the production process. In such a context, a general material handling task is an operation involving the employment of a transporter in the movement of a sheet, a panel, a block, or any other raw material or semi-finished part from one department to another or to the dry-dock within the shipyard area. A general logistic task is thus an operation which originates from a service request from a manufacturing department and involves the assignment of an adequate transporter and the establishment of a trajectory to the destination department taking into account the available pathways, the other transportation operations currently being performed and the presence of obstacles that may lead to intersection or safety risks during transportation. In the DT framework proposed, once the service request has been issued by a department, specifying all the related attributes (source dept., destination dept., weight and dimensions of the part to be transported, etc.), the service workflow starts with the task assignment process which involves the definition of the mission goals related for the logistic service requested, the assignment of the start and destination pairs as well as the temporal constraints (i.e., when the vehicle is expected to be in the designated start and destination areas). The subsequent motion planning phase aims at defining how the vehicle should displace across the shipyard area considering the presence of obstacles conditioning the viability thus defining the trajectory envelopes containing the kinematically feasible paths. The coordination phase is responsible for further refining the trajectory envelopes by avoiding deadlocks and collisions with other vehicles and obstacles in the operating area. The trajectories thus obtained are finally transferred to the vehicle control system controlling the movement of the vehicle according to the trajectory planned. The vehicle control system is also capable of sensing the nearby environment and transmitting relevant elements (e.g. presence of obstacles) to the central system which updates the routes coherently. In the proposed DT framework, hence, operations related to the planning of the overall logistic system are carried out by a centralized management system, while the transporters participate in the process as decentralized units capable of autonomous decisions, and seamlessly interconnected with the central system in a common digital space. With such features, the transporter system configures as a cyber-physical system as described within the Industry 4.0 paradigm (Jeong et al., 2020), which is constantly interconnected to the central system and involves a sensing and actuation layer as well as decentralized decision-making capabilities. Based on such consideration, and coherently with the recent research (Agnusdei et al., 2021) the DT here proposed involves three fundamental elements: the digital space, the physical space and the interconnection layer, as depicted in Figure 30.1.

30.3 The DT Digital Space

The digital space is the virtual ecosystem where the digital counterparts of the physical objects coexist and interact with the centralized system for top-level decision-making and management. In the logistic system considered here, the digital representation of the transporter allows for the visibility of its position, operational status and task assigned. Based on such information, a state vector is assigned to each transporter, which includes its GPS coordinates, speed and service condition. At each new service request, the central decision-making system assigns the task to a specific resource and generates its motion path taking into account the efficiency of the system and the safety of operations. The path-planning phase relates to the problem of finding a geometric path from an initial position to a terminating position such that each intermediate state on the path is feasible (no collisions and violation

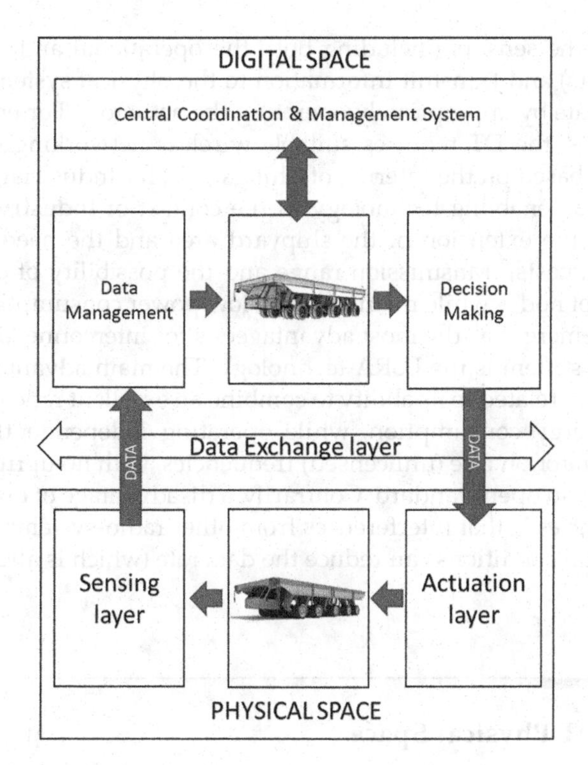

FIGURE 30.1
Structure of the Digital Twin of the shipyard logistic system.

of motion constraints such as road boundaries, etc.). Subsequently, a trajectory is determined, represented by a sequence of states visited by the vehicle, parameterized by time and velocity. The trajectory planning problem is thus concerned with the real-time planning of the actual vehicle's transition from one feasible state to the next, as well as the corresponding acceleration (or velocity). At each planning cycle, hence, the path planner module generates a set of possible trajectories from the vehicle's current location, and evaluates the optimal trajectory with respect to their length, duration and safety score referred to the probability of collisions with other vehicles, workers or obstacles.

30.4 The DT Data Exchange Layer

The data exchange layer is the middle layer of the DT model proposed, designed to bridge the physical and digital spaces by implementing suitable networking technologies and enabling the possibility to receive the data collected by the sensors (including both the operational and environmental parameters) and transmit information to the physical system in order to modify its state by appropriately operating the actuators. For enabling such functionalities, the DT requires suitable wireless networking communication systems based on the Internet of Things (IoT) for Industrial automation, which is a key enabling technology in the context of Industry 4.0. Taking into account the extension of the shipyard area and the need to combine performance, costs, transmission range and the possibility of connecting a multiplicity of nodes while maintaining a low power consumption, the technology that emerges as the most advantageous for interconnecting the shipyard logistic system is the LoRA technology. The main advantages of LoRa technology are related to its ability to combine an excellent ratio performance with low energy consumption, while operating independently from any external operator on free (unlicensed) frequencies, with no upfront licensing cost and with an open standard. Contrarily, a disadvantage of operating with open frequencies is that interferences from other radio-systems and adverse environmental conditions can reduce the data rate (which is already low).

30.5 The DT Physical Space

This section discusses the design of the autonomous material handling vehicle for the shipbuilding logistic operations to be employed within the DT framework described. The transporter considered is based on the manually operated multi-wheeled hauling vehicle equipped with a hydraulically operated lifting facility with load capacity of approx. 200 tons, and is able to reach a speed of 20 km/h when unloaded and 10 km/h at the maximum payload. The transporter is constituted by a frame structure with an engine power unit, a hydraulic lifting facility, a computer-controlled manoeuvring device, an air braking system and driving/control system with one or two driving cabins. The manoeuvring system involves a hydraulic turning system operated through a computerized controller enabling various manoeuvring modes such as conventional turning, oblique marching, tilting head or tail and circling from any radius. The system is also self-propelled by an Internal Combustion Engine (ICE), and, in more recent models, with a hybrid/electric propulsion. Coherently with the above described overall framework of the DT, the traditional transporter must be equipped with an autonomous guidance system and an automation layer to allow interaction with the nearby environment and communication with the digital space of the DT. According to the functionalities of guidance systems, their automation level is generally classified (Flemisch et al., 2010) into three levels. In the driver-assisted level, the vehicle is manually controlled by the driver and the automation system only provides visual, acoustic or haptic warnings. The semi-automated guidance level introduces the "longitudinal" control of the vehicle through an automated speed adaptation module, which is highly automated in which the automation controls the longitudinal and lateral movements of the vehicle. In this research, for economical, legal and psychological reasons, the most basic system has been chosen, and the functional architecture considered is depicted in Figure 30.2.

The perception module gathers data from the sensors and provides information concerning the nearby environment to the vehicle control module (VCM), which involves the supervision and decision-making functions of the system. The VCM, keeps track of the movement of the vehicle, checking the actual position of the vehicle against the pre-established trajectory received from the centralized control system and raises a warning when significant differences are detected. In addition, the VCM constantly gathers the alerts from the perception module concerning the presence of obstacles on the way, receives updates about the trajectory the centralized traffic control system and interacts with the driver (through the human–machine interface) who directly operates the vehicle. The most relevant specific functionalities of the modules implemented in the autonomous transporter are discussed below.

30.5.1 Vehicle Interconnection Module (VIM)

The VIM allows information exchange between the vehicle and the centralized control system of the DT enabling a bidirectional data flow between the vehicle and the centralized traffic control system, to dynamically adapt to the changes of the operating environment. The interconnection module involves a microcontroller and an integrated LoRa/GPS Shield. The microcontroller chosen is an Arduino Uno R3 based on ATmega328 AVR with 30 digital inputs/outputs (six of which can be used as PWM outputs and six can be used as analogue inputs). The LoRa/GPS shield employed is Lora BEE equipped with the SX1276/SX1278 transceiver with integrated GPS based on MTK MT3339 module. The embedded system allows to continuously retrieve GPS position of the vehicle and can provide up to three different frequency bands; 433, 868 and 915 MHz.

30.5.2 Human–Machine Interface (HMI)

The HMI enables the interaction between the driver and the vehicle, thus directly impacting the cognitive load and attention requested to the driver, and ultimately determining the resulting safety performance. In particular, the design of an HMI for a transport vehicle should allow the driver interaction with the system, while both his hands are engaged in manoeuvring and his sight is directed to the front. With such requirements, the design choice has been to implement a double interaction system involving a touchscreen and a voice control interface capable of translating the trajectory information from the centralized traffic control system into spoken messages broadcasted

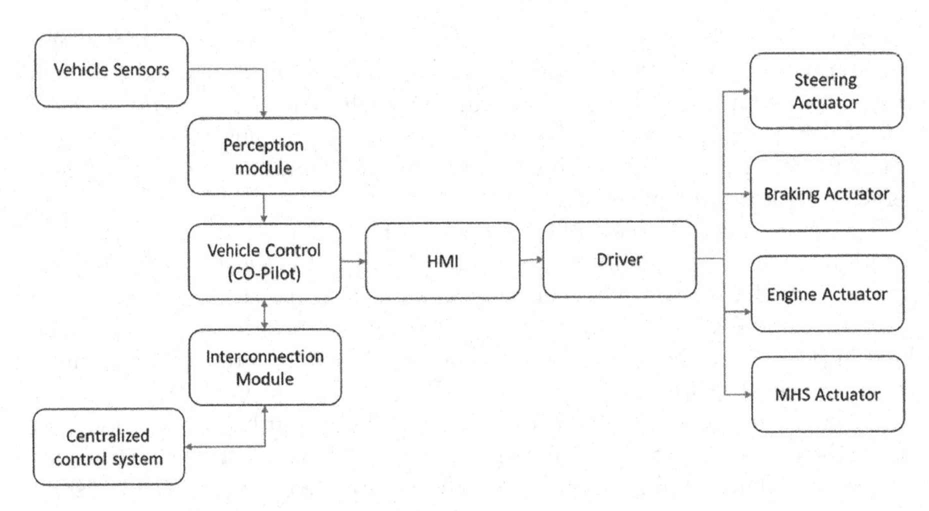

FIGURE 30.2
Architecture of the vehicle's assisted guidance system.

through the cabin loudspeakers and translating the voice messages from the driver into text messages and transferring their relevant information to the centralized system. Vocal interfaces have undergone substantial improvements in the last decade and have now reached a commercial maturity level due to the spread of assisted living and info-tainment systems for the automotive industry (Pfleging et al., 2012). Voice interfaces are also a very promising application field in industrial logistics and interesting research has been conducted for example on voice-controlled forklifts (Walter et al., 2015). While the reliability of voice recognition interfaces is now well recognized, a substantial drawback is their low accuracy in noisy environments.

The main technologies involved are speech-to-text (STT) (a.k.a. speech recognition) for the input part, and text-to-speech (TTS) translation for the feedback part. The information about the trajectory of the vehicle generated by the centralized system is thus visualized on the touchscreen in the driver's cabin, and also communicated to the driver via TTS. A TTS synthesizer is a system generating speech (acoustic waveforms) from text by breaking the input text into sentences subdividing them into tokens (such as words, numbers, dates and other types). These tokens are then converted to a sequence of natural words that can be pronounced by synthesizer. The STT process is designed to transform the acoustic signal of a spoken message into text. The system developed in this research employs the Sphinx-4 system for speech recognition, while the vocal synthesis is enabled by a TTS system freely available. Sphinx is an open source speech recognition system jointly designed by Carnegie Mellon University, Sun Microsystems Laboratories and Mitsubishi Electric Research Laboratories. Other TTS and STT systems are currently being developed by tech giants such as Google, Microsoft and IBM as free or paid services.

30.5.3 Vehicle Control Module (VCM)

The VCM is the system designed to supervise/control the vehicle operations and to ensure it is appropriately manoeuvred to follow the trajectory computed by the central planning algorithm. As discussed before, the specific functions of the VCM are strictly related to the level of automation of the vehicle: in low-automation scenarios the control of the vehicle is left to the driver, and the automation system only provides information about the nearby environment, while in high automated systems the control of the vehicle is entirely demanded to the automation and no human intervention is required. Despite the technological progress made in recent years, even in the automotive sector where the investments from commercial companies are more consistent, fully automated driving systems are not yet commercially mature. The most used classification was developed by Society of Automotive Engineers (SAE) and includes five levels from simple driving assistance (co-pilot function) to completely automated vehicle control (Harner, 2019). This research refers to the first development of a Variable Control System(VCS) for industrial applications in the shipbuilding industry where the system acts a co-pilot, which supervises the manoeuvres of the driver and provides information in order to improve the efficiency and safety of his operations. The scope of this research is thus limited to the development of a vehicle controller capable of receiving a trajectory from the centralized traffic management system and generating a sequence of steering and acceleration operations to be translated into voice commands for the driver. The VCM also involves a decision support system to raise specific warnings when:

1. Changes to the trajectory are communicated from the central system.
2. The vehicle significantly deviates from its assigned trajectory.
3. Significant safety issues are detected (e.g. obstacle avoidance).
4. Vehicle-related critical events are detected.

The first warning is triggered by a new communication event from the centralized system, and involves the VCM to update the current trajectory and to issue a message for the HMI module to inform the driver. To enable the second warning, the VCM module constantly compares the local GPS data with the assigned trajectory, and when the difference exceeds a pre-established threshold, the warning event is triggered and sent to the HMI. The third warning is instead triggered by a specific obstacle detection function based upon the definition of a "warning" and a "safety" zone around the vehicle. The presence of obstacles in the safety zone also triggers an autonomous "safe vehicle stop" function. The last early warning message refers to the vehicle state and is related to the detection of an anomaly or a failure in a critical component of the vehicle. The trigger event, in this case, is the activation of an alarm in the vehicle cockpit. Further developments of the VCM module are mainly addressed to the introduction of predictive/forecasting functions capable of detecting incipient risks and originating different operational states (e.g. regular, cautious/degraded and emergency/critical) as in similar referenced systems (Xu et al 2021).

30.5.4 Sensing Layer and Perception Module

The sensing layer provides the vehicle (and the driver) with real-time information about the operating environment. It is designed to improve or augment the operator's capabilities by means of sensors such as: CCTV cameras, ultrasonic, radar, Radio Frequency Identification (RFID), etc. The sensing layer implemented in the vehicle considered in this research involves the systems described below.

GPS tracking system: The GPS tracking system is based on the commercial GPS module previously described and is designed to constantly track the vehicle trajectory through the determination of the instant position and the current speed vector. Such information is constantly transferred to the VCM module to detect divergences from the trajectory assigned to the vehicle.

Vision system: The vision system implements four FLIR Blackfly colour cameras for rear, right, front and left directions. The cameras are equipped with a 1/1.2″ sensor for a maximum resolution of 1,900×1,200 at 41 frames per second. F2.4 lenses are coupled to the cameras, yielding a horizontal field-of-view of 75°. The information from the vision system can be transferred to the centralized traffic control system when an object is detected that prevents following the assigned trajectory.

Obstacle detection: Obstacle detection plays a critical role in enhancing the safety of the transport system. The system is based on ultrasonic obstacle detection technology which is now mature and widely employed in the automotive and industrial sector and consists in transmitting acoustic waves at a frequency between 25 and 50 kHz (above the human hearing range) at a pre-established time length and spacing. The presence of an obstacle is then detected through the variations in the reflected pulses which are converted into electrical signals and passed to a signal processing system for their interpretation.

30.5.5 Actuation Module

The actuation module controls all the actuators of the vehicle and has the fundamental function of translating the manoeuvres requested by the driver acting on the pedals and steering wheel into actual movements of the transporter. Automated steering is operated through the generation of an electronic signal similar to the one generated by the encoder connected to the steering wheel in manual systems. Each wheel of the carrier has an angular sensor, allowing the electronic control of the target steering angles by a Programmable Logic Controller(PLC). The throttle pedal is the actuator controlling the power request to the engine through a PLC, the pedal is an accelerator position sensor (APS) which is part of a closed loop throttle position system which reads the feedback for the Throttle position sensor (TPS) and directs the power request to the electric or the ICE engine, both PLC controlled.

The shipyard transporter is finally equipped with a compressed air braking system constituted by a braking cylinder, a pedal and a brake drum. The main element of the system is the braking air cylinder, which ensures safety braking if a compressor breakdown should occur.

30.6 System Validation

This section discusses the results of an experimental study where the functionalities of the semi-autonomous transporter described above have been validated through a sequence of use-cases each one involving the generation of a logistic service request and corresponding trajectory. The study took place at the shipyard of Palermo (Italy), which is constituted by 11 functional areas distributed on a surface of approx. 1,000 m^2. In each use case analysed, a real service request has been initially considered and a corresponding trajectory has been generated and transmitted to the vehicle control system. The driver then executed the service according to the spoken instructions received by the VCM through the HMI. An example of the trajectory generated referring to the transport request for a sheet metal from the raw materials storage area to the mechanical workshop at the shipyard of Palermo (Italy) is represented in Figure 30.3.

The use case allowed verifying if the VCM correctly translated the trajectory into a list of commands, and that the commands were correctly spoken by the HMI. Also, the correct functioning of the interconnection module was tested and off-track warnings as well as obstacle detection were successfully validated. The automation system installed also allowed analysing the utilization profiles during vehicle operations in 11 tests for a maximum duration of 1 working day (560 minutes), and the corresponding results are reported in Figure 30.4 in terms of travelled distance as a function of time. This analysis showed that the maximum travelled distance per working day is around 25 km, and the minimum is approx. 10 km, and the utilization profiles can actually vary substantially.

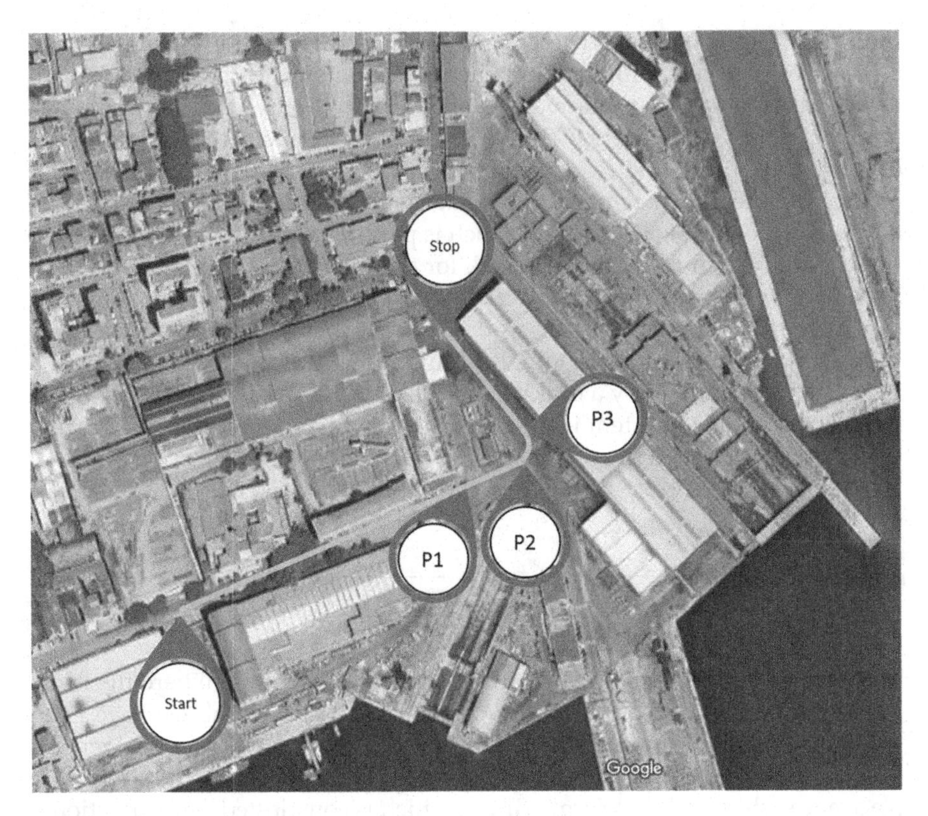

FIGURE 30.3
The generated trajectory representation.

30.7 Conclusions

As the competition among manufacturing organizations increases, the efficiency of logistic operations is necessary for achieving superior supply chain performance. The DT technology adopted within the industry 4.0 paradigm can be in such regard a fundamental game changer for the development of more efficient smart production systems while preserving the safety of operations. This is particularly true in Engineering-to-Order(ETO) and Open and Closed Loop Production(OKP) systems such as the shipbuilding industry where an integrated production environment with collaborative manufacturing and logistic resources represents a fundamental goal for the achievement of Shipyard 4.0. This research focuses on the development of a DT of a logistic system for the shipbuilding industry, based on the employment of autonomous moving vehicles and consisted in conceptualizing and formulating

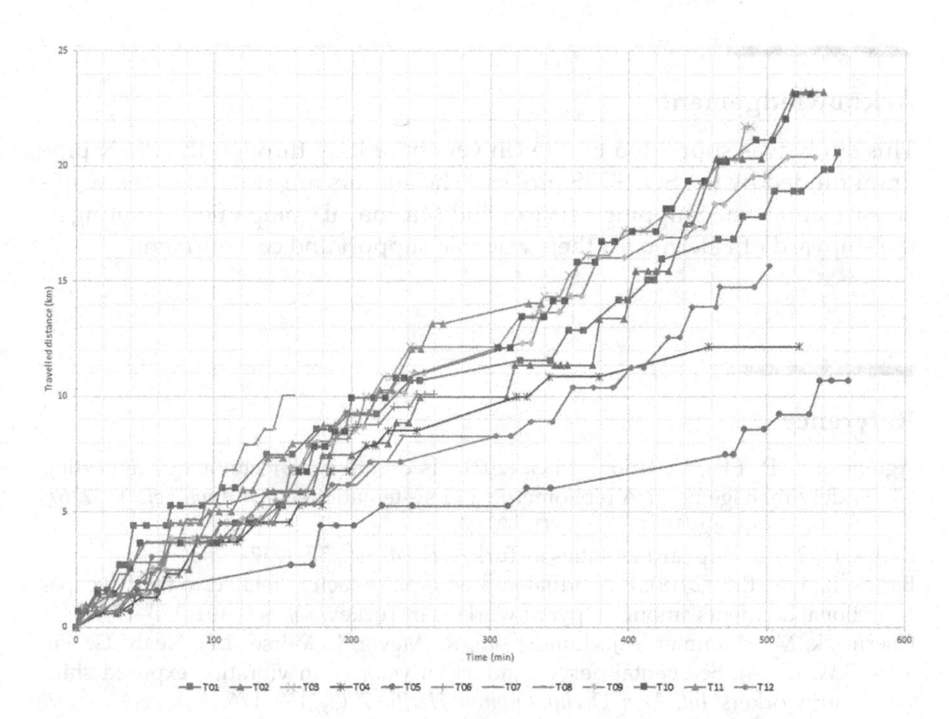

FIGURE 30.4
Utilization profile analysis based on the travelled distance.

a DT framework coherent with the relevant literature, and in developing a prototype starting from traditional manned heavy-duty transporter typically employed in shipbuilding. The experimental analysis allowed to validate the design assumptions and to gather significant data for determining the utilization profile of the vehicle and the characteristics of missions performed. Despite the relevance of the achieved results, the validation of the final objectives is still linked to the ability to integrate the proposed model of the smart shipyard logistic management system into the general framework of Shipyard 4.0, considering the lack of a standardized reference framework. With such substantial elements of uncertainty, future developments will focus on the deployment of suitable artificial intelligence and machine learning methods for analysing the data gathered by the vehicles (position, speed, load, etc.) in order to optimize the trajectory envelopes taking into account spatial and temporal constraints and to plan the manoeuvres of all the vehicles with the objective of minimizing the safety risks related to the material handling operations. Further implementations will also extend the scope of the DT framework to the maintenance operations and to the interaction with other smart resources in the shipyard.

Acknowledgements

This study was supported by the University of Italy through the PRIN program funded by the SO4SIMS project.. The authors are grateful to the workers of Tecnonaval, Shipping Services and Marinaval Companies operating in the shipyard of Palermo for their valuable support and commitment.

References

Agnusdei, G.P., Elia, V., Gnoni, M.G. (2021). Is digital twin technology supporting safety management? A bibliometric and systematic review. *Appl. Sci.*, 11, 2767. https://doi.org/10.3390/app11062767.

Barlas, B. (2012). Shipyard fatalities in Turkey. *J. Saf. Sci.*, 50, 1247–1252.

Barlas, B., Izci, B.F. (2018). Individual and workplace factors related to fatal occupational accidents among shipyard workers in Turkey. *Saf. Sci.*, 101, 173–179.

Cherniack, M., Brammer, A.J., Lundstrom, R., Meyer, J., Morse, T.F., Nealy, G., Fu, R.W. (2004). Segmental nerve conduction velocity in vibration-exposed shipyard workers. *Int. Arch. Occup. Environ. Health*, 77(3), 159–176.

Coggon, D., Palmer, K.T. (2016). Are welders more at risk of respiratory infections? *Thorax*, 71(7), 581–582.

Correa, A., Walter, M.R., Fletcher, L., Glass, J., Teller, S., Davis, R. (2010). Multimodal interaction with an autonomous forklift. In: *Proceedings of the ACM/IEEE International Conference on Human-Robot Interaction*, 2010, pp. 243–250.

Crispim, J., Fernandes, J., Rego, N. (2020). Customized risk assessment in military shipbuilding. *Reliab. Eng. Syst. Saf.*, 197, 106809.

Efe, B. (2019). Analysis of operational safety risks in shipbuilding using failure mode and effect analysis approach. *Ocean Eng.*, 187, 106214.

European Commission, Directorate-General for Internal Market, Industry, Entrepreneurship and SMEs, A vision for the European industry until 2030: Final report of the Industry 2030 high level industrial roundtable, Publications Office, 2019. https://data.europa.eu/doi/10.2873/102179.

Flemisch, F., Nashashibi, F., Rauch, N., Schieben, A., Glaser, S., Gerald, T., Resende, P., Vanholme, B., Löper, C., Thomaidis, G., Mosebach, H., Schomerus, J., Hima, S., Kaussner, A. (2010). Towards highly automated driving: Intermediate report on the HAVEit-joint system. In: *Proceedings of the 3rd European Road Transport Research Arena*, Brussels, Belgium, November 2010, pp. 1–12.

Gamberi, M., Manzini, R., Regattieri, A. (2009). A new approach for the automatic analysis and control of material handling systems: Integrated layout flow analysis (ILFA). *Int. J. Adv. Manuf. Technol.*, 41, 156–167. https://doi.org/10.1007/s00170-008-1466-9.

Gillibrand, S., Ntani, G., Coggon, D. (2016). Do exposure limits for hand-transmitted vibration prevent carpal tunnel syndrome? *Occup. Med.*, 66(5), 399–402.

Harner, I. (2019). Venture for America Fellow. The 5 Autonomous Driving Levels Explained. Retrieved from https://www.iotforall.com/5-autonomous-driving-levels-explained/. (Accessed 11/12/2021).

Jeong, Y., Flores-Garcia, E., Wiktorsson, M. (2020). A design of digital twins for supporting decision-making in production logistics. In: *Proceedngs of Winter Simulation Conference (WSC)*, 2020, pp. 2683–2694. https://doi.org/10.1109/WSC48552.2020.9383863.

Lee, J., Cameron, I., Hassall, M. (2019). Improving process safety: What roles for digitization and Industry4.0? *Process Saf. Environ. Prot.*, 132, 325–339.

Liu, Y., Ma, X., Qiao, W., Luo, H., He, P. (2022). Human factor risk modeling for shipyard operation by mapping fuzzy fault tree into Bayesian network. *Int. J. Environ. Res. Public Health*, 19, 297. https://doi.org/10.3390/ijerph19010297.

Longo, F., Padovano, A., Nicoletti, L., Elbasheer, M., Diaz, R. (2021). Digital twins for manufacturing and logistics systems: Is simulation practice ready? In: *Proceedings of the 33rd European Modeling & Simulation Symposium (EMSS 2021)*, pp. 435–442. https://doi.org/10.46354/i3m.2021.emss.062.

Malherbe, L., Mandin C. (2007). VOC emissions during outdoor ship painting and health-risk assessment. *Atmos. Environ.*, 41(30), 6322–6330.

Pagès, J., Armangué, X., Salvi, J., Freixenet, J., Martí, J. (2001). Computer vision system for autonomous forklift vehicles in industrial environments. In: *9th Mediterranean Conference on Control and Automation*, MED'2001, Dubrovnik, Croatia.

Parrot, A., Warshaw, L. (2017). Industry 4.0 and the digital twin: Manufacturing meets its match. In *A Deloitte Series on Industry 4.0, Digital Manufacturing Enterprises, and Digital Supply Networks*. Deloitte University Press, New York, pp. 1–17.

Pfleging, B., Schneegass, S., Schrnidt, A. (2012). Multimodal interaction in the car-combining speech and gestures on the steering wheel.Proceedings of the 4th international conference on automotive user interfaces and interactive vehicular applications. 2012: 155–162.

Pradnya, T.C., Ganesh, R. (2014). An autonomous industrial load carrying vehicle. *Adv. Electr. Electron. Eng.*, 4(2), 169–178.

Ramirez-Peña, M., Abad Fraga, F.J., Sánchez Sotano, A.J., Batista, M. (2019). Shipbuilding 4.0 index approaching supply chain. *Materials (Basel, Switzerland)*, 12(24), 4129. https://doi.org/10.3390/ma12244129.

Ramirez-Peña, M., Sánchez Sotano, A., Pérez-Fernandez, V., Salguero, J., Abad, F., Gomez-Parra, A., Batista, M. (2020). 1. Supply chain 4.0 in shipbuilding industry. In J. Davim (Ed.), *Manufacturing in Digital Industries: Prospects for Industry 4.0*. De Gruyter, Berlin, Boston, MA, pp. 1–22. https://doi.org/10.1515/9783110575422-001.

Shao, G., Jain, S., Laroque, C., Lee, L.H., Lendermann, P., Rose, O. (2019). Digital twin for smart manufacturing: The simulation aspect. In: *Proceedings Winter Simulation Conference*, December 2019 (Bolton 2016), pp. 2085–2098.

Sujono, S., Lashkari, R.S. (2007). A multi-objective model of operation allocation and material handling system selection in FMS design. *Int. J. Prod. Econ.*, 105(1), 116–133.

Tamba, T.A., Hong, B., Hong, K.S. (2009). A path following control of an unmanned autonomous forklift. *Int. J. Control Autom. Syst.*, 7(1), 113–122.

Tao, F., Qi, Q., Wang, L., Nee, A.Y.C. (2019). Digital twins and cyber-physical systems toward smart manufacturing and Industry 4.0: Correlation and comparison. *Engineering*, 5(4), 653–661.

Tao, F., Zhang, M. (2017). Digital twin shop-floor: A new shop-floor paradigm towards smart manufacturing. *IEEE Access*, 5, 20418–20427.

Tompkins, J.A., White, J.A., Bozer, Y.A., Frazelle, E.H., Tanchoco, J.M.A., Trevino, J. (2003). *Facilities Planning*, 3rd edn. Wiley, New York.

Tsoukalas, V.D., Fragiadakis, N.G. (2016). Prediction of occupational risk in the shipbuilding industry using multivariable linear regression and genetic algorithm analysis. *Saf. Sci.*, 83, 12–22.

Walter, M.R., Antone, M., Chuangsuwanich, E., Correa, A., Davis, R., Fletcher, L., et al. (2015). A situationally aware voice-commandable robotic forklift working alongside people in unstructured outdoor environments. *J. Field Robot.* 32, 590–628. https://doi.org/10.1002/rob.21539.

Xu, W., Sainct, R., Gruyer, D., Orfila, O. (2021). Safe vehicle trajectory planning in an autonomous decision support framework for emergency situations. *Appl. Sci.*, 11, 6373. https://doi.org/10.3390/app11146373.

31

Digital Twin Applications in Electrical Machines Diagnostics

Georgios Falekas, Ilias Palaiologou, Zafeirios Kolidakis, and Athanasios Karlis
Democritus University of Thrace

Digital Twins are a reimagined concept newly explored in many areas of research, as it is a broad context of virtualization of a product or procedure for online monitoring and control. This idea is closely expressed by the industrial condition-monitoring sector and especially that of electrical machines. To explain its usage in this department, a closer look at the definition is taken, to highlight parallel goals and attempt to aid the establishment of a concrete definition in literature and industry.

31.1 Digital Twin Definition in Maintenance

The original definition expanded to include not only products but also any process with a physical or real-world aspect inside the concept of the DT from the core in maintenance. This idea took shape when it was reinvestigated by the National Aeronautics and Space Administration (NASA) in 2012, birthing the DT concept as it is known to date (Glaessgen and Stargel 2012). Important review work (Tao et al. 2019) and literature consensus (Fuller et al. 2020; M. Liu et al. 2021) define five dimensions into the concurrent DT: Physical, Virtual, Data, Service, and Connection.

Representation of the physical system in the virtual space is realized via a process called twinning or mirroring. A physical element, a virtual representation of it, and the bi-directional data connections that feed data from the physical to the virtual representation and information from the virtual representation to the physical comprise the system. The distinction between "data" and "information" is important to distinguish between the two. The two representations are linked throughout their entire lifecycle, meaning

DOI: 10.1201/9781003425724-37

that the virtual system is not a static representation. A complete lifecycle consists of:

- Creation or Design;
- Production or Manufacture;
- Operation and Sustainment or Service;
- Disposal or Retirement.

This core concept of the Digital Twin envisaged a system that couples physical entities to virtual counterparts, leveraging the benefits of both the virtual and physical environments to the benefit of the entire system. Product information is captured, stored, and evaluated in what is called product-embedded information (PEI). This enables learning techniques, such as machine learning and pattern recognition, to be applied to the current, as well as future products of similar or varying nature. This process in essence enables the application of a knowledgeable, data-driven approach to the monitoring, management, and improvement of a product throughout its lifecycle.

ISO 23247-2:2021 is one of the few standards pertinent to DTs and has recently been published, incorporating concurrent concepts. In literature, various definitions are given according to the authors' insight and within the scope of their prevalent work domain. A number of definitive, inclusive, and synergistic definitions are repeated here, with notable contributions in bold:

> Tao et al.: *"the digital twin is a* **multiphysics, multiscale, probabilistic, ultrafidelity** *simulation that reflects, in a* **timely manner**, *the state of the corresponding twin based on the* **historical data, real time sensor data**, *and physical model"* (Tao et al. 2019).
>
> Gabor et al.: *"the digital twin is a special simulation, built based on the* **expert knowledge** *and real data collected from the existing system, to realize a more accurate simulation in* **different scales of time and space**" (Gabor et al. 2016).
>
> Zhuang et al.: *"virtual, dynamic model in the virtual world, that is fully consistent with its corresponding physical entity in the real world and can* **simulate** *its physical counterpart's* **characteristics, behaviour, life, and performance**, *in a timely fashion"* (Zhuang, Liu, and Xiong 2018).
>
> Liu et al.: *"the digital twin is a* **living model** *of the physical asset or system,* **which continually adapts** *to operational changes based on the collected online data and information and can* **forecast** *the future of the corresponding physical counterpart"* (M. Liu et al. 2021).

Contributions of these definitions to the industrial DT paradigm are discussed along with the characteristics. The authors proposed a focused definition in the context of electrical machine condition monitoring utilizing the previous definitions' key points as follows:

> The Digital Twin is an organic multiphysics, multiscale, probabilistic simulation that can represent the physical counterpart of a system in

real-time, based on the bi-directional flow and complete volume of PEI, encapsulating the full lifecycle data to facilitate knowledge sharing and integration.

(Falekas and Karlis 2021)

31.2 Digital Twin Characteristics

Definition of the basic idea of the DT branches into the different characteristics an approach should include to be considered as a DT. These characteristics are and should be further discussed, since they comprise the base tools and processes the DT framework (DTF) provides, with the aim of staying true to the definition as it has been embraced by industry.

Four major keywords frequently seen in DT literature, along with their definitions, include:

- Multiphysics – different system descriptions must cooperate with each other;
- Multiscale – simulation must adapt to required accuracy of the application;
- Probabilistic – core purpose is to predict the future of a system, given an initial state;
- Ultrafidelity – offer the application-required precision down to the best achievable level, given the hardware.

These four keywords represent the pillars of any engineering procedure. It is important to keep in mind that representation of any real-world system will never be perfect, owing to physical limitations, and the DT is no exception to the rule. Rather, the DT concept offers a promising paradigm to become the best representor to date. The classic trade-off between time, cost, and precision remains. One novel concept offered is that this trade-off is included under the multiscale and ultrafidelity characteristics of the DT and is able to be included as an input, optimizable parameter to the application.

Information is the basis of the DT. It is retrieved from the physical entity and processed to create the virtual twin through the twinning process. Then, it is constantly, bi-directionally exchanged to maintain the link between the two spaces, enable control of the physical entity, and update the virtual twin. This information can be tapped into at any time to realise ancillary procedures such as condition monitoring or making predictions under varying assumptions. In addition, information is separated and filtered to provide services to end users of varying specialization, while preserving confidentiality according to industry standards. Therefore, sources, handling, and storage are of paramount importance. Information is extracted from raw data retrieved from the physical entity and is stored as structured data in hardware. This data comes from three major sources:

- Historical data;
- Real-time sensors;
- Models.

Historical data temper expectations and aid in setting the information retrieval experiment, calibrating sensory equipment, and making basic predictions. These data are currently included in maintenance reports, time series, or tables, but are aimed to be included in the DT of systems to aid in the advancement of future DTs. Sensors applied on the physical entity feed the DT with real-time information, and constitute the actual links forming one part of the Connections' dimension. Finally, information simulated from tried-and-tested models can be considered as actual information embedded in the DT.

DTs are ultimately made of components, down to the lowest possible level. Each physical system is analysed to a fidelity. For example, a permanent magnet synchronous electrical machine model may consist of the copper wiring, stator and rotor laminations, permanent magnet blocks, screws and miscellaneous components. These are characterized by their geometric orientation in 3D space, the material properties, physical interactions, and environmental conditions. Depending on the method or quality of information retrieval and modelling, the Black–Grey–White Box paradigm aids in understanding the various approaches in twinning these components in the virtual space, as depicted in Figure 31.1.

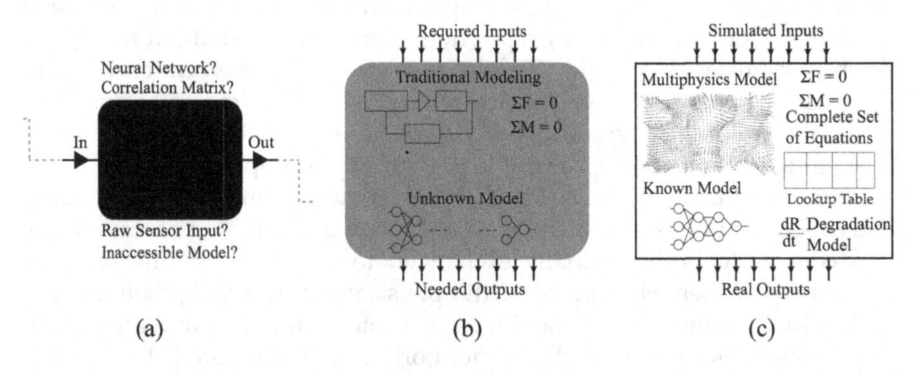

FIGURE 31.1
The Box paradigm illustration on describing physical components in the virtual space: (a) Black Box, (b) Grey Box, and (c) White Box.

a. Black Boxes include only inputs and outputs. Inner workings of the component are unknown.
b. Grey Boxes can model the component's effects and results, with theoretical knowledge.
c. White Boxes contain knowledge of all the functionalities and inner workings of the model.

The advancement of technology has improved the amount of information that can be gathered from the virtual space. Today's simulators can test the virtual model's performance capabilities to the desired degree, without carrying around unnecessary details. On the physical side, sensory, memory, and bandwidth advancements increase both the availability and quality of data.

To deliver the essential benefits gained from the linkage between virtual and physical models, a Unified Repository that links the two models together is required. Its main purpose is to enable a two-way connection between the virtual and physical models.

On the virtual side, a unique mark in specific areas (such as the centre of the rotor of a machine) can serve as a unique tag for the physical model. These marks can create a lightweight model while highlighting specific areas.

31.2.1 Internet of Things in Industry and Effects on Digital Twinning

"*Internet of Things* (IoT) is a term given to devices connected to the internet, with a sense of intelligence and the ability to collect information on their environment". This connection between all devices gives their user the ability to watch everything synergistically, and is intrinsically intertwined with the idea of the DT. Furthermore, when considering the DT of a single product or process, the IoT can connect different DTs together and enable data exchange and history comparison, enabling specialized decisions to achieve a specific goal.

A specialized subcategory, the Industrial IoT, is similar but refers to industrial processes. Like its broader counterpart, it can have a huge impact on improving manufacturing processes by providing a specialized scope on the amount of data that can be used for the construction and development of the virtual model. In this way, the performance, production rate, costs, waste, and many other critical deliverables within the industry setting can be improved significantly.

Simplistically, the IoT facilitates data aggregation and filtering. One primary IoT challenge is the effort to control the flow of data in the system, leveraging unstructured data volume. Data filtering is essential since part of it may be useless, slowing and convoluting PEI. Another challenge is the existing infrastructure, especially in smaller industries and less developed countries, where upgrading is even more important. The infrastructure is currently technologically behind the vision of IoT due to its rapid growth in the last decade.

The Digital Twin supports three of the most powerful tools in the human knowledge, Conceptualization, Comparison and Collaboration, as put by Grieves (2014).

When humans look at a problem, they **conceptualize** its context. DT can create different scenarios produced from the virtual model and pinpoint the actual development of the problem and most importantly, provide insight on issues that may have been lost on traditional problem formulation, owing to difficult to discern variables.

Comparison of desired and actual results is important in both problem solving and prototype engineering. The DT is surmised to be the best tool in evaluating an ideal characteristic, the tolerance limits around that ideal measurement, actual results, and future predictions. This process can save precious time in the physical space, where failed efforts are both costly and time-consuming.

Collaboration between both different parties and approaches is important to achieve highly specific goals. The DT not only facilitates sharing of ideas and existing endeavours but also intelligent approaches such as a human theory and an Artificial Intelligence (AI) machine learning approach. DTs can communicate between each other and selectively extract data and use them as an input in their virtual model. In this way, they share "experience" in a never-before-seen scale and use it to predict more quickly and accurately.

31.2.2 Data Sharing Issues in Industrial Internet of Things

One can surmise that the DT depends almost solely on the integrity of the data provided to be built correctly and thus be useful in industry, where immediate results and returns on investment are required. One glaring issue in this chapter's authors' relevant work is the low availability of proper form of industrial data when a singular application is concerned. For example, synchronous generator condition monitoring during shutdown is typically made once a year, even less when sensors and previous measurements indicate healthy operation. Furthermore, signals regularly received often are integrated under large intervals, obscuring transient effects where most of the information regarding the inner workings of the system is stored. Finally, high-fidelity monitoring (small interval measurements) is done on a very limited basis and does not cover an indicative portion of the product lifecycle. These factors render case-by-case digital twinning largely ineffective compared to traditional condition-monitoring techniques, especially for smaller industries where usually no predictive maintenance is done at all.

This issue is one not to be taken lightly, as according to a pertinent research, a budget of $2bn. with an estimated 12% annual growth is spent on maintenance of the large electrical machines (EMs) annually, $1.4bn of which is spent on plants without a diagnostic procedure. State-of-the-art paradigms are expected to reduce these costs by up to 60%, making the DT approach an even more lucrative investment (Sadeghi et al. 2017). A promising solution to this problem comes in the form of dataspaces, as proposed to be highly collaborative with the DT by Usländer et al. (2022).

Industries are concerned with two major issues associated with data sharing:

1. What is the benefit of participating in sharing important business and equipment data with potential rivals?
2. How do owners maintain control and ownership over their data, as well as precisely tracking what is shared?

Both of these issues are not to be taken lightly, as information is the commodity of the century and potential leaks on the internet can be extremely costly. The dataspace solution provides a sort of "in-between" space where these data are appropriately exchanged. To date, this information is not a part of the DTF until it is incorporated in its Data dimension. The DTF immediately benefits large players in industry due to pre-existing data abundance, especially when considering similar applications. Smaller players require an outside data source to be able to use the DTF to its full extent. An incentive possibility is a data broker, able to exchange information between players.

Intertwined with data sharing is the concept of privacy and security. He et al. (2022) investigate the matter in an application of a vehicular DT aimed at performing the massive calculation required for autonomous driving in the cloud instead of the limited local capabilities of the vehicle. In this case, the danger of an attack is twofold:

- Communication impairment may render the vehicle unusable or unsafe for operation;
- Communication retrieval can expose critically private information to malicious parties.

Both of these issues extend to any serious application of the DT, as growth incentive for businesses, and as safety concerns for users and the environment. However, tackling this problem is no different than social networks, e-banking, and blockchains, but one could argue its increased criticality. Figure 31.2 presents an early concept of the in-between dataspace, with data transactions and typical possible inter-twin malicious attacks.

Intra-twin malicious protection is up to the user, while a summary of these attacks can be found in the reference. The issue at hand when discussing the assignment of a broker, and thus an independent data handler, is inter-twin attacks, most of which can be summarized in three categories. Direct Denial of Service (DDoS) attacks clog the information pipeline preventing normal traffic. This kind of attack is of little consequence in this scheme, since exchanged information is processed for utility at a later date. Sybil attacks,

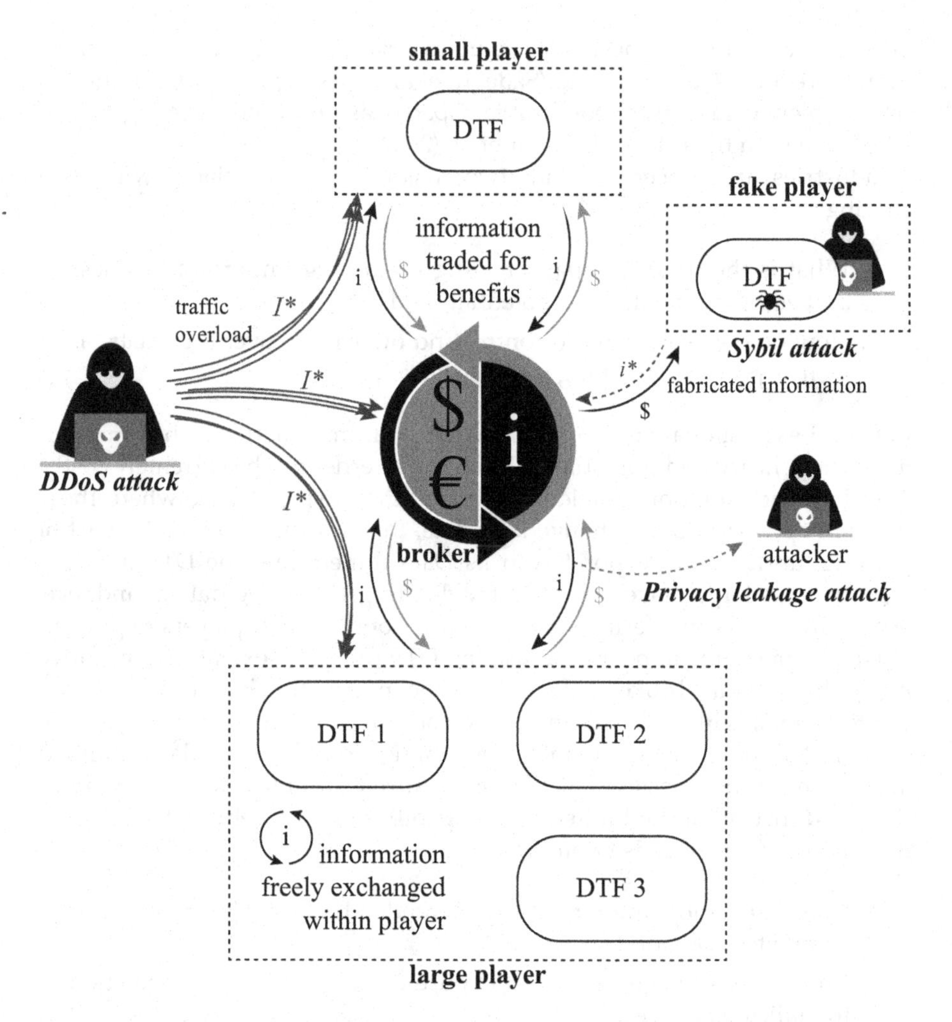

FIGURE 31.2
Two digital twin frameworks co-existing in an area of the global dataspace enacting in data transactions, with indicative malicious intent players.

injecting false information in the dataspace, can be tackled with a rigorous certification procedure and do not pose an imminent threat. On the contrary, privacy leakage attacks are the biggest concern since the exchanged information can be worth up to a significant amount of currency.

31.2.3 Misconceptions and Their Usage in Diagnostics

Digital Twin is a not yet established guideline and therefore used as an umbrella term. Following the best-received definitions, when considering the necessary attributes pertaining to the DT, the term is often used wrongly leading to misconceptions. Investigation of examples containing this issue leads research to create two terms in the subspace of DT; digital model and digital shadow. They differ in the level of integration between the physical and digital counterparts, while following the main pillars of the DT paradigm. Examples and criticism can be found in Falekas and Karlis (2021). The three terms are sequential subclasses.

The main difference is in the data flow automation, resulting in differing capabilities. A DT is both necessarily a shadow and a model, while a Digital Shadow is also a Digital Model.

- Digital Model: there is no automatic data exchange between the physical model and virtual model. Once the digital model is created, a change in the physical model will not affect the virtual model and vice versa.
- Digital Shadow: it has one-way flow between the physical and the virtual objects. A change in the physical object leads to one-way change in the virtual object. This results in an incapability to control the physical aspect, therefore the creation is not a DT.
- Digital Twin: data flow between the physical object and the virtual model bidirectionally. A change made on the physical object leads to a change on the virtual and vice versa, allowing control.

It is important to note that these definitions are themselves under scrutiny in the broader DT predicament and therefore not final, but generally well received. On the contrary, many new works keep referring a digital shadow as a twin. It is not yet clear whether this separation is beneficial or not when discussing the advancement of DT research. The authors surmise that it may aid in guiding interested parties more clearly.

This misconception is not presented here as a preventative measure, but rather as an encouraging suggestion for researchers to employ the correct term when naming their work. In reality, most condition-monitoring applications in industry require digital shadows while research into a possible future endeavour for the project can be completely made with only a snapshot of the physical aspect, namely, a digital model. The importance is that these DT subcategories still encompass the aspects that make the framework possibly the best candidate.

31.3 Digital Twin in Industrial Condition Monitoring

Employment of the technology in the industrial predictive maintenance sector, state-of-the-art in condition monitoring, is explored. Special focus is given to electrical machines (EMs), since they are both the primary mobilization force of industry and its electricity supplier. Around 95% of global energy includes synchronous generators in its production chain (Ritchie, Roser, and Rosado 2020), while around 50% is consumed by induction motors in the United States (McCoy and Douglass 2014).

Hence, proper operation with minimal losses and downtime is paramount. EMs should be closely monitored, as sub-optimal operation results in considerable energy waste. Furthermore, accurate lifespan prediction minimizes downtime and therefore operation costs. EMs consist of a multitude of electromechanical (multiphysics) parts which can be affected by a number of faults, with varying degrees of severity. These necessities of monitoring, prediction, assessment, learning and causality analysis perfectly align with the DT paradigm. Perceived benefits in the industry are listed below, along with their references (Jones et al. 2020):

- Cost reduction;
- Reduction of risk and design time;
- Reduction of complexity and reconfiguration time;
- Improved after-sales service;
- Benefit management and decision-making;
- Improved security and efficiency;
- Increased safety and reliability;
- Enhanced flexibility and competitiveness.

Consideration of experts is strongly advised in specialized domains. Therefore, the DT for condition-monitoring applications ought to be designed with two interfaces in mind; design engineer and specialized end-user, or back- and front-end. A complete DTF for usage in predictive maintenance is presented in Figure 31.3 (Falekas and Karlis 2021). The five dimensions may contain both front- and back-end utilities. For example, AI is employed both in the machine learning mode in the virtual space, and as information processing for front-end in the service dimension.

31.3.1 Design and Manufacturing Stage

Despite the potential of smart manufacturing, confusion about the DT concept and its implementation in real-life systems, mainly in small to medium enterprises, still exists. Lacking the expertise or resources required to implement DTs effectively is commonly encountered in industry. Further complications can be added by differing standards, technologies, and procedures used. Understanding the ideal concept is essential for reducing constraints and costs in manufacturing, making the understanding of the following factors necessary (Shao and Helu 2020):

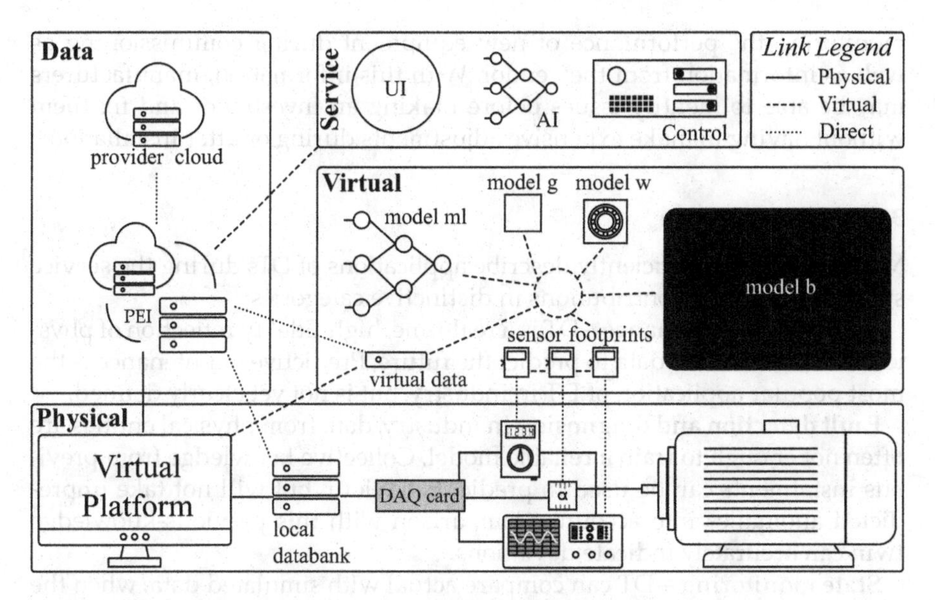

FIGURE 31.3
Indication of a complete proposition of the DT framework in the context of industrial predictive maintenance, highlighting aspects of interest. Abbreviations: ml: machine learning, w: white, g: grey, b: black, indicating the Box paradigm.

Application – identifies the fidelity required to enable the desired decision or control action in relation to its specific needs, as well as the need for real-time or offline updates.

Viewpoint – of the desired decision or control action determines whether a product, process, or system twin is needed. The viewpoint decides the methods required and whether the DT should mimic observable behaviour or model the system state.

Context – determines how information should be provided, and can be influenced by the needs of the viewpoint.

Potential **benefits** provided by the DT implementation in manufacturing are:

Minimization of equipment downtime – reduces equipment downtime by using a "machine health twin" to track, troubleshoot, identify, and forecast equipment faults.

Optimizing production planning and scheduling – a "scheduling and routing twin" can gather information to analyse the production system's current state and any changes in customer demand, inventory, and resources. Following that, this information can facilitate demand-driven, on-time delivery, resource optimization, cycle-time reduction, and inventory-cost reduction.

Enabling virtual commissioning – to enable system optimization and continual improvement, a "commissioning twin" can leverage data from monitoring the performance of new equipment during commissioning as well as information from the vendor. With this information, manufacturers may be able to identify issues before making an investment and fix them without having to make expensive adjustments during or after installation.

31.3.2 Service Stage

M. Liu et al. (2021) efficiently describe applications of DTs during the service stage, highlighting contributions in distinctive categories:

Predictive maintenance – DT is a real-time, high-fidelity reflection of physical entities and uses data to predict the future. Predictive maintenance is the most popular application of DT in industry, but is not yet clearly defined.

Fault detection and diagnosis – in industry, data from physical entities are often not enough to train a reliable model. Collective knowledge from previous instalments can be used to predict behaviour, but will not take unpredicted anomalies into account. Comparison with this previous-knowledge twin can accurately indicate deviations.

State monitoring – DT can compare actual with simulated data, when the aforementioned knowledge is unavailable. The virtual model provided by a DT is very accurate and keeps updating through the product lifecycle.

Performance optimization – initial design may not be suitable, indicated by changes taking place as the machine operates. Data from the lifecycle of every component of the product make DTs aware of the physical twin's performance and as a result optimize the product and predict its performance.

Virtual test – as a close representation of the physical entity, DT can test certain system parts under assumed conditions that may lead to failure of the physical part, but may be necessary to prepare for, finding limitations without destructive tests.

31.3.3 Retirement Stage

Information aggregated during the previous stages of the DTF is not only stored but also processed to a high degree of quality, meaning that contained information is invaluable in industry in all aspects, be it new designs evolved from previous, term data incorporated in small industries to aid twinning, or as a monetary value when sold through the broker. This stage currently can be surmised to not actually exist, as this information is scattered in various forms and is of little use. Few literature works are noted.

31.3.4 Applications in Predictive Maintenance

Current research efforts are made to introduce the DT technology in predictive maintenance and further improve its performance. Apart from the financial incentive, other benefits include improved real-time diagnostics, higher equipment efficiency, implementation of advanced algorithms, and high-fidelity models. Finally, a plethora of problems with insufficient datasets can be solved as the DT generates numerous synthetic data (You et al. 2022).

At present, three methods with unique characteristics and specific applicable fields are used in predictive maintenance (Luo et al. 2020).

31.3.4.1 Reliability Statistics Method

Data regarding historical fault events are used to predict relevant future faults. This method requires less detailed information and no mathematical approach for the DT model. For the prediction, information is contained in series of different probability density functions. Methods are based on reliability statistics, such as the Weibull distribution, Bayesian method, and fuzzy logic. Despite their simplicity, those methods do not take the environment of the product and its complexity or the performance degradation of equipment into consideration, resulting in low prediction accuracy. It is the preferred solution for simple, commercial products in large batches.

31.3.4.2 Physical Model–Based Method

The operation principle of the target system is described via mathematical and physical models that reflect its degradation process. Adequate description of the physical entity provides accurate fault predictions. Complex mechanical and electrical systems are best described by this method thanks to captured correlations.

31.3.4.3 Data-Driven Method

Precise fault and performance evaluation is not needed as necessary data are collected in real time. Utilization of big datasets is done via tools such as autoregressive models, artificial neural networks, and support vector machines. Information regarding system degradation is obtained, stored in historical data, and then used to extract certain features, providing further knowledge. As the installation of sensors is not always feasible, advanced algorithms for data acquisition and analysis are constantly needed.

Due to the complexity of the DT model, it is necessary to realize quantitative and qualitative data transformation into one singular dataset. This can be achieved with *super-networks*.

> Super-network is a multilayer network with large scale, complex connections and various nodes and they are superior to the existing networks and of higher order and if we compare them to ordinary complex networks, super-network has the characteristics of multi-level, multi-attribute or multi-criterion.
>
> *(Z. Liu et al. 2019)*

The purpose of building a data super-network model is to analyse the data we have from the DT to predict when a fault may occur and programme the suitable repairs. By using this technology, data from the super-network model of physical, virtual, and service layer are votary. The data needed to create the service layer come from the physical and virtual dimensions at the same time. During the manufacturing process, preliminary instructions can be given to the models.

31.3.5 Examples in Literature and Industry

Werner, Zimmermann, and Lentes (2019) present an example application of the DTF in condition monitoring, a short breakdown of which is presented here as an indicative step-by-step guide for the reader. Different key aspects are highlighted in bold:

To predict the remaining useful life of a machine, one can gather information from other machines that have been used for similar applications in conjunction with **sensors** located on the actual machine. The data are transported via Ethernet and W-Lan to the virtual model for predictions.

The input data in the **prediction model** are prepared according to the chosen algorithms. Preparation is done via filters and analysis components, whose main purpose is to reduce data complexity. Model output is an estimation of the remaining useful life.

Concerning the **simulation model**, the input data are typically sourced from computer-aided engineering data. These are CAD and data from previous simulation phases. The difference between the simulation model and the prediction model is that the former output can be used as input for the latter.

The **decision support system** uses the output data from the predictive and the simulation models as input to propose maintenance actions, optimizing time and cost of application.

Step 1: state analysis and definition of objectives.

Existing industries already use data-driven and/or physics-based approaches, but do not necessarily take maximum advantage of their data and therefore cannot use them to perform advanced monitoring. State analysis is the initial step in identifying goals and appropriate actions. State analysis can be performed front-to-back or vice versa; start from proposed actions and work out the end goals; or decide the thresholds for end goals and choose actions accordingly. Identification of critical components is also important.

Step 2: checking the availability of data, information, and knowledge.

Previous knowledge and aggregate information are to be thoroughly examined and appropriately filtered before initializing the process of predictive maintenance. Possible information gaps can be identified in this stage, meaning insufficient data for a future update of the DTF. Appropriate strategies are formulated in this stage, typically resulting in additional sensory instalments or configuration changes.

Step 3: integration of data-driven modelling.

Data analysis uses machine learning algorithms and statistical methods to connect acquired data with causal links. These algorithms can describe the actual model. After appropriate training, the data-driven model has the ability to detect trends across historical and newly acquired data and make behaviour predictions. Data scientists typically opt for a combination of algorithms to overcome a situation where no improvement is noted in pre-existing methodologies. The algorithm with the most accurate results isn't always the best choice; industries operate on strict time and monetary budgets, often a trade-off for fidelity and a "better" scientific method.

Step 4: integration of physics-based models.

The physics-based model typically follows data-driven modelling to assess for missed dependencies or correlations, since it provides an additional barrier to check for possible mistakes. The appropriate simulation software or combination thereof is then chosen based on the simulation technique, the necessary accuracy, and the allocated time.

This simulation can then be extrapolated and distributed for usage as an early version of the DTF to present to clients, end users, and engineers from different departments.

Step 5: hybridization of modelling approaches.

Once the data-driven and physics-based models have been successfully created, they are combined to formulate the hybrid model. Simulation model output is theoretical since it is not produced from real data. These data can be combined with newly attained sensor measurements and used as predictive model input. The experiment is complimented with further simulated data, which may not occur in the real world, i.e., data about a failure which have not appeared in any previous iterations or experiments. The 3D model of the machine is produced, providing an upgraded version of the DTF.

Sensor and previous experiment data come a long way in creating a realistic virtual model. This process is iteratively performed, resulting in a state-of-the-art approach following the pillar of any engineering procedure in approaching a perfect representation.

31.4 Conclusion and Complete Guideline

This chapter aims to introduce the concept of digital twinning under the scope of electrical machine diagnostics. DTF is a flexible concept able to process a multitude of applications, each with its own requirements and specifications. This presentation is not at all different in its paradigms from the general DTF concept. Rather, the example given to better guide interested reads pertains to electrical machine predictive maintenance. The complete guideline is present in Figure 31.4.

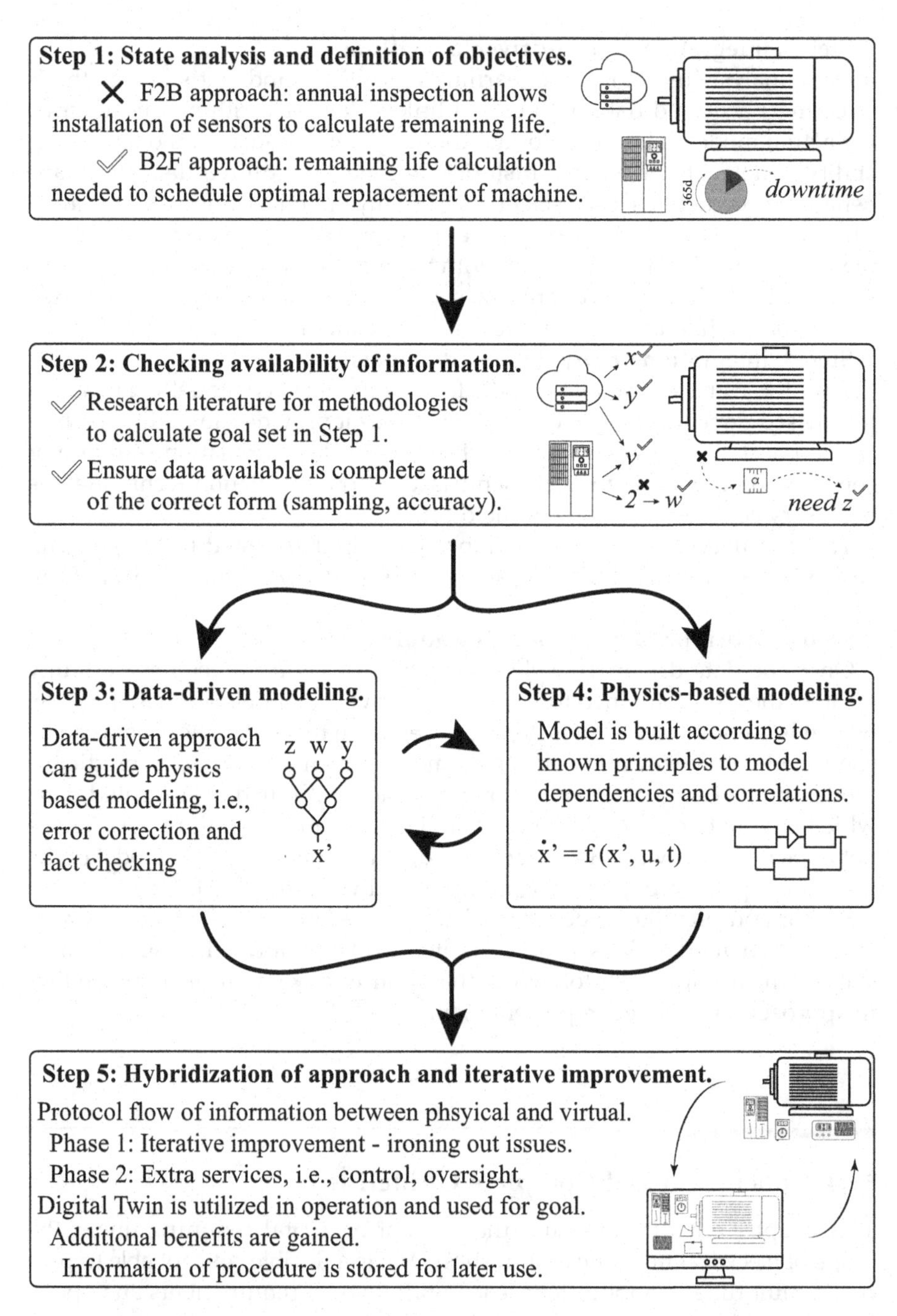

FIGURE 31.4
Guideline for creation of a DT framework to be employed in industrial predictive maintenance of electrical machine systems.

The DTFs created across the industry are as of now "isolated"; while researchers may opt for information exchange and initial runs of dataspaces are witnessed, no definite protocol exists. However, no robust applications of the DTF exist either. A common guideline is surmised to aid better definition of these new paradigms. The given example of electrical machine maintenance indicates core aspects that, should they be covered, applications are robust enough to warrant a deep inspection by experts.

References

Falekas, Georgios, and Athanasios Karlis. 2021. "Digital Twin in Electrical Machine Control and Predictive Maintenance: State-of-the-Art and Future Prospects." *Energies* 14 (18). https://doi.org/10.3390/en14185933.

Fuller, Aidan, Zhong Fan, Charles Day, and Chris Barlow. 2020. "Digital Twin: Enabling Technologies, Challenges and Open Research." *IEEE Access* 8: 108952–71. https://doi.org/10.1109/ACCESS.2020.2998358.

Gabor, Thomas, Lenz Belzner, Marie Kiermeier, Michael Till Beck, and Alexander Neitz. 2016. "A Simulation-Based Architecture for Smart Cyber-Physical Systems." In *2016 IEEE International Conference on Autonomic Computing (ICAC)*, 374–79. IEEE. Wuerzburg, Germany.

Glaessgen, Edward, and David Stargel. 2012. "The Digital Twin Paradigm for Future NASA and U.S. Air Force Vehicles." In: *Collection of Technical Papers - AIAA/ASME/ASCE/AHS/ASC Structures, Structural Dynamics and Materials Conference*, no. April. https://doi.org/10.2514/6.2012-1818.

Grieves, Michael. 2014. "Digital Twin: Manufacturing Excellence through Virtual Factory Replication." *White Paper* 1: 1–7.

He, Chao, Tom H Luan, Rongxing Lu, Zhou Su, and Mianxiong Dong. 2022 "Security and Privacy in Vehicular Digital Twin Networks: Challenges and Solutions." in IEEE Wireless Communications, vol. 30, no. 4, pp. 154-160, August 2023, doi: 10.1109/MWC.002.2200015.

Jones, David, Chris Snider, Aydin Nassehi, Jason Yon, and Ben Hicks. 2020. "Characterising the Digital Twin: A Systematic Literature Review." *CIRP Journal of Manufacturing Science and Technology* 29: 36–52. https://doi.org/10.1016/j.cirpj.2020.02.002.

Liu, Mengnan, Shuiliang Fang, Huiyue Dong, and Cunzhi Xu. 2021. "Review of Digital Twin about Concepts, Technologies, and Industrial Applications." *Journal of Manufacturing Systems* 58 (PB): 346–61. https://doi.org/10.1016/j.jmsy.2020.06.017.

Liu, Zhifeng, Wei Chen, Caixia Zhang, Congbin Yang, and Hongyan Chu. 2019. "Data Super-Network Fault Prediction Model and Maintenance Strategy for Mechanical Product Based on Digital Twin." *IEEE Access* 7: 177284–96. https://doi.org/10.1109/ACCESS.2019.2957202.

Luo, Weichao, Tianliang Hu, Yingxin Ye, Chengrui Zhang, and Yongli Wei. 2020. "A Hybrid Predictive Maintenance Approach for CNC Machine Tool Driven by Digital Twin." *Robotics and Computer-Integrated Manufacturing* 65 (September 2019): 101974. https://doi.org/10.1016/j.rcim.2020.101974.

McCoy, Gilbert A., and John G. Douglass. 2014. *"Premium Efficiency Motor Selection and Application Guide-A Handbook for Industry."* US Department of Energy: Washington, DC, USA.

Ritchie, Hannah, Max Roser, and Pablo Rosado. 2020. "Energy." *Our World in Data*. https://ourworldindata.org/energy.

Sadeghi, Iman, Hossein Ehya, Jawad Faiz, and Hossein Ostovar. 2017. "Online Fault Diagnosis of Large Electrical Machines Using Vibration Signal-A Review." In: *Proceedings - 2017 International Conference on Optimization of Electrical and Electronic Equipment, OPTIM 2017 and 2017 Intl Aegean Conference on Electrical Machines and Power Electronics, ACEMP 2017*, 470–75. https://doi.org/10.1109/OPTIM.2017.7975013.

Shao, Guodong, and Moneer Helu. 2020. "Framework for a Digital Twin in Manufacturing: Scope and Requirements." *Manufacturing Letters* 24: 105–7. https://doi.org/10.1016/j.mfglet.2020.04.004.

Tao, Fei, He Zhang, Ang Liu, and Andrew Y. C. Nee. 2019. "Digital Twin in Industry: State-of-the-Art." *IEEE Transactions on Industrial Informatics* 15 (4): 2405–15. https://doi.org/10.1109/TII.2018.2873186.

Usländer, Thomas, Michael Baumann, Stefan Boschert, Roland Rosen, Olaf Sauer, Ljiljana Stojanovic, Jan Christoph Wehrstedt, and Fraunhofer Iosb. 2022. "Symbiotic Evolution of Digital Twin Systems and Dataspaces." *Automation* 3(3): 378–98.

Werner, Andreas, Nikolas Zimmermann, and Joachim Lentes. 2019. "Approach for a Holistic Predictive Maintenance Strategy by Incorporating a Digital Twin." *Procedia Manufacturing* 39: 1743–51.

You, Yingchao, Chong Chen, Fu Hu, Ying Liu, and Ze Ji. 2022. "Advances of Digital Twins for Predictive Maintenance." *Procedia Computer Science* 200: 1471–80.

Zhuang, Cunbo, Jianhua Liu, and Hui Xiong. 2018. "Digital Twin-Based Smart Production Management and Control Framework for the Complex Product Assembly Shop-Floor." *The International Journal of Advanced Manufacturing Technology* 96 (1): 1149–63.

32

Building a Digital Twin – Features for Veneer Production Lines – Observations on the Discrepancies between Theory and Practice

Jyrki Savolainen and Ahsan Muneer
LUT University

32.1 Introduction

The utilization of digital twin (DT) in large-scale industrial applications has been extensively discussed in academic literature. However, only a few publications provide real-life case descriptions on building a digital twin model from scratch. This chapter aims to fill that gap by providing a case example from the veneer and LVL (laminated veneer lumber) industry, discussing the applicability of the DT concept, and identifying central pain points observed from simulation and data perspectives that are likely to be generalizable for many process industrial organizations undertaking DT projects.

The brief history on the idea of DT can be traced back to aeronautics and the need to forecast the performance and structural life of individual operating aircrafts (see discussion, e.g., Madni et al., 2019; Rosen et al., 2015; Stark et al., 2017; Tuegel et al., 2011). Some scholars, such as Min et al. (2019), Negri et al. (2017), and Qi and Tao (2018), say that the DT of the whole industrial system is still at its beginning and in many ways conceptual. For the purposes of this paper, we use digital twins to mean, stated by Alam and Saddik (2017), highly detailed computer models that interact with physical reality.

The production of veneer involves various physical phenomena that require adequate consideration in the DT model. In the past, one of the most widely used ways has been randomly selecting sheets and evaluating them manually. Today, the quality inspection is done online in a separate processing equipment by automated imaging and the physical flows of material are steered accordingly. Nevertheless, plant-wide optimization solutions remain a challenge due to the complexity of the production process, which involves multiple pieces of equipment interconnected in a veneer production line that requires dynamic optimization. To address this challenge, the veneer/LVL

DOI: 10.1201/9781003425724-38

manufacturing industries are investing in process automation and DT of manufacturing lines to enhance product quality standards, reliability, and efficiency (Urbonas et al., 2019).

Despite the advances in computerization of industrial processes since the 1950s, the idea of using a digital twin model for automated communication and operation of multiple intelligent machines in concert is relatively new. That is, the previous generations of industry automation have relied on human operator wisdom in the overall production line optimization while focusing on the computerized optimization of single equipment at a time. In the case of complex and multi-phase industrial processes, such as veneer and LVL introduced below, this sometimes leads to situations where the production line optimum is not reached due to lack of automated communication between equipment and the inability of humans to continuously change production settings "on the fly" within the timeframe of minutes or hours. In this chapter, this gap of real-time optimization is addressed with DT.

The chapter continues with a background on Industry 4.0 followed by the case description. Then key findings of the project are given and their generalizability to other applications is evaluated within the I4.0/DT framework. The chapter closes with conclusions and discussion.

32.2 Background

In 2013, a German initiative labeled the fourth industrial revolution as *Industry 4.0* (I4.0) that has later emerged as a prospective technological paradigm. One of the areas most impacted by I4.0 is manufacturing, with a focus on enhancing production processes to optimize operational performance, product or service development, and supply chain planning (Zheng et al., 2020). The underlying technology behind I4.0 is Cyber-Physical Systems (CPS), which makes manufacturing systems modular and flexible, allowing mass production of highly customized products (see, e.g., Veza et al., 2015; Li, 2018).

Ishwarappa and Anuradha (2015) write that the integration of information technology with the industrial systems and enterprise data has become increasingly rich consisting of various types of datasets. In addition, the amount of data generated by industrial systems has been rapidly increasing. Zhong et al. (2017) say that IoT-enabled manufacturing transforms traditional manufacturing devices into smart manufacturing objects (SMOs) that can connect, communicate, and interact with one another to execute manufacturing logic automatically and adaptively.

The interconnectedness of equipment and mutual communication brings with it new challenges. Therefore, as observed by Wang et al. (2022), data analytics as a whole is the vital component in the development of intelligent industrial systems. One of the objectives of I4.0 paradigm can be considered to include the gathering of all the available data to make traditional factories and manufacturing processes more intelligent and thus achieve higher levels of operational efficiency and productivity. By gradually incorporating additional sensors, autonomous systems, and actuators into the manufacturing process, the factories are made smarter, more dynamic, and adaptive enabling machines and equipment to self-optimize their actions (see, e.g., Roblek et al., 2016).

In our view, self-optimization requires that the optimization software (or agent) is aware of *what happens next* and we propose that DT could be the solution for accurate forecasting, once it is connected with the ability to simulate production process into the future which (Grieves & Vickers, 2017) refers to as "simulation front run". The word "agent" is a process or entity designed to accomplish a task constantly and independently in a non-deterministic context with other processes and elements. The case example of the digital twin dealt within this paper is a veneer production line with a vast amount of data from multiple individual processes that have stochastic elements. The simulation process should be able to weigh different courses of action with regard to maximizing a given value function of the digital twin model using policies that are either prescribed or explored by the agent during simulation. In the intelligent manufacturing context, the agent paradigm is observed as one of the most effective ways of optimization (Lu, 2017) when the agent(s) are left to perform tasks in a condition from which they are separated and have their own knowledge and understanding of their surrounding environment; they employ preference in interacting within their environment, developing plans, making autonomous choices, and performing actions to change the environment (see discussion, e.g., Adeyeri et al., 2015).

DT of industrial processes requires, first, a deep understanding of the physical processes and, second, the ability to analyze the data produced in these physical processes. This "model-driven" approach, based on the rigorous understanding of the scientifically proven relationships, enables one to extrapolate process values into the future by introducing simulated, and possibly even unforeseen, variable combinations into the digital twin model. On the other hand, a "model-free" approach may also be used, when the correlation or other numerical interactions between system parameters are to be unrevealed with data mining techniques which in general refer to the processes of determining correlations in large databases by applying multiple levels of analysis.

Machine learning means that no specific model is given to the computer to apply for the dataset but rather it should discover the patterns independently. A review of the commonly applied Machine Learning (ML) methods is provided in Kotsiantis et al. (2006). Before applying data mining methods in sensor data, outliers should be removed from the data as these can influence the model parameter estimation results and the data analysis process. These outliers can arise due to multiple reasons, including sensor failure and improper processing of missing data. Outliers are sometimes helpful providing information that can lead to the unearthing of new information (Cheng et al., 2018; Xu et al., 2015).

Data analysis has become much more complicated due to recent trends in gathering and utilizing diverse datasets. One of the characteristics of such large datasets is that they contain significant levels of redundancy, and the likelihood of unrelated data entities is high (Houari et al., 2016). The process of transforming a multidimensional data representation into a low-dimensional representation is known as dimensionality reduction. The idea of dimensionality reduction techniques is to transform a multidimensional dataset into a low-dimensional dataset while preserving as much of the data's original meaning. Data in low dimensions are simple to examine, process, visualize, and interpret (Zebari et al., 2020). Principal component analysis (PCA) and partial least squares are some of the widely known multivariate statistical projection methods that can handle enormous volumes of data and compress the information into low-dimensional latent variable components, making monitoring and interpreting results more manageable (Kourti et al., 1996). In this chapter, we discuss the aspects of data availability, data types, dimensionality reduction, and simulation data analysis in the context of digital twins using the example case of veneer production line.

32.3 Data and Methodology

Any company's digital transformation is not only about investing in new technology, but it should be based on the business strategy and both internal and external business processes. Rajnai and Kocsis (2018) describe digitalization as an incremental advancement through steps including technological and organizational changes. In the case of the veneer industry, companies are aligning their strategies toward extracting value from manufacturing processes through advanced data analytics and DT.

The digital twin model of veneer production should have enough generalized, yet flexible, structure that enables not only the optimization of the original case factory but also could be extended to other plants with new data and minor changes in the structure – such as adding/removing pieces of equipment from the process. Ideally, it will be a dynamic simulation model, which links the production data of different process equipment with the physical and computational models of these equipment, so that the veneer production line can be coordinated by using different quality raw materials. In other words, the ultimate aim is to produce on-target quality product from the available material stock and not spend the highest quality veneer sheets to meet order book targets at a specific timeframe given. The DT project was started with a prototypical flowsheet of the veneer production line shown in Figure 32.1.

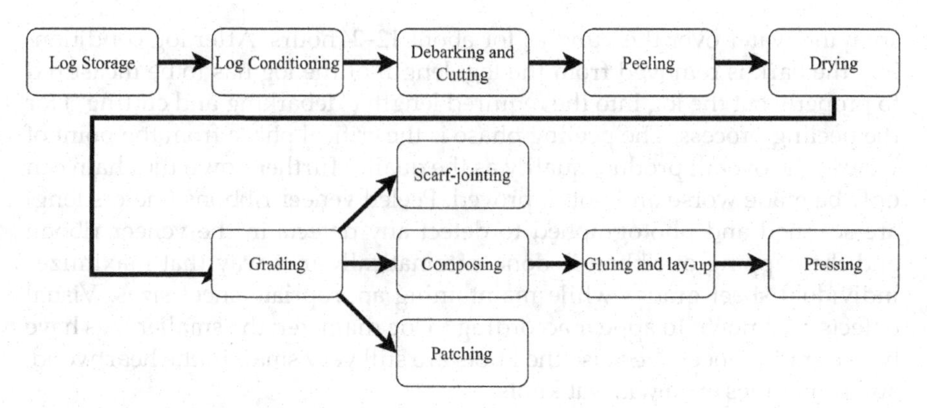

FIGURE 32.1
General process flowchart for the plywood and LVL manufacturing.

Building a fully applicable digital model is a huge technological and organizational effort. Therefore, the focus of the study was further narrowed down to the peeling and drying processes as a starting point of the modeling. It was figured that if these critical processes could be modeled adequately then, later on, the scope could be extended to cover the downstream processes as well using similar principles and level of modeling required for the DT. For the sake of comprehensiveness, the process of veneer production is described below in full.

The production of plywood and LVL is based on multiple production stages illustrated in Figure 32.1. The above shown set of processes is required to convert wooden logs into veneer sheets and, further, to plywood or LVL as needed where veneer sheets are being glued together. In gluing, the wood grain directions of sheets vary by 90°, making it strong in all directions and various dimensions of products – most commonly sizes 4 ft×8 ft and 8 ft×4 ft – are being made according to the customer requirements where the limiting factor is the woodblock length in the initial peeling process.

One of the major challenges from the production optimization point of view in veneer manufacturing is the varying wood quality which by nature varies a lot depending on the factors that are partly beyond the control of a veneer line operator, such as the habitat and growing conditions of trees. To control the quality of veneer sheets, multiple automated measurements are taken from each sheet such as the moisture, temperature, and density. The main defining characteristic is the visual image of the sheet after the main process phases which inherently dictate its quality class are often indicated by letters, e.g., grades A–D and zero ("0") for waste. The quality is decided by computer imaging where the defects such as knots, holes, and splits reduce the grade evaluation.

The first step of the process is the "log conditioning" (see Figure 32.1) where the logs are conditioned by soaking them in a hot water pool and spraying water over the bundles for about 12–24 hours. After log conditioning, the bark is removed from the log, length of the log has to be measured to properly cut the log into the required length ("debarking and cutting") for the peeling process. The peeling phase is the critical phase from the point of view of the overall product quality as the quality further down the chain can only be made worse and not improved. Peeled veneer ribbons (meters long) are scanned and photographed to detect any defects in the veneer ribbon and the clipping of ribbon is done automatically in a way that maximizes individual sheet grades while maintaining appropriate sheet sizes. Visual defects are known to appear according to log diameter: the smaller logs have better grade veneers because the knots are still very small in the heartwood, and sometimes even without knots.

Peeling sheets are stored in stacks of sheets (few hundred sheets each) for hours to few days before they are fed to the dryer. The veneer sheet quality is ultimately defined by the effectiveness of the drying process where the drying parameters (such as speed and temperature) should be adjusted according to the input material (peeling sheet) quality, e.g., moisture and thickness. During the drying process, the width of the individual veneer sheets shrink up to 10% due to varying shrinking properties on their surfaces. After the drying process, full-size sheets are stacked and transported directly to the lay-up or scarf jointing. Broken or defective veneers are stacked for the composing process and proceed to the further process of gluing, sanding, and trimming.

Some of the studies dealing with the technical aspects of the veneer process include a research of Çolak et al. (2007) who investigate log steaming and veneer drying conditions on technical properties and durability of LVL and solid sawn lumber suggesting partial least squares method as an effective method of quality control. Han et al. (2015) apply non-linear programming and operational research theory to optimize the energy consumption during veneer drying. Ahmed et al. (2020) provide a detailed study on the effect of drying temperature on the veneer sheet quality. They list several process parameters that must be controlled, including airflow, gas usage, drying speeds (time), dryer zone temperatures, and chain side temperature. Demirkir et al. (2013) presented a method by applying ANN to predict the intermediate bonding strength values based on the peeling and drying temperature of the veneer sheets finding positive and negative correlations with the bonding strength depending on the bonding agent applied. Within the scope of the digital twin model here, the physical details were kept close to minimum and the main efforts were spent on image analysis which are discussed in detail later in this chapter.

32.3.1 Model Description

Matlab Simulink was used for the purpose of model building as it provides a graphical, hierarchical model building environment with dynamic

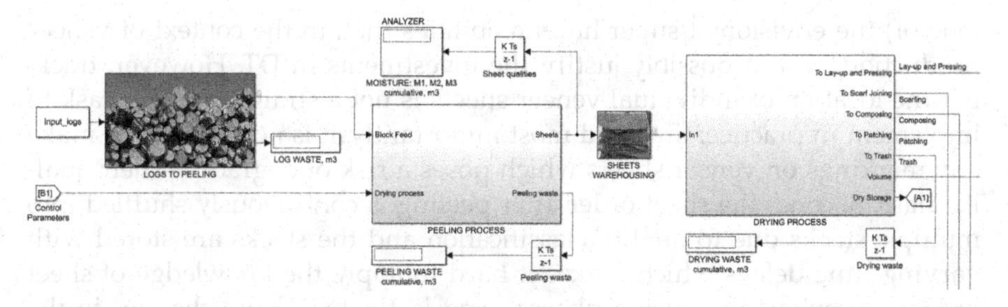

FIGURE 32.2
Model flowsheet in Matlab Simulink where each equipment block contains a detailed flowsheet diagram of the process.

simulation where programmatic ("command line") use of model objects is possible for the purposes of, e.g., optimization and random simulations. In the project, an Excel-flowsheet was constructed that enables an average model user can run the Simulink without any programming skills. Later, the Excel interface could be converted into a web-version (HTML) making the digital twin available over the internet. The flowsheet of the system model is provided in Figure 32.2.

32.4 Results

Despite the ongoing general discussion on the availability and accessibility of data sources, a lot of real-life issues still persist regarding the data availability from a legal aspect as well as from a technical point of view. A key question common to any DT-project is the data and result ownership between the digital twin model developer and its customers: the sole ownership of raw data is seen by the data owners as something valuable whereas in the digital twin setting this is not completely accurate. In fact, the data become valuable if and only if one is allowed to utilize it inside the digital twin model, whereas the digital twin model cannot exist without the initial data for testing. This could be regarded as a *chicken and egg* problem where it is not clear which one comes first: DT or the data. It is evident, however, that neither the data nor DT model alone is able to produce economic value and, therefore, we suggest ensuring the data availability as far as possible first and secondly starting the model building which is the most time-consuming task.

In veneer production, the tracking of individual sheets is not possible by humans which means that efficiency losses are incurred in veneer production daily due to inadequate and/or ill-timed material flow planning. The sheet tracking with route planning and scheduling can be considered as (one of) the envisioned super-human abilities that, in the context of veneer production, would possibly justify the investments in DT. However, tracking the location of individual veneer sheets is not a straightforward task to implement in practice: first, and most importantly, it is not allowed to make any markings on veneer sheets which poses a risk of degrading their quality class. Second, the sheet order after peeling is continuously shuffled into multiple stacks due to quality classification and the stacks are stored with varying time delays which makes it hard to apply the knowledge of sheet order in previous processing phases. And lastly, the visual changes in the sheets induced by the certain processing stages (such as drying) can be significant in terms of altering the colors, producing defects and shrinking the dimensions of the sheets.

Due to the constraints in marking or labeling individual sheets for tracking, the most important data for matching the sheets are the images taken at the different phases of veneer production. Based on matching images, the DT model should be able to improve the control of both the inbound logistics and physical product quality: that is, if one was able to know the location and qualities of different sheets on a sheet-by-sheet basis, it would open up new types of optimization possibilities to the current practice where some of the processes are mostly viewed as black boxes where no statistical models exist on how different qualities of wood behave under given process conditions of the equipment.

What makes the attempt of using sheet tracking exceptionally challenging from the computational perspective is that raw color images are unstructured data that need to be somehow structured for the purposes of analysis by the computers. In other words, the question is how to reduce the data dimensionality in a way that enables identifying the whole variety of individual sheets while keeping the data size of the representation at the minimum. A trade-off between the presentation size and recognizability can be stated: it should be able to simplify the image representation so that it is compact enough and easy to analyze, but at the same time, it should maintain the difference from all other images in the data. In practice, storing and using full images sized tens of megabytes for analysis is not an option as thousands of new images are being produced daily.

Admittedly, many advances on Artificial Neural Networks (ANN) in the current scientific literature have been reported that utilize raw image data but, noticeably, they are based on training image datasets of millions of labeled images. In the case of veneer production, manually labeling thousands of individual color images into training dataset pairs showing peeling input (x) and drying output (y) is laborious – if impossible as tens of thousands of sheets may be processed daily in the largest operations. As an extra twist, there is a high level of randomness on how and where the new defects appear on individual sheets which makes it questionable whether even advanced deep neural networks would be able to generalize veneer image data.

One documented effort on the use of ANN in matching veneer images is provided by Jalonen et al. (2021) who use Siamese ANNs to match peeling sheets with drying. Their model is fitted on the data of approximately 2,500 sheet pairs of manually labeled spruce sheets showing that the performance of matching is inadequate with realistically sized datasets. During the DT project, an alternative, much simpler, approach with gray-level correlation matrices (GLCMs; see Haralick et al., 1973) was adopted which is based on the textural differences in the raw images. The GLCM-method produced an excellent ~95% accuracy with the test dataset of ~2,000 sheet images of spruce but it could not be verified if the method would generalize to less textured wood qualities such as birch.

From the software point of view, digital twins in the literature present themselves in the literature as computer programs, more or less, without further specifics on the actual hardware/software platform. Today, there is no off-the-shelf software for building digital twin models – or some exist but they are usually tailored and restricted to specific equipment or industries. The unavailability of platforms is probably due to the immense list of requirements on what the DT should be able to do. In this study, the Matlab Simulink® was selected on the basis of the problem characteristics such as the need for dynamic simulation and ability to optimize. Matlab Simulink® is unfortunately a proprietary software and its licensing fees seldom justify its use for applications where the economic optimization potential is small or more uncertain than the case described here. From a scientific advancement point of view, we regard the unavailability of a software platform as a serious handicap for developing generalized digital twin frameworks for industry applications.

The related, and higher level, question in the choice of software is the philosophical standpoint on the detail-level of the model which we see as a decision between top-down or bottom-up approach on the basis of what is the ultimate goal of DT. In the first option, top-down, the (managerial) interest may be in having an overall presentation of the production process with a model timestep of hours or days which should not be confused as a DT project but rather it should be kept as a simulation model exercise due to the absence of real-time interplay with the real data feed. On the other hand, if the customer is genuinely interested in having a deeper, data-based insight to the process on a minute or second basis, then the window of opportunity of DT implementation exists. Based on our research, the properties of modern simulation software (such as the one selected here) do have adequate capabilities for both cases as graphical interfaces, computational capabilities, and scalability (cloud computing) are available. However, utilizing modern simulation software requires at least a team of one or two highly skilled experts in the organization who are able to update, maintain, and run the digital twin model.

Instead of software capabilities, the main problems of DT seem to lie in the availability of data and technical details of physical processes. When it comes to data gathering, there are seldom too much digital measurements available that could be potentially beneficial for the DT model. Probably, the default case in the heavy industry cases is such that a variety of measurement history data exit in the process automation system (DCS/digital control system) but the amount of metadata might be sparse. The metadata, in our perspective, is often something that exists outside the sight of the digital world that might affect, or even discard, the measurements taken from a given period of time.

As an example of missing datapoints from the veneer industry, the stacks of quality-controlled sheets are often manually, i.e., with a forklift, transferred from peeling to a storage from where they are later on moved to the dryer based on the needs of the current order book of the production line. Therefore, there is no explicit knowledge on how the stacks are being moved on the factory floor as the decisions on feeding the subsequent processes are left for the human operator. A second example is the raw material feed which consists of logs that are stored outside in piles in the factory premises. Due to the sheer size of material stock, there are limited possibilities to optimize what type of size or quality distribution of logs is currently going into the process and if it were, there is still uncertainty of the real material quality not being revealed until the logs arrive to the peeling process. In this project, the missing knowledge of the input material quality was dealt with the idea that the previous logs (in peeling) represent the currently best guess on what is coming next and that would enable planning ahead.

Admittedly, it would be technically possible to automate some blind spots of the log storage-peeling-veneer storage-dryer logistics by, e.g., providing peeling stacks with IDs for improved traceability. From an economic perspective, however, this makes no sense unless the proof of monetary benefit can be given ex-ante to the management. In the case of heavy industries, many of the equipment have been installed decades ago before the emergence of a new industrial automation paradigm. This means that the software is mainly built for running the machine's own proprietary pieces of code to perform the desired task as efficiently as possible and "if it works, don't touch it". In other words, in many cases there are no in-built machine-to-machine communication abilities in the legacy equipment and even the question of synchronizing the clock of such a self-reliant machine regularly with the other processes may be a practical issue that needs resolving at the first place.

32.5 Conclusions and Discussion

This chapter discussed the applicability of the DT paradigm in the context of veneer industry identifying several practical issues in the model building, data availability, and the use of unstructured data. It was suggested that the key concerns in being able to build and implement digital twins are related to the data availability and how to utilize it efficiently especially in the case of unstructured datasets that are traditionally utilized only by the human operators for high-level decisions. A second topic that the scientific literature is not addressing adequately is the absence of computational models of physical processes that are used for day-to-day operations on the factory floor that might be fitted as components of the larger, system-wide digital twin. The summary of insights is provided in Table 32.1 based on the project. It is suggested that many of the points raised are in common with other, established heavy industries as well.

TABLE 32.1

Generalized Take-Aways on Digital Twin Implementation Based on the Case-Project in the Veneer Industry

Topic / Issue	Proposed Solution
Philosophical approach to modeling	Clarify whether the customer is looking for a top-down (fast to build; not completely accurate) solution or a bottom-up (accurate; slow to build) digital twin
Choice of software	Prefer hierarchical and flexible simulation software that does not create a bottleneck for the model properties; ensure reasonable costs of licensing (vs. expected benefits) and the customer commitment to the software platform in the long term
Data/results ownership	Agree on who owns the (potential) results of the digital twin built on data and how the model can be utilized to cover similar applications with new sets of data
Lack of formalization of physical processes	Even though many industrial processes have a solid physical theory and calculations but once taken into action they are treated as black boxes by the operators which means that the computational models may be non-existent at the start of the DT project
Unavailability of digital measurements (and data) in critical parts of the process	Evaluate trade-off between the effort of automation versus human effort and accept the fact that certain things are better to be left for human operators; be critical on assessing whether the case data available has the pre-requisites for DT
Legacy equipment that are incompatible with the machine-to-machine communication paradigm	Start your DT-project using a bottom-up approach with the equipment that are sophisticated enough for model building. Proven results with tangible value serve as a motivation to upgrade weak spots of the overall process
Unstructured data formats (such as images) serve as a basis for humans' decision-making	At the beginning of the project, explore ways to formalize unstructured critical data (assumed to be available) while preserving its distinctiveness and reasonable size. It is unlikely that ready-made solutions exist for context-specific industrial cases and therefore the data pre-processing and dimensionality reduction can take time more than budgeted

(Continued)

TABLE 32.1 (*Continued*)

Generalized Take-Aways on Digital Twin Implementation Based on the Case-Project in the Veneer Industry

Topic / Issue	Proposed Solution
Data representativeness and data labeling	Be aware of the data gathering restrictions and (un)availability of metadata for historical data. Even big datasets have little to no value to algorithm development if they lack proper timestamps (or other metadata) and labeling
Reliability and validity	Due to uncertainty in the manufacturing process, dynamic changes in real-time data often cause volatility in the models and even declare the earlier obtained model invalid once new data are being introduced
Dimensionality reduction and de-noising	In the industrial manufacturing process, multidimensional data is ubiquitous. It provides ample information but also poses significant challenges to data mining and pattern recognition methods due to its sparseness and redundancy

Despite the practical challenges, the paradigm of DT is worthy of further research and (attempts of) application. It should be highlighted that within this research, the focus of model building was on a prototypical veneer production line with an amount of automation that is typical for such systems. As a topic of future research or application, one could cover a state-of-the-art veneer production line that would be, from the beginning, designed to meet the requirements of Industry 4.0 using the latest technological equipment available, thus circumventing some of the foundational issues identified in this chapter.

Acknowledgment

This research acknowledges the support from the Finnish Strategic Research Council project #335980 and #335990.

References

Adeyeri, M. K., Mpofu, K., & Adenuga Olukorede, T. (2015). Integration of agent technology into manufacturing enterprise: A review and platform for industry 4.0. *IEOM 2015-5th International Conference on Industrial Engineering and Operations Management, Proceeding*. https://doi.org/10.1109/IEOM.2015.7093910.

Ahmed, S. S., Cool, J., & Karim, M. E. (2020). Application of decision tree-based techniques to veneer processing. *Journal of Wood Science, 66*(1). https://doi.org/10.1186/S10086-020-01904-0.

Alam, K. M., & Saddik, A. El. (2017). C2PS: A digital twin architecture reference model for the cloud-based cyber-physical systems. *IEEE Access*, *5*, 2050–2062. https://doi.org/10.1109/ACCESS.2017.2657006.

Cheng, Y., Chen, K., Sun, H., Zhang, Y., & Tao, F. (2018). Data and knowledge mining with big data towards smart production. *Journal of Industrial Information Integration*, *9*, 1–13. https://doi.org/10.1016/J.JII.2017.08.001.

Çolak, S., Çolakoğlu, G., & Aydin, I. (2007). Effects of logs steaming, veneer drying and aging on the mechanical properties of laminated veneer lumber (LVL). *Building and Environment*, *42*(1), 93–98. https://doi.org/10.1016/J.BUILDENV.2005.08.008.

Demirkir, C., Özsahin, Ş., Aydin, I., & Colakoglu, G. (2013). Optimization of some panel manufacturing parameters for the best bonding strength of plywood. *International Journal of Adhesion and Adhesives*, *46*, 14–20. https://doi.org/10.1016/J.IJADHADH.2013.05.007.

Grieves, M., & Vickers, J. (2017). Digital twin: Mitigating unpredictable, undesirable emergent behavior in complex systems. In F. J. Kahlen, S. Flumerfelt, & A. Alves (Eds.), *Transdisciplinary Perspectives on Complex Systems: New Findings and Approaches* (pp. 85–113). Springer International Publishing. https://doi.org/10.1007/978-3-319-38756-7_4.

Han, C., Zhan, T., Xu, J., Jiang, J., & Lu, J. (2015). Process optimization for multi-veneer hot-press drying. *Drying Technology*, *33*(6), 735–741. https://doi.org/10.1080/07373937.2014.983243.

Haralick, R. M., Shanmugam, K., & Dinstein, I. (1973). Textural features for image classification. *IEEE Transactions on Systems, Man, and Cybernetics*, *SMC-3*(6), 610–621. https://doi.org/10.1109/TSMC.1973.4309314.

Houari, R., Bounceur, A., Kechadi, M. T., Tari, A. K., & Euler, R. (2016). Dimensionality reduction in data mining: A Copula approach. *Expert Systems with Applications*, *64*, 247–260. https://doi.org/10.1016/J.ESWA.2016.07.041.

Ishwarappa, & Anuradha, J. (2015). A brief introduction on big data 5Vs characteristics and hadoop technology. *Procedia Computer Science*, *48*(C), 319–324. https://doi.org/10.1016/J.PROCS.2015.04.188.

Jalonen, T., Laakom, F., Gabbouj, M., & Puoskari, T. (2021). Visual product tracking system using siamese neural networks. *IEEE Access*, *9*, 76796–76805.

Kotsiantis, S. B., Zaharakis, I. D., & Pintelas, P. E. (2006). Machine learning: A review of classification and combining techniques. *Artificial Intelligence Review*, *26*(3), 159–190. https://doi.org/10.1007/s10462-007-9052-3.

Kourti, T., Lee, J., & Macgregor, J. F. (1996). Experiences with industrial applications of projection methods for multivariate statistical process control. *Computers & Chemical Engineering*, *20*(SUPPL.1), S745–S750. https://doi.org/10.1016/0098-1354(96)00132-9.

Li, L. (2018). China's manufacturing locus in 2025: With a comparison of "Made-in-China 2025" and "Industry 4.0." *Technological Forecasting and Social Change*, *135*, 66–74. https://doi.org/10.1016/j.techfore.2017.05.028.

Lu, Y. (2017). Industry 4.0: A survey on technologies, applications and open research issues. *Journal of Industrial Information Integration*, *6*, 1–10. https://doi.org/10.1016/J.JII.2017.04.005.

Madni, M. A., Madni, C. C., & Lucero, D. S. (2019). Leveraging digital twin technology in model-based systems engineering. *Systems*, 7(1), 7. https://doi.org/10.3390/systems7010007.

Min, Q., Lu, Y., Liu, Z., Su, C., & Wang, B. (2019). Machine learning based digital twin framework for production optimization in petrochemical industry. *International Journal of Information Management, 49*, 502–519. https://doi.org/10.1016/j.ijinfomgt.2019.05.020.

Negri, E., Fumagalli, L., & Macchi, M. (2017). A review of the roles of digital twin in CPS-based production systems. *Procedia Manufacturing, 11*, 939–948. https://doi.org/10.1016/j.promfg.2017.07.198.

Qi, Q., & Tao, F. (2018). Digital twin and big data towards smart manufacturing and Industry 4.0: 360 degree comparison. *IEEE Access, 6*, 3585–3593. https://doi.org/10.1109/ACCESS.2018.2793265.

Rajnai, Z., & Kocsis, I. (2018). Assessing Industry 4.0 readiness of enterprises. *SAMI 2018- IEEE 16th World Symposium on Applied Machine Intelligence and Informatics Dedicated to the Memory of Pioneer of Robotics Antal (Tony) K. Bejczy, Proceedings*, 2018-February. https://doi.org/10.1109/SAMI.2018.8324844.

Roblek, V., Meško, M., & Krapež, A. (2016). A complex view of Industry 4.0. *SAGE Open, 6*(2), 215824401665398. https://doi.org/10.1177/2158244016653987.

Rosen, R., von Wichert, G., Lo, G., & Bettenhausen, K. D. (2015). About the importance of autonomy and digital twins for the future of manufacturing. *IFAC-PapersOnLine, 48*(3), 567–572. https://doi.org/10.1016/j.ifacol.2015.06.141.

Stark, R., Kind, S., & Neumeyer, S. (2017). Innovations in digital modelling for next generation manufacturing system design. *CIRP Annals, 66*, 169–172. https://doi.org/10.1016/j.cirp.2017.04.045.

Tuegel, E. J., Ingraffea, A. R., Eason, T. G., & Spottswood, M. S. (2011). Reengineering aircraft structural life prediction using a digital twin. *International Journal of Aerospace Engineering*, 14. https://doi.org/10.1155/2011/154798.

Urbonas, A., Raudonis, V., Maskeliunas, R., & Damaševičius, R. (2019). Automated identification of wood veneer surface defects using faster region-based convolutional neural network with data augmentation and transfer learning. *Applied Sciences, 9*(22), 4898. https://doi.org/10.3390/APP9224898.

Veza, I., Mladineo, M., & Gjeldum, N. (2015). Managing innovative production network of smart factories. *IFAC-PapersOnLine, 28*(3), 555–560. https://doi.org/10.1016/J.IFACOL.2015.06.139.

Wang, J., Xu, C., Zhang, J., & Zhong, R. (2022). Big data analytics for intelligent manufacturing systems: A review. *Journal of Manufacturing Systems, 62*, 738–752. https://doi.org/10.1016/J.JMSY.2021.03.005.

Xu, S., Lu, B., Baldea, M., Edgar, T. F., Wojsznis, W., Blevins, T., & Nixon, M. (2015). Data cleaning in the process industries. *Reviews in Chemical Engineering, 31*(5), 453–490. https://doi.org/10.1515/REVCE-2015-0022.

Zebari, R. R., Mohsin Abdulazeez, A., Zeebaree, D. Q., Zebari, D. A., & Saeed, J. N. (2020). A Comprehensive Review of Dimensionality Reduction Techniques for Feature Selection and Feature Extraction. *Journal of Applied Science and Technology Trends, 1*(2), 56–70. https://doi.org/10.38094/jastt1224.

Zheng, T., Ardolino, M., Bacchetti, A., & Perona, M. (2020). The applications of Industry 4.0 technologies in manufacturing context: A systematic literature review. *International Journal of Production Research, 59*(6), 1922–1954. https://doi.org/10.1080/00207543.2020.1824085.

Zhong, R. Y., Xu, X., Klotz, E., & Newman, S. T. (2017). Intelligent manufacturing in the context of Industry 4.0: A review. *Engineering, 3*(5), 616–630. https://doi.org/10.1016/J.ENG.2017.05.015.

33

Experiments as DTs

Jascha Grübel
ETH Zürich

DTs (DTs) have become the buzzword of the decade, permeating research, industry and even governance as indicated by Gartner's hype measure [47]. At the same time, DTs are highly under-specified and each DT is based on a slightly different conceptualisation. So far, most DTs have focused on the digitalisation of crucial components of a system [18]. More recent work has unveiled a methodology to describe the five core components to construct effective DTs [21,47] but other conceptualisations still prevail [49]. The DT is said to arise by combining five environments that take on different roles in the formation of a DT (see Figure 33.1 and Ref. [21]). First, the Physical Environment captures and changes the reality of the Physical Twin (PT) through sensors and actuators.[1] Second, the Data Environment stores information required to form the DT. This can range from cloud services and subscription to local sensor networks to databases and data warehouses. Third, the Analytical Environment is taking the information as input to some modelling, simulation or analysis. Fourth, the Virtual Environment enables the interaction of users with content in all environments through dashboards [5] and immersive analytics [32] such as Fused Twins [20]. Lastly, the Connection Environment brings together the other Environments by providing APIs and Access Control and standardisation [46]. All connections are bi-directional and allow DTs to interact with the PT. This conceptual framework is generalised to accommodate all kinds of DTs regardless of actual implementation.

[1] While an *Ideal DT* [21] would perfectly match the PT, real-word applications usually limited themselves to a *Practical DT* [21], which only captures important aspects of the PT relevant to the task.

DOI: 10.1201/9781003425724-39

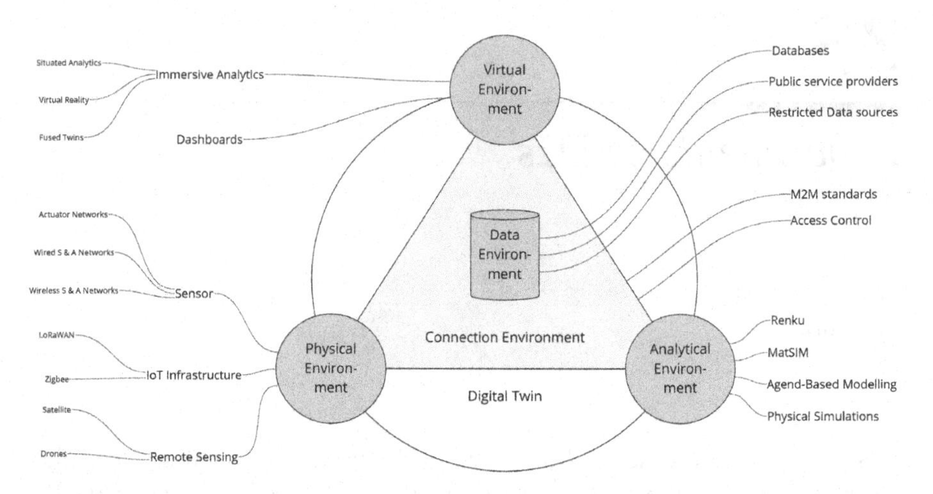

FIGURE 33.1
Overview of Digital Twin components. Five environments (shades of grey) form the Digital Twin (white circle). Example content/implementations of the respective components are shown in the mind map. The Physical Environment captures the reality of the PT to the required degree. The Data Environment stores information necessary for the Digital Twin. The Analytical Environment uses Machine Learning platforms (e.g. Renku) and simulation platforms (e.g. MatSIM) to obtain information on higher order processes. The Virtual Environment provides access to the Digital Twin through Immersive Analytics or Dashboards. The Connection Environment.

IDEAL VS PRACTICAL DTs [21]

DTs have not been theorised on much and "in the wild" are mostly *practical* solutions to problems. As far as theoretisation goes, DTs have been looked at from the perspective of similitude [16] which postulates that an *ideal* Digital Twin conforms to a multi-physics, multi-scale, probabilistic, ultra-fidelity simulation that mirrors a PT based on historical data, real-time sensor data and a physical model. However, except for very specific cases such as air plane turbines ideal DTs have been impractical. Universal ideal DTs implementing all aspects of similitude do not exist and are only planned in the near future for air planes [48]. Partial practical DTs are more common that cover some aspect of similitude and have been labelled according to their degree of completeness from digital models and digital shadows to (ideal) DTs [42,44]. However, instead of hard classifications, this chapter instead promotes to look at the environments of a Digital Twin and assess the implementation according to the completeness of each environment.

However, there is another side to consider beyond representation and interaction where DTs are considered as the ultimate platform for experiments. Experiments are conducted in research to understand mechanisms, in industry to test components and in policy to comprehend trade-offs in alternatives. Experiments are often conducted *ad hoc*[2] and the experiment process itself is not part of the design [1]. Consequently, it is often difficult to replicate the outcome of experiments [9] which can have severe implications in research, industry and governance. Defining experiments as DTs can change this and make experiments in all contexts more transparent and reproducible. Furthermore, the theoretical frame of experiments allows gaining better insights into what DTs are and how they differ from their cognates (i.e. systems with similar properties but different names [10]).

33.1 DTs

The concept of DTs was coined in 2003 [18] to conceptualise the matching of the physical world with a digital representation through some kind of data collection process in an industrial context. The idea stuck around and slowly emancipated itself from its industrial roots expanding to various other fields and disciplinary contexts where the simple description makes sense to abstract processes like in smart cities [6]. While DTs have found wide applications, they have remained under-specified and carry a very different meaning in different communities. For example, in aviation, DTs stand for similitude of the digital representation to the highest possible physical degree [16]. On the other hand, for Smart Cities, DTs only require eventual consistency [6], i.e., representing some past abstract state of the city correctly such as traffic load or the population in contrast to properly simulating the physical movement of cars and people as similitude would require.

DTs (DTs) are both innovative and at the same time just an iteration of what has come before. This relation is characterised in the term cognates [10] which originally describes words of similar origins and here is transferred to technologies from similar origins. DTs reflect aspects of previous cognate systems [21] such as Cyber-Physical-Systems (CPS) [52], the Internet of Things (IoT) [4], wireless sensor and actuator networks (WSAN) [2], machine-to-machine (M2M) [3], remote sensing (RS) [8], ubiquitous and pervasive computing (UPC) [51], smart object (SO) [41] and computation-on-the-wire (COW). These systems often partially implement DTs (see Table 33.1) and together seem to share all properties of a Digital Twin. Ultimately, a Digital Twin goes beyond these systems by combining many key features.

[2] Meaning that the experiment is conducted for its own sake rather than as part of triangulation or systematic replication [37].

TABLE 33.1
Cognates of DTs [21]

	CPS	IoT	WSAN	M2M	RS	UPC	SO	COW
Scope								
System	Cl	Op	Op	Op	Op	Op	Cl	Op
Location	Sta	Both	Both	Both	Sta	Sta	Mob	Both
Environment								
Physical	Yes	Yes	Yes	-	Yes	Yes	Yes	-
Data	Yes	*	-	-	-	Yes	Yes	Yes
Analytical	Yes	-	-	-	-	*	*	Yes
Virtual	*	-	-	-	-	*	Yes	*
Connection	Yes	*	*	Yes	*	*	*	Yes

Note: -Not defined; *Possible but not strictly necessary
Abbrevations: Sta, Stationary; Mob, Mobile; Op, Open; Cl, Close.

CPSs [52] are the oldest term and the most general which is why CPS best matches a Digital Twin most closely in terms if not in scope. The interaction between the physical and the cyber (digital) is at the focus and an emphasis is put on control theory and the management of closed stationary systems such as factories or machines [21]. However, CPS never had the aspiration to scale up global systems or represent open systems. Nonetheless, CPSs have been extended to Cyper-Physical-Social Systems (CPSS) where the human component is given more space and open systems are also considered.

The Internet of Things [4] was always considered as an open system that mimics the Internet itself. The comparison to the Digital Twin appears obvious at first but breaks down under scrutiny. The IoT is mostly occupied with how to extract information from the physical environment. Newer IoT stack also contain data processing but that can be understood in the context of Connection Environment as necessary transformations of the data. Ultimately, IoT is agnostic to its use and therefore more a practical implementation of the physical environment for a Digital Twin than a model for Digital Twin itself. Nevertheless, many IoT solutions are actually DT when adding data processing, data analysis, and data visualization capabilities..

Wireless Sensor and Actuator Networks [2] are the main stay of any IoT and CPS solution. As such they are deeply connected to how DTs work. They provide the foundation on which data are extracted from the physical environment. While they are integral to producing an IoT, they are a mere component and should not be mistaken for a Digital Twin.

Machine-to-machine communication [3] is an integral component of a Digital Twin as the complex composition of machines needs to coordinate themselves. A centralised approach is both too complex and does not scale automatically as the number of sensor devices increases. M2M does not have any inherent model for data, analysis or visualisation and therefore cannot be equated with a Digital Twin.

Remote Sensing [8] is a difficult term because it has earned two separate meanings over time. In a geographic context, Remote Sensing is mostly about camera-based sensors on drones or satellites whereas in IoT and WSAN the word is often meant as readings from 'far away' sensors. This conflagration of terms isn't very useful and here we focus on the camera-based version as the other is technically covered by general IoT. Tracking large-scale systematic changes with local sensing equipment can be difficult. Remote Sensing offers an opportunity to capture the open system around the physical environment of a Digital Twin. While Remote Sensing is too coarse to produce fine-tuned DTs, it can be an important tool to complete DTs.

> The most profound technologies are those that disappear. They weave themselves into the fabric of everyday life until they are indistinguishable from it.
>
> **Mark Weiser**
> *Former CTO at Xerox PARC*

UPC [51] idealised a Digital Twin before the term was coined by Michael Grieves [18]. UPC imagines technology to be disappearing into the background of everyday life while facilitating common activities. In its original formulation, it was aimed as a call for action rather than a particular set of technologies. It is often also called Ambient Intelligence [14] pointing at a clever environment that supports users in their everyday life. The parallels to the Digital Twin are obvious but there is no explicit data-orientation.

SOs [41] follow in the footsteps of UPC by making everyday objects intelligent. However, SOs focus on a encapsulated Digital Twin within the PT. It is also a Augmented Virtuality [21,34] representation of Fused Twins [20] embedding the Digital Twin *in situ* of the PT.

Computations-on-the-wire [7] implies the processing of data as they are transferred from a source to a destination. It also summarises hydrological paradigms [21] such as cloud, fog, mist and dew computing into a continuum that describes the optimal computing location according to the subsidiarity principle [21]. In the context of computing, the subsidiarity principle can be understood as aiming to perform the computation as close as physically possible and as far away as technically necessary while transporting data from the source to the user. While computations-on-the-wire are not directly equated with DTs, they feature prominently in the discourse how to create DTs and are often used as implicit standins for the Connection Environment but also as implementations for Data, Analytical and Virtual Environments. When talking about the infrastructure required to build a Digital Twin, the fuzzy distinction between hydrologically inspired terms requires a more abstract view for the trade-offs encapsulated in the subsidiarity principle and computations-on-the-wire (Figure 33.2).

In this section, we developed a deeper understanding for the components of DTs. We also reviewed cognates of DTs to see nuanced differences between

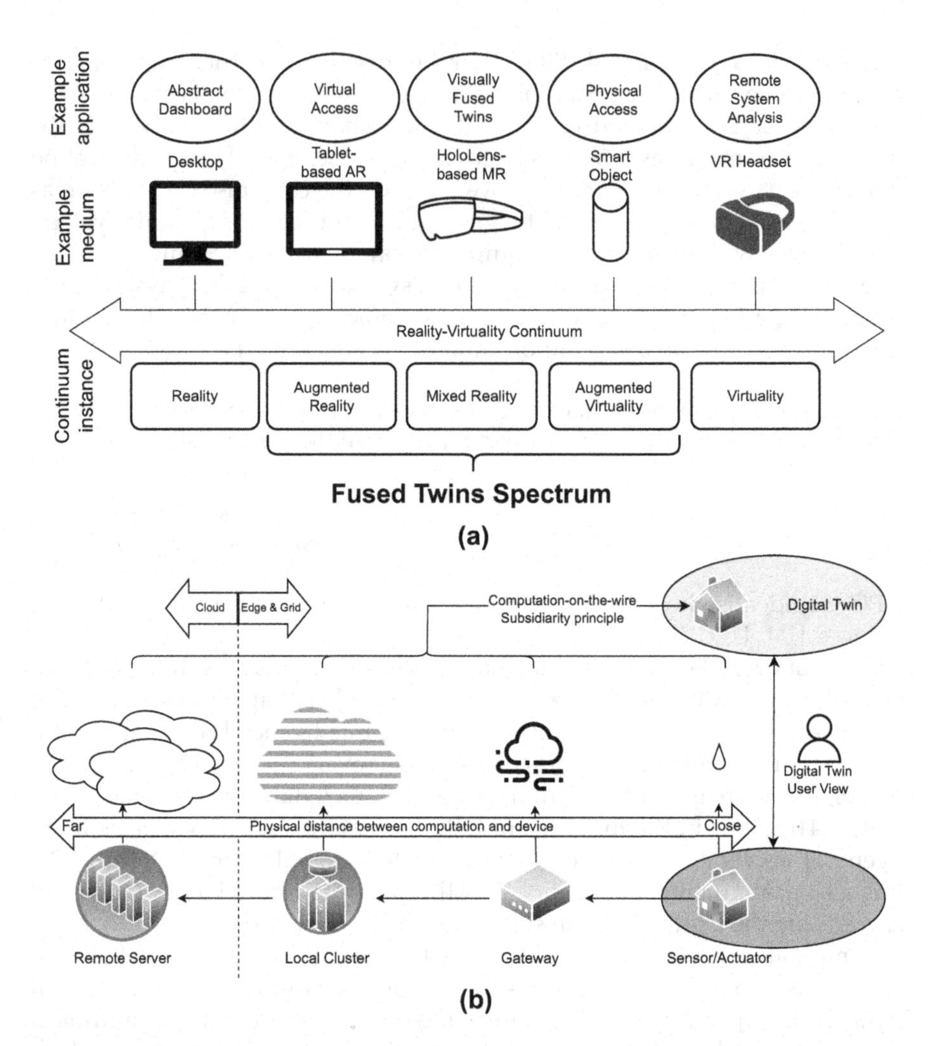

FIGURE 33.2
(a) DTs' Interaction [21] on the Reality–Virtuality–Continuum [34]. The continuum spans from reality via different modes of Extended Reality (XR) to Virtuality or Virtual Reality. Different input modalities are associated with different forms of XR from Augmented Reality (AR) on tablets via Mixed Reality (MR) through head-mounted holographic displays to Augmented Virtuality (AV) in Smart Objects. In full reality, access to DTs focuses on abstract dashboards, whereas in full virtuality, access focuses on remote analysis of the Digital Twin. Image licensed under Create Commons [21]. (b) Computation-on-the-wire. Computations form the basis for any Digital Twin and can be performed "on-the-wire" between the producer and the consumer of the data. The computing location can be locally at the sensor/actuator or more remotely at the gateway, a local cluster or remote server. Trade-offs in computational intensity, traffic, latency, accuracy, energy consumption and security [14,40] map to a hydrologically inspired terminology (i.e., cloud, fog, mist and dew). The hydrological terms encapsulate trade-offs according to the subsidiarity principle to decide where to place the computation physically. DTs hide computational complexity from the end users' view through abstraction while relying on many different data and computing sources. (Images licensed under Create Commons [21].)

a Digital Twin and its technological siblings or standins. Equipped with a detailed understanding of DTs, we also study experiments in more detail before we come to the synthesis of both.

33.2 Experiments

DTs exist for humans to be able to interact with digital representations of reality through Human–Computer Interaction (HCI) [28,30]. At the same time, HCI is also used to understand how humans behave through behavioural experiments [13,50]. Experimental psychology has developed three main methodologies of assessing behaviour: observational measures, physiological measures and subjective measures [13]. In common across all methodologies is to collect data. Experiments require rigorous data collection which has been shown to be a major difficulty through replication attempts [9]. In recent times, Virtual Reality (VR) has often been used to produce a fully controlled environment and collected data [24]. Several generations of frameworks have been developed for the task [1]. First-generation frameworks focused on enabling the hardware for VR experiments. Second-generation frameworks enriched experimental control by focusing on setting up experiments. Recently, a third generation of frameworks is emerging that puts reproducibility and automation at the forefront. To understand the requirements for implementing experiments that "cannot with advantage be omitted" [38], the six pillar taxonomy of behavioural experiments was developed [1]; see Figure 33.3.

> Science may be described as the art of systematic oversimplification – the art of discerning what we may with advantage omit.
>
> **Karl Popper**
> *Philosopher of Science*

The first pillar covers documentation of an experiment and includes information on the design, registration and management of the experiment. Typical documents include the experimental protocol, a pre-registration of the design, an analysis plan, readmes and a data management plan. A major cause for the replication crisis [29] is the inability to reproduce the exact conditions of an experiment because knowledge is transferred *viva voce*[3] [27].

Proper documentation enables subsequent repetitions of an experiment without personal knowledge of the original experiment.

The second pillar covers the infrastructure and environment of an experiment and includes the equipment, servers and virtual machines. It contains

[3] Latin: literally translated "with living voice", commonly translated "by word of mouth".

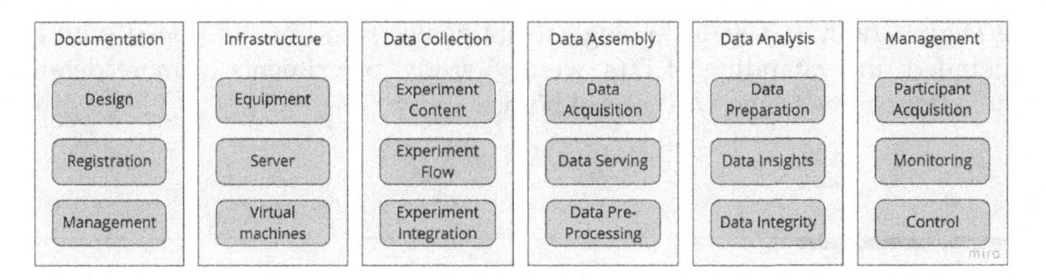

FIGURE 33.3
Six pillars taxonomy of behavioural experimentation. The pillars are documentation, infrastructure and environment, data collection, data assembly, data analysis and management. Each pillar consists of multiple essential tasks that need to be completed [1].

a precise definition of data processing equipment, content delivery systems, participant equipment, analysis equipment, researcher equipment and images and containers for virtual machines. Preproducibility [45] considers experiments to be "recipes" and in that context, infrastructure can be considered the list of required ingredients.

The third pillar covers the data collection of an experiment and includes experiment content, experiment flow and experiment integration. Data need to be collected from different sources describing the content of an experiment like tasks, questionnaires, scores, sensor measurements but also the flow of the experiment like order, timing and interactions. These data then should be served through some kind of Application Programming Interface (API) to manage the data collection and transfer data. Some efforts to implement this already happened in second-generation frameworks [24] but generally, data collection has not been conducted with rigour and was often limited to key experimental variables instead of the overall experiment resulting in gaps in the documentation that ultimately reduced reproducibility.

The fourth pillar covers data assembly of an experiment and includes data acquisition, data serving and data pre-processing. The collected data are distributed across different machines (experiment data itself, the infrastructure used to run the experiment, physical sensors connected to participants and metadata) and need to be integrated in the data acquisition process. A data API must be available to serve the data for analysis and visualisation. Lastly, data pre-processing often must take place because data may be too large to store, not anonymous yet or badly structured because of proprietary sources. Data assembly is really an orphaned aspect of an experiment as data are often dumped in text files with little order and no thoughts for FAIR data [53].

The manifesto for FAIR data states that all data should be findable, accessible, interoperable and reusable. Therefore, FAIR data required a data base, indexing, metadata and a standardised format, which is usually lacking in experiment implementations [1].

The fifth pillar covers data analysis of an experiment and includes data preparation, data insights and data integrity. Data need to be inspected and validated before it can be wrangled and edited. Often, data visualisation is a useful tool when preparing and checking the data. Lastly, insights are gained with statistical analysis, machine learning and simulations. Data analysis is often decoupled from the data collection process, which can lead to problematic outcomes [17,35]. Pre-registration is supposed to lead to planning the analysis before conducting a study. However, the technical infrastructure also plays a role because often analysis is conducted as black boxes to reviewers and readers and end up not being replicable. Data processing platforms such as Renku[4] allow for a systematic analysis and documentation of analysis that is required to bring experiments to the next level.

PRE-REGISTRATION [17]

Performing analysis after the fact (of collecting the data) enables wrong interpretation of information. When data are interpreted to come to a conclusion, the results may not hold under replication. Typical actions include p-hacking, publication bias, data dredging and inappropriate forms of post hoc analysis. Hypothesising after the Results are Known (HARKing) [31] sums up these behaviours. Pre-registration provides an alternative by defining the analysis before it is applied reducing errors of the following kind [31]:

- Integrating Type I errors into accepted theory
- Proposing results that do not hold under replication
- Pretending post hoc explanations are *a priori* explanations
- Loosing valuable information about what did not work
- The application error of statistical method
- Producing an inaccurate model of science
- Encouraging 'fudging' in grey areas
- Reducing serendipitous findings
- Encouraging adoption of too narrow and context-bound theory
- Encouraging retention of old (wrong) theory
- Denying plausible alternative hypotheses
- Violating basic ethical principles

Pre-registration is storing the proposed experiment and analysis plan in a public repository and therefore allowing for clear expectations of the original data collection. Note that post hoc analysis is not per se wrong but it needs to be clearly declared as such to enable proper understanding and ultimately replication of experiments.

[4] https://renkulab.io/.

The sixth pillar covers the management of an experiment and includes participant acquisition, monitoring and control. Participants need to be recruited according to the criteria of the experiment, their payment has to be coordinated and participants have to be connected to measuring hardware and possibly be immersed in a VR setup. These steps require monitoring especially in an automated online context [1]. During an experiment, additional monitoring of an experiment is required for quality control. It is important to oversee progress, enable live observation of the ongoing experiment and check the status of the infrastructure while the experiment is ongoing. In general, these kinds of applications are neither reported about nor documented well. Consequently, the quality of data acquisition is often hard to judge and may be another reason why experiments are not reproducible.

The "Experiments as Code" (ExaC) Paradigm was proposed to address these issues in VR experiments [1] by implementing a framework that conforms to the six pillars. The resulting framework is the first third-generation VR framework although it is not feature complete [19]. It is too early to tell whether the scientific community is up to adopt these methods because they require a certain overhead but the options are now available.

33.3 Synthesis

The previous two sections have given us a quick introduction into both DTs and Experiments in VR. There are strong parallels between both and some slight differences, see Figure 33.4. In the following, we will equate participants of an experiment with the PT of a Digital Twin. It follows that an Experiment and a Digital Twin are similar if not the same. In this section, we will compare both and gain new insights on each informed by the other. Ultimately, we can learn how to combine both to gain more out of Experiments and more out of DTs.

First, we observe the similarity in the collection and assembly of data between Experiments and DTs. The taxonomy's pillar of Data Collection in an experiment equates to the physical environment of a Digital Twin. Data collection in an experiment is tasked with gathering measurements of key properties of a participant to test a hypothesis. In contrast, or rather parallel, the Physical Environment is gathering key properties of a PT to represent a Digital Twin. We can re-imagine the experiment to construct a Digital Twin of the participant. However, the Physical Environment is more than just the measurements that we ultimately store. It contains the full PT including all the possible measurements that we forego for practical reasons. Transferring this imagery to Experiments opens up our view for triangulation [37] by being able to include new variables and measurements and models. If we understand an Experiment as a Digital Twin and conduct an Experiment in

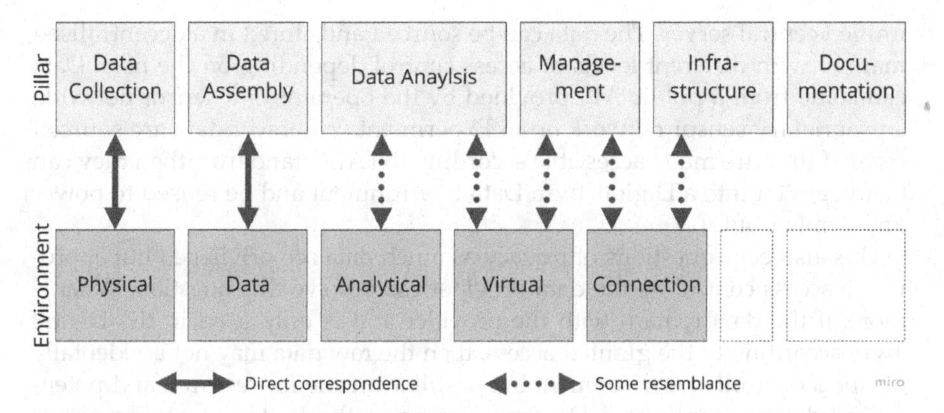

FIGURE 33.4
Comparison of DTs and Experiments as Code. There is a direct correspondence between Physical and Data Environment and Data Collection and Assembly, respectively. There is some overlap between Data Analysis and Analytical and Virtual Environment as well as between Management and Virtual and Connection Environment. Lastly, Infrastructure overlaps with the Connection Environment but also contains definitions of the hardware used which is absent in the Digital Twin model. The Digital Twin has no category for Documentation.

a Digital Twin, then we can easily implement triangulations and reuse DTs for reproduction.

TRIANGULATION [37]

Triangulation is an alternative to replication. Replication implies to exactly repeat the same experiment and come to the same conclusion in the sense of Popper [38]. In order to obtain proper copy preproductivity [45], documentation is required at each step, for example in the code method [1]. However, there are internal biases that can appear through HARKing [31] that suggest a successful replication without actually proving a theory. Triangulation overcomes these issues by not only replicating the exact same experiment but also thinking up alternative experiments that show the same outcome on different variables and through different measurements and models. Ultimately, triangulation allows developing a more robust proof for hypotheses.

Second, having established data collection, we can discuss data assembly. Again, there is a match between the pillar of Data Assembly and the Data Environment. In an Experiment, it is crucial to structure data according to the FAIR principles. However, DTs also benefit from a FAIR structure [43]. A Digital Twin may inspire the comparison to a digital avatar of the PT but the comparison is lacking. Data storage for a Digital Twin does not require a unified central server. The data can be sourced and stored in a decentralised manner with different levels of access control depending on the user. Data can come from a public API provided by the operator of a sensor network, a proprietary sensor network or an Experiment. Wherever data are sourced from, if they are made accessible according to FAIR standards, then they can be integrated into a Digital Twin Data Environment and be reused to power any number of DTs.

This also begs questions of privacy as much data are privileged but appropriate access control by the data provider can resolve this question. What is more, if the data remain with the provider and is only used in the Digital Twin according to the granted access, then the raw data may not accidentally escape a controlled environment but is still available for research and potentially industrial application. Data can even be synthesised [36] for wider access to allow some operations on otherwise restricted data. Such a Digital Twin model would be beneficial in research, industry and government to use the best possible data sources without damaging the integrity of the data provider by maintaining privacy, anonymity and restricted access where necessary.

Third, we can look at the partial match of Data Analysis in the six pillars with the Analytical and Virtual Environment. The Analytical Environment can be fully represented with the Data Analysis pillar. In recent experiment methodologies, analysis remains separated from the conducting of the experiment. However, the Experiments as Code Paradigm are more similar to DTs where the analysis of data is an integral component of the overall system. Integrated analysis helps with issues of reproducibility and can provide high-quality statistics by automatically ensuring the validity of data.

However, there is crucial difference in the temporal dimension between DTs and Experiments. In DTs, the analytics are an automated ongoing procedure to maintain a representation of the PT. In Experiments, the analysis is a snapshot in time that is (often) manually analysed for a singular purpose (hypothesis testing). Here, both sides can learn from one another. DTs have the capacity for hypothesis testing and can continuously evaluate hypotheses allowing for temporal triangulation through regular repetition. At the same time, Experiments can benefit from regularised analytics. While there may not be typically continuous recordings (but see Sea Hero Quest [12] and SPACE [11]), the Analytical Environment allows to maintain typical tests for DTs of similar experiments. Properly applied, DTs could help Experiments to be conducted at a more rigorous standard.

Fourth, the Virtual Environment finds some resemblance in the Data Analysis and the Management pillar which in turn also has overlap with the Connection Environment which in turn has an overlap with Infrastructure. The split can be explained by a difference in focus of the Digital Twin and Experiments. In DTs, Analytics are automated and the Virtual Environment provides visual access and interactions for the analytics. In Experiments, Analysis often includes visualisation and therefore overlaps with the Virtual Environment. Similar observations apply for Management. Here, in DTs,

User Interaction is focused on the Virtual Environment and as such Management functionality is exposed to the user there. However, the actual management is conducted in the Connection Environment. In Experiments, Infrastructure, Management and Analysis are clearly separated because neither setting up the experiment nor managing the experiment is considered part of the experiment analysis. In DTs, (automated) analytics, (virtual) visualisations and connections between environments are viewed separately because modular exchangeable units are used for analytics and virtual environments, whereas the Connection Environment is the invisible glue (management) of the Digital Twin. The different foci set modularity against semantic boundaries when designing tasks. However, these differences can easily be overcome by delineating clearly between management and analytics in DTs and between modelling and visualising in experiments.

Fifth and last, there are certain pillars such as Documentation and parts of Infrastructure that find no correspondence in the Digital Twin. This is a severe lacking in definition. The Experiments as Code paradigm postulates that provisioning and deploying are critical steps for any open research data practice. A Digital Twin therefore should not only be well-defined internally but also provide instructions on how to provision for it and how to deploy it. Obviously, there are automated solutions to this [1], but documentation is critical to enable new users to engage with the technology.

Summarising this comparison, three main points should be remembered. First, DTs and Experiments as Code have a near identical understanding of how to collect and curate data. Second, DTs have a stronger focus on modularity of components, whereas Experiments more clearly delineate semantic boundaries between pillars and both can profit from taking the considerations of the other more into account. Third, DTs have a less strict requirement for documentation and information of provisioning and deployment, which ultimately hurts open research data practices. We can bring together these paradigms through running Experiments in DTs.

The VR used in many experiment applications also highlights the problem of the difference of real-world information and purely virtual information in DTs. Often, pure virtual information is considered to be an internal representation of the Digital Twin [47,49], but here I argue that the Experiments as DTs perspective gives another understanding. Using the Virtual Twin (i.e., the purely virtual information fed to a Digital Twin) as a distinct component allows us to compare it to PT; see Figure 33.5. By switching between a Virtual Twin and a PT as input for the Digital Twin, it becomes possible to implement Digital Threads in a less self-referential way. The Digital Twin does not maintain its own simulated existence but is receiving input either from the real world or a virtual world that represents the real world sufficiently for the functioning of the Digital Twin. In practise, this can mean to have a virtual world of a future building and running tests/experiments on its energy efficiency, user experience and more based on the same Digital Twin that later-on provides these services in the real world.

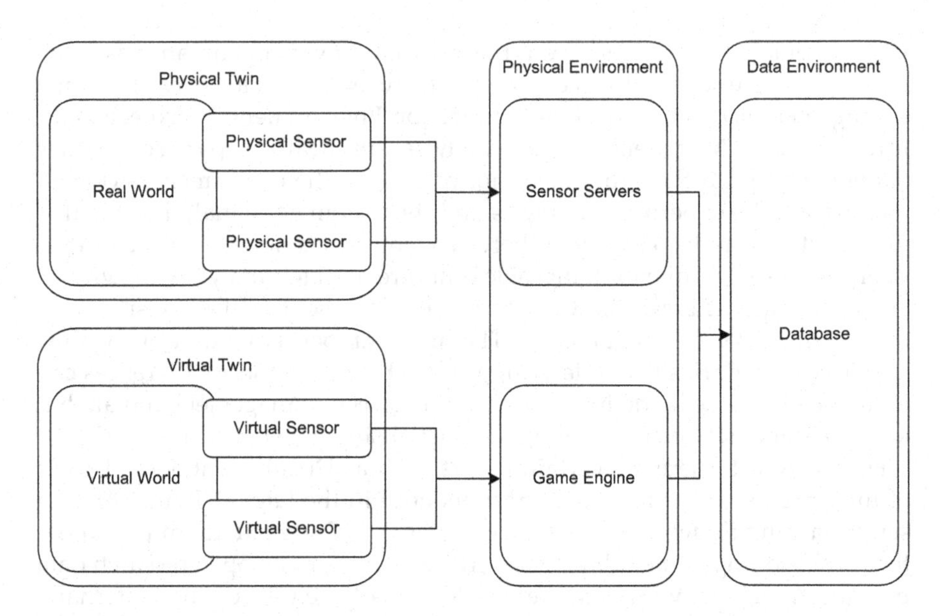

FIGURE 33.5
Virtual Twins and Physical Twins (PTs) as input for DTs. DTs are also used in Digital Threads [33] to develop the representation of a system through its life cycle from design over use to de-commission. The separation in Virtual and PT allows the use of a Digital Twin before the PT is present and beyond the use of the PT when it has already been decommissioned.

33.4 Case Study: A Building Digital Twin

In this section, we look at an example of a Digital Twin for a building at a university with a focus on understanding human behaviour in buildings [20,22,23]. The implementation of each Environment for the Digital Twin is explored in detail to give insights regarding how a system can be implemented that satisfies both the requirements of DTs and Experiments as Code. A Dense Indoor Sensor Network (DISN) [22,23] has been used in the Physical Environment. The Fused Twins paradigm [20,21] is used for the Virtual Environment. R and cogARCH [15] are used for the Analytical Environment. The Data Environment was created as an EVEREST database [22,23] and the Connection Environment is not closer defined but consists of LoRaWAN networks, virtual servers and Unity-Based Apps for XR devices [22,23].

33.4.1 The Digital Twin

The Physical Environment of the Digital Twin consists of a DISN [22,23] with 390 sensors that is deployed in the public areas of a building. The sensors capture environmental characteristics of the building, some of which are co-varying with human presence. Passive sensing [21] is a way to implement privacy by design. Here, no direct data of human activity is collected and only correlational information is available. Furthermore, the focus is on public areas to maintain privacy and anonymity of the building's users in their offices and other regularly used spaces that would help to identify individuals.

The Data Environment is implemented as the EVEREST database [23,25] consisting a POSTGRESQL database with extensions for temporal (timescaleDB) and spatial (postgis) data. The database contains both transmission metadata as well as measurements from the Physical Environment.

The Analytical Environment of the Digital Twin consists of the BIM model of the PT [23,25] that is embedded in the simulation engine cogARCH [15] and the Experiments in Virtual Environments (EVE) framework [24]. Furthermore, for management purposes, R-scripts are run regularly to understand and manage the DISN [26].

The Virtual Environment takes the BIM model and presents it in Mixed Reality [20]. The Fused Twins approach [21] is taken to embed the Digital Twin *in situ* in its PT. Furthermore, a classical VR model is available to explore the Digital Twin on a desktop [23].

The Connection Environment consists of the LoRaWAN infrastructure, a flask server that implements REST-HAL [39] and in the Unity Game Engine in the EVE framework [24]. The flask server provides a standardised interface to retrieve information from the Data Environment and pipe it into any other environment for further processing. The REST-HAL extension to the REST protocol is used to allow for automated traversal of the REST interface without prior knowledge of the API. If users acquire access to the API, then they can freely explore the data according to the FAIR principles. A game engine is used as a standardised platform to conduct spatial computations, visualise information and provide users with access to the system.

33.4.2 The Experiment

The introduced Digital Twin has been collecting and managing data since March 2020, just before the COVID pandemic's first lockdown. From the Physical Environment, a total of 183 million data points across 119.3 million transmissions have been collected and stored in the Data Environment, see Fig. 33.1. Consequently, the sensor data provide insight into a natural experiment due to the differences in measurement between the lockdown and other phases of the pandemic response [22,23]. Switzerland experienced two lockdowns and we compare the lockdown with the subsequent time of re-opening, see Figure 33.6. The lockdown lasted from March 16 2020 to April 22 2020 and from January 15 2021 to April 23 2021. The re-opening had several stages and we take the stages that allow open access to the building but permits building users to work from home which lasted from August 3 2020 to October 15 2020 and April 23 2021 to September 6 2021.

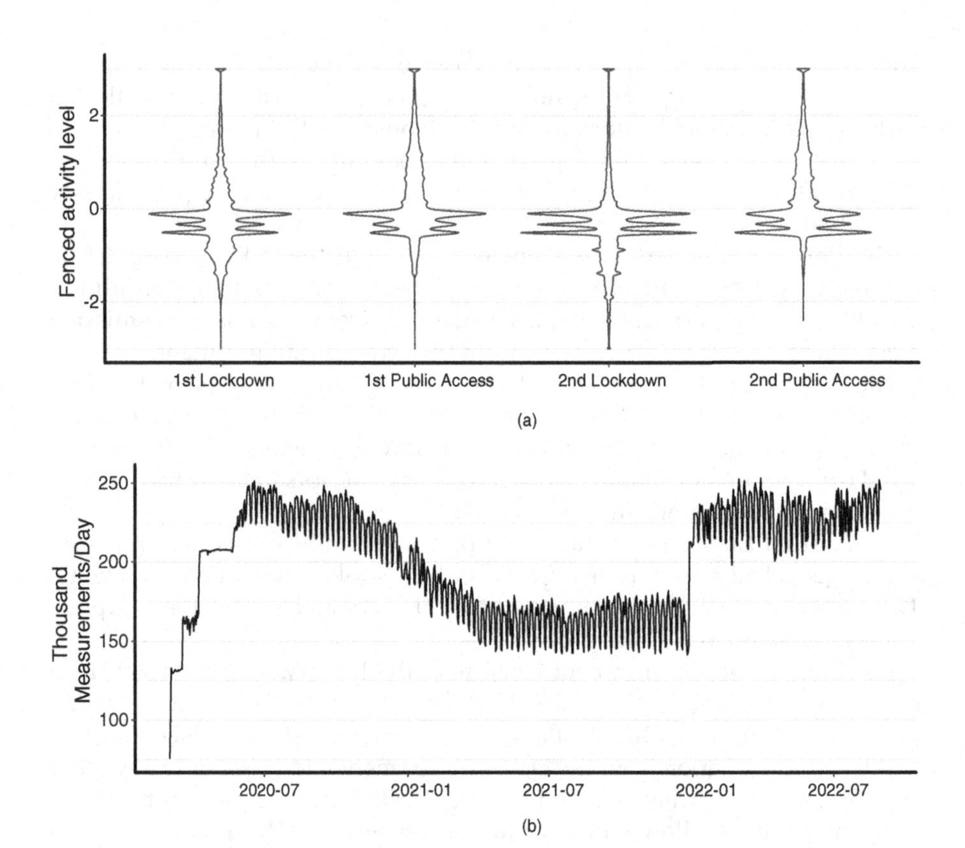

FIGURE 33.6
(a) Data Volume of Digital Twin. The DISN [22,23] in the Physical Environment has produced 183 million data points, which are stored in the Data Environment. The reduction in volume in 2021 is due to empty batteries which have been replaced in early 2022. (b) Lockdown Comparison. Activity is represented as the variation in sensors reported on a scale of three standard deviations [22,23] from the average activity over the whole period. The violin plot indicates how many measurements were done at each level of activity throughout the four observed phases. First, the daily rhythm is between 0 and –0.5, which appears independent of the phases. The width patterns seem to be wider during Lockdowns indicating some baseline state. Second, the main body of the violin is above 0 for Public Access indicating overall more activity than average, whereas the main body of the violin is below 0 for the Lockdowns indicating reduced activity.

During the lockdowns, average activity decreases to –0.198 and –0.380, respectively, whereas during public access the activity reaches 0.157 and 0.177, respectively.

Overall, the sensor system allowed us to investigate usage in the building. Initial research focused on the impact of COVID but we will continue to use the system to get more insights into spatial behaviour within the building. The Digital Twin approach makes it easy for us to add data in real time to our analysis and update our visualisations accordingly.

33.5 Conclusion

The Digital Twin is the new kid on the block and while it may appear very similar to old-timers, these cognates appear sufficiently different on closer inspection [21], which merits the attention. Furthermore, DTs have a lot of untapped potential, especially when they are combined with rigorous practices from experiments. In this chapter, the parallels between Experiments as Code [1] and DTs [21] were drawn to give insights regarding how both can profit from one another. Experiments focus on replicability and robustness through triangulation which are underpinned by a FAIR data management. DTs can also be FAIR [43] but would benefit even more from not only managing the data well but also focusing on best practices regarding how to procure data and how to describe themselves on a meta level.

DTs are still not standardised and this book is a testimony to the width and depth in which the topic is currently explored. This chapter has offered new insights regarding how to explore DTs from the perspective of Experiments. It is important to note that a technology is not developed in a vacuum but derives from other technologies and often is concurrent with its cognates. Whether a system is a new technology or just an instance of another technology is often not decided by the contemporaries but by historians. As of today, the Digital Twin seems to separate itself from its cognates as a technology of its own. By acknowledging its roots and at the same time cherishing its differences, we can form the technology towards a tool that enables a more systematic, effective and reliable tool for experimentation and testing in all kinds of context from academia to industry and government. Experimentation is *a priori* not necessarily a question of technology but more of procedure. Replication crisis [29] and more automated answers to improve replicability [1] show how automation technology can improve programs.

Experiments are conducted everywhere, all the time. However, more often than not, this is not acknowledged. Any form of testing is an experiment and could benefit from rigorous conducting. DTs as an automation platform are an optimal tool to ensure that any test is also an experiment. In this chapter, the similarity between Experiments and DTs has been made evident. An example study of a Digital Twin of a Building has been shown, which is also used to conduct experiments [20,22,23]. An outlook was also given with a second case study regarding how DTs at a national level could be designed in the near future. Ultimately, by thinking of Experiments as DTs and DTs as Experimentation platforms, we will be able to get more out of these technologies and make use of their full potential.

References

1. Leonel Aguilar, Michal Gath-Morad, Jascha Grübel, Jasper Ermatinger, Hantao Zhao, Stefan Wehrli, Robert W Sumner, Ce Zhang, Dirk Helbing, and Christoph Hölscher. Experiments as Code. Available on arXiv., 2022.
2. Ian F. Akyildiz, Weilian Su, Yogesh Sankarasubramaniam, and Erdal Cayirci. Wireless sensor networks: A survey. *Computer Networks*, 38(4):393–422, 2002.
3. Sergey Andreev, Olga Galinina, Alexander Pyattaev, Mikhail Gerasimenko, Tuomas Tirronen, Johan Torsner, Joachim Sachs, Mischa Dohler, and Yevgeni Koucheryavy. Understanding the iot connectivity landscape: A contemporary M2M radio technology roadmap. *IEEE Communications Magazine*, 53(9):32–40, 2015.
4. Luigi Atzori, Antonio Iera, and Giacomo Morabito. The internet of things: A survey. *Computer Networks*, 54(15):2787–2805, 2010.
5. Michael Batty. A perspective on city dashboards. *Regional Studies, Regional Science*, 2(1):29–32, 2015.
6. Michael Batty. DTs. *Environment and Planning B*, 45(5):817–820, 2018.
7. Shilpi Bhattacharyya, Dimitrios Katramatos, and Shinjae Yoo. Why wait? Let us start computing while the data is still on the wire. *Future Generation Computer Systems*, 89:563–574, 2018.
8. Diego M. Botín-Sanabria, Adriana-Simona Mihaita, Rodrigo E. Peimbert-García, Mauricio A. Ramírez-Moreno, Ricardo A. Ramírez-Mendoza, and Jorge de J. Lozoya-Santos. Digital twin technology challenges and applications: A comprehensive review. *Remote Sensing*, 14(6):1335, 2022.
9. Colin F. Camerer, Anna Dreber, Felix Holzmeister, Teck-Hua Ho, Jürgen Huber, Magnus Johannesson, Michael Kirchler, Gideon Nave, Brian A. Nosek, Thomas Pfeiffer, et al. Evaluating the replicability of social science experiments in nature and science between 2010 and 2015. *Nature Human Behaviour*, 2(9):637–644, 2018.
10. Susanne E. Carroll. On cognates. *Interlanguage Studies Bulletin (Utrecht)*, 8(2):93–119, 1992.
11. Giorgio Colombo, Jascha Grübel, Karolina Minta, Jan M. Wiener, Marios Avraamides, Christoph Hölscher, and Victor R. Schinazi. Spatial performance assessment for cognitive evaluation (space): A novel tablet-based tool to detect cognitive impairment. In *4th Interdisciplinary Navigation Symposium (iNAV 2022)*. Merano, Italy. Virtual Meeting. June 14–16, 2022; Poster abstract.
12. Antoine Coutrot, Ed Manley, Sarah Goodroe, Christoffer Gahnstrom, Gabriele Filomena, Demet Yesiltepe, Ruth C. Dalton, Jan M. Wiener, Christian Hölscher, Michael Hornberger, et al. Entropy of city street networks linked to future spatial navigation ability. *Nature*, 604(7904):104–110, 2022.
13. Stephen F. Davis. *Handbook of Research Methods in Experimental Psychology*. Stephen F. Davis. John Wiley & Sons, 2008.
14. Christian Flügel and Volker Gehrmann. Scientific workshop 4: Intelligent objects for the internet of things: Internet of things-application of sensor networks in logistics. In *European Conference on Constructing Ambient Intelligence*, Berlin, Heidelberg, pages 16–26, 2009.

15. Michal Gath-Morad, Leonel Aguilar, Ruth Conroy Dalton, and Christoph Hölscher. cogARCH: Simulating wayfinding by architecture in multilevel buildings. In *Proceedings of the 11th Annual Symposium on Simulation for Architecture and Urban Design*, Society for Computer Simulation International, PO Box 17900, San Diego, CA, United States, pages 1–8, 2020.
16. Edward Glaessgen and David Stargel. The digital twin paradigm for future nasa and us air force vehicles. In *53rd AIAA/ASME/ASCE/AHS/ASC Structures, Structural Dynamics and Materials Conference 20th AIAA/ASME/AHS Adaptive Structures Conference 14th AIAA*, Honolulu, Hawaii, page 1818, 2012.
17. Joseph E. Gonzales and Corbin A. Cunningham. The promise of preregistration in psychological research. *Psychological Science Agenda*, 29(8):2014–2017, 2015.
18. Michael Grieves and John Vickers. Digital twin: Mitigating unpredictable, undesirable emergent behavior in complex systems. In Kahlen, J., Flumerfelt, S., Alves, A. (eds) Transdisciplinary Perspectives on Complex Systems, pages 85–113. Springer, Heidelberg, 2017.
19. Jascha Grübel. The design, experiment, analyse, and reproduce principle for experimentation in virtual reality. *Frontiers in Virtual Reality*, 4:1069423, 2023.
20. Jascha Grübel, Michal Gath-Morad, Leonel Aguilar, Tyler Thrash, Robert W. Sumner, Christoph Hölscher, and Victor R. Schinazi. Fused twins: A cognitive approach to augmented reality media architecture. In *MAB '20: Proceedings of the 5th Media Architecture Biennale Conference*, Amsterdam and Utrecht Netherlands, 2021.
21. Jascha Grübel, Tyler Thrash, Leonel Aguilar, Michal Gath-Morad, Julia Chatain, Robert W. Sumner, Christoph Hölscher, and Victor R. Schinazi. The hitchhiker's guide to fused twins: A review of access to DTs in situ in smart cities. *Remote Sensing*, 14(13):3095, 2022.
22. Jascha Grübel, Tyler Thrash, Leonel Aguilar, Michal Gath-Morad, Didier Hélal, Robert W. Sumner, Christph Hölscher, and Victor R. Schinazi. Dense indoor sensor networks: Towards passively sensing human presence with lorawan. *Pervasive and Mobile Computing*, 84:101640, 2022.
23. Jascha Grübel, Tyler Thrash, Didier Hélal, Robert W. Sumner, Christoph Hölscher, and Victor R. Schinazi. The feasibility of dense indoor lorawan towards passively sensing human presence. In *2021 IEEE International Conference on Pervasive Computing and Communications (PerCom)*,Kassel, Germany, pages 1–11. IEEE, 2021.
24. Jascha Grübel, Raphael Weibel, Mike Hao Jiang, Christoph Hölscher, Daniel A. Hackman, and Victor R. Schinazi. Eve: A framework for experiments in virtual environments. In *Spatial Cognition* X, Philadelphia, PA, USA, pages 159–176. Springer, 2016. Editors:Thomas Barkowsky, Heather Burte, Christoph Hölscher, Holger Schultheis.
25. Jascha Grübel and Michal Gath-Morad. Fused twin base (github), March 2022.
26. Jascha Grübel, Tyler Thrash, Didier Hélal, Robert W. Sumner, Christoph Hölscher, and Victor R. Schinazi. LoRaWAN DISN Transmission Meta Data, January 2021. The accompanying research is presented at IEEE International Conference on Pervasive Computing and Communications 2021 (PerCom'21). The research that produced this data set is funded by ETH Zürich under the grant ETH-15 16-2. We thank Michal Gath-Morad for the BIM used for distance computations.

27. Rachida Hassani and Younes El Bouzekri El Idrissi. Communication and software project management in the era of digital transformation. In *Proceedings of the International Conference on Geoinformatics and Data Analysis*,Marseille France, pages 22–26, 2018.
28. Martin G. Helander. *Handbook of Human-Computer Interaction*. Editors: Jean Vanderdonckt, Philippe Palanque, Marco Winckler. Elsevier, 2014.
29. John P. A. Ioannidis. Why most published research findings are false. *PLoS Medicine*, 2(8):e124, 2005.
30. Julie A. Jacko. *Human Computer Interaction Handbook: Fundamentals, Evolving Technologies, and Emerging Applications*. ·CRC Press, Inc. Subs. of Times Mirror 2000 Corporate Blvd. NW Boca Raton, FL, United States, 2012.
31. Norbert L. Kerr. Harking: Hypothesizing after the results are known. *Personality and Social Psychology Review*, 2(3):196–217, 1998.
32. Kim Marriott, Falk Schreiber, Tim Dwyer, Karsten Klein, Nathalie Henry Riche, Takayuki Itoh, Wolfgang Stuerzlinger, and Bruce H. Thomas. *Immersive Analytics*, volume 11190. Springer Cham, 2018.
33. Deborah Mies, Will Marsden, and Stephen Warde. Overview of additive manufacturing informatics:"a digital thread". *Integrating Materials and Manufacturing Innovation*, 5(1):114–142, 2016.
34. Paul Milgram and Fumio Kishino. A taxonomy of mixed reality visual displays. *IEICE Transactions on Information and Systems*, 77(12):1321–1329, 1994.
35. Mistler S. Planning your analyses: Advice for avoiding analysis problems in your research. *Psychological Science Agenda*, 26(11), 2012.
36. Kirill Müller and Kay W. Axhausen. Hierarchical IPF: Generating a synthetic population for Switzerland. *Arbeitsberichte Verkehrs-und Raumplanung*, 718, 2011.
37. Marcus R. Munafò and George Davey Smith. Robust research needs many lines of evidence. *Nature*, 553(7689):399–401, 2018.
38. Karl R. Popper. *The Open Universe: An Argument for Indeterminism*, Totowa, N.J. : Rowman and Littlefield, volume 2. Psychology Press, 1992.
39. Leonard Richardson, Mike Amundsen, Michael Amundsen, and Sam Ruby. *RESTful Web APIs: Services for a Changing World*. O'Reilly Media, Inc., 2013.
40. Rodrigo Roman, Pablo Najera, and Javier Lopez. Securing the internet of things. *Computer*, 44(9):51–58, 2011.
41. Tomás Sánchez López, Damith C. Ranasinghe, Mark Harrison, and Duncan McFarlane. Adding sense to the Internet of Things. *Personal and Ubiquitous Computing*, 16(3):291–308, 2012.
42. Michael Schluse, Marc Priggemeyer, Linus Atorf, and Juergen Rossmann. Experimentable DTs-streamlining simulation-based systems engineering for industry 4.0. *IEEE Transactions on industrial informatics*, 14(4):1722–1731, 2018.
43. Erik Schultes, Marco Roos, Luiz Olavo Bonino da Silva Santos, Giancarlo Guizzardi, Jildau Bouwman, Thomas Hankemeier, Arie Baak, and Barend Mons. Fair DTs for data-intensive research. *Frontiers in Big Data*, 5:883341, 2022.
44. Samad M. E. Sepasgozar. Differentiating digital twin from digital shadow: Elucidating a paradigm shift to expedite a smart, sustainable built environment. *Buildings*, 11(4):151, 2021.
45. Philip B. Stark. Before reproducibility must come preproducibility. *Nature*, 557(7706):613–614, 2018.

46. Ronak Sutaria and Raghunath Govindachari. Making sense of interoperability: Protocols and standardization initiatives in iot. In *2nd International Workshop on Computing and Networking for Internet of Things*, Aveiro, Portugal, page 7, 2013.
47. Fei Tao, He Zhang, Ang Liu, and AYC Nee. Digital twin in industry: State-of-the-art. *IEEE Trans. Ind. Informat.*, 15:2405–2415, 2018.
48. The Economist. DTs in cockpits will help planes look after themselves, 05 2022.
49. Adam Thelen, Xiaoge Zhang, Olga Fink, Yan Lu, Sayan Ghosh, Byeng D Youn, Michael D Todd, Sankaran Mahadevan, Chao Hu, and Zhen Hu. A comprehensive review of digital twin-part 1: Modeling and twinning enabling technologies. *Structural and Multidisciplinary Optimization*, 65(12):354, 2022.
50. David Waller and Lynn Nadel. *Handbook of Spatial Cognition*. American Psychological Association, Washington, DC, 2013.
51. Mark Weiser. The computer for the 21st century. *Scientific American*, 265(3):94–105, 1991.
52. Norbert Wiener. *Cybernetics or Control and Communication in the Animal and the Machine*, volume 25. MIT Press, 1961. **DOI:** https://doi.org/10.7551/mitpress/11810.001.0001
53. Mark D. Wilkinson, Michel Dumontier, IJsbrand Jan Aalbersberg, Gabrielle Appleton, Myles Axton, Arie Baak, Niklas Blomberg, Jan-Willem Boiten, Luiz Bonino da Silva Santos, Philip E. Bourne, et al. The fair guiding principles for scientific data management and stewardship. *Scientific Data*, 3(1):1–9, 2016.

34

Digital Twins–Enabled Smart Control Engineering and Smart Predictive Maintenance

Jairo Viola, Furkan Guc, and YangQuan Chen
University of California

34.1 Introduction

Groundbreaking technologies in Industry 4.0 like artificial intelligence, Big Data, Real-Time Data Analytics, and machine/deep learning have been incorporated along the multiple layers of industrial processes. Thus, the added value from a process can be generated to increase its knowledge and develop novel capabilities that make these systems smarter. It means systems that are more efficient and resilient to the variability are produced not only by their changes but also by their environment interaction composed of several multiple interacting agents, which can be understood as Digital Transformation.

One crucial technology for Digital Transformation is the DT. It can be described as a virtual representation of a physical system, replicating its behavior through real-time updatable models based on the real-time asset datastreams. Therefore, a DT enables novel capabilities into Industry 4.0 systems like failure detection, self-optimizing control (SOC), real-time analytics, or fault-aware warnings based on extensive and real-time knowledge of the system.

This chapter introduces a novel framework for DT design and implementation applied to Smart Control Engineering and Smart Predictive Maintenance development. First, the definition of a smart system and its relationship with DTs for developing the enabling capabilities is presented. Then, the difference between what is and is not a DT is established. After that, the sequential methodological framework for DT applications is introduced with Smart Control Engineering and Smart Predictive Maintenance concepts. Two case studies are developed in this chapter to show the use of the DT systematic development framework. One uses the DT and Smart Control Engineering for temperature uniformity control. The other employs the DT to enable

 DOI: 10.1201/9781003425724-40

the Smart Predictive Maintenance framework used for fault diagnosis on a mechatronic testbed.

34.1.1 What Is A Smart System?

Defining a smart system is required to talk about Smart Control and Engineering and Smart Predictive Maintenance. It can be defined as a system with the capabilities of sensing, actuation, and control, able to make decisions based on real-time and historic knowledge in order to satisfy a set of desired performance specifications in the presence of uncertainties and external disturbances.

A smart system incorporates the sensing, control, and actuation features as well as the interaction with other smart systems. Besides, according to the Smart and Autonomous Systems (S&AS) program of NSF (Natural Sciences Foundation) [10], the characteristics of a smart system are:

- *Cognizant*: The system is aware of capabilities and limitations to face the dynamic changes and variability.
- *Taskable*: The system is capable of handling high-level, often vague instructions based on the automated stimulus or human commands.
- *Reflective*: A smart system is able to learn from previous experience to improve its performance.
- *Knowledge rich*: All the reasoning processes made by a smart control system are performed over a diverse body of knowledge from a rich environment and sensor-based information.
- *Ethical*: The smart behavior of a system adheres to a system of societal and legal norms.

Some applications of smart systems includes smart transportation [17,27], smart grid [2,13], unmanned autonomous vehicles, and robotic systems [4]. However many of these smart systems use to operate inside controlled environments with low variability, making the smart systems unable to handle unanticipated changes or dynamics given not only by a variable environment but also for the system non-modeled behaviors. In the context of Industry 4.0, the industrial processes are just a part of a more complex CPS, equipped with multiple sensors that produce big data as well as the interaction among its agents.

34.1.2 What Is Smart Control Engineering?

The concept of Smart Control Engineering begins with the Cognitive Process Control (CPC) proposed by Chen in Ref. [26]. It can be defined as:

> A branch of control engineering that leverage groundbreaking technologies like Edge Computing, IoT, AI, Big Data, Data Analytics to enhance the smartness and performance of a control system.
>
> *MESALab (2021)*

Likewise, the Smart Control Engineering introduces a set of new characteristics like:

- Each system incorporates smartness capabilities given by groundbreaking technologies and supported by edge computing devices, performing real-time health awareness status of the system.
- Involves one or multiple real-time optimization and artificial intelligence decision-based stages that seek the optimal performance of the system, based on economic, environmental, health, or fault conditions, operating as a single system or into an interconnected environment.
- Requires the use of virtual representation of each single element composing the system in order to perform analysis of the system behavior under different conditions that may not be feasible for evaluation during real operation due to the cost or risk associated.
- Introduces resilient behavior on the smart system being able to adapt to unknown changes on the system based on evolving controllers and its control actions according to the system's current health status.

These features of Smart Control Engineering are supported by three core technologies corresponding to the industrial artificial intelligence (IAI), DT, and SOC.

The IAI is defined as any application of AI related to the physical operations or systems of an enterprise [7,12]. It is focused on helping an enterprise monitor, optimize, or control the behavior of these operations and systems to improve their efficiency and performance [5].

In the case of the DT, it can be interpreted as a virtual, digital equivalent of a physical product, usually across multiple stages of its lifecycle [6]. Under the scope of Smart Control Engineering, the DT became a relevant tool to represent fault behaviors and undesired situations of a system that in real life cannot be evaluated due to the high risk and economic cost.

Besides, SOC is a control strategy to find the optimal response of a system based on an economical cost function with multiple optimization stages at different timescales [1,3,15]. In the case of Smart Control Engineering, the SOC acts as that upper layer of control and optimization that takes the real-time analytics provided by IAI and the exhaustive analysis performed via real-time updated DT to generate a smart response of the system.

34.1.3 What Is a Digital Twin?

A DT can be defined as a virtual mirroring representation or a copy of a physical system. It is fed with real-time data coming from the system and is able to reflect any change present in the system. The virtual model is composed by multiple physics-based models of the system that is complemented with data-driven models.

A DT begins with a physical system composed of many different subsystems or elements that performs a specific task. The datastreams from the physical system are sent in real-time to the DT environment using IoT and edge computing devices. Inside the DT environment, a DT prototype describes the physical behavior of each component of the physical system and its interaction using multidomain physics tools and data-driven models incorporated in the DT environment. Thus, many DT instances can be created in the DT environment based on the DT prototype. The main purpose of the DT instance is to create an individual representation of the physical system for specific tasks such as controller design, health and prediction of system components, and revenue analysis.. Each instance has its own attributes defined at the time when the instance is created, working also as a time capsule to represent previous behaviors of the systems.

34.1.4 What Is Not a Digital Twin?

The term "Digital Twin" is employed in different ways with substantial conceptual variations among researchers and engineers. For example, in engineering, the DT is considered as a tool with benefits in system design, optimization, process control, virtual testing, predictive maintenance, or lifetime estimation. However, the difference between a model and a DT is not made clear among other contexts. The most important components of a DT are:

- A model of the physical system.
- A real-time varying dataset of the system.
- A real-time update or adjustment of the system model based on the varying dataset.

Thus, one way to differentiate between a DT and simulation model is using the fact provided by Ref. [22]. "A Digital Twin without a physical twin is just a model". It means that a DT has to be associated with an object that physically exists. So, the virtual representation can enhance new analysis further than real-life experimentation and assessment, based on an updatable model supported by a real-time changing dataset.

Therefore, it is essential to differentiate between a simulation model and a DT. A traditional model-based design (MBD) involves the verification and validation of models, which can be used to optimize the design or operation of a device or process. The model is usually validated by comparing experimental results with the model results and performing parameter estimation, but this operation is not performed in real-time since MBD can exist even without the presence of a physical counterpart; therefore, simulation is not a DT, but it is an essential part of itself.

34.2 Digital Twin Development Framework

The methodological framework developed in Ref. [18] is used to perform successful implementation of a DT. The framework is composed by five steps corresponding to the target system definition, system documentation, multidomain simulation, DT assembly and behavioral matching (BM), and DT evaluation and deployment.

Step 1 focuses on recognizing the current status of the physical system to be replicated via DT with two possible scenarios. The first scenario is that the physical system is in the conceptual design stage, and the second scenario is to perform the required tasks when the physical system is running..

Step 2 consists of collecting all the available information of the system to create the most accurate representation. It includes the control algorithms, sensors and actuators datasheets, troubleshooting and problem records, cumulative experience of the system engineers and operators, and the system data streams.

In Step 3, the definition and configuration of the simulation models for representing the real system behavior are performed. In this case, defining the simulation domains related to the system is required including governing equations and data-driven models to derive model representations of the different subsystems.

In Step 4, based on a stable and operative multidomain simulation model, the system response is calibrated via BM, defined as an optimization based on system real-data procedure to find the parameters of each subsystem composing the DT in order to fit its complete system dynamics, coinciding with the real state of the physical system.

Finally, Step 5 is the DT implementation, performing the validation and simultaneous deployment with the real system. The DT response is monitored via a supervisory interface to perform the DT offline and simultaneous execution of the system.

34.3 Smart Control Engineering Enabled by Digital Twin Case Study: Self-Optimizing Control for Temperature Uniformity

The temperature uniformity control system using a real-time thermal infrared (TIR) vision feedback presented in Figure 34.1a is employed as a case study for the DT using the systematic development framework. The first and second steps correspond to the target system definition and system documentation. All the details regarding these steps for the proposed system can be found in Ref. [20,21]. The third step consists in defining the multidomain simulation model. It is divided in three simulation domains, namely, electrical, thermal, and digital, corresponding to the Peltier electric and thermal components as well as the digital controller implementation as shown in Figure 34.1b. The multidomain simulation model is built on Matlab/Simulink/Simscape and can be downloaded from https://www.theedgeai.com/dtandscebook.

The fourth step corresponds to the BM. In this case, the Peltier parameters Seebeck coefficient α, internal resistance R, heat capacity C, and heat conductance K are the most critical due to the nonlinear nature of this device. Thus, several real datasets are acquired by applying different step reference signals to evaluate its dynamic behavior for four different setpoints, 30°C, 50°C,70°C and 90°C. Based on the system responses, the BM is performed using the Simulink Design Optimization (SLDO) toolbox by minimizing the cost function 0.1 obtaining the family of parameters shown in Table 34.1. From these results, a DT model discrimination is performed. It consists in finding the nominal model on the family of plants applying information gain metrics including Akaike Information Criteria (AIC) or ν-gap metric in order to determine the set of parameters with the best likelihood that ensure the stability of the system in open and closed-loop setups. So, the set of parameters for 70°C response corresponds to the nominal model of the DT. All the details on the DT model discrimination can be found in Ref. [19].

$$F(x) = \sum_{k=0}^{N} e^2(k). \tag{34.1}$$

Finally, Step 5 corresponds to the system deployment. It consists of a supervisory interface and a communication architecture to connect and monitor the physical asset with its DT. Thus, the DT is fed in real-time with the same control action, output temperature, and environmental temperature, among other variables of the physical system. The monitoring interface is implemented in Matlab using the appdesigner tool and its operation can be visualized at https://youtu.be/acXTNmcCIYs.

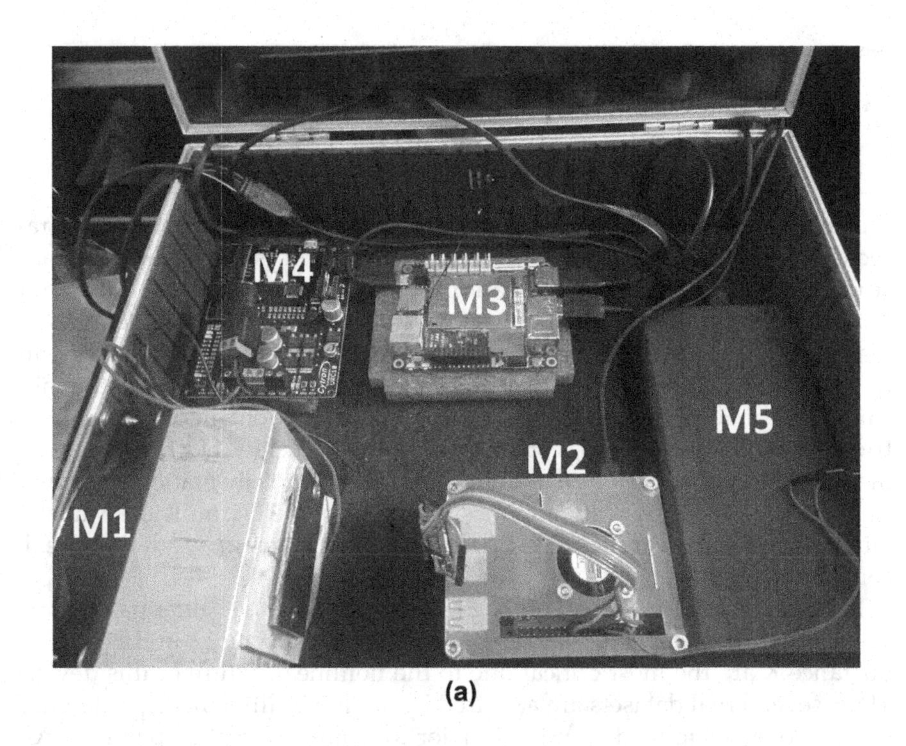

(a)

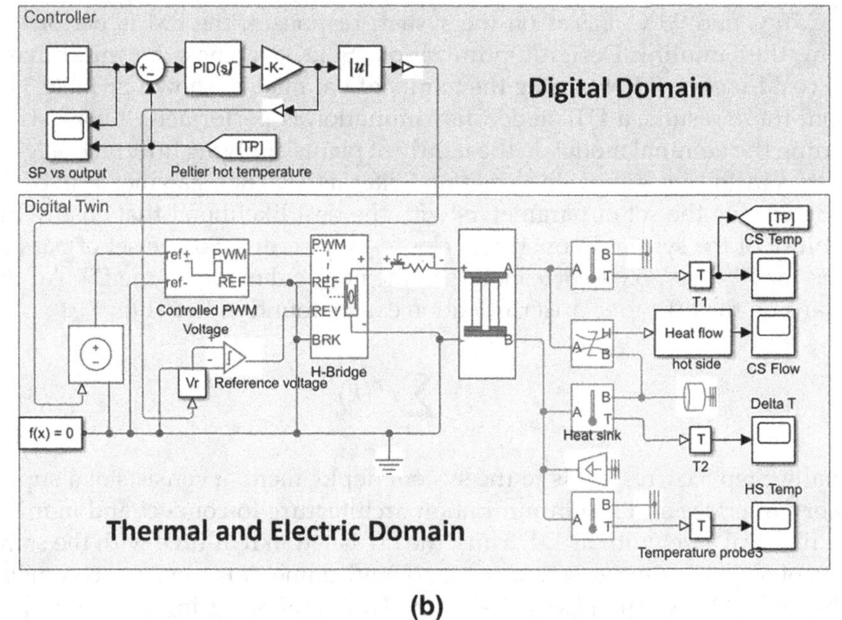

(b)

FIGURE 34.1
Digital Twin (DT) showcase system: (a) real-time vision feedback infrared temperature uniformity control and (b) DT.

TABLE 34.1

Behavioral Matching Results for Different Setpoints

	Setpoint			
Parameter	**30°C**	**50°C**	**70°C**	**90°C**
α	96.3 mV	82.5 mV	21.1 mV	29.5 mV
R	3.3 Ω	3.3 Ω	3.3 Ω	3.3 Ω
K	0.3 K/W	0.35 K/W	0.286 K/W	0.38 K/W
C	34.9 J/K	31.93 J/K	11.1 J/K	13.7 J/K

34.3.1 Digital Twin–Enabled Capability: Self-Optimizing Control Framework

The Real-Time Self-Optimizing Control architecture is shown in Figure 34.2a. As can be observed, the Self-Optimizing Control acts as an upper optimization layer which takes the reference r, error signal e, and output y of the system to find the optimal values of the controller $c(s)$, corresponding to a PID controller (34.2) for a plant $p(s)$, modeled as a FOPDT system given by Eq. (34.3), where K, τ, and L correspond to the system gain, time constant, and delay, respectively.

$$c(s) = k_p + \frac{k_i}{s} + k_d s \tag{34.2}$$

$$p(s) = \frac{K}{\tau s + 1} e^{-Ls} \tag{34.3}$$

The SOC control problem is stated by Eqs. (34.4) and (34.5), where T_s is the system settling time, OV is the overshoot percentage, $\mu = k_p, k_i, K_d$ are the proportional, integral and derivative gains of the PID controller, A and B correspond to the maximum overshoot and settling time, $k_{p_{\min,\max}}$, $k_{i_{\min,\max}}$, and $k_{d_{\min,\max}}$ are the minimum and maximum values for the PID gains, and $W_{1,2,3}$ are the weights for the Overshoot, Settling time and the Integral Square Error index, respectively. In this case, the SOC controller executes an optimization step after one cycle of the periodic reference signal r, computing a performance cost function for that period.

$$\min_{\mu \in \mathbb{R}} J = W_1 OV(\mu) + W_2 T_s(\mu) + W_3 \int_0^t e(t,\mu)^2 \, dt, \tag{34.4}$$

subject to the constraints

$$\begin{aligned} & OV(\mu) < A;\ T_s(\mu) < B, \\ & k_{p_{\min}} \le k_p \le k_{p_{\max}}, \\ & k_{i_{\min}} \le k_i \le k_{i_{\max}}, \\ & k_{d_{\min}} \le k_d \le k_{d_{\max}}. \end{aligned} \tag{34.5}$$

The proposed SOC benchmark capabilities can be extended by replacing the FOPDT system proposed in the benchmark problem for the DT of a real system. Thus, the performance of the SOC controller with GCNM optimization algorithm can be evaluated for a real system through its DT.

An FOPDT model of the thermal system is identified as Eq. (34.6) to determine if the system has a lag dominant, balanced, or delay dominant dynamics and sets the right optimization parameters (k_p, k_i, OV, T_s) for the SOC controller, corresponding to a lag dominated system. Thus, the design specifications for the SOC PI controller are $0.01 \le k_p, k_i \le 1$, 5 seconds settling time, 2% overshoot, and reference signal period of 100 seconds. The only difference is the initial conditions for k_p and k_i, which are calculated using the ZN method proposed in Ref. [25] based on Eq. (34.6). The obtained values are $k_p = 102$ and $k_i = 85$. For the SOC controller application, the thermal parameters of the Peltier thermoelectric module selected for the DT are $C = 38.15$ J/K, $\alpha = 0.097989$ V/K, $R = 3.3\ \Omega$, and $K = 0.2207$ K/W.

$$g(s) = \frac{0.63}{32.53s + 1} e^{-0.6s}. \tag{34.6}$$

The performance of the SOC controller with the DT of the Peltier system to improve its closed-loop response is shown in Figure 34.2. As can be observed, the SOC controller is able to improve the closed-loop response of the system about after 10,000 seconds. Likewise, the settling time and overshoot performance specifications are satisfied when the SOC PI terms reach the optimal value of $k_p = 162$ and $k_i = 82$. In the case of the cost function evolution, its value has small variations due to the presence of random noise which has no significant impact on the final values of k_p and k_i. Thus, we can say that using the DT in combination with the proposed SOC architecture is the first step toward the Smart Control Engineering implementation on real systems by allowing the performance evaluation of the control strategy as well as the repeatability of the system behavior under repetitive tasks.

34.4 Smart Predictive Maintenance

Classical approach for the predictive maintenance should be defined before introducing the concept of smartness enabled by DTs. As one of the

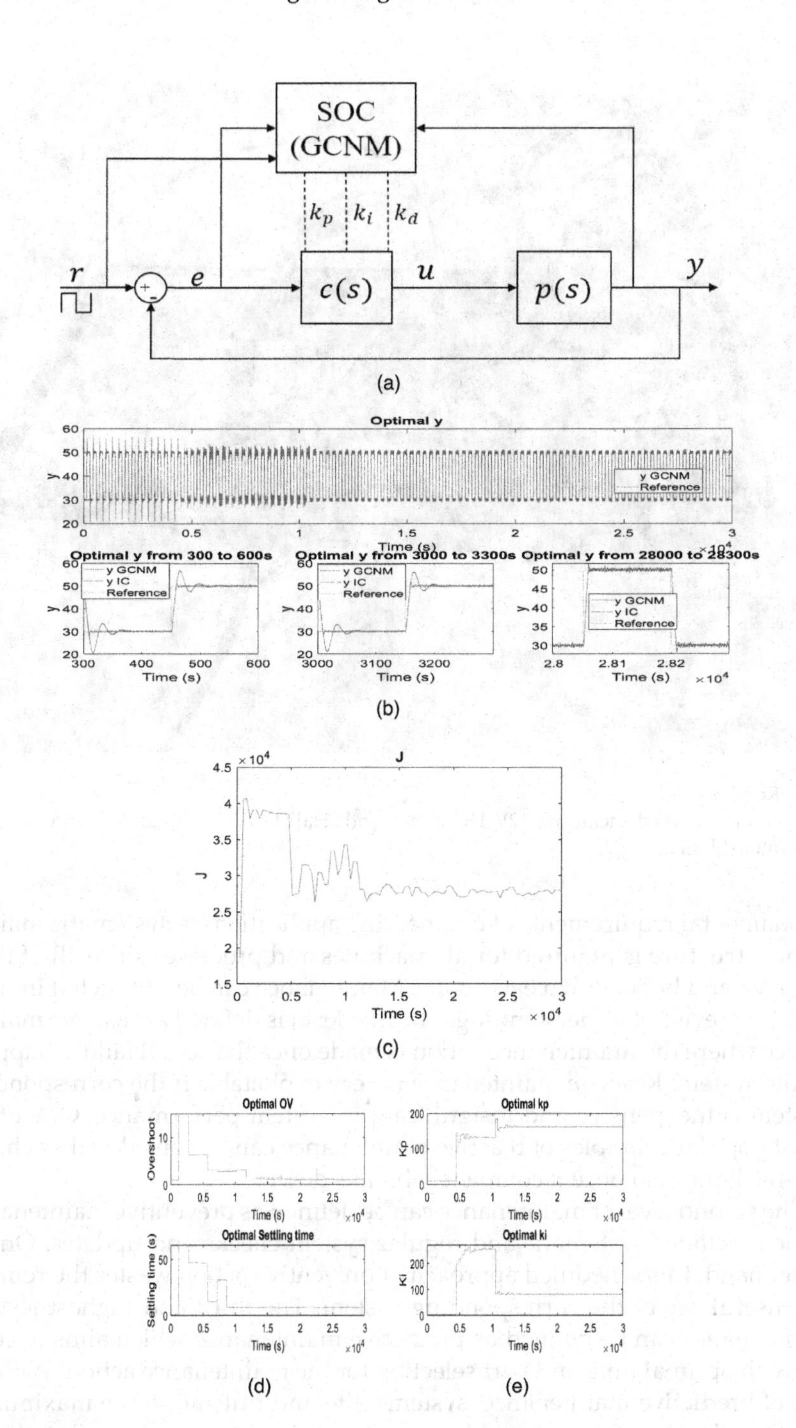

FIGURE 34.2
Self-optimizing controller: (a) architecture, (b) time response, (c) cost function, (d) overshoot and settling time, and (e) PI gains evolution for the Peltier thermal system Digital Twin.

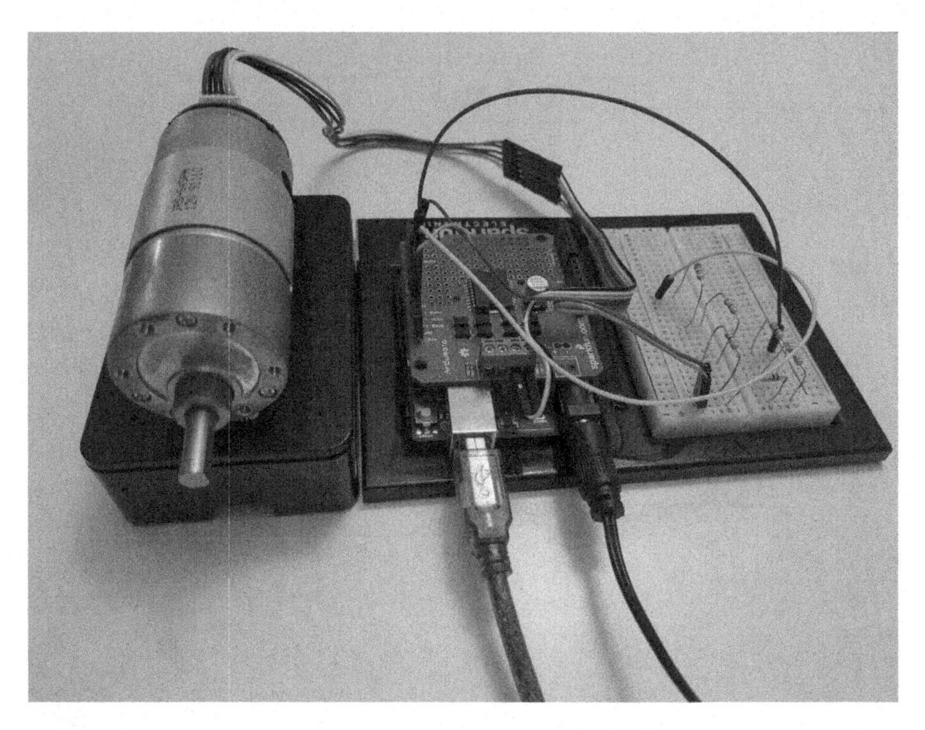

FIGURE 34.3
Mechatronic testbed including 12V DC motor with Hall Effect encoder, Arduino Uno, and Ardumoto Shield.

fundamental requirements of engineering applications, a systematic maintenance structure is required for all machines and processes since all of them degrade and break down eventually. Maintenance can be conducted in three different levels of understanding. The first level is defined as reactive maintenance where the maintenance action is made once the actual failure happens in the system. Reactive maintenance is very exploitable if the corresponding system is inexpensive and insignificant in system performance. One of the most popular examples of reactive maintenance can be considered as changing the light bulb once it completely breaks down.

The second level of maintenance can be defined as preventive maintenance, which includes systematic and regular system checks and updates. On the other hand, this scheduled approach of preventive action wastes the remaining useful life of the corresponding system. The third and highest level of maintenance can be defined as predictive maintenance which aims to come up with optimal time and part selection for the maintenance action. With the aid of predictive maintenance, systems' life and utilization are maximized. On the other hand, best possible usage condition can be achieved with predictive maintenance before the actual failure happens. Besides, the severity and the location can be indicated with predictive maintenance algorithms within complex system architecture. In this context, predictive maintenance enables users to reduce the downtime, optimize the spare parts inventory, and maximize the equipment lifetime [11].

Classical workflow for the predictive maintenance can be defined in three steps. The first step starts with the data acquisition for all healthy and faulty cases for the corresponding system. In this step, having data with various environmental conditions has a dramatic influence on the overall performance of the predictive maintenance. Therefore, this step requires collection a variety of data for the corresponding system in different cases. As the last part of the first step, data should be pre-processed for next steps such as converting time domain signals into frequency domain. The second part of the classical approach is the heart of the predictive maintenance which can be defined as extracting the condition indicator from raw data and development of predictive maintenance models. Selection of the proper signature fingerprints enables users to have an effective predictive maintenance structure. Users can start with the simplest approach of the predictive maintenance models where it only differentiates between healthy and faulty case which can be considered as anomaly detection. The second approach is to have both healthy and faulty data detection among different classes of fault cases. This approach enables to achieve fault classification. The third approach predicts the remaining useful life of the corresponding system depending on the transition from healthy and faulty states. Finally, the last step of the classical workflow is to have deployment and integration of the predictive maintenance model. This final part can utilize various combinations of edge computing devices and cloud services [9,8,14].

34.4.1 SPM Framework Enabled by the Digital Twin

Due to the fundamental requirement of the predictive maintenance models, all methods require a high-volume data set collected with a variety of labels and environmental conditions. As it can be observed from most of the engineering applications, real system data may not be available in the desired level of volume and variability. In that context, mathematical models and simulations can be used to synthetically generate a data set. However, mathematical models introduce undesired mismatch errors which lead to huge effects on the predictive maintenance accuracy and performance, especially for complex systems. Therefore, synthetic data generation can be leveraged by key contributions of the DT framework. Since DTs are behavioral copies of the real systems, synthetic data generated by DTs are reliable and high volume in the sense of predictive maintenance. In this sense, key concepts of DTs like multidomain simulation and BM can be leveraged in terms of Smart Predictive Maintenance [18].

The second part of the Smart Predictive Maintenance is the utilization of the Smart Big Data. Predictive maintenance algorithms mostly rely on the high-volume data. However, how to use existing data is also an important concept in the performance of the predictive maintenance algorithm. The concept of Smart Big Data is defined as the utilization of the existing knowledge or underlying dynamics of the corresponding system. The concept can be extended to smart applications like physics informed experiment design and complexity aware advanced data analytics.

34.4.2 Mechatronic Testbed Case Study

A case study of the Mechatronic Testbed is utilized in order to demonstrate various steps and approaches of Smart Predictive Maintenance framework. Position and speed control of a rotating shaft is one of the key systems in both literature and industry. It directly provides rotary motion and, coupled with wheels or drums and cables, can provide translational motion. On the other hand, one of the most common sensor–actuator pairs in control systems are the encoder and DC motor. Corresponding Mechatronic Testbed has 12V DC motor with Hall Effect encoder, along with Arduino Uno and Ardumoto Shield, as shown in Figure 34.3.

34.4.2.1 Physical Twin and Digital Twin

Position and speed control of a rotating shaft problem is considered as one of the complex and comprehensive systems to perform Smart Predictive Maintenance. In this context, both real physical twin and DT should be defined explicitly. In this case study, the mechatronic testbed given in Figure 34.3 is utilized as the physical twin that can be considered as the real system in Smart Predictive Maintenance sense. A fault case is introduced to the case study including motor misalignment as harmonic torque disturbance periodic with the motor shaft position. The misalignment torque disturbances are injected to the real system with a variable severity that can be considered as more realistic in the sense of predictive maintenance. Finally, fault injected mechatronic testbed is considered as the physical twin.

Before generating the DT of the system, target system definition should be performed. If one recalls the basics of the rotating shaft problem, couple of steps should be clarified. First, the generated motor torque T is proportional to the armature current i by a constant factor K_t.

$$T = K_t i \tag{34.7}$$

Secondly, the back emf, e, is proportional to the angular velocity of the shaft $\theta^{\cdot}$ by a constant factor K_b.

$$e = K_b \dot{\theta} \tag{34.8}$$

In SI units, the motor torque and back emf constants can be considered as equal, that is, $K = K_t = K_b$ for further simplification of the parameters.

By using Newton's second law and Kirchhoff's voltage law, mathematical model for the testbed can be defined as:

$$J\dot{\theta} + b\dot{\theta} = Ki \tag{34.9}$$

$$L\frac{di}{dt} + Ri = V - K\dot{\theta} \tag{34.10}$$

where J and b are the rotor inertia and damping, R and L are armature resistance and inductance, and K is the back emf and motor torque constant. At the end, model parameters are limited to the number of five in the BM of the DT. In terms of the DT, a simulation model is generated using the mathematical model and manufacturer values for the model parameters as shown in Figure 34.4a. Then, the BM is performed to update DT parameters using SLDO as visualized in Figure 34.4b. Initial and updated parameters of the DT are demonstrated in Table 34.2. After the validation of the DT along with the physical twin output, a misalignment fault is also injected to the DT as harmonic torque disturbance periodic with the motor shaft position with various severity levels. When the DT is ready, a high-volume synthetic

TABLE 34.2
Behavioral Matching Results for the Mechatronic Testbed Digital Twin

Parameter	Initial Value	Final Value
Armature resistance R [Ω]	4	2.2667
Armature inductance L [H] [Ω]	2.75E-6	2.75E-6
Back emf constant K [$V/(rad/s)$]	0.0274	0.022322
Rotor inertia J [$kg\ m^2$]	3.2284E-6	1.2305E-6
Rotor damping b [$N\ m/(rad/s)$]	3.5077E-6	3.4568E-6

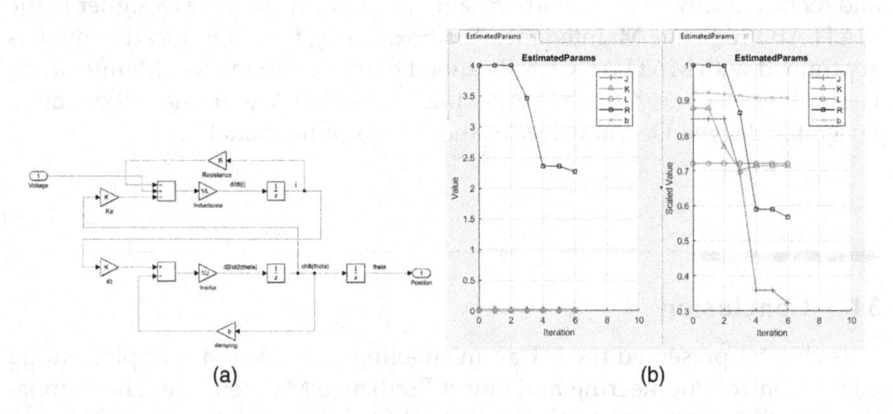

FIGURE 34.4
Mechatronic testbed: (a) multidomain Simulink model and (b) behavioral matching iterations.

data is generated including various healthy and faulty case scenarios under different environmental conditions.

34.4.2.2 Smart Big Data

Smart Big Data is defined as the utilization of the existing system information in the corresponding process. For both DT generation and Predictive Maintenance development, parameter identification is one of the key required processes. Traditional identification processes are costly most of the time and introduce a certain level of error into the final product. With the aid of the concept of Smart Big Data, the underlying dynamics and knowledge of the system should be utilized.

In the mechatronic testbed case study, rotating shaft properties are highly dominant in terms of mechanical faults and vibrations [23,24]. Understanding the effect of the misalignment is the first step of performing Smart Predictive Maintenance. In this sense, existing knowledge on the misalignment should be included in terms of Smart Big Data. With the aid of this approach, the user will be enabled to utilize the existing system information and underlying dynamics of the system as much as possible. Classical identification methods for the misalignment are costly and challenging since the problem itself is highly complex. In this context, misalignment effect can be considered as a harmonic torque disturbance periodic with respect to the shaft position [16].

34.4.2.3 Fault Diagnosis

In the final step of the mechatronic testbed case study, predictive maintenance fault classification model is generated which is enabled by key technologies of D T and Smart Big Data. For healthy and three different faulty cases, four classes of labeled data are generated as healthy, low, medium, and high misalignment severity levels. Signature condition indicators are extracted and hierarchically ordered with the aid of Diagnostic Feature Designer in the MATLAB Predictive Maintenance Toolbox. Lastly, fault classification models are trained with MATLAB Classification Learner for Predictive Maintenance. Highest test accuracy is achieved in the level of 85.0% with the utilization of Ensemble Bagged Decision Tree as the classification model.

34.5 Conclusion

This chapter presented the DT as an enabling technology for implementing Smart Control Engineering and Smart Predictive Maintenance. The foundations for the mentioned technologies and two case studies were developed: a self-optimizing control for Peltier thermal systems and a smart predictive fault detection method for mechatronic testbed. The results show that the DT is a crucial tool enabling smartness for different processes. Likewise, using DTs, it was possible to evaluate the performance of the proposed systems under hazardous conditions that could be costly to reproduce on the physical setup, which increases the knowledge of the system. Thus, future applications based on DTs like prognosis, remaining useful life, real-time data-driven supported control, or real-time data analytics can be developed toward its implementation using edge computing devices.

References

1. Hyo Sung Ahn, Yang Quan Chen, and Kevin L. Moore. Iterative learning control: Brief survey and categorization. *IEEE Transactions on Systems, Man and Cybernetics Part C: Applications and Reviews*, 37(6):1099–1121, 2007.
2. Amam Hossain Bagdadee, Zahirul Hoque, Li Zhang, Amam Hossain Bagdadee, Zahirul Hoque, and Li Zhang. IoT based wireless sensor network for power quality control in IoT based wireless sensor smart network for power quality control in grid smart grid. *Procedia Computer Science*, 167:1148–1160, 2020.
3. Kartitk Bariyur and Miroslav Krstic. *Real-Time Optimization by Extremumseeking Control*. John Wiley & Sons, Inc. 605 Third Ave. New York, NY, United States, 2003.
4. Ben Jye Chang and Jhih Ming Chiou. Cloud computing-based analyses to predict vehicle driving shockwave for active safe driving in intelligent transportation system. *IEEE Transactions on Intelligent Transportation Systems*, 21(2):852–866, 2020.
5. Cloudpulse Strategies. Artificial Intelligence for Industrial Applications. Technical Report, CloudPulse Strategies, 2017.
6. Michael Grieves. Digital twin: Manufacturing excellence through virtual factory replication. *White Paper*, 1:1–7, 2014.
7. Meng Hao, Hongwei Li, Xizhao Luo, Guowen Xu, Haomiao Yang, and Sen Liu. Efficient and Privacy-Enhanced Federated Learning for Industrial Artificial Intelligence. *IEEE Transactions on Industrial Informatics*, 16(10):6532–6542, 2020.
8. Hashem M. Hashemian. State-of-the-art predictive maintenance techniques. *IEEE Transactions on Instrumentation and Measurement*, 60(1):226–236, 2010.
9. R. Keith Mobley. *An Introduction to Predictive Maintenance*. A volume in Plant Engineering, Elsevier, 2002. https://doi.org/10.1016/B978-0-7506-7531-4.X5000-3
10. NSF. Smart and Autonomous Systems (S&AS), 2018.
11. Paul Peeling. Big data and machine learning for predictive maintenance. *Matlab Expo*, 2017.
12. Ricardo Silva Peres, Xiaodong Jia, Jay Lee, Keyi Sun, Armando Walter Colombo, and Jose Barata. Industrial artificial intelligence in industry 4.0-systematic review, challenges and outlook. *IEEE Access*, 8:220121–220139, 2020.
13. Morteza Rahimi, Maryam Songhorabadi, and Mostafa Haghi Kashani. Fogbased smart homes: A systematic review. *Journal of Network and Computer Applications*, 153:102531, 2020.

14. Yongyi Ran, Xin Zhou, Pengfeng Lin, Yonggang Wen, and Ruilong Deng. A survey of predictive maintenance: Systems, purposes and approaches. arXiv preprint arXiv:1912.07383, 2019.
15. Sigurd Skogestad. Plantwide control: The search for the self-optimizing control structure. *Journal of Process Control*, 10(5):487–507, 2000.
16. Alok Kumar Verma, Somnath Sarangi, and Mahesh Kolekar. Misalignment faults detection in an induction motor based on multi-scale entropy and artificial neural network. *Electric Power Components and Systems*, 44(8):916–927, 2016.
17. Patricio Vicuna, Sandeep Mudigonda, Camille Kamga, Kyriacos Mouskos, and Charles Ukegbu. A generic and flexible geospatial data warehousing and analysis framework for transportation performance measurement in smart connected cities. *Procedia Computer Science*, 155(2018):226–233, 2019.
18. Jairo Viola and YangQuan Chen. Digital twin enabled smart control engineering as an industrial ai: A new framework and case study. In *2020 2nd International Conference on Industrial Artificial Intelligence (IAI)*, Shenyang, China, pages 1–6, 2020.
19. Jairo Viola, YangQuan Chen, and Jing Wang. Information-based model discrimination for digital twin behavioral matching. In *2020 2nd International Conference on Industrial Artificial Intelligence (IAI)*, Shenyang, China, pages 1–6, 2020.
20. Jairo Viola, Piotr Oziablo, and Yang Quan Chen. An Experimental Networked Control System with Fractional Order Delay Dynamics. In *Proceedings of the 2019 IEEE 7th International Conference on Control*, Delft, Netherlands, *Mechatronics and Automation, ICCMA 2019*, pages 226–231, 2019.
21. Jairo Viola, Alberto Radici, Sina Dehghan, and YangQuan Chen. Low-cost real-time vision platform for spatial temperature control research education developments.International Design Engineering Technical Conferences and Computers and Information in Engineering Conference. American Society of Mechanical Engineers, 2019, 59292: V009T12A030.
22. Louise Wright and Stuart Davidson. How to tell the difference between a model and a digital twin. *Advanced Modeling and Simulation in Engineering Sciences*, 7(1):1–13, 2020.
23. M. Xu and R. D. Marangoni. Vibration analysis of a motor-flexible coupling-rotor system subject to misalignment and unbalance, part i: Theoretical model and analysis. *Journal of Sound and Vibration*, 176(5):663–679, 1994.
24. M. Xu and R. D. Marangoni. Vibration analysis of a motor-flexible coupling-rotor system subject to misalignment and unbalance, part ii: Experimental validation. *Journal of Sound and Vibration*, 176(5):681–691, 1994.
25. Dingyü Xue, YanQuan Chen, and Derek Atherton. *Linear Feedback Control, Analysis and design with MATLAB - Advances in Design and Control*. Society for Industrial and Applied Mathematics, 3600 University City Science Center Philadelphia, PA, United States, 2007.
26. YangQuan Chen. Cognitive Process Control (ppt), 2012.
27. Fenghua Zhu, Zhenjiang Li, Songhang Chen, and Gang Xiong. Parallel transportation management and control system and its applications in building smart cities. *IEEE Transactions on Intelligent Transportation Systems*, 17(6):1576–1585, 2016.

Part 7

Digital Twins in Building

35

3D City Models in Planning Activities: From a Theoretical Study to an Innovative Practical Application

Gabriele Garnero
Università degli Studi di Torino

Gloria Tarantino
Politecnico di Torino

35.1 Introduction

During the last few years, digital 3D city models have achieved a high presence as valuable planning tools used by a large number of public administrations and private firms spread all over the world. Initially, the early use of 3D city models has been dominated by visualization only, and the main purpose was providing public access to users for an attractive view of the urban environment and all its geographic elements in a certain area, taking advantage of 3D models for tourism and marketing tasks. In recent times, by virtue of new software and new modeling technologies, 3D spatial and non-spatial information has been implemented in several cities. Consequently, 3D city models have become estimable for various domains beyond visualization and have been extended to larger number of tasks, such as urban planning, disaster simulation, virtual-heritage conservation and many others. However, on the one hand, the increasing number of different applications that employ 3D city models, where each of them requires its own specific LoD, and on the other hand, the complexity of 3D model generation process, have led to a fuzzy vision about the real possibilities of utilization that 3D city models have.

35.1.1 Outline

In light of that, the first part of this chapter (Section 35.2) provides a comprehensive inventory of use cases, where specific 3D data requirements are

DOI: 10.1201/9781003425724-42

classified for certain applications, hence delineating which 3D models with their specific LoDs fit-for-purpose. Since visualization seems to be the only criterion that is suitable and can cover almost all fields of application, a wide range of use cases that employ 3D city models has been chosen and categorized into two groups. The first regards non-visualization use cases, where the visualization of 3D city models is not required and the results of the spatial operations can be just stored in a database. The second group concerns visualization-based use cases, where the visualization of 3D city model is essential and without it the use cases part of this group would not make much sense. Successively, the use of 3D models as support tools throughout the planning process is investigated in-depth. Indeed, 3D city models are commonly used to display virtually existing cities as well as publicly divulge three-dimensional visualization about future developments, plans and projects in a 3D environment. Furthermore, these visualizations can be used in two-way communication where citizens can comment back and propose better alternatives after having inspected the plan, providing either positive or negative feedbacks to local authorities, information to citizens about hypothetical new developments in a 3D environment. Subsequently, to better understand how much attention the opinion of diverse actors deserves throughout the whole process, two different approaches widespread among municipalities are described, which use virtual city modeling to make engagement of different stakeholders faster and easier during the decisions-making process: the User-oriented approach and the E-participation technology.

The second part (Section 35.3) regards the description of a project developed at Vaxholms Stad, the municipality office of Vaxholm, a small town of the Stockholm County in Sweden. The project aimed to create a system that allows 3D visualizations of Vaxön's urban structure, which is the most populated island of the whole municipality chosen as the area of study. By virtue of the software FME, and the combination of data source with different formats, a geo database has been created to represent and manage a virtual LoD2 3D buildings model of Vaxholm, which has been visualized on Google Earth, allowing citizens to easily read and view the 3D model into a free web-based mapping platform as Google Earth. In addition, the detailed planning and project of Vauxhall are added to the three-dimensional model, and the planning work is still in progress. Indeed, citizens had the possibility not only to inspect the 3D city model of Vaxön on GE but also, from the plans mapped on the earth surface, to have access directly by the model to the interested area and to all the detailed information about a certain plan on the municipality web page. Furthermore, thanks to Google Earth functions, citizens can leave a feedback by e-mail to the city administration, claiming for their needs about a specific plan or just providing their points of view about what has been mapped on the 3D model. Finally, based on the results obtained, the possible improvements of the adopted program are predicted, and other applications of the 3D building model that may be extended in the future are prospected.

35.1.2 Materials and Methodologies

The objective of this article is to investigate what the real potential of 3D city models can be for planning activities as well as for many other domains related to the city's development, in order to identify an efficient procedure to create 3D city models for municipalities, and which advantages they can get if equipped with this planning tool. To gather such information, the main methodology used in this essay was based on a survey and literature review of, mostly, online resources publicized in the last two decades, such as scientific journals, academic articles, theses and project reports. These documents were related in some cases, to the current application and utilization of 3D city models in diverse domains and in others, to the broad number of different approaches used today to create 3D city models at various levels of details. Most of the literature found about the topic debated includes cases of study taken as examples, in which cities were already equipped with 3D city models, and that were mainly chosen in Austria, Kenya and Sweden, where the 3D city model of Vaxholm has been developed. This study area was selected because it presents a mixture of historic and modern buildings with homogeneous architecture styles, which made easier the reconstruction of the 3D buildings, and where the highest building was about 20 m. Once all the documents were retrieved, a comprehensive and systematic synthesis was delineated through the sections of this essay, aiming to sort out the objectives aforementioned. First, an overview was reported about which application fields the use of 3D city models could be convenient for. Successively, a hypothetical procedure was proposed, already implemented in Sweden, to create a 3D city model, which in turn will be used as a tool for citizens' engagement throughout the decision-making process.

35.2 Theoretical Overview

35.2.1 Geometrics and Semantic Properties of 3D City Models

3D city models are digital representations of certain objects in the urban environment, which include earth surface, vegetation, infrastructure, landscape elements and buildings, created through a modeling process [1]. The third dimension indicates 3D GIS (Geographic Information System) data, where each dimension is geometrically defined. However, besides geometry, an important component of 3D city models is semantic information, which can be described as any information that is not visible as the geometry is, e.g., the land use of a building. Both geometric and semantic properties of 3D models are stored in diverse 'levels-of-detail' (LoD). The LoD approach, defined by the CityGML[1] standard, is a coherent modeling of geometric and topological properties at each level, where geometric objects get assigned to semantic objects. The LoD definition is mainly used for the details of the buildings, which are the most important items in a 3D city model, and it is a fundamental concept in 3D city modeling since it defines the degree of abstraction of real-world's elements [1].

35.2.2 Related Works

Nowadays, the development of new technologies has allowed the implementation of 3D geo-information data in several cities, which is rapidly growing and expanding in different research fields. The early use of 3D city models has been dominated by visualization only. However, by virtue of new software and new modeling methodologies, 3D city models have become estimable for several domains beyond visualization, and are currently used in a large number of purposes. During the past years, some researchers have studied the applicability of 3D geo-information, focusing on solving industrial and experimentation problems. For instance, Ross in 2010 [2], proposed a general taxonomy of 3D use cases, which relies on the type of data that each model contains: applications based on geometry (e.g. estimation of the shadow); applications based on geometry and semantic information (e.g. estimation of the solar potential); applications based on domain-specific extensions and external data (e.g. noise emission calculation). However, it is important to note that such categories are not 'exclusive' in all cases, but some applications might fit in more than one category. For example, to estimate the propagation of noise in an urban environment, only the geometry of buildings is needed. Furthermore, if hypothetically semantic information of geometries is also known, such as inhabitants or building's material, these data may represent important improvements for predictions and better assessment of noise consequences [3].

Biljecki et al. [3] conducted another study in 2015, where they argued that visualization might be considered as the only criterion that is suitable and can cover almost all categories of applications. Hence, the taxonomy of use cases mentioned above can be further categorized into two groups. The first concerns non-visualization use cases, where the visualization of the 3D model and the results of the 3D spatial operations can be visualized, but it is not essential to achieve the task of the use case since the results can be stored in a database. The second group regards visualization-based use cases, where instead, visualization is essential and the use cases would not make much sense without it. Based on these two groups, distinct use cases have been identified in several application domains.

---

1 *City GML:* Today, the number of cities that are representing their 3D city models according to the CityGML standard is growing. This standard has been issued by the OGC (Open Geospatial Consortium) to decompose articulated objects [1].

35.2.2.1 Non-visualization Use Cases

Environmental simulations part of this group are, for instance, the estimation of solar irradiation, where 3D data are used to evaluate the solar potential on rooftops in urban areas, and thus, to assess how much a building is exposed to the sunlight, in order to evaluate the suitability of a roof surface for installing photovoltaic panels above it. Another analysis concerns the energy demand estimation, where 3D city models are used to combine the data of the building's volume, number of floors, type of buildings, and other features to predict the energy demand for heating or cooling. A further estimation regards the distribution and shape of a building type in a neighborhood, which may help marketing and real estate management fields, forecasting its potential for taxation and valuation of buildings [3].

35.2.2.2 Visualization-Based Use Cases

Visualization is considered as one of the most used applications of 3D city models since it is able to provide panoramic views, web visualizations, profiling and other related works. Furthermore, it is generally used for an attractive presentation of the results from such analyses, which can be related to GIS, as in a visibility analysis, or which are not necessarily related to GIS, as economic activities [3]. Indeed, 3D city models are fundamental for various kinds of visibility analyses.

In light of the classification above mentioned, it is clear that 3D city models are currently used in a lot of domains for several purposes. The second group related to visualization-based applications is broader than the first one. This suggests that visualization is a fundamental feature of the contemporary workflows involving 3D city models. Therefore, this analysis has revealed some interesting patterns about the development and large utilization of 3D city models, and how a lot of use cases already prove the valuable role and their growth over time.

35.2.3 A Support Tool for Urban Planning and Facility Management

In recent times, 3D city models have been widely used by designer and urban planners as decision-making tools employed to explore, plan and actively act on cities. For instance, a visualization application of 3D city models can virtually display existing cities as well as may provide urban information to citizens about hypothetical developments in a 3D environment [4]. Furthermore, today's web technologies and availability of 3D city models, at different LoDs, enables local governments to communicate spatial plans to their citizens, but also it can be used in two-way communication where citizens can comment back and propose better alternatives, submitting either positive or negative feedbacks to local authorities [5]. This means that 3D city models can be useful to investigate local dynamics and best fitting urban indicators for a future enhancement [3].

35.2.3.1 User-Oriented

A useful method to achieve what is mentioned above might be a solution-oriented approach. This process starts with the collection of citizens' issues and needs, then understanding them carefully and subsequently, trying to figure out what the optimal solutions that solve citizens' problems might be. Therefore, user-oriented requirements may become helpful tools for a more transparent communication and a better decision-making process, which can improve the quality of the planning process. In Salzburg, Austria, a study has been accomplished thanks to the research project *Digital Cities*,[2] which Autodesk has conducted with Z_GIS and the City of Salzburg as a pilot city. The research was focused on the analysis of the impact that a future urban development could have in Salzburg. To do that, a combination of city data with realistic visualizations and simulation tools allowed Salzburg authorities to view and interact with the city landscape and analyze the impact of future urban planning, tourism, and economic development projects before they are built [6]. Successively, aiming to create an integrated tool for the working processes, a user-oriented approach was implemented. Indeed, users could express their needs that, in a second moment, have been structured in detailed requirements for digital cities. These requirements covered all the components of the digital cities working environment and aimed to embrace all the tasks that had value for the City of Salzburg [7]. Thus, this kind of analysis conducted, in order to be user-oriented, was context-oriented as well. At the end, two application areas have been selected: the first area concerns visual communication of planned development and navigation in the city model, considered at a different scale, from building modifications to planning the whole city districts, which represents the communication basis to involve stakeholders of different areas of expertise. The second application area regards the management of geographic objects that need to be represented in three-dimensions and that are spread over a big area of the city. In both application areas, respectively, planned modifications and geographic objects were integrated into their surroundings and could be analyzed by their spatial relations with other geographic elements [7]. In light of that, it can be argued that the digital cities environment may help cities like Salzburg to visualize and communicate proposed changes in urban areas to inhabitants. The data collected in Salzburg, especially about urban planning tasks, can be suitable for other urban contexts, such as cities with equal size, number of inhabitants, levels of development, social environment, and historical buildings structure.

---

[2] Digital Cities: 'The Digital City initiative is Autodesk's unique technology designed to provide a collaborative environment for visualizing, analyzing and simulating the future impact of urban design and development at a city-wide scale'. http://www.autodesk.com/digitalcities.

35.2.3.2 E-Participation

During the past years, E-participation has been frequently used to involve citizens in urban planning and management [5]. Indeed, web technologies, today, facilitate the communication of citizens' feedbacks about development plans promoted by authorities in order to eliminate the need for citizens to gather together in a certain place and in a specific moment. Through these technologies, a citizen may choose how, when and where to take part, even anonymously if he wants, to the decision-making process simply using a web portal. Aiming to prove the feasibility of the E-participation approach in developing countries, a 3D model with a web portal access has been created for the city of Kismu, in Kenya. Experiments have been held to measure the ability of groups of people from different backgrounds. The 3D model created was visualized in a web portal, provided by ArcGIS online, and each participant had the possibility to take part remotely, without meetings to take part in the whole process. Essentially, each citizen could easily create an account, log in, view, navigate through and leave comments in the portal. Subsequently, opinions, proposals and various alternatives gathered have been discussed throughout plenary workshops [5]. To verify participants' abilities coping with the designed 3D model, two tasks have been identified: the first regards 2D maps on A3 sheets with road networks and a list of feature names, where each participant had to pick the name of a feature in the list, locate it and mark it in the 2D map within a time limit. The second task had the same process, but with a 3D city model stored in a web-based portal. The participants' performance has been measured calculating the time needed to complete the tasks and counting the number of correct objects identified. Considering these factors, the results have shown that for all groups, the 3D task has taken less time than the first one and also appeared in more correct answers than the 2D task [5].

35.3 A Practical Application of 3D City Models

35.3.1 Public Use of the 3D City Models in the Swedish Environment

Since 1970 in Sweden, after a decentralization process that has brought about a considerable power transition to counties and especially to municipalities, the fields of territorial and urban planning are managed by the public sector and municipalities, who are considered the main actors [8]. In light of that, the 3D industry in Sweden is increasingly developing thanks to the recurring employment of 3D models in public use and particularly in city planning applications widespread in many municipalities. Indeed, many Swedish cities are currently engaged in projects that aim to reinvent the

FIGURE 35.1
From the left: Stockholm Royal Sea Port; *Min Stad* in Gothenburg; touchscreen of Linköping's 3D city model [10].

use of 3D city models for the promotion of the city development and public participation [9].

As shown in Figure 35.1, in projects like Stockholm Royal Sea Port, *Min Stad* and the Linköping's touch screen, the digital dialogue's structure consists essentially of a web portal or a touch screen, which uses an interactive 3D city model as background and encourages citizens to post ideas, according to their needs, that can be useful for the development and enhancement of several projects in certain contexts [10]. Therefore, the recent broad employment of 3D city models for public uses and stakeholders engagement have allowed many cities in Sweden to digitally promote the implementation of new projects in their territory as well as inviting an increasing number of citizens to take part in the decision-making process.

35.3.2 Study Area

The generation of a 3D city model for the city of Vaxholm will be carried out throughout this chapter as a case study to get firsthand experience with some of the available methods to create 3D city models. The whole process to create the model, the results obtained and the potential future applications and improvements that the model may reach will be presented in this chapter as well. The geographic area used in this case study, as aforesaid, is the municipality of Vaxholm, which is located in northwest of Stockholm and is often referred as the capital of its Archipelago [11].

35.3.3 The Generation Process

Throughout the next paragraphs will be presented the workflow generated to reconstruct a 3D building model with textured roofs at LoD2. The project developed aimed to create a system that allows 3D visualizations of Vaxön's urban structure, the main and most populated island of the whole municipality chosen as study area. Through the use of the software FME, and the combination of different data formats, LIDAR point clouds, 2D footprints in shapefiles and ortho images in ECW (Enhanced Compression Wavelet), a geo database handled in a FME workflow has been created to represent and manage the virtual 3D building models of Vaxholm, as shown in Figure 35.2.

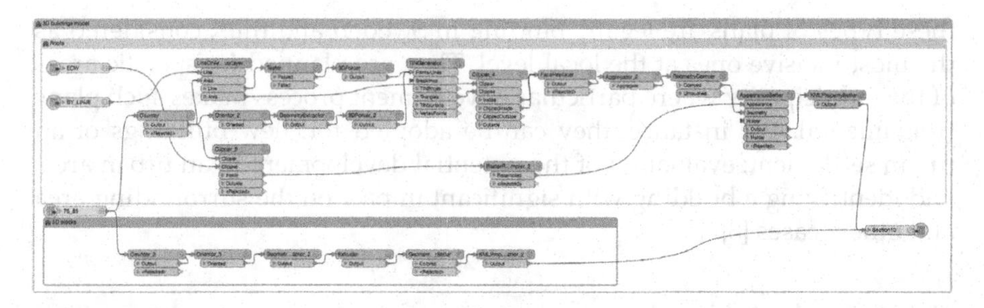

FIGURE 35.2
Workspace assembled to shape integrated 3D buildings models (Own elaboration from FMEDesktop).

The KML format was chosen because it is compatible with GE, and thus, allowing citizens to easily read and view the 3D model in a free widespread web-based mapping application as GE. Furthermore, the KML model has been spread out on the earth surface in GE, which, as in reality, presents the earth slopes according to the sea level, as shown in Figure 35.3.

Furthermore, *Detaljplaner* (Detailed plans) and *Projekt* (Projects) *pågående planarbeten* (with an ongoing plan work) have been added to the 3D buildings model, representing a zoning spread on the island of Vaxön in proximity to those buildings' part of the plans. Therefore, adding these plans has improved the usability of the 3D model, not only as a mere 3D visualization of the municipality, but rather as an effective urban planning tool. Indeed,

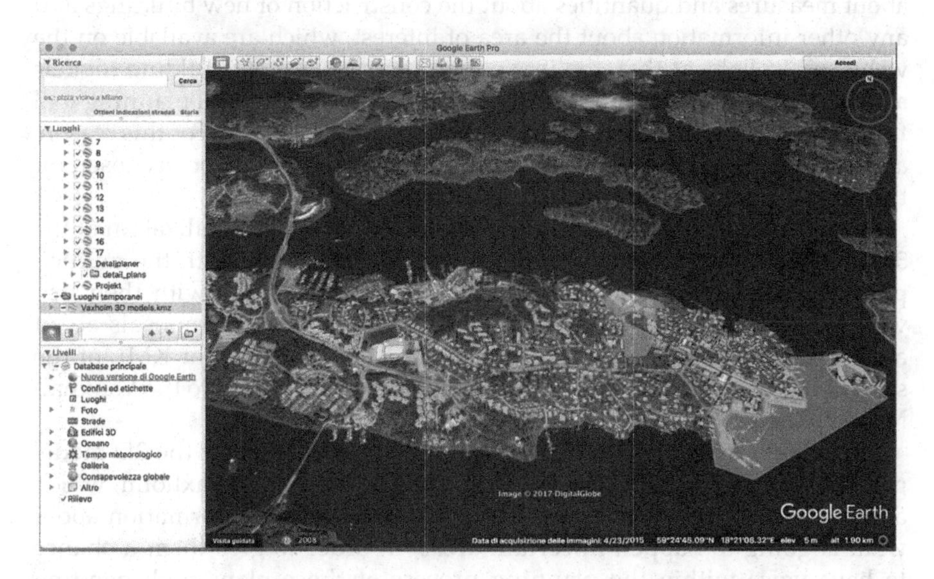

FIGURE 35.3
Visualization on Google Earth of the 3D models, detailed plans and projects (Own elaboration from GE).

these types of plans are legally binding in Sweden and thus, considered as the most incisive ones at the local level. They are planned for specific areas of the municipality where particular development process makes such plans fundamental. For instance, they can be adopted for: new buildings of an urban settlement; evaluation of the potential development in an urban area; and identifying a building with significant impact on the surrounding area and others cases [8].

35.3.4 Results Achieved

As aforementioned, only the plans where the planning process was still ongoing have been selected and added to the 3D model. The reason was to allow the municipality of Vaxholm, throughout the *iter* of approval, to take into consideration different alternatives proposed by citizens about those plans, and eventually, modifying some features before the approval of a certain plan. Indeed the plans mapped on GE can be easily switched on and switched off from the legend on the left side, according to the users' interest, and also present an URL link in the attributes content that from a clickable pop-up window connects each user directly to the plan information, published on Vaxholm municipality's webpage, as shown in Figure 35.4. In this way, citizens have the possibility not only to inspect the 3D city model of Vaxön on a GE but also to have access directly from the model, specifically on the interested area, to all the documents about analysis and studies conducted before drawing up the plan, planimetries and different types of maps about measures and quantities about the construction of new buildings and any other information about the area of interest, which are available on the webpage. In light of that, the integration between the 3D building models and the plans with pending approval may represent a helpful planning tool for those citizens who are not familiar with a web-GIS portal. For this reason, a free widespread program with an easy interface as GE has been chosen for the visualization of an interactive 3D building model.

Furthermore, the inhabitants of Vaxholm, besides the visualization of the 3D models and the connection that detailed plans have with the municipality's webpage, have also the possibility to login in GE with their private e-mail account and leave opinions, complaints and proposals about a building or a whole area, In order to attach a screenshot or KML of the subject they are interested in to the email, an email was sent to the Vauxhall Municipality.

Therefore, Google Earth, besides a simple visualization of the 3D model generated, provides a user-friendly tool to the citizens of Vaxholm, which allows them to have a clear and quick access to all the information about what is going to happen in certain areas, in order to give them a chance to have voice within the planning process of those plans with pending approval.

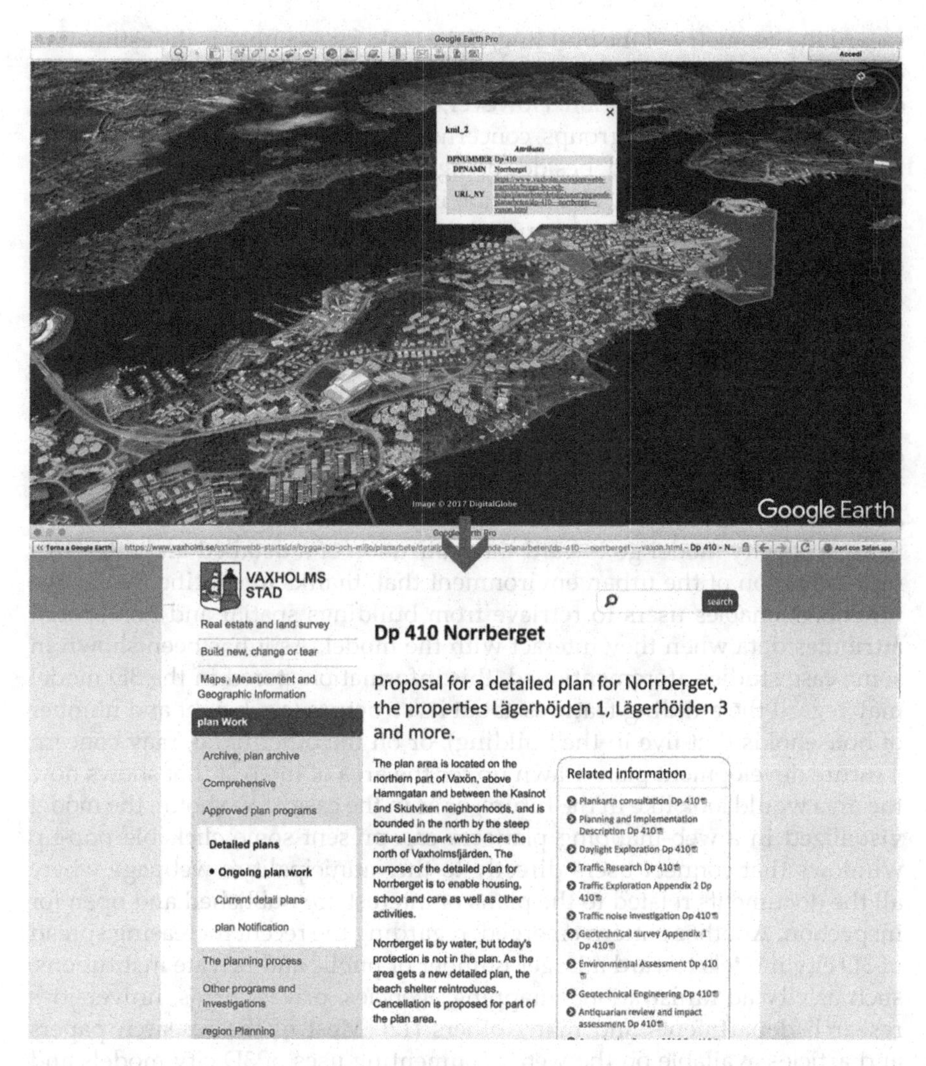

FIGURE 35.4
On the top: the pop-up window with the URL link (Own elaboration from GE) that connects users directly to the detailed plan information on the municipality's website, on the bottom [11].

35.4 Conclusions and Future Improvements

The evidence presented in this article clearly indicates that 3D city models have proved to be estimable for a large number of domains during the last few years and thus have been recently used in large number of application ambits and for diverse purposes related to cities' development that, throughout this chapter, have been classified according diverse criteria. These principles can regard the geometry of the building, if the task, for example, is the estimation of shadow; or semantic information stored in each building, if we refer to the estimation of solar irradiation. However, the most relevant criterion followed to individuate two main groups, concerns the visualization or non-visualization of 3D models. Indeed, besides being the first early-use domain of application for 3D city models, visualization is arguably an indivisible part of the workflows that involve 3D city models and it can be considered as the only criterion that is adaptable to almost all categories of applications mentioned in the last chapters. Successively, in examining how the utilization of 3D city models in purposes related to urban analyses is growing, some interesting patterns have been delineated. For instance, it has been revealed that, by virtue of new technologies and methods for data acquisition and processing that have enhanced the efficiency of 3D city models, the requirements of 3D city models have changed direction: from a mere realistic geographical representation of cities that provides to users public access for the exploration of 3D elements, currently, the goals that a city administration wants to achieve using a 3D city model go toward the realization of a detailed and attractive representation of the urban environment that, thanks to specific interactive functions, enables users to retrieve from buildings spatial and non-spatial attributes data when they interact with the model. As it has been shown in some case studies aforementioned, this information stored in the 3D model may regard the building features about its structure (e.g. height and number of households that live in the building), or on the other hand, may concern a future development plan drawn up for the area of interest that shows how the area would look like in the future, or as in the case of Vaxholm, the model visualized in a web-mapping platform can present some clickable pop-up windows that connect users directly to the municipality's webpage where all the documents related to the plans of interest are published and open for inspection. Another pattern emerged, regarding the recent increasing spread of 3D city models toward a larger number of public and private institutions, such as city administrations, mapping agencies, private firms, universities research departments and many others [12]. Most of the research papers and articles available on the web, documenting uses of 3D city models and integration of them as planning tools employed by several municipalities, have been published during the last decade. Hence, this fundamental support tool has been already adopted by every municipality that has a dense urban structure and a copious population or, for the cities where it is not yet, they are in the process of setting up 3D city models as part of their planning tools [13].

Given the results achieved, this research might be helpful for all stakeholders involved in 3D city modeling community with different level of decision-making, in the way that they may use it to make improvements to their product or at least understand the range of applications that 3D geo-information can offer today. Also, it might be beneficial as a reference that provides a detailed insight defining use case scenarios and then according to the purposes that have to be achieved, setting the suitable requirements when procuring the 3D datasets. Even though the large number of cases of study that have been mentioned proves already the estimable role and the high demand of 3D city models, further technologies' improvements, new scenarios and cases of application, are expected in the following years. One of the biggest tasks related to the field of 3D city models is to find cost-effective and avoid time-consuming approaches to create models rich in semantic information. For instance, improvements toward the integration of computer graphics, GIS and BIM would allow on the one side, the realization of more detailed 3D city models, and on the other side, rich information stored in the model would allow new types of applications to be planned thereby increasing the possibilities associated with 3D city models.

The integration of 3D modeling with dynamic visibility analysis is being developed in order to allow users (citizens, planners, members of landscape commissions, ... up to students) to analyze the urban proposals presented using this key [14,15].

In conclusion, in light of the all above-mentioned considerations, it can be asserted that virtual city modeling for many municipalities is a new approach to manage cities' development and to encourage participation of the public within the planning process. A 3D city model can help cities, both similar to and different from Vaxholm, to visualize, inspect, and communicate proposed developments and changes of their urban environment. The approach followed to generate the model can be applicable in other contexts with similar characteristics, such as equal size, number of inhabitants, historical background and building structure, as a support tool for more transparent communication with the citizens that have the potentiality to improve the quality of the planning process. However, there is a lot more to investigate in this field that unluckily is outside the scope of this dissertation, but it is clear that the more detailed is the model and rich of semantic information, the greater are the possibilities to include additional applications with the model as a basis.

The tools developed are used in academic activities and represent a significant example to be promoted as best practices at various administrative levels.

Acknowledgments

This chapter is the result of a Master of Science Thesis handed-in to the Politecnico of Turin. The authors would like to acknowledge the support of the Building Law and GIS Unit (Bygglov- och GIS-enhet) of Vaxholms Stad, Sweden, for the development of the 3D city model used as case study.

References

1. Biljecki, F. The concept of level of detail in 3D city models. GISt Report No. 62. Delft University of Technology. **2013**. pp. 1–25.
2. Ross, L. *Virtual 3D City Models in Urban Land Management-Technologies and Applications*. Ph.D. Thesis, Technische Universität Berlin (TUB), Berlin, Germany, **2010**.
3. Biljecki, F.; Stoter, J.; Ledoux, H.; Zlatanova, S.; Çöltekin, A. Applications of 3D city models: State of the art review. *ISPRS Int. J. Geo-Inf.* **2015**, 4, pp. 2842–2889.
4. Buhur, S.; Ross, L.; Büyüksalih, G.; Baz, I. 3D city modeling for planning activities, case study: Haydarpasa train station, haydarpasa port and surrounding backside zones. *Istanbul. Int. Arch. Photogramm. Remote Sens. Spat. Inf. Sci.* **2009**, 38. pp. 1–6.
5. Onyimbi, J.R.; Koeva, M.; Flacke, J. Public participation using 3D city models. E-participation opportunities in Kenya. *GIM International*, **2017**, 31. pp. 29–31.
6. Autodesk. *Autodesk Announces Salzburg, Austria, as First Pilot City of Its Digital Cities Initiative*. Press Room Archieve, **2008**. Available online: https://www.autodesk.com/digitalcities (Accessed the 23rd September 2017).
7. Albrecht, F.; Moser, J. Potential of 3D City Models for Municipalities - The User-Oriented Case Study of Salzburg. **2008**. pp. 2–8.
8. Solly, A. *The Europeanization of Spatial Planning: The Case of Sweden*. Master Thesis in Territorial, Urban, Environmental and Landscape Planning at Politecnico of Turin, **2013**.
9. GIM International. *Sweden Excels in Public Use of 3D in Smart City Applications.* Mapping the world. **2015**. Available online: https://www.gim-international.com/content/news/swedish-cities-excel-in-public-use-of-3d-in-smart-city-applications (Accessed the 1st October 2017).
10. Agency9. *Swedish Cities Innovate 3D Use in Smart City applications.* **2015**. Available online: https://agency9.com/swedish-cities-innovate-3d-use-in-smart-city-applications/ (Accessed the 14th of November 2017).
11. Vaxholms Stad. *Vaxholm - skärgårdens huvudstad.* **2017**. Available online: https://www.vaxholm.se/turistwebb-startsida.html (Accessed the 3rd of December 2017).
12. Salata, S.; Garnero, G.; Barbieri, C. A.; Giaimo, C. The integration of ecosystem services in planning: An evaluation of the nutrient retention model using InVEST software. *Land*, **2017**, 6. 48p [DOI: 10.3390/land6030048].
13. Minucciani, V.; Garnero, G.: Available and implementable technologies for virtual tourism: A prototypal station project. In B. Murgante et al. (Eds.): *ICCSA 2013*, Part IV, published in Lecture Notes in Computer Science (including subseries Lecture Notes in Artificial Intelligence and Lecture Notes in Bioinformatics), LNCS 7974 (ISSN: 0302-9743, ISBN: 978-3-642-39649-6), pp. 193–204. Springer, Heidelberg (**2013**) [DOI: 10.1007/978-3-642-39649-6-14].
14. Garnero, G.; Fabrizio, E. Visibility analysis in urban spaces: A raster-based approach and case studies. *Environment and Planning B-Planning & Design*, **2015**, 42. pp. 688–707 [DOI: 10.1068/b130119p].
15. Fabrizio, E.; Garnero, G. Visual impact in the urban environment: The case of out-of-scale buildings. *Journal of Land Use, Mobility and Environment*, **2014**. pp. 377–388 [DOI: 10.6092/1970-9870/2477].

36

Exploiting Virtual Reality to Dynamically Assess Sustainability of Buildings through Digital Twin

Muhammad Shoaib
Politecnico di Milano

Lavinia Chiara Tagliabue
University of Turin

Stefano Rinaldi
University of Brescia

36.1 Introduction

Considering the global impacts of climate change, sustainability is becoming the major concern area for governments, policymakers, industries, and researchers. There is a global consensus that collective efforts are required at all levels to mitigate the negative impacts of climate change. Governments are making policies, researchers are contributing through innovation, and industries are familiarizing with environment-friendly approaches. Similarly, the construction industry is implementing strategies to reduce its carbon footprint. It is important to mention that the construction sector is one of the major carbon emitters as many industries run in tandem with the construction sector. The goal of sustainable built environment is to reduce its environmental impact. The introduction of green building rating systems has standardized the sustainability assessment process. According to the World Green Building Council, *Green building rating and certification systems require an integrated design process to create projects that are environmentally responsible and resource-efficient throughout a building's lifecycle: from siting to design, construction, operation, maintenance, renovation, and demolition*. On the other hand, Building Information Modeling (BIM) has transformed the industry showing its value throughout the project lifecycle. The digital design through BIM can help in reducing the carbon footprint of the construction industry.

DOI: 10.1201/9781003425724-43

However, further innovation is required to optimize the built environment through dynamic sustainability assessment techniques based on the digital twin (DT) technology. Research studies regarding the sustainable built environment, green buildings rating systems, BIM, and DT are significant. The gist of the research studies is that green buildings are an effective way to lower the environmental impacts associated with construction. BIM has transformed the construction lifecycle and the DT domain in the built environment is relatively new. The DT studies concluded that like other industries DT is immensely suitable for the construction sector too. However, establishing the connection of the cyber-physical world and the use of that data is quite complex. The important takeaway from the literature is that the use of Internet of things (IoT), sensors, and other data collection methods can provide a huge amount of data on the built environment which can then be used for a whole slew of different purposes. The establishment of Green Digital Twin (GDT) is to take the leverage of the DT to assess sustainability. The core objective of this research study is to provide the mechanism for the dynamic sustainability assessment by harnessing the potential of BIM, IoT, Artificial Intelligence (AI), and VR. In the actual reality (AR) / VR environment, the complex system that establishes the connection between the information-physical world, collects data through various Internet of Things, and processes the data to obtain the output of the results and visualization will be GDT.. This research is based on a case study and also proposes a framework that provides the basis of the GDT. The proposed framework is the amalgamation of different DT models available and it is specific to the built environment but it can be applied to the other domains with some customization. The main research purpose is to check the parameters related to Air and Materials from WELL and LEED building standards, respectively. Although a large number of parameters can be checked through the GDT, this study focuses on two parameters checked to test the model validity.

36.2 Literature Review

A literature review was conducted to understand the existing research and what is being implemented in the market. The interdisciplinary nature of the project requires a complete understanding of the knowledge and research status in different fields. Firstly, sustainable construction, which is a vast topic. However, the scope of this study is limited to the avenues related to green buildings, their rating systems, and their evaluation methodologies. Secondly, the role of BIM in sustainable construction. Finally, the literature review discusses the DT technologies in detail, their role in green buildings, and facility management.

In order to explain sustainability in construction, it is crucial to understand its importance in the construction sector. Researchers and practitioners are concerned about the environmental issues that resulted due to the unprecedented urbanization in the last decade and developments associated with it (Chuai et al., 2021). The construction sector anywhere in the world is one of the most important sectors with a handsome contribution to the global and local economy. According to the studies by McKinsey & Company, the construction sector contributes approximately 14% to the global gross domestic product (GDP) (McKinsey & Company, 2020). According to the estimates, the construction industry is one of the greatest CO_2 emitters accounting for 36% of total greenhouse gas (GHG) emissions and consuming 40% of the global energy (Li et al., 2021). Furthermore, the energy consumption in the use phase of the buildings is estimated to be one quarter of total CO_2 emissions (Huang et al., 2018). Sustainable construction takes inspiration from sustainable development as it aspires to achieve the core values of sustainability.

There are different attributes that a building should have to be considered as a "Green Building". According to the World Green Building Council, following attributes must be met by a building to be considered as Green Building (WGBC, 2015)

- Improved indoor environment quality
- Efficient use of land, energy, and water resources
- Incorporation of renewable energy resources
- Responsible use of materials and nontoxic products/materials
- Adoption of circular approach in project and product life cycles.

These are the most common attributes that need to be assessed and verified before considering a building green. The core work of the thesis is centered around the assessment of these attributes and the process is improved further. The benefits of green buildings are proven and various research studies have stated the fact. Balaban and Puppim de Oliveira (2017) concluded that the green buildings based on the principles of sustainability are effective in lowering the environmental impacts and ultimately contribute to healthier and clean urban development. Green buildings are considered to improve human health, enhance indoor air quality, water, energy, and be resource-efficient (Chi et al., 2020).

BIM adoption has transformed the built environment lifecycle in many ways for all the stakeholders. On the other hand, technological innovation especially the IoT implementation in different industries has provided better insights into the systems and enhanced data quality for decision-making. Construction projects are data-intensive and the whole lifecycle of any project is dependent on data input from various sources and stakeholders. The influence of tech-based solutions is ubiquitous in different industries. Digital transformations provided new avenues to enhance the economic, social, and environmental resilience of systems. Similarly, the integration of sensors, IoT, cloud computing, AI, and other supporting technologies that are reshaping the construction industry have led to the emergence of DT in construction (Lu et al., 2020). The concept of the DT is new and yet in nascent stages for the construction industry. However, other industries especially the manufacturing sector is using this quite efficiently for the last two decades.

The use of DTs in a built environment is inextricably linked with the BIM adoption as only the BIM is considered as the foundational block of the DT. Irrespective of the industry domain, "Data" is the key to the development of the DT. Different research studies have explored the potential of the DT in the built environment and presented a diverse set of DT models. According to the Ernst and Young (EY) white paper, DTs have four major applications within the built environment. The four applications are:

i. Building maintenance and operations
ii. Health and wellness
iii. Enhanced occupant's experience and data collection for decision-making
iv. Sustainability evaluation.

There are several frameworks for DT presented by researchers over the years in the built environment domain. Most of these frameworks provide their approach to connect the physical object with the virtual and then leverage the data from there.

In general, there are four types of DT models:

- *Interface-oriented*: Interface-oriented DTs are mainly used for the monitoring of the objects or assets lifecycle, and data for manufacturing or re-engineering purposes. The implementation of it is difficult in the AEC industry as the data or information exchange is quite abstract and the layers of information aren't clearly defined. It provides an interface between the physical and virtual objects, but the explicit link between these two is not clear. These types of DTs are not suitable for individual asset information recovery or data input.
- *Prototypical*: These types of DTs provide more detailed data from the physical assets and can be linked to the Building Management Services or Systems (BMS). Apart from the usual capability of sharing the operations and maintenance data, these DTs can also provide periodic data for events or occurrences as defined by the use or DT owner. Prototypical DTs take into account the asset's lifecycle as well as use-case descriptions. Regarding Built environment applications, these DTs can be deployed in parallel with BMS solutions and monitoring protocols (Wang & Luo, 2021). All sorts of stakeholders (operators, owners, authorities, and end-users) can interact with the

DTs. Contextual visualizations are included in the list of DT services because the context in which an item is located is extremely important for built assets.

- *Model-based*: Model-based DTs are quite like the BIM concepts and are used for assets that require a higher degree of a digital replica of complex physical objects/structures. To effectively depict the changing conditions of the physical asset, proper interaction between the various models is required. It is important to note that the Model-based DTs require a higher degree of precision, communication, and interoperability protocols, and every four nodes mentioned in the below model are fully integrated. The DT model will have four different nodes enabling its functionality, the function of the four nodes is as follows:
    i. Configurator node: It is a user interface that allows various categories of users in the built environment to customize the DT's structure, functionalities, and control rules.
    ii. Ontologies node: It is a collection of components and resources used to customize certain DTs to meet the needs of the user.
    iii. Simulation node: Various Algorithms play their role here. This process or node evaluates and simulates the performance of various DTs.
    iv. Visualization node: This is a visual environment that provides a visual presentation of simulations and evaluations that have been completed.
- *Service-based*: These types of DTs are usually developed to manage the construction and operation of large-scale complex assets. For example, nuclear facilities, manufacturing units, oil, and gas refineries. These DTs aim to automate the complex workflows in which the user takes the central position. Through this real-time monitoring, control and maintenance services of a complex system can be handled manually or through automation. Service-based DTs enable the user to handle an array of isolated or interconnected DTs. These DTs are modeled in an explicit manner in which data workflows are rather complex but provide a better virtual representation of the physical asset.

Boje et al. (2020) concluded that the data information exchange has been made easy by the introduction of IFC but still a lot of barriers are there to harness the true potential of the DT. However, Bilal et al. (2016) concluded that most of the studies are related to operations and maintenance of few key parameters for facility management. A research study (Ye et al., 2019) proposed a DT model for structural health monitoring. The DT concept is also getting popularity for the management of urban assets. A research study conducted by ETH Zurich (Schrotter & Hürzeler, 2020) proposed a DT model for urban planning. In this study, extensive use of spatial data and the data from the physical objects were synced to create a DT for Zurich. The use of DT is also getting to measure and evaluate parameters related to the indoor environment quality, occupant's behavior, and facility management to predict the behavior of parameters. An occupancy-oriented DT study was conducted to evaluate the factors and optimize the operational phase of the buildings (Seghezzi et al., 2021). This study is aimed to produce a BMS for facility management based on occupancy DTs. IoT, cameras, sensory devices, and advanced data modeling techniques were used to obtain the results from the DT. Indoor air quality predictions for the educational building were made by using sensory data (Tagliabue, Re Cecconi, et al., 2021). These predictions can be used to further optimize the building operations by adjusting the HVAC systems. The approach adopted in this research can be used to measure various other factors of the indoor air quality ranging from temperature to measuring the accumulation of volatile organic compound (VOC) accumulation. Some parameters have already been measured in this research. Similarly, by harnessing the potential of forecast and anticipation through data processing an approach for the smart built environment was presented in the research (Tagliabue, 2021). Furthermore, a case study–based research (Tagliabue, Cecconi, et al., 2021) presented an approach in which the concepts of Green BIM along with the DTs were used to conduct the sustainability assessment for an educational building. It concluded that the development of DTs and the data analytics during the use phase allows for better building management and real-time sustainability evaluation. DT enhances the decision-making capability of the stakeholders and facility managers. It allows them to perform what-if scenarios on the operational phase of the building and then optimize the operational energy of the building (Khajavi et al., 2019). The data of the DT are quite useful not only for the building for which they were developed but these data can also be used for big data analytics to study the behavior of future projects (Qi & Tao, 2018).

36.3 Methodology

There are various reasons to have a dynamic DT for sustainability assessment. The need arises from the initiatives to achieve SDGs globally and locally. The European Union (EU) Green Deal is a comprehensive ambitious plan to reduce carbon emissions. The role of the built environment in reducing emissions is immense. Sustainability in buildings is measured on a range of parameters and factors. Green building rating systems serve as a foundational block to quantify the sustainability parameters. Each category is further divided into points and each point has its criteria to be fulfilled. Based on the points achieved, a certification is awarded to establish the measures taken to make a building more sustainable. However, the green building rating systems are diverse and the categories cover a range of building characteristics. All these characteristics are parameters that can be made dynamic through the use of sensors and other techniques. The need for the development of a dynamic DT is evident considering,

- Building, either certified or not certified, is a living element or object producing a lot of data that can be used for various purposes.
- Parameters of the green building need to be recorded and reported periodically, and maintained over the lifecycle of the building.
- Effectively support the built environment control and monitoring to durably evolve toward a green and sustainable dimension.
- Provide a decision support system to stakeholders based on the real performance of the buildings.
- DT's results or outputs can provide useful insights for future planning of the building for designers, owners, researchers, and building regulatory authorities.
- DTs can make the green building certification, recertification, and performance process efficient which is currently based on documentation review and sometimes on-field testing. For example, the certification process in LEED and WELL involves documentation review. The time, after credit submittals, in reviews, comments, etc. can vary from 45 to 90 days which further delays the whole process.

The foundation of a green or sustainable built environment is based on the principles of sustainability which is a bigger domain. It aims to achieve the SDGs and every industry has its methods and ways to achieve that. Similarly, the green buildings at the MACRO level consider the same attributes. The rating systems are developed in a way that upon implementation they provide environmentally friendly, economic, and socially viable solutions. The green DT is focusing on the parameters of the green buildings. Various green building rating systems are there and each has its own set of parameters that needs to be evaluated.

It is very important to establish the Key Performance indicators as identifying them can help to develop strategies and mechanisms for data input, processing, and output. Considering the complexity and diversity of the built environment parameters, the KPIs have been divided into three layers:

- *Primary indicators*: These are the MACRO elements and define a major field. For example, all the elements that can impact the environment by any means will come under the environment domain and the environmental domain will become the primary KPI. Similarly, all the elements related to human health will come under the wellness primary KPI.

- *Secondary indicators*: These are the ones that can be grouped under the primary indicators or the subset of primary indicators. For example, energy generation and use directly impact the environment due to the carbon emissions associated with it.
- *Tertiary indicators*: These indicators are the subset of secondary indicators and include various criteria. Tertiary indicators are the foundational block of any DT. For example, all the credits related to energy are part of tertiary indicators. The relation between the KPIs is shown as follows (Figure 36.1).

In the case study for green DT, the tertiary indicators are of prime importance and they will be the focus. The values for the tertiary indicators will be taken from the standards (LEED, WELL, LBC, BREEAM, etc.).

After the review of different studies, it is evident that the concept of the DT in the sustainable built environment or overall in the construction industry is not mature. It is getting popular and research studies have already proved the potential of the DTs. A uniform approach for the development of "Digital Twins" is required. Considering the factors from the built environment perspective, the DT paradigm intends to improve existing construction processes and models (nD BIMs) and their accompanying semantics (e.g. IFC, COBie) within the scope of a cyber-physical world. The scope of these digital models will be that they will reflect the constructed physical asset at any time. Hence, the concept of DT requires that the integration of construction data models with new technologies is essential to fully harness the potential of DTs. In the previous section,

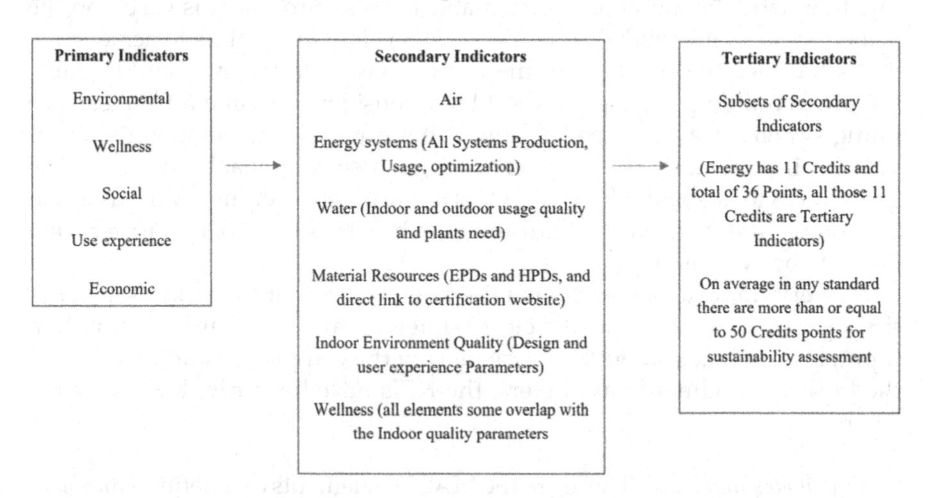

FIGURE 36.1
KPIs and their connections.

a separate discussion about the frameworks and types of DTs has been done. However, due to the complexity and diversity of built environment parameters and associated interoperability issues, a single model of DTs can't be adopted. Amalgamation and customization of various DT models are required to develop a model for the DTs. A generic framework for the DT is presented below which is an effort to provide an idea about the different layers a DT can have.

A DT is a combination of several modules, such as a computer model, a physical model, communication services, and data analytics. These modules work in synchronization to monitor, learn, and optimize the complete system operation. However, the implementation of the DT concept may require new processes, methods, and novel platforms to interact with each of these modules. GDT for dynamic sustainability assessment can only be done by integrating the data from various sources for different sustainability parameters. The sources of information collection can be any but not limited to the real-time sensor, BMS, feedback devices at the public service use areas, cloud services, and asset management software. The asset information model is the key in the whole process that can lead to the development of the DT. However, the development of a federated model is highly dependent on the quality and quantity of the data that can be generated. Data are the lifeline for any DT. Considering the nature of this study in which the assessment of the sustainability parameters needs to be done, it is important to identify what kind of databases are required. Usually, there are two types of databases, static and dynamic data.

Static data: Static data represent the information related to the parameters that are fixed and will not change over time or for a longer period. Commonly static data include the characteristics of the material data, physical features like windows, walls, and door information. Furthermore, the metadata in the BIM model can incorporate a lot of information that comes under the static data for the DT. For example, warranty information, material sound absorption value, transmissivity, fire rating, U-value, operation and maintenance schedule, operation manuals, etc. If we further expand the type of static data from a sustainability assessment point of view, then it will be regarding the site information, type of plants on the site, materials HPDs, and EPDs.

Dynamic data: Dynamic data refer to all sorts of data from the building that are not constant and keep changing over time, randomly or periodically. Mainly dynamic data are from the IoT, sensors, and other monitoring devices placed in the buildings. For example, all the data coming from the sensor's data to monitor the carbon dioxide level in any room, lighting levels monitoring sensors, occupancy sensors, HVAC speed controlling devices data, humidity monitoring data, and energy consumption data at any given time frame. Usually, this dynamic data are processed by simple algorithms or AI for simulations to extrapolate and predict the behavior of certain parameters. The proposed framework for the green DT is shown below. It is a comprehensive framework that defines each layer separately and a bottom-up approach. The framework starts from the basic BIM data or model and relies mostly on the data that is fed to either the BIM model or directly to the server available. It is highly dependent on the nature of the study or the data collection method that can vary from study to study. Furthermore, sometimes the nature of the parameters under consideration can create problems regarding data interoperability, data recording, security, quality, and quantity of the data. It is a proposed model and has the provision for the nth number of DTs but in a real-case scenario, it depends on the number of credits and parameters associated with those credits are being measured (Figure 36.2).

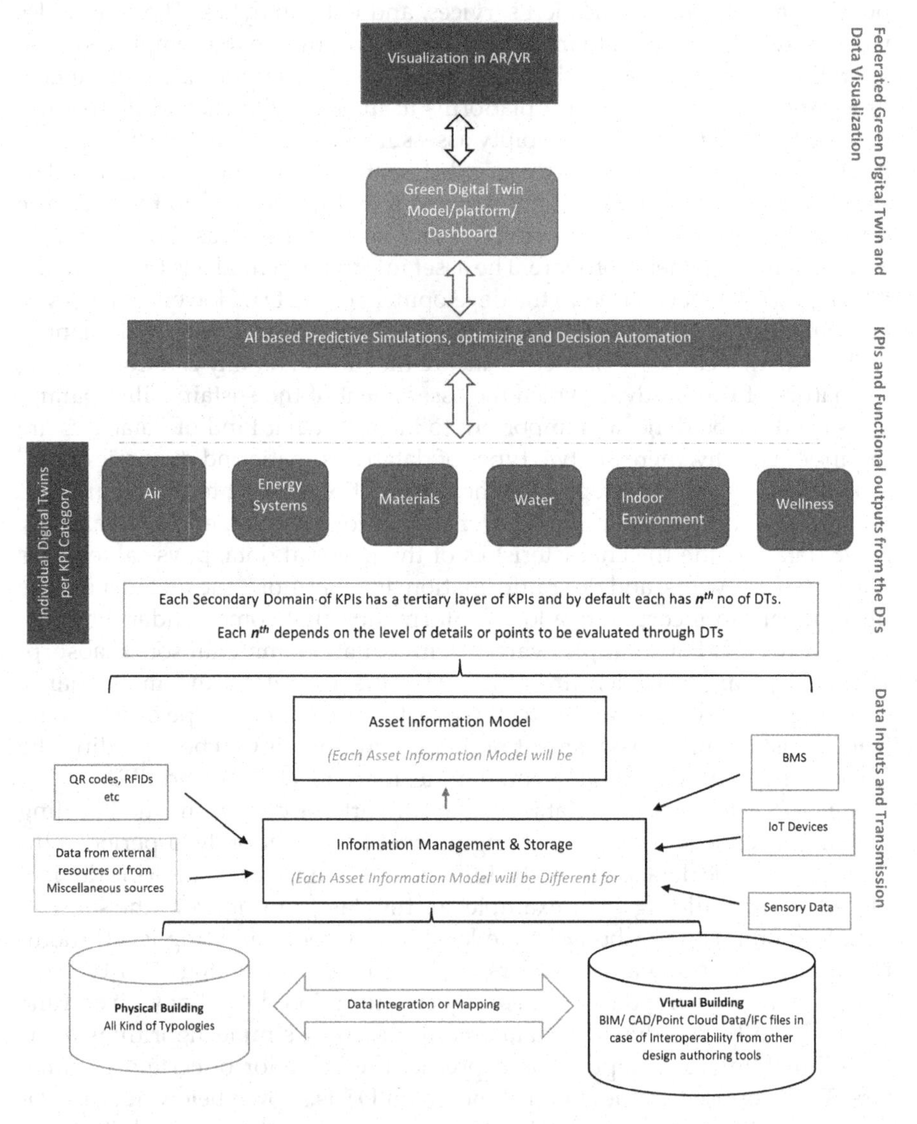

FIGURE 36.2
Proposed green digital twin framework.

36.4 Case Study

To verify the sustainability assessment through DT, a case study was conducted on the eLUX lab building at the University of Brescia, Italy. There are many parameters related to sustainability. However, for this case study, one parameter from LEED and WELL are tested through DT and then the data is visualized in the virtual reality (VR). The parameters that are measured are related to

- The Material and Resources Category from the LEED V4.1 rating system
- AIR credit category from the WELL V2 building standard.

Recycled content and EPDs belong to the static data category while the CO concentration is a value that is from the dynamic data category. The Revit model for the eLUX lab is used in conjunction with the Escape VR engine to visualize the result. The complete process is explained in this section.

Our target is the furniture, chairs to be precise, and walls, to measure the two parameters from the material credit category. A Revit family of the chair was generated and placed in the 3D model.

36.5 Results

In the next step, an EPD parameter of the Chair Revit element was added and the EPD was linked in the model. The authentic EPD for the chair was selected from the international EPD Portal. In this step, BIM data are linked with the Encapse and turned the models into immersive 3D experiences for the verification of the EPD. For example, the color of the selected chair is yellow and the data confirm that this product is EPD verified.

Similarly, the staircase wall has recycled content and the recycled material is defined as a parameter in the BIM model for that wall. The VR selection of the image confirms the recycled content. The amount of percentage of the recycled material is unknown that's why it is not showing any value. If we choose the other elements that don't have EPDs, the data will not show that they are EPD verified.

This is how we conduct the sustainability assessment of the building components by harnessing the potential of the BIM data and then using the VR engine. It is important to remember that the BIM data are the basis of the sustainability assessment. Now moving on to the next scenario in which the same EPD verification is being done but in the BIM environment. Furthermore, by selecting the object in the BIM and using the filer option in BIM schedules, we can see in the following pictures that all the chairs with the EPDs are selected. The purpose of filtering, selecting, or verification within the BIM environment is to highlight the fact that the metadata of the BIM model, objects, or elements is of prime importance. Figure 36.3 shows the EPD metadata of the chair in the BIM model, and then the selection of all chairs that are EPD verified.

From the verification of the material EPD and recycled content, we can say that the establishment of a cyber-physical world connection can help verify the sustainability parameters, hence validating the green DT for the material credit category.

The WELL building standard air category parameter for carbon monooxide concentration was monitored, analyzed, and shown in the BIM model. According to the standard, the concentration of the CO_2 must be less than 1,000 PPM in any given space. These data can be monitored in various ways, analyzed, and results can be shown in the model. However, here the results are only shown in the changing colors to show how the color scheme of the spaces will change based on the CO_2 concentration levels. In further advanced

Environmental Product Declaration - Ovo Lounge Chair						
TYPE	QR code	Cost	EPD	Design country	Manufacturer name	Manufacturer country
Ovo Lounge Chair	https://www.environdec.com/library/epd3806	2145.00	https://www.environdec.com/library/epd3806	Foster + Partners, UK	Benchmark Woodworking Ltd	UK & Europe
Ovo Lounge Chair	https://www.environdec.com/library/epd3806	2145.00	https://www.environdec.com/library/epd3806	Foster + Partners, UK	Benchmark Woodworking Ltd	UK & Europe
Ovo Lounge Chair	https://www.environdec.com/library/epd3806	2145.00	https://www.environdec.com/library/epd3806	Foster + Partners, UK	Benchmark Woodworking Ltd	UK & Europe
Ovo Lounge Chair	https://www.environdec.com/library/epd3806	2145.00	https://www.environdec.com/library/epd3806	Foster + Partners, UK	Benchmark Woodworking Ltd	UK & Europe
Ovo Lounge Chair	https://www.environdec.com/library/epd3806	2145.00	https://www.environdec.com/library/epd3806	Foster + Partners, UK	Benchmark Woodworking Ltd	UK & Europe
Ovo Lounge Chair	https://www.environdec.com/library/epd3806	2145.00	https://www.environdec.com/library/epd3806	Foster + Partners, UK	Benchmark Woodworking Ltd	UK & Europe
Ovo Lounge Chair	https://www.environdec.com/library/epd3806	2145.00	https://www.environdec.com/library/epd3806	Foster + Partners, UK	Benchmark Woodworking Ltd	UK & Europe
Ovo Lounge Chair	https://www.environdec.com/library/epd3806	2145.00	https://www.environdec.com/library/epd3806	Foster + Partners, UK	Benchmark Woodworking Ltd	UK & Europe
Ovo Lounge Chair	https://www.environdec.com/library/epd3806	2145.00	https://www.environdec.com/library/epd3806	Foster + Partners, UK	Benchmark Woodworking Ltd	UK & Europe

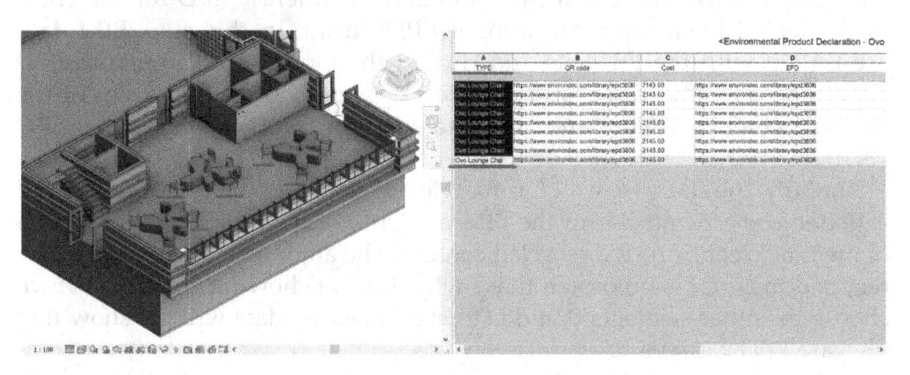

FIGURE 36.3
Selection of the elements with EPDs and recycled content in the cyber-physical world.

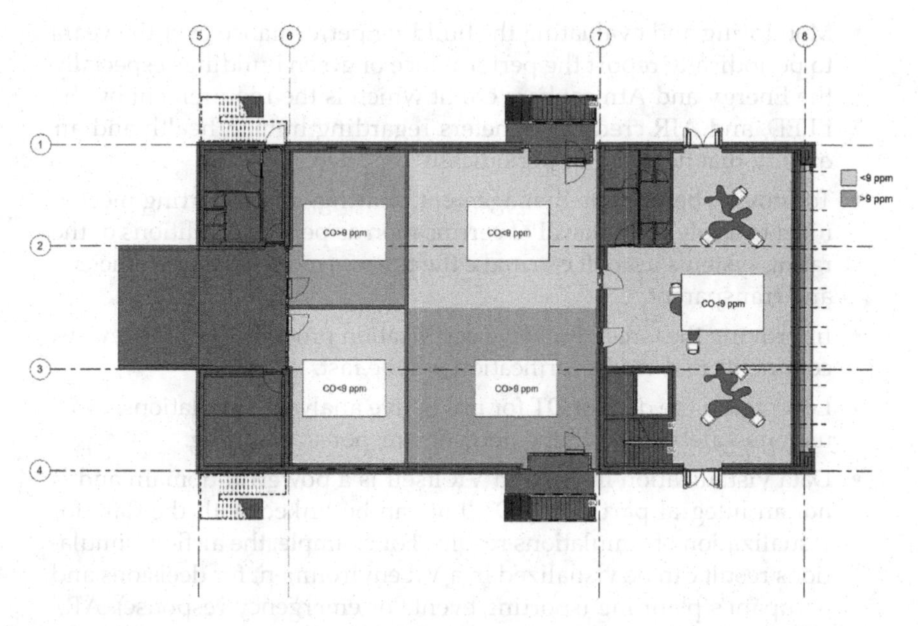

FIGURE 36.4
Indoor condition verification in the cyber-physical world.

studies, it can be linked with the alarms or HVAC systems to induce the fresh air or exhaust mechanism to keep the CO concentration in range (Figure 36.4).

36.6 Conclusions

Based on the case study and the proposed green DT framework it can be concluded that the process of sustainability assessment through DT is highly dependent on the BIM and other input data. The sustainability parameters assessment is quite efficient, fast, and transparent through the DT. Each parameter of sustainability has its requirement for the DT. Hence, each DT will be unique in its structure and might have the same or different components and layers of the DT framework with respect to the other DTs. Considering the wide applications of the green DT, the following conclusions can be inferred from the research study. The development and implementation of green DT can be used.

- To facilitate the transition of sustainability assessment from the static approach to the dynamic approach. With sensors, IoT devices, and other tools, various parameters can be digitalized.

- Monitoring and evaluating the building performance over the years to periodically report the performance of green buildings especially the Energy and Atmosphere credit which is the requirement by the LEED, and AIR credit parameters regarding human health and air quality that needs to be periodically reported.
- To provide better data management, sharing, and reporting mechanism to apply for renewal of certification as per the conditions of the rating systems used. It can make the process more easy, fast, efficient, and transparent.
- Improving the Green Building certification process. With DT review, comments and credit verification will be fast.
- Leveraging the data of DT for predictive analysis, simulations, evacuations, safety, and other emergency response planning.
- Data visualization in AR and VR itself is a powerful domain and is not an integral part of the DT, but can be linked with the data for visualization of simulations results. For example, the airflow simulations result can be visualized in a VR environment for decisions and occupant's planning (sporting events or emergency response). AR/VR can also be used for WELL and LEED features (lighting, thermal comfort, airflow modeling, and quality views) verification.

References

Balaban, O., & Puppim de Oliveira, J. A. (2017). Sustainable buildings for healthier cities: Assessing the co-benefits of green buildings in Japan. *Journal of Cleaner Production*, 163, S68–S78. https://doi.org/10.1016/J.JCLEPRO.2016.01.086.

Bilal, M., Oyedele, L. O., Qadir, J., Munir, K., Ajayi, S. O., Akinade, O. O., Owolabi, H. A., Alaka, H. A., & Pasha, M. (2016). Big Data in the construction industry: A review of present status, opportunities, and future trends. *Advanced Engineering Informatics*, 30(3), 500–521. https://doi.org/10.1016/J.AEI.2016.07.001.

Boje, C., Guerriero, A., Kubicki, S., & Rezgui, Y. (2020). Towards a semantic Construction Digital Twin: Directions for future research. Automation in Construction, 114, 103179. https://doi.org/10.1016/J.AUTCON.2020.103179.

Chi, B., Lu, W., Ye, M., Bao, Z., & Zhang, X. (2020). Construction waste minimization in green building: A comparative analysis of LEED-NC 2009 certified projects in the US and China. *Journal of Cleaner Production*, 256, 120749. https://doi.org/10.1016/J.JCLEPRO.2020.120749.

Chuai, X., Lu, Q., Huang, X., Gao, R., & Zhao, R. (2021). China's construction industry-linked economy-resources-environment flow in international trade. *Journal of Cleaner Production*, 278, 123990. https://doi.org/10.1016/j.jclepro.2020.123990.

Huang, L., Krigsvoll, G., Johansen, F., Liu, Y., & Zhang, X. (2018). Carbon emission of global construction sector. *Renewable and Sustainable Energy Reviews*, 81, 1906–1916. https://doi.org/10.1016/j.rser.2017.06.001.

Khajavi, S. H., Motlagh, N. H., Jaribion, A., Werner, L. C., & Holmstrom, J. (2019). Digital twin: Vision, benefits, boundaries, and creation for buildings. *IEEE Access*, 7, 147406–147419. https://doi.org/10.1109/ACCESS.2019.2946515.

Li, Q., Zhang, L., Zhang, L., & Jha, S. (2021). Exploring multi-level motivations towards green design practices: A system dynamics approach. *Sustainable Cities and Society*, 64, 102490. Retrieved August 25, 2021, from https://www.sciencedirect.com/science/article/pii/S2210670720307101.

Lu, Q., Xie, X., Parlikad, A. K., & Schooling, J. M. (2020). Digital twin-enabled anomaly detection for built asset monitoring in operation and maintenance. *Automation in Construction*, 118, 103277. https://doi.org/10.1016/J.AUTCON.2020.103277.

Pierre-Ignace Bernard and Frédéric Remond. The next normal in construction. McKinsey & Company. (2020).

Qi, Q., & Tao, F. (2018). Digital twin and big data towards smart manufacturing and industry 4.0: 360 degree comparison. *IEEE Access*, 6, 3585–3593. https://doi.org/10.1109/ACCESS.2018.2793265.

Schrotter, G., & Hürzeler, C. (2020). The digital twin of the city of Zurich for urban planning. *PFG - Journal of Photogrammetry, Remote Sensing and Geoinformation Science*, 88(1), 99–112. https://doi.org/10.1007/S41064-020-00092-2.

Seghezzi, E., Locatelli, M., Pellegrini, L., Pattini, G., Giuda, G. M. di, Tagliabue, L. C., & Boella, G. (2021). Towards an occupancy-oriented digital twin for facility management: Test campaign and sensors assessment. *Applied Sciences*, 11(7), 3108. https://doi.org/10.3390/APP11073108.

Tagliabue, L. C. (2021). eLUX: The case study of cognitive building in the smart campus at the University of Brescia. *BIM-Enabled Cognitive Computing for Smart Built Environment*, 190–224. https://doi.org/10.1201/9781003017547-8.

Tagliabue, L. C., Cecconi, F. R., Maltese, S., Rinaldi, S., Ciribini, A. L. C., & Flammini, A. (2021). Leveraging digital twin for sustainability assessment of an educational building. *Sustainability*, 13(2), 480. https://doi.org/10.3390/SU13020480.

Tagliabue, L. C., Re Cecconi, F., Rinaldi, S., & Ciribini, A. L. C. (2021). Data driven indoor air quality prediction in educational facilities based on IoT network. *Energy and Buildings*, 236, 110782. https://doi.org/10.1016/J.ENBUILD.2021.110782.

Wang, P., & Luo, M. (2021). A digital twin-based big data virtual and real fusion learning reference framework supported by industrial internet towards smart manufacturing. *Journal of Manufacturing Systems*, 58, 16–32. https://www.sciencedirect.com/science/article/pii/S0278612520301990?casa_token=fu2D4MArED4AAAAA:CsPwVXFyENu9sVbz5DmSvttGB_7adHtP41uyTzf3NVQGLUXtnHmahoLIU5AFX53kR3L2a5ltJU

WGBC. (2015). About Green Building. https://www.worldgbc.org/what-green-building.

Ye, C., Butler, L., Calka, B., Iangurazov, M., Lu, Q., Gregory, A., Girolami, M., & Middleton, C. (2019). A digital twin of bridges for structural health monitoring. *Structural Health Monitoring 2019: Enabling Intelligent Life-Cycle Health Management for Industry Internet of Things (IIOT) - Proceedings of the 12th International Workshop on Structural Health Monitoring*, 1, 1619–1626. https://doi.org/10.12783/SHM2019/32287.

37

Riding the Waves of Digital Transformation in Construction – Chances and Challenges Using Digital Twins

Bianca Weber-Lewerenz
RWTH Aachen University

37.1 Introduction

The digital transformation is a key competence in the Construction Industry 4.0 and enables a more economical, more efficient construction life cycle. Taking into account the negative effects of digitization requires the responsible use of innovative technologies. This is a signpost for the design of the construction branch with sustainable growth that is fair for people, society and the environment. With increasing technical feasibility, the human need for security and the desire to handle and use new technologies responsibly increases (Weber-Lewerenz 2021a, b, c). Thus, it is a prerequisite to deal with its ethical implications by shaping a reliable and legally protective framework that provides both orientation and safety.

Practical experiences, shared in expert interviews over the period of 2019–2021, provide signposts for the use of the latest digital technologies for the cities of tomorrow. Overall, German companies were able to achieve sales of almost €60 billion (BMWi 2019a) in 2019 with products and services that directly use AI. In 2019, 30% of the companies in the German economy that used AI in the company had vacancies in the field of AI (BMWi 2019b). BIM and new technologies such as AI, AIoT, IoT are not only part of the strategy for a sustainable digital transformation in all areas of buildings, mobility and smart cities (Gomez-Trujillo and Gonzalez-Perez 2021) but also contribute – in recognition of ethical digital responsibility – to a successful social transformation and resilient ecosystems at (Feroz et al. 2021). New technologies are important catalysts in the life cycle of buildings, in the reduction of the ecological footprint, in the participation and co-creation of all (Weber-Lewerenz and Vasiliu-Feltus 2022). They enable a more structured, more transparent basis for decision-making and for forecasting increases in

 DOI: 10.1201/9781003425724-44

efficiency by visualizing the planned end product with its technical operating equipment, building usage data and operating data. Malfunctions, risk hazards, environmental impacts, user behaviour, energy consumption and a holistic life cycle of the building can be simulated. Digital twins enable holistic platforms merging all relevant data and data interfaces. BIM delivers considerable added value (Lo et al. 2021). BIM is rising towards the next consolidated methodology: Building Lifecycle Management. In this way, the digital transformation can be designed collaboratively and environmental protection-related challenges can be consciously tackled (Ye et al. 2020).

New innovative technologies such as artificial intelligence (AI), virtual reality (VR), and metaverse support human work with even more sophisticated technology simulate the planned project in design and use with all technical interfaces. This not only allows the assessment of the necessary resources of completion and operator costs, time and quality, and efficiency but also offers a consistent data structuring without data loss, and a visualized representation of scenarios (Ernstsen et al. 2021). Such technologies provide a strong foundation for sensible, responsible use, resource conservation and environmental protection and are being researched and already applied at various German locations. Innovation needs innovation champions, i.e., role models who inspire others and get them to innovate their own future (Leiringer and Cardellino 2010). Accordingly, this chapter sums up the status quo of the development of digital twin and quintessence of the look behind the scenes shared in expert interviews conducted with innovation champions in construction.

37.2 Best Practices

The diverse digital methods such as digital twins and AI applications already implemented in practice contribute to an eye-opening effect. The experience in user practice with digitization and AI ranges from structural and civil engineering, Technical Building Automation, real estate management, monument preservation, formwork technology, tunnel technology, timber construction, to fire protection, intelligent buildings, smart cities and the latest software developments for the construction industry. There are no limits to the variety of fields in which AI can be and is already used in construction, as surveys by construction contractors and research institutions have shown. The crux lies in the development of a digital agenda that suits the company.

Digitized processes simplify construction, registration, measurement and reconstruction work, component sections and measurements; they enable more efficient municipal management (firefighting, policing, sanitation, waste management). Thus, they send a strong signal to the construction industry to deal with potential and already implemented areas of application of AI. Digital online construction site inspection with VR glasses, simulation of renovation concept and building in stock, acoustic and fire protection concepts, AI sensors for detecting defects/damage, AI monitoring and controlling in the construction progress with continuously generated invoicing, forecast and strategy data systems as a basis for business decisions (predictive technologies) become even more efficient through AI. 3D to 10D – twins with simulation, reality and forecast models provide real-time mapping of construction progress, time and costs. The effects of planning changes can be shown to the customer virtually, so that changes are incorporated into the planning as early as possible and not later, when things get really expensive. The new technologies help to increase customer trust and corporate compliance through transparency and illustration; legal disputes and supplements can be avoided. The entire administrative effort on the part of the company is reduced. In particular, the innovative technologies offer opportunities to work more sustainably in construction, i.e., more resource-saving, energy-efficient and environmentally friendly, and to significantly expand value chains throughout the entire construction life cycle.

Integrating existing buildings into Smart Cities by help of visualizing by digital twins (Figure 37.1.) and setting up fire protection systems that enable the intelligent communication between buildings represent significant challenges. In order to ensure the efficient operation and safe functioning of networked buildings in Smart Cities, e.g., and to protect them from crises that threaten the existence of the company, sophisticated, highly modern fire protection is required (Zhansheng et al. 2020; Fan et al. 2019).

The digital twin seems to be the all-purpose tool of the moment. In the construction industry in particular, BIM is the most efficient method to date, to merge technical interfaces on a single data platform and digitally map a building or an urban environment. The digital twin created in such process not only visualizes the building's shell and interior but also all of the technical lines that run through the building and penetrate parts of the building. The digital twin offers to everyone involved one common exchange platform, in real time and without data loss.

Smart cities not only include new construction but also the upgrading of the comparatively larger building stock to intelligent building automation systems in accordance with standards, SDGs, climate and ecological strategies (Weber-Lewerenz 2022). The building sector causes almost 40% of global CO_2 emissions and is therefore a key to achieving climate neutrality by 2050. Innovative building technologies in existing buildings enable efficient construction planning through visual representation of all trades involved (Figure 37.2) and support independent monitoring and operation (Figure 37.3), control, analysis and optimization of energy consumption.

Emerging technologies allow decision-makers to understand the impacts resulting from policy recommendations or new proposals for urban systems. On the user side, they enable better design and analysis of city-scale interventions (Figure 37.4).

FIGURE 37.1
Digital twin visualizing the integration of a new building into existing urban environments. (University of Lucerne.)

Smart technologies in the building stock enable a much more efficient, economical and safer form of testing, building operation and maintenance (building automation, communication between buildings). In practice, compliance control systems are essential, especially in building automation, as well as methods supporting the operator and user in the automated check. It is thus essential to structure complex data and interfaces and merge to a single data platform for efficient collaboration of all participants. In construction practice, this includes approval and permit processes with the authorities and the overall digital communication; digital platforms and digital twins not only speed up approval and permit procedures but also ensure transparency and quality management for all involved. Also, compliance with occupational health and safety on construction sites can be monitored using BIM: by integrating crane information models into BIM, compliance with lifting plan requirements can be checked. Either through the BIM-based performance-based parking lot ventilation concept compliance test, or through the BIM-based automated code compliance inspection system..

FIGURE 37.2
Digital twin of wall and ceiling penetrations in buildings. (www.cadsys.de.)

Best practices applying digital twins can be allocated especially in the field of monument preservation. With the help of a proof-of-concept GIS application, geometric and semantic data can be stored and queried. Using this data model for automated code review is an example of how GIS data can seamlessly complement BIM data. It demonstrates how it supplements other important data in order to create an even more holistic digital twin.

Monument preservation, duplication of severely damaged parts of historic buildings within smart cities mean applying modern technologies, creating digital twins and connecting the concerned modernized buildings with surrounding intelligent buildings and infrastructure.

Here, designated digital technologies and AI methods apply, especially in measuring and reconstructing works, cuts and measurements of building components, fitting of re-manufactured components into the existing building, production of building materials, Forecast Modelling (investment planning and control) and simulation of construction models (3–7D, e.g., with construction-accompanying continuous recording of the actual construction time, costs and quality compared to the original planning). Virtual Reality

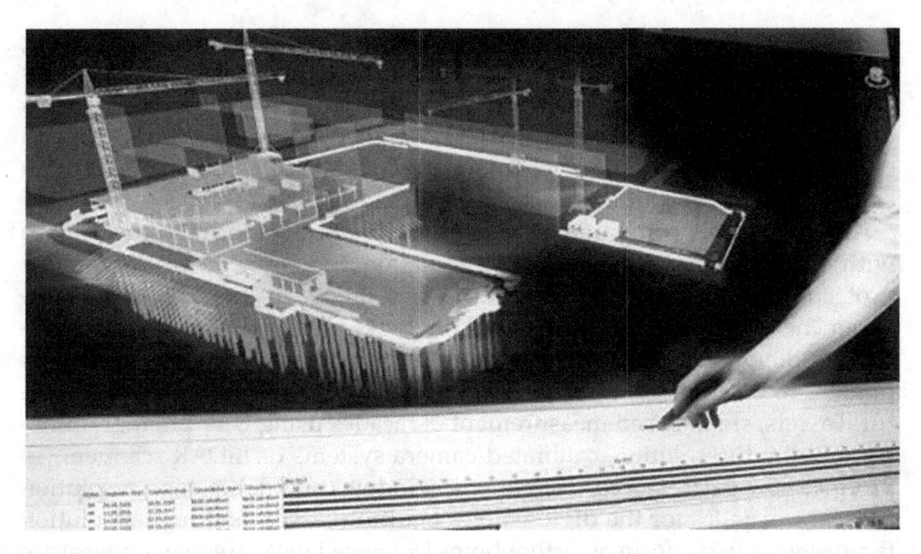

FIGURE 37.3
Digital twin building and operation (Siemens).

FIGURE 37.4
Digital twins for planning and implementing smart cities, e.g., in Singapore. (publish-industry Printing (https://www.industr.com/de/digital-twins-bilden-die-basis-fuer-smart-city-konzepte-2616454).)

and Augmented Reality are used to create virtual models of churches, castles and historical buildings, and facilitate walk-throughs without individual presence. Object recognition using Machine Learning enables the identification of building types, classes, damage or anomalies in images.

Defective areas, often in hard-to-reach locations, can be allocated, repairing and maintenance can be planned. Digital online site visits with VR glasses get routine, Simulation of Renovation, Acoustic and Fire Protection Concepts can be realized, the pros and cons be weighed while researching design changes. AI-based sensors recognize and track defects and damages with changes of material, air humidity and temperature.

AI technologies compare new data with stored data to track compliance with structural requirements and regulations (monument protection, fire protection, environmental protection) when taking measures. Via 3D measurements and using drones for UAV photogrammetry and UAV laser scanning in the area of monument protection digital twins can be created. It is possible to record the inventory, and 3D documentation helps to digitize cultural assets, stone-based measurement of facades using UAV photogrammetry with high-resolution, calibrated camera systems or LiDAR scanners. To produce new parts, stucco, decorations, windows and doors, high-resolution 3D scans of relics for the digitization of cultural assets and high-resolution façade views in the form of Orthophotos (corrected measurement images) are the basis for designing and for recording and documenting damages.

Architectural monuments, ground and area monuments can be used for digital tours in public relations with the help of digital twin models. In this way, the high accuracy requirements of monument preservation and monument protection can be met, (damage) documentation in the area and monument protection can be performed in a highly efficient manner, as well as the entire documentation, cataloguing and archiving, planning of renovation measures and reconstruction. Digital twins have become indispensable methods both for highly sensible restoration and their use in public relations.

The French global player and Construction Company VINCI is an innovation champion of a special kind: Europe is characterized by a rich building culture. Maintenance with the help of reproductions that are as identical as possible and the restoration of existing buildings are performed by using AI methods beyond the digital twin. The search for integrative options is part of Monument Preservation focusing on the reconstruction of the Cathedral of Notre Dame in Paris (Figure 37.5) or digitizing and renovating the Mount Saint Michel in Normandy, France.

After the fire in 2019, 3D scans were carried out to create the digital twin. The Point Cloud–based model captures the original structures and digitally preserves them for future generations. VINCI uses Sixense Software created by VINCI Construction Group and its ScanLab. High-precision 3D and photogrammetric measurements enable automated risk detection. The company's own AI programme is called Leonard VINCI. Recordings of the façade automatically map the technical floor plans. The advantages and experience consist in the intervention times; scaffolding construction working period are drastically reduced. This results in a lot more efficient and shorter restoration time, so that tourist masterpieces can be accessed within a short period of time. A classical example from practice is Mont Saint Michel in Normandy, France.

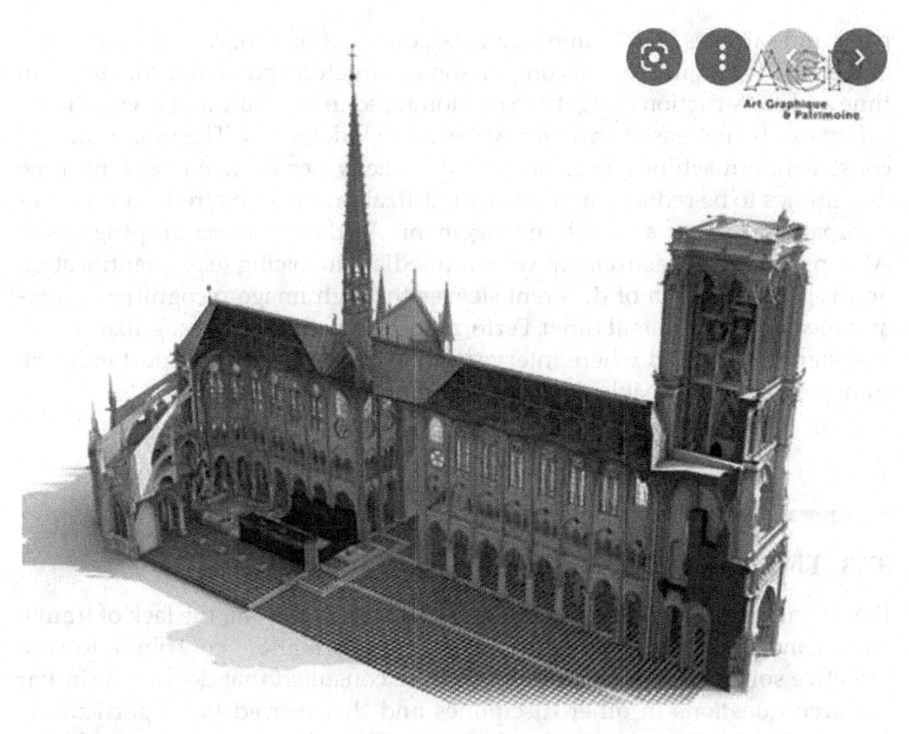

FIGURE 37.5
Reconstruction of Notre Dame Cathedral Paris, France, via digital twin. (AGP Art Graphique & Patrimoine.)

Another Best Practice represents the research project SDaC – Smart Design and Construction (Wolber and Steuer 2022). In their research project, which is funded with EUR 9 million by the Federal Ministry for Economic Affairs and Energy and runs over a period from April 2020 to March 2023, Jan Wolber and Dominik Steuer show practical applications that address the greatest and most complex challenges enter, such as B, through the strong fragmentation of specialization, there is an urgent need for innovative technologies to meet the quality and speed requirements of data collection while reducing resource consumption.. The mission: *We are developing a platform that enables organizations from the construction industry to access information easy to access and use them intelligently with added value.* AI is recognized as a key solution, because for it the advantages include: the automation of repetitive processes, feedback from the user increases the quality of forecasts, reduction of errors and increase in quality, time freed up for creativity and other services, support for decisions and reduction of complexity and new revenue streams through business models.

The recognized application possibilities of AI range from enabling through image and object recognition, awarding, supply chains to planning automation, shortage prediction, identification of hazards, identification of identical parts and samples to deadline forecasts, construction progress and construction diary. For example, the comparison of targeted and actual construction time and construction budget can be monitored in real time and occupational safety can be increased through AI-based risk detection. The monitoring of construction machines, utilization and capacity per trade enables machine downtimes to be reduced and efficient utilization to be ensured. AI is part of occupational safety and risk management. As the construction progresses, AI supports the measurement with immediate invoicing (e.g. quantification and type recognition of different sleeves through image recognition generates measurement in real time). Performance deviations are recognized by AI and signal when and where intervention is necessary and support efficient, transparent and traceable documentation for supplements.

37.3 The Ethical Implications

Empirical values and observations, in particular regarding the lack of framework conditions for a successful digital transformation, contribute to constructive solutions. Literature sources were consulted that deal with similar research questions in other disciplines and that proved to be particularly informative for the construction industry. These help to transfer problems and questions that other departments are also dealing with to the construction industry. Secondary data were collected to locate the innovation discourses. As this understanding develops, it becomes clear that the social, economic and political influencing factors that innovative technologies bring with them have so far been disregarded in the critical debate, especially in the construction industry, but due to their considerable impact on people, the common good, nature and the environment, new approaches are required. Technical feasibility also requires more social responsibility, as the study shows. The debate must be conducted at all political, economic and social levels (Xue et al. 2014). The entrepreneurially responsible use of digitization and AI, CDR for short, is one of the most important approaches. The construction industry is still hesitant to address ethical, social issues, but this is being pushed by such a new research field and the knowledge transfer between research and practice, increasingly recognized as a key lever for the success and sustainability of digital transformation.

The increasing awareness in the construction industry for a balanced, sustainable human–machine interaction is noticeable, as confirmed by the experts surveyed. The research findings from technology ethics (Grunwald and Hillerbrand 2021; Grunwald 2020; Jonas 1987) and technology assessment by Armin Grunwald (Grunwald 2010) are helpful here. New technologies are increasingly playing a key design role. However, they require a new culture of thinking and value-based action – not limited to corporate

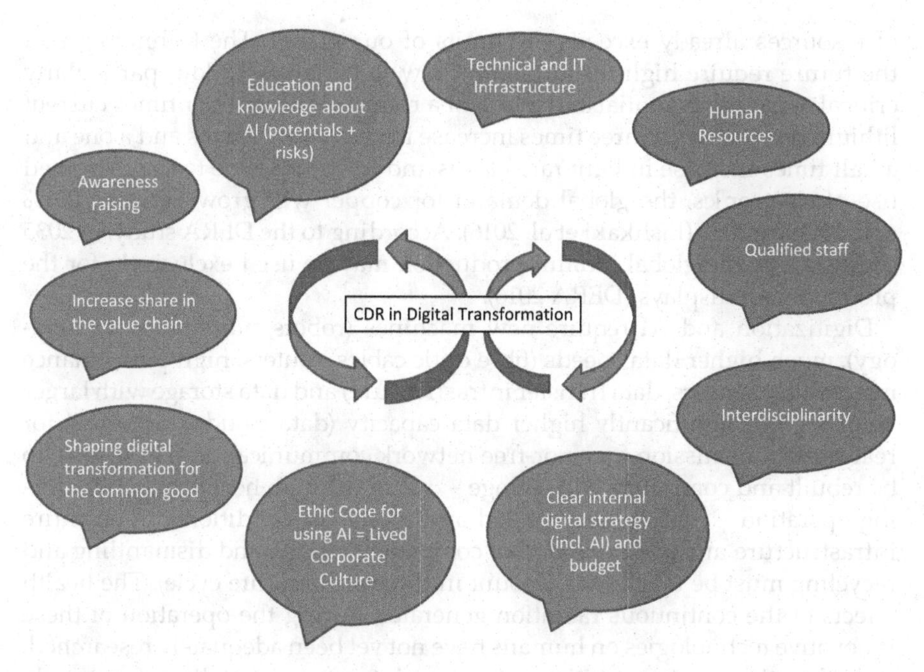

FIGURE 37.6
Mind map digital transformation strategy – interaction of the key elements. (Bianca Weber-Lewerenz.)

environments. The study identified key success factors for shaping digital change in Construction (Figure 37.6). These factors include enterprise digital infrastructure, communication skills, expansion of knowledge through appropriate education and training, academic teaching skills, cross-disciplinary close interface work, creative awareness, etc.. These factors are mutually dependent, form the prerequisite for the next step, the respective success depends on the other and only enables the next stage of the transformation.

37.4 The Dark Side of Digitization

When it comes to designing intelligent cities of tomorrow, e.g., the question immediately arises as to how intelligent digital standards can be achieved. And how can smart cities not only be smart but also sustainable and resilient? In this respect, Building 4.0 can certainly be seen as a catalyst for the digital transformation. However, strong negative effects of digital methods and AI are referred to as the "dark side" of digitization (Pilgrim 2017) because they have social and ecological effects on the raw materials sector. Economic growth is difficult to decouple from resource consumption. The consumption of resources already exceeds the limits of our planet. The technologies of the future require high resources and raw materials with low, particularly critical recycling potential and a lack of a return strategy. Four times current lithium production, a three times increase in heavy rare earths and a one and a half times increase in light rare earths and tantalum. Due to the increased use of electronics, the global demand for copper will grow between 231% and 341% by 2050 (Elshkaki et al. 2016). According to the DERA study, in 2035 up to 34% of the global indium production may be used exclusively for the production of displays (DERA 2016).

Digitization and AI require new machines (robots, automation technology), much higher data speeds (fibre optic cables, routers, high-performance microchips, sensors, data transfer infrastructure) and data storage with larger volumes, i.e., significantly higher data capacity (data clouds, IoT, AIoT) for real-time transmission for error-free network communication. These have to be rebuilt and computers and storage – due to the high heat production during operation – continuously cooled and rooms air-conditioned. The entire infrastructure must be expanded or completely rebuilt, and dismantling and recycling must be taken into account in the planning (life cycle). The health effects of the continuous radiation generated during the operation of these innovative technologies on humans have not yet been adequately researched: fully functioning smart cities require uninterrupted, intelligent networking of buildings, e-mobility, smartphones, etc. with data clouds, IoT and the expansion of a multiple of base stations ahead. This can only work with a high, uninterruptible power supply and acceptance of the additional energy consumption, the high intensity of electromagnetic radiation (networking among the devices, radio) and CO_2 production. The end-to-end full automation of buildings requires hardware such as machines and sensors, and thus an increasing production of the necessary inventory materials and with the corresponding consumption of resources. The fifth generation of mobile data transmission (5G) requires broadband expansion using fast fibre optic networks. 5G combines the previous mobile communications standards, WiFi, satellite and landline networks into a holistic communication network. From this, it can be seen that digitization can become an energy guzzler, because in 2025 data centres will account for around 4%–11% of global energy consumption; at the same time, high energy saving and waste heat utilization potentials are forecast and localized (Höfer et al. 2020).

The digital value chain is being taken ad absurdum. When raw materials are increasingly used, and when inhumane mining work, abuse of people and the environment, and long-term damage to health are accepted, how should we reduce the ecological footprint and conduct sustainable management in order to develop and apply better, more economical, and more innovative technologies? (Rüttinger 2016). The danger lies not only in resource consumption, but in increasingly complex risk areas of data misuse. Protection is one of the success criteria for sustainability. This also applies to data communication: Municipalities are largely administratively unable to handle approval processes digitally, since many companies lack a digital infrastructure (Fiedler 2022). Declarations of intent by the German government on digitization do not correspond to the current situation and the practice of public administration.

One result of the study is that the design of the digital transformation must include where innovative technologies make sense, conserve resources and support people in their work. A clear distinction must be made between empty promises, the concealment of major security risks or exploitation. There is a great danger of becoming part of so-called greenwashing, i.e., propagating superficial advantages and high benefits, which, however, come at the expense of violations of personal and data rights, environmental depletion and resource consumption that are harmful to people and society. The research "CDR in Construction 4.0" investigates the question of how such a countermeasure can succeed; the excellence initiative for sustainable, human-led AI in construction wants to strengthen this awareness. The Scientific Advisory Council of the Federal Government and the German Federal Office for Radiation Protection (BfS) also warn against unchecked digitization that is not aligned with sustainability criteria.

37.5 Résumé and Outlook

BIM and AI technologies offer essential support for structuring complex data and interfaces merging to a single data platform such as digital twins to achieve efficient collaboration of all participants. Theoretical principles, guidelines and Lessons Learned lay foundation for a trustful, human-led, sustainable construction practice. It is not limited to approval and permit processes with the authorities, so that fast, smooth, structured and quality-enhancing building project procedures and the holistic life cycle succeed and strengthen sustainability.

When it comes to designing intelligent cities of tomorrow, the question raises immediately on how Smart Cities, e.g., shall be designed or new building structure shall be integrated into the existing urban environment to achieve smart digital standards. Furthermore, how can future Smart Cities not only be smart – but also sustainable and resilient? In this respect, Construction 4.0 can be considered as a catalyst for digital transformation and therefore requires further research. It also represents key competences: it carries high potential for an economic more efficient construction life cycle, while requiring responsible handling with innovative technologies.

Digital methods and AI have high potential in many areas, such as competitive, innovative business models. The ethical orientation framework with value-based engineering adds value on the technical, economic, social and ecological level. It leads to the conclusion that Germany has an important role model function for other countries leading into the new technology century. Particularly surprising in the course of research was and is the growing realization in the construction industry that innovative and technical progress can only be made possible with orientation-giving and confidence-building measures as the lynchpin.

For the success and sustainability of modern digital technologies and methods from AI, as proven by best practices, a responsible corporate design of the digital transformation is essential. The study comes to the conclusion that the pioneers of the construction industry, best practices and innovation champions are important role models that offer orientation when navigating innovative technologies in Construction 4.0 and additionally strengthen the courage to tackle them. Digital twins are the best efficient method to start riding the waves of Digital Transformation in Construction, use diverse chances, master unique challenges and set new standards in the Digital Era.

Acknowledgements/Other Declarations

This chapter is to encourage the full use of the potential of the Digital Era in Construction. The unexploited reservoir of digital twin and AI potential is to increase the share in the value chain for more sustainability and success at all levels.

I would like to thank all of the experts interviewed in this research. They not only shared valuable insights but contributed to the founding of the *Excellence Initiative for Sustainable, human-led AI in Construction.*

Statistics and reports were of great support with data and facts to underlay the statements and recommendations made in this paper. The huge interest in this paper by experts, companies, decision-makers and activists in the Digital Era demonstrates the key role of women leadership.

This research received no specific grant from any funding agency in the public, commercial or not-for-profit sectors.

References

BMWi. 2019a. Einsatz von Künstlicher Intelligenz in der Deutschen Wirtschaft. Stand der KI-Nutzung im Jahr 2019, Studie des Bundesministeriums für Wirtschaft und Energie (BMWi). [Application if AI in German Economy, Status in 2019, Study by the German Federal Ministery for Economy and Energy]. p.20. https://www.bmwk.de/Redaktion/DE/Publikationen/Wirtschaft/einsatz-von-ki-deutsche-wirtschaft.pdf?__blob=publicationFile&v=8#:~:text=Im%20Jahr%202019%20setzten%20rund17%2C8%20%25%20erheblich%20h%C3%B6her. [accessed May 06, 2022].

BMWi. 2019b. Einsatz von Künstlicher Intelligenz in der Deutschen Wirtschaft Stand der KI-Nutzung im Jahr 2019, Studie des Bundesministeriums für Wirtschaft und Energie (BMWi). [Application if AI in German Economy, Status in 2019, Study by the German Federal Ministery for Economy and Energy]. p.28. https://www.bmwk.de/Redaktion/DE/Publikationen/Wirtschaft/einsatz-von-ki-deutsche-wirtschaft.pdf?__blob=publicationFile&v=8#:~:text=Im%20Jahr%202019%20setzten%20rund17%2C8%20%25%20erheblich%20h%C3%B6her. [accessed May 06, 2022].

DERA Deutsche Rohstoffagentur 2016. Rohstoffe für Zukunftstechnologien 2016. Bundesanstalt für Geowissenschaften und Rohstoffe (BGR). [Raw materials for emerging technologies 2016. Federal Institute for Geosciences and Raw Materials]. https://www.deutsche-rohstoffagentur.de/DERA/DE/Downloads/Studie_Zukunftstechnologien-2016.html. [accessed May 06, 2022].

Elshkaki, A. et al. 2016. Copper demand, supply, and associated energy use to 2050. *Journal of Global Environmental Change*, Vol. 39, pp. 305–315. https://doi.org/10.1016/j.gloenvcha.2016.06.006.

Ernstsen, S.N. et al. 2021. How innovation champions frame the future: Three visions for digital transformation of construction. *Journal of Construction Engineering and Management*, Vol. 147, Iss. 1. https://doi.org/10.1061/(ASCE)CO.1943-7862.0001928.

Fan, C. et al. 2019. Disaster city digital twin: A vision for integrating artificial and human intelligence for disaster management. *International Journal of Information Management*. https://doi.org/10.1016/j.ijinfomgt.2019.102049.

Feroz, A.K. et al. 2021. Digital transformation and environmental sustainability: A review and research agenda. *Journal for Sustainability*, Vol. 13, Iss. 3, p. 1530. https://doi.org/10.3390/su13031530.

Fiedler, M. 2022. BIM and Construction Authorities - A report. Structured Data for digital collaboration in the field of infrastructural construction: BIMSTRUCT. 2022. https://www.bsdplus.de/fachartikel/von-einem-der-auszog-das-fuerchten-zu-lernen-bim-und-baubehoerde-ein-erfahrungsbericht.html. [accessed May 06, 2022].

Gomez-Trujillo, A.M. and Gonzalez-Perez, M.A. 2021. Digital transformation as a strategy to reach sustainability. *Journal for Smart and Sustainable Built Environment*, Vol. 11, Iss. 4, pp. 1137–1162. https://doi.org/10.1108/SASBE-01-2021-0011.

Grunwald, A. 2010. Technikfolgenabschätzung: Eine Einführung. [*Technology Assessment: An Introduction*]. Vol. 1. edition sigma.

Grunwald, A. 2020. Verantwortung und Technik: zum Wandel des Verantwortungsbegriffs in der Technikethik. [Responsibility and Technic]. In: Seibert-Fohr A. (ed). *Entgrenzte Verantwortung*. [Unlimited Responsibility]. Springer, Berlin, Heidelberg. https://doi.org/10.1007/978-3-662-60564-6_13.

Grunwald, A. and Hillerbrand, R. 2021. Überblick über die Technikethik. [Overview Technical Ethics]. In: Grunwald A., Hillerbrand R. (eds). *Handbuch Technikethik*. J.B. Metzler, Stuttgart. https://doi.org/10.1007/978-3-476-04901-8_1.

Höfer, T. et al. 2020. Additional energy consumption in data centers when the 5G standard is introduced. Conference Paper Presented at the Conference "*Sustainable Data Centers - Chances and Development Possibilities in Baden-Württemberg*". https://www.nachhaltige-rechenzentren.de/wp-content/uploads/2020/03/1-Madlener_5G-Standard-und-Rechenzentren.pdf [accessed May 06, 2022].

Jonas, H. 1987. Warum die Technik ein Gegenstand für die Ethik ist: fünf Gründe. Technik und Ethik. [Why Technology is a subject of ethics: five reasons. *Technic and Ethics*]. Vol. 2, pp. 81–91.

Leiringer, R. and Cardellino, P. 2010. Tales of the expected: Investigating the rhetorical strategies of innovation champions. *Journal of Construction Management and Economics*. Vol. 26, Iss. 10, pp. 1043–1054. https://doi.org/10.1080/01446190802389394.

Lo, C.K. et al. 2021. A review of digital twin in product design and development. *Journal of Advanced Engineering Informatics*, Vol. 48, p. 101297. https://doi.org/10.1016/j.aei.2021.101297.

Pilgrim, H. 2017. *The Dark Side of Digitalization: Will Industry 4.0 Create New Raw Materials Demands?* PowerShift e.V. Publishing, Berlin.

Rüttinger, L. 2016. *Case Studies on Environmental and Social Impacts of Bauxite Mining and Processing in the Boké and Kindia Region, Guinea*. Adelphi Publishing, Berlin. [accessed May 06, 2022].

Weber-Lewerenz, B. (2021a). Corporate digital responsibility in construction engineering. *International Journal of Responsible Leadership and Ethical Decision-Making (IJRLEDM)*. 2(1). https://doi.org/10.4018/ijrledm.2020010103.

Weber-Lewerenz, B. (2021b). Corporate digital responsibility (CDR) in construction engineering - Ethical guidelines for the application of digital transformation and artificial intelligence (AI) in user practice. *SN Applied Sciences*. 3(10). https://doi.org/10.1007/s42452-021-04776-1.

Weber-Lewerenz, B. (2021c). Ethical Aspects in AI in Construction. *Conference Presentation and Proceedings of Reser Conference on 14oct2021*, Heilbronn. The Disruptive Role of Data, AI and Ecosystems in Services. Bernd Bienzeisler, Katrin Peters, Alexander Schletz (eds). Fraunhofer Institute Publishing. pp. 247 et sqq. https://publica.fraunhofer.de/dokumente/N-642928.html.

Weber-Lewerenz, B. 2022. Thinking otherwise: Integrating Existing Buildings and Monument Protection in Smart Cities -Experience shared from user practice. In: *Impact of Digital Twins in Smart Cities Development*. Ingrid Vasilliu-Feltes. IGI Global. https://doi.org/10.4018/978-1-6684-3833-6.

Weber-Lewerenz, B. and Vasiliu-Feltus, I. 2022. Empowering digital innovation by diverse leadership in ICT - A roadmap to a better value system in computer algorithms. *Humanistic Management Journal*, Vol. 7, Iss. 1. pp. 117–134. https://doi.org/10.1007/s41463-022-00123-7.

Wolber, J. and Steuer, D. 2022. AI in Construction. *Online-Presentation of the Research Project Group SDaC for the 17th meeting of the Regional Working Group Karlsruhe* as of March 30, 2022, German Lean Construction Institute - GLCI e.V.

Xue, X. et al. 2014. Innovation in construction: A critical review and future research. *International Journal of Innovation Science*. Vol. 6, Iss. 2. pp. 111–126. https://doi.org/10.1260/1757-2223.6.2.111.

Ye, Z. et al. 2020. Tackling environmental challenges in pollution controls using artificial intelligence: A review. *Journal of Science of the Total Environment*, Vol. 699, p. 134279. https://doi.org/10.1016/j.scitotenv.2019.134279.

Zhansheng, L. et al. 2020. A framework for an indoor safety management system based on digital twin. *Journal of Sensors*. Vol. 20. https://doi.org/10.3390/s20205771.

38

A Framework for the Definition of Built Heritage Digital Twins

Marianna Crognale, Melissa De Iuliis, and Vincenzo Gattulli
Sapienza University of Rome

38.1 Introduction

The management and use of data obtained from multidisciplinary approaches for the conservation of Built Cultural Heritage (BCH) is a crucial issue to tackle and a particular goal of academic and industrial research. The protection of historic structures is paramount due to the vulnerability of this type of building, whose value is further improved by the artworks therein placed, e.g., sculptures, paintings, etc. In the last decade, the construction industry has made remarkable advancements in preserving the BCH structural integrity due to the technology push (e.g., integrated sensing, Industrial Internet of Things (IoT)), and demand growth. Conservation relates to the periodic maintenance and use of heritage. It is not an exceptional event but an open-ended process of knowledge, understanding, maintenance, management, and enhancement, where sustainability, participation, and education are essential. Conservation and transformation also imply attention to the environment because heritage is related to natural, anthropic, cultural, and historical contexts. In the European Union, 35% of buildings are over 50 years old, and almost 75% of buildings are structurally and energy inefficient. Therefore, improving historical buildings' structural maintenance and energy efficiency contributes to reducing energy consumption and lowering greenhouse gas emissions. Current tools and workflows for digital representation, information, and management enables new perspectives in terms of geometry acquisition and data dissemination (Gattulli et al. 2016; Masciotta et al. 2017; Cavalagli et al. 2018; Clementi et al. 2018; Hoskere et al. 2019; Spencer Jr et al. 2019). In civil engineering, 3D models have increasingly been used for enhancing project management. That is, 3D models, in particular "capturing" 3D models or "point cloud data" of existing structures, include important data and information related to the building projects that are essential for Building Information Modeling (BIM) processes (e.g., building geometry,

DOI: 10.1201/9781003425724-45

material properties, construction typology). Compared to the way projects and interventions are managed through CAD and storage files generated by them, BIM systems allow a more efficient and dynamic management mode (Lauria et al. 2022).

Within this context, the concept of Digital Twin (DT) has gradually attracted the attention of the building sector. DT defined as "a model of the physical object or system, which connects the physical system in real-time" (Grieves and Vickers 2017), is becoming popular as a comprehensive approach to managing, planning, and predicting the building infrastructure (Hou et al. 2020; Borowski 2021; Lee and Lee 2021). DTs are an attractive novel paradigm that could become an essential tool in the battle against climate change (Errandonea et al. 2020; Opoku et al. 2021), as they could help to understand how a building works either in its structural performance during the service life or in the energy management to ensure the hydro-thermal optimal condition. DT can be seen as a digital replica or a virtual counterpart of a physical entity that serves as a "living" digital simulation model. Generally, a DT is developed using a great deal of data, such as operational data acquired from the monitored structure, advanced data processing, and interpretation rotes. Experiencing the behavior measuring the structural response of complex damaged constructions allows having the background necessary for designing reliable DTs (Potenza et al. 2015). Angjeliu et al. (2020) developed a procedure for creating an accurate digital model, which integrates the experimental physical reality to study the structural response of Milan Cathedral. Due to the improvement of geomatics methodologies, several solutions are available in the literature for the generation of refined models of real-world structures through meshing point clouds (Barazzetti et al. 2015). This technology, known as Scan-to-FEM, is typically obtained using remote sensing techniques, i.e., laser scanning and digital photogrammetry (Yastikli 2007; Remondino 2011). Geometric data obtained from 3D laser scanning are largely used for structural purposes. For instance, a two-step methodology to convert the point cloud to a BIM model was proposed by Barazzetti et al. (2015). They imported the BIM model into FEM software, demonstrating how the approach could be applied to achieve structural analysis. Castellazzi et al. (2015) developed a new semi-automatic procedure to transform three-dimensional point clouds into three-dimensional finite element models. The procedure included the construction of a refined discretized geometry with a reduced amount of time to be employed for structural analyses.

Similarly, a Scan-to-FEM procedure for historic structures characterized by different geometrical irregularities was proposed by Fortunato et al. (2017); Pepe et al. (2020). Nagakura and Sung (2014) realized a 3D model by capturing building textures and materials through photogrammetry. The process consisted of three main steps: (1) collection of spatial and documentary data, (2) data processing and dense surface model (DSM) creation, and (3) Heritage Building Information Modeling (HBIM) modeling.

Despite the number of studies on implementing point cloud into a continuum model, developing Scan-to-FEM tools suitable for BCH is still an ongoing process. The application of the DT concept to accurately simulate the structural response of the BCH over time can be considered a proper solution. The core of features compared to previous generations of models (e.g., BIM) is determined by the bidirectional seamless connection with the physical world and the computer intelligence provided by the DT. Heesom et al. (2020) developed a systematic collaborative HBIM of a 19th-century multi-building industrial site in the United Kingdom. The development of HBIM is aimed at bringing together tangible and intangible data to provide a model for future work and BIM data sets used during the asset lifecycle. The need to investigate the integration of heritage assets in the HBIM process and the application of the DT concept based on HBIM digital models to propose an alternative and more efficient method for preventive conservation was highlighted by Jouan and Hallot (2019). Angjeliu et al. (2020) proposed a strategy for developing the simulation model for DT applications in historic masonry buildings. The DT model was hierarchically organized in their work and can be updated continuously when new information on historic buildings is available. Piaia et al. (2021) introduced a software solution for on-site condition assessment and management of assets with embedded BIM software. HBIM software, combined with results from non-destructive testing and geometric surveying, was presented by Mol et al. (2020) to perform modeling, storage, and analysis of geometric data.

Furthermore, two DTs of the Ballroom and St. Francis of Assisi Church were proposed by Dezen-Kempter et al. (2020) through Terrestrial Laser Scanning (TSL) and a low-cost uncrewed aerial vehicle (UAV). Funari et al. (2021) presented a novel parametric Scan-to-FEM approach suitable for architectural heritage. The approach started from the 3D survey of a Portuguese monument and culminated with the definition of a detailed FEM that can be used for predicting future scenarios.

The present chapter describes a Scan-to-BIM framework for the DT generation of a monumental building in Rome. A set of registered laser scanning point clouds is used to create a BIM model that, in turn, allows the design of the structural FEM for the structural analysis (BIM to FEM). The BIM model is also implemented to develop a common data platform for visualizing the structural response under environmental noise. The platform is developed through the integration of IoT sensors and the Revit model to:

i. obtain a dynamic and automated exchange of data between sensors and the BIM model,
ii. test the monitoring system's performance,
iii. include and visualize sensor data directly on the BIM model.

38.2 Digital Twin General Concept

The DT concept was first proposed during Michael Grieves' 2003 presentation on product lifecycle management. Grieves (2016) wanted to shift from the predominantly paper-based and manual product data to a digital model of the product. Similarly, Cyber-Physical Systems (CPS) (Lee and Seshia 2016) and the IoT (Mukhopadhyay and Suryadevara 2014) focus on the connection of a physical system to data collection, and computational and communication systems. The main difference stands in different perspectives the techniques approach from: CPS concept from the system engineering and control perspective, IoT from networking and IT perspective, and DT from a computation modeling (e.g., machine learning, artificial intelligence) perspective. According to Grieves (2014), the DT concept consists of three components (see Figure 38.1):

a. Physical product in Real space or experimental reality.
b. Virtual representation of that product in the virtual space.
c. Connections of data and information tie the virtual and real products together.

The triad (physical model, virtual simulation model, and big data) is known as the DT. Although in the DT, the data and information collected in the real space represent only a part of the existing structure, they realized the basis for developing the virtual space. Therefore, virtual space is an idealized version of the collected data into a mathematical model (Angjeliu et al. 2020).

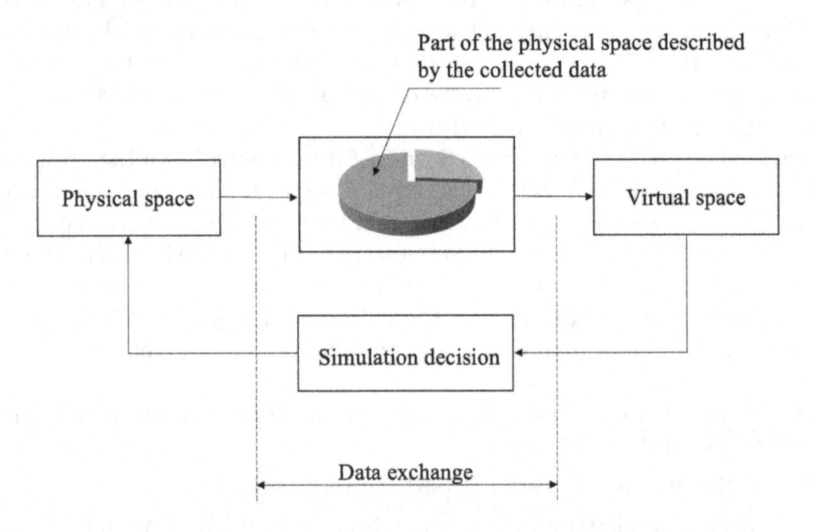

FIGURE 38.1
Schematic representation of the Digital Twin concept. (Adapted from Angjeliu et al. 2020.)

A DT is a living model that can be used for numerical simulations, visualizations, or complex data management. Inspection data get layered onto this model and then analyzed.

38.3 Data Collection

38.3.1 Site Description

The project designed by the architect Carlo Aymonino has been interpreted and studied by Trait D'Union, who has faithfully accomplished the architect's will. The architectural project of Aymonino started in 1993, shortly after the return of the Capitoline hill of the statue of Marco Aurelio, at the end of a restoration that lasted nine years that highlighted the impossibility of exposing it again to external atmospheric agents. The Marco Aurelio Esedra Hall is the new grand glass hall built inside what was called the "Giardino Romano" in Palazzo dei Conservatori. The hall represents a prestigious piece of modern architecture within the ambit of this Municipal Museum complex that links the historic part of Palazzo dei Conservatori to those parts of the museum that have been more recently constructed. The large and bright hall was built in an open area that historically marked the boundary between the properties of the Conservators and the Caffarellis. The new space shows an elliptical plan that recalls the shape of Michelangelo's place (see Figure 38.2a) (Pärn and Edwards 2017). A thin glassy skin covers the steel structure, shaped like a frustum of an ellipse. To support and anchor it, new supporting disposal for glasses has been designed, innovative compared with others used for applications of suspending glass and able to solve problems of constraint and support. The panel of structural glass, with 91 different triangular measures, consists of a 12 mm thick, light float, supporting slab, and a security slab with a selective couche to optimize the sun-protection function of the transparent panel. Panels are united in the upper part through special silicone glues on a specific gasket to safeguard expansion movements. The natural ventilation in the inner surface of monolithic glasses is guaranteed by a labyrinth in which the air circulates constantly.

The elliptical metal-and-glass shed was to be supported by six 75 cm diameter sheer steel pillars without capitals and bases. The six columns support the massive box section girders at two different altitudes that hold up the perimetrical glass enclosures, the horizontal elliptical roof, and the lateral surface, as shown in Figure 38.2b. Thus, the glazed roof is divided into two levels connected by a glass "drum" that rests on the lower elliptical beam. The drum emerges from the lower roof of the space between the ellipse and the walls of the Giardino Romano. The upper ellipse has a structural mesh of 2 cm thick steel lamellar beams arranged in a grid of 1.94 ×1.94 m in support

(a)

(b)

FIGURE 38.2
Images of the Hall of Marco Aurelio (a) top view of the roof and (b) elliptical structure of the Hall. (© Sovrintendenza Capitolina.)

of the glass above. The latter comprises laminated triangular glass panels of different sizes, consisting of a 12 mm transparent float plate tempered flat and a 6 mm selective anti-solar safety glass.

38.3.2 Laser Scanner Acquisition for Geometry Reconstruction

An essential ingredient of the DT is the correct geometry of the analyzed structure. It can be obtained from measurement of the real buildings, or existing models developed for other purposes may be used. The complexity of historical constructions, with irregular geometry, inhomogeneous materials, variable morphology, alterations, and damages, poses numerous challenges in the digital modeling and simulation of structural behavior. Existing technologies for measurements Scanning (with laser scanning or digital photogrammetry) and for modeling CAD or BIM must be accounted for. Collecting the best and most accurate data about real-world conditions, often known as the "as-built", is defined as Reality Capture RC. As a RC technique, 3D laser scanning is one of the most widely used 3D imaging techniques, and the captured data using 3D laser scanners is also known as the "point cloud". The result is a 3D file that can be saved, edited, and even 3D-printed. Each 3D scanning technology has its pros and cons, and costs which need to be considered. Laser scanning is used for many processes, including surveying, manufacturing inspection, building construction quality control, and heritage building restoration. Laser scanning devices are now integrated within numerous built environment applications (Díaz-Vilariño et al. 2018). They are a consolidated technology for collecting, documenting, and analyzing 3D data on the as-built status of buildings and infrastructures (Barazzetti et al. 2015). The reversed 3D model will be created in this framework through the 3D scanning procedure. The whole planned procedure is described in the following. The geometrical reconstruction of Marco Aurelio Esedra's Hall will start with the acquisition with a Leica BLK360 laser scanner. 3D-point accuracy in the instrumental reference system is expected to be better than 4 mm. All the geometrical information derived from the laser scan acquisition will be accurately used to create the "as-built" BIM model. A diagnostic analysis would be necessary to combine material properties and geometrical information. The analysis helps to consider the materials' temporal deterioration and the structure's historical evolution with modifications and restorations. Finally, this BIM model will be compared with a second BIM model directly derived from the original technical drawings. Details of the model are provided in the following section.

38.3.3 BIM Description

The generation of historic BIM is a challenging task as the available tools are not always sufficient to represent the irregularities of historic buildings (Barazzetti et al. 2015). Indeed, BIM software and object libraries are mainly developed for modern constructions and for the design phase. One of the advantages of BIM technology is the interoperability for the different professional operators (Gholizadeh et al. 2018). The BIM model of the structure of Aymonino was developed following the original projects of the structure with the commercial package Revit. As the BIM will be converted into a 3D model for finite element analysis, the work must integrate all the requirements for the structural analysis, such as the first modeling phases, definition of material properties, loads, and boundary conditions. Structural elements were classified into category, family, type, and instance. As mentioned before, the 3D model was obtained from the 2D drawings provided by the laser cloud. The structure evolves in height following three levels: ground floor, lower cover, and upper cover. On the ground floor, there is the exhibition area with all horizontal connections, stairs, ramps, and connecting paths to the two lower levels leading to the wall of the temple of Jupiter (see Figure 38.3a). In the BIM model, the topographic coordinates have been inserted to have all the environmental data at hand in case analyses related to the soil type, wind, and the impact of the lighting are performed. The constructive elements of the structure have been identified using BIM families to get a digital model as close as possible to the existing one. The main steel structure is composed of different columns and all the connections are bolted (Figure 38.3b). The lower roof is formed by steel blades that fit into the

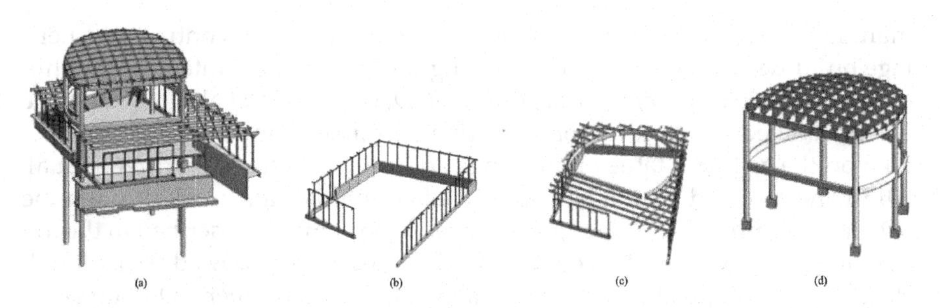

FIGURE 38.3
Images of: (a) BIM model, (b) system of columns, (c) blades' system, and (d) elliptical dome. (© Sovrintendenza Capitolina.)

outer masonry of the museum. Figure 38.3c depicts the lower cover that has elements of horizontal closure of the glass plates applied to the blades' system. The top cover is dome type, formed by semi-elliptical steel blades, and connected with steel joints through welding and bolting. The beam system is finally joined to the upper half-elliptical joist connected to the central circular pillars. Rectangular windows are designed between the first and the second half-elliptical steel truss, as shown in Figure 38.3d. After the main structural elements have been added to the model, architectural elements have been designed, such as plasterboard walls covering the main structure, the false ceiling, the amphitheater staircase, the stairs, and the connecting ramp to the northern part of the museum complex. To preserve BIM interoperability, objects have been then imported following the parametric modeling principles with a categorizations based on Revit families (Steel et al. 2012).

38.4 Interoperability: from BIM to FEM

BIM interoperability, which is the ability of two or more systems to exchange information, is one of the main tasks of this framework. Software available on the commercial market allow the integration between BIM technology and finite element analysis without losing the high level of geometric detail obtained from the original point clouds. The hall of Marco Aurelio has also been modeled through a numerical finite element model directly derived from the BIM model. The 3D FEM model was exported from Autodesk Revit to SAP2000. When exporting from Revit to SAP2000, the following issues are identified: the CSIxrevit translation program does not recognize the connections between some connected components, such as columns decentralized from the axis of the beams or beams with different heights. Furthermore, some elements modeled in BIM are not recognized by SAP2000. The FEM model presents a combination of beam elements and zero-mass shell elements for the two horizontal glazed roofs. Shell elements are used to model the roofs' weight, as they are zero-mass and zero stiffness elements, with an applying distributed surface load equal to the roofs' dead load. An elastic and linear constitutive law has been adopted for the material, (S450), using the following nominal parameters: yield strength f_{yk}=450 MPa, Young's modulus E=210,000 MPa, and specific weight ρ=7,850 kg/m^3.

Commonly, the purposes of a FE analysis seek to facilitate the development of a final design of structural monitoring system that will be installed in the coming months, and the visualization of the monitoring data.

38.5 Long-Term Monitoring System

The objectives of the structural monitoring systems should be considered in this chapter. The following actions are usually taken to develop a preliminary design of a structural monitoring system: first, to define the structural behaviors to be monitored, e.g., accelerations; second, to define the monitoring approach and the structural elements (in this step, the required devices to monitor the defined behaviors are selected). Usually, three types of devices are needed in a monitoring system: sensors, a communication network, and a processing unit. In the last step, physical installation process of the monitoring system is defined. The phase described below concerns the continuous monitoring of the health state of the selected structure through a network of accelerometric sensors deployed following the detailed study of the complex through surveys and model analysis. Inspections carried out in recent months have made it possible to define the sites and methods of installation of the sensor nodes and the gateway node for the whole monitoring system of the Capitoline Museums complex. Moreover, the installations will be minimally invasive, and will not compromise the structural elements or the surface finishes of the sites where they will be placed. At the end of the project, the situation prior to the installation of the sensors network will be restored. The project proposes the installation of six sensor nodes, two for a single column that supports the hall's dome, thus involving three of the six alternating columns. The future sensors' position for the monitoring of Marco Aurelio's Hall is shown in Figure 38.4a and b. The sensors to be installed in the future are temperature and triaxial accelerometers. Since the six sensor nodes shall be installed outdoors in direct contact with current external weather conditions, they will be housed in suitable IP68 certified containers, as shown in Figure 38.4c. These containers will be installed directly on the cover of the drum and externally at the base of the selected columns using silicone-based adhesives that will be entirely removed at the end of the monitoring. For installation operations, we will be using the walkway for maintenance and cleaning machine outside the structure from which the installation sites of

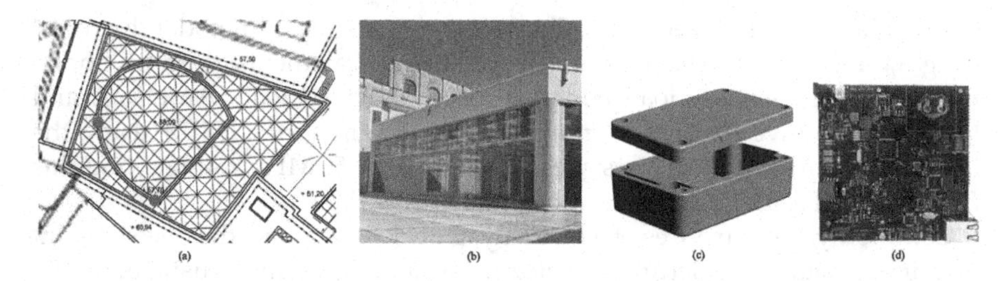

FIGURE 38.4
Sensor's layout: (a) top view, (b) detail of the external positioning, (c) IP68 container, and (d) SHM board.

the sensor nodes can be easily accessed, for both positioning and maintenance of the same. Each sensor device communicates directly with a central device, called a controller/gateway. The controller/gateway node, the real heart of the sensor network, is a radio device that manages the configuration of the network and the transmission of the detected data (environmental and acceleration measurements). The data collected is then made available on cloud platforms and remotely accessible through smartphones, tablets, and computers. The gateway for the sensor network will be placed in the cabin at the top of Palazzo Caffarelli. Concerning the sensor board, SHM-Board v2 has been designed as a highly versatile and high-performance device for real-time SHM of structures and infrastructures, able to promptly intervene to avoid or limit structural-related risks and direct or indirect consequences for people or things. The board, illustrated in Figure 38.4d, is based on an ultra-low-power micro-controller (MCU) of the STM32F4 family, produced by ST Microelectronic: through its 12-bit ADC, it is connected to external low-noise MEMS accelerometer (Kionix KXRB5-2050), mounted directly on the board. SHM-Board v2 is equipped with 4-channel 24-bit ADC able to sample at 100Hz and on-board data processing to minimize the total amount of data to be transmitted. From the point of view of communications, SHM-Board v2 has an Ethernet interface (from which it can be powered via PoE) and other interfaces to connect more peripherals. The board also provides the capability to communicate wirelessly (RF) with other nodes of the monitoring network through W-MBUS protocol at 169 or 868 MHZ that allows communication over long distances and is much less exposed to obstacle attenuation. The integrated temperature sensor allows the evaluation of the thermal effect on the structure and on the sensor thus allowing to distinguish seasonal variations from real inclinations. The MEMS operating principle guarantees good thermal stability and excellent linearity.

38.5.1 Monitoring System within the BIM Environment

This section presents a general approach to model structural performance monitoring systems and to include sensor data directly onto the BIM model.

It enables the creation of BIM models enriched with sensor data that can be exchanged using IFC specifications. The objectives of the monitoring system are quantifying the acceleration in various structural elements, providing baseline measurements for future monitoring, and developing an intuitive 3D-visualization of the monitoring data for preliminary assessment of structural behavior. Generally, the objectives of most structural monitoring systems can be classified into (1) anomaly detection, (2) sensor deployment studies, and lastly (3) damage detection. Therefore, the first step in modeling a structural monitoring system is the definition of the purposes of the developed system. Later, a preliminary design of the structural monitoring system should be defined before developing the modeled system. In this step, the structural behaviors, the monitoring approach, and the required devices to monitor are selected. In the last step, the monitoring system's physical installation process is determined, related to level of detail, accuracy, frequencies of recorded data, and the retrieved data (e.g., data are stored directly on the BIM model or linked from databases). To design a system that closely represents the real-life situations, data are represented by entities (sensor or a structural element, etc.). The different entities are aggregated into "sensor systems", which in turn are aggregated into "monitoring assemblies" (Davila Delgado et al. 2017). Every entity is instantiated at its respective location and populated with its corresponding acquired data. A graphical script has been developed to using the visual programming plugin Dynamo for Autodesk Revit and it works as follows: (1) the file path for the external data files is selected, in which data from the monitoring system are stored. All the data are compiled into a single list and the number of required sensors is inferred from the data; (2) the BIM element to be used and the lines along which the sensors will be populated are selected; and (3) the corresponding values for the analyzed measurements are assigned to each sensor. Before using data from the structural monitoring system, they must be processed into the correct physical quantity and units. This process can be automated through additional software that uses Application Programming Interface (API). Finally, data acquired by the proposed monitoring system are presented directly on the BIM model facilitating their interpretation. The result is a BIM model that has been augmented with sensor data that represents the state and performance of the constructed asset. The BIM model can be automatically updated to represent dynamic data, and new sensor data can be imported as it becomes available. The solution proposed for the real-time measurement and monitoring of environmental parameters is one innovative part of this paper, especially for existing buildings with significant architectural constraints. In these cases, the monitoring systems must be easily integrable into the building, from the installation/architectural point of view, low-cost, easy to find on the market, and easily manageable even by a non-expert user.

38.6 Conclusions

This study presents a data-driven BIM approach to leverage structural monitoring data of historical buildings. A strategy has been presented for the development of the simulation model for DT applications in historical buildings. The key features of the methodology are the correspondence of geometry to structural components and their materials. The work develops an approach that integrates a 3D information model and IoT systems to generate a detailed BIM which is then used for structural simulation via finite element analysis. A parametric BIM model of structural monitoring system is generated to enable the visualization of the monitoring data in a 3D environment in a dynamic and interactive manner. Most information is extracted from the BIM, preserving the consistency of BIM logic for structural simulation. That is, the finite element model reflects the geometric complexity of the construction without an excessive simplification of structural elements. The DT model can be updated continuously in line with the acquired knowledge of historical buildings. The proposed methodology aims at providing the possibility of developing several operations related to assessment and maintenance. It helps to understand the structural action in the different construction stages, study the damage of parts during the construction life, interpret the causes, and find solutions.

Future developments will require the implementation of the proposed framework to the analyzed case study to understand the behavior of elements of the construction as well as to check the stress level that is a preliminary information for the detailing of the restoration process to be planned. Furthermore, future work will focus on an investigation of the analyzed structure under analysis (in terms of composition and structural behavior). Different techniques such as non-destructive (sonic tests, geo-radar), minor destructive (flat jack, coring), will be planned to characterize construction materials and techniques, mechanical properties, and foundation levels. Investigation under different actions, such as extreme event of rainfall due to climatic change, wind and earthquake, will be accounted for in new studies to propose cloud-to-BIM-FEM into a tool for simulation pre- and post-restoration, risk management, and disaster prevention.

Acknowledgments

This research work was part of ERIS Project supported by LAZIO INNOVA (n. G09493-PO FESR LAZIO 2014/2020)The authors are gratefully to the Capitoline Museums of Rome.

References

Angjeliu, G., Coronelli, D., and Cardani, G. (2020). "Development of the simulation model for Digital Twin applications in historical masonry buildings: The integration between numerical and experimental reality." *Computers & Structures*, 238, 106282.

Barazzetti, L., Banfi, F., Brumana, R., Gusmeroli, G., Previtali, M., and Schiantarelli, G. (2015). "Cloud-to-BIM-to-FEM: Structural simulation with accurate historic BIM from laser scans." *Simulation Modelling Practice and Theory*, 57, 71–87.

Borowski, P. F. (2021). "Digitization, digital twins, blockchain, and industry 4.0 as elements of management process in enterprises in the energy sector." *Energies*, 14(7), 1885.

Castellazzi, G., D'Altri, A. M., Bitelli, G., Selvaggi, I., and Lambertini, A. (2015). "From laser scanning to finite element analysis of complex buildings by using a semi-automatic procedure." *Sensors*, 15(8), 18360–18380.

Cavalagli, N., Comanducci, G., and Ubertini, F. (2018). "Earthquake-induced damage detection in a monumental masonry bell-tower using long-term dynamic monitoring data." *Journal of Earthquake Engineering*, 22(sup1), 96–119.

Clementi, F., Gazzani, V., Poiani, M., Antonio Mezzapelle, P., and Lenci, S. (2018). "Seismic assessment of a monumental building through nonlinear analyses of a 3D solid model." *Journal of Earthquake Engineering*, 22(sup1), 35–61.

Davila Delgado, J. M., Butler, L. J., Gibbons, N., Brilakis, I., Elshafie, M. Z., and Middleton, C. "Management of structural monitoring data of bridges using BIM." Proceedings of the Institution of Civil Engineers-Bridge Engineering. Thomas Telford Ltd, 2017, 170(3): 204–218.

Dezen-Kempter, E., Mezencio, D. L., Miranda, E. D. M., De Sá, D. P., and Dias, U. (2020). "Towards a Digital Twin for Heritage Interpretation-from HBIM to AR visualization." RE Anthr. Des. Age Humans: Proc. 25th Int. Conf. Comput. Archit. Des. Res. Asia, CAADRIA 2020. 2020, 2: 183–191.

Díaz-Vilariño, L., Frías, E., Balado, J., and González-Jorge, H. (2018). "Scan planning and route optimization for control of execution of as-designed BIM." *International Archives of the Photogrammetry, Remote Sensing & Spatial Information Sciences*, 42(4): 143–148.

Errandonea, I., Beltrán, S., and Arrizabalaga, S. (2020). "Digital Twin for maintenance: A literature review." *Computers in Industry*, 123, 103316.

Fortunato, G., Funari, M. F., and Lonetti, P. (2017). "Survey and seismic vulnerability assessment of the Baptistery of San Giovanni in Tumba (Italy)." *Journal of Cultural Heritage*, 26, 64–78.

Funari, M. F., Hajjat, A. E., Masciotta, M. G., Oliveira, D. V., and Lourenço, P. B. (2021). "A parametric scan-to-FEM framework for the digital twin generation of historic masonry structures." *Sustainability*, 13(19), 11088.

Gattulli, V., Lepidi, M., and Potenza, F. (2016). "Dynamic testing and health monitoring of historic and modern civil structures in Italy." *Structural Monitoring and Maintenance*, 3(1), 71–90.

Gholizadeh, P., Esmaeili, B., and Goodrum, P. (2018). "Diffusion of building information modeling functions in the construction industry." *Journal of Management in Engineering*, 34(2), 04017060.

Grieves, M. (2014). "Digital twin: Manufacturing excellence through virtual factory replication." *White Paper*, 1(2014), 1–7.
Grieves, M. (2016). "Origins of the Digital Twin Concept." DOI: 10.13140/RG.2.2.26367.61609.
Grieves, M., and Vickers, J. (2017). "Digital twin: Mitigating unpredictable, undesirable emergent behavior in complex systems." In: Kahlen, J., Flumerfelt, S., Alves, A. (eds) *Transdisciplinary Perspectives on Complex Systems*, Springer, Cham, pp. 85–113.
Heesom, D., Boden, P., Hatfield, A., Rooble, S., Andrews, K., and Berwari, H. (2020). "Developing a collaborative HBIM to integrate tangible and intangible cultural heritage." *International Journal of Building Pathology and Adaptation*, 39(1), 72–95.
Hoskere, V., Park, J.-W., Yoon, H., and Spencer Jr, B. F. (2019). "Vision-based modal survey of civil infrastructure using unmanned aerial vehicles." *Journal of Structural Engineering*, 145(7), 04019062.
Hou, L., Wu, S., Zhang, G., Tan, Y., and Wang, X. (2020). "Literature review of digital twins applications in construction workforce safety." *Applied Sciences*, 11(1), 339.
Jouan, P.-A., and Hallot, P. (2019). "Digital twin: A HBIM-based methodology to support preventive conservation of historic assets through heritage significance awareness." *International Archives of the Photogrammetry, Remote Sensing and Spatial Information Sciences*, 42(2019).
Lauria, M., Mussinelli, E., and Tucci, F. (2022). "Producing Project." Published by Maggioli Editore in Open Access with Creative Commons License Attribution-NonCommercial-NoDerivatives 4.0 International (CC BY-NC-ND 4.0).
Lee, D., and Lee, S. (2021). "Digital twin for supply chain coordination in modular construction." *Applied Sciences*, 11(13), 5909.
Edward A. Lee and Sanjit A. Seshia, Introduction to Embedded Systems, A Cyber-Physical Systems Approach, Second Edition, MIT Press, ISBN 978-0-262-53381-2, 2017.
Masciotta, M.-G., Ramos, L. F., and Lourenço, P. B. (2017). "The importance of structural monitoring as a diagnosis and control tool in the restoration process of heritage structures: A case study in Portugal." *Journal of Cultural Heritage*, 27, 36–47.
Mol, A., Cabaleiro, M., Sousa, H. S., and Branco, J. M. (2020). "HBIM for storing life-cycle data regarding decay and damage in existing timber structures." *Automation in Construction*, 117, 103262.
Mukhopadhyay, S. C., and Suryadevara, N. K. (2014). *Internet of things: Challenges and Opportunities*. Part of the Smart Sensors, Measurement and Instrumentation book series (SSMI,volume 9):1–17.
Nagakura, T., and Sung, W. "Ramalytique: Augmented reality in architectural exhibitions." *Conference on Cultural Heritage and New Technologies 19th Prooceedings*, Vienna, Austria, 2014, pp. 3–5.
Opoku, D.-G. J., Perera, S., Osei-Kyei, R., and Rashidi, M. (2021). "Digital twin application in the construction industry: A literature review." *Journal of Building Engineering*, 40, 102726.
Pärn, E., and Edwards, D. (2017). "Conceptualising the FinDD API plug-in: A study of BIM-FM integration." *Automation in Construction*, 80, 11–21. s
Pepe, M., Costantino, D., and Restuccia Garofalo, A. (2020). "An efficient pipeline to obtain 3D model for HBIM and structural analysis purposes from 3D point clouds." *Applied Sciences*, 10(4), 1235.

Piaia, E., Maietti, F., Di Giulio, R., Schippers-Trifan, O., Van Delft, A., Bruinenberg, S., and Olivadese, R. (2021). "BIM-based cultural heritage asset management tool. Innovative solution to orient the preservation and valorization of historic buildings." *International Journal of Architectural Heritage*, 15(6), 897–920.

Potenza, F., Federici, F., Lepidi, M., Gattulli, V., Graziosi, F., and Colarieti, A. (2015). "Long-term structural monitoring of the damaged Basilica S. Maria di Collemaggio through a low-cost wireless sensor network." *Journal of Civil Structural Health Monitoring*, 5(5), 655–676.

Remondino, F. (2011). "Heritage recording and 3D modeling with photogrammetry and 3D scanning." *Remote Sensing*, 3(6), 1104–1138.

Spencer Jr, B. F., Hoskere, V., and Narazaki, Y. (2019). "Advances in computer vision-based civil infrastructure inspection and monitoring." *Engineering*, 5(2), 199–222.

Steel, J., Drogemuller, R., and Toth, B. (2012). "Model interoperability in building information modelling." *Software & Systems Modeling*, 11(1), 99–109.

Yastikli, N. (2007). "Documentation of cultural heritage using digital photogrammetry and laser scanning." *Journal of Cultural Heritage*, 8(4), 423–427.

39

Digital Twins in Architecture: An Ecology of Practices and Understandings

Anca-Simona Horvath
Aalborg University

Panagiota Pouliou
The Royal Danish Academy of Fine Arts (KADK)

39.1 Introduction

Lately, the term digital twin has been growing in popularity in diverse fields, from manufacturing to healthcare, and from aerospace to the architecture, engineering, and construction (AEC) industry. The concept was first introduced in the early 2000s by Grieves in the context of industrial design [26,27] and defined as *a digital informational construct about a physical system that is created as an entity on its own. This digital information would be a "twin" of the information that was embedded within the physical system itself and be linked with that physical system through the entire life cycle of the system*. Further, in Grieves' definition, the digital twin has three parts: the physical products in real space, the virtual replicas in virtual space, and the connections of data and information that tie the virtual and physical products together [26].

Al-Sehrawy&Kumar conducted a background analysis in all published research between 2012 and 2019 in the manufacturing, aerospace, and AEC industries, and concluded that a digital twin is *the concept of connecting a physical system to its virtual representation via bidirectional communication* [4].

While conceptually the notion of the digital twin was embraced by architecture as early as the 2000s, the field faces unique challenges making it one of the least digitized industries. Among the most important differences between manufacturing and architecture we note: in architecture each design is different, the construction sites of these designs are different (even when part of the construction is pre-fabricated, there are unique problems of transportation and assembly). The automation level on construction sites is still minimal, and the projects need to be integrated within existing natural or human-built habitats that are not digitized. In other words, as Ref. [80]

 DOI: 10.1201/9781003425724-46

states: the construction process is far from being standardized. Moreover, the prospected life of construction projects spans several decades, while technological artifacts (the digital twin being one such artifact) are typically designed for shorter lifespans.

Therefore, in spite of a number of studies and research papers connecting architecture with digital twins, there is no consensus as to how such a twin should be built, by whom, who should operate it, or even to what extent it can be useful. Through this paper, we attempt to map the diverse practices and understandings reported in research on digital twins for architecture. We understand architecture in its broad sense and analyze papers that have to do with matters of concern to the architectural field, including built heritage, design, construction, facility management and post-occupancy, and urban planning. We ask:

- What is understood by the concept of digital twin in architecture?
- What are their use cases and how is the concept of digital twin distributed across the different subfields that concern architecture?

The rest of the chapter is structured as follows: in the next section, we describe the materials and methods used to conduct a systematic literature review. Section 39.3 details the main findings of the study, while in Section 39.4, we discuss these findings and give an account of the current ecosystem of understandings, tools, and practices of digital twins in architecture.

39.2 Materials and Method

We started our data collection by performing a search for the string *digital twin* in the *Web of Science* (WoS) database. This search and subsequent data collection were conducted by one of the authors in August 2022. We refined the results to only include entries from the following *WoS* categories: computer science interdisciplinary applications (598), engineering civil (290), construction building technology (171), multidisciplinary sciences (107), architecture (17), urban studies (15), social sciences interdisciplinary (14), humanities multidisciplinary (10), and cultural studies (4). This resulted in 1,226 papers. We screened the titles and abstracts of these papers and only kept papers dealing with the field of digital twins for architectural applications. This left us with 119 papers, 6 of which were in German, and thus removed from the analysis. Finally, we performed the review on 113 papers: 30 papers in conference proceedings, 6 editorials, and 77 research articles in journals.

We analyzed this corpus using a mixed-method strategy conducting first a quantitative analysis, and later a qualitative analysis.

In the first stage, we put the abstracts of the papers through K-means [43], an unsupervised machine learning algorithm introduced by Ref. [2]. K-means is useful in finding the words with high frequency in a text. Before using K-means, we removed symbols and connection words such as conjunctions and articles. Next, we vectorized the texts using a technique called Term Frequency (TF)–Inverse Dense Frequency (IDF) [2]. The words were then separated into clusters, as proposed by Ref. [17]. The K-means algorithm initializes k random positions in the vector plane, equal to the number of requested clusters, in our case $k=3$, $k=10$, and assigns the data points to the nearest k position. The algorithm calculates the average position of the data points in a cluster and offsets the respective k position to it. When the data points have the minimum distance from their respective k positions, the algorithm stops and exports the clusters. In order to visualize the correlations between the abstracts and the clusters, we used the Principal Components Analysis (PCA) [49] algorithm. PCA is a technique for dimension reduction and visualization of clusters. PCA was used to reduce the dimensions of the TF-IDF and K-means results, allowing us to present the analysis as a 2D graphical representation.

Following the quantitative analysis of the abstracts and complementary to it, we conducted a thematic qualitative analysis based on the grounded theory approach [23]. We analyzed the abstracts and conclusions of all papers included in the review, together with the results from the clustering from the quantitative analysis. We started with a period of familiarization with the literature followed by a period of coding and generating themes. Once generated, the themes were reviewed, defined, and renamed over several iterations.

39.3 Findings

In this section, we start by presenting the keywords associated with the papers included in the review, followed by the results from the quantitative clustering of the abstracts. We then summarize the different types of digital twin research under the main themes that emerged from the qualitative analysis.

Figure 39.1 shows the percentages of *WoS* keywords associated with the papers included in this review, some papers had multiple keywords associated with them, while others only have one. We visualized the percentages of all the keywords (giving the same weight to each) and noticed that most keywords relate to the fields of civil engineering (37.3%) and the construction and building industry (27.1%). A significant proportion of work also relates to computer science (16.5%). Urban planning and architecture form only 3.4% and 2.5% of the keywords, respectively, while sustainable science &

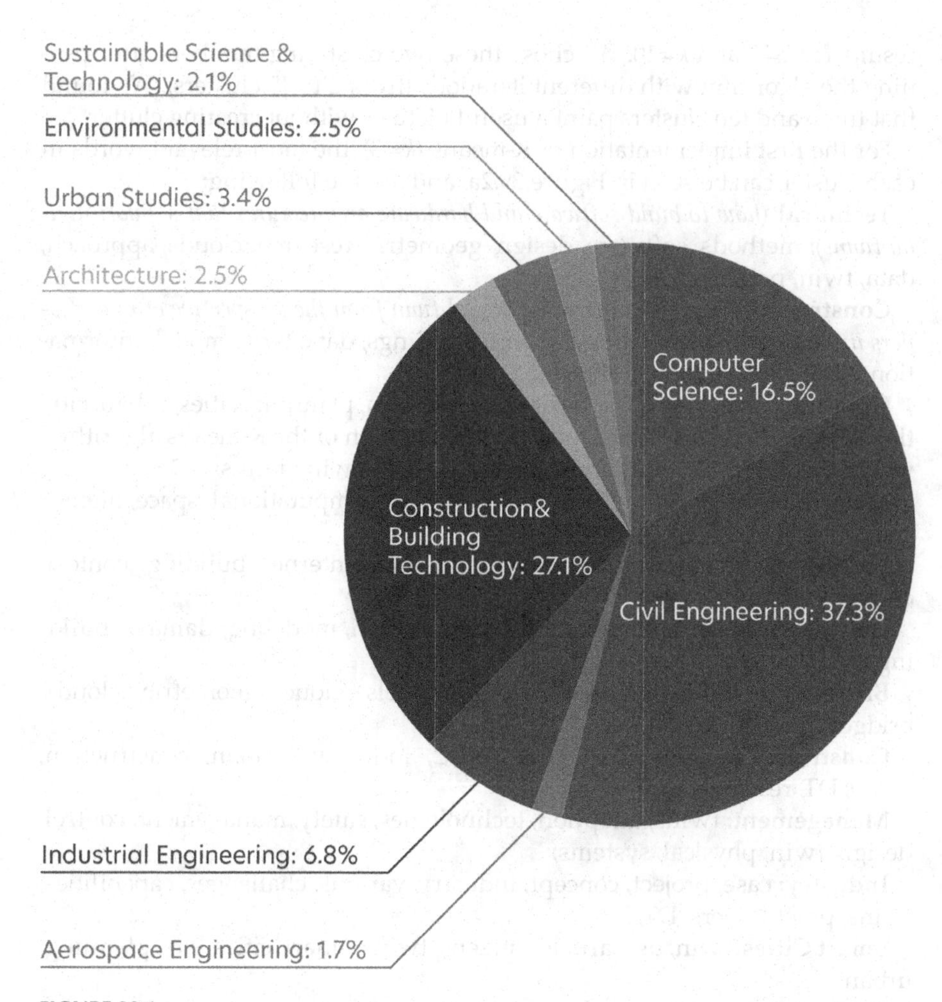

FIGURE 39.1
Percentages of keywords associated with the reviewed papers as per Web of Science categories.

technology and environmental studies make up for 2.1% and 2.5% each. One limitation of our study is that we only look at *WoS* as a source, which does not include all humanities and social sciences entries, that sometimes make up for architecture and urban planning research. However, the distribution of the keywords confirms what others have noted previously [57], namely, that digital twins are mostly discussed in the construction and management of buildings rather than in the conceptual and design phases of architecture.

39.3.1 Quantitative Clustering of Abstracts

In this subsection, we present the results from using the K-means clustering algorithm on the abstracts of the reviewed papers. We showcase the clustering results for $k=3$ and $k=10$. We chose these two clustering numbers after running the algorithm with different iterations (from 2 to 15 clusters). We found that three and ten clusters paint a useful picture, without creating clutter.

For the first implementation of K-means ($k=3$), the most relevant words in each cluster can be seen in Figure 39.2a, and are the following:

Technical (*how to build software and hardware architectures that support digital twins*): methods, software, design, geometric, research, clouds, approach, data, twin, point;

Construction (*how to construct a digital twin from the perspective of practitioners in architecture*): modeling, system, buildings, data, twin, model, information, BIM, construction, building;

Urban: DT, based, twins, physical, smart, twin, planning, cities, urban, city; the clusters formed in the second implementation of the K-means algorithm, where $k=10$ (Figure 39.2b) correspond to the following terms:

Material: enables, model, architectural, site, computational, space, materials, construction, design, physical;

Building: IoT, buildings, twin, things, data, internet, building, context, comfort, energy;

Heritage: traditional, model, BIM, information, modeling, damage, building, heritage, buildings, structural;

Bridges: methods, method, based, models, cloud, geometric, clouds, bridges, point, bridge;

Construction: technology, twin, AEC, industry, human, construction, study, DT, research;

Management: twins, adoption, technologies, safety, management, control, design, twin, physical, systems;

Industry: case, project, concept, industry, various, challenges, capabilities, twins, practitioners, DT;

Smart Cities: twin, use, article, citizens, twins, smart, cities, city, planning, urban;

BIM: twin, model, based, modeling, process, framework, information, building, construction, BIM;

Virtual Reality: twin, urban, based, world, physical, data, reality, virtual, cities, city.

The clustering results show that the concept of the digital twin is connected to and includes concepts such as BIM, VR, and smart cities. There are several areas of concern in the lifecycle of a building: from heritage and conservation to structure and materials, and construction and management.

39.3.2 Qualitative Thematic Analysis

Next, we conducted a thematic analysis of the abstracts and conclusions of the papers. The outcome of the analysis was refined in maps of what digital twins for architecture mean across two scales: time (the moment in a built object's life) and space (the scale of the built object).

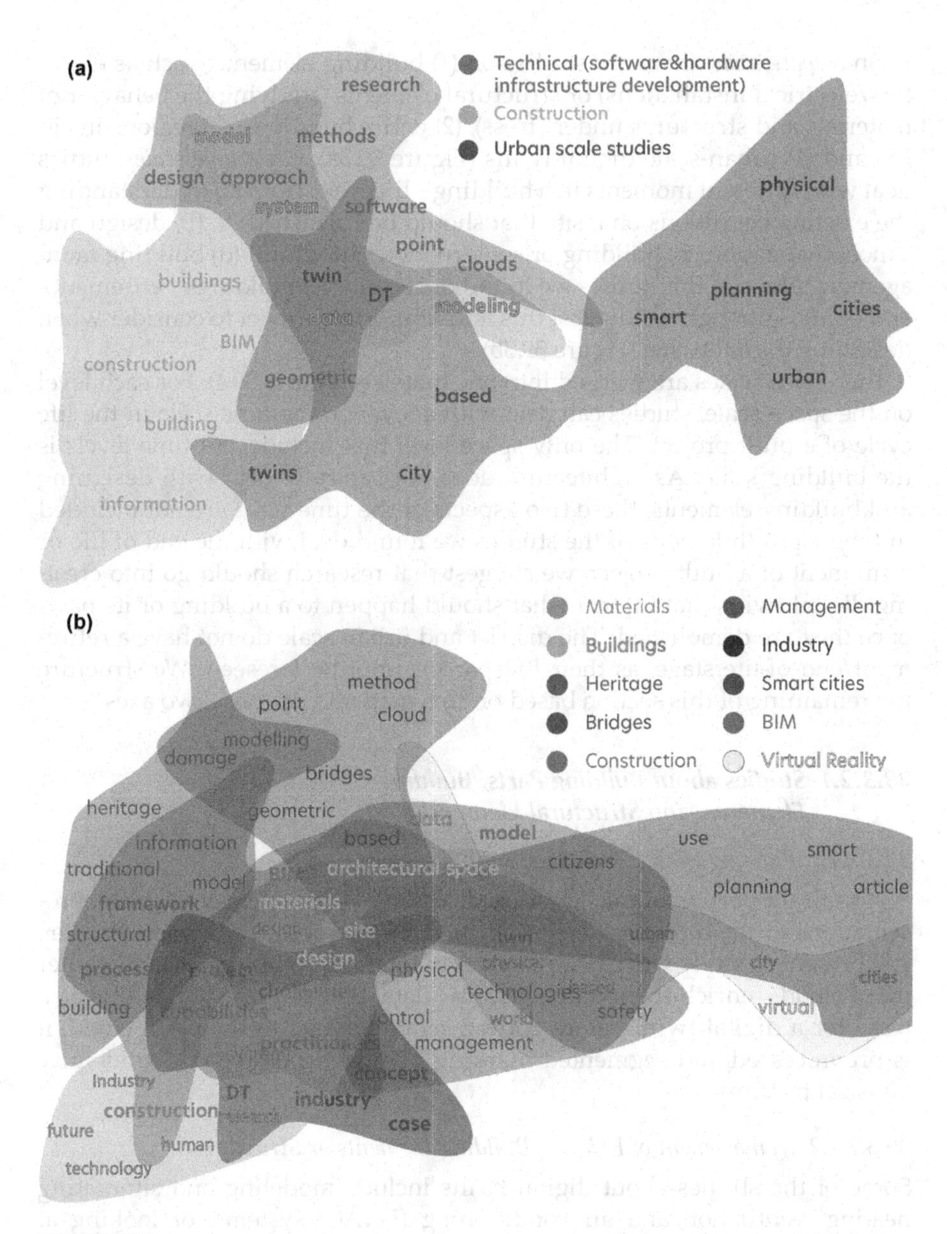

FIGURE 39.2
(a) The three clusters and most representative words attached to each cluster for the k=3 implementation of the K-means algorithm applied to the abstracts of the 113 reviewed papers. (b) The ten clusters and most representative words attached to each cluster for the k=10 implementation of the K-means algorithm applied to the abstracts of the 113 reviewed papers.

On a *space scale*, there are studies on (1) building elements (such as elevators/electrical installations) or structural elements (studying the behavior of materials and structures under stress), (2) entire buildings, (3) regions in cities, and (4) urban-scale digital twins (Figure 39.3a). On a *time scale*, studies deal with different moments in a building's lifecycle: from (a) understanding the existing conditions on a site that should be furnished, to (b) design and conceptualization, (c) building or construction phase, and (d) building management. None of the studies we found deals with (e) buildings' retirement/end of life, although we suggest this is an important aspect to consider when designing digital twins (Figure 39.3b).

These two scales are merged into one map (see Figure 39.4). For each level on the space scale, studies can deal with a level on the time scale in the life cycle of a built project. The only space level that includes all time levels is the building scale. As architecture does not typically deal with designing and building elements, these two aspects of the time scale are not included in the map. While none of the studies we found deal with the end of life or retirement of a built project, we suggest that research should go into creating digital twins that inform what should happen to a building or its parts once they are demolished. The district and urban scale do not have a retirement/end-of-life stage, as their lifespans cannot be foreseen. We structure the remaining of this section based on this map and on these two axes.

39.3.2.1 Studies about Building Parts, Building Elements, and Structural Elements

39.3.2.1.1 Existing Built Conditions/Built Heritage

A significant portion of the literature deals with ways to create digital representations of the built heritage or existing infrastructure. For example, Ref. [69] presents a study where they laser scan interior environments and gather point clouds, enrich them with semantic data, and propose to use this as a basis for a digital twin. Before adding semantic data to the point cloud, it is preprocessed and segmented marking homogeneous regions to identify physical features.

39.3.2.1.2 Management of Existing Building Elements or Structures

Some of the studies about digital twins include modeling and simulating heating, ventilation and air conditioning (HVAC) systems, or looking at building components such as elevators or escalators usage. For example, Ref. [76] develops a digital twin using the software Modelica [55] for an energy recovery ventilation unit to predict physical system behavior. Reference [12] reports on a post-occupancy study analyzing elevator usage. They establish an elevator operation simulation model, predict elevator usage under different conditions, and optimize operation measures.

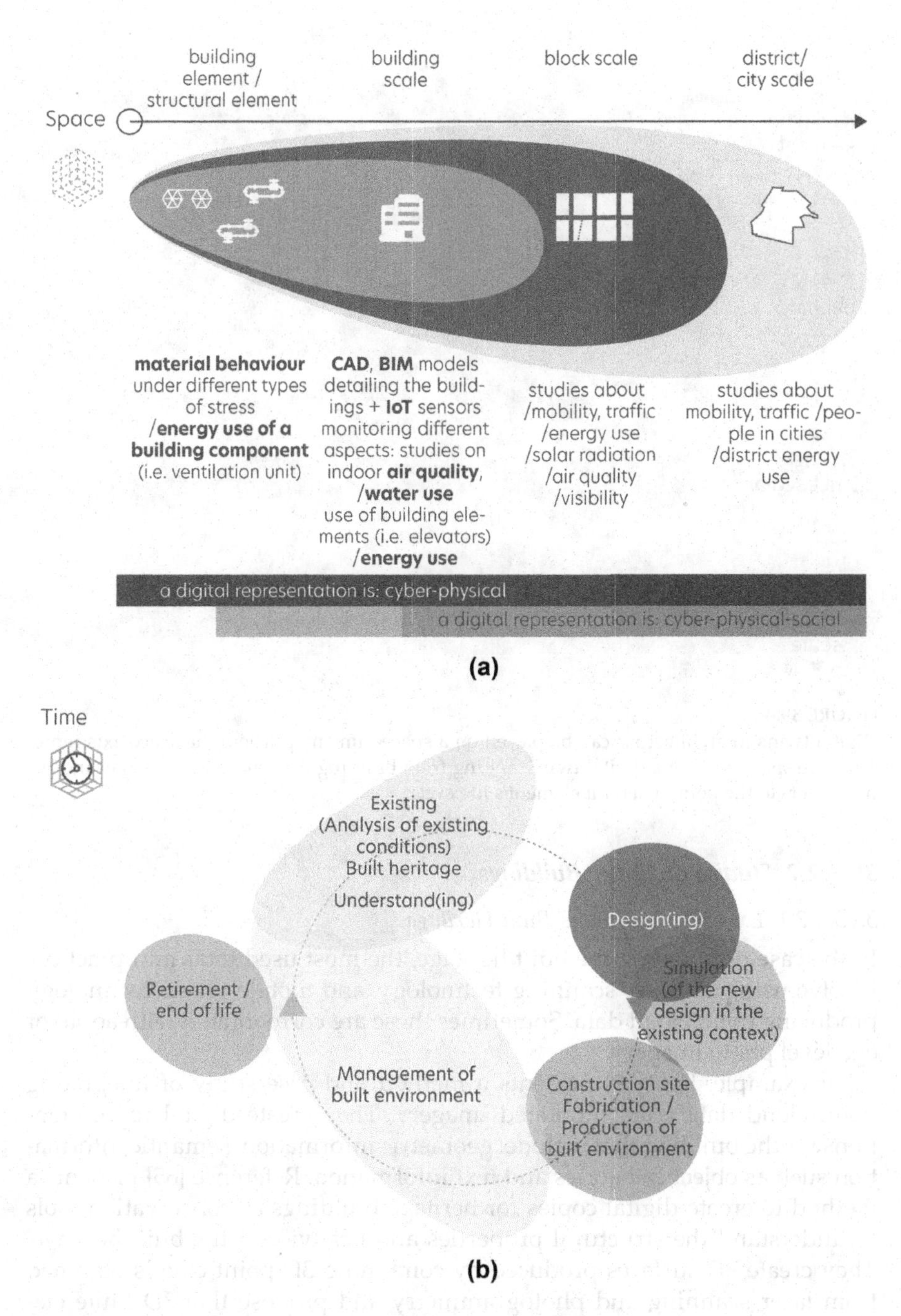

FIGURE 39.3
(a) On a space scale, projects range from creating digital twins of buildings or structural elements, to twins for entire buildings, to studies on block scale, district, or city scale. (b) Studies can deal with different moments in a built project's life time: from understanding existing conditions about a site, to designing new buildings, construction, and facility management.

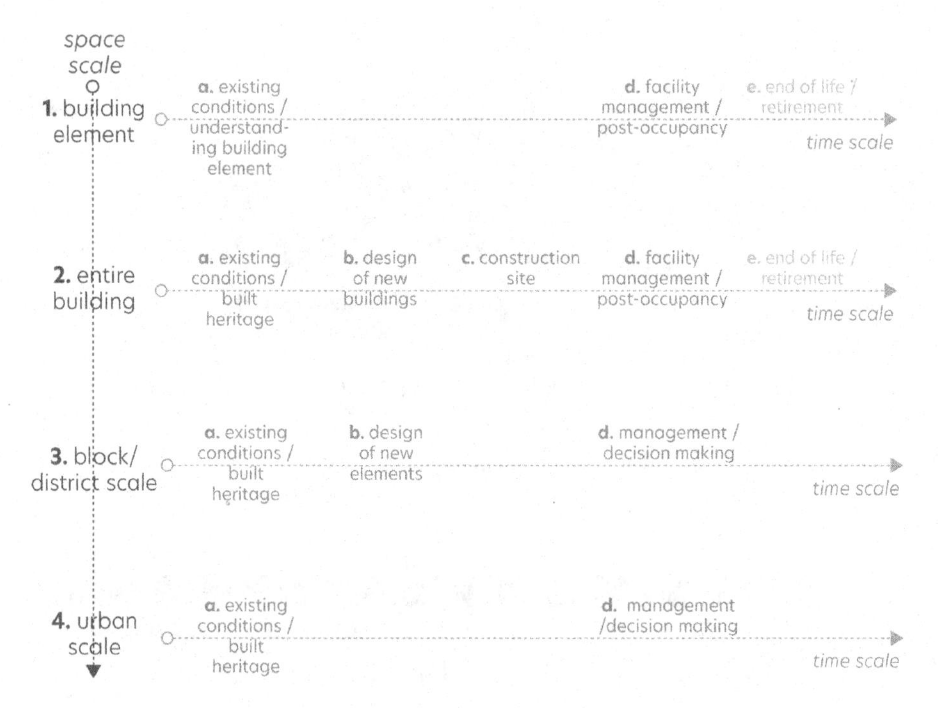

FIGURE 39.4
Digital twins in architecture can be placed on a space–time map where the space axis represents the space scale of a digital twin, ranging from building element to urban scale, and the time refers to the point in a built element's life cycle.

39.3.2.2 Studies on Entire Buildings

39.3.2.2.1 Existing Conditions/Built Heritage

In the case of mapping the built heritage, the most used tools and practices involve using 3D laser scanning technology, and mobile phone technology producing point cloud data. Sometimes these are corroborated with aerial or eye-level photo imagery.

For example, Ref. [59] presents a method and case study of integrating point cloud data with annotated imagery. They create digital representations for the building that include: geometric information, semantic information such as object categories and text information. Reference [65] presents a method to create digital copies for heritage buildings as conservation tools to understand the structural properties and behavior of the built heritage. They create 3D surfaces produced by combining 3D point clouds obtained from laser scanning and photogrammetry and propose that 3D finite element models (FEM) can be developed based on the geometries derived from the 3D models.

In the case of structural maintenance, the most used tools include aerial imagery, open map data, and finite element analysis (FEA) tools. Reference

[47] presents a method to automate the creation of a digital twin for an existing bridge, using annotated point clouds. The method follows a slicing strategy to generate 3D shapes using the Industry Foundation Classes (IFC), followed by fitting them to the labeled bridge point clusters.

Reference [37] proposes a systematic approach for making a digital representation of a highway based on Open Street Map data and their own engineering expertise. Reference [68] presents a study on a digital twin for a bridge that is used for preventive maintenance containing two elements: a maintenance information management system based on a 3D information model and a digital inspection system using image processing. Reference [1] presents a case study of a digital twin of an automated highway maintenance project. Based on interviews, they find that AEC practitioners face difficulties in (1) creating a shared understanding due to the lack of consensus on what a digital twin is, (2) adapting and investing in digital twin due to inability to exhaustively evaluate and select the appropriate capabilities in a digital twin, and (3) allocation of resources for digital twin development due to the inability to assess the impact of digital twin on the organizational conditions and processes. Practitioners used terms such as BIM, IoT, or BIM+IoT when asked to describe what a digital twin is and concluded that *the reality is that in the current practice and the foreseeable future, digital twins would still have limited technological capabilities*. Further, finding out what changes should be made in the organization in order to create, maintain, and generate value from a digital twin is still unclear.

39.3.2.2.2 Design of New Buildings

BIM acts as the shared platform for architecture, structural engineering, and HVAC work and is often referred as the basis for digital twins [6,31,38,41,79]. In the design of new buildings, the BIM is one of the first steps in creating a digital twin, after the digitalization of existing conditions for where a building should be inserted. However, there are sometimes problems and discrepancies between a building as designed (as it should be) and the building-as built, as pointed out by Ref. [38].

BIM can be easily integrated with VR systems, yet it still faces integration problems with other building management systems (BMS), geographic information systems (GIS), and IoT sensor data. BIM adoption is still strongly correlated to country GDP, with countries from the Global North leading the way, but it is expected that BIM and the IFC will become mandatory in public contracts for countries around the world in the future [3,78,82].

Apart from BIM creation as a basis for digital twins, many studies describe projects where the behavior of a structure or a part of a structure is simulated to understand material behavior under different types of stress. In Ref. [18], a digital twin for a structure containing a scaled physical model and its digital simulation is presented and automated decision-making for structural integrity is discussed. Similarly, Ref. [44] presents a case study for a digital twin of a long-span cable-stayed bridge. They assess the collapse fragility under strong earthquakes by using a scaled physical model of the bridge and compare the behavior of the physical model to a FEM analysis. Reference [35] describes a digital twin of a steel bridge designed to understand the behaviors of the material under fatigue and understand material properties of the structure under (environmental) stress. The study highlights the following challenges related to digital twins: (1) the lack of understanding of steel bridge fatigue and (2) the insufficiency of the present technologies. Similarly, Ref. [36] presents a case study of a road extension project. They start by suggesting a method for building road digital twins from online map data and report using: ArcGIS10.3 to process digital surface models and digital terrain model data, Autodesk Civil 3D assisted in the road modeling process, and Autodesk Navisworks Manage for clash detection.

39.3.2.2.3 Construction Sites

The next types of digital twins on the time scale in a building's life are digital twins of construction sites. Here, studies present management systems of construction sites and logistics [25] as well as management systems for prefabrication of building elements. For example, studies look at management of material and energy flows to and from the construction site and management [51] but also on training and safety of workers. Reference [73] proposes to define a subsection of the digital twin: digital twin for construction safety.

Reference [71] describes a method of using a robotic hand to train a construction worker to do complex woodwork. They work with an experienced woodworker, who should inspect wood for its quality and try to digitize some of this knowledge by using video and voice capturing and explain that the usage for the system would be to create meaningful digital material profiles for a CNC machine and to help less skilled wood-workers tap into the knowledge of more experienced ones.

39.3.2.2.4 Facility Management and Post-occupancy

The design and construction phases are relatively short in a building's life cycle compared to the overall life span of a building, which should exceed a few decades [38]. Facility management and post-occupancy play a critical role in this life cycle, and much of the literature on digital twins deals with these stages. Apart from studies that monitor the energy or water usage for entire buildings, there are studies that look at indoor air quality and building use and occupancy.

The notion of automated BMS is connected to the concept of the smart home and has to do with the management layer of the data collected from a building through sensors. BMS is currently used for energy and climate control, and energy reduction in buildings [32]. In facility management the functionalities of a digital twin are currently exercised by various combinations of systems such as BMS and building automation system (BAS). However, both currently lack integration with the existing BIM [70]. Reference [83] presents a case study of a digital twin used for the operation and management of a building. Based on the physical entity of the building and its BIM model, this framework combines IoT with algorithms such as neural networks and decision trees and includes data about: the building structure, the equipment in the building, and the building's energy consumption. The data are collected by computer maintenance management systems and a BAS and the behavior model should have the ability to 'self-evolve' by analyzing the relevant supplemented data.

One of the challenges identified relating to BMS and BAS is cybersecurity [16]: current digital twin systems might be vulnerable to cyber-threats, and this should be taken into account in the design stage. Connected to this are privacy issues.

Reference [24] describes a study that applies digital twin to indoor air quality management. The study monitors CO_2 levels, temperature, and humidity values of rooms within a building and claim the twin can be used to improve work productivity and reduce the risk for virus infections. The study highlights how using digital twins poses problems on existing software engineering practices. Similarly, Ref. [62] presents a model of the interior comfort for a heritage building and report using the Dragonfly and Honeybee plugins for Grasshopper in Rhino3D [52,63,64].

Reference [9] presents a practical example of a digital twin controlling people flow in a heritage building using an agent-based simulation on top of a BIM, which receives data from physical sensors installed around the building (people inflow) and changes the state of physical actuators (retractable barriers controlling outflow). The authors use a custom Dynamo script, as an addition to Revit and use Netlogo to get input from the sensors.

To summarize, studies of digital twins in this category include (1) physical aspects of how a building looks like, (2) how HVAC and other automated building systems are used, (3) FEA and FEM models about structures, (4) data about light, environment, and air quality, and (5) how people use a built space (gathered from: cameras, mobile phone data, entry points/check points in buildings, for example, entries using cards).

39.3.2.3 Block/District-Level Digital Twins

39.3.2.3.1 Existing Conditions

In Ref. [8], a digital twin for a 170 hectare district-scale university campus is presented detailing how the twin was built. The twin has a web user interface including visualizations of the 3D campus, real-time data from sensors, energy demand simulation results from city energy analyst, and occupancy rates from WiFi data. This twin is used to create scenarios of use of the built environment that can then help to understand energy needs in various instances.

39.3.2.3.2 Management and Decision-Making at a Block/District Scale

Reference [53] presents a conceptual model of a city, based on a metaphor of organic physiology with a model including: the environment, nature, society, and economy. Later, in Ref. [54], this conceptual framework is used in a case study for a digital twin for the Camp Nou Stadium in Barcelona claiming that for the purpose of the stadium, only two layers from the city physiology are necessary, namely, the *built domain* and *society*. They use data from three sources: mobile phone data, Foursquare data, and ticketing information which is then modeled using agent-based algorithms. The purpose of the digital twin is to understand possible behavior of visitors to the stadium, and to inform facility management (i.e. hotspots and bottlenecks).

39.3.2.4 Urban-Scale Studies

Many cities try to build digital twins that can help with data-informed decision-making both for the design of new buildings or areas, and in the management of the city's transportation systems, energy and waste management systems, or disaster response. Examples of cities that have digital twins include Helsinki, Singapore, Hong Kong, and London. The larger the scale, the more disciplines and stakeholders are involved in the design and decision-making processes regarding the built environment. As we saw in the quantitative analysis of the abstracts, the concept of digital twin overlaps with that of the smart city.

39.3.2.4.1 Existing Conditions

At the city scale, similar to the smaller scales, much of the work on digital twins focuses on data gathering. This data are gathered either about the physical elements of the city, about its social or socio-economic elements, or a combination of both (i.e. citizens reporting potholes to inform municipalities about construction work). Reference [34] analyzes the currently available data that can be used in digital twins for smart cities and identify: datasets related to the (1) environment, (2) infrastructure, (3) healthcare, (4) education, and (4) government. In other words, there are data on flows of energy, materials, transportation, data on people and relationships between them, data on governance and laws, and data on the weather and the environment.

Reference [39] presents a study of using drones imagery to create a 3D representation of a section of a city. The system consists of a drone that acquires both first person and overhead views at the building construction site, a controller that operates the drone, and a PC on a server that performs AR rendering with occlusion handling. Similarly, Ref. [45] presents a method for building inspection that merges aerial drone animations and BIM and visualizes this using AR technology.

39.3.2.4.2 Management and Decision-Making at City Scale

Here studies focus on how to deal with the different data types including data processing and integration and case studies about how digital twins were used in decision-making. Reference [16] proposes the data needs, data standards, and data sources to develop city building datasets for urban building energy modeling stating that urban building energy modeling is becoming a proven tool to support energy efficiency programs for buildings in cities and that the development of a city-scale dataset of the existing building stock is a critical step. They report the use of the following tools and standards: CityGML, Geo-JSON, FileGDB, ArcGIS pro, IoT, webGL, Unity, and VR/AR, agent-based modeling and developed methods to integrate city building datasets using common standards.

Reference [20] describes a digital twin for the small German city Herrenberg. The twin includes data about: (1) the 3D representation of the built environment; (2) a street network model using the theory and method of space syntax; (3) an urban mobility simulation with SUMO and wind flow simulation; (4) qualitative and quantitative data using Volunteered Geographic Information with a mobile application developed for this purpose; and (5) a pollution simulation using an empirical data set from a sensor network. Their study involved visualization of the twin through VR and was meant to support participatory processes from citizens that should inform city planning. The study does not report on how the twin is used to inform decision-making but rather on how the visualization of the twin is developed and shown to different stakeholders that should participate in decision-making regarding the city.

The study in Ref. [81] reports on deploying a digital twin online allowing for citizen feedback on urban planning. The system is used to provide feedback on proposed buildings and green spaces by allowing users to tag problems in urban areas such as potholes.

A very different example of a digital twin on the urban scale comes from research from the fields of computer science and mathematics. This deals with software and hardware architectures for a city-level digital twin, and how to integrate the different data streams. Reference [33] describes a concept for software architecture for a digital twin of a city and proposes that digital twins of individual elements of the urban environment should be consistently built on a single hardware and software platform. Reference [5] proposes a conceptual software architecture of a digital twin that could be applied to a smart city that includes: (1) a physical layer, (2) a cyber layer (stores data about the physical objects), (3) an integration layer, (4) a service layer, and (5) an interaction layer (where users can see 3D models, graphs of the data, etc.).

To summarize, on an urban scale, digital twins become highly complex, including data about physical, environmental, and social elements. There is on-going work on mapping the built environment and systems dubbed digital twins used in decision-making (of traffic, energy, or water use) and simulation (of environmental disasters, traffic, energy use) at city levels.

39.4 Discussion

In this section, we discuss four aspects of digital twins that emerge from our analysis. First, the concept of digital twin is understood differently in the various disciplines connected to architecture. For the field of architecture, the digital twin represents an ecology of understandings, practices, tools, and data types. These are all fragmented and the concept is in its infancy both from a theoretical perspective (i.e. what is it, how should it be defined) and from a practical perspective (i.e. how should one be designed, with what tools, by whom, and why) in spite of many studies that try to define what a digital twin is [10,60,66]. Next, we discuss the concept of the digital twin itself: as digital twins rely on digital representations and abstractions, digital twins will never be twins. Third, the twins rely on large amounts of data, and their relationship to these data needs to be addressed critically. Finally, we end by discussing the expected life and life span of a digital twin for an architectural project.

39.4.1 Digital Twins in Architecture: An Ecology of Practices and Understandings

From the quantitative analysis we see that a digital twin is connected to technologies such as BIM, VR, and IoT, but also concepts such as smart city and smart building. While some studies claim that a digital representation of existing aspects of the built environment constitutes a digital twin, other studies claim that a digital twin is more than CAD, BIM, point clouds, or VR, and includes sensor data streams that can support maintenance work and decision-making. This ambiguity is not unusual, as most new terminology and technology is loosely defined in the initial stages of its implementation [29] and it is likely to persist in the future, as the concept gains maturity.

A digital twin can be understood rather as an ecology of tools, practices, and understandings. Not only are there different understandings of what a digital twin is across different disciplines [74], but there are also different approaches to creating digital data about the same physical entities across these different disciplines. For example, multiple disciplines deal with urban 3D modeling in different ways [40]. A twin only exists when there is a corresponding physical entity connected to it. The twin contains geometric and non-geometric information about the physical entity and supports decision-making about its physical counterpart.

Building digital twins requires expertise from fields such as mathematics, hardware engineering, and computer science that were not traditionally associated to architectural practice. A large portion of the literature we reviewed look at how to build technology that supports digital twins for architecture and increasingly experts from these fields contribute to architectural related fields through designing these technologies.

In the design stage of an architectural project, a digital twin is based on BIM and CAD data. In the structural design phase, the twin is based on FEA, FEM and computer fluid simulations. In the construction phase, a digital twin is based on site management and logistics, coupled with data extracted from the BIM model, although some work also describes digital twins about digital manufacturing for construction (this includes material profiles that can feed in digital fabrication tools). In the management and operation phase, the digital twin is based on BMS and BAS. In the care for the built heritage, the digital twin is mostly based on segmented and annotated point cloud data. The IFC and the CityGML are both standards that solve some interoperability issues between all these types of representations.

All of the above technologies can come together to make a digital twin for architecture, but it is important to differentiate the notion of the digital twin from BIM, VR, or IoT. A digital twin always refers to an existing physical system and represents that system as accurately as possible, while a CAD, BIM or VR model can all exist without their physical counterparts.

39.4.2 Digital Twins Will Never Be Twins

However, no matter how accurate the data that make a digital twin are, the concept ultimately relates to digital representation and abstraction.

The idea of creating digital mirrors of physical worlds was suggested as early as 1993 [22]. Yet, the physical world is not completely understood, and therefore, cannot be mirrored. This lack of a complete understanding of the physical world is often mentioned in studies from structural engineering. For example, in the design of a digital twin for a structure, Ref. [18] highlights how their twin is a model, abstraction, based on current scientific understanding of physical entities: *the virtual environment provides an idealization of the physical environment under a specific level of abstraction through mathematical models based on laws of physics, data, or both.*

The foundation of a digital twin is digital representation [82] – representation here being the keyword. There are different ways to represent physical elements (i.e. a brick wall can be represented through the 3D model of its geometry, through a mathematical formulation, through the numbers of bricks and quantity of mortar needed to build it, or through all of these together). The same brick wall will have different representations for different stakeholders that contribute to the design, building, or management of the building, in its different stages. This has led some to challenge the term itself [15], with Ref. [10] arguing that in the end the digital twin can only function as a model, or an abstraction of the real world. Reference [75] states that the digital twin is embodied and immersed in what it is supposed to mirror, and thus is no longer an independent representation. Other terms connected to digital twin have been proposed, such as smart city or building, but also cyber-physical system, cyber–physical–social system [75], or cyber–physical–social ecosystems [15].

The term digital twin should not be understood as a complete representation of reality, as this is not possible. Instead, it is a model that can be useful in: informing new designs, maintenance of existing infrastructure, resource consumption optimization on a construction site, water use and energy use optimization at district and city scale, and in understanding how people use the built environment. The different fields that contribute to architecture each contribute with parts to this model.

39.4.3 Digital Twins Should Engage Critically with Data

Regardless of how the term is defined, digital twins rely on data, and both the theory and the practice surrounding digital twins for architecture should engage critically with these data.

Data fusion, defined as integration of various sources and types of data streams, and especially the integration of physical and virtual data, is still a major concern in digital twin research [72]. Together with this, there is a lack of common data standards and tools [28,30,60,67,77,80].

Creating and managing a digital twin can be seen as a twofold problem: a tool problem (where tools include algorithms and software such as BIM, CAD, GIS, and IoT technology) and a data problem, including the infrastructures of data collection and storage.

While finding ways to solve problems around data integration and interoperability is important, it is equally important to look at data from a critical standpoint.

Data are bounded in space and time [14], and data collection, processing, and visualization are always subjective. Reference [46] makes a case about data locality describing data as *cultural artifacts created by people, at a time, in a place and with the instruments at hand.* The environment of data collection, who collects it, and for what purpose are all important.. This is especially important in cases when digital twins use or rely on data about people, in other words, when the digital twin corresponds to a cyber–physical–social ecosystem (see Figure 39.3). Reference [56] argues that moving from a purely technical toward a socio-technical understanding of technology for smart cities has profound implications on the conceptualizations of technology. The General Data Protection Regulation (GDPR) directive in Europe has fundamentally reshaped the way data is handled [11], what kind of data can be stored, how it can be stored and used and for how long. Companies have even started discarding data after GDPR was enforced [61]. The directive tries to address ethical aspects related to data science, namely, monitoring, surveillance, and discrimination, because as Ref. [21] says: *today, data science is a form of power.* The GDPR directive is one of the first to regulate digital infrastructures. As digital technologies become part of critical infrastructure, more regulations of how these technologies should be designed will be implemented, similar to regulations about the built environment.

Digital twins involve information systems [58], data science, and data governance, and need to discuss not only how to extract value from the data but also how to ensure transparency of algorithms that facilitate data-informed decision-making, especially when these decisions involve people.

To summarize, the data infrastructures needed to support a digital twin relevant for architecture face two main challenges, one conceptual, and the other practical and ethical.

First: data about a physical entity come always with a certain granularity over space and time and this granularity reflects how accurate the data about a physical system can be. No matter how detailed, the data are always incomplete, as our current understanding of the physical world is incomplete. A voxel (defined as 3D pixel or three-dimensional unit of digital space) can serve as a good metaphor: one can ask – what does a voxel of real space include? How accurate *can* the voxel be? How should these data be represented (i.e. what kind of data types and abstractions should we employ)? Importantly, what is the size of this voxel/or the granularity we need, and what data are actually necessary and valuable, when, and for whom?

The second challenge is practical and relates to issues of both cybersecurity and ethics. Reference [7] proposes to create established standards for a cybersecurity layer for the built environment. Breaches in a digital twin system can incapacitate a building's functioning, and, as Ref. [7] states, hardly any research today focuses on this issue. Furthermore, ethical issues about the types of data and algorithms used for decision-making need to be taken into account.

39.4.4 Can a Digital Twin Live as Long as a Building or a City?

What should the life and the life span of a digital twin be? The answer to this question seems straightforward, as in the definition given by Grieves; the life of a digital twin corresponds to the life of a physical product. Therefore, the development of the digital twin starts at the design stage, and its existence continues throughout the building's life cycle. However, life spans of buildings typically exceed several decades. Digital twins are based on technological artifacts that typically are designed with shorter life spans in mind. In addition, digital twins should continuously collect data about the physical environment and help with decision-making about it. One might ask: how should a digital twin be planned, in order for that technological infrastructure to still be relevant, and maintainable 50 or 100 years into the future? Would some of the data be deleted periodically, and only 'relevant summaries' be recorded?

The built environment is responsible for 40% of total energy use, over 30% of CO_2 emissions, and 25% of the generated waste in Europe on an annual basis [13,48] and research within architecture is increasingly looking at ways to improve sustainability in buildings and cities. While only a small percentage of the *WoS* keywords associated to the articles we analyze are 'environmental studies' (2.5%) or 'sustainable science and technology' (2.1%), making for a total of under 5% (see Figure 39.1), many studies on digital twins explain that some of the uses of a twin can be energy use optimization and reduction. References [42,50,77] suggest that information can be extracted throughout the construction life cycle from conception to operation, and propose that these data may improve the sustainability aspects of projects (for example, by reducing project waste). Nonetheless, this might be unlikely to happen with growing privacy issues surrounding data governance.

However, as Ref. [18] puts it, when describing a structural engineering study: *there also exist computational and energy constraints, hence the need for careful planning during the design stage.* It then becomes important to consider which data are stored, by whom, and for how long. Further, having more data does not necessarily translate into more value.

The anatomy of an AI system is a map of the Amazon's Echo service [19] showing complex infrastructures of data collection and processing that feed into an AI system unfolding a multi-layered ecosystem of servers, data, and people behind the digital assistant. Considering how much it costs to collect, store, and process the data needed to create digital twins at scale, what are the material costs that go into the data banks and how much energy they consume could be issues of future inquiry in digital twin research related to architecture and the built environment. The digital twin should conceptually include the infrastructures of data collection and processing that feed the twin. This is important when designing a new twin as these infrastructures can be energy consuming, and involve material resources themselves which is why the concept of infinite information and processing power needs to be challenged, as this is ultimately a sustainability concern.

39.5 Conclusion

In this chapter, we analyzed the concept of digital twin as it relates to architecture through a systematic literature review of 113 papers. We analyzed the papers through a mixed-method technique, first by conducting a quantitative clustering analysis on the abstracts and then a qualitative thematic analysis on the abstracts and conclusions of the papers. We understood architecture in its broad sense, involving all its matters of concern, from built heritage and conservation, to designing new buildings, construction, management and post-occupancy studies, and end of life. We present the findings on a space–time map that shows the studies about digital twins in architecture deal with different scales (from building element to entire buildings, districts, and cities) and different moments in a built project's life cycle. The concept of the digital twin is still in its infancy both from a theoretical perspective (i.e. how should the notion be defined) and from a practical perspective (how should a twin be built, by whom, who should use it, and when). The twins are always representations of reality based on current technology and scientific understanding of the world, and so digital twins will never be twins, but such representations can be useful. In addition, digital twins should engage critically with the data they make use of, and should consider the infrastructures of data storage and processing they would need throughout their life span. Finally, digital twins as technological artifacts will need to consider how their lifespan correlates to that of the built environment they mirror.

References

1. Ashwin Agrawal, Vishal Singh, Robert Thiel, Michael Pillsbury, Harrison Knoll, Jay Puckett, and Martin Fischer. Digital twin in practice: Emergent insights from an ethnographic-action research study. In *Construction Research Congress 2022*, Arlington, Virginia, March 2022. American Society of Civil Engineers.
2. Akiko Aizawa. An information-theoretic perspective of tf-idf measures. *Information Processing & Management*, 39(1):45–65, 2003.
3. Eisa Al Hammoud. Comparing bim adoption around the world, Syria's current status and future. *International Journal of BIM and Engineering Science*, 4:64–78, 2021.
4. Ramy Al-Sehrawy and Bimal Kumar. Digital twins in architecture, engineering, construction and operations. A brief review and analysis. In Eduardo Toledo Santos and Sergio Scheer, editors, *Proceedings of the 18th International Conference on Computing in Civil and Building Engineering*, pages 924–939, Cham, 2021. Springer International Publishing.
5. Kazi Masudul Alam and Abdulmotaleb El Saddik. C2ps: A digital twin architecture reference model for the cloud-based cyber-physical systems. *IEEE Access*, 5:2050–2062, 2017.
6. Sepehr Alizadehsalehi and Ibrahim Yitmen. Digital twin-based progress monitoring management model through reality capture to extended reality technologies (DRX). *Smart and Sustainable Built Environment*, 12(1):200–236, 2021.
7. Kaznah Alshammari, Thomas Beach, and Yacine Rezgui. Cybersecurity for digital twins in the built environment: Research landscape, industry attitudes and future direction. *International Journal of Civil and Environmental Engineering*, 15(8):382–387, 2021.
8. Pradeep Alva, Martin Mosteiro-Romero, Clayton Miller, and Rudi Stouffs. Digital twin-based resilience evaluation of district-scale archetypes. In *International Conference for the Association for Computer-Aided Architectural Design Research in Asia*, pages 525–534, Sydney, Australia, 2022.
9. Ugo Maria Coraglia Armando Trento, Gabriel Wurzer. A digital twin for directing people flow in preserved heritage buildings, Conference Paper· December. 2019.
10. Michael Batty. Digital twins. *Environment and Planning B: Urban Analytics and City Science*, 45(5):817–820, 2018.

11. Olivia Benfeldt, John Stouby Persson, and Sabine Madsen. Data governance as a collective action problem. *Information Systems Frontiers*, 22:299–313, 2020
12. Gabriele Bernardini, Elisa Di Giuseppe, Marco D'Orazio, and Enrico Quagliarini. Occupants' behavioral analysis for the optimization of building operation and maintenance: A case study to improve the use of elevators in a university building. In John Littlewood, Robert J. Howlett, and Lakhmi C. Jain, editors, *Sustainability in Energy and Buildings 2020*, pages 207–217. Springer, Singapore, 2021.
13. José Pedro Carvalho, Luís Bragança, and Ricardo Mateus. Optimising building sustainability assessment using bim. *Automation in Construction*, 102:170–182, 2019.
14. Silvia Casini. *Giving Bodies Back to Data: Image Makers, Bricolage, and Reinvention in Magentic Resonance Technology*. MIT Press, 2021.
15. Marianna Charitonidou. Urban scale digital twins in data-driven society: Challenging digital universalism in urban planning decision-making. *International Journal of Architectural Computing*, 20(2):238–253, 2022.
16. Yixing Chen, Tianzhen Hong, Xuan Luo, and Barry Hooper. Development of city buildings dataset for urban building energy modeling. *Energy and Buildings*, 183:252–265, 2019.
17. Jung Hee Cheon, Duhyeong Kim, and Jai Hyun Park. Towards a practical cluster analysis over encrypted data. Cryptology ePrint Archive, Paper 2019/465, 2019. https://eprint.iacr.org/2019/465.
18. Manuel Chiachío, María Megía, Juan Chiachío, Juan Fernandez, and María L. Jalón. Structural digital twin framework: Formulation and technology integration. *Automation in Construction*, 140:104333, 2022.
19. Kate Crawford and Vladan Joler. Anatomy of an ai system: The amazon echo as an anatomical map of human labor, data and planetary resources, 2018. https://anatomyof.ai/.
20. Fabian Dembski, Uwe Wössner, and Mike Letzgus. The digital twin tackling urban challenges with models, spatial analysis and numerical simulations in immersive virtual environments, pages 795–804. Blucher, São Paulo, 2019.
21. Catherine D'Ignazio and Lauren F. Klein. *Data Feminism*. Publisher: The MIT Press, 2020.
22. David Gelernter. *Mirror Worlds: Or the Day Software Puts the Universe in a Shoebox...How It Will Happen and What It Will Mean.*(NY, 1991; online edn, Oxford Academic, 12 Nov. 2020), https://doi.org/10.1093/oso/9780195068122.003.0008, a ccessed 8 Dec. 2023..
23. Barney Glaser and Anselm Strauss. *The Discovery of Grounded Theory: Strategies for Qualitative Research*. New York, Routledge, 1999.
24. Hari Shankar Govindasamy, Ramya Jayaraman, Burcu Taspinar, Daniel Lehner, and Manuel Wimmer. Air quality management: An exemplar for model-driven digital twin engineering. In *2021 ACM/IEEE International Conference on Model Driven Engineering Languages and Systems Companion (MODELS-C)*, Fukuoka, Japan, pages 229–232, 2021.
25. Toni Greif, Nikolai Stein, and Christoph M. Flath. Peeking into the void: Digital twins for construction site logistics. *Computers in Industry*, 121:103264, 2020.
26. Michael Grieves. Digital twin: Manufacturing excellence through virtual factory replication. *White Paper*, 1(2014):1–7, 2014.

27. Michael Grieves. Origins of the digital twin concept. *Florida Institute of Technology*, 8:3–20, 2016.
28. Anca-Simona Horvath. Assessing site-geometry for architectural design using graph theory. In Cosmin Chiorean, editor, *Proceedings of the Second International Conference for PhD students in Civil Engineering and Architecture*, Cluj-Napoca, Rumunsko, elektronický zdroj, pages 611–619, 2014. U.T. Press. Conference date: 10-12-2014 through 12-12-2014.
29. Anca-Simona Horvath. How we talk(ed) about it: Ways of speaking about computational architecture. *International Journal of Architectural Computing*, 20(2):150–175, 2022.
30. Anca-Simona Horvath and Radu Becus. Cluj Geoweb. In *1st International Edition of Cadet Inova for Young Inventors*, page 221, 2015. Conference date: 15-04-2015 through 17-04-2015.
31. Anca-Simona Horvath, Clara Vite, Naja L. Holten Møller, and Gina Neff. Messybim: Augmenting a building information model with messy talk to improve a buildings' design process. In María Menéndez-Blanco, Seçil Uğur Yavuz, Jennifer Schubert, Daniela Fogli, and Fabio Paternò, editors, *CHItaly 2021 Joint Proceedings of Interactive Experiences and Doctoral Consortium*, Bolzano-Bozen, Italy, volume 2892, pages 7–14. *CEUR Workshop Proceedings*, 2021. Conference date: 11-07-2021 through 13-07-2021.
32. Md. Faruque Hossain. Chapter seven - best management practices. In Md. Faruque Hossain, editor, *Sustainable Design and Build*, pages 419–431. Butterworth-Heinemann, 2019. https://doi.org/10.1016/C2017-0-02236-X.
33. Sergey Ivanov, Ksenia Nikolskaya, Gleb Radchenko, Leonid Sokolinsky, and Mikhail Zymbler. Digital twin of city: Concept overview. In *2020 Global Smart Industry Conference (GloSIC)*, pages 178–186, 2020.
34. Michael Jacobellis and Mohammad Ilbeigi. Digital twin cities: Data availability and systematic data collection. In *Construction Research Congress 2022*, Arlington, Virginia,pages 437–444, 2022.
35. Fei Jiang, Youliang Ding, Yongsheng Song, Fangfang Geng, and Zhiwen Wang. An architecture of lifecycle fatigue management of steel bridges driven by digital twin. *Structural Monitoring and Maintenance*, 8(2):187–201, 2021.
36. Feng Jiang, Ling Ma, Tim Broyd, Ke Chen, and Hanbin Luo. Underpass clearance checking in highway widening projects using digital twins. *Automation in Construction*, 141:104406, 2022.
37. Feng Jiang, Ling Ma, Tim Broyd, Weiya Chen, and Hanbin Luo. Building digital twins of existing highways using map data based on engineering expertise. *Automation in Construction*, 134:104081, 2022.
38. Karen Kensek. Bim guidelines inform facilities management databases: A case study over time. *Buildings*, 5(3):899–916, 2015.
39. Naoki Kikuchi, Tomohiro Fukuda, and Nobuyoshi Yabuki. Future landscape visualization using a city digital twin: Integration of augmented reality and drones with implementation of 3D model-based occlusion handling. *Journal of Computational Design and Engineering*, 9(2):837–856, 2022.
40. Thomas H. Kolbe and Andreas Donaubauer. *Semantic 3D City Modeling and BIM*, pages 609–636. Springer, Singapore, 2021.
41. Mergen Kor, Ibrahim Yitmen, and Sepehr Alizadehsalehi. An investigation for integration of deep learning and digital twins towards construction 4.0. *Smart and Sustainable Built Environment*, 12(3):461–487, 2023.

42. Alexander Koutamanis, Boukje van Reijn, and Ellen van Bueren. Urban mining and buildings: A review of possibilities and limitations. *Resources, Conservation and Recycling*, 138:32–39, 2018.
43. K. Krishna and M. Narasimha Murty. Genetic k-means algorithm. *IEEE Transactions on Systems, Man, and Cybernetics, Part B (Cybernetics)*, 29(3):433–439, 1999.
44. Kaiqi Lin, You-Lin Xu, Xinzheng Lu, Zhongguo Guan, and Jianzhong Li. Digital twin-based collapse fragility assessment of a long-span cablestayed bridge under strong earthquakes. *Automation in Construction*, 123:103547, 2021.
45. Donghai Liu, Xietian Xia, Junjie Chen, and Shuai Li. Integrating building information model and augmented reality for drone-based building inspection. *Journal of Computing in Civil Engineering*, 35(2):04020073, 2021.
46. Yanni Alexander Loukissas. *All Data Are Local: Thinking Critically in a Data-Driven Society*. MIT Press, 2019. https://ieeexplore.ieee.org/servlet/opac?bknumber=8709328.
47. Ruodan Lu and Ioannis Brilakis. Digital twinning of existing reinforced concrete bridges from labelled point clusters. *Automation in Construction*, 105:102837, 2019.
48. Zhihan Lyu, Yang Liu, Yuhui Sun, Ang Yang, and Jing Gao. Digital twin-based ecogreen building design. *Complexity*, 2021, 2021: 1–10.
49. Andrzej Maćkiewicz and Waldemar Ratajczak. Principal components analysis (pca). *Computers & Geosciences*, 19(3):303–342, 1993.
50. Azad M. Madni, Carla C. Madni, and Scott D. Lucero. Leveraging digital twin technology in model-based systems engineering. *Systems*, 7(1):7, 2019.
51. Oleg Maryasin. Home automation system ontology for digital building twin. In *2019 XXI International Conference Complex Systems: Control and Modeling Problems (CSCMP)*, Samara, Russia, pages 70–74, 2019.
52. McNeel&Associates. Rhinoceros 3d. https://www.rhino3d.com/, 2022.
53. Irene Meta, Fernando M. Cucchietti, Diego Navarro, Eduardo Graells-Garrido, and Vicente Guallart. A physiology-inspired framework for holistic city simulations. *Cities*, 126:103553, 2021.
54. Irene Meta, Feliu Serra-Burriel, José C. Carrasco-Jiménez, Fernando M. Cucchietti, Carla Diví-Cuesta, Carlos García Calatrava, David García, Eduardo Graells-Garrido, Germ´an Navarro, Quim Lázaro, Patricio Reyes, Diego Navarro-Mateu, Alex Gil Julian, and Imanol Eguskiza Martínez. The camp nou stadium as a testbed for city physiology: A modular framework for urban digital twins. *Complexity*, 2021:1–15, 2021.
55. Modelica. https://modelica.org/, 2022.
56. Timea Nochta, Li Wan, Jennifer Mary Schooling, and Ajith Kumar Parlikad. A socio-technical perspective on urban analytics: The case of city-scale digital twins. *Journal of Urban Technology*, 28(1–2):263–287, 2021.
57. De-Graft Joe Opoku, Srinath Perera, Robert Osei-Kyei, and Maria Rashidi. Digital twin application in the construction industry: A literature review. *Journal of Building Engineering*, 40:102726, 2021.
58. Karen S. Osmundsen, Christian Meske, and Devinder Thapa. Familiarity with digital twin totality: Exploring the relation and perception of affordances through a heideggerian perspective. *Information Systems Journal*, 32(5):1064–1091, 2022.

59. Yuandong Pan, Alexander Braun, Ioannis Brilakis, and André Borrmann. Enriching geometric digital twins of buildings with small objects by fusing laser scanning and ai-based image recognition. *Automation in Construction*, 140:104375, 2022.
60. Dessislava Petrova-Antonova and Sylvia Ilieva. Methodological framework for digital transition and performance assessment of smart cities. In *2019 4th International Conference on Smart and Sustainable Technologies (SpliTech)*, Island of Brac, Croatia, pages 1–6, 2019.
61. Alison B. Powell. *Undoing Optimization: Civic Action in Smart Cities*. Yale University Press, New Haven, CT 06511-8909, 2021.
62. Federico Mario La Russa and Cettina Santagati. From the cognitive to the sentient building - machine learning for the preservation of museum collections in historical architecture, /Machine Learning for the preservation of museum collections in historical architecture. In Proceedings of the 38th eCAADe Conference on Education and Research in Computer Aided Architectural Design in Europe, Berlin, Germany. 2020: 16–17.
63. David Rutten. Grasshopper. https://www.grasshopper3d.com/page/download-1, 2022.
64. Mostapha Sadeghipour Roudsari and Michelle Pak. Ladybug: A parametric environmental plugin for grasshopper to help designers create an environmentally-conscious design, 2013.
65. Amirhosein Shabani, Margarita Skamantzari, Sevasti Tapinaki, Andreas Georgopoulos, Vagelis Plevris, and Mahdi Kioumarsi. 3d simulation models for developing digital twins of heritage structures: Challenges and strategies. *Procedia Structural Integrity*, 37:314–320, 2022.
66. Muhammad Shahzad, Muhammad Tariq Shafiq, Dean Douglas, and Mohamad Kassem. Digital twins in built environments: An investigation of the characteristics, applications, and challenges. *Buildings*, 12(2):120, 2022.
67. Guodong Shao and Moneer Helu. Framework for a digital twin in manufacturing: Scope and requirements. *Manufacturing Letters*, 24:105–107, 2020.
68. Chang-Su Shim, Ngoc-Son Dang, Sokanya Lon, and Chi-Ho Jeon. Development of a bridge maintenance system for prestressed concrete bridges using 3d digital twin model. *Structure and Infrastructure Engineering*, 15(10):1319–1332, 2019.
69. Vladeta Stojanovic, Matthias Trapp, Rico Richter, Benjamin Hagedorn, and Jürgen Döllner. Semantic enrichment of indoor point clouds an overview of progress towards digital twinning. *Blucher Design Proceedings*, 7(1):809–818, 2019.
70. Vladeta Stojanovic, Matthias Trapp, Rico Richter, Benjamin Hagedorn, and Jürgen Döllner. Towards the generation of digital twins for facility management based on 3d point clouds. *Management*, 270:279, 2018.
71. Chika Sukegawa, Arastoo Khajehee, Takuya Kawakami, Syunsuke Someya, Yuji Hirano, Masako Shibuya, Koki Ito, Yoshiaki Watanabe, Qiang Wang, Tooru Inaba, Alric Lee, Kensuke Hotta, Mikita Miyaguchi, and Yasushi Ikeda. Smart hand for digital twin timber work: The interactive procedural scanning by industrial arm robot.SMART HAND FOR DIGITAL TWIN TIMBER WORK, 2:131–140, 2022.
72. Fei Tao, Meng Zhang, and Andrew Yeh Chris. Nee. *Digital Twin Driven Smart Manufacturing*. Academic Press, Elsevier, 2019. https://doi.org/10.1016/C2018-0-02206-9.

73. Teizer J, Johansen K W, Schultz C. The concept of digital twin for construction safety. In *Construction Research Congress 2022*, Arlington, Virginia, pages 1156–1165, 2022.
74. Joeran Tesse, Ulrich Baldauf, Ingrid Schirmer, Paul Drews, and Sebastian Saxe. Extending internet of things enterprise architectures by digital twins exemplified in the context of the Hamburg port authority. In The 27th annual Americas Conference on Information Systems, online, 2021.
75. Martin Tomko and Stephan Winter. Beyond digital twins - A commentary. *Environment and Planning B: Urban Analytics and City Science*, 46(2):395–399, 2019.
76. Christian Vering, Philipp Mehrfeld, Markus Nürenberg, Daniel Coakley, Moritz Lauster, and Dirk Müller. Unlocking potentials of building energy systems' operational efficiency: Application of digital twin design for HVAC systems. In *16th International Building Performance Simulation Association (IBPSA)*, University College Dublin, 2019.
77. Ahmed Vian, Aziz Zeeshan, Tezel Algan, and Riaz Zainab. Challenges and drivers for data mining in the aec sector. *Engineering, Construction and Architectural Management*, 25:1436–1453, 2018.
78. Christos Vidalakis, Fonbeyin Henry Abanda, and Akponanabofa Henry Oti. Bim adoption and implementation: Focusing on smes. *Construction Innovation*, 20:128–147, 2019.
79. Clara Vite, Anca-Simona Horvath, Gina Neff, and Naja L. Holten Møller. Bringing human-centredness to technologies for buildings: An agenda for linking new types of data to the challenge of sustainability. In Antonella De Angeli, Luca Chittaro, Rosella Gennari, Maria De Marsico, Alessandra Melonio, Cristina Gena, Luigi De Russis, and Lucio Davide Spano, editors, *CHItaly '21: 14th Biannual Conference of the Italian SIGCHI Chapter*, pages 1–8, United States, 2021. Association for Computing Machinery. Conference date: 11-07-2021 through 13-07-2021.
80. Yuxi Wei, Zhen Lei, and Sadiq Altaf. An off-site construction digital twin assessment framework using wood panelized construction as a case study. *Buildings*, 12(5):566, 2022.
81. Gary White, Anna Zink, Lara Codecá, and Siobhán Clarke. A digital twin smart city for citizen feedback. *Cities*, 110:103064, 2021.
82. Jiaying Zhang, Jack C. P. Cheng, Weiwei Chen, and Keyu Chen. Digital twins for construction sites: Concepts, lod definition, and applications. *Journal of Management in Engineering*, 38(2):04021094, 2021.
83. Yuhong Zhao, Naiqiang Wang, Zhansheng Liu, and Enyi Mu. Construction theory for a building intelligent operation and maintenance system based on digital twins and machine learning. *Buildings*, 12(2):87, 2022.

40

Developing a Construction Digital Twin for Bridges: A Case Study of Construction Control of Long-Span Rigid Skeleton Arch Bridge

Chunli Ying and Long Chen
Loughborough University

Daguang Han
Southeast University

Kaixin Hu
Chongqing Jiaotong University

Yu Zhang
Shenyang Jianzhu University

Guoqian Ren
Tongji University

Yanhui Liu
Southwest Jiaotong University

Yongquan Dong
Chongqing Jiaotong University

Yatong Yuan
China Construction Fifth Engineering Bureau

40.1 Introduction

Digital twins, as dynamic asset representations that mimic real-world behaviors, offer promising solutions for managing, planning, predicting, and demonstrating assets in the built environment (Chen et al. 2021).

DOI: 10.1201/9781003425724-47

These digital twins, based on data from physical assets or systems, facilitate improved decision-making, providing positive feedback to their physical counterparts. Various modeling approaches enable accurate representations of physical reality at different scales. Although numerous digital twins exist for the built environment, few can connect or share data across organizations, industries, or geographies due to limited interoperability.

The digital twin concept has been increasingly applied to various fields, overlapping with virtual simulation and numerical simulation in civil engineering. Building information modeling (BIM), which carries extensive information about buildings, is widely utilized as a digital twin carrier in engineering. Developed in the 1970s, BIM was defined by experts like Charles Eastman (Eastman et al. 2011). BIM technology and its applications are more advanced in the United States and some European countries than in China. European and American scholars have extensively studied BIM in the construction industry, with Caiyun Wan (Wan et al. 2004) providing recommendations for using BIM standard structural analysis models for prestressed loads and load combinations. Bridge Information Modeling (BrIM) is a BIM extension in the bridge domain (Costin et al. 2018).

As BIM-related standards and software expand in Europe, the United States, and Asia, BIM models become ideal data carriers for digital twin models (DTMs) in civil engineering. The digital twin BIM model (DT-BIM), built and updated on digital twin concepts, offers enhanced information processing capabilities and application prospects, serving as a precise digital description and information carrier for real-world physical objects. This chapter demonstrates the application of digital twin theory during the construction of a reinforced concrete arch bridge. Reinforced concrete arch bridges have proliferated since the 1800s, with nearly 80 worldwide featuring span diameters over 200 m. Monier built the first concrete arch bridge in 1875, and Australian engineer Josef Melan improved the design by using steel as a skeleton in 1890. Continuous advancements have increased span lengths, exemplified by the Hoover Bridge in 2010, which utilized steel–concrete composite construction. By 2007, China led the world with 199 arch bridges featuring spans of at least 100 m (Wang 2018).

The Rigid Skeleton Construction Approach is the predominant method for building large-span reinforced concrete arch bridges. The structural steel rigid skeleton (SSRS) is built first, followed by sectional concrete pouring to form the final arch structure. The steel arch skeleton's position determines the bridge's shape, making the accurate fabrication and installation of SSRS sections critical for construction quality. One challenge is obtaining precise SSRS section dimensions in a timely manner and controlling on-site installation.

DT-BIM is a digital model that manages and simulates the life cycle of physical objects. To maintain synchronization with real-world objects, DTM must integrate IoT technology for real-time updates. This approach enables predictive simulation, analysis, and visualization of physical models. Digital twin models have proven effective in constructing large-span reinforced concrete arch bridges. Ruodan Lu employed a slice-based object fitting method for automated modeling of point clouds in reinforced concrete bridges (Lu & Brilakis 2019), generating geometric digital twin models (GDTM). Their experiments showed an average modeling distance of 7.05 cm and a modeling time of 37.8 seconds. Masoud Mohammadi used the digital twin concept with UAV photogrammetry and TLS to evaluate and monitor a steel truss bridge in Australia, resulting in accurate data capture and 3D reconstruction models (Mohammadi et al. 2021).

To summarize, the digital twin concept is less widely used in bridge construction quality control. In this chapter, the GDTM is created using a series of algorithmic processes that begin with establishing point cloud data of stiff skeleton segments and end with this GDTM. The stiff skeleton's construction control is realized on this basis by combining mechanical information. Compared with previous methods, the developed GDTM can (1) not only provide precise 3D dimensions but also reduce the impact of human factors on data processing, and (2) be useful to control each key step to achieve a better overall force state based on the inherent mechanical information.

40.2 Literature Review

Creating a construction DTM involves three steps: (1) developing a GDTM; (2) predicting a DTM; and (3) generating a mechanical analysis DTM model by providing mechanical information, construction process analysis, and force control of the completed bridge. The key to realizing a DTM is acquiring consistent, comprehensive, and accurate object information through reality capture technology, which primarily utilizes 3D scanning and digital image processing methods.

Research into digital image processing for concrete structure detection has advanced significantly in the 21st century. Abdel-Qader compared four crack detection techniques, finding the fast Hal transform (FHT) to be the most reliable (Abdel-Qader et al. 2003). Yiyang proposed an algorithm for glass crack detection using segmentation and feature extraction (Yiyang 2014). In conclusion, digital image processing techniques effectively acquire and process object texture information but are unsuitable for precise geometry acquisition in complex steel structures like stiff skeleton segments.

The 3D laser scanning is widely used in capturing surface shapes and reverse modeling of complex geometries, such as body surface organ reconstruction (Zhang 2007) and steel bar corrosion pattern identification (Kashani et al. 2013). Combined with BIM technology and photogrammetry, 3D scanning manages project progress and quality inspection. S. Yoon explored optimal scanning arrangements (Yoon et al. 2018) for identifying positioning errors in prefabricated steel beam shear nails (Bosché & Guenet 2014). In addition, 3D scanning was employed to measure stone arch bridge surface alignment and analyze finite element simulations (Lubowiecka et al. 2009). C. Andreotti (Andreotti et al. 2015) used 3D scanning to detect minor masonry structure anomalies post-earthquake. These case studies confirm 3D laser scanning as a reliable method for obtaining GDTMs of bridge-like structures.

Prediction is crucial in DTM for structural objects. Case demonstrated a successful virtual assembly method for complex steel structures using generalized Procrustes analysis (Case et al. 2014). Zhou developed a mesh-based parameter extraction method for bolt hole features, enabling accurate prediction of alignment and stress in assembled steel structures (Zhou et al. 2021). To accelerate cable-stayed bridge construction, Donggun Kima devised a 3D laser point cloud-based method for measuring and predicting temporal variations in displacement between welded precast segments, reducing Cheonsa Bridge construction time by 10 days (D. Kim et al. 2020).

These cases indicate that predicting assembly behavior and actual construction of complex steel structures using 3D laser point clouds is feasible. However, bolted stiff skeletons require more stringent accuracy and prediction thresholds compared to welded prefabricated sections. Further verification of the method's feasibility in actual engineering is needed.

Final quality of constructed targets relies on construction control. Uncertainties and risk factors arise during construction as the structure's size and height increase. Seungho Kim focused on a five-stage construction process, demonstrating a reliable method for tracking construction progress via point clouds (S. Kim et al. 2020). Nisha Puria et al. proposed a semi-automated bridge monitoring method, with results indicating sensitivity to factors such as point cloud density (Puri & Turkan 2020). Yawei Qin combined 3D laser scanning and BIM technology for integrated engineering quality control but did not include mechanical behavior control in arch bridge construction (Qin et al. 2019). Research on point clouds for constructed object control has mostly focused on construction progress and quality assessment. However, none of the aforementioned cases address the mechanical properties of the constructed object. For complex rigid skeletons, structural behavior during construction directly impacts safety and final quality. Therefore, providing rigid skeleton DTM mechanical information is essential for constructing mechanical analysis DTM models and performing force control for completed bridges.

40.3 Methodology

40.3.1 Framework for Generating Digital Twin Models in Construction

This study aims to address the limitations in current research on DTM for complex steel structures, using a complex rigid skeleton bridge as a case study. The process consists of three steps: (1) creating the GDTM using 3D laser scanning and feature extraction algorithms; (2) predicting the behavior of the DTM to understand the structural installation; and (3) controlling the construction process with a mechanically endowed DTM, informed by bridge-related control theory. This approach aims to enhance the application of digital twin technology in complex steel structures and improve construction progress, quality (Figure 40.1).

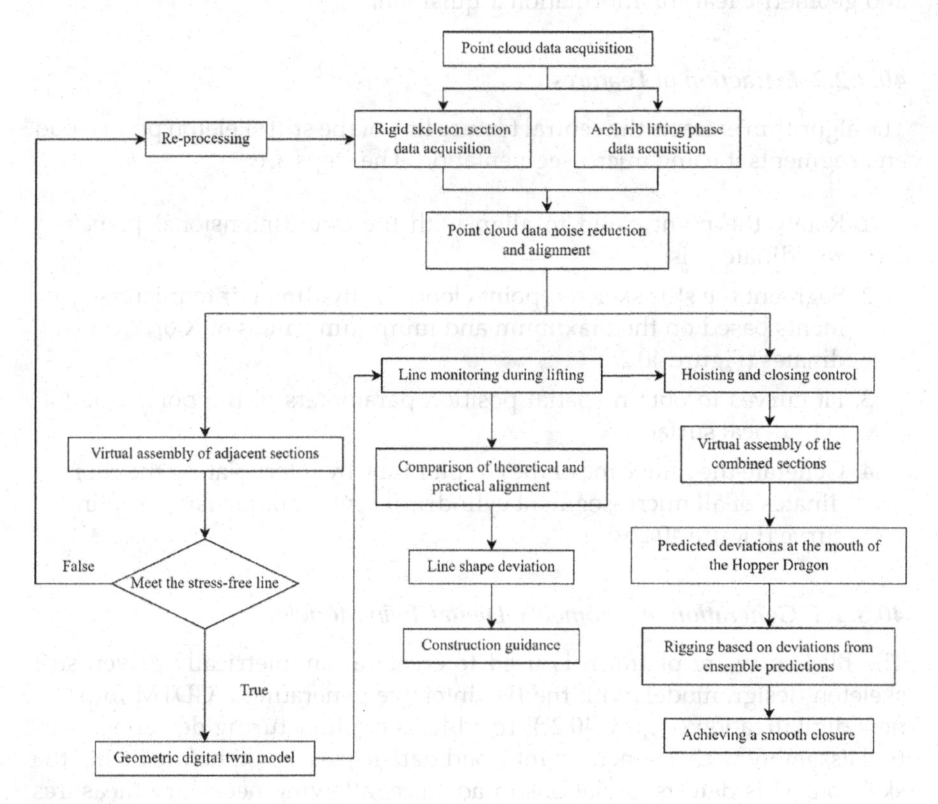

FIGURE 40.1
Generation process of digital twin model.

40.3.2 Generation of Geometric Digital Twin Models

40.3.2.1 Point Cloud Pre-processing

Point cloud pre-processing is crucial for ensuring the accuracy and efficiency of post-processing in large-sized skeleton data. This involves noise reduction and point cloud alignment, which can be addressed through a robust filtering algorithm targeting the noise points and an improved ICP algorithm for fine alignment. The improved ICP algorithm consists of three main steps: (1) utilizing the KD-tree algorithm to find neighboring points in both directions; (2) determining the normal vector of the point set and filtering based on a set queue value for the angle cosine; and (3) employing the SVD decomposition method to calculate rotation and translation vectors between the point sets, iterating until the condition is satisfied. This approach effectively mitigates noise and aligns point clouds, improving data analysis accuracy and geometric feature information acquisition.

40.3.2.2 Extraction of Features

The algorithm extracts the central feature line of the stiff skeleton point cloud and segments it using micro-segmentation. The steps are:

1. Rotate the point cloud to align with the two-dimensional plane's coordinate axis.
2. Segment the stiff skeleton point cloud by dividing it into micro-segments based on the maximum and minimum values of X or Y coordinates (Figure 40.2a).
3. Fit curves to obtain spatial position parameters of the point cloud cylindrical surface.
4. Generate the centerline of the circular tube by interpolating the coordinates of all micro-segment cylindrical center points using a spline curve (Figure 40.2b).

40.3.2.3 Generation of Geometric Digital Twin Models

The BIM software platform is used to create a parametrically driven stiff skeleton design model, with the IFC interface generating a GDTM for stiffness digitalization (Figure 40.2c). To address manufacturing deviations, virtual assembly with segment point cloud data is performed before lifting the skeleton. This detects deviations in advance, allowing necessary measures to be taken. The splicing process aligns bolt hole coordinates on adjacent flanges using the Prandtl analysis. The algorithm calculates optimal rotation translation parameters to align skeleton segments, ensuring precise assembly of all sections (Figure 40.2d).

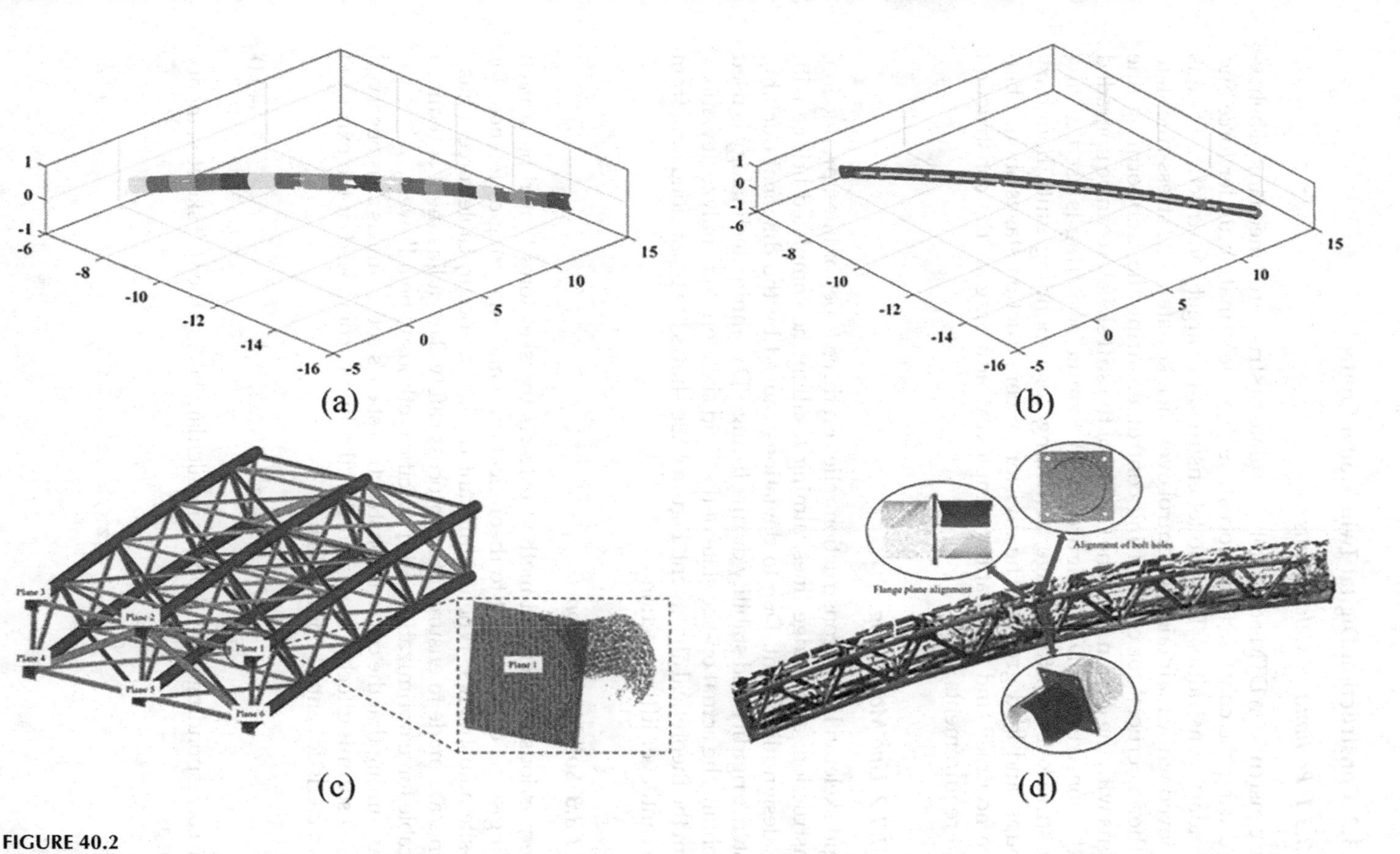

FIGURE 40.2
Generation of geometric digital twin models: (a) the effect of point cloud segmentation; (b) generation of stiff skeleton centerlines; (c) model of a geometric digital twin; and (d) virtual skeleton assembly.

40.3.3 Construction Digital Twin Model Control

40.3.3.1 Preliminary Calculations

The construction DTM accurately simulates a structure's mechanical behavior during the construction process, such as deformation under force and boundary constraints. This model ensures the target construction state is achieved while maintaining control over each key step. The stress-free state method is crucial for cable lifting control, ensuring the arch forming line aligns with the design line. The GDTM of the stiff skeleton can be upgraded to the construction DTM (Figure 40.3a). Symmetric lifting is used during the stiff skeleton lifting stage, activating corresponding units, boundary groups, and load groups. The division of construction stages should balance accuracy and calculation complexity, adhering to the construction plan requirements.

40.3.3.2 Line Monitoring

Rigid skeleton lifting construction relies on three types of lines: production, construction, and bridge lines, aiming to align the completed bridge with the design alignment. Due to deviations caused by the diagonal buckling system's rigidity and stability during lifting, 3D scanning technology is used to obtain the actual on-site state of the rigid skeleton and analyze deviations from the theoretical alignment. Figure 40.3b shows the point cloud data from the cantilever lifting section.

40.3.3.3 Soli Modification

The previously described method extracts the skeleton's centerline at each lifting stage, comparing it to theoretical lines and analyzing differences for on-site adjustments. When significant deviations occur, cable force adjustments are made to ensure lifting process safety. This adjustment is reduced to cable force optimization, with specific methods to be followed.

Assuming that the current construction stage's actual line is z^1 as observed by 3D scanning technology, and that the line is composed of the extracted characteristic control points, then

$$Z^1 = \left(z^1{}_1, z^1{}_2, z^1{}_3, \ldots, z^1{}_i\right) \tag{40.1}$$

The corresponding control points is calculated using the stiff skeleton theory line as

$$Z^2 = \left(z^2{}_1, z^2{}_2, z^2{}_3, \ldots, z^2{}_i\right) \tag{40.2}$$

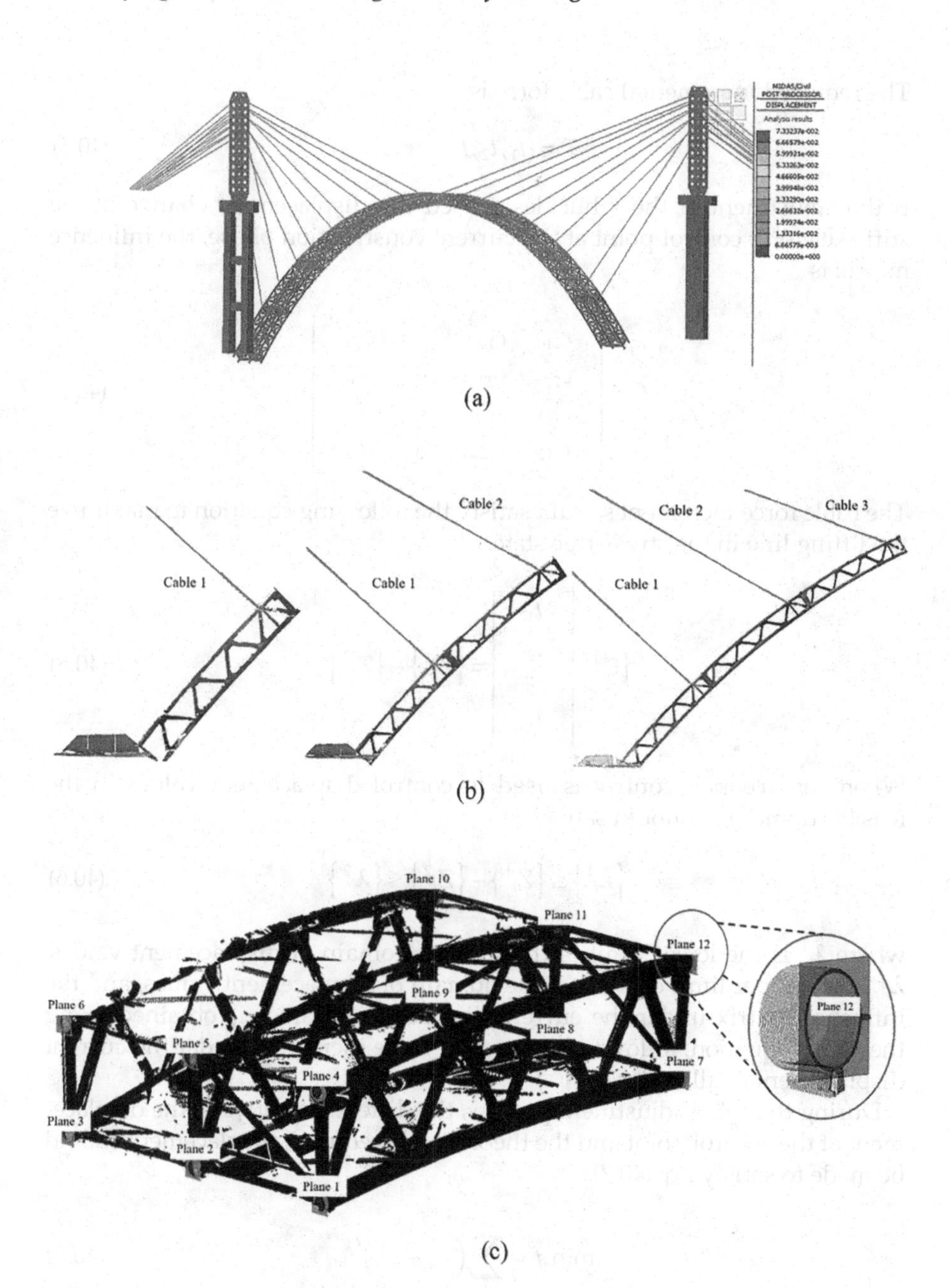

FIGURE 40.3
Construction digital twin model control: (a) analysis of lifting construction model; (b) point cloud data for lifting construction; and (c) port plane of the combined section.

The required incremental cable force is

$$T = (t_1, t_2, t_3, \ldots t_n) \tag{40.3}$$

If the adjustment of the # buckles caused the displacement change of the stiff skeleton's control point at the current construction phase, the influence matrix is

$$C = \begin{bmatrix} c_{11} & c_{12} & \cdots & c_{1n} \\ c_{21} & c_{22} & \cdots & c_{2n} \\ \vdots & \vdots & \ddots & \vdots \\ c_{i1} & c_{i2} & c_{i3} & c_{in} \end{bmatrix} \tag{40.4}$$

The cable force increment should satisfy the following equation to maximize the lifting line in the stress-free state.

$$[C]* \begin{bmatrix} t_1 \\ t_2 \\ \vdots \\ t_n \end{bmatrix} = \{Z_i^1\} - \{Z_i^2\} \tag{40.5}$$

When construction control is used to control displacement values in the feasible domain, Z should satisfy:

$$\{\lambda_i^1\} \leq \{Z_i^1\} - \{Z_i^2\} \leq \{\lambda_i^2\} \tag{40.6}$$

where λ_i^1 is the lower limit of the feasible domain of displacement values, λ_i^2 is the upper limit of the feasible domain of displacement values, and the influence matrix under the current displacement values is obtained using the above equation, followed by the cable force increment under the current displacement of all nodes.

During the cable adjustment process, the difference between the displacement of the control point and the theoretically required displacement should be made to satisfy Eq. (40.7).

$$\min f = \sum_{i=1}^{n} \left(z^1{}_i - z^2{}_i\right)^2 \tag{40.7}$$

Using point cloud data analysis, the line deviation determines the displacement increment for buckling. Buckles interact at each stage in the system. Ensuring stiffness and stability, on-site cable adjustments require calculating cable force increment with finite element modeling for the current construction stage.

40.3.3.4 Closure of Stiff Skeleton

Skeleton closure is critical for the completed bridge alignment. If stress-free lifting alignment control isn't met, the jointed section length is typically adjusted on-site, making construction harder and causing alignment errors. Ensuring no deviation in the joint section fabrication and installing the cantilever in a stress-free state allows smooth jointing and stress-free lifting of the entire rigid skeleton.

Utilizing 3D scanning technology, point cloud data are gathered for each section before lifting and pre-assembly completion, determining the feasibility of successful skeleton joining. Cable force adjustments are made based on end deviation values, ensuring a seamless arch bridge closure.

Assuming that the deviations of the jointing opening in the stiff skeleton's maximum cantilever state are dx_i and ry_i, the influence matrix of the unit variation of the initial tensile force on the deviations of the jointing opening is C. The optimization of the cable force can be attributed to the cable force adjustment, and the specific adjustment methods are as follows:

$$[C]\begin{Bmatrix} T_1 \\ T_2 \\ \vdots \\ T_n \end{Bmatrix} = \begin{Bmatrix} dx_1 \\ ry_1 \\ \vdots \\ dx_i \\ ry_i \end{Bmatrix} \tag{40.8}$$

where $T_1, T_2, \ldots, T_n$ denotes the nth buckler's unit change.

Control of the jointed end deviation from the embedded section is crucial for smooth installation and achieving designed alignment in the completed bridge. Hinge end deviation is derived from assembly results, while distance deviation between cantilever and hinge section ports is obtained through point cloud data–based hinge port plane fitting, utilizing the plane's central coordinate point (Figure 40.3c).

The objective function is the minimum sum of squared deviations or the minimum maximum deviation to achieve collocation with the optimal attitude. The following attitude adjustment equation produces a better collocation attitude and optimal results:

$$R_{xyz} = R_{\alpha} R_{\beta} R_{\gamma} T_{xyz} \tag{40.9}$$

where α, β, and γ are the rotation angles around the X, Y, and Z axes during the merging section's attitude adjustment, and x, y, and z are the translation distances along the X, Y, and Z directions. The following is the splicing error function:

$$f(\alpha, \beta, \gamma, x, y, z) = \sum_{i=1}^{n} \sqrt{(a_1 - b_1)^2 + (a_2 - b_2)^2 + \cdots (a_n - b_n)^2} \tag{40.10}$$

where a_n, b_n is the corresponding coordinate point after splicing, if there is a deviation in the splice regardless of the splice's attitude, the cable force must be adjusted to compensate for the deviation.

40.4 Validation and Discussion

40.4.1 Data Collection

This chapter is based on a 155 m main span upper-bearing stiff skeleton arch bridge. The bridge has 11 lifting sections, with a maximum lifting weight of 120.5t. The skeleton's steel pipe material is Q235, and the bridge is built with cable-lift diagonal suspension.

For the stiff skeleton from the prefabricated plant, the instrument-target distance should not exceed 20 m. To enable point cloud data stitching and feature extraction, at least three target spheres are used as reference features between data from two stitched stations.

The collected point clouds attempt to cover all the stiff skeletons of the arch bridge for the stiff skeleton lifting construction. Due to the distance between the two banks where the arch bridge is located, the complete data cannot be obtained by single side scanning and must be collected separately on both banks.

40.4.2 Analysis and Results

The stiff skeleton point cloud data was preliminarily aligned and then fine-aligned using an improved ICP method. The deviations of the prefabricated plant's fine-aligned stiff skeleton point cloud were compared, resulting in average deviations of 0.00043, 0.0004, and 0.00017 m in the X, Y, and Z directions, respectively.

In the construction stage, the same method is used to finely align the point cloud data. The result shows that the deviation of the corresponding planes of the bridge pier after fine alignment is less than 1 mm.

The alignment accuracy of the point cloud reaches the millimeter level, and the deviation amount is basically within 1 mm. It shows that the method's alignment accuracy can achieve high-precision alignment of the point cloud, which can be used for later deformation analysis.

The extracted stiff skeleton center curve is used to generate the solid stiff skeleton based on the extracted stiff skeleton center curve, and the deviation of the solid stiff skeleton from the point cloud after the inverse is analyzed.

The accuracy of the stiff skeleton's centerline obtained by this method is good and can meet the requirements of use. The deviations of the coordinates fit by this algorithm are all in the millimeter range. The stiff skeleton line obtained by virtual assembling is the actual line of the stress-free state method, which should be theoretically consistent with the manufactured line.

Using the stress-free state method for lifting control in this case, the pre-lift value of each stage guides site construction. The lifting process must ensure cable lifting meets the required elevation.

The lifting process has a significant impact on the alignment of the stiff skeleton, particularly due to the buckling rope's effect on the main arch ring. To study this, the upper main chords of a specific section were examined, and the actual and theoretical alignments were compared after accounting for lifting system conversion.

Deviations in position points were analyzed by comparing measured chord lines to theoretical lines using 3D scanning. The actual and theoretical alignments were found to be similar in trend, with minor deviations in the main chords. After completing a hoisting stage, the extracted line was compared to the theoretical one, revealing gradually increasing deviations. A displacement matrix was obtained based on control points in the deviation line, and the load coefficient and influence matrix of cable force were determined by applying unit force to the buckling cable. This allows the deviation of the downline used to calculate the cable force increment..

The calculated cable force values are analyzed in the model, and the construction is guided by the pre-lift values obtained from the later analysis.

The alignment was adjusted on-site based on the obtained pre-lift value and buckling force, and the results were obtained by re-comparing the theoretical alignment with the measured alignment.

The overall line deviation is obviously reduced after the cable force is adjusted, and the deviation of each point is unevenly distributed due to the processing error of the main chord, load distribution, and point cloud error, but the deviation is small and fluctuates within millimeters, and the overall effect is good.

This method is used to adjust the cable force in order to improve the accuracy of the influence matrix and bring the obtained cable force closer to the actual one. The adjusted stiff skeleton line is closer to the theoretical line with the deviation only in millimeters, indicating that the stiff skeleton line can be hoisted for good control by using this method.

To obtain the splicing results of the ports, extract the port plane coordinates of the maximum cantilever section and the port plane coordinates of the merging section and perform the optimal attitude alignment by the coordinates.

According to the assembling results, the lower chord's end deviation in the Y direction is large, while the upper chord's end deviation is small, with the end deviation of 1, 2, 3 and 7, 8, 9 being between 1 and 2 cm and the end deviation of 4, 5, 6 and 10, 11, 12 being between 2 and 3 cm.

All stiff skeleton point cloud sections can achieve the requirement of stress-free alignment using the previous virtual assembly method. The manufacturing deviation of the jointed section is essentially unaffected, indicating that the deviation of the port is caused by the stability of the diagonal tension buckling system, construction load, and temperature during the construction process. So, the deviation of the jointed port can be eliminated by adjusting the stiff skeleton alignment to achieve smooth stiff skeleton jointing.

If the rigid skeleton lifting process ensures that the rigid skeleton alignment meets the stress-free alignment, the final alignment of the bridge will be consistent with the design alignment, according to the stress-free state method. As a result, the buckling cable force can be adjusted prior to the jointing to ensure a smooth installation of the jointed section. Because the actual manufacturing line of the stiff skeleton section meets the requirement of stress-free line, and the entire lifting is controlled according to the stress-free line, the jointing requirement of stress-free length and curvature can be met as long as the horizontal and vertical deviation of the jointed end is zero, and the cable force adjustment model is shown in Figure 40.4a.

The influence matrices C_1 and C_2 for the upstream and downstream under this merging deviation by adjusting the six cable forces at the cantilever end using the method described in the previous section.

Field adjustments were made based on the deviation results and calculation results of the cable force to obtain the actual deviation value of the adjusted hinge opening, as shown in Figure 40.4a.

Figure 40.4 shows that the deviation of the adjusted joint port has been reduced to within 1–2 mm. This method can predict whether the dimensional deviation of the joint section meets the requirements of the joint in advance. On the one hand, the joint port can be adjusted according to the deviation value to meet the requirements of the joint, while on the other hand, it can improve efficiency and reduce construction difficulty.

The results shown in Figure 40.4b were obtained by comparing the theoretical and actual alignment of the completed bridge by studying the three lower chord rod stiffening skeletons at the bottom of the arches.

According to Figure 40.4c, the deviation of the bare arch alignment in the cross-bridge direction after the bridge is completed is 2.5 mm at most, indicating anti-symmetry; the deviation of the alignment in the elevation direction is 1.4 cm at most, indicating symmetry, and the maximum displacement difference accounts for 7/8,000 of the calculated span, which is within the specification requirement. The deviation of the three main chords is nearly identical, and the overall deviation is small, indicating that the line control effect is effective.

40.4.3 Discussion

A DTM detects SSRS section processing accuracy and installation quality using key data from: (1) stiff skeleton point cloud collocation station accuracy; (2) algorithm extraction accuracy for stiff skeleton features (bolt holes, alignment); (3) incomplete flange fit impact on alignment in virtual pre-assembly;

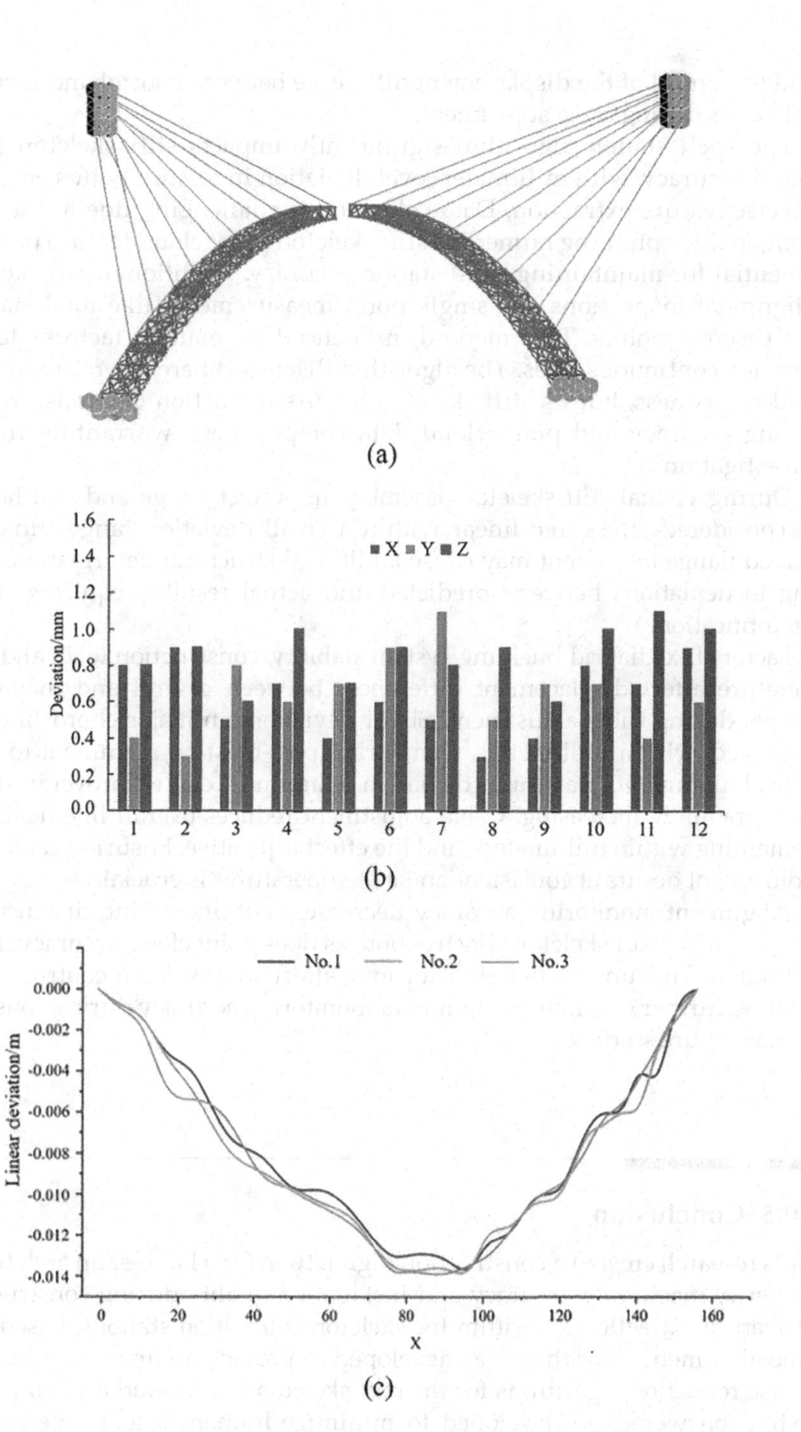

FIGURE 40.4
Data results: (a) combined lifting rigging force adjustment model; (b) adjusted deviation of hinge opening; and (c) deviation of main chord elevation line after completion of the bridge.

and (4) control of the displacement difference between control and theoretical points during cable adjustment.

The spell station algorithm significantly impacts stiff skeleton point cloud accuracy, with millimeter-level deviation in feature points ensuring precise feature extraction. Data collection is challenging due to the site's complexity; obtaining numerous stiff skeleton point cloud feature points is essential for maintaining spell station accuracy. Traditional stiff skeleton alignment inspections use single point measurements like total stations for discrete points. This method, influenced by manual factors, doesn't provide continuous lines. The algorithm discussed here offers automation and robustness, but its stiff skeleton feature extraction depends on both fitting accuracy and point cloud data completeness, warranting further investigation.

During virtual stiff skeleton assembly, imperfect flange and bolt hole fit is considered stress-free linear within a small deviation range. However, forced flange alignment may cause additional structural deformation, leading to deviations between predicted and actual results, requiring future quantification.

Factors like diagrid buckling system stability, construction load, and temperature affect displacement differences between control and theoretical points during cable adjustment. Slight deviations in major chord lines are observed, within millimeters. Comparing post-hoisting alignment to theoretical alignment, maximum deviation is around 2 cm, with overall deviation gradually increasing. Cable adjustment reduces overall line deviation, remaining within millimeters, and the effect is positive. Ensuring cable force adjustment occurs at consistent on-site temperatures is crucial.

Alignment monitoring accuracy decreases with increasing distance and angle during stiff skeleton construction, as does point cloud accuracy due to vibration. This unconsidered factor may affect stiff skeleton control, necessitating further research on alignment monitoring accuracy during construction in future studies.

40.5 Conclusion

This research created a construction digital twin for a large-span arch bridge to verify machining accuracy and installation quality during construction. A feature extraction algorithm for skeleton collocation stations, based on a fine alignment algorithm, was developed to provide accurate 3D geometry. Noise reduction algorithms for the stiff skeleton point cloud and alignment extraction were also developed to minimize human factors' interference. The virtual pre-assembly technique was used to simulate the assembly of adjacent stiff skeletons, with the developed GDTM accurately predicting installation positions and quality. Post-closure, the arch bridge nears design alignment, with improved alignment control compared to prior methods. Future research will develop construction digital twins with semantic info to simulate operational bridge behavior based on its digital twin.

References

Abdel-Qader, I., O. Abudayyeh et al. 2003. "Analysis of Edge-Detection Techniques for Crack Identification in Bridges." *Journal of Computing in Civil Engineering* 17 (4): 255–63. https://doi.org/10.1061/(ASCE)0887-3801(2003)17:4(255).

Andreotti, C., D. Liberatore, and L. Sorrentino. 2015. "Identifying Seismic Local Collapse Mechanisms in Unreinforced Masonry Buildings through 3D Laser Scanning." *Key Engineering Materials*, 628: 79–84.

Bosché, F., and E. Guenet. 2014. "Automating Surface Flatness Control Using Terrestrial Laser Scanning and Building Information Models." *Automation in Construction* 44 (aug.): 212–26. https://doi.org/10.1016/j.autcon.2014.03.028.

Case, F., A. Beinat, F. Crosilla et al. 2014. "Virtual Trial Assembly of a Complex Steel Structure by Generalized Procrustes Analysis Techniques." *Automation in Construction* 37 (jan.): 155–65. https://doi.org/10.1016/j.autcon.2013.10.013.

Chen, L., Xie, X., Lu, Q., Parlikad, A. K., Pitt, M., Yang, J. 2021. "Gemini Principles-Based Digital Twin Maturity Model for Asset Management." *Sustainability* 13: 8224. https://doi.org/10.3390/su13158224Eastman.

Costin, A., A. Adibfar et al. 2018. "Building Information Modeling (BIM) for Transportation Infrastructure - Literature Review, Applications, Challenges, and Recommendations." *Automation in Construction* 94 (OCT.): 257–81. https://doi.org/10.1016/j.autcon.2018.07.001.

Eastman, C. M. et al. 2011. *BIM Handbook: A Guide to Building Information Modeling for Owners, Managers, Designers, Engineers and Contractors*. John Wiley & Sons. DOI:10.1002/9781119287568

Kashani, M. M., A. J. Crewe et al. 2013. "Use of a 3D Optical Measurement Technique for Stochastic Corrosion Pattern Analysis of Reinforcing Bars Subjected to Accelerated Corrosion." *Corrosion Science* 73 (aug.): 208–21. https://doi.org/10.1016/j.corsci.2013.03.037.

Kim, D., Y. Kwak, and H. Sohn. 2020. "Accelerated Cable-Stayed Bridge Construction Using Terrestrial Laser Scanning." *Automation in Construction* 117: 103269. https://doi.org/10.1016/j.autcon.2020.103269.

Kim, S., S. Kim, and D. E. Lee. 2020. "Sustainable Application of Hybrid Point Cloud and BIM Method for Tracking Construction Progress." *Sustainability* 12: 4106.

Lu, R., and I. Brilakis. 2019. "Digital Twinning of Existing Reinforced Concrete Bridges from Labelled Point Clusters." *Automation in Construction* 105 (SEP.): 102837.1–102837.16.

Lubowiecka, I., J. Armesto, P. Arias et al. 2009. "Historic Bridge Modelling Using Laser Scanning, Ground Penetrating Radar and Finite Element Methods in the Context of Structural Dynamics." *Engineering Structures* 31 (11): 2667–76. https://doi.org/10.1016/j.engstruct.2009.06.018.

Mohammadi, M., M. Rashidi et al. 2021. "Quality Evaluation of Digital Twins Generated Based on UAV Photogrammetry and TLS: Bridge Case Study." *Remote Sensing* 13 (17): 3499. https://doi.org/10.3390/rs13173499.

Puri, N., and Y. Turkan. 2020. "Bridge Construction Progress Monitoring Using Lidar and 4D Design Models." *Automation in Construction* 109 (Jan.): 102961.1–102961.15.

Qin, Y., W. Shi, M. Xiao et al. 2019. "Quality control of bridge steel component engineering based on BIM+3D laser scanning technology." *Journal of Civil Engineering and Management* 36 (4): 7. https://doi.org/10.3969/j.issn.2095-0985.2019.04.019

Wan, C., P. H. Chen, and R. L. K. Tiong. 2004. "Assessment of IFCS for Structural Analysis Domain." *Journal of Information Technology in Construction (ITcon)* 9(5): 75–95.

Wang, Z. 2018. "Research on the mechanical performance of the main arch of the cable-hoisted concrete arch bridge." PhD Thesis, Chongqing Jiaotong University.

Yiyang, Z. 2014. "The Design of Glass Crack Detection System Based on Image Preprocessing Technology." In *2014 IEEE 7th Joint International Information Technology and Artificial Intelligence Conference*, Chongqing, China, 39–42. IEEE.

Yoon, S., W. Qian, and H. Sohn. 2018. "Optimal Placement of Precast Bridge Deck Slabs with Respect to Precast Girders Using 3D Laser Scanning." *Automation in Construction* 86 (FEB.): 81–98. https://doi.org/10.1016/j.autcon.2017.11.004.

Zhang, H. 2007. "Research on simulation repair and reconstruction of body surface organs based on reverse engineering technology. " PhD Thesis, Peking Union Medical College.

Zhou, Y., D. Han, K. Hu et al. 2021. "Accurate Virtual Trial Assembly Method of Prefabricated Steel Components Using Terrestrial Laser Scanning." *Advances in Civil Engineering* 2021. https://doi.org/10.1155/2021/9916859.

41

Urban-Scale Digital Twins and Sustainable Environmental Design: Mobility Justice and Big Data

Marianna Charitonidou
Athens School of Fine Arts

41.1 Introduction: How Urban-Scale Digital Twins Can Enhance Sustainable Environmental Design

This chapter examines the critiques of 'digital universalism', reflecting upon the role of urban-scale digital twins in data-driven decision-making concerning urban policies and urban planning. According to Gerhard Schrotter and Christian Hürzeler, an urban-scale digital twin can contribute to the "lifecycle management of the individual components as well as the entire data inventory".[1] Manuel Castells' theory concerning what he calls "network society" is useful for comprehending how big data can be used for urban governance.[2] Moreover, at the core of this chapter is the exploration of how urban-scale digital twins can help us use big data to enhance social advocacy. Incorporating urban-scale digital twins in the decision-making processes concerning urban planning, urban planners can shape new participatory design methods. However, they should not neglect or underestimate the risks of 'digital universalism'.[3] As Giorgio Caprari remarks in "Digital Twin for Urban Planning in the Green Deal Era: A State of the Art and Future Perspectives", the European Union has set the following goals regarding sustainable urban planning strategies: firstly, the empowerment of urban actors towards common goals; secondly, the development of people-oriented urban planning strategies that aim to contribute to the social equity of communities; thirdly, the development of digital platforms and other digital tools that intend to enhance interactive and proactive approaches in urban planning decision-making, and "the creation of integrated, open, and functional technological infrastructures for the development of programmes and the provision of services (data-driven planning)".[4]

DOI: 10.1201/9781003425724-48

To realise the central role of Europe within the framework of the endeavours to incorporate urban-scale digital twins in decision-making concerning urban planning, we should take into account the fact that "Europe is emerging as the main centre of development of urban digital twins, with over 60% of the existing"[5] urban-scale digital twins. As Jaume Ferré-Bigorra, Miquel Casals and Marta Gangolells remark in "The adoption of urban digital twins", among existing urban-scale digital twins that are either in operation or under development are the twins of the following cities or districts: that of Athens in Greece, that of Plzeň in the Czech Republic, that of Dublin Docklands in Ireland, that of Herrenberg in Germany, that of Vienna in Austria, that of Zurich in Switzerland, that of New York in United States of America, that of London in the United Kingdom, and that of Helsinki in Finland.[6] Other cases of urban-scale digital twins are those of Cambridge, Gothenburg, Munich, Newcastle, Paris, Rennes and Rotterdam.[7]

Two programmes that play a major role in shaping sustainable urban planning methods are the European New Green Deal, the Agenda for Sustainable Development and its Sustainable Development Goals, which is also known as SDGs. The former – the European Green Deal – is based on the intention to achieve zero net emissions by 2050. Moreover, this programme places particular emphasis "on achieving a circular economy by 2050, creating a sustainable food system and protecting biodiversity and pollinators".[8] As John Hatcher remarks, in "Digital twins can help sustainability", which was published in June 2022 in *Smart Building Magazine*, "60% of organizations across major sectors are leaning on digital twins as a catalyst [...] to fulfil their sustainability agenda".[9] Hatcher also highlights, in the same article, that "digital twin implementations are set to increase by 36% on average over the next five years".[10] Giorgio Caprari, in "Digital Twin for Urban Planning the Green Deal Era: A State of the Art and Future Perspectives", discerns the following main characteristics of urban-scale digital twins: firstly, their "scalability"; secondly, their "predictability", which becomes possible thanks to the use of simulation algorithms; and thirdly, their capacity to integrate new elements thanks to the use of IoT sensors, and data updated concerning *in situ* real-time data, and, finally, their capacity to enhance cooperation due to the fact that they can be broadly accessible. Caprari also underscores the fact that citizens can download and upload data enhancing in this way social equity and participatory design methods.[11] Gordon S. Blair, in "Save Share Reprints Request Digital twins of the natural environment", distinguishes three challenges concerning the creation of digital twins: firstly, the challenge of "bringing the environmental assets together in one logical place, including both data assets and modelling assets"; secondly, the challenge of allowing different assets to work together as part of a larger digital twin architecture", and, thirdly, the challenge of ensuring "that the necessary storage and processing capacity is available when it is needed, especially given the sizes of the challenges and the associated potentially very large datasets".[12]

To enhance social equity when we introduce big data in urban planning decision-making, it is pivotal to bear in mind that thinking locally means thinking critically.[13] This goes hand in hand with the recognition of the significance of shaping approaches that aim to enable us to reveal the specificities and implications of the local contexts in which data are created. Some key questions in the field of critical data studies that are connected to the issues addressed in the paper are the following: How big data are collected? Which is the impact of the local conditions of the collection and creation of data on research methods? To what extent sense data are operational part of economic systems? Which social groups take advantage of the creation of big data?[14] The point of departure of this paper is the intention to investigate how data are collected and instrumentalised when urban-scale digital twins are used for urban planning decision-making. Useful for responding to these questions is Christine L. Borgman's remark that "entities become data only when someone uses them as evidence of a phenomenon, and the same entities can be evidence of multiple phenomena".[15]

Of great importance for the reflections developed here is Manuel Castells's theory, which would contribute to a better understanding of the relationship between big data and urban planning in a data-driven society and the new kind of temporality in the so-called 'network society'.[16] Michael Batty and Castells' work are related to the transition from spatial perspectives concerning the investigation of urban data to topological perspectives.[17] Dietmar Offenhuber and Carlo Ratti, in *Decoding the City: Urbanism in the Age of Big Data*, mention that the "term big data refers to the availability of massive amount of machine-readable information".[18] Yanni Alexander Loukissas departs from the following principles: all data are local; data have complex attachments to place; data are collected from heterogeneous sources; data and algorithms are inextricably entangled; interfaces recontextualise data; and data are indexes to local knowledge.[19] A notion that is of great significance for this paper is that of 'local reading'.[20] Digital twins enhance evidence-based operational decisions and experimentation on urban policies. The current state of research concerning the role of digital twins in shaping urban policies is characterised by a dichotomy between scholars that focus on the technological and sustainable benefits of the use of urban-scale digital twins and researchers that criticise 'digital universalism'. This chapter intends to challenge this dichotomy, shaping methods based on a socio-technical perspective of using urban-scale digital twins, and combining the technical, sustainable and social advantages of their use.

Criticising 'digital universalism' goes hand in hand with realising that the creation of digital twins is based on the use of a limited set of variables and processes. The myth of 'digital universalism' is based on the belief that "once online, all users could be granted the same agencies on a single network, all differences could dissolve, and everyone could be treated alike".[21] As Loukissas highlights, in *All Data Are Local: Thinking Critically in a Data-Driven Society*, "[i]f left unchallenged, digital universalism could become a new kind of colonialism in which practitioners at the 'periphery' are made to conform to the expectations of a dominant technological culture".[22] According to Loukissas, "[a]spiring to the ideology of big data means seeking to collect everything on a subject, downplaying the importance of data's origins, and assuming that data alone can entirely supplant other ways of knowing".[23] Stefania Milan and Emiliano Treré call for a "de-Westernization of critical data studies".[24]

An important shift within the field of smart cities is the shift from technical to socio-technical perspectives.[25] This shift is related to the idea that the concept of smart city should be related to the endeavour to reveal "multiple dimensions beyond an infrastructure-technology focus".[26] Useful for understanding that the evangelism that accompanies the discourse around smart cities is not something new, but has a long history is the remark of Benjamin H. Bratton that "[w]ell before smart cities evangelism, the modernist call for a more intense technologization of design's disciplinary doxa, blending urban and cybernetic programs, was a predominant discourse".[27] According to Stefania Milan and Emiliano Treré, the myth of 'data universalism' refers to "the tendency to assimilate the cultural diversity of technological developments in the Global South to Silicon Valley's principles". Milan and Treré criticise the "hyperbolic narratives of the 'big data revolution'", arguing that "the main problem with data universalism is that it is asocial and ahistorical, presenting technology [...] as something operating outside of history and of specific sociopolitical, cultural, and economic contexts".[28] A key question concerning the myth of "data universalism" is the following: "how does datafication unfold in countries with fragile democracies, flimsy economies, impending poverty?" The uneven access to the technologies and data that make possible smart cities and urban-scale digital twins should be seriously taken into account if we wish to go beyond the myth of 'data universalism'. On the unevenness of technology and data, Simon Joss, Frans Sengers, Daan Schraven, Federico Caprotti, and Youri Dayot have shed light on the "competitive dynamics created between world cities posited as 'model' smart cities and various second- and third-tier 'follower' cities".[29]

41.2 Urban-Scale Digital Twins and Socio-Technical Perspectives

The term 'digital twin' refers to the digital representation enabling comprehensive data exchange and can contain models, simulations and algorithms describing their counterpart and its features and behaviour in the real world. A 'digital twin' is a digital representation of a physical process, person, place, system or device. The term 'digital twin' firstly emerged in the field of manufacturing sector to refer to digital simulation models that run alongside real-time processes. 'Digital twins' are digital replicas of physical entities. Their creation is based on the use of advanced technological applications, such as sensing, processing and data transmission. Digital twins are used in the field of urban analytics, as well as in the field of computational social sciences. ABI Research forecasts that urban digital twin deployments will exceed 500 by 2025.[30] According to Michael Batty, "[t]he idea of the digital twin [...] has emerged from the representation of the city in terms of its physical assets".[31] The digital twins are able to get updated following the changes of the physical equivalents thanks to the pairing between the virtual and the physical world. To understand what is the main idea behind the creation of digital twins, we should bear in mind that "[a]n ideal digital twin would be identical to its physical counter-part and have a complete, real-time dataset of all information on the object/system".[32]

Recently, within the domain of urban planning and, more particularly, within the field of smart cities, the notion of urban-scale digital twin has acquired a central place. Li Deren, Yu Wenbo and Shao Zhenfeng define the 'digital twin' as a "simulation process that makes full use of physical models, sensors, historical data of operation, etc., to integrate information of multi-discipline, multi-physical quantities, multi-scale, and multi-probability". They also highlight the fact that the current debates concerning the notion of digital twin are characterised by plurality of how this concept is understood. They remark that "a consensus definition has not yet been formed". The common denominator of the different definitions of the term is the shared interest in the "bi-directional mapping relationship that exists between physical space and virtual space". The creation of digital twins is based on the intention to establish "real-time connection[s] between the virtual and the real". In the case of digital twins, the digital models, apart from "observing, recognizing, and understanding"[33] the physical world, they also aim to control and transform it. Martin Mayfield has emphasised the role of urban-scale digital twins in providing a holistic approach to urban and infrastructure design.[34] Anah Boyd and Kate Crawford, in "Critical Questions for Big Data: Provocations for a cultural, technological, and scholarly phenomenon", analyse critically the role of big data within the current cultural and technological context of data-driven societies.[35] Li Deren, Yu Wenbo, Shao Zhenfeng argue that at the core of the development of urban-scale digital twins is the creation of "a complex giant system between the physical world and the virtual space that can map each other and interact with each other in both directions".[36] They also underscore that the continuous generation of massive urban big data and the use of sensors within the cities for which the digital twins are created are necessary for the construction of urban-scale digital twins.

41.3 How 'Digital Twins' Affect Urban Planning Decision-Making

Among the challenges of data-driven approaches are the measurement errors, the biases, the existence of false positives and false negatives, the undesired discrimination effects, the complexity, the network effects, the non-linear dynamics, the wicked problems and an ensemble of convergence issues. Apart from the aforementioned issues, a problem that should be highlighted is the fact that, in general, the digital twin approaches have been largely ignorant of people and what relates to them. This means that the ways in which urban-scale digital twins function often neglect the importance of social interactions, competition and cooperation, social norms, laws and regulations, culture, history, politics, democracy, human rights, ethics, and essential non-material qualities. It is, therefore, indispensable to develop approaches that aim to incorporate questions related to the aforementioned aspects in the ways in which urban-scale digital twins are created and used. The fact that the role of urban-scale digital twins in the decision-making processes concerning urban planning will become even more important during the next years makes the incorporation of aspects related to democracy, human dignity, and solidarity in how the urban-scale digital twins function even more necessary. Clare Wildfire distinguishes two categories of benefits of the city-scale digital twins: the reactive benefits, on the one hand, and the predictive benefits, on the other. Wildfire relates the first category of benefits to the capacity of enhancing "real-time or near real-time interventions and improve the smooth day-to-day running of the city or asset", and the second category to the use of data for the improvement of "longer-term scenario planning to steer appropriate (and equitable) investment decisions".[37] Li Deren, Yu Wenbo, and Shao Zhenfeng analyse the application "Smart City Traffic Brain", which is based on the use of digital twins and collects the big data concerning travel trajectories and "real-time dynamic traffic information".[38] Michael Batty has remarked that "one of the quests in city modelling is to merge social and economic processes with the built environment and to link functional and physical processes to socio-economic representations".[39] The shift from technical to socio-technical perspectives goes hand in hand with the effort to construct urban-scale digital twins that aim to "reflect the specifics of the urban and socio-political context".[40] According to Martin Tomko and Stephan Winter, "[t]he term 'digital twin' has been applied to representations of buildings and aggregations thereof such as precincts or entire cities – as long as these representations preserved aspects of temporal dynamics and self-updating ('4D')". Tomko and Winter have criticised the term 'digital twin'.[41] Their critique departs from Batty's remark that "a computer model of a physical system can never be the basis of a digital twin [i.e., 'mirror'] for many elements of the real system are ignored in any such abstraction".[42] Tomko and Winter, in contrast with Batty, argue that the notion of 'digital twin' should be replaced. They suggest that the notion that should replace the term 'digital twin' should be 'cyber–physical–social system with coupled properties'. They claim that this shift in the description of this phenomenon goes hand in hand with a recognition of the fact that digital models do not function exclusively as a "passive reflection of a mirror", but most importantly as systems serving to establish methods of action. They also mention that "[t]he coupling also implies that the system to describe is not a purely digital one".[43]

Martin Tomko and Stephan Winter, to render explicit why the term "digital twin" is problematic, remark that "the 'digital twin' is embodied and immersed in what it is supposed to mirror, and thus is no longer an independent representation". To understand how "digital twins" can affect urban planning methods we could bring to mind that "[t]he digital side of [the] [...] coupled system, however, can react, predict, and act". An example of how they can serve for predicting and acting is their use for "controlling the traffic lights according to traffic, guide by digital signage". As Tomko and Winter highlight, "the digital side of the coupled systems (the 'digital twin') morphs into the physical environment by communication and control, a phenomenon studied by cybernetics". What makes 'digital twins' operative is the "bi-directional coupling between the physical artefact and their digital counterpart",[44] as well as the "bi-directional coupling across the digital, physical, and social spheres".[45] Tomko and Winter, to render explicit the importance of "bi-directional coupling", use the term 'coupled ecosystem' and 'cyber–physical–social ecosystem' to refer to the coupling of the physical and the digital system. The bi-directional coupling between the real and the digital artefacts becomes possible thanks to the use of "snapshot[s] of the current or past representations", which serve "to predict by extrapolation".[46]

A worth-mentioning urban-scale digital twin is that of London, which was created by AccuCities. The urban-scale digital twin of London is accessible through the application named Plan.City. The datasets of the aforementioned urban-scale digital twin of London are publicly available. The users of Plan.City can build a maximum building envelope using a 3D Builder Tool. Unreal Engine 4 (UE4) is used to assist decision-making concerning the urban environment of London[47] (Figures 41.1–41.3). In July 2022, the Barcelona Supercomputing Center – Centro Nacional de Supercomputación (BSC-CNS), the CINECA Consortium of Universities, the University of Bologna, the city council of Barcelona and the city council of Bologna signed an agreement concerning the creation of a project focusing on urban governance. This project is based on the development of urban-scale digital twin applications. At the core of this collaborative project between Bologna and Barcelona is the intention to shape evidence-based decision-making strategies concerning not only public policies, but also impact assessment. This collaborative project aims to place particular emphasis on developing urban governance methods concerning parameters related to urban mobility and

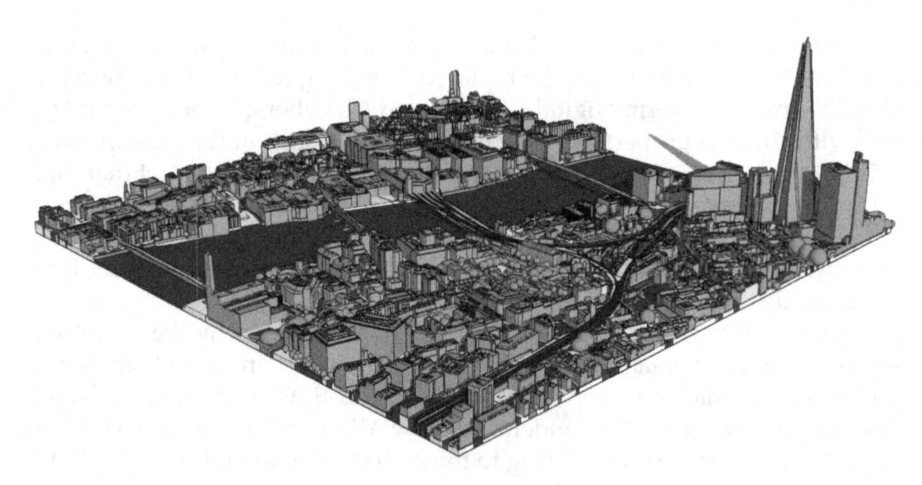

FIGURE 41.1
Image from 3D London | AccuCities. Licensed under a Creative Commons Attribution-ShareAlike 4.0 International License. (https://www.accucities.com/product/tq3280-free-3d-london-samples/ (accessed 7 January 2023))

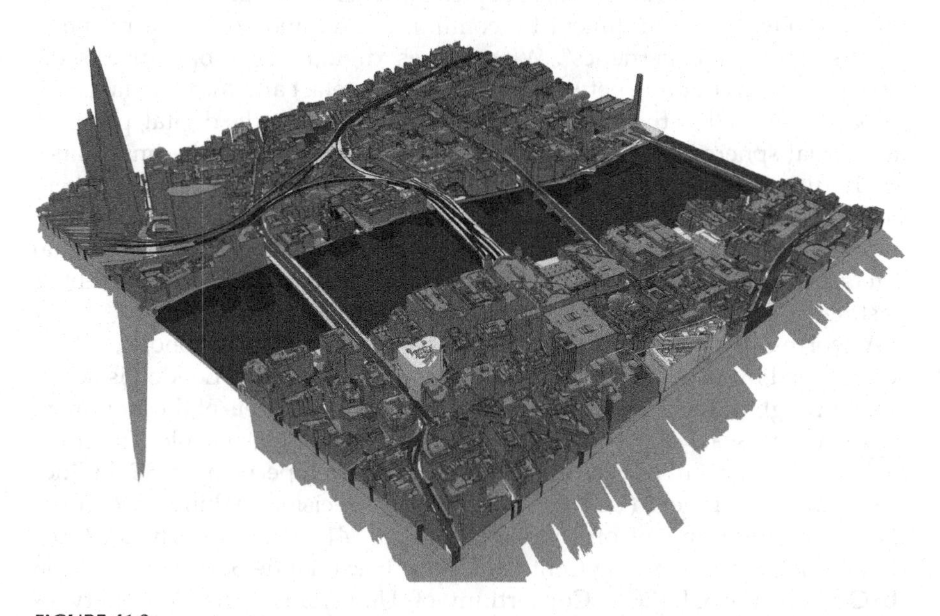

FIGURE 41.2
Image from 3D London | AccuCities. Licensed under a Creative Commons Attribution-ShareAlike 4.0 International License. (https://www.accucities.com/product/tq3280-free-3d-london-samples/ (accessed 7 January 2023))

sustainable environmental design. More specifically, special attention will be paid to urban planning policies intending to eliminate greenhouse-gas emissions. Moreover, this project aims at the development of urban-scale digital twins to be used by the municipality of Bologna as well as the municipality

FIGURE 41.3
Towers of the Kalasatama Center from the northwest. The Figure 41.3 shows a sectional view of the airflow coming from the right edge (south) of the image at a height of about 50 m. The wind weakens as it hits the tower blocks and causes slow (black) airflows. Whirlwind phenomena occur next to the tower buildings. (https://www.hel.fi/static/liitteet-2019/Kaupunginkanslia/Helsinki3D_Kalasatama_Digital_Twins.pdf (accessed 7 January 2023))

of Barcelona for implementing ecological public policy models in both cities. Aitor Hernández-Morales, in his article entitled "Barcelona bets on 'digital twin' as future of city planning", refers to the MareNostrum supercomputer, which is housed in Torre Girona chapel in Barcelona and is among the most powerful data processors internationally.[48] This supercomputer is used for the aforementioned collaborative project between Barcelona and Bologna. He also mentions that this supercomputer is expected to process data with the objective of improving urban policies.

41.4 The Case of the Urban-Scale Digital Twin of Kalasatama District in Helsinki

In "Urban development with dynamic digital twins in Helsinki city", Mervi Hämäläinen examines the case of the digital twin of the Kalasatama district (Figure 41.3). The urban-scale digital twin of Kalasatama district was developed in 2018 and became available to citizens' mobile devices thanks to the browser-based OpenCities Planner application. Citizens were asked to give their feedback through this platform. Among the technologies on which CityPlanner is based are 5G network, and urban model updating through robotics and extended reality. B. Green, which is funded by the Interreg Central Baltic Programme, collaborated with the City of Helsinki's environmental services and Kalasatama's urban planning. Hämäläinen analyses Open Cities Planner, an application that was developed in the framework of this project "to complement and reinforce the usage of the Kalasatama digital twin platform".[49] This application aimed to familiarise citizens with the use of the data of the digital twin and to enhance "participation and interaction among Kalasatama residents". A case in which the aforementioned application was used to promote the participation of citizens is the "public participation GIS (PPGIS) poll" that intended to explore, through the use of the application, what locations the residents of Kalasatama would recommend to the visitors of the district.

Mervi Hämäläinen's remark that "Digital twin platforms were also exploited to integrate citizens into urban development initiatives and activities in Helsinki" makes us wonder to what extent the use of urban-scale digital twins can contribute to establishing participatory design methods. Despite the fact that Hämäläinen argues that the digital twin of Helsinki made "design processes more transparent and open",[50] in many cases the access of the general public to the digital twins is not possible. This does not help to make the design processes more transparent and open. Efforts should be made during the years to come in order to achieve the transparency and openness that Hämäläinen describes. The digital twin of Kalasatama district also aimed to provide a platform to simulate cases related to wind and solar parameters, including data concerning airflows, shadows, solar rays and air pressure. The purpose of these simulations is to provide sustainable urban planning solutions, testing the impact of the aforementioned data on the built artefacts.

41.5 The Case of 'Digital Urban European Twins'

Another interesting case that is based on the use of urban-scale digital twins is the project entitled 'Digital Urban European Twins' (DUET), which is "a cooperative endeavour, involving 15 different partners from across Europe"[51] (Figure 41.4). Another ongoing project that also focuses on the use of digital twins is Low-Emission Adaptive last mile logistics (LEAD).[52] Among the cities that are involved in this project are Madrid, The Hague, Budapest, Lyon, Oslo, Porto, Athens, Antwerp and Pilsen. DUET is based on the intention to experiment with new decision-making processes, using on-demand logistics operations. Through the use of urban-scale digital twins, it explores the potentials of new strategies as far as low-emission logistics operations are concerned. Urban-scale digital twins render possible the integration of data and the modelling of possible strategies, a wide variety of solutions for shared, connected and low-emission logistics operations. Among the

FIGURE 41.4
DUET. Simulation of traffic-induced air pollution in Pilsen. (https://www.digitalurbantwins.com/post/major-cities-of-europe-learn-how-to-build-local-digital-twins-from-duet (accessed 10 October 2022))

parameters that the urban-scale digital twins of the DUET project simulate are those concerning different traffic, air and noise. Particular attention is paid to the exploration of traffic models, including static, dynamic and a local mobility models (Cityflows). In order to simulate air quality data concerning traffic volume, road network, wind speed and wind direction are used. Despite the interest of the project in taking into account how citizens perceive urban landscape, and in trying to render the urban-scale digital twins "citizen-centric", the efforts to "model citizens by looking at their emotional state" entail risks given that they are based on significant abstraction and simplifications. While urban-scale digital twins seem very efficient in simulating parameters related to sustainable environmental design, they are rather problematic when they intend to simulate "how people experience their built environment [in order] [...] to develop smart cities in line with evolving patterns and preferences".[53]

41.6 Towards "Mobility Justice" Perspective: Commoning Practices and Big Data

Manuel Castells's approach is useful for deciphering the tension between the real and the ideal at stake during the process of abstracting sets of variables and processes in the case of urban-scale digital twins. Castells argues that the societal system corresponding to the digital era is based on informationalism and globalism. He also claims that societal processes cannot be understood or represented without the underlying technology. At the core of Castells' *Communication Power* is the following question: "where does power lie in the global network society?".[54] Castells distinguishes four categories of power in the network society: networking power, network power, networked power and network-making power. He argues that the network-making power is the "paramount form of power in the network society",[55] and that the network society is organised around the following three concepts: 'space of flows', 'space of places' and 'timeless time'.[56] Castells, through these concepts, intends to render explicit that the "incorporation of the impact of advanced forms of networked communication"[57] calls for a new understanding of societies. According to Castells, in network society there are no boundaries, and, for this reason, contemporary urbanisation and networking dynamics should be studied conjointly.

A remark of Castells that can help us better understand why 'digital universalism' is not compatible with the intention to challenge inequalities is the following: "the network of decision-making and generation of initiatives, ideas and innovation is a micro network operated by face-to-face communication concentrated in certain places".[58] Castells' informational city emphasizes the significance of the "incessant flows of information, goods, and people".[59] According to Castells, cities should be understood as processes and not as places. His approach can help us better understand how the creation of urban-scale digital twins influences how the public sphere is conceived. A concept that is useful for addressing the issue of unequal access to urban-scale digital twins' data is that of 'mobility justice',[60] that Mimmi Sheller uses to suggest a new way of understanding inequality and uneven accessibility to the mobility commons.[61] This concept is useful for analysing how urban-scale digital twins can be used for traffic simulations. The main idea behind the use of the term 'mobility justice' is the intention to render explicit that while mobility is a fundamental right for everyone, it is experienced unequally along the lines of gender, class, ethnicity, race, religion and age. As Mimmi Sheller argues, in "Mobility justice in urban studies", "a strong theorization of mobility justice is the best way to bridge these various dimensions of urban inequalities".[62] Sheller also highlights the fact that shaping urban planning methods that aim to promote a sustainable future for the cities should go hand in hand with using less destructive modes of urban mobility. To shape approaches that promote the use of big data for urban analytics without neglecting the social aspects involved in the strategies of formation of urban policies, it is important to bear in mind the weaknesses of 'digital universalism' and the assumptions on which the creation of urban-scale digital twins. In order to do so, pivotal is the epistemological shift from technical to socio-technical perspectives.

Of great importance for understanding the connections between the ongoing debates around the concepts of commons and commoning and the history of advocacy planning movement in the late 1960s are the debates concerning the critiques of urban renewal in Philadelphia during the late 1950s. Denise Scott Brown has commented on advocacy planners' critique of urban renewal programme, highlighting that it "derived from the problem that urban renewal had become 'human removal'".[63] She has also "underscored that the main argument of advocacy planners was that architects and urban planners' leadership had diverted urban renewal from a community support to a socially coercive boondoggle".[64] A distinction that is at the core of the debates around participation-oriented strategies is that between the collaborative and the co-production approaches. The distinction between the collaborative and the co-production approaches is examined by Vanessa Watson, in "Co-production and collaboration in planning: The difference", where she remarks that "co-production" and "collaborative or communicative planning", despite their shared concern "with how state and society can engage in order to improve the quality of life of populations [...] with an emphasis on the poor and marginalized", differ in the sense that co-production "works outside (and sometimes against) established rules and procedures of governance in terms of engagement with the state, while this is much less usual (although not impossible) in collaborative and communicative planning processes".[65]

The reflections developed here regarding the role of urban commons in shaping urban planning strategies aimed to render explicit the relevance and importance of establishing methods intending to examine the actual practices of citizens while making decisions. Within the framework if this endeavour to better grasp the role of urban commons in shaping urban planning strategies, a concept that is of great importance is that of negotiated planning, which "focuses less on normative expressions of how planning should be (i.e., informed by evidence and participation) and more on the actual practices evident in cities".[66] As Vanessa Watson has highlighted, 'negotiated planning' strategies should be based on a close analysis of "the difficulties of [...] [the] processes as well as to the range of contexts and conditions within which participation takes place",[67] which would save them (the 'negotiated planning' strategies) from the traps of an idealised image of collaborative planning based on the Habermasian model.[68] The shift from 'collaborative' towards 'negotiated planning' is related to the intensification of the interest not only in the commoning practices, but, most importantly, in the actual "actors and power dynamics [...] involved", as well as in "the 'virtuous cycle' of planning, infrastructure, and land".[69] A characteristic of "negotiated planning" that is note-worthy is the attention it pays to the actual "power-laden compromises, contests [...] among various arms of the state, civil society, and the local and international private sector".[70] More specifically, 'negotiated planning' approaches place particular emphasis on "the actions and agendas of a whole range of stakeholders who together work to conFigure 41 4 fragile system which is constituted through and co-constitutive of each urban context".[71] 'Negotiated planning' strategies in urban planning and architecture are based on the idea that design should be shaped in close dialogue with the actual commoning practices. Within such a perspective, architecture and urban planning act as actors connecting planning, infrastructure and land.[72]

At the core of this paper is the conviction that the ways in which urban commons can inform our understanding of the shared codes and conventions characterising the production of co-housing practices during the process of urban planning. As David Harvey has highlighted, our understanding of commons should go beyond natural resources given that all resources are socially defined.[73] According to John Bingham-Hall, the notion of common "suggests a community of commoners that actively utilise and upkeep whatever it is that is being commoned".[74] In Patrick Bresnihan's perspective, "[t]he noun 'commons' has been expanded into the continuous verb 'commoning' to denote the continuous making and remaking of the commons through shared practice".[75] The role of urban commons in spatial planning decision-making is at the core of the current debates. At the core of the reflections developed in this paper is the exploration of ways in which we could combine the use of advanced digital tools such as digital twins and civic-oriented participatory design methods. However, despite the aspirations of urban-scale digital twins to enhance the participation of citizens in the decision-making processes relayed to urban planning strategies, the fact that they are based on a limited set of variables and processes makes them problematic. During the coming years, given the galloping development of urban-scale digital twins applications globally, it will be of pivotal importance to shape methodological tools offering the possibility to develop new forms of social advocacy around big data, bringing together the reflection on smart cities and the debates around urban commons.

The research project was supported by the Hellenic Foundation for Research and Innovation (H.F.R.I.) under the "3rd Call for H.F.R.I. Research Projects to support Post-Doctoral Researchers" (Project Number: 7833)

Notes

1 G. Schrotter, C. Hürzeler, 2020, 'The digital twin of the city of Zurich for urban planning', *PFG – Journal of Photogrammetry, Remote Sensing and Geoinformation Science*, 88: 99–112. https://doi.org/10.1007/s41064-020-00092-2.

2 M. Castells, 2010, *The Rise of the Network Society. Second edition with a new preface*, Chichester, West Sussex: Wiley-Blackwell.

3 M. Charitonidou, 2021. 'Public spaces in our data-driven society: The myths of digital universalism', in *Deep City – Climate crisis, democracy and the digital. Proceedings of the International Latsis Symposium 2021*, Lausanne, Switzerland: EPFL. https://doi.org/10.3929/ethz-b-000465249.

4 G. Caprari, 2022. 'Digital twin for urban planning in the green deal era: A state of the art and future perspectives', *Sustainability*, 14(10): 6263. https://doi.org/10.3390/su14106263.
5 J. Ferré-Bigorra, M. Casals, M. Gangolells, 2022, 'The adoption of urban digital twins', *Cities*, 131, 103905. https://doi.org/10.1016/j.cities.2022.103905.
6 Ibid.
7 Caprari, 'Digital twin for urban planning in the green deal era: A state of the art and future perspectives'.
8 'EU responses to climate change Society'. Available at: https://www.europarl.europa.eu/news/en/headlines/society/20180703STO07129/eu-responses-to-climate-change (accessed on 1 August 2022).
9 J. Hatcher, 2022, 'Digital twins can help sustainability', *Smart Building Magazine*. Available at: https://smartbuildingsmagazine.com/news/digital-twins-can-help-sustainability.
10 Ibid.
11 Caprari, 'Digital twin for urban planning in the green deal era: A state of the art and future perspectives'.
12 G. S. Blair, 2021, 'Digital twins of the natural environment', *Patterns*, 2(10). https://doi.org/10.1016/j.patter.2021.100359.
13 M. Charitonidou, 2022, 'Urban scale digital twins in data-driven society: Challenging digital universalism in urban planning decision-making', *International Journal of Architectural Computing*, 20(2): 238–253. https://doi.org/10.1177/14780771211070005.
14 Loukissas, Y. A., 2019, *All Data Are Local: Thinking Critically in a Data-Driven Society*, Cambridge, MA: The MIT Press; Charitonidou, 'Urban scale digital twins in data-driven society: Challenging digital universalism in urban planning decision-making'.
15 C. L. Borgman, 2015, *Big Data, Little Data, No Data: Scholarship in the Networked World*, Cambridge, MA: The MIT Press, 28.
16 Castells, *The Rise of the Network Society. Second edition with a new preface*, Castells, 1989, *The Informational City: Information Technology, Economic Restructuring and the Urban-Regional Process*, Oxford: Basil Blackwell; Castells, 2009, *Communication Power*, Oxford: Oxford University Press. Castells, 2010, 'Globalisation, networking, urbanisation: Reflections on the spatial dynamics of the information age', *Urban Studies*, 47(13): 2737–2745. Permanent link: https://www.jstor.org/stable/43079956; Castells, 2021, 'From cities to networks: Power rules', *Journal of Classical Sociology*, 21(3–4): 260–262. https://doi.org/10.1177/1468795X211022054.
17 M. Batty, 2018, 'Digital twins', *Environment and Planning B: Urban Analytics and City Science*, 45(5): https://doi.org/10.1177/2399808318796416.
18 D. Offenhuber, C. Ratti, (eds.), 2014, *Decoding the City: Urbanism in the Age of Big Data*, Basel: Birkhäuser, 7.
19 Loukissas, *All Data Are Local: Thinking Critically in a Data-Driven Society*.
20 Loukissas, Y. A., 2017, 'Taking Big Data apart: Local readings of composite media collections', *Information, Communication & Society*, 20(5): 651–664. https://doi.org/10.1080/1369118X.2016.1211722.
21 A. S. Chan, 2014, *Networking Peripheries Technological Futures and the Myth of Digital Universalism*, Cambridge, MA: The MIT Press, 7.
22 Loukissas, *All Data Are Local: Thinking Critically in a Data-Driven Society*, 10.
23 Ibid., 16.

24 S. Milan, E. Treré, 2019, 'Big Data from the south(s): Beyond data universalism', *Television & New Media*, 20(4), 321. https://doi.org/10.1177/1527476419837739.
25 T. Nochta, L. Wan, J. M. Schooling, A. K. Parlikad, 2021, 'A socio-technical perspective on urban analytics: The case of city- scale digital twins', *Journal of Urban Technology*, 28(1–2): 263–287. https://doi.org/10.1080/10630732.2020.1798177.
26 S. Joss, F. Sengers, D. Schraven, F. Caprotti, Y. Dayot, 2019, 'The smart city as global discourse: Storylines and critical junctures across 27 cities', *Journal of Urban Technology*, 26(1), 24, https://doi.org/10.1080/10630732.2018.1558387.
27 B. H. Bratton, 2016, *The Stack: On Software and Sovereignty*, Cambridge, MA: The MIT Press, 172.
28 Milan, Treré, 'Big Data from the south(s): Beyond data universalism', 324.
29 Joss et al., 'The smart city as global discourse: Storylines and critical junctures across 27 cities'.
30 ABI Research, 2021, 'New Urban Use Cases Drive over 500 Cities to Adopt Digital Twins by 2025', 5 January 2021. Available at: https://www.abiresearch.com/press/new-urban-use-cases-drive-over-500-cities-adopt-digital-twins-2025/ (accessed 3 March 2021).
31 Batty, 'Digital twins', 818.
32 G. White, A. Zink, L. Codecá, S. Clarke, 'A digital twin smart city for citizen feedback', *Cities*, 110 (2021), 103064. https://doi.org/10.1016/j.cities.2020.103064.
33 L. Deren, Y. Wenbo, S. Zhenfeng, 2021, 'Smart city based on digital twins', *Computational Urban Science*, 1(4), 1. https://doi.org/10.1007/s43762-021-00005-y.
34 M. Mayfield, 2020, 'Can urban scale digital twins address climate adaptation?' *Buildings and Cities*, 28 January 2020. Available at: https://www.buildingsandcities.org/insights/commentaries/urban-scale-digital-twins-climate-adaptation.html (accessed 3 April 2021).
35 A. Boyd, K. Crawford, 2012, 'Critical questions for Big Data: Provocations for a cultural, technological, and scholarly phenomenon', *Information, Communication & Society*, 15(5): 662–679. https://doi.org/10.1080/1369118X.2012.678878.
36 Deren, Wenbo, Zhenfeng, 'Smart city based on digital twins', 1.
37 C. Wildfire, 2021, 'How can we spearhead city-scale digital twins?' *Infrastructure Intelligence*, 9 May 2018. Available at: http://www.infrastructure-intelligence.com/article/may-2018/how-can-we-spearhead-city-scale-digital-twins (accessed 3 April 2021).
38 Deren, Wenbo, Zhenfeng, 'Smart city based on digital twins', 6.
39 Batty, 'Digital twins', 819.
40 Nochta, Wan, Schooling, Parlikad, 'A socio-technical perspective on urban analytics: The case of city- scale digital twins', 263.
41 M. Tomko, S. Winter, 2019, 'Beyond digital twins – A commentary', *Environment and Planning B: Urban Analytics and City Science*, 46(2), 396. https://doi.org/10.1177/2399808318816992.
42 Batty, 'Digital twins', 817.
43 Tomko, Winter, 'Beyond digital twins – A commentary', 396.
44 Ibid., 397.
45 Nochta, Wan, Schooling, Parlikad, 'A socio-technical perspective on urban analytics: The case of city-scale digital twins', 268.
46 Tomko, Winter, 'Beyond digital twins – A commentary', 397.
47 https://www.ribacpd.com/accucities/230545/3d-city-models-overview/411006/movie/ (accessed 3 April 2021).

48 A. Hernández-Morales, 2022, 'Barcelona bets on 'digital twin' as future of city planning', *Politico.* Available at: https://www.politico.eu/article/barcelona-digital-twin-future-city-planning/ (accessed 1 June 2022)
49 M. Hämäläinen, 2021, 'Urban development with dynamic digital twins in Helsinki city', *IET Smart Cities.* https://doi.org/10.1049/smc2.12015.
50 Ibid.
51 'Digital urban European twins'. Available at: https://www.digitalurbantwins.com (accessed 3 April 2021).
52 'LEAD Strategies for shared-connected and low-emission logistics operations. Available at: https://www.leadproject.eu (accessed 3 April 2021).
53 L. Raes, P. Michiels, T. Adolphi, C. Tampere, T. Dalianis, S. McAleer, P. Kogut, 2021, 'DUET: A framework for building secure and trusted digital twins of smart cities', *IEEE Internet Computing.* https://doi.org/10.1109/MIC.2021.3060962.
54 Castells, *Communication Power,* 42.
55 Ibid., 47.
56 A. White, 2016, 'Manuel Castells's trilogy the information age: economy, society, and culture, Information', *Communication & Society,* 19(12), 1674. https://doi.org/10.1080/1369118X.2016.1151066.
57 Ibid., 1673–1674.
58 Castells, 'Globalisation, networking, urbanisation: Reflections on the spatial dynamics of the information age', 2742.
59 F. Stalder, 2008, *Manuel Castells. The Theory of the Network Society,* Cambridge, MA: Polity Press, 163.
60 Charitonidou, 2021, 'Mobility and migration as constituting elements of urban society: Migration as a gendered process and how to challenge digital universalism', in *Proceedings of the ACSA/EAAE Teachers Conference. Curriculum for Climate Agency: Design (in)Action.* ACSA/EAAE. https://doi.org/10.3929/ethz-b-000503331.
61 M. Sheller, 2018, *Mobility Justice: The Politics of Movement in an Age of Extremes,* London: Verso.
62 Sheller, 2020, 'Mobility justice in urban studies', in Jensen, O. B. et al, *Handbook of Urban Mobilities,* London; New York: Routledge, 13–22.
63 D. Scott Brown, 2009, 'Towards an 'Active Socioplastics'', in idem., *Architecture words 4: Having words,* London: Architectural Association, 32; Charitonidou, 2021, 'The 1968 effects and civic responsibility in architecture and urban planning in the USA and Italy: Challenging 'nuova dimensione' and 'urban renewal'' *Urban, Planning and Transport Research,* 9(1): 549–578. https://doi.org/10.1080/21650020.2021.2001365; Charitonidou, 2022, 'Denise Scott Brown's active socioplastics and urban sociology: From Learning from West End to Learning from Levittown', *Urban, Planning and Transport Research,* 10(1): 131–158. https://doi.org/10.1080/21650020.2022.2063939.
64 Scott Brown, 'Towards an 'Active Socioplastics'', 33.
65 V. Watson, 2014, 'Co-production and collaboration in planning: The difference', *Planning Theory & Practice,* 15(1), 71. https://doi.org/10.1080/14649357.2013.866266.
66 L. R. Cirolia, S. Berrisford, 2017, 'Negotiated planning: Diverse trajectories of implementation in Nairobi, Addis Ababa, and Harare', *Habitat International,* 59 (2017): 71–79. https://doi.org/10.1016/j.habitatint.2016.11.005.
67 Watson, 'Co-production and collaboration in planning: The difference', 63.

68 M. Tewdwr-Jones, P. Allmendinger, 1998, 'Deconstructing communicative rationality: A critique of Habermasian collaborative Planning', *Environment and Planning A: Economy and Space*, 30(11): 1975–1989. https://doi.org/10.1068/a301975.
69 Cirolia, Berrisford, 'Negotiated planning: Diverse trajectories of implementation in Nairobi, Addis Ababa, and Harare', 77.
70 Ibid., 71.
71 Ibid.
72 M. Charitonidou, 2022, 'Housing programs for the poor in Addis Ababa: Urban commons as a bridge between spatial and social', *Journal of Urban History*, 48(6): 1345–1364. https://doi.org/10.1177/0096144221989975.
73 D. Harvey, 2011, 'The future of the commons', *Radical History Review*, 109: 101–102.
74 J. Bingham-Hall, 2016, *Future of cities: Commoning and collective approaches to urban space*. Government Office for Science.
75 P. Bresnihan, 2016, 'The more-than-human commons: From commons to commoning', in S. Kirwan, L. Dawney, J. Brigstocke, eds., *Space, Power and the Commons: The Struggle for Alternative Futures*. London; New York: Routledge, 96.

Part 8

Digital Twins in Transportation

42

Digital Twins in Transportation and Logistics

Mariusz Kostrzewski
Warsaw University of Technology

42.1 Introduction

The initial conceptual approaches focused on digital twin applications are primarily associated with industrial production (manufacturing). This is noted by both researchers as Singh et al. (2022) and online industry websites such as Newton (2021) who stated that there was "no better example of a digital twin application than (...) the manufacturing industry." A digital twin concept is beginning to penetrate more and more spheres of everyday social and economic life. There is a growing interest in this technological concept both in the world of science and in various branches of industry, from manufacturing through architecture, engineering and construction industry, urban design, consulting up to logistics and transport, i.e., to all the branches wherever direct or indirect contact with technological processes occurs.

The digital twin concept has recently raised interest in most of the research disciplines and economy sectors, including the sector of transportation. All sectors of an economy cannot proceed without transportation. Therefore, whenever new technology is applied, transportation absorbs, develops, and transforms it necessarily to be improved itself as a service sector of the economy.

As currently, Digital twin technology has attracted great attention in transportation. Therefore, NASA (Shafto et al., 2010) set a peculiar circle as the beginning of digital twin, which directly includes vehicles. ("an integrated multi-physics, multi-scale, probabilistic simulation of a vehicle or system that uses the best available physical models, sensor updates, fleet history, etc., to mirror the life of its flying twin" Shafto et al. (2010, 2012) after Negri et al. 2017). On the other hand, Michael W. Grieves is believed to be the father of the term digital twin, and Grieves (2005) is the author who set the assumption to a paradigm of digital twins.

This chapter aims to consider applications of the digital twin in application together with logistics as an integral part of transportation. The chapter is divided into several sections. The next section discusses a general view of

DOI: 10.1201/9781003425724-50

transport research on the digital twins' concept, and it states some research questions. To answer them, a research methodology is given in Section 42.3. In Section 42.4, special attention is paid to the considerations on the application of digital twin accrued to the various branches of transportation. In addition, the digital twin nomenclature and definitions are investigated. Such an approach allows for answering research questions (Section 42.5).

42.2 State-of-the-Art

Typically, the state-of-the-art of any research area can be identified as a result of the review papers' analysis, hereby the state-of-the-art of digital twins in the transportation area of interest. Ríos et al. (2015) focused on the literature review of the digital twin in the aerospace sector. They investigated the terms of digital twin together with its synonyms (product avatar, product digital twin, and digital counterpart) as well as current approaches related to digital twins and product lifecycle management. Kritzinger et al. (2018) focused on terminology as well. Nevertheless, they aimed to develop a classification between digital model, digital shadow, and digital twin varied by the ways of data flow between virtual and physical objects, either manual or automatic. Uninterrupted fashionable considerations concern intelligent transport systems; therefore, Bao et al. (2021) reviewed the application of digital twin for such systems, including differences between traffic simulation approaches with digital twin ones.

A more general approach was developed by Rasheed et al. (2020), who reviewed the modeling perspective of digital twins, including applications in the case of transportation among other economic sectors. Meanwhile, Barricelli et al. (2019) apart from definitions, comments, and statements underlined the necessity of focusing on the socio-technical design aspects of digital twins' lifecycle.

Based on the brief analysis of the state-of-the-art, it seems that there is no digital twin review paper treating transportation as a whole sector of the economy. Certain areas of transportation as the aerospace sector, internal transport and intralogistics, and intelligent transport systems were discussed separately. Therefore, this lack of digital twin consideration for all the branches of transport at once can be treated as a research gap. Consequently, two research questions are proposed to be considered, namely:

RQ1: How does the transport deal with absorbing the technology of digital twins and replicating it for its use?

RQ2: What topics and research interests, taken up in the consideration of digital twins, accrue to the various branches of transportation?

42.3 Methodology

The main aim of this chapter is to present a literature review together with certain authors' opinions, reflections, and insights on digital twins developed around transport and logistics. The high-quality publications were collected based on the Scopus database. The following pseudocode was applied: "digital twin" AND "transport" AND (LIMIT-TO (SUBJAREA, "ENGI")) AND (LIMIT-TO (LANGUAGE, "English")) AND (LIMIT-TO (EXACTKEYWORD, "Digital Twin")). Such pseudocode triggering led to complete 284 publications (collected in July 2022).

The details of the methodology are given in Figure 42.1. As can be observed in Stage 6 of Figure 42.1, many publications were excluded as they were not related to the transport area of interest. In the case of most of these publications, the term "transportation" or "transport" appears in the names of the affiliating institution, titles of references, or as a single word in the body of the publication. Also, the publications with digital twin listed only as a keyword, with no results mentioned, were omitted. Therefore, in addition to conducting a panoramic review (a superficial review of the literature), a thorough analysis of the content was necessary; therefore, consequently a systematic review was conducted. During the thorough review, special attention was paid to the core subject matter of the paper, the terms or synonyms that appeared in the context of the digital twin nomenclature, and to which branch of transportation the research was assigned. Such an approach allows for answering research questions.

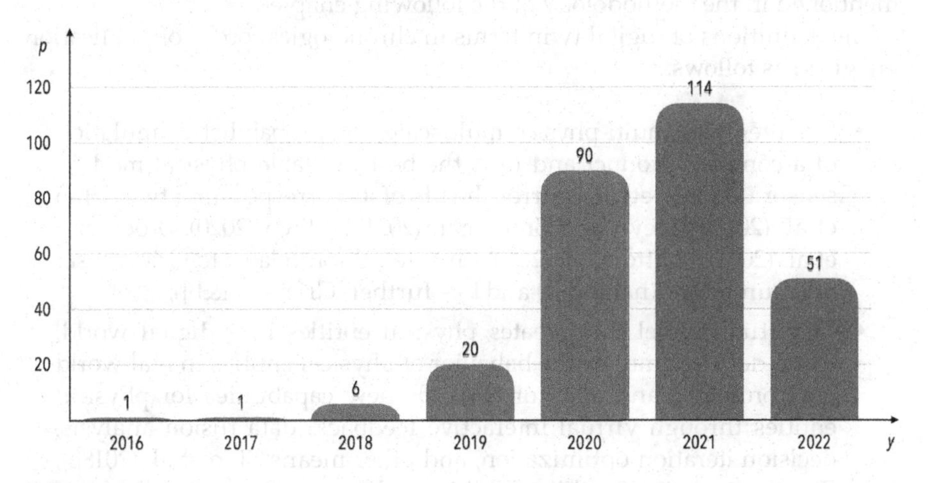

FIGURE 42.1
The methodology applied to the research.

42.4 Results and Discussion

The concept of the digital twin refers to the cooperation between the software and the physical object. In the case of a digital twin, the software is a specific "entity" that collects data on its own, processes them, and then transfers the results directly to physical objects, equipment, and means of transport, which would be automated or autonomous. In the case of the digital twin, the interaction between the physical object and its digital model is mutual, fully integrated into both directions mentioned above. When a change in any state of a physical object occurs, this automatically leads to a change in a digital model (and vice versa). If an organization's processes are not fully automated and require an employee to be present and perform actions, a digital shadow enters the considerations. A digital shadow is characterized by a one-way dataflow between a physical object and a digital model. A change in any state of a physical object is reflected in a digital model of a physical object, while a reverse situation does not occur (Fuller et al. 2020). A digital shadow can be understood de facto as a simulation model.

According to Deuter and Pethig (2019), many heterogeneous definitions of a digital twin were developed; every single digital twin may be supported by its definition assuming that the digital twin paradigm is "a) physical products in Real Space, b) virtual products in Virtual Space, and c) the connections of data and information that ties the virtual and real products together" (Grieves 2014). In the definition part of the chapter, these a), b), and c) points are referenced as a set of Grieve's assumptions.

The definitions of digital twin refer mainly to the publications that are mentioned in the methodology of the following chapter.

The definitions of digital twin terms in chronological order of publication are given as follows:

- "an integrated multi-physics, multiscale, and probabilistic simulation of a complex product and uses the best available physical models, sensor updates, etc., to mirror the life of its corresponding twin" Tao et al. (2018a), Steyn and Broekman (2021b), Steyn (2020), Broekman et al. (2021), Shafto et al. (2010) (this definition relates to Grieves' set of assumptions, namely a) and b) – further: Grieves {a&b}),
- "a virtual model that creates physical entities in a digital world, using data to simulate the behavior of physical entities in real-world environments, and adds or expands new capabilities for physical entities through virtual interactive feedback, data fusion analysis, decision iteration optimization, and other means" Tao et al. (2018b), Zhang et al. (2020); this definition relates to Grieves' full set of assumptions – further Grieves {a&b&c},

- "virtual and computerised counterpart of a physical system, used to simulate it, exploiting real-time synchronisation of data" Kritzinger et al. (2018), Hofmann and Branding (2019) (the definition is given in the context of manufacturing, nevertheless it is sufficiently generalized to be applied to other areas as well; Grieves {a&b&c}),
- "a digital representation of a physical entity, which can simulate the entire life cycle of the operating system and synchronize the mapping with the physical twin" Souza et al. (2019) (Grieves {a&b}),
- "[d]igital twins can be defined as digital representations of physical entities that employ real-time data to enable understanding of the operating conditions of these entities", Moya et al. (2022) (Grieves {a&b}),
- "a digital replica of a real-world asset" Lu and Brilakis (2019), Parrott and Lane (2017) (Grieves {a&b}),
- "[a] computerized models that represents the network state for any given moment in real time (Ivanov and Dolgui 2020)", Dolgui et al. (2020) (Grieves {a&b}),
- "a virtual representation of a physical asset enabled through data and simulators for real-time prediction, optimization, monitoring, controlling, and improved decision making", Rasheed et al. (2020) (Grieves {a&b&c}),
- "a digital representation of key elements of the autonomous ship as a key tool for simulation-based testing", Pedersen et al. (2020) (Grieves {a&b}, yet the authors suggest replicating solely key elements of a physical structure),
- "the combination of specification languages and visual block programming languages to enable industrial users to test and/or build their own [d]igital [t]win models at a suitable abstraction level and with low entry barriers", Pi i Palomés et al. (2021) (this definition does not relate openly to Grieves' set of assumptions; it is a rare definition by the necessary subjects and operations to be performed during digital twin design); however, the authors mentioned as well that a digital twin "is defined as a virtual reproduction of a system based on simulations, either with real-time or historical data, that allows representing, understanding, and predicting scenarios of the past, present, and future, with verified and validated models", based on Fonseca i Casas et al. (2021) (Grieves {a&b&c}),
- "the digital twin should include four important parts: a model of the object built based on an integrated multiphysics, multiscale, probabilistic simulation approach; an evolving set of data relating to the object; a means of dynamically updating or adjusting the model in

accordance with the data; mirror and predict activities/performance over the life of its corresponding physical twin", West and Blackburn (2017), Wright and Davidson (2020) after Spiryagin et al. (2021) (Grieves {a&b&c}),

- "a virtual mapping of a physical product (...) always associated with a singular, physical product instance based on a customer's specific order", Eigner et al. (2021); and also: "a bi-directional data exchange between the physical product and its digital counterpart" (Eigner et al. 2021) (Grieves {a&b&c}).

Eigner et al. (2021) studied various definitions of digital twin (concerning Product Lifecycle Management – PLM) and finally proposed their definition of the digital thread, which is a result of digital twin approaches studying together with the digital model for and is called as follows (Helu et al. 2017; Singh and Willcox 2018; Eigner et al. 2021):

> [d]igital thread connects the configuration items (...) of the lifecycle-phase specific partial models of a digital model as well as the digital twin and, furthermore, also its physical twin over the entire product lifecycle. In addition to information from the PLM, team data management (TDM), and the authoring systems, it can also integrate further product-relevant information from other IT legacy systems. The objective of the digital thread is the traceability of the development, procurement, and production history of a component at any time.

Therefore, this context treats the digital twin in a more complex way by setting it as a factor related to the management of the organization and establishing it as a complex digital thread. This interesting perspective may become one of the future agendas of digital twin development.

Some of the above-given definitions or semi-definitions are characterized by simplicity, e.g.: Kritzinger et al. (2018), Hofmann and Branding (2019), Souza et al. (2019), Moya et al. (2022), Lu and Brilakis (2019), Parrott and Lane (2017), Ivanov and Dolgui (2020), Dolgui et al. (2020), and Pedersen et al. (2020), whereas the other are more complex as e.g.: Tao et al. (2018a), Rasheed et al. (2020), Pi i Palomés et al. (2021), Fonseca i Casas et al. (2021), and Spiryagin et al. (2021). With the succession of years, researchers are beginning to adhere more and more to the assumptions pointed out by Grieves. Over time, the complexity of the digital twin definitions has increased.

Of interest is the definitional subjectivity expressed by the terms applied to call the physical objects being under digitization into digital twins. They relate to:

- System, as in: Kritzinger et al. (2018), Hofmann and Branding (2019), Pi i Palomés et al. (2021), Fonseca i Casas et al. (2021),
- Entity, as in: Moya et al. (2022), Souza et al. (2019),

- Asset, as in: Parrott and Lane (2017), Lu and Brilakis (2019), Rasheed et al. (2020),
- Key elements of a physical structure, as in: Pedersen et al. (2020),
- Product, as in: Shafto et al., (2010), Tao et al. (2018a), Steyn (2020), Steyn and Broekman (2021b), Broekman et al. (2021), Eigner et al. (2021),
- Network, as in: Ivanov and Dolgui (2020), Dolgui et al. (2020).

The transition from the digital twin paradigm to the development of a compact definition requires increased attention, time, and in-depth study of practical applications, which may take the next few years.

The main employment of digital twin technology within reviewed publications' sample concerning certain transport branches or application areas are given in Table 42.1: air transport, rail transport, road transport, maritime, and inland water transport.

According to Figure 42.1 and Table 42.1, most of the research on the digital twin in air, rail, road, inland transport was constituted in 2021. As de Paula Ferreira et al. (2020) mentioned, the quantity of digital twin publications has recently increased exponentially, even more reason to expect an increase in interest in the second half of 2022 and beyond.

This section focused on the considerations of digital twin nomenclature, both in the case of terms that are applied to recall this virtual object in research consideration and the terms applied to call the physical objects being under digitization into the digital twin. Many of the publications considered are a

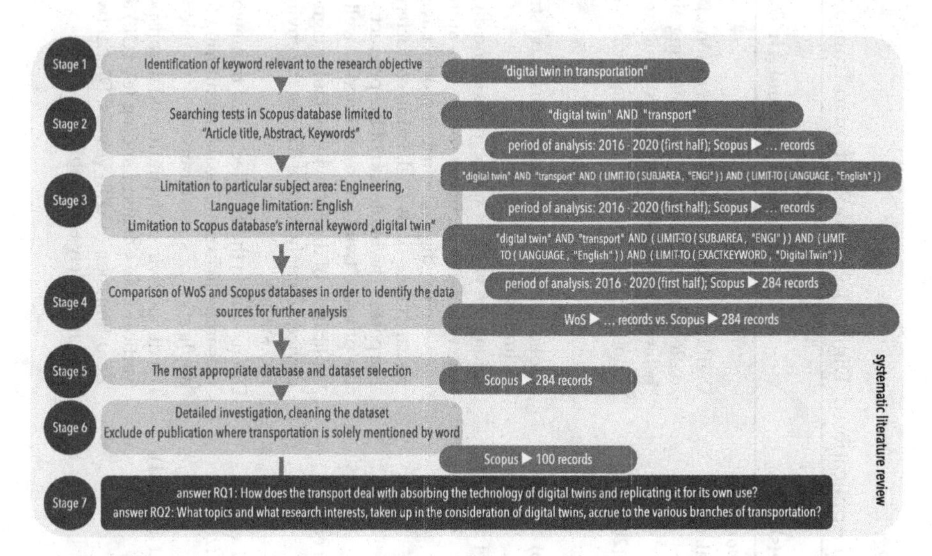

FIGURE 42.2
The nomenclature of digital twin and synonymous terms per year.

TABLE 42.1

The Main Application of Digital Twins in Relation to Transport Branches

Reference(s)	Digital Twins and Air Transport's Relation	How Is Digital Twin Called?	Publication's Type
Air Transport			
Tuegel et al. (2011)	Digital twin applied for the integration of computation related to structural deflections and temperatures affecting aircraft conditions to predict the behavior of the physical objects' structure.	Airframe digital twin	Journal
Ríos et al. (2015)	Various applications of the digital twins in the aerospace sector.	Product avatar, product digital twin, and digital counterpart	Review
Ross (2016)	Application of digital twins focused on structural health monitoring (SHM) technologies.	Digital twin	Chapter
Barricelli et al. (2019)	Among the various applications of the digital twin in the aerospace sector in line with other economy sectors, the authors mentioned the differences between the digital twin and product avatar as deriving from two different research lines (Ríos et al. 2015), characterized by different capabilities and purposes. The limitation of the product avatar as a perfect virtual replica of the physical twin was mentioned as well.	Digital twin, product avatar, intelligent product, and smart product	Review
Ezhilarasu et al. (2021)	Application of digital twins in fault diagnostics of aircraft systems and devices, namely, for the engine, the environmental control system, and the fuel system.	Digital twin	Journal
Ezhilarasu and Jennions (2021)	Discussion on the architecture of framework for air transport vehicles reasoning to explore health monitoring for mentioned vehicles by application of such a framework coupled with digital twins.	Digital twin	Journal
Portapas et al. (2021)	The digital twin of the multimodal transport system consisting of electric Vertical Take-Off and Landing (eVTOL) aircraft and electric Autonomous Zero-Emission (eAZE) vessels. Digital twin applied for studies on the system's safety and viability.	Digital twin	Conference paper

(Continued)

TABLE 42.1 (*Continued*)

The Main Application of Digital Twins in Relation to Transport Branches

Reference(s)	Digital Twins and Air Transport's Relation	How Is Digital Twin Called?	Publication's Type
Saifutdinov and Tolujevs (2021)	Consideration of the ground traffic controller's replacement with the digital twin dedicated to the centralized control system of the airport ground transportation.	Digital twin	Journal
Wang et al. (2021)	The digital twin–based decision-supports system which constitutes the analysis and strategy layer of local airspace.	Digital twin	Conference paper
Yiu et al. (2021)	Digital twin approach provides real-time collaborative decision-making training for a pilot (at least one) and an air traffic control officer (at least one).	Digital twin	Journal
Railway Transport			
Rasheed et al. (2020)	An integrated approach between vehicles and mobile cloud computing (passengers).	Indirect as cyber-physical system	Journal
Kaewunruen and Lian (2019)	Railway turnouts or (switches and crossings) with the application of Building Information Modeling (BIM) to improve the overall information flow of the turnout planning and design, manufacturing pre-assembly and logistics, construction and installation, operation and management, and demolition, thereby achieving better project performance and quality.	Digital twin	Journal
Ariyachandra and Brilakis (2020)	Digital twin for railway overhead line equipment, which supports analyzing different catenary configurations, infrastructure.	Digital twin, geometric digital twin	Conference paper
Bustos et al. (2021)	Digital twin as one of the four-stage methodology of high-speed train adaptation to Industry 4.0, among physical systems, information, and communication technology infrastructure, and diagnosis.	Digital twin	Journal
Broo and Schooling (2021)	The framework of digital twin–based system application integrating several assets, such as bridges, railways, and transport systems which aim in aiding climate-conscious, sustainable decision-making.	Digital twin	Journal

(*Continued*)

TABLE 42.1 (*Continued*)

The Main Application of Digital Twins in Relation to Transport Branches

Reference(s)	Digital Twins and Air Transport's Relation	How Is Digital Twin Called?	Publication's Type
Jeschke and Grassmann (2021)	Considering how digital twins in the logistics systems of German rail transport can be designed; the authors treat their work as the instrument for achieving the operational optimization and intelligent management of rail transport assets.	Digital twin	Journal
Kampczyk and Dybeł (2021)	The diagnostic application of digital twins to railway turnouts structural condition together with consideration of weather conditions' influence – the authors focused on monitoring temperature and other atmospheric conditions, infrastructure.	Digital twin	Journal
Spiryagin et al. (2021)	Full-scale locomotive design together with its digital twin was the main interest in this research.	Digital twin	Journal
Zhang et al. (2021a)	Framework for digital twin–assisted fault diagnosis for railway point machine.	Digital twin	Conference paper
Road Transport			
Besselink et al. (2016) quoted after Rasheed et al. (2020)	Control of large-scale road freight transportation problems with the application of cyber-physical system (fuel consumption at a minimal state with integrated routing and transport planning was included).	Indirect as cyber-physical system	Conference paper
Hofmann and Branding (2019)	Coordination of traffic density of the vehicles' road infrastructure and shipments. As truck congestion leads to heavy traffic in the case of large vessels, the digital twin can be applied for alternatives in performance forecasts based on the current system status.	Digital twin	Journal
Lu and Brilakis (2019)	Geometric digital twin applied in concrete bridges identification and analyses, maintenance; generate geometric digital twins in industry foundation classes format using four types of point engineering/ architectural clusters.	Digital twin, digital twinning, and geometric digital twin	Journal

(*Continued*)

TABLE 42.1 (*Continued*)

The Main Application of Digital Twins in Relation to Transport Branches

Reference(s)	Digital Twins and Air Transport's Relation	How Is Digital Twin Called?	Publication's Type
Omer et al. (2019)	Bridges inspection was considered via their digitization with LIDAR and virtual reality VR technology into digital twin; the importance of the solutions is highlighted by bridge collapses which have affected local economies, traffic congestion, and a loss of life.	Digital twin	Journal
Shao et al. (2020)	Fundamental analysis of bridges to develop digital twins for large-scale structures (bridges' mockups), their condition assessment, and intelligent damage identification.	Digital twin	Journal
Varga et al. (2020)	Synchronization of real and virtual car for digital twin with Sensor Interface Specifications communication standard.	Digital twin	Journal
Broekman et al. (2021)	Applied a mini edge computing platform for real-time edge processing, which serves as a digital twin of a multi-lane freeway; video data were ensured with the application of Unmanned Aerial Vehicle (UAV) filming above the freeway. This is an example of road transport supported by UAV.	Digital twin	Journal
Chen et al. (2021)	Vehicle-to-everything technology (V2X), coupled with a digital twin platform, is applied for dynamic analyses of the intelligent network safety in the case of autonomous vehicles moving in this network.	Digital twin	Journal
Charissin et al. (2021)	The digital twin of motorways is applied as a part of a virtual simulator for investigation of the effectiveness in the case of a system ensuring driving safety.	Digital twin	Journal
Deng et al. (2021)	The digital twin model is introduced to construct the individual trip chain and the bus operation chain, respectively, for demand-responsive transit; the solution can be applied to transport poverty.	Digital twin	Conference paper
Jiang et al. (2021)	The digital twin–based solution for non-deterministic fatigue life prediction of steel bridges.	Digital twin	Journal
Meža et al. (2021)	The digital twin for a road construction project in the case of road constructed with the use of secondary raw materials.	Digital twin	Journal

(*Continued*)

TABLE 42.1 (*Continued*)

The Main Application of Digital Twins in Relation to Transport Branches

Reference(s)	Digital Twins and Air Transport's Relation	How Is Digital Twin Called?	Publication's Type
Niaz et al. (2021)	Applied for autonomous driving tests with V2X technology scenarios.	Digital twin	Conference paper
Shadrin et al. (2021)	Applied of the digital twin for autonomous vehicles indicates an increasing degree of adequacy of the digital twin to an actual vehicle; it ensures minimizing the discrepancy between results of field tests and software modeling and applying this technology in practice – environmental conditions should be identical, therefore clarifying road sensors are necessary, etc.	Digital twin	Conference paper
Van Mierlo et al. (2021)	Digital twin opens a new door for the next generation of electric vehicles via new terms and technology applications: a digital twin for design (DT4D), a digital twin for control design (DT4CD), a digital twin for virtual validation (DT4VV), and a digital twin for reliability (DT4R).	Digital twin	Journal
Saroj et al. (2021)	A digital twin for an intelligent transportation system to analyze its impact on connected corridor traffic and environmental performance.	Digital twin	Journal
Temkin et al. (2021)	Constructed a dynamic 3D model (digital twin) of infrastructure and technological system elements using large arrays of telemetric data in order to analyze open-pit technological roads in the mining industry (continuous, stochastic, and nonstationary modeling object).	Digital twin	Journal
Zhang et al. (2021b)	Built a digital twin model–based on a vehicle real-time monitoring platform, which aims to predict the energy consumption of electric vehicles.	Digital twin	Journal
Hu et al. (2022)	Internet of vehicle coupling with digital twin may support traffic scheduling and alleviate traffic jams (traffic prediction method).	Digital twin	Journal
Liao et al. (2022)	The digital twin approach based on a vehicle-to-cloud communication of data is sent from vehicles, processed in digital twin, and sent back.	Digital twin	Conference paper
Sell et al. (2022)	Safety evaluation and validation for autonomous vehicles.	Digital twin, digital model	Journal

(*Continued*)

TABLE 42.1 ***(Continued)***

The Main Application of Digital Twins in Relation to Transport Branches

Reference(s)	Digital Twins and Air Transport's Relation	How Is Digital Twin Called?	Publication's Type
Maritime and Inland Water Transport			
Hofmann and Branding (2019)	Mentioned in the second row of road transport part in this table.	Digital twin	Journal
Hatledal et al. (2020)	Preliminary results of the vessel's digital twin with co-simulation support.	Digital twin	Journal
Nikolopoulos and Boulougouris (2020)	Simulation of the response in variations of the geometrical, design variables of the vessel under uncertainty.	Digital twin	Journal
Pedersen et al. (2020)	Digital twin for autonomous vessels navigation system.	Digital twin	Journal
Chu et al. (2021)	Responses of the vessel's motion coupled with the freight loading crane were analyzed with the application of digital twin, including multi-body dynamics.	Digital twin	Journal
Coraddu et al. (2021)	The authors investigated a digital twin of the dual fuel engine in vessels.	Digital twin	Journal
Taskar and Andersen (2021)	Consideration of separate wave spectra for wind waves and swell on performance prediction of the models of two ships called digital twins.	Digital twin	Journal
Wu et al. (2021)	The digital twin of infrastructural facilities allocated along inland water paths to improve the efficiency of daily monitoring, evidence collection, and emergency response.	Digital twin	Journal
Agostinelli et al. (2022)	The digital twin for a maritime port integrated with real-time data occurred with sensors to improve data management.	Digital twin	Journal
Zhou et al. (2022)	Condition monitoring port rail-mounted gantry crane system based on digital twin to collect operational data.	Digital twin	Journal

preview of a fully functioning system, however, proven and proven purely scientific and utilitarian results will have to wait several years.

Numerous publications, not necessarily included in this chapter, consider digitization and recall a particular system, device, or tool as a digital twin. Digitization certainly is one of the solid bases and conditions for an object to be substituted as a digital twin, yet it is not a sufficient condition; familiarizing with the publications shows among their many examples the lack of distinction between digital twin, digital shadow, and digital model. The author's intention was not to indicate which of the publications are given erroneously, and which correctly, so it is applied to "trust" the authors of the publications as to the legitimacy of calling solutions by the name of "digital twin". This was observed by other researchers as well. Also, Baalsrud et al. (2021) mentioned, "[f]urthermore, the analysis revealed that quite a few applications are more to be characterized as [digital shadows] than [digital twins]."

Contradictions are related not solely to the nomenclature. The application of the digital twin accrued to the various branches of transportation was also a result of this section and reading the publications indicates some contradictions in applications of the digital twin in various branches of transport as well. For example, Yiu et al. (2021) mentioned that "the potential of digital twin-based automation technologies is yet to be unleashed in numerous industries, such as for air traffic operations in the aviation industry", However, in previously published studies, Saifutdinov and Tolujevs (2021) specifically considered air traffic operations. "[as] a result [of airplanes catastrophe], the stocks about the digital twinning all rose by the daily limit, and digital twinning technology also attracted more and more attention (Chen 2020)" Zhang et al. (2021b). Admittedly, the difference between the publication of these research papers is just over a quarter. However, it should also be mentioned here that most of the presented research for air transport considers condition monitoring or faults diagnostics of the aircraft. Considering, for example, rail transport is a similar matter of contradictions, according to Bustos et al. (2021) "The railway industry is starting to apply the Digital Twin, but it is mainly oriented towards infrastructure (track, signaling, catenary, etc.) and traffic management ('In-Depth Focus: Digital Twins', 2021)." This is often true; however, researchers' interest evolves in vehicles and broader systems as Spiryagin et al. (2021) and Rasheed et al. (2020).

42.5 Conclusion

This chapter aims to find answers to research questions. Firstly, how does the transport deal with absorbing the technology of digital twins and replicating it for its use? Despite the rather modest number sample of the publications' pool, it turns out that the digital twin has penetrated all branches of transport.

What is additionally intriguing is that the topics extend beyond strictly defined transportation industries. The digital twin is being developed in transportation in such a way that it contributes to combining various transportation media: road transport is coupled with air transport via Unmanned Aerial Vehicle (UAV) mapping the territory as, e.g., in Broekman et al. (2021), road transport and maritime transport are combined as, e.g., in the case of consideration of traffic difficulties around ports given in Hofmann and Branding (2019), to mention only a few.

The second research question concerns topics and research interests, related to digital twins, and accrued to the various branches of transportation. The investigation of publications allowed to specify the following list of such topics:

- Support for means of transport and transport devices (e.g., for their condition analyses):
  - air transport: Tuegel et al. (2011), Ezhilarasu et al. (2021), Ezhilarasu and Jennions (2021);
  - rail transport: Spiryagin et al. (2021);
  - road transport: Varga et al. (2020), Shadrin et al. (2021), Van Mierlo et al. (2021), Zhang et al. (2021b), Sell et al. (2022);
  - maritime transport: Hatledal et al. (2020), Nikolopoulos and Boulougouris (2020), Coraddu et al. (2021), Taskar and Andersen (2021), Zhou et al. (2022).
- Support for infrastructure:
  - rail transport: Kaewunruen and Lian (2019), Ariyachandra and Brilakis (2020), Broo and Schooling (2021), Kampczyk and Dybeł (2021), Zhang et al. (2021a);
  - road transport: Lu and Brilakis (2019), Omer et al. (2019), Shao et al. (2020), Jiang et al. (2021), Meža et al. (2021), Steyn and Broekman (2021b), Temkin et al. (2021);
  - inland transport: Wu et al. (2021);
- Applications in control systems and decision-support systems, planning support systems, or training systems:
  - air transport: Saifutdinov and Tolujevs (2021), Wang et al. (2021), Yiu et al. (2021);
  - rail transport: Bustos et al. (2021), Jeschke and Grassmann (2021);
  - road transport: Hofmann and Branding (2019), Broekman et al. (2021), Chen et al. (2021), Deng et al. (2021), Saroj et al. (2021), Hu et al. (2022), Liao et al. (2022);
  - maritime transport: Pedersen et al. (2020), Agostinelli et al. (2022).

The above list can provide important guidance for researchers who would be interested in considering research topics not yet fully explored. It is also wholeheartedly recommended to focus more on a human factor in digital twin investigation, especially that – as Petzhold et al. (2020) mentioned – it leads to a new role of humans in the increasing digitization of the world: "the role of operator changes from that of a manual worker to that of a supervisor and problem solver."

References

Agostinelli, S., Cumo, F., Nezhad, M.M., Orsini, G., and Piras, G. 2022. Renewable energy system controlled by open-source tools and digital twin model: Zero energy port area in Italy. *Energies* 15:1817.

Ariyachandra, M.R.M.F., and Brilakis, I. 2020. Digital twinning of railway overhead line equipment from airborne lidar data. *Proceedings of the 37th International Symposium on Automation and Robotics in Construction, ISARC 2020: From Demonstration to Practical Use - To New Stage of Construction Robot*; pp. 1270–1277.

Baalsrud, H.J., Zafarzadeh, M., Jeong, Y., Li, Y., Ali Khilji, W., Larsen, C., and Wiktorsson, M. 2021. Digital twin testbed and practical applications in production logistics with real-time location data. *International Journal of Industrial Engineering and Management* 12(2):129–140.

Bao, L., Wang, Q., and Jiang, Y. 2021. Review of digital twin for intelligent transportation system. *Proceedings - 2021 International Conference on Information Control, Electrical Engineering and Rail Transit, ICEERT 2021*; pp. 309–315.

Barricelli, B.R., Casiraghi, E. and Fogli, D. 2019. A survey on digital twin: Definitions, characteristics, applications, and design implications. *IEEE Access* 7:167653–167671.

Besselink, B., Turri, V., van De Hoef, S.H., Liang, K.-Y., Alam, A., Mårtensson, J., and Johansson, K.H. 2016. Cyber-physical control of road freight transport. *Proceeding of the IEEE* 104(5):1128–1141.

Broekman, A., Gräbe, P.J., and Steyn, W.J.vd M. 2021. Real-time traffic quantization using a mini edge artificial intelligence platform. *Transportation Engineering* 4:100068.

Broo, D.G., and Schooling, J. 2021. A framework for using data as an engineering tool for sustainable cyber-physical systems. *IEEE Access* 9:22876–22882.

Bustos, A., Rubio, H., Soriano-Heras, E., and Castejon, E. 2021. Methodology for the integration of a high-speed train in Maintenance 4.0. *Journal of Computational Design and Engineering* 8(6):1605–1621.

Proceedings - 2020 IEEE International Conference on Engineering, Technology and Innovation, ICE/ITMC 2020; pp. 1–8.

Chen, G. 2020. *Data Twinning*; Publishing House of Electronics Industry: Beijing, China.

Chen, X., Min, X., Li, N., Cao, W., Xiao, S., Du, G., and Zhang, P. 2021. Dynamic safety measurement-control technology for intelligent connected vehicles based on digital twin system. *Vibroengineering Procedia* 37:78–85.

Chu, Y., Li, G., Hatledal, L.I., Holmeset, F.T., and Zhang, H. 2021. Coupling of dynamic reaction forces of a heavy load crane and ship motion responses in waves. *Ships and Offshore Structures* 16(supl):58–67.

Coraddu, A., Oneto, L., Ilardi, D., Stoumpos, S., and Theotokatos, G. 2021. Marine dual fuel engines monitoring in the wild through weakly supervised data analytics. *Engineering Applications of Artificial Intelligence* 100:104179.

de Paula Ferreira, W., Armellini, F., and de Santa-Eulalia, L.A. 2020. Simulation in industry 4.0: A state-of-the-art review. *Computers and Industrial Engineering* 149:106868.

Deng, S., Zhong, J., Chen, S., and He, Z. 2021. Digital twin modeling for demand responsive transit. *Proceedings 2021 IEEE 1st International Conference on Digital Twins and Parallel Intelligence, DTPI 2021*; pp. 410–413.

Deuter, A., and Pethig, F. 2019. The Digital Twin Theory, Eine neue Sicht auf ein Modewort. *Industrie 4.0 Management* 35 (1):29–30.

Dolgui, A., Ivanov, D., and Sokolov, B. 2020. Reconfigurable supply chain: the X-network. *International Journal of Production Research* 58(13):4138–4163.

Eigner, M., Detzner, A., Schmidt, P.H., and Tharma, R. 2021. Holistic definition of the digital twin. *International Journal of Product Lifecycle Management* 13(4):343–357.

Ezhilarasu, C.M., and Jennions, I.K. 2021. Development and implementation of a Framework for Aerospace Vehicle Reasoning (FAVER). *IEEE Access* 9(9500228):108028–108048.

Ezhilarasu, C.M., Skaf, Z., and Jennions, I.K. 2021. A generalised methodology for the diagnosis of aircraft systems. *IEEE Access* 9(9319676):11437–11454.

Fonseca i Casas, P., Garcia i Subirana, J., García i Carrasco, V., and Pi i Palomés, X. 2021. Sars-cov-2 spread forecast dynamic model validation through digital twin approach, Catalonia case study. *Mathematics* 9(14):1660.

Fuller, A., Fan, Z. Day, C., and Barlow, C. 2020. Digital twin: Enabling technologies, Challenges and Open Research. *IEEE Access*, 8:108952–108971.

Grieves, M., 2014. Digital twin, manufacturing excellence through virtual factory replication. *White Paper* 1, 1–7. https://www.3ds.com/fileadmin/PRODUCTS-SERVICES/DELMIA/PDF/Whitepaper/DELMIA-APRISO-Digital-Twin-Whitepaper.pdf (accessed June 19, 2022).

Grieves, M.W. 2005. Product lifecycle management: The new paradigm for enterprises. *International Journal of Product Development* 2(1–2):71–84.

Hatledal, L.I., Skulstad, R., Li, G., Styve, A., and Zhang, H. 2020. Co-simulation as a fundamental technology for twin ships. *Modeling, Identification and Control* 41:297–311.

Helu, M., Hedberg, T., and Feeney, A.B. 2017. Reference architecture to integrate heterogeneous manufacturing systems for the digital thread. *CIRP Journal of Manufacturing Science and Technology* 19:191–195.

Hofmann, W., and Branding, F., 2019. Implementation of an IoT- and cloud-based digital twin for real-time decision support in port operations. *IFAC-PapersOnLine* 52(13):2104–2109.

Hu, C., Fan, W., Zeng, E., Hang, Z., Wang, F., Qi, L., and Bhuiyan, M.Z.A. 2022. Digital twin-assisted real-time traffic data prediction method for 5G-enabled internet of vehicles. *IEEE Transactions on Industrial Informatics* 18(4):2811–2819.

In-Depth Focus: Digital Twins. 2021. In-depth focus: Digital twins. *Global Railway Review* 27(02):19. https://www.globalrailwayreview.com/article/120531/digital-twins-in-depth-focus/ (accessed July 26, 2022).

Ivanov, D., and Dolgui, A. 2020. A digital supply chain twin for managing the disruption risks and resilience in the era of Industry 4.0. *Production Planning and Control* 32(9):775–788.

Jeschke, S., and Grassmann, R. 2021. Development of a generic implementation strategy of digital twins in logistics systems under consideration of the german rail transport. *Applied Sciences* 11:10289.

Jiang, F., Ding, Y., Song, Y., Geng, Y., and Wang, Z. 2021. Digital Twin-driven framework for fatigue life prediction of steel bridges using a probabilistic multiscale model: Application to segmental orthotropic steel deck specimen. *Engineering Structures* 241:112461.

Kaewunruen, S., and Lian, A. 2019. Digital twin aided sustainability-based lifecycle management for railway turnout systems. *Journal of Cleaner Production* 228:1537–1551.

Kampczyk, A., and Dybeł, K. 2021. The fundamental approach of the digital twin application in railway turnouts with innovative monitoring of weather conditions. *Sensors,* 21:5757.

Kritzinger, W., Karner, M., Traar, G., Henjes, J., and Sihn, W. 2018. Digital Twin in manufacturing: A categorical literature review and classification. *IFAC PapersOnLine* 51(11):1016–1022.

Liao, X., Wang, Z., Zhao, X., Han, K. Tiwari, P., Barth, M.J., and Wu, G. 2022. Cooperative ramp merging design and field implementation: A digital twin approach based on vehicle-to-cloud communication. *IEEE Transactions on Intelligent Transportation Systems* 23(5):4490–4500.

Lu, R., and Brilakis, I. 2019. Digital twinning of existing reinforced concrete bridges from labelled point clusters. *Automation in Construction* 105(102837):1–16.

Meža, S., Mauko Pranjić, A., Vezočnik, R., Osmokrović, I., and Lenart, S. 2021. Digital twins and road construction using secondary raw materials. Journal of Advanced Transportation, 2021, 2021: 1-12.

Moya, B., Badías, A., Alfaro, I., Chinesta, F., and Cueto, E. 2022. Digital twins that learn and correct themselves. *International Journal for Numerical Methods in Engineering* 123(13):3034–3044.

Negri, E., Fumagalli, L., and Macchi, M. 2017. A review of the roles of digital twin in CPS-based production systems. *Procedia Manufacturing* 11:939–948.

Newton, E. 2021. These 7 industrial sectors are seeing remarkable results from digital twins. *Revolutionized.* https://revolutionized.com/industrial-sector-digital-twins/ (accessed July 26, 2022).

Niaz, A., Shoukat, M.U., Jia, Y., Khan, S., Niaz, F., and Raza, M.U. 2021. Autonomous driving test method based on digital twin: A survey. *2021 International Conference on Computing, Electronic and Electrical Engineering, ICE Cube 2021 – Proceedings;* pp. 1–7.

Nikolopoulos, L., and Boulougouris, E., 2020, A novel method for the holistic, simulation driven ship design optimization under uncertainty in the big data era. *Ocean Engineering* 218:107634.

Omer, M., Margetts, L., Mosleh, M.H., Hewitt, S., and Parwaiz, M. 2019. Use of gaming technology to bring bridge inspection to the office. *Structure and Infrastructure Engineering* 15(10):1292–1307.

Parrott, A., and Lane, W. 2017. Industry 4.0 and the digital twin. https://www2.deloitte.com/insights/us/en/focus/industry-4-0/digital-twintechnology-smart-factory.html (accessed May 2, 2019).

Pedersen, T.A., Glomsrud, J.A., Ruud, E.-L., Simonsen, A., Sandrib, J., and Eriksen, B.-O.H. 2020. Towards simulation-based verification of autonomous navigation systems. *Safety Science* 129:104799.

Pi i Palomés, X., Tuset-Peiro, P., and i Casas, P.F. 2021. Combining low-code programming and SDL-based modeling with snap in the industry 4.0 context. *Companion Proceedings -24th International Conference on Model-Driven Engineering Languages and Systems, MODELS-C 2021*; pp. 741–750.

Portapas, V., Zaidi, Y., Bakunowicz, J., Paddeu, D., Valera-Medina, A., and Didey, A. 2021. Targeting global environmental challenges by the means of novel multimodal transport: Concept of operations. *Proceedings of the 2021 5th World Conference on Smart Trends in Systems Security and Sustainability, WorldS4 2021*; pp. 101–106.

Rasheed, A., San, O., and Kvamsdal, T. 2020. Digital twin: Values, challenges and enablers from a modeling perspective. *IEEE Access* 8:21980–22012.

Ríos, J., Hernández, J.C., Oliva, M., and Mas, F. 2015. Product avatar as digital counterpart of a physical individual product: Literature review and implications in an aircraft. In *Transdisciplinary Lifecycle Analysis of Systems*, eds. R. Curran, M. Wognum, and M. Borsato, 657–666. Amsterdam, the Netherlands: IOS Press BV.

Ross, R.W. 2016. Integrated vehicle health management in aerospace structures. In *Structural Health Monitoring (SHM) in Aerospace Structures*, ed. F. G. Yuan, 3–31. Amsterdam, the Netherlands: Elsevier Ltd.

Saifutdinov, F., and Tolujevs, J. 2021. Time and space discretization in the digital twin of the airport transport network. *Transport and Telecommunication* 22(3):257–265.

Saroj, A.J., Roy, S., Guin, A., and Hunter, M. 2021. Development of a connected corridor real-time data-driven traffic digital twin simulation model. *Journal of Transportation Engineering Part A: Systems* 147(12):4021096.

Sell, R., Malayjerdi, E., Malayjerdi, M., and Baykara, B.C. 2022. Safety toolkit for automated vehicle shuttle - Practical implementation of digital twin. *ICCVE 2022 - IEEE International Conference on Connected Vehicles and Expo.*

Shadrin, S.S., Makarova, D.A., Ivanov, A.M., and Maklakov, N.A. Safety assessment of highly automated vehicles using digital twin technology. *2021 Intelligent Technologies and Electronic Devices in Vehicle and Road Transport Complex (TIRVED)*, 11–12 November 2021. Moscow, Russian Federation; pp. 1–5.

Shafto, M., Conroy, M., Doyle, R., Glaessgen, E., Kemp, C., LeMoigne, J., and Wang, L., 2010. DRAFT modeling, simulation, information technology & processing roadmap. *Technology Area* 11. https://www.nasa.gov/pdf/501321main_TA11-MSITP-DRAFT-Nov2010-A1.pdf (accessed July 26, 2022).

Shafto, M., Conroy, M., Doyle, R., Glaessgen, E., Kemp, C., LeMoigne, J., and Wang, L., 2012. Modeling, simulation, information technology & processing roadmap. *Technology Area* 11. https://fdocuments.in/document/modeling-simulation-information-technology-processing-r-the-modeling-simulation.html?page=1, (accessed July 26, 2022).

Shao, S., Zhou, Z., Deng, G., Du, P., Jian, C., and Yu, Z. 2020. Experiment of structural geometric morphology monitoring for bridges using holographic visual sensor. *Sensors* 20:1187.

Singh, M., Srivastava, R., Fuenmayor, E., Kuts, V., Qiao, Y., Murray, N., and Devine, D. 2022. Applications of digital twin across industries: A review. *Applied Sciences* 12:5727.

Singh, V., and Willcox, K.E. 2018. Engineering design with digital thread. *AIAA Journal* 56(11):4515–4528.

Souza, V., Cruz, R., Silva, W., Lins, S., and Lucena, V. 2019. A digital twin architecture based on the industrial internet of things technologies. *IEEE International Conference on Consumer Electronics;* pp. 1–2.

Spiryagin, M., Wu, Q., Polach, O., Thorburn, J., Chua, W., Spiryagin, V., Stichel, S., Shrestha, S., Bernal, E., Ahmad, S., Cole, C., and McSweeney, T. 2021. Problems, assumptions and solutions in locomotive design, traction and operational studies. *Railway Engineering Science* 30(3):265–288.

Steyn, W.J.M. 2020. Selected implications of a hyper-connected world on pavement engineering. *International Journal of Pavement Research and Technology* 13(6):673–678.

Steyn, W.J.V.D.M., and Broekman, A. 2021a. Process for the development of multiscale digital twins of local roads - A case study. *GeoChina 2021 Conference Theme: Civil and Transportation Infrastructures: From Engineering to Smart and Green Life Cycle Solutions,* NanChang.

Steyn, W.J.V.D.M., and Broekman, A. 2021b. Development of a digital twin of a local road network: A case study. *Journal of Testing and Evaluation, 2021,* 51(1).

Tao, F., Cheng, K., Qi, Q., Zhang, M., Zhang, H., and Sui, F. 2018a. Digital twin-driven product design, manufacturing and service with big data. *The International Journal of Advanced Manufacturing Technology* 94: 3563–3576.

Tao, F., Qi, Q., Wang, L., and Nee, A.Y.C. 2018b. Digital twin and its potential application exploration. *Computer Integrated Manufacturing Systems* 24(1):1–18.

Taskar, B., and Andersen, P. 2021. Comparison of added resistance methods using digital twin and full-scale data. *Ocean Engineering* 229:108710.

Temkin, I., Myaskov, A., Deryabin, S., Konov, I., and Ivannikov, A. 2021. Design of a digital 3D model of transport-technological environment of open-pit mines based on the common use of telemetric and geospatial information. *Sensors* 21, 6277.

Tuegel, E.J., Ingraffea, A.R., Eason, T.G., and Spottswood, S.M. 2011. Reengineering aircraft structural life prediction using a digital twin. *International Journal of Aerospace Engineering* 154798:1–14.

Van Mierlo, J., Berecibar, M., El Baghdadi, M., De Cauwer, C., Messagie, M., Coosemans, T., Jacobs, V.A., and Hegazy, O. 2021. Beyond the state of the art of electric vehicles: A fact-based paper of the current and prospective electric vehicle technologies. *World Electric Vehicle Journal* 12(20).

Varga, B., Szalai, M., Fehér, Á., Aradi, S., and Tettamanti, T. 2020. Mixed-reality automotive testing with sensoris. *Periodica Polytechnica Transportation Engineering* 48(4):357–362.

Wang, C., Deng, T., Wang, X., Zhao, P., and Wu, Q. 2021. Local airspace traffic prediction and flow control strategy recommendation system. *Proceedings 2021 IEEE 1st International Conference on Digital Twins and Parallel Intelligence, DTPI 2021;* pp. 465–468.

West, T., and Blackburn, M. 2017. Is digital thread/digital twin affordable? A systemic assessment of the cost of DoD's latest Manhattan project. *Procedia Computer Science* 114:47–56.

Wright, L., and Davidson, S. 2020. How to tell the difference between a model and a digital twin. *Advanced Modeling and Simulation in Engineering Sciences* 7(13):1–13.

Wu, Z., Ren, C., Wu, X., Wang, L., Zhu, L., Lyu, Z., 2021. Research on digital twin construction and safety management application of inland waterway based on 3D video fusion. *IEEE Access* 9(9502678):109144–109156.

Yiu, C.Y., Ng, K.K.H., Lee, C.-H., Chow, C.T., Chan, T.C., Li, K.C., and Wong, K.Y. 2021. A digital twin-based platform towards intelligent automation with virtual counterparts of flight and air traffic control operations. *Applied Sciences* 11:10923.

Zhang, K., Qu, T., Zhou, D., Jiang, H., Lin, Y., Li, P., Guo, H., Liu, Y., Li, C., and Huang, G.Q. 2020. Digital twin-based opti-state control method for a synchronized production operation system. *Robotics and Computer-Integrated Manufacturing* 63:101892.

Zhang, S., Dong, H., Maschek, U., and Song, H. 2021a. A digital-twin-assisted fault diagnosis of railway point machine. *Proceedings 2021 IEEE 1st International Conference on Digital Twins and Parallel Intelligence, DTPI 2021*; pp. 430–433.

Zhang, Z., Zou, Y., Zhou, T., Zhang, X., and Xu, Z. 2021b. Energy consumption prediction of electric vehicles based on digital twin technology. *World Electric Vehicle Journal* 12:160.

Zhou, Y., Fu, Z., Zhang, J., Li, W., and Gao, C. 2022. A digital twin-based operation status monitoring system for port cranes. *Sensors* 22, 3216.

43

Digital Twin–Driven Damage Diagnosis and Prognosis of Complex Aircraft Structures

Xuan Zhou and Leiting Dong
Beihang University

43.1 Introduction

43.1.1 Historical Overview

Most of the structural failures in aircrafts are due to fatigue fracture, in which cyclic fatigue loads cause cracks to initiate in a fatigue critical component and grow eventually, affecting the structural integrity. The philosophy to ensure aircraft structural integrity has been evolving from the safe-life to the damage tolerance with the development of engineering practice, as shown in Figure 43.1. The damage tolerance relies on the assumption that inevitably there are initial damages which will subsequently grow over a period of time prior to the catastrophic failure. Fracture mechanics methodology is used to predict the fatigue crack growth in the structure, and inspection intervals are devised to ensure that the crack does not grow to a critical size. However, there are many uncertainties in the crack growth process. For example, loads, structural geometry, and crack growth behaviors may have some level of dispersion between various aircraft and flight missions. Therefore, it is desirable to carry out damage diagnosis and prognosis to improve decision-making in a condition-based maintenance (CBM) framework. In comparison to fixed-interval inspections, this could reduce unexpected downs and also unnecessary scheduled maintenance.

Individual aircraft tracking (IAT) programs have been widely implemented on many types of aircraft in recent years [1–3]. With the recorded flight data from the installed data acquisition unit, it aims to track the potential fatigue damage growth and life consumption for each aircraft in a fleet. However, most IAT systems in engineering practice only monitor load data such as aircraft overload, while epistemic uncertainties that also lead to discrepancies in damage states, such as geometric and material parameters, are not considered. The U.S. Air Force has been funding research on the airframe digital

 DOI: 10.1201/9781003425724-51

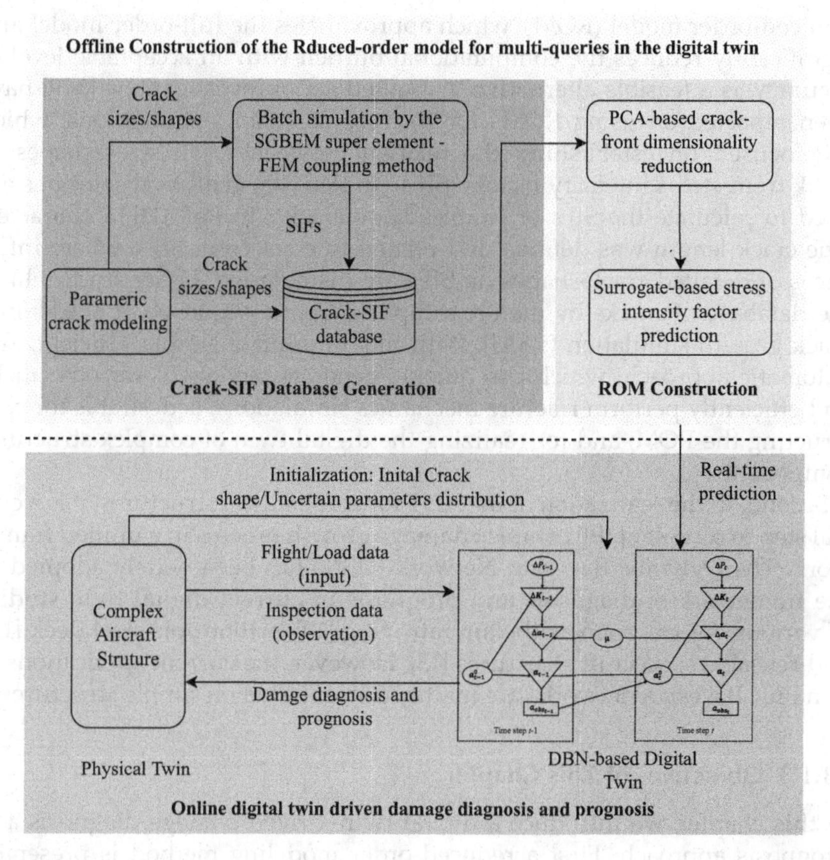

FIGURE 43.1
Framework of the digital twin–driven damage diagnosis and prognosis.

twin (ADT), which is an extension of the IAT, also known as Probabilistic and Predictive Individual Aircraft Tracking (P²IAT) in Spiral 1 of the project. ADT enables structural damage diagnosis and prognosis by creating a multiphysics, multiscale, and probabilistic virtual model [4] of an as-built system that can integrate multiple heterogeneous and uncertain sources of information from models and data to support decisions for proactive fleet maintenance.

43.1.2 Challenges in the Damage Diagnosis and Prognosis

However, there are still some challenges to be addressed for the damage diagnosis and prognosis of complex aircraft structures utilizing the capability of the digital twin.

First, a rapid crack growth simulation method is desired to handle multi-queries in the digital twin. However, full-order fracture mechanics simulation methods cannot fulfill this requirement. In this regard, the reduced-order model (ROM), which approximates the full-order model and significantly reduces the computational burden with an acceptable level of accuracy, is a feasible alternative. A limited set of investigations [5–9] have been reported on using ROMs for fracture mechanics simulations, which are focused on establishing the relation between crack sizes/shapes to crack-front stress intensity factors (SIFs). In Ref. [10], analytical solutions are used to calculate the SIFs of simple-shape cracks. In Ref. [11], a characteristic crack length was defined and related to crack-front SIFs, where influences of detailed crack-shapes on SIFs are disregarded. Other studies limit the database of cracks by merely sampling from a sequence of cracks in a crack growth simulation [5,8,11]. With this in mind, a simple, efficient, and automatic approach, which can quickly generate models of various cracks and efficiently perform fracture mechanics simulations, is desirable for constructing the ROM, and for realizing the digital twin of complex structural components.

Second, in the realization of the ADT for aeronautical structures, it is a critical step to represent the complex damage growth process in a unified framework. The Dynamic Bayesian Network (DBN) has been widely adopted as the framework of diagnosis and prognosis in current digital twin studies of various objects, such as the aircraft wing [10], orthotropic steel deck [12], and reusable spacecraft structures [13]. However, the current applications of dynamic Bayesian networks are mainly concentrated on simple structures.

43.1.3 Objectives of This Chapter

In this chapter, we introduce a digital twin–driven damage diagnosis and prognosis approach. First a reduced-order modeling method is presented to address the timeliness challenge [14], which adopts a fast and efficient fracture mechanics simulation method, the SGBEM (Symmetric Galerkin Boundary Element Method)–FEM (Finite Element Method) coupling method to generate the fracture mechanics database of cracks with various sizes and shapes in a complex structural component, and then the ROM is constructed by the PCA-based dimensionality reduction of crack-front coordinates and a surrogate-based SIF computation. Several regression methods are explored to establish the surrogate model, and their performance is compared. Then a DBN-based digital twin is built to mirror the physical crack growth process and carry out the damage diagnosis and prognosis of complex aircraft structures, in which the ROM is used to provide the capability of fast multi-queries. Finally, the proposed method is applied to the crack growth process of a round-robin helicopter component. The results show that the digital twin can be updated with the crack measurement and improve the capability of damage diagnosis and prognosis.

In the rest of this chapter, Section 43.2 introduces the method of constructing a fracture mechanics database based on the "SGBEM super element–FEM coupling method". Section 43.3 presents a data-driven reduced-order modeling approach using the constructed database, and further compares the SIF predictions using different regression methods. In Section 43.4, the deterministic crack growth predictions of the round-robin helicopter component is conducted to demonstrate the capability of the constructed ROM. Section 43.5 shows the construction of the digital twin and conducts the damage diagnosis and prognosis. In Section 43.6, some summary completed with some summaries (Figure 43.1).

43.2 Crack-SIF Database Generation by the SGEBM–FEM Method

In general, the reduced-order modeling of the crack growth can be divided into two stages: the offline stage and the online stage. In the offline stage, a database containing various cracks is established by full-order fracture mechanics simulations, and the ROM is constructed using the database; while in the online stage, the ROM is used in the crack growth prediction and the remaining life assessment as a surrogate of the full-order simulation. During this procedure, constructing the database of fracture mechanics simulations is crucial, which nevertheless requires significant computational burden, especially when cracks in complex three-dimensional structural components are considered.

To construct the database of SIFs for various cracks, there are generally three steps: (1) Characterize cracks of interests with several parameters, and sample in the parameter space to generate cracks of various locations/shapes/sizes. (2) For each sample crack, construct the CAE model with meshes, material parameters, loads, etc. (3) Conduct batch fracture mechanics simulations to obtain the SIFs for each sample crack. In this process, the human labor of mesh generation in step (2) and the computational burden to simulate the structure with various cracks can be significantly reduced by exploring the advantages of the "SGBEM super element–FEM coupling method" [15,16].

In the SGEBM super element–FEM coupling method [15], modeling the uncracked structure (by FEM) and the cracked local sub-domain (by SGBEM) with different methods can significantly simplify and accelerate the process of constructing the database of SIFs for various cracks. On the one hand, because of the independence of the FEM mesh and crack-surface mesh in the preprocessing stage, step (2) can be significantly simplified. It is worth noticing that the FEM of the uncracked global structure needs to be constructed only once, and the surface mesh of crack with various shapes/sizes can be automatically done by parameterized modeling tools, as shown in the next subsection. On the other hand, due to the accuracy of SGBEM for modeling a small cracked region and the efficiency of FEM for modeling uncracked complex structure, such a combined simulation approach can significantly reduce the computational time in step (3). The computational accuracy and simulation efficiency of the coupled SGBEM–FEM method for fracture and fatigue analyses of 2D and 3D complex structures have also been systematically evaluated and demonstrated in Refs. [17,18]. In this section, we demonstrate the procedure of constructing the database of various cracks by this approach using a round-robin helicopter component, as shown in the next two subsections.

43.2.1 Round-Robin Helicopter Component and the FEM Model

A round-robin helicopter component, the lift frame described in Cansdale and Perrett [19] and Irving, Lin and Bristow [20], is used for demonstration.

The geometry of the component is a flanged plate with a central lightning hole, where a 2mm corner crack is initiated (see Figure 43.2), according to which the FEM was constructed. Then the geometry model is constructed in SOLIDWORKS and imported in MSC/PATRAN to generate the FEM of the uncracked structure, where uniform tensile loads are applied at both ends, and the materials and load spectrum are given as follows.

As shown in Ref. [21], the helicopter component was made of AL 7010-T73651, whose average room temperature fracture toughness in the (L-T) orientation is $33.4\,\text{MPa}\sqrt{\text{m}}$. It is assumed to have a Young's modulus of 70,000 MPa and a Poisson's ratio of 0.3 in this study.

The component was subjected to the Asterix spectrum, a representative helicopter spectrum loading, which is also shown in Figure 43.2. The Asterix spectrum was derived from the strain data measured on a helicopter lift frame. It has an average R value of 0.82 and a limited number of negative R ratio excursions [22]. There are very few load cycles with an R ratio of less than $R=0.7$. The Asterix spectrum, which represents a single load block, corresponds to 190.5 flight hours with 371,610 load cycles. In this paper, the largest far field stress in the spectrum is set to 130 MPa.

43.2.2 Constructing Sample of Various Cracks Based on Parametric Modeling

The SGBEM super element–FEM coupling method takes two inputs for the fracture mechanics simulation, the FEM of the uncracked structure and the crack-surface mesh. In order to construct the database of SIFs with various cracks, a large number of cracks with various sizes/shapes need to be generated and meshed, which can be time-consuming and labor-intensive if done manually. In this paper, a PATRAN-based parametric modeling tool is implemented to automate and speed up the process.

The process of crack growth in the helicopter component can be divided into several stages. At each stage, the PATRAN parameterized modeling script is generated by an in-house MATLAB code, which is then run automatically to

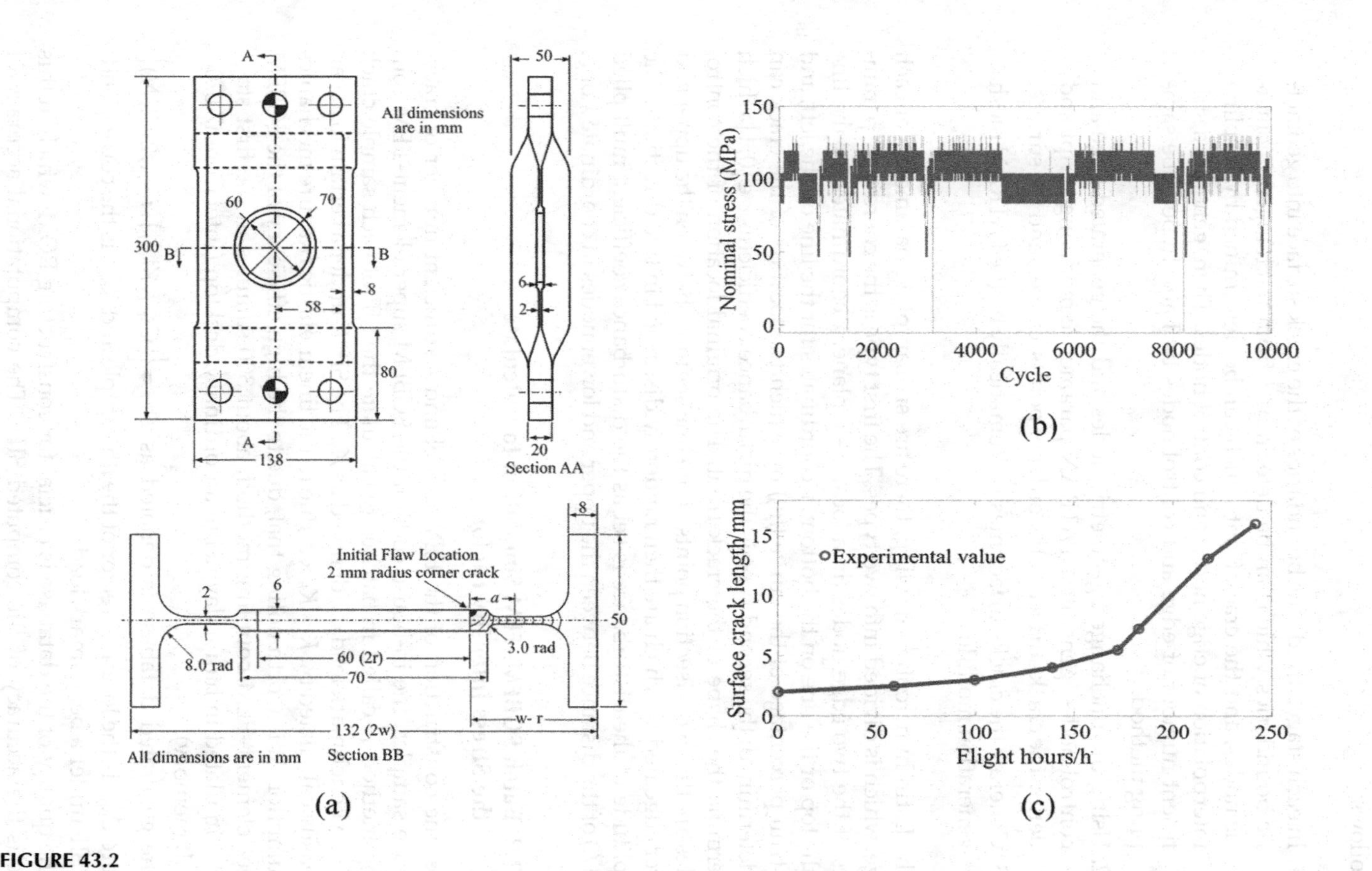

FIGURE 43.2
Geometry, load spectra, and test data in the round-robin test: (a) geometry of the helicopter lifting frame, (b) the Asterix load spectrum, and (c) the test data of crack growth.

generate a meshed crack surface. Such a parameterized modeling process is as follows:

1. In each stage, divide the boundaries of the crack surface into geometric boundaries which can be determined from unchanged geometric nodes, and the crack front which can be determined by B-spline interpolation of edge nodes and control nodes. Then, generate a sufficient number of edge and control nodes by Latin hypercube sampling method.
2. Using the unchanged geometric nodes and the generated edge and control nodes, generate a PATRAN parametric modeling script and define the crack surface in the order of points, curves, and the surface.
3. Generate the crack-surface mesh automatically by the PATRAN mesh generation module.

In the Latin hypercube sampling, there are six sampling parameters at each stage, which is divided into two types. The first type is the coordinate parameters of the two edge nodes (in a local $x-y$ plane, x coordinates for the line on the top or the line on the bottom, y coordinates for the line on the left, and the trim percentage of the curve (tp)) by which the location of the crack can be determined; the second type is the disturbance coefficients $\epsilon \in [0,1]$ which determine the shape of the crack front. The original location of the control nodes are the two trisection points of the line segment between the upper and lower edge nodes, which are then randomly disturbed in two directions. The amplitude of the disturbance is set as the disturbance coefficient multiplied by 1/3 of the distance between the upper and lower nodes in coordinate x or y.

43.2.3 Batch SGBEM–FEM Simulation to Calculate the Stress Intensity Factor

After the construction of the FEM model and a sufficient number of crack-surface samples, the in-house code of the SGBEM super element–FEM coupling method is called in batches to calculate the SIF of each sample crack under a benchmark load σ_b, denoted as K_b. The SIF K under a certain load can be easily calculated by $K = K_b \times \sigma$ due to the linear elastic fracture mechanics assumption. It is worth noting that, due to the high efficiency of the SGBEM super element–FEM coupling method, about 300 samples in the first stage were simulated in half a day using an ordinary desktop (Intel Core i7-6700, 16 GB memory).

The generated database is denoted as $\mathcal{D}_{\text{ori}} = \left\{ \left(\mathbf{C}_{\text{ori},i}, K^b_{\text{ori},i} \right) : i = 1,2,\ldots,N) \right\}$, where $\mathbf{C}_{\text{ori},i}$ is the location vector of the ith sample and $K^b_{\text{ori},i}$ is the corresponding SIF under a benchmark load.

The quality of the database is crucial for constructing ROM, which in this case is the accuracy of the computed SIFs. The computational accuracy of the SGBEM–FEM method has been systematically verified in our two-part paper in 2013 [17,18], where there are many examples of 2D through-thickness cracks and 3D embedded/surface cracks.

43.2.4 Nodal Spacing Equalization

In the parametric crack-surface modeling process, the generated number of nodes at the crack front may be different for various cracks, and the nodal spacing may also be non-uniform. Therefore, to establish the relation between nodal coordinates and the SIFs at crack fronts for different cracks, a nodal spacing equalization procedure in Ref. [5] is used here to set the same number of equidistant nodes along the crack front for a consistently parameterized representation of the crack fronts between different cracks.

The node spacing equalization aims to set the same number of equidistant nodes on the front curve to generate a consistently parameterized representation of the crack fronts between different cracks. In this paper, the method in Ref. [5] is used, but corresponding improvements are made for specific problems.

For a crack-front curve with non-equispaced nodes, we can represent the nodes with coordinates:

$$\mathbf{X} = \{\mathbf{X}_1, \mathbf{X}_2, \ldots, \mathbf{X}_m\} \tag{43.1}$$

where m is the number of the nodes and $\mathbf{X}_i$ is the location vector for the ith node.

A cubic spline interpolation is used to generate a spline representation $f_1(\xi)$ of the crack front, where $\xi \in [0,1]$ is the spline's natural parameter. Assuming that $p+1$ equispaced nodes along the crack front are to be generated, the spline parameter for these nodes are:

$$\mathbf{s} = \left\{0, \frac{1}{p}, \frac{2}{p}, \ldots, \frac{p-1}{p}, 1\right\} \tag{43.2}$$

and the location vector of the generated equispaced nodes along the crack-front can be calculated by $\mathbf{x} = f_1(\xi)$:

$$\mathbf{x} = \{\mathbf{x}_0, \mathbf{x}_2, \ldots, \mathbf{x}_p\} \tag{43.3}$$

For each node $\mathbf{x}_i$, we can have two representations: Cartesian coordinate and corresponding spline parameter.

Meanwhile, the spline parameter values corresponding to the original non-equispaced nodes $\mathbf{X}_i$ can be obtained by $\xi = f_1^{-1}(\mathbf{X})$, which can be expressed as:

$$\mathbf{S} = \{0, \xi_1, \xi_2, \ldots, \xi_{m-1}, 1\} \tag{43.4}$$

The cubic spline interpolation is also used to fit the SIF.

$$\mathbf{K}^b = f_2(\xi) \tag{43.5}$$

Then the corresponding SIF at the newly generated equispaced nodes $\mathbf{K}^b_{\text{equi}}$ can be obtained by the spline function f_2.

$$\mathbf{K}^b_{\text{equi}} = \{K_0, K_1, K_2, \ldots, K_p\} \tag{43.6}$$

By equispacing all samples, a set of crack-front curves with a consistently parameterized representation can be obtained, which can be represented as a matrix $\mathbf{C}_{\text{equi}}$. Then the database with equispaced nodes is denoted as $\mathcal{D}_{\text{equi}} = \left\{\left(\mathbf{C}_{\text{equi},i}, K^b_{\text{equi},i}\right) : i = 1,2,\ldots,N\right)\right\}$.

43.3 Reduced-Order Fracture Mechanics Modeling of Cracked Complex Structure

In this paper, a ROM $\mathcal{M}^r$ is used to replace the time-consuming fracture mechanics simulation model $\mathcal{M}$. In this model, the input consists of parameterized parameters of the crack-front, and the output is the corresponding SIFs along the crack front under a benchmark load $\mathbf{P}^b$. The ROM can be expressed as:

$$\mathbf{K}^b = \mathcal{M}^r(\mathbf{a}) \tag{43.7}$$

where $\mathbf{a}$ are crack parameters, and $\mathbf{K}^b$ are SIFs along the crack front under a benchmark load $\mathbf{P}_b$, which is set to be 130 MPa.

The ROM consists of two parts: the PCA-based dimensionality reduction of crack-front nodal coordinates and the surrogate-based SIF prediction, as described below.

43.3.1 PCA-Based Dimensionality Reduction of Crack-Front Nodal Coordinates

In the simulation database, the vector $\mathbf{C}_{\text{equi},i}$, which contains the Cartesian coordinates of equispaced nodes at the crack front, is a high-dimensional representation of the crack front. Directly using $\mathbf{C}_{\text{equi},i}$ to construct the surrogate model would affect the computational burden and the prediction accuracy of the surrogate model. In this study, the Principal Component Analysis (PCA) is firstly conducted for the dimensionality reduction of nodal coordinates at the crack front.

In the matrix $\mathbf{C}_{\text{equi}}$, each column represents a sample crack, the total number of samples is denoted as N, and the number of rows $M = 2p$ in which p represents the number nodes along the crack-front or original features. The PCA is conducted to reduce the dimension of the database, in the following process.

First, the $\mathbf{C}_{\text{equi}}$ is subtracted by the mean value of the total samples $\bar{\mathbf{C}}$.

$$\mathbf{A} = \mathbf{C}_{\text{equi}} - \bar{\mathbf{C}}_{\text{equi}} \tag{43.8}$$

Then the Singular Value Decomposition (SVD) is executed for **A**.

$$\mathbf{A}_{m\times n} = \mathbf{U}_{M\times M}\Sigma_{M\times N}\mathbf{V}_{N\times N}^{T} \tag{43.9}$$

where $\mathbf{U} \in \mathbb{R}^{M\times M}$ and $\mathbf{V} \in \mathbb{R}^{N\times N}$ are orthogonal matrices, and $\Sigma = \text{diag}_{M\times N}\{\sigma_1, \sigma_2, \ldots, \sigma_N\}$ contains the singular values $\sigma_1 \geq \sigma_2 \geq \cdots \geq \sigma_M \geq 0$.

After the SVD of **A**, the first q singular values can be used to approximate the matrix **A**.

$$\mathbf{A} \approx \mathbf{U}_{M\times q}\Sigma_{q\times q}\mathbf{V}_{q\times N}^{T} \tag{43.10}$$

The value of q is determined so that the cumulative variance contribution rate of principal components β, or the retained cumulative information, should exceed a threshold r. In this paper, r is set to be 0.99.

$$\beta(q) = \frac{\sum_{i=1}^{q}\sigma_i^2}{\sum_{j=1}^{N}\sigma_j^2} >= r \tag{43.11}$$

With this approximation, a crack front $\mathbf{C}_i$ can be projected into a q-dimensional subspace by

$$\mathbf{W}_i = \mathbf{U}_{M\times q}^{T}\left(\mathbf{C}_i - \bar{\mathbf{C}}\right) \tag{43.12}$$

where $\mathbf{W}_i$ is the principal component coefficients corresponding to the first q singular values, which can also be regarded as a q-dimensional latent space representation of the crack-front nodal coordinates $\mathbf{C}_{\text{equi},i}$.

Conversely, the formula below can be used to restore the $\mathbf{W}_i$ to the full-order space.

$$\mathbf{C}_i = \mathbf{U}_{M\times q}\mathbf{W}_i + \bar{\mathbf{C}} \tag{43.13}$$

For all samples in the database, the principal component coefficients of each sample can be combined to form a matrix **W**, which is a low-order representation of $\mathbf{C}_{\text{equi}}$ in the database $\mathcal{D}_{\text{equi}}$. Thus, the final database can be denoted as $\mathcal{D}=\left\{\left(\mathbf{W}_i, K_i^b\right): i=1,2,\ldots,N)\right\}$.

43.3.2 Surrogate-Based Stress Intensity Factor Prediction

The database $\mathcal{D}$ is used to construct the surrogate model, in which the input is the principal component coefficient **W** and output is the corresponding SIF. In this paper, four widely used methods are adopted to construct the models, which are least square support vector model (LS-SVM), Gaussian process regression (GPR), multi-variable polynomial regression (MPR), and artificial neural network (ANN), respectively. For succinctness the four methods are not discussed in detail, interested readers can refer to Refs. [23–25] for more information.

Here, the accuracy of the four surrogate models trained with the database of the first stage (273 samples) are shown. The training set consists of 205 samples and the test set consists of 68 samples. The error is defined as the Mean Absolute Percentage Error (MAPE):

$$\varepsilon = \frac{1}{q \times n_p} \sum_{i=1}^{q} \sum_{j=1}^{n_p} \frac{\left|\hat{y}_{ij} - y_{ij}\right|}{y_{ij}} \times 100\% \tag{43.14}$$

where N is the number of samples in the training or test set. q is the number of component coefficients, $\hat{y}$ is the prediction by the surrogate model, and y is the SIFs obtained by the original full-order simulations.

The training error and test error of the four ROMs are shown in Table 44.1. As can be seen, in this stage, the accuracy of the ROMs are similar, which are all about 3%–5%. In addition, as a surrogate to the high-fidelity model, the computational efficiency of the ROM is much higher. The computation time per call of the ROM is about 0.015 seconds, which is roughly four orders of magnitude smaller than the time needed for the high-fidelity simulation.

TABLE 44.1

Comparison of Training and Test Error of Four Reduced-Order Models

Model	LS-SVR	LS-SVR	LS-SVR	LS-SVR
Training error	0.0277	0.0222	0.0287	0.0324
Test error	0.0307	0.0287	0.0434	0.0508

43.4 Validation of the Reduced-Order Crack Growth Prediction

In this section, we demonstrate how to embed the ROM into a deterministic crack growth prediction algorithm, which can take online monitored load data and give a real-time prediction of crack growth. This provides the multi-query capability for the digital twin accounting for various uncertainties in the crack growth. Since the four regression models show comparable performance in the prediction of crack-front SIFs, the ROM trained by the LS-SVR method is chosen for demonstration.

43.4.1 Fatigue Crack Growth Law

Fatigue crack growth laws, which relate the crack growth rate da/dN to fracture mechanics parameters (such as the range of SIFs ΔK), can significantly influence the crack growth predictions. In this paper, as adopted in Ref. [21], a generalized version of the Frost–Dugdale law is used as the crack growth model:

$$\frac{da}{dN} = Ca^{1-\gamma/2}\Delta K^{\gamma} \tag{43.15}$$

where C and γ are material constants, $C = 1.28 \times 10^{-11}$, $\gamma = 3$.

43.4.2 Crack Growth Simulation Process

The simulation process of fatigue crack growth is illustrated as follows. The crack growth is calculated for a number of load cycles in each step. Within the step, for a crack front with equispaced nodes, it is first determined which stage it belongs to, and then projected to the PCA subspace by Eq. (43.12) first to obtain principal component coefficients, which are then fed into the reduced-order fracture mechanics model to calculate the SIFs under the reference stress $\hat{K}^b$. Then, according to the load cycle in the current step, the normalized stress range $\Delta\sigma$ can be extracted, and the SIF range ΔK can be obtained by Eq. (43.16). It should be noted that, at stages 3 and 5, the lower end of the crack front is located in the lower line segment, and the two cracks may be separated. Therefore, interpolation between the two leading edges is required.

$$\Delta\hat{K} = \Delta\sigma \cdot \hat{K}^b \tag{43.16}$$

The crack growth increment can be calculated by the crack growth model shown in Eq. (43.15). By adding the increment to the current crack front, a new crack front can be obtained. The boundary of the newly computed crack-front may not fall on the boundary of the structure, and the new nodal spacing may not be equal. However, the boundary of the newly determined crack front may not fall exactly on the boundary of the structural geometry. In such cases, linear interpolation and extrapolation are used to correct the boundary of the after-growth crack-front curve, which follows the node equispacing. Such a process is repeated for each load step, until the critical crack size is reached. In this study, the critical crack size is set to be 20 mm.

43.4.3 Results and Discussion

The predicted path of crack growths from 2 to 20 mm by the ROM is shown in Figure 43.3a. It can be seen that, starting from a 2 mm corner crack, the

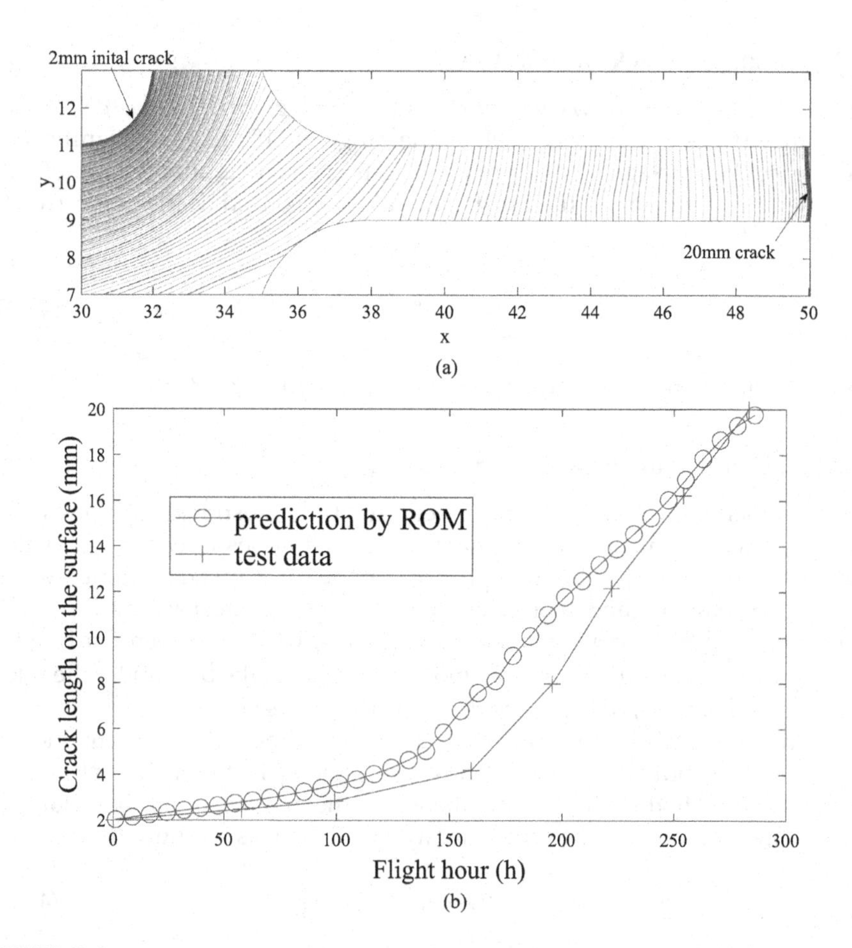

FIGURE 43.3
Crack growth path and time from 2 to 20 mm. The initial crack is a 2 mm corner crack in the upper left corner of the figure a (curve at the beginning), and the 20 mm crack is on the right side of the figure a (line at the end) [14]. (a) Crack growth path and (b) crack growth time, compared with the round-robin test data [19].

crack gradually grows across the thickness and then transforms into a through-thickness crack.

The comparison of the crack growth simulation and the round-robin test data is shown in Figure 43.3b, from which we can see that fairly good agreements are obtained. In the first 150 hours, the predicted growth is slightly faster than that in the test, while after that, the predicted crack growth is slightly slower.

43.5 Digital Twin–Driven Damage Diagnosis and Prognosis

In traditional aircraft structural management, the influence of various uncertainties on fatigue crack growth is not considered, which affects the damage state of each aircraft significantly. In order to effectively track the damage evolution to support the arrangement of the inspection and maintenance, it is necessary to comprehensively consider the influence of various uncertainties on the damage evolution. In this section, a digital twin in the forms of DBN is constructed, which can integrate various epistemic and aleatory uncertainties. (Epistemic uncertainty is produced by lack of knowledge and is reducible, whereas aleatory uncertainty is a naturally occurring variable that is irreducible [10].) Based on the digital twin model, the diagnosis and prognosis of the damaged structures are conducted, where the measured crack lengths are used as observations to update the digital twin to improve the accuracy of damage prognosis and remaining useful life prediction.

43.5.1 Constructing the Digital Twin by the Dynamic Bayesian Network

For the crack growth problem, the state of interest is the crack length vector **a**. The fatigue crack growth law is as follows:

$$\frac{da}{dN} = f(\mathbf{a}, \Delta\mathbf{K}; \mu) \tag{43.17}$$

where $\frac{da}{dN}$ are the increments of crack lengths per cycle, **a** are the crack lengths, $\Delta\mathbf{K}$ are the ranges of SIFs, and μ are the material parameters, which are considered as uncertain parameters.

Appending unknown material parameters μ to crack lengths **a** at time step t, the augmented state vector is defined as $\mathbf{X}_t = [\mathbf{a}_t, \mu_t]^T$. The functional relationship of crack lengths between two adjacent time steps can be obtained as follows:

$$\mathbf{a}_t = \mathbf{a}_{t-1} + \frac{d\mathbf{a}_{t-1}}{dN}\Delta N \tag{43.18}$$

Then the augmented state function can be obtained:

$$\mathbf{X}_t = \begin{bmatrix} \mathbf{a}_t \\ \mu_t \end{bmatrix} = \begin{bmatrix} \mathbf{a}_{t-1} + \dfrac{d\mathbf{a}_{t-1}}{dN}\Delta N \\ \mu_{t-1} \end{bmatrix} \tag{43.19}$$

The measurement of the crack lengths $\mathbf{a}_t$ is assumed as $\mathbf{Z}_t \sim N\left(\mathbf{a}_t, \sigma_\varepsilon^2\right)$ to consider the measurement noise. However, due to the limitation of measurement tools, the full shape of the crack front cannot be accurately obtained. Sometimes only the crack length at the surface of the structure is available. Then the corresponding observation function is obtained:

$$z_t = a_t^{\text{surf:enumerate}} + \varepsilon_t \tag{43.20}$$

where a_t^{surf} is the crack length at the node where the crack front intersects the structural surface, and random variable ε_t is the standard deviation of measurement results, which is determined by the accuracy of measurement tools.

Further, the crack growth process is represented by the DBN. The DBN is an extension of the Bayesian network in the time domain, which is effective for integrating various sources of uncertainty and heterogeneous information and tracking the changes of state of time-variant systems. When the observation of any child node is acquired, the DBN is updated by Bayesian inference. In DBN, the following two tasks need to be accomplished by Bayesian inference:

1. Forward propagation, i.e., obtaining the state variables $\mathbf{X}_t$ according to the state variables $\mathbf{X}_{t-1}$ at the previous time step and the conditional probability distribution (CPD) between the networks at two adjacent time steps;
2. Backward inference, i.e., updating the joint probability distribution of the state variables $\mathbf{X}_t$ when any child node is observed.

The definition of diagnosis step and prognosis step in Ref. [10] is adopted here, i.e., a time step with only forward propagation is defined as a "prognosis step", which is implemented by the ROM constructed in Section 43.3; and a time step with both forward propagation and backward inference is defined as a "diagnosis step". The diagnosis step is performed only when damage is observed, and the prognosis step is performed in other cases. At the diagnosis step, the Bayesian inference algorithm is used to update the joint probability distribution of $\mathbf{X}_t$ based on the observation result.

At time step 1, the distribution of random variables in the BN is the prior distribution provided by the user. The prior distribution of BN at other time steps is obtained by propagating the posterior distribution of BN at the previous time step.

The particle filter is widely used as the inference algorithm for DBN because of its ability to deal with non-Gaussian state variables [10]. There are two challenges in the PF: (1) degeneracy, meaning that all but one particle will have negligible weights after a few iterations and (2) sample impoverishment, meaning the loss of sample diversity. In this chapter, RPF is used as the Bayesian inference algorithm for DBN.

43.5.2 Digital Twin Initialization and Relevant Parameter Setting

In the adopted generalized Frost–Dugdale law, two parameters, i.e., C and γ, affect the accuracy of diagnosis and prognosis. In this paper, both C and γ are considered as uncertain parameters to be estimated. Due to the influence of measurement errors, the initial size of the crack in the component may not be exactly 2mm but a certain deviation from it. In this paper, the initial size of the crack is also regarded as an uncertain parameter. Based on the deterministic crack growth model Eq. (43.15), in order to consider the influence of process noise, the random crack growth process is described as Ref. [26]:

$$\mathbf{a}_t = \mathbf{a}_{t-1} + \exp(\omega_{t-1}) \left.\frac{d\mathbf{a}}{dN}\right|_{t-1} \Delta N \tag{43.21}$$

where $\mathbf{a}_t$ are the crack lengths at time step t, ΔN is the number of load cycles experienced from time step $t-1$ to time step t, and ω_{t-1} is a noise term following Gaussian distribution $N\left(-\frac{\sigma_\omega^2}{2}, \sigma_\omega^2\right)$, leading to $E(\exp(\omega_{t-1})) = 1$ [27]. The crack growth rate $\left.\frac{da}{dN}\right|_{t-1}$ is expressed as follows:

$$\left.\frac{d\mathbf{a}}{dN}\right|_{t-1} = C_{t-1}\mathbf{a}_{t-1}^{1-\gamma_{t-1}/2}\Delta\mathbf{K}_{t-1}^{\gamma_{t-1}} \tag{43.22}$$

In Eq. (43.22), model parameters C and γ are replaced by C_{t-1} and γ_{t-1}, respectively, to consider the uncertainty of model parameters. The prior distribution of C is set to $\log(C_t|_{t=0}) \sim U(-10.9, -10.8)$ due to insufficient knowledge. Here $\log(C_t)$ is used instead C which is a very small value compared to other quantities. The prior distribution of γ is set to $\gamma_t|_{t=0} \sim N(3, 0.1^2)$, with the expectation equal to 3 as defined in Ref. [21]. The prior distribution of the corner crack length is adopted as $a_t|_{t=0} \sim N(2, 0.1^2)$ to consider the uncertainty of the initial crack size. The measurement error is set as $\varepsilon_t \sim N(0, 0.05)$ to consider the measurement uncertainty, and σ_ω^2 is set as 0.1 to consider the effect of process noise on the crack growth.

The number of particles is set as $N_s = 2{,}000$ to balance the computational cost and simulation accuracy. At the initial time step $t = 0$, the particle set $\left\{\mathbf{x}_{t=0}^{(i)}\right\}_{i=1}^{N_s} = \left\{\mathbf{a}_{t=0}^{(i)}, \log C_{t=0}^{(i)}, \gamma_{t=0}^{(i)}\right\}_{i=1}^{N_s}$ is randomly sampled from the prior probability density functions (PDFs) mentioned above. After that, each particle $\mathbf{x}_t^{(i)} = \left[\mathbf{a}_t^{(i)}, \log C_t^{(i)}, \gamma_t^{(i)}\right](i = 1, 2, \ldots, N_s)$ evolves with time step. The increment of loading cycle in Eq. (43.21) is set as $\Delta N = 1{,}000$ to balance the computational cost and the accuracy. When the time step corresponding to the experimental observation (i.e., the diagnosis step) is reached, the measurement is input into the DBN to obtain the posterior PDF for each parameter in state vector $X_t = \left[\mathbf{a}_t, \log C_t, \gamma_t\right]$.

43.5.3 Discussion of the Diagnosis and Prognosis Results

Figure 43.4a and b shows the comparison of probabilistic crack growth with and without diagnoses. Compared with the case without diagnoses, the case with diagnoses can effectively reduce the uncertainty of crack growth. The uncertainty of crack size distribution approximately decreases to the measurement error at each inspection and increases gradually between two adjacent inspections. The results confirm that the proposed digital twin model can improve the diagnosis and prognosis of crack growth using DBN and observation data.

Figure 43.4c and d shows the updated PDFs of logC and γ with inspections, respectively. After several updates, the two PDFs become narrower and taller, and their expectations are updated to specific values that reflect the real crack growth process. In the original set $\left\{\mathbf{x}_{t=0}^{(i)}\right\}_{i=1}^{N_s}$, the particles that do not conform to the real process are eliminated. It is noted that there are multiple peaks in Figure 43.4c and d. It is speculated that different parameter combinations will make the crack growth prediction results in the previous interval tend to the same inspection result, especially in the case of partial observations.

43.6 Summary

This chapter introduced a damage diagnosis and prognosis approach for complex aeronautical structures, which consists of two parts: the real-time crack growth prediction of complex structures based on PCA and regression methods, and the DBN-based damage diagnosis and prognosis of complex structures. In the first part, a real-time simulation approach to conduct deterministic/probabilistic predictions of crack growth is developed, by combining the advantages of the SGBEM super element–FEM coupling method for fracture mechanics simulations to construct an offline database and the advantages of the reduced-order modeling approach to accelerate the online simulation efficiently. The proposed method is advantageous for

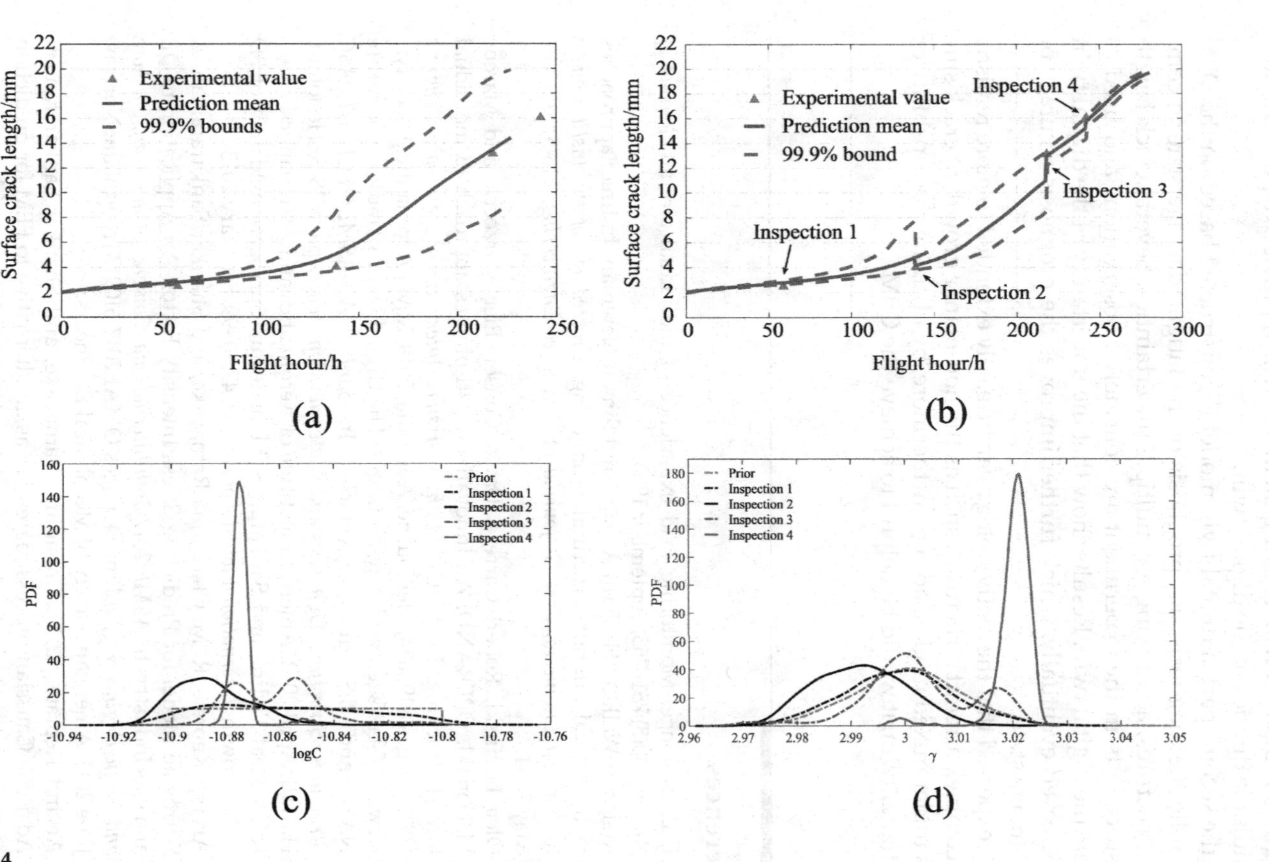

FIGURE 43.4
Comparison of probabilistic crack growth with and without diagnosis: (a) without diagnosis, (b) with diagnosis, (c) updating of the log C, and (d) updating of the γ.

the real-time simulation of the growth of complex-shaped cracks in complex structures compared with existing methods, which provides a powerful tool for multiple queries in the digital twin.

In the second part, a digital twin model is constructed based on the DBN to handle the diagnosis and prognosis of the fatigue damage state in complex structures, accounting for multiple uncertainties. Several crack length measurements in the experiment are considered observations and used to update the digital twin. Results show that the uncertainty of the digital twin is decreased eventually, which further improves the prognosis accuracy of the damage state.

It is expected that the methodology can be easily extended to more realistic and complex aircraft structures, such as the non-planar growth of crack surfaces under mixed mode loading. Furthermore, by utilizing the capability of the digital twin, it would be helpful to achieve the CBM.

References

1. J. B. de Jonge. Monitoring load experience of individual aircraft. *Journal of Aircraft*, 30(5):751–755, September 1993.
2. Malcolm Wallace, Hesham Azzam, and Simon Newman. Indirect approaches to individual aircraft structural monitoring. *Proceedings of the Institution of Mechanical Engineers, Part G: Journal of Aerospace Engineering*, 218(5):329–346, May 2004.
3. Oleg Levinski, Robert Carrese, David Conser, Pier Marzocca, and Marcus McDonald. OPERAND: An innovative multi-physics approach to individual aircraft tracking. In *AIAC18: 18th Australian International Aerospace Congress (2019)*, pages 849–854, Melbourne, Australia, 2019. Royal Aeronautical Society.
4. Edward Glaessgen and David Stargel. The digital twin paradigm for future NASA and U.S. air force vehicles. In *53rd AIAA/ASME/ASCE/AHS/ASC Structures, Structural Dynamics and Materials Conference*, page 1818, Honolulu, Hawaii, April 2012. American Institute of Aeronautics and Astronautics.
5. K. Hombal Vadiraj and Sankaran S. Mahadevan. Surrogate modeling of 3D crack growth. *International Journal of Fatigue*, 47:90–99, February 2013.
6. Arvind Keprate, R. M. Chandima Ratnayake, and Shankar Sankararaman. A Surrogate Model for Predicting Stress Intensity Factor: An Application to Oil and Gas Industry. In *ASME 2017 36th International Conference on Ocean, Offshore and Arctic Engineering*, volume 4, pages OMAE2017-61091, Trondheim, Norway, June 2017. American Society of Mechanical Engineers.
7. Arvind Keprate, R. M. Chandima Ratnayake, and Shankar Sankararaman. Adaptive Gaussian process regression as an alternative to FEM for prediction of stress intensity factor to assess fatigue degradation in offshore pipeline. *International Journal of Pressure Vessels and Piping*, 153:45–58, June 2017.

8. Patrick E. Leser, Jacob D. Hochhalter, James E. Warner, John A. Newman, William P. Leser, Paul A. Wawrzynek, and Fuh-Gwo Yuan. Probabilistic fatigue damage prognosis using surrogate models trained via three-dimensional finite element analysis. *Structural Health Monitoring*, 16(3):291–308, May 2017.
9. Shankar Sankararaman, You Ling, Christopher Shantz, and Sankaran Mahadevan. Uncertainty quantification in fatigue crack growth prognosis. *International Journal of Prognostics and Health Management*, 2(1):1–15, 2011.
10. Chenzhao Li, Sankaran Mahadevan, You Ling, Sergio Choze, and Liping Wang. Dynamic Bayesian network for aircraft wing health monitoring digital twin. *AIAA Journal*, 55(3):930–941, March 2017.
11. Patrick E. Leser, James E. Warner, William P. Leser, Geoffrey F. Bomarito, John A. Newman, and Jacob D. Hochhalter. A digital twin feasibility study (Part II): Non-deterministic predictions of fatigue life using in-situ diagnostics and prognostics. *Engineering Fracture Mechanics*, 229:106903, April 2020.
12. Jin Zhu, Wei Zhang, and Xuan Li. Fatigue damage assessment of orthotropic steel deck using dynamic Bayesian networks. *International Journal of Fatigue*, 118:44–53, January 2019.
13. Yumei Ye, Qiang Yang, Fan Yang, Yanyan Huo, and Songhe Meng. Digital twin for the structural health management of reusable spacecraft: A case study. *Engineering Fracture Mechanics*, 234:107076, July 2020.
14. Xuan Zhou, Shuangxin He, Leiting Dong, and Satya N. Atluri. Real-time prediction of probabilistic crack growth with a helicopter component digital twin. *AIAA Journal*, 60(4):2555–2567, April 2022.
15. Leiting Dong and Satya N. Atluri. SGBEM (using non-hyper-singular traction BIE), and super elements, for non-collinear fatigue-growth analyses of cracks in stiffened panels with composite-patch repairs. *CMES: Computer Modeling in Engineering & Sciences*, 89(5):415–456, 2012.
16. Shuangxin He, Chaoyang Wang, Xuan Zhou, Leiting Dong, and Satya N. Atluri. Weakly singular symmetric galerkin boundary element method for fracture analysis of three-dimensional structures considering rotational inertia and gravitational forces. *Computer Modeling in Engineering & Sciences*, 131(3):1857–1882, 2022.
17. Leiting Dong and Satya N. Atluri. Fracture & fatigue analyses: SGBEM- FEM or XFEM? Part 1: 2D structures. *CMES: Computer Modeling in Engineering & Sciences*, 90(2):91–146, 2013.
18. Leiting Dong and Satya N. Atluri. Fracture & fatigue analyses: SGBEM- FEM or XFEM? Part 2: 3D solids. *CMES: Computer Modeling in Engineering & Sciences*, 90(5):379–413, 2013.
19. R. Cansdale and B. Perrett. The helicopter damage tolerance round robin challenge. In *Workshop on Fatigue Damage of Helicopters*, volume 12, pages 99–128, Pisa, Italy, 2002. University of Pisa.
20. P. E. Irving, J. Lin, and J. W. Bristow. Damage tolerance in helicopters report on the round Robin challenge. In *59th American Helicopter Society Annual Forum*, pages 1642–1652, Fairfax, VA, 2003. Vertical Flight Society.
21. Ung Hing Tiong and Rhys Jones. Damage tolerance analysis of a helicopter component. *International Journal of Fatigue*, 31(6):1046–1053, June 2009.
22. Robert E. Vaughan and Jung Hua Chang. Life prediction for high cycle dynamic components using damage tolerance and small threshold cracks. In *59th American Helicopter Society Annual Forum*, volume 3, pages 1712–1720, Phoenix, Arizona, May 2003. Vertical Flight Society.

23. Johan A. K. Suykens, Tony Van Gestel, Joseph De Brabanter, Bart De Moor, and Joos P. L. Vandewalle. *Least Squares Support Vector Machines*. World scientific, River Edge, NJ, 2002.
24. Carl Edward Rasmussen and Christopher K. I. Williams. *Gaussian Processes for Machine Learning*. Adaptive Computation and Machine Learning. MIT Press, Cambridge, MA, 2006.
25. R. M. V. Pidaparti and M. J. Palakal. Neural network approach to fatigue-crack-growth predictions under aircraft spectrum loadings. *Journal of Aircraft*, 32(4):825–831, July 1995.
26. Jann N. Yang and Sherrell D. Manning. Stochastic crack growth analysis methodologies for metallic structures. *Engineering Fracture Mechanics*, 37(5):1105–1124, January 1990.
27. Jian Chen, Shenfang Yuan, and Xin Jin. On-line prognosis of fatigue cracking via a regularized particle filter and guided wave monitoring. *Mechanical Systems and Signal Processing*, 131:1–17, September 2019.

44

Digital Twins and Path Planning for Aerial Inspections

Antonio Bono, Luigi D'Alfonso, Giuseppe Fedele, and Anselmo Filice
University of Calabria

44.1 Introduction

Infrastructure inspection and monitoring has become increasingly important over the past decade as awareness of the perishable nature of construction materials has grown. Dam and bridge failures are among the worst infrastructure disasters, both in terms of loss of life and economic impact [1]. One of the most fatal cases in recent years is the failure of the Morandi Bridge in Genoa, Italy.

The causes of such phenomenon are diverse and are usually classified into two main categories: natural factors such as floods, scour, earthquakes, landslides and wind and human factors such as inadequate design and construction, overloading of vehicles, fire, terrorist attacks, and lack of inspection and maintenance. This last aspect is of critical importance according to Ref. [2]. Statistics on Chinese bridge failures between 2010 and 2019 have shown that lack of maintenance is one of the main causes of collapses.

The main drawback of infrastructure inspections is their cost and disruption to travelers. Unmanned aerial vehicles (UAVs) are one of the most promising low-cost alternatives for this task, with many benefits in terms of operator safety, time required, and traffic management (see Ref. [2] for an excellent cost–benefit analysis).

At a similar level is the monitoring of landfills to control and manage municipal waste, especially for information related to spatial and volumetric characteristics, or to monitor gas emissions and air quality. Again, the use of these technologies has only increased in the last 5 years.

Currently, most research on automated inspection of infrastructure using UAVs focuses on missions conducted by a single drone [3,4]. However, as is well known in the robotics community, single-robot systems are outperformed by Multi-Robot Systems (MRS) in several aspects:

DOI: 10.1201/9781003425724-52

- *Parallelism*: multiple agents can collaborate on the same task or on different tasks.
- *Robustness*: if one robot fails or is attacked, the group can still perform the task after a possible reconfiguration phase. In a monolithic solution, the failure of one robot means the failure of the mission.
- *Simplicity*: in general, agents in MRS are less complex than in monolithic solutions, where greater physical and computational capabilities are required to accomplish the same task. This leads to a lower probability of software/hardware failures.
- *Cost*: The simplicity of the agents often makes building a MRS cheaper than building a single, more skilled robot capable of the same tasks.

Of the main challenges faced in automating a multi-UAV inspection, we have focused on two in this work:

1. A fleet path planning that allows all components to perform the necessary analysis while avoiding collisions with infrastructure and other aircraft.
2. A control of the UAVs that guarantees the tracking of the generated trajectories despite the inevitable error in the actual position of the aircraft through GNSS localization systems.

Solving such problems is only the first step toward real applications.

Indeed, programmatic management of environmental and civil infrastructure inspections is managed by ad hoc information systems called management systems. However, these systems are not yet configured for a MRS and Dynamic Building Information Modeling BIM-augmented or Digital Twin approach to inspection [5,6]. In addition, BIM, a user-friendly and data-driven information platform, has not been connected to the current safety inspection process to better understand the risk issues identified for the infrastructures [7]. Also, the use of Digital Twins and Heritage Building Information Modeling (HBIM) in current World Heritage management systems has not yet been clarified.

The use of Digital Twins in a specific workflow must necessarily go hand in hand with other processes used in the environment and civil infrastructure field for whole life cycle management: BIM and specific management systems such as BMS, WDS (Waste Disposal Sites) management, which we will refer to as WMS (Waste Management System), DCMS (Dam Control Management System), or HMS (Heritage Management System) in this paper. By using the software BIM, the digital aspects of a construction project are managed with the appropriate digital software. In this work, BIM, photogrammetry, and simulation software were used. The increasing use of BIM and photogrammetric software, as we will see below, leads to innovative solutions in the areas of interest for this research work and promotes interoperability to actually enable the solution of multidisciplinary problems.

In an attempt to address the lack of integration between innovative management systems and multi-robot inspections, starting from the results obtained in Ref. [20], in this chapter we propose a workflow that combines the latest techniques used by management systems with MRS-based inspection. In particular, we show how the digital representation of civil and environmental infrastructures through digital twins is a very useful tool. Thanks to this information, the end user can easily use a digital twin representation to define the waypoints of interest for the inspection, which can later be followed by the real drones without further measurements and on-site inspections.

44.2 Proposed Methodology

The methodology proposes a generic workflow based on the intelligent use of Digital Twins and UAVs. The idea aims to integrate environmental and civil infrastructure monitoring platforms (BMS, WMS, and HMS), which are now based on algorithms for prioritizing interventions and generating inspection reports, with UAVs to achieve rapid update of the digital twin and safe inspection of infrastructure. In particular, the use of a digital twin to efficiently locate infrastructure points of interest and a methodology for automatic control and trajectory planning of a UAV is proposed. Using this new approach, in the future the use of a single manually or semi-automatically controlled drone will be replaced by a swarm of UAVs efficiently controlled using control laws tested and developed in a numerical environment using a digital twin for the infrastructure in question. Moreover, it will be possible to schedule inspections at all stages of the infrastructure's life, using additional machine learning, artificial intelligence tools, and dynamic BIM to ensure timely, fast, and safe inspection of the infrastructure, even with GPS demented. Thanks to dynamic and advanced BIM, we can visualize and manage issues and tasks and instantly assign them to project stakeholders. By collecting photos and videos and linking them to precise locations within the Digital Twin, it is possible to add a variety of on-site information to dynamic and augmented-BIM [7].

The workflow proposed in Figure 44.1 provides an overview of the steps to be followed in the case of deploying a swarm of UAVs in automatic mode, with modifications when we work with regular or articulated infrastructures. Simulations of the controller and reference trajectories were performed based on two digital twins.

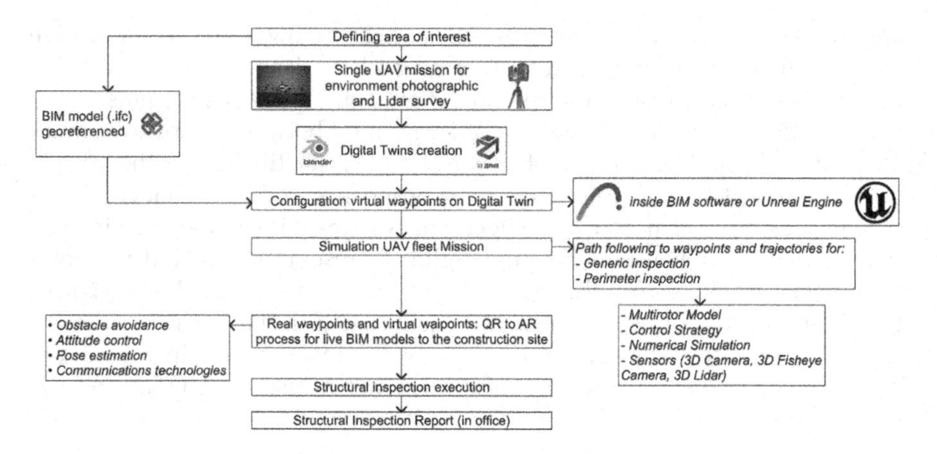

FIGURE 44.1
Workflow of the proposed method.

44.2.1 Digital Twins Creation

Regarding the methodology proposed in this document to obtain a digital twin representation of the infrastructure, two alternatives can be distinguished depending on whether a digital infrastructure model is available:

1. If a 3D model such as BIM exists, it is possible to work with an interoperable file of Industry Foundation Class (IFC) type. To import this model into Unreal Engine, it must be georeferenced and transformed. Integrating the BIM model into the 3D territorial context means using coordinates, GIS data, cartographic maps, orthophotos, topographic databases, and technical maps. With this basic information, we can realistically reconstruct the three-dimensional territorial context in which our 3D model will be located. This alternative concerns new structures designed according to the philosophy of BIM.
2. There are two options for twinning when the only available information is the paper project, which is always the case for old structures: either create the model BIM using the information on paper, or perform a new survey using an UAV with LiDAR and photographs to create a georeferenced photogrammetric model. In the first case, it is necessary to manually create a 3D model from which a BIM can be created. In the second case, photogrammetry can be performed instead with a single UAV mission equipped with LiDAR and supported by a laser scanner total station if needed. Dedicated software such as 3DF Zephyr can then be used to create an already georeferenced network. Finally, using BIM, software such as Archicad, a

FIGURE 44.2
Santa Liberata church (Santo Stefano di Rogliano, Cosenza – Italy) digital twin creation, mesh with texture.

BIM, is created from the mesh file. To simulate the real drone in this realistic digital model, one can create a 3D model of the aircraft itself using 3D modeling software such as Blender.

In this work, we have created two Digital Twins for simulation purposes (Figures 44.2 and 44.3), the Church of Santa Liberata with a monumental facade in the province of Cosenza (Italy) and a Digital Twin of a facility with a landfill for demolition materials in the province of Cosenza (Italy). These are two examples of case studies of civil infrastructure and environment.

44.2.2 Virtual Waypoints

The methodology of this document recommends basing the selection of waypoints directly on the digital twins. Using the software BIM, this process can be performed on the georeferenced 3D model during the infrastructure design phase. Alternatively, you can use the Unreal Engine to set up waypoints by simply specifying targets. Once these locations of interest have been established, the proposed methodology allows for the autonomous creation of UAV trajectories to facilitate data collection and avoid aircraft collisions. The selected waypoints for the proposed simulation test are shown in Figure 44.3.

FIGURE 44.3
Landfill for demolition materials (Cosenza, Italy) point cloud: trajectory and virtual waypoints in Unreal Engine. Perspective view.

44.2.3 Multirotor Model

For infrastructure inspection, multirotor UAVs are usually preferred over fixed-wing UAVs due to their high maneuverability and hovering capability. Therefore we considered in this work the common commercial model Asctec® hexacopter NEO and the relative identification procedure used in Ref. [9]. The related dynamic model and state representation, necessary for the design of the control architecture, are reported in detail in our previous work [20].

44.2.4 Trajectory Generation

To allow a fleet of UAVs to inspect a general infrastructure, we need to define a set of reference trajectories for each UAV that will allow the aircraft to approach points of interest and avoid collisions with other members of the fleet. Since trajectory planning is a core problem in robotics and control, there are many opportunities in this regard. However, in the context of automated inspections, certain aspects such as ease of use for the operator and flexibility in the spatial arrangement of UAVs depending on the data to be collected are of great importance. With these goals in mind, we have developed two ad hoc kinematic path planners suitable for different types of inspections. They are described in detail in the next two sections.

44.2.4.1 Generic Inspection

This first path planner considers the case of a general inspection where UAVs are to fly near a set of points of interest. The generation of such references is done in two steps:

1. The generation of a fleet trajectory: first, the user selects a set of waypoints on a digital twin (Figure 44.3) corresponding to the points of interest to be inspected, as well as the average speed to be maintained during the inspection. With these data, a standard interpolation method or path planning algorithm can be used to generate a smooth reference trajectory for the centroid (geometric center) of the entire fleet.
2. Generation of trajectories for the individual UAVs: The kinematic model of the fleet presented in Ref. [10] generates non-colliding trajectories for the individual UAVs around the previously calculated centroid trajectory.

To understand the reasons for choosing the model presented in Ref. [10], we briefly report its main features here.

Considering n UAVs with one smooth fleet trajectory, that is, $\underline{p}^{(F)}: R \rightarrow R^3, \underline{p}^{(F)} \in C^1$, the fleet kinematic model consists of a combination of an attractive term to $\underline{p}^{(F)}(t)$ and a hard limiting repulsion function:

$$\underline{\dot{p}}(t) = -\alpha\left(\underline{p}^{(i)}(t) - \underline{p}^{(F)}(t)\right) + \underline{\dot{p}}^{(F)}(t) - M\sum_{j=1}^{n} \frac{\underline{p}^{(i)}(t) - \underline{p}^{(j)}(t)}{\left(\left\|\underline{p}^{(i)}(t) - \underline{p}^{(j)}(t)\right\|^2 - 4\mu^2\right)^2}, i = 1,2,N \tag{44.1}$$

with $\underline{p}^{(i)}(t) = \left[p_1^{(i)}(t), p_2^{(i)}(t), p_3^{(i)}(t)\right]^T = \left[\underline{x}^{(i)}, \underline{y}^{(i)}, \underline{z}^{(i)}\right]^T$ the position reference for the ith UAV; $M \in R^{3\times3}$ a full rank matrix; and α, $\mu \in R_{>0}$. The reasons for choosing this model are twofold. The first is its ability to generate trajectories that, at each time instant, are confined into non-intersecting hyper-balls centered at the corresponding current positions and having the same radius $\mu \in R_{>0}$. In Ref. [10], this property is formally stated as follows:

Statement: If $\left\|\underline{p}^{(i)}(t) - \underline{p}^{(j)}(t)\right\| > 2\mu$, $\forall(i,j)$, $i \neq j$ then $\left\|\underline{p}^{(i)}(t) - \underline{p}^{(j)}(t)\right\| > 2\mu$ holds true $\forall t$ and $\forall(i,j)$, $i \neq j$.

The second reason is the ability of the M matrix to change the geometric shape of the fleet, allowing the user to choose between different ellipsoidal formations that can be transformed into rectilinear formations (for more details, see Ref. [10]). Unlike other formation control algorithms, in this case, it is not necessary to specify the distances that the agents must maintain between each other in order to achieve the desired formation. This benefits both the scalability of the proposed strategy and the user, who does not have to worry about specifying all the inter-distances.

Along with the position reference, we also provide the UAVs with an attitude reference $\underline{\eta}^{(i)} = \left[\underline{\phi}^{(i)}, \underline{\theta}^{(i)}, \underline{\psi}^{(i)}\right]^{\mathsf{T}}$. Since the purpose of the mission is to acquire data via various sensors, the flat attitude, i.e., zero roll and pitch, is the most advantageous. As far as yaw is concerned, it was chosen to use the angle of the tangent to the trajectory in the x–y plane. Therefore, the attitude reference results in

$$\underline{\eta}^{(i)}(t) = \left[0, 0, atan2\left(\underline{y}^{i}(t), \underline{x}^{i}(t)\right)\right]^{\mathsf{T}} \tag{44.2}$$

where $atan2(\cdot)$ is the four quadrant version of the inverse tangent function. Combining the pose and attitude reference in one vector we obtain the reference vector

$$r^{i}(t) = \left[r_1^{(i)}(t), r_2^{(i)}(t), \dots, r_6^{(i)}(t)\right]^{\mathsf{T}} = \left[\underline{p}^{(i)}(t)^{\mathsf{T}}, \underline{\eta}^{(i)}(t)^{\mathsf{T}}\right]^{\mathsf{T}} \in R^6. \tag{44.3}$$

In order to use such a reference with the discrete-time position controller, we sample it according to the sampling time. For the sake of readability, with an abuse of notation, we denote the sampled trajectory as:

$$r^{(i)}(k) \equiv r^{(i)}(k\ Ts). \tag{44.4}$$

44.2.4.2 Perimeter Inspection

The path planner just presented leverages the operator's experience and expertise to define waypoints of interest and plan an effective inspection of the artifact under consideration or even specific parts of it. This freedom of choice, however, is little exploited in the very many cases of buildings with well-defined floor plans (residential buildings, architectural heritage, etc.). In these cases, in fact, inspection consists of following the perimeter of the plan (perhaps at different heights) so as to collect data relative to the entire structure. In order to facilitate the end user, we have developed a specific path planner for these cases that requires only the building plan as information and allows MAVs to rotate on the perimeter of the building floor plan at a desired height.

Such a path planner stems from solving a more general problem addressed in Ref. [16]: the arrangement of kinematic agents inside a desired (possibly moving) region, a problem also known as region following/tracking [17]. Such a region is defined by a *star-shaped set*, i.e., a set $S \in R^2$ that has at least one point $x_0 \in S$ such that for all $x \in S$ the line segment joining x and x_0 is contained in S. Since buildings have usually big dimension and many sides, the exact perimeter of the floor may not define a star-shaped set in the planar space. It is always possible, however, to define a star-shaped region that embeds the original perimeter in a way that is effective for inspection (see Figure 44.4).

A fundamental characteristic of the proposed solution is the use of a reference frame mapping that is able to translate the star-shaped region following problem into a simpler circular region following task. The original star-shaped region S, thus, is transformed into a ball centered on the vantage point x_0 with radius ρ, i.e., $B(x_0, \rho)$. The control law is designed in such a circular auxiliary reference frame (where all computations are simplified) and then converted to the original frame. Defining $\zeta^{(i)}(t) \in R^2$ as the agent position at time t in the original frame and $\xi^{(i)}(t)$ as the same position in the circular reference frame, the kinematic model presented in [cit] can be expressed as follows:

FIGURE 44.4
The Church of Santa Liberata seen from above. The cyan rectangle is the reference perimeter for the MAVs, the white arrows shows the origin of the reference frame.

$$\dot{\xi}^{(i)}(t) = -\beta\xi^{(i)}(t) + \gamma^2 \frac{\xi^{(i)}(t)}{\xi^{(i)}(t)^T \xi^{(i)}(t) - \gamma^2\rho^2}$$

$$+ \sum_{j=1, j\neq i}^{N} g_I\left(\left\|\xi^{(i)}(t) - \xi^{(j)}(t)\right\|\right)\left(\xi^{(i)}(t) - \xi^{(j)}(t)\right) \tag{44.5}$$

where

- $g_I(\cdot) : R \to R$ is a function that models the interaction between agents that prevents them from colliding with each other.
- $\gamma \in [0,1)$ is the parameter used to drive the agent into a circular ring (later mapped in a strip of the original star-shaped set). If $\gamma = 0$, the entire ball is reachable while for values close to one only a very thin ring near the border is reachable ($\gamma = 1$ would imply no space available at all). For our application we will use this latter setting.
- β is a control knob. Its value comes out from the main result of Ref. [16].

It is worth noting that even if the summation considers all the agents of the team, this does not mean that they are all connected to each other. The interaction function $g_I(\cdot)$, in fact, is null when any two agents are at a distance greater than the detection range of the agents, namely, r_d. Such a range, obviously, depends on the perception sensor mounted on the MAVs.

One feature of the model that is critical for automatic inspection is the ability of agents to rotate around the vantage point of the region. To achieve such behavior, the following term must be added to the control law (44.3):

$$\lambda A\xi^{(i)}(t), A = \begin{bmatrix} 0 & -110 \end{bmatrix}, \lambda > 0 \tag{44.6}$$

where the positive scalar λ defines the rotational speed (rad/s) for the ith agent along the curve and A is a skew-symmetric matrix.

The model presented above is confined to the planar space while the MAVs during a real inspection move in a 3D space. This discrepancy is not an issue anyway. It is always possible, in fact, to use any kind of interpolation method to generate the MAV trajectory along the z-axis and use it along with 2D path given by Ref. [8]. In such a way, we can define the actual MAVs path reference as $\underline{p}^{(i)}(t) = \left[\zeta^{(i)}(t)^\top, \underline{z}^{(i)}(t)^\top\right]^\top$. Finally, for what concerns the attitude reference $\underline{\eta}^{(i)} = \left[\underline{\phi}^{(i)}, \underline{\theta}^{(i)}, \underline{\psi}^{(i)}\right]^\top$, the considerations done for the previous path planner in Section 2.4.1 still hold.

44.2.4.3 Position Controller

To fulfill the inspection mission, the position controller must be able to meet two basic requirements:

1. Maintain a low tracking error to account for the collision avoidance characteristics of the two proposed path planners.
2. Cope with the hardware constraints of the considered UAV (maximum thrust, etc.).

For this purpose, we have chosen a Nonlinear Model Predictive Controller (NMPC) [18]. The advantages of this approach are as follows:

- By using an appropriate cost function in the formulation of the Optimal Control Problem (OCP), the maximum tracking error never exceeds μ, so that the UAVs are always within the safety hyperball defined in Section 2.4.1 (generic path planner).
- The hardware constraints of the considered platform are handled formally. This prevents nonlinearities caused by instruction saturations from affecting performance.
- Using a nonlinear model, the tracking performance is better than the linearized version, as shown in Ref. [9].

The assumed OCP for the *i*-th drone is as follows:

$$\min q^{(i)}(k)\sum_{j=1}^{n_y}\sum_{l=1}^{h}\left\{\left\{w_j\left[r_j^{(i)}(k+l|k)-r_j^{(i)}(k+l|k\right]\right\}^2\right\} \tag{44.7}$$

$$\text{s.t } s^{(i)}(k+1)=f\left(s^{(i)}(k),u^{(i)}(k)\right) \tag{44.8}$$

$$s^{(i)}(0)=s^{(i)}(t_0) \tag{44.9}$$

$$u^{(i)}(k)\in U \tag{44.10}$$

where:

- k is the current control interval;
- h is the prediction horizon;
- $n_y=6$ the number of system output variables [20];
- $q^{(i)}(k)=\left[u^{(i)}(k|k)^{\top},u^{(i)}(k+1|k)^{\top},\ldots,u^{(i)}(k+h|k)^{\top}\right]^{\top}$ is the vector of decision variables;

- $w = [w_1, w_2, \ldots, w_6]^{\mathrm{T}}$ is the tuning weight vector for the system outputs, constant for all predictions;
- $U = \left\{ u \in R^4 \mid [T_{\min}, \phi_{\min}, \theta_{\min}, \dot{\psi}_{\min}]^{\mathrm{T}} \leq u \leq [T_{\max}, \phi_{\max}, \theta_{\max}, \dot{\psi}_{\max}]^{\mathrm{T}} \right\}$ is the control input admissible set.

Only the first control input $u(k \mid k)$ is applied to the system, and the process is repeated the next time step in a receding horizon fashion.

44.2.4.4 State Estimation

Considering that like in Ref. [20], the output of the system does not correspond to the whole state, a state estimator is needed to use the proposed NMPC. Since the considered system is nonlinear, following Ref. [11], we propose a Square Root Unscented Kalman Filter (SR-UKF) to estimate the state of the UAV. Compared to the classical version of the UKF presented in Ref. [12], this filter is computationally more efficient.

The process model of the filter is

$$\tilde{s}(k+1) = f\left(s(k), u(k), T_s\right) + d(k) \sim (0, D) \tag{44.11}$$

with $D \in R^{9\times 9}$ the constant process noise covariance matrix, which is a filter tuning parameter. Exploiting the knowledge about the measurement errors (see Ref. [20]), the filter measurement model is the same as the system output reported in Ref. [20]. The linearity of the measurement model, moreover, further reduces the computational load of the algorithm [13].

Using the SR-UKF, the position controller, at each time step, is provided with an estimate of the entire state, namely, $\hat{s}^{(i)}(k \mid k)$. This quantity is actually used instead of the true state $s^{(i)}(k)$ in Eq. (44.6).

44.2.5 Numerical Simulation

Finally, in this section, we report the performance of the proposed inspection strategy by means of two numerical simulations. In the first one, the inspection concerns a dump, a changing environment where the operator's expertise is supported by the first type of proposed path planner (see Section 2.4.1). The second, on the other hand, concerns a church, an architectural asset with a well-defined plan where the second type of path planner is more efficient.

The software tool used for this aim is the MATLAB® programming and numeric computing platform. An important aspect in the choice of such an environment is the opportunity to generate executable code on the on-board computers of real UAVs from the high-level implementation of the numerical simulation (see in this regard the Model Predictive Control Toolbox™ and the EMBOTECH® FORCESPRO solver [14]). The chosen UAV is the Asctec® NEO hexacopter, whose physical and dynamic parameters are retrieved from Ref. [9] and reported in Table 1 in Ref. [20]. In both simulations, we consider a fleet of three UAVs ($n = 3$).

44.2.5.1 Position Controller Setup

In order to highlight the fleet behaviors with the two different path planners, we decided to use a single controller setup that could provide satisfactory trajectory tracking performance. The OCP reported in Section 44.2.4.3 is performed at 20 Hz ($T_s = 0.05$ seconds), with a prediction horizon $h = 10$ and a tuning weight vector $w = [10,10,10,5,5,5]^\top$ that gives priority to the position tracking error. The hard constraints that define the input admissible set U are reported in Table 1 in Ref. [20].

Regarding the setup of the SR-UKF, the constant process noise covariance matrix, which is a filter tuning parameter, is chosen as

$$D = \text{diag}\left(1_3^\top \varepsilon_p^2, 1_3^\top \varepsilon_\eta^2, 1_3^\top \varepsilon_{\dot{p}}^2\right), \varepsilon_p = \varepsilon_\eta = 10^{-2}, \varepsilon_{\dot{p}} = 10^{-3} \tag{44.12}$$

and the constant process noise covariance matrix is

$$V = \text{diag}\left(1_3^\top \sigma_p^2, 1_3^\top \sigma_\eta^2\right), \sigma_p = 0.05 \text{ m}, \sigma_\eta = 2\frac{\pi}{180} \text{ rad} \tag{44.13}$$

where σ_p and σ_η are the position and attitude measurement standard deviations, respectively.

44.2.5.2 Generic Path Planner Simulation

In order to generate the fleet trajectory $r^{(F)}$, Akima splines [15] are used to interpolate the chosen waypoints with an average speed of 0.5 m/s.

Next, we considered the generation of the individual UAV position reference trajectory through the fleet kinematic model (44.1). In order to avoid collision between the aircrafts, the chosen safety radius is $\mu = 0.8$ m. Please note that, with such a radius and the UAV initial positions in meters (MAV 1: $x = 10.5$, $y = 7.5$, $z = 9$, yaw = 0; MAV 2: $x = -70$, $y = -68.5$, $z = -71.5$, yaw = 0; MAV 3: $x = -2$, $y = -2$, $z = -2$, yaw = 0), the necessary condition of Statement is satisfied. Since the desired approximate shape of the fleet is a vertical line, the interaction matrix is chosen as $M = -0.01$ diag(1, 2, 3) where the biggest eigenvalue is relative to the z-axis. The speed convergence parameter, finally, is $\alpha = 0.2$. Once the position trajectories $\underline{p}^{(i)}$ are obtained, using (44.2), it is possible to compute the complete reference signals $r^{(i)}$.

As expected, during the simulation the UAVs track their references despite the measurement noise. Finally, it is possible to appreciate how the NMPC ensures very good tracking performance while respecting the hard constraints on the inputs end. Thanks to the MATLAB® UAV Toolbox Interface for Unreal Engine® Projects, we created a 3D video to better show the behavior of the UAVs during the inspection: [YouTube video: https://youtu.be/JwT2XR8zgq8].

The operation ends with the automatic writing of the inspection report and the updating of the priority index for predictive maintenance operations as is the case in the current management systems.

44.2.5.3 Perimeter Inspection Simulation

To assess the performance of the perimeter inspection strategy, we chose as case of study the Church of Santa Liberata near Cosenza (Italy). A plan of the church along with the surrounding region that MAVs take as reference is shown in Figure 44.4 and the UAV initial positions in meters (MAV 1: $x = -2$, $y = 17$, $z = 0$, yaw=0; MAV 2: $x = 0$, $y = 17$, $z = 0$, yaw=0; MAV 3: $x = 2$, $y = 17$, $z = 0$, yaw=0). The user has only to set the coordinates of the four vertexes of the rectangle according to a chosen reference frame (in this case it is on the center of the plan but in a real inspection it could be anywhere, e.g., the position of a GCS) and choose the angular velocity of the MAVs. In this example such a velocity is chosen very low ($\lambda = 1.3$) to allow drones to make in depth analysis but of course higher velocity is possible when only a check is necessary. A suitable choice for the other knobs of the kinematic planner (see Ref. [16]) that can be considered fixed for any perimeter inspection are: $d = 1$, $h = 0.9$, $\delta = 1, \gamma = 0.99$. The last parameter, in particular, allows the tracking region to be only a thin strip, i.e., the perimeter.

Finally, it is possible to appreciate how the NMPC ensures very good tracking performance while respecting the hard constraints on the inputs. Finally, it is possible to appreciate how the NMPC ensures very good tracking performance while respecting the hard constraints on the inputs end. Thanks to the MATLAB® UAV Toolbox Interface for Unreal Engine® Projects, we created a 3D video to better show the behavior of the UAVs during the inspection: [YouTube video: https://youtu.be/81K12EU967Y].

44.2.6 Co-simulation Sensors

The update of the Digital twin, as closely as possible corresponding to reality, requires the use of various types of sensors: 4k or higher cameras, LiDAR to detect the dimensions and displacements of the infrastructure, multispectral cameras to identify particular degradation phenomena (armor corrosion) and thermal imaging cameras. By sharing these sensors on several drones, the computational load is further lightened and consequently smaller, cheaper and lighter drones can be used.

The quality of the data collected from the sensors is therefore crucial for effective use of the digital twin. In this regard, some simulation environments, such as Simulink, offer the ability to simulate not only the physical behavior of the drones during the mission but also of the sensors mounted on board [19]. This possibility, of course, ensures that only one real inspection is needed, thus reducing cost and time.

44.2.7 Structural Inspection Execution and Final Report

The planned trajectory on the georeferenced digital twin, consisting of way-points (x, y, z), is transferred to the flight plan (geographic coordinates) of the drone fleet, which automatically tracks it while the operator monitors the mission from the GCS.

The collected data are downloaded to the cloud and used to update the digital twin, which enables the automatic identification of defects in infra-structures by comparing the initial state and the state of the site, exploiting the full potential of innovative AI algorithms that enable the detection and classification of different types of defects with high reliability [21].

The process ends with the automatic generation of the inspection report and the updating of the priority index for predictive maintenance actions.

44.3 Conclusions

In this chapter, we presented a comprehensive strategy to manage automatic infrastructure inspection through the use of a fleet of UAVs and digital twin. Indeed, the latter allows the most recent state of the building to be known and this information is to be used for proper inspection planning. In this regard, we have shown two different algorithms for trajectory planning depending on whether the infrastructure under inspection has a regular structure or not. In the following, we have proposed a UAV control strategy capable of coping with the non-perfect location information such as those normally found when using a GNSS. The generality of the proposed approach makes it possible to consider very different civil and environmental infrastructures and to integrate the information obtained into dedicated management systems.

Acknowledgments

Special thanks for support to: 3Dflow S.r.l., Piano Lago Calcestruzzi S.r.l. and Cosmo S.r.l.

References

1. Deng, L.; Wang, W.; Yu, Y. State of the art review on the causes and mechanisms of bridge collapse. *Journal of Performance of Constructed Facilities*. **2016**, *30*, 04015005.
2. Tan, J.S.; Elbaz, K.; Wang, Z.F.; Shen, J.S.; Chen, J. Lessons learnt from bridge collapse: A view of sustainable management. *Sustainability*. **2020**, *12*, 1205.
3. Akanmu, A. Towards Cyber-Physical Systems Integration in Construction. Ph.D. Thesis, The Pennsylvania State University, State College, PA, USA, 2012.
4. Ham, Y.; Han, K.K.; Lin, J.J.; Golparvar-Fard, M. Visual monitoring of civil infrastructure systems via camera-equipped unmanned aerial vehicles (UAVs): A review of related works. *Visualization in Engineering*. **2016**, *4*, 1.
5. de Freitas Bello, V.S.; Popescu, C.; Blanksvärd, T. Bridge Management Systems: Overview and framework for smart management. In *Proceedings of the IABSE Congress Ghent 2021-Structural Engineering for Future Societal Needs*, Ghent, Belgium, 22–24 September 2021.
6. Helmerich, R.; Niederleithinger, E.; Algernon, D.; Streicher, D.; Wiggenhauser, H. Bridge inspection and condition assessment in Europe. *Transportation Research Record*. **2008**, *2044*, 31–38.
7. Liu, D.; Chen, J.; Hu, D.; Zhang, Z. Dynamic BIM-augmented UAV safety inspection for water diversion project. *Computers in Industry*. 2019, *108*, 163–177.
8. Ciampa, E.; De Vito, L.; Pecce, M.R. Practical issues on the use of drones for construction inspections. *Journal of Physics. Conference Series*. **2019**, *1249*, 012016v.
9. Kamel, M.; Burri, M.; Siegwart, R. Linear vs nonlinear mpc for trajectory tracking applied to rotary wing micro aerial vehicles. *IFAC PapersOnline*. **2017**, *50*, 3463–3469.
10. D'Alfonso, L.; Fedele, G.; Franzè, G. Path tracking and coordination control of multi-agent systems: A robust tube-based mpc scheme. *IFAC PapersOnline*. **2020**, *53*, 6950–6980.
11. Van der Merwe, R.; Wan, E.A. The square-root unscented kalman filter for state and parameter-estimation. In *Proceedings of the 2001 IEEE International Conference on Acoustics, Speech, and Signal Processing*, Salt Lake City, UT, USA, 7–11 May 2001; Volume 6, pp. 3461–3464.
12. Julier, S.J.; Uhlmann, J.K. New extension of the Kalman filter to nonlinear systems. In *Proceedings of the AEROSENSE '97: Signal Processing, Sensor Fusion, and Target Recognition VI*, Orlando, FL, USA, 21–25 April 1997; International Society for Optics and Photonics: Bellingham, WA, USA, 1997; Volume 3068, pp. 182–193.
13. Tagliabue, A.; Kamel, M.; Verling, S.; Siegwart, R.; Nieto, J. Collaborative transportation using MAVs via passive force control. In *Proceedings of the 2017 IEEE International Conference on Robotics and Automation (ICRA)*, Singapore, 29 May–3 June 2017; pp. 5766–5773.
14. Domahidi, A.; Jerez, J. Forces Professional. Embotech AG, 2014–2019. Available online: https://www.embotech.com/ (accessed on 7 April 2022).
15. Akima, H. A new method of interpolation and smooth curve fitting based on local procedures. *Journal of the ACM* **1970**, *17*, 589–602.
16. D'Alfonso, L.; Fedele, G.; Bono, A. Distributed region following and perimeter surveillance tasks in star-shaped sets. *Systems & Control Letters*, **2023**, *172*, 105437.

17. Chen, F.; Ren, W. A connection between dynamic region-following formation control and distributed average tracking. *IEEE Transactions on Cybernetics.* **2017**, *48*(6), 1760–1772.
18. Wang, D., et al. Efficient nonlinear model predictive control for quadrotor trajectory tracking: Algorithms and experiment. *IEEE Transactions on Cybernetics.* **2021**, *51*(10), 5057–5068.
19. https://it.mathworks.com/products/uav.html.
20. Bono, A.; D'Alfonso, L.; Fedele, G.; Filice, A.; Natalizio, E. Path planning and control of a UAV fleet in bridge management systems. *Remote Sensing.* **2022**, *14*(8), 1858.
21. Tropea, M.; Fedele, G.; De Luca, R.; Miriello, D.; De Rango, F. Automatic stones classification through a CNN-based approach. *Sensors.* **2022**, *22*(16), 6292.

Part 9

Digital Twins in Energy

45

Digital Twin Security of the Cyber-Physical Water Supply System

Nikolai Fomin and Roman V. Meshcheryakov
V. A. Trapeznikov Institute of Control Sciences of Russian Academy of Sciences

45.1 Cyber-Physical Water Supply System as a Digital Twin of CPS

Consider the cyber-physical water supply system (CPWSS) as the digital twin (DT) of the cyber-physical system (CPS). Modern smart cities are a set of CPSs – the CPWSS is one of them. CPWSS is considered to be a digital water supply system (DWSS) of a modern city (Bartos and Kerkez 2021). The CPWSS is created using specialized equipment and software in order to centralize information about the quality and quantity of water resources, monitor the state of systems and increase the level of public safety. Data exchange takes place in a secure, centralized and automated way at each production stage. The chemical and biological composition, resource losses, equipment condition and wear level, quantitative indicators of water resource consumption are analysed (Torfs et al. 2022). Based on the international strategies of control DTs analysis, all countries have a lot of water problems like high wear of water supply systems, security problems, anthropogenic impact on water resources and dependence on other countries (Fomin and Meshcheryakov 2020). The incorporation of the Digital Twin in water distribution networks contributes to creating a more sustainable, efficient and smartwater grids (Conejos et al. 2019). The contrast of DWSS is an analogue water supply system (AWSS) – an enterprise in the water supply industry that does not use centralized systems for accounting and monitoring of resource, equipment condition, operational monitoring of resource losses and predictive analytics. Both types of DT enterprises will be used in the study to assess vulnerability levels and control features.

According to the terminology of DT, mandatory attributes are a description of the functioning of a real object in digital form (Negri et al. 2017). The description is based on the twin's business process. Dividing the work

DOI: 10.1201/9781003425724-54

of CPWSS into the main functionality, we have four stages: Extraction; Preparation, Transportation and Consumption:

- *Extraction*: Extraction of a water resource from sources (ground, underground) for the purpose of further use. The possibility of operational analysis of quantitative and qualitative characteristics of a water resource.
- *Preparation*: Purification and normalization of the water resource to the maximum permissible concentrations of substances. Water becomes suitable for consumption. Normalization of the water resource is carried out in an automated way. The human factor and potential risks of deliberate sabotage by personnel are reduced.
- *Transportation*: Delivery of water resources from water treatment stations to the consumer. Systems for monitoring qualitative and quantitative indicators during transportation with the provision of reliable operational information about losses have been installed.
- *Consumption*: Water use by the consumer. The ability to track the quality of the water delivered and the amount of water consumed by the consumer, improving the level of safety and health.

45.2 Security Analysis of Digital Twins CPWSS Models

As any CPS, the DT is subject to interference in its functioning. The security of DT is considered in Eckhart and Ekelhart (2019) and represents a set of measures to ensure the reliability, integrity and availability of information and the stability of equipment in the CPS. In recent years, the threat to the critical infrastructure of water systems has increased. Preventive security mechanisms are often insufficient to identify and neutralize the actions of intruders. Ensuring the safety of the functioning of the CPS of a Smart City is a complex task (Habibzadeh et al. 2019; Estévez 2020; Iskhakov et al. 2018). Analysing the existing trends in the functioning of the CPS of modern cities (da Silva et al. 2022), it is possible to identify a list of current vulnerabilities and threats to control security (Vielberth et al. 2021). The basic terms of CPS security were formulated by the authors in previous studies (Fomin and Meshcheryakov 2021):

- CPS security threat is a set of conditions and factors that create a potential or real danger of a violation of the safety of the functioning of the CPS.
- Vulnerability of the CPS is a lack (weakness) of the infrastructure, software (software–technical), software–hardware levels and the

influence of the human factor on the functioning of the CFS as a whole, which (which) can be used to implement security threats.
- CPS security risk is the product of the degree of probability of a threat to the functioning of the CFS by the size (magnitude) of potential consequences.

Cyber-physical attacks on the water supply infrastructure have increased with an increase in the number of connected equipment and automation of water supply – transformation into "smart water" control. Identifying and neutralizing potential vulnerabilities is an important process that allows you to safely and effectively use the benefits of digital equipment in the water supply and water security industry of Smart Cities (Yang et al. 2019; Su et al. 2020) creating the support and decision-making CPS (Shcherbakov et al. 2020; Cao et al. 2020). One of the ways is to improve the algorithms for detecting cyber-physical attacks of water supply control (Taormina et al. 2018). An important aspect is the procedure for detecting an attack on the CPS. There are various mechanisms for a comprehensive assessment of the state of systems, including the implementation of sustainable and safe remote monitoring (Iskhakov and Meshcheryakov 2019).

A set of measures to assess the state of the system at each time step, including with the artificial intelligence, contributes to the timely detection of certain attacks (Zeadally et al. 2020). Ge et al. (2020) have created a numerically efficient algorithm for achieving stability and detecting attacks on remote monitoring systems. In Ramotsoela et al. (2019), several traditional methods of detecting anomalies are considered and evaluated in the context of detecting attacks in water distribution systems. These algorithms were centrally trained on the entire feature space and compared with multi-stage detection methods that were developed to isolate both local and global anomalies.

The infrastructure of the CPWSS and wastewater disposal is socially significant. When modelling the WSS (Mishra et al. 2019), a universal agent-based structure was used to assess the collective behaviour of the control system during cyberattacks launched against the implementation of this model. The model of urban water cycle control in real time is considered in Sun et al. (2020) divided into water supply and urban drainage systems necessary for the functioning of urban agglomeration. The paper by Christodoulou et al. (2017) is devoted to the automatic detection of water losses in water distribution networks by dynamic analysis of time series associated with water consumption within the network and the use of a classifier for wavelet detection of points of change to identify anomalies in the consumption structure. A critical review of the disclosed documented and malicious cybersecurity incidents in the water sector described in Hassanzadeh et al. (2020) confirms the frequent cases of external interference. The results indicate an increase in the frequency, diversity and complexity of cyber threats in the water sector. Ensuring security both at the level of the Internet of Things and at the network level is crucial for the operation of the CPS as a whole. We need to comprehensively consider the vulnerabilities of CPS and supplement the OSI model with the details of CPS.

Methods have been developed for identifying and classifying CPS vulnerabilities (Fomin et al. 2021). In particular, the CPS security assessment method (M_1) is an element of the security model of the socio-economic system of the digital water utility. Method M_1 is included in the list of measures and mechanisms for forming a policy to ensure information and cyber security of the facility's control system. The input of the Model M_1 international management security issues, standards, resource information, and best practices in water resource management. The output is a classified list of potential vulnerabilities, threats and risks of the CPWSS.

Stage Grouping of vulnerabilities: In this step, we classify potential vulnerabilities, threats and risks of functioning of the CPWSS. The described vulnerabilities and threats will be the basis for building new models for controlling water supply systems. It is important to note that risk management will increase the level of strategic security of cities. When assessing the vulnerabilities of digital water utilities, the degree of threat impact on the stability of the water utility control system should be divided into several circuits:

1. Operation of IoT and IIoT terminal equipment;
2. Integration the CPSs to the Smart City;
3. Centralization of data into a single information system of the country.

The presented paper does not address the issues of the third circuit – the possibility of centralizing the exchange of information from several Smart Cities. Such systems of combining information from several cities and regions represent a higher-level system. Large-scale systems of this kind require additional analysis of vulnerabilities and threats (Chow 2017), comparison of architecture options for building DT of smart cities (Hyunbum and Ben-Othman 2020; Zhao et al. 2019; Chatterjee et al. 2018).

Let's group the issues that control the CPWSS:

- limited water resources;
- increased wear of water supply systems;
- increase in the urban population;
- degradation of water sources;
- lack of backup water supply sources;
- non-compliance of equipment at water treatment plants with modern types of pollution – both chemical and biological: antibiotics, microplastics and biologically active bacteria.

Based on the conducted research and the introduced term of vulnerability of the CPS, we group the vulnerabilities of the CPWSS:

- Group 1 – Infrastructure level (water supply systems, equipment, water resources);
- Group 2 – Software level;
- Group 3 – Hardware level (digital equipment);
- Group 4 – Human factor level (human influence on the system).

Stage Risk calculation (RL_i): The calculation consists of several sub-steps. The primary need for each vulnerability (Vul_i) calculates the level of risk (RL_i) and specifies the probability ($Prob_i$) and the potential implications of vulnerability (Imp_i) according to formula (45.1).

$$RL_i = Prob_i * Imp_i \tag{45.1}$$

The tables were expertly filled in, the calculation is based on a scale from 0 to 1. The total values are shown below:

- Group 1 – Infrastructure level (water supply systems, equipment, water sources): Probability $\left(\sum Prob_i\right)$ AWSS=0.78; DWSS=0.17; Implications $\left(\sum Imp_i\right)$ AWSS=0.78; DWSS=0.70;
- Group 2 – Software level: Probability $\left(\sum Prob_i\right)$ AWSS=0.60; DWSS=0.69; Implications $\left(\sum Imp_i\right)$ AWSS=0.68; DWSS=0.73;
- Group 3 – Hardware level (digital equipment): Probability $\left(\sum Prob_i\right)$ AWSS=0.62; DWSS=0.62; Implications $\left(\sum Imp_i\right)$ AWSS= 0.80; DWSS=0.79;
- Group 4 – Human factor level (human influence on the system): Probability $\left(\sum Prob_i\right)$ AWSS=0.46; DWSS=0.36; Implications $\left(\sum Imp_i\right)$ AWSS=0.69; DWSS=0.75.

The total values of the level of risk (RL_i) for each Vul_i are presented below (AWSS/DWSS):

- The use of outdated technologies to detect the degree of water pollution ($Vul_{1.1}$)=Medium/Low;
- High level of wear of the water supply infrastructure ($Vul_{1.2}$) = Medium/Low;
- Lack of backup water supply sources ($Vul_{1.3}$)=Medium/Low;
- Lack of fresh water storage facilities ($Vul_{1.4}$)=Medium/Low;
- Lack of network health monitoring systems ($Vul_{1.5}$)=High/Low;
- Absence of main systems for monitoring the condition of water supplied to the consumer ($Vul_{1.6}$)=High/Low;
- Program code, undeclared features ($Vul_{2.1}$)=Medium/Medium;
- Software injections ($Vul_{2.2}$)=Medium/Medium;
- External accessibility (potential hacker intrusion) ($Vul_{2.3}$)=Medium/Medium;
- Backdoors of imported software ($Vul_{2.4}$)=High/High;
- Availability for connecting equipment ($Vul_{3.1}$)=Medium/Medium;
- The possibility of signal interception, suppression, distortion ($Vul_{3.2}$)=Medium/Medium;
- External accessibility (potential hacker intrusion) ($Vul_{3.3}$)=High/Medium;
- Backdoors of imported hardware ($Vul_{3.4}$)=High/High;
- Remote control of equipment ($Vul_{3.5}$)=Medium/Medium;
- Accidental incorrect actions of the staff ($Vul_{4.1}$)=Low/Low;
- Deliberate negative actions of personnel on the functioning of systems (sabotage) ($Vul_{4.2}$)=Low/Low;
- Potential data leaks ($Vul_{4.3}$)=Medium/Medium;
- Incorrect actions of personnel in case of an emergency (due to poor training) ($Vul_{4.4}$)=Medium/Medium.

Present the results of degree risk Vul_i in the form of a graph (Figure 45.1).

The security of the functioning of the CPS as a socio-economic system of a Smart City cannot be ensured only by identifying vulnerabilities, threats and calculating risks. A theoretical and practical tool is needed aimed at checking the identified vulnerabilities, threats and risks to the stability potential of the CPS. This will make it possible to carry out a practical test of the stability of the CPS to potential vulnerabilities in laboratory conditions. Based on the method of increasing the stability of the CPS to potential vulnerabilities (M_2), the stability of the system is checked by implementing the n-th number of scenarios (impacts on the system). The M_2 method is based on disabling the system and determining its stability. The purpose of the method is to form a list of requirements for the CPS. These requirements are the feedback loop of

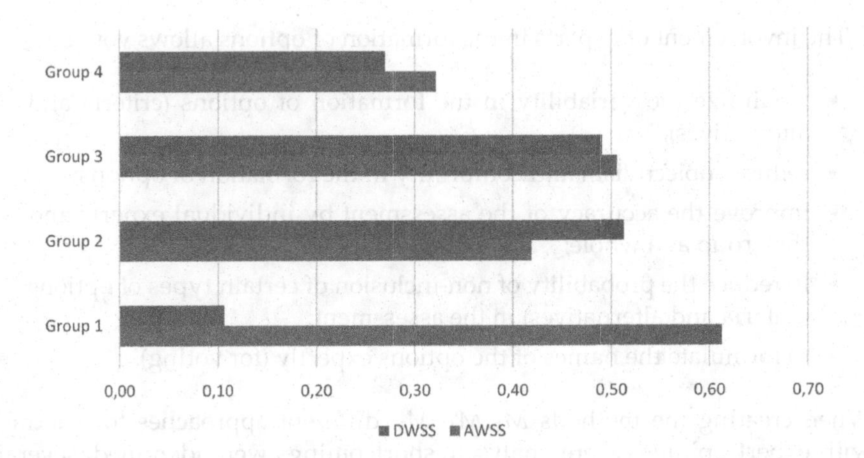

FIGURE 45.1
Degree of risk of the CPWSS vulnerability groups (analogue and digital type).

the CPS model. It is possible to improve the security level of the functioning of the CPS and increase the strategic security of Smart Cities.

45.3 Methods of Security Assessment CPWSS and Forming Control CPWSS Model

The formation of the CPWSS control model is based on the DT safety assessment methods created by the authors: Method M_3 – Method of forming criteria for assessing the safety of alternatives to control the CPWSS; Method M_4 – Method of forming alternatives safety control to the CPWSS; Method M_5 – Method of choosing an alternative control of the CPWSS.

The research was carried out with the involvement of an expert group of different professional groups and specialties. A single group of experts from 50 people of 5 profiles was divided into mini groups of 5 people (k) 4 times during the work in order to neutralize potential collusion and deliberate changes in the results. Each expert is assigned a unique identifier ID. As a result, we get a sample without repetitions from each mini-group calculated according to the formula (45.2):

$$C_n^k = \frac{n!}{(n-k)!^* k!} \tag{45.2}$$

The fourfold grouping into mini groups must be divided into two parts: the first two times – to select criteria, the other two times – to select alternatives.

The involvement of experts in the formation of options allows you to:

- maximize the variability in the formation of options (criteria and alternatives);
- reduce subjectivism and conformity in the formation of options;
- improve the accuracy of the assessment by individual experts and the group as a whole;
- to reduce the probability of non-inclusion of certain types of options (criteria and alternatives) in the assessment;
- to formulate the names of the options expertly (for voting).

When creating the methods M_3, M_4, M_5, different approaches to working with expert opinions were analysed, shortcomings were identified, several methods were compared:

- *Delphi method*: Main idea is dividing into two groups – experts and analysts, who gradually conduct discussions and analysis on choosing the best option in the direction under study. *Disadvantages of the method:* A large number of steps at each of the three stages, the work of experts is opaque, the search for a solution to a specific problem, rather than a set of problems and their dependencies, the influence of conformity, the possibility of manipulation by an expert and analytical group.
- *Analytic hierarchy process (AHP)*: The goal, the formation of evaluation criteria and a list of management alternatives. There is a pairwise comparison of criteria and alternatives. The alternative with the maximum value of the weight coefficients is the most preferred. *Disadvantages of the method:* Dependence on the selected criteria and alternatives, often formed abstractly; the ability to influence the results with a small sample.

The basis of the created methods M_3, M_4 is the modified Delphi method. M_5 is based on the AHP method:

- *Method M_3*: Method of forming criteria for assessing the safety of alternatives control the CPWSS. Each expert is provided with a primary list of criteria from the author-researcher. In Round 1, the expert gives his assessment to each of the criteria, or offers his own variants of criteria, analysing the negative factors of the functioning of the CPS. The results of Round 1 are processed by the author-researcher and a list of criteria is formed for evaluation by a group of experts in Rounds 2 and 3. According to the results of

Round 3, the author-researcher forms a final list of criteria for M_5. Differs from the well-known expert assessment of the negative factors of the functioning of the CPWSS.

- *Method* M_4: Method of forming alternatives safety control the CPWSS. Each expert is provided with a primary list of control alternatives from the author-researcher. In Round 1, the expert gives his assessment of each of the alternatives, or offers his options for alternative management, analysing the negative factors of the functioning of the CPS. The results of Round 1 are processed by the author-researcher and a list of alternatives to the management of CPS is formed for evaluation by a group of experts in Rounds 2 and 3. According to the results of Round 3, the author-researcher forms a final list of alternatives control of CPS for M_5. Differs from the well-known expert assessment of the negative factors of the functioning of the CPWSS.
- *Method* M_5: Method of choosing an alternative control of the CPWSS. The method is based on the AHP method. The obtained criteria based on M_3 and control alternatives based on M_4 are used to evaluate the choice of control alternatives by a group of experts.

The received answers of experts are checked for: *Verification* – Methods of verification; *Adequacy* – Comparison of the primary author's assessment and the responses received in different combinations of mini groups; *Reliability* – Expert opinions are obtained in an automated way through user verification; *Consistency* – During the calculations, the consistency of the results is checked.

The correlation of the methods M_3, M_4, M_5 is presented in Figure 45.2.

Based on the identified vulnerabilities, threats and international issues of water resources control, a primary list of criteria for assessing the safety of control alternatives (CrR_i), is proposed. The general description of the criteria is presented in the form of formula (45.3)

$$\{CrR_1, \ldots, CrR_n\}, i = \overline{1,n} \tag{45.3}$$

where $n = 4$. Finally we have:

1. The ability to detect chemical infections at water treatment plants and water supply networks (CrR_1).
2. The ability to detect biological infections at water treatment plants and water supply networks (CrR_2).
3. The stability of control systems to cyberattacks (CrR_3).
4. Centralization of data on the production and consumption of water resources stages (CrR_4).

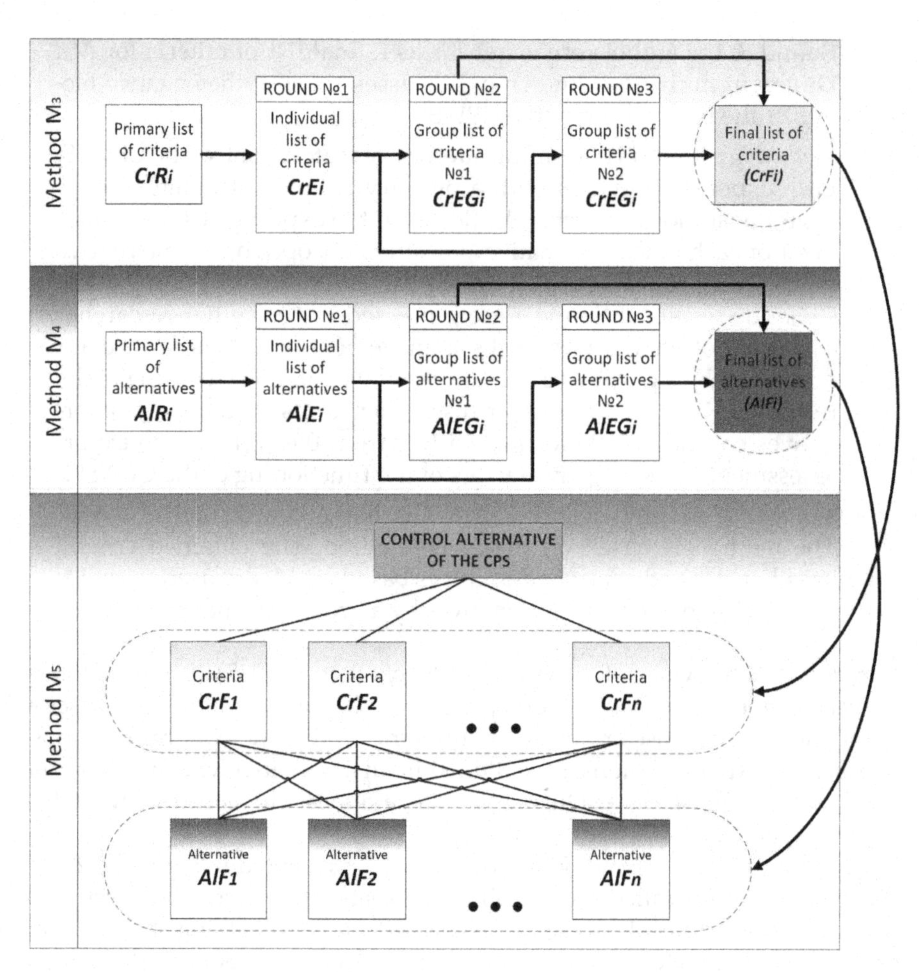

FIGURE 45.2
The correlation of the methods M_3, M_4, M_5.

The list of criteria (CrR_i) is provided to each expert in Round 1. In Round 1, the expert gives his assessment to each of the criteria, or offers his own variants of criteria, analysing the negative factors of the functioning of the CPS. The expert answers 15 questions – 2 questions with examples $Coun_i$ and 3 general questions on water supply control. Each of the case-questions describes the current control problems of the CPS. The results are processed in an automated way using the author's software. The author's software provides a data loading module – a list of control alternatives (AlR_i) and a list of evaluation criteria (CrR_i). The software was developed using web technologies, based on the PHP programming language and MySQL database. At the same time as loading the data, the software also adds text and images related to the problem In order to preserve the variability of the choice, each interviewed expert is allowed to submit his own version (CrE_i). These proposed options can be included in the group list ($CrEG_i$) for expert evaluation. When starting work with software, the expert receives a message about the expert's consent to the processing of the Personal Data Protection Policy. After confirmation, the expert enters his full name and selects the field of activity (expertise in the field). In case of refusal to sign the consent, the expert is not allowed to assess. After that, the expert receives a message about the generated list (CrR_i). Next, he is presented with 15 questions to select suitable criteria for evaluating the control alternative. Based on the results of work with the software of stage 1, a conclusion is formed. It contains the full name of the expert, the results of the evaluation criteria in the form of a list, the number of these answers and the final list is processed and formed. Each of the 50 experts (experts specialization: water supply, cybersecurity, economy, civil servants, IT specialists) gave an average of 2 of their own CrE_i. In total, after Round 1, the total number of criteria variants, let's call it *CrAl*, was more than 100 copies (45.4).

$$\sum CrAl = \sum CrR_i + \sum CrE_i \tag{45.4}$$

Finally,

$$\text{The set of } \; CrAl = \{CrAl_1, CrAl_2, \ldots, CrAl_n\}, \tag{45.5}$$

Based on the results of the analysis of the results of Round 1, a list of criteria was formed for evaluation by a group of experts ($CrEG_i$) for Rounds 2 and 3 by combining elements of a set of criteria similar in meaning. An additional condition is $10 \le CrEG_i \le 20$. This condition will allow preserving the variability of the choice at the stage of group selection.

Round 2 – Group Criteria List №1

According to the conditions, a single group of experts from 50 people of 5 profiles was divided into mini groups of 5 people (k) 4 times during the work in order to neutralize potential collusion and deliberate changes in the results. Each expert is assigned a unique identifier ID. As a result, we get a sample without repetitions from each mini-group calculated according to the formula (45.6)

$$C_n^k = \frac{n!}{(n-k)!^* \, k!} \tag{45.6}$$

The fourfold grouping into mini groups must be divided into two parts: the first two times – to select criteria, the other two times – to select alternatives.

Each group was presented with 10 questions with a list of the formed variants of the $CrEG_i$, in our case of 15 options. The questions reflect the features and control problems of the CPWSS. To answer the question, the mini-group must select 7 from the $CrEG_i$, rank in order of priority, where 1 is the highest priority with a score of 10 points, priority 2=9 points, and priority 7=4 points. The result of the group's work is answers to 10 questions with priorities.

We get the matrix $CrGroup_x$ (45.7), where x is the group number, m is the question number and n is $CrEG_i$.

$$CrGroup_x = \begin{pmatrix} CrEG_{11} & CrEG_{12} & CrEG_{1n} \\ \dots & \dots & \dots \\ CrEG_{m1} & CrEG_{m2} & CrEG_{mn} \end{pmatrix} \tag{45.7}$$

For each of the m questions from the n answer options of the mini-group, the grades given form a rating matrix. The sum of points per question = 49 points, the total sum of $CrGroup_x$ = 490 points.

According to the results of Round 2, the values of each of the max($CrEG_i$) are ranked in ascending order and the one who scored the maximum is assigned a rating of 1 and then ratings are assigned in descending order of the values of $CrEG_i$.

Round 3 – Group Criteria List №2

The calculation takes place in the same way with Round 2; the difference is the change (rearrangement) of the mini-group (45.6). This is necessary to minimize lobbying of the opinions of some participants in relation to others. The result of the work of the Round 3 group is answers to 10 questions with priorities.

Round 4 – Final list of criteria

After counting Round 3, we form the final list of criteria (CrF_i). It is calculated by the formula (45.8)

$$CrF_i = \frac{CrEG_{\text{round1 } i} + CrEG_{\text{round2 } i}}{2} \tag{45.8}$$

Expert responses of Rounds 2 and 3 and the final values of the points received by each of the criteria are shown in Figure 45.3.

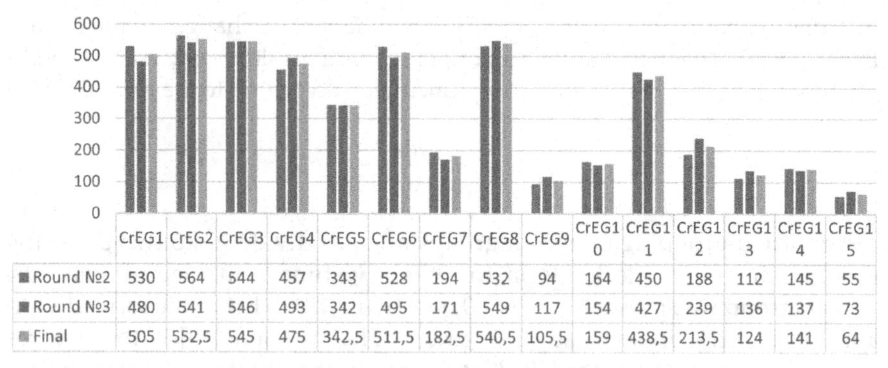

	CrEG1	CrEG2	CrEG3	CrEG4	CrEG5	CrEG6	CrEG7	CrEG8	CrEG9	CrEG10	CrEG11	CrEG12	CrEG13	CrEG14	CrEG15
Round №2	530	564	544	457	343	528	194	532	94	164	450	188	112	145	55
Round №3	480	541	546	493	342	495	171	549	117	154	427	239	136	137	73
Final	505	552,5	545	475	342,5	511,5	182,5	540,5	105,5	159	438,5	213,5	124	141	64

FIGURE 45.3
Expert responses $CrEG_i$.

Based on the results of M_3, a list of criteria for evaluating the safety of alternatives to the control of the CPWSS was formed. Criteria max($CrEG_i$) has CrF_i that was provided a rating $CrEG_i \leq 7$. The obtained criteria with the maximum rating (CrF_i) will be used in the calculations of M_5.

45.4 Security Model of the DT CPWSS

The security model of the socio-economic system for creating a DT of CPWSS was practically tested in St. Petersburg, Russia. The scientific community, under the leadership of Doctor of Economics, foreign member of the Russian Academy of Sciences V.L. Kvint, carried out comprehensive work on the analysis of the functioning of the water supply company "Vodokanal of St. Petersburg" – the second largest water supply company after Moscow. The result of the work was the "Development Strategy of the Vodokanal of St. Petersburg until 2035 and for a longer term". The document was successfully adopted by the management of the State Unitary Enterprise "Vodokanal of St. Petersburg". The strategy developers include one of this chapter's author – Nikolai Fomin, who was responsible for creating the section on strategic security and technological superiority of the Vodokanal of St. Petersburg (Fomin and Danilov 2019). The strategy pays special attention to water quality as a source of life and improving the efficiency of water supply systems control. At the heart of the security model of the CPWSS control and DT is the security model of the socio-economic system of the digital water utility (Figure 45.4).

45.5 Conclusion

1. Digital water supply system is CPS. CPS is a DT of smart cities. International strategies of control DTs tell as these countries have a lot of water problems like high wear of water supply systems, security problems, anthropogenic impact on water resources, and dependence on other countries.
2. According the security analysis of DTs CPWSS models we know that cyber-physical attacks on the water supply infrastructure have increased with an increase in the number of connected equipment and automation of water supply – transformation into "smart water" control. Identifying and neutralizing potential vulnerabilities is an important process that allows you to safely and effectively use the

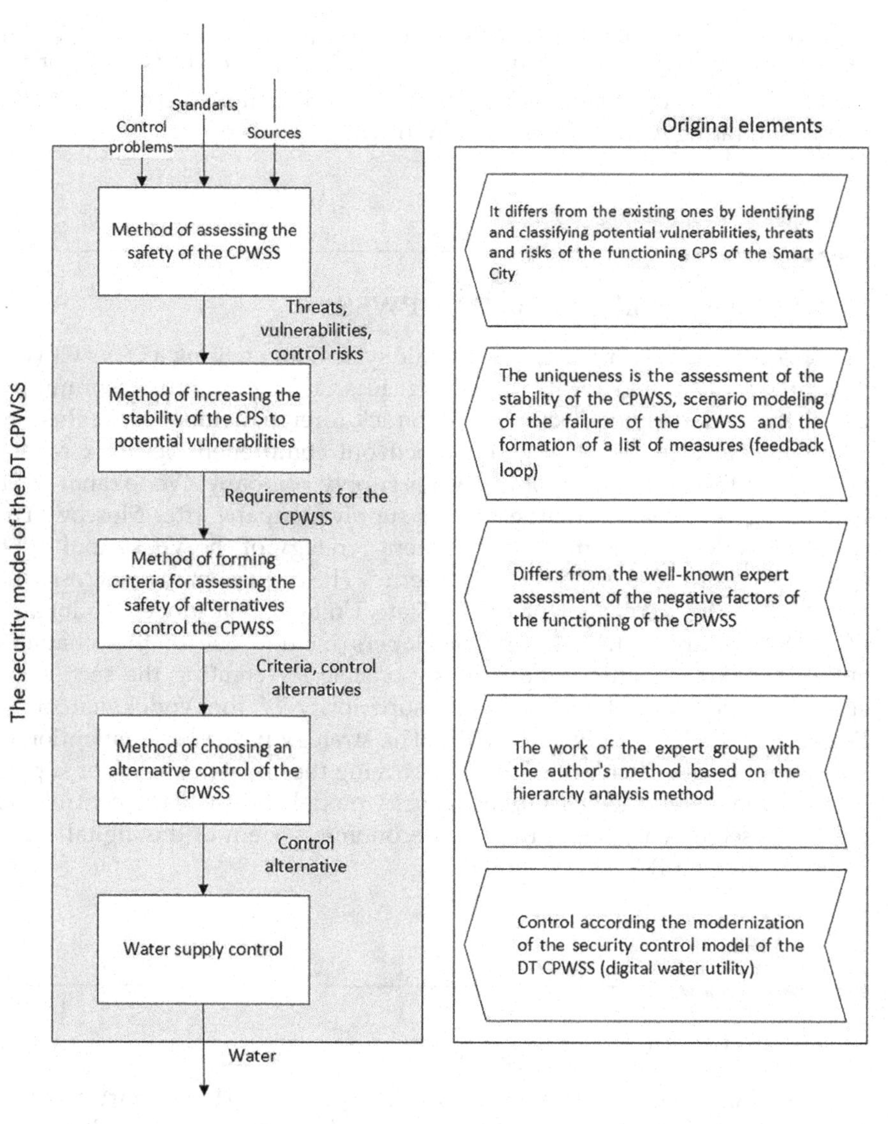

FIGURE 45.4
Security model of the DT CPWSS.

benefits of digital equipment in the water supply and water security industry of Smart Cities.

3. DT – has a lot of vulnerabilities, threats and functionality risks. The range of CPS vulnerabilities is wider than the OSI model. To assess the vulnerabilities of the CPS, it is necessary to supplement the system elements related to the features of the CPS and important to add extra security layers: exploitation layer, intersystem layer, external

layer and control layer. The calculation of risks/vulnerabilities using M_1, compared for AWSS and DWSS, presents the results of degree risk Vul_i of each group of vulnerabilities, like №1 – Infrastructure level (water supply systems, equipment, water sources) or №2 – Software level has different levels for AWSS and DWSS.

4. Security of DT is a complex task. A security model for DT CPWSS was established using the author's security assessment which was practically tested in the water supply company "Vodokanal of St. Petersburg" (St. Petersburg, Russia) – the second largest water supply company after Moscow.

References

Bartos, M., & Kerkez, B. (2021). Pipedream: An interactive digital twin model for natural and urban drainage systems. *Environmental Modelling & Software*, 144, 105120.

Cao, R., Hao, L., Gao, Q., Deng, J., & Chen, J. (2020). Modeling and decision-making methods for a class of cyber-physical systems based on modified hybrid stochastic timed petri net. *IEEE Systems Journal*, 14(4), 4684–4693.

Chatterjee, S. et al. (2018). Prevention of cybercrimes in smart cities of India: From a citizen's perspective. *Information Technology & People*, 32(5), 1153–1183.

Chow, R. (2017). The last mile for IoT privacy. *IEEE Security & Privacy*, 15(6), 73–76.

Christodoulou, S. E., Kourti, E., & Agathokleous, A. (2017). Waterloss detection in water distribution networks using wavelet change-point detection. *Water Resources Management*, 31(3), 979–994.

Conejos, P., Martínez Alzamora, F., Hervás, M., & Alonso Campos, J. C. (2019). Development and use of a digital twin for the water supply and distribution network of Valencia (Spain). In *17th International Conference, CCWI*. 1-4 September 2019, Exeter, United Kingdom.

da Silva, A. C. F., Wagner, S., Lazebnik, E., & Traitel, E. (2022). Using a cyber digital twin for continuous automotive security requirements verification. *IEEE Software*, 40(1), 69–76.

Eckhart, M., & Ekelhart, A. (2019). Digital twins for cyber-physical systems security: State of the art and outlook. *Security and quality in cyber-physical systems engineering*, 383–412.

Estévez, A. T. (2020). The fifth element: Biodigital & genetics. *Environmental Management of Air, Water, Agriculture, and Energy*, 95–212. ISBN 978-0-367-18484-1. https://doi.org/10.1201/9780429196607

Fomin, N. A., & Danilov, A. N. (2019). Digitalization of water supply infrastructure control: Problems, solutions. In *Proceedings of the 13th All-Russian Meeting on Control Problems (VSPU XIII, Moscow, 2019)*, pp. 1832–1835. Institute of Control Sciences of Russian Academy of Sciences, Moscow.

Fomin, N. A., & Meshcheryakov, R. V. (2020). Features of controlling the large-scale cyber-physical water supply systems in cities of different countries. In *Proceedings of the 13th International Conference "Management of Large-Scale System Development" (MLSD)*, pp. 1–4. IEEE, Piscataway.

Fomin, N. A., Meshcheryakov, R. V., Iskhakov, A. Y., & Gromov, Y. Y. (2021). Smart city: Cyber-physical systems modeling features. In *Society 5.0: Cyberspace for Advanced Human-Centered Society*, Alla G. Kravets, Alexander A. Bolshakov, and Maxim Shcherbakov (eds.) pp. 75–90. Springer, Cham.

Fomin, N., & Meshcheryakov, R. (2021). Modelling smart city cyber-physical water supply systems: Vulnerabilities, threats and risks. In *Futuristic Trends in Network and Communication Technologies. FTNCT 2020. Communications in Computer and Information Science*, vol 1395. Springer, Singapore.

Ge, X. H., Han, Q. L., Zhang, X. M., Ding, D. R., & Yang, F. W. (2020). Resilient and secure remote monitoring for a class of cyber-physical systems against attacks. *Information Sciences*, 512, 1592–1605.

Habibzadeh, H., et al. (2019). A survey on cybersecurity, data privacy, and policy issues in cyber-physical system deployments in smart cities. *Sustainable Cities and Society*, 50, 101660.

Hassanzadeh, A., Rasekh, A., Galelli, S., Aghashahi, M., Taormina, R., Ostfeld, A., et al. (2020). A review of cybersecurity incidents in the water sector. *Journal of Environmental Engineering*, 146(5), 13.

Hyunbum, K., & Ben-Othman, J. (2020). Toward integrated virtual emotion system with AI applicability for secure CPS-enabled smart cities: AI-based research challenges and security issues. *IEEE Network*, 34(3), 30–36.

Iskhakov, A. Y., Iskhakova, A. O., Meshcheryakov, R. V., Bendraou, R., & Melekhova, O. (2018). Application of user behavior thermal maps for identification of information security incident. *SPIIRAS Proceedings*, 6(61), 141–171.

Iskhakov, A., & Meshcheryakov, R. (2019). Intelligent system of environment monitoring on the basis of a set of IOT-sensors. In *2019 International Siberian Conference on Control and Communications (SIBCON)*, pp. 1–5. IEEE, Tomsk, Russia.

Mishra, V. K., Palleti, V. R., & Mathur, A. (2019). A modeling framework for critical infrastructure and its application in detecting cyber-attacks on a water distribution system. *International Journal of Critical Infrastructure Protection*, 26, 19.

Negri, E., Fumagalli, L., & Macchi, M. (2017). A review of the roles of digital twin in CPS-based production systems. *Procedia Manufacturing*, 11, 939–948.

Ramotsoela, D. T., Hancke, G. P., & Abu-Mahfouz, A. M. (2019). Attack detection in water distribution systems using machine learning. *Human-Centric Computing and Information Sciences*, 9, 22.

Shcherbakov, M. V., Glotov, A. V., & Cheremisinov, S. V. (2020). Proactive and predictive maintenance of cyber-physical systems. In *Cyber-Physical Systems: Advances in Design & Modelling*, Alla G. Kravets, Alexander A. Bolshakov and Maxim V. Shcherbakov (eds.) pp. 263–278. Springer, Cham.

Su, Y., Gao, W., & Guan, D. (2020). Achieving urban water security: A review of water management approach from technology perspective. *Water Resources Management*, 34, 4163–4179.

Sun, C. C., Puig, V., & Cembrano, G. (2020). Real-time control of urban water cycle under cyber-physical systems framework. *Water*, 12(2), 17.

Taormina, R., Galelli, S., Tippenhauer, N. O., Salomons, E., Ostfeld, A., Eliades, D. G., et al. (2018). Battle of the attack detection algorithms: Disclosing cyber attacks on water distribution networks. *Journal of Water Resources Planning and Management*, 144(8), 11.

Torfs, E., Nicolaï, N., Daneshgar, S., Copp, J. B., Haimi, H., Ikumi, D., … & Nopens, I. (2022). The transition of WRRF models to digital twin applications. *Water Science and Technology*, 85(10), 2840–2853.
Vielberth, M., Glas, M., Dietz, M., Karagiannis, S., Magkos, E., & Pernul, G. (2021). A digital twin-based cyber range for SOC analysts. In *IFIP Annual Conference on Data and Applications Security and Privacy*, pp. 293–311. Springer, Cham.
Yang, L., Elisa, N., & Eliot, N. (2019). Privacy and security aspects of E-government in smart cities. In *Smart Cities Cybersecurity and Privacy*, Danda B. Rawat, Kayhan Zrar Ghafoor (eds.) pp. 89–102. Elsevier.
Zeadally, S., Adi, E., Baig, Z., & Khan, I. A. (2020). Harnessing artificial intelligence capabilities to improve cybersecurity. *IEEE Access*, 8, 23817–23837.
Zhao, L. et al. (2019). Optimal edge resource allocation in IoT-based smart cities. *IEEE Network*, 33(2), 30–35.

46

Digital Twin in Smart Grid

Hui Cai, Xinya Song, and Dirk Westermann
Ilmenau University of Technology

46.1 Development of Digital Twin (DT) in Smart Grid

The foundation of DT technology is the joint usage of the spreading information communication, embedded sensors collecting descriptive high-dimensional data, artificial intelligence improvements in data processing and high-performance computing technology. It is the integrated solution of multi-technologies. Since the first official artificial intelligence (AI) declaration in 1956 [1,2], several similar thoughts to the current DT concept have sprouted up. In the initial definition of the AI concept, various descriptions like thinking machines, being human-like rather than becoming human, a replica of the human mind, emulation of human reasoning, etc., were proposed for its broad sense. With its development and the advancement of computation technology, AI applications benefit several fields. About 10 years after the first declaration of AI, the first DT concept–based system, called mirror system, was created to monitor unreachable physical spacecraft in the emission process by NASA in 1970 [1,3]. The engineers utilised the simulation environment to develop the operation of Apollo 13 during the emission. According to the simulation test, they found out that an improvised air purifier could be the possible solution to get the crew of Apollo 13 back to earth alive [1].

The success of Apollo 13 motivated the precursors of the DT concept with consideration of the potential applications with the mirror system. The simulation model could play a role in bridging physical and digital spaces. However, this model lacks a seamless connection and real-time data exchange, which could not be the proper DT. The first milestone of the DT concept was in 2002 [4]. The concept named Product Lifecycle Management (PLM) was informally introduced with the title "Conceptual Ideal for PLM". In this concept, three primary elements were proposed:

 DOI: 10.1201/9781003425724-55

i. Physical space (also called real space in Ref. [4])
ii. Virtual space
iii. The link for data flow between physical space and virtual space

The link in the figure enables the data exchange and allows the convergence and synchronisation of the virtual and physical systems. The simulation in virtual space can be applied to optimise the physical system. The connection between them would be linked throughout the entire lifecycle of the system.

The discussion of the proposed elements in the PLM concept continued in the research report from Framling [5]. It exploited the seamless connection provided by the spread of Internet technologies between the envisioned agent and the physical counterpart. The definition of the DT in aerospace was described as an integrated multi-physics, multi-scale, probabilistic simulation of a system that uses the available physical models, sensor devices, fleet history, etc., to replicate the life of its flying twin [6]. In the same year, the conceptual Airframe DT (ADT) model was proposed [7] to investigate applications like predicting the life of aircraft structure and management of it over the entire lifecycle. This model is described as a computational model of individual aircraft. It was expected to be the possible way to improve the management effectiveness for the U.S. Air Force [8].

The successful application of DT in aerospace has motivated research and implementation to some extent in manufacturing and industrial fields. In these areas, DT is defined as a computer-based digital model or set of digital models that can mirror its physical counterpart, receives information from the physical system, accumulates valuable information and helps in decision-making and the execution of processes. DT in the manufacturing area uses computer-based digital models to monitor the production process, and with the assistance of AI algorithms, an autonomous and intelligent manufacturing approach is executed with minimised human intervention. It can respond to failures or unexpected contingency with automated decision-making among a set of alternative actions to prevent the damage on the whole process at the supervisory level. Also, the connectivity between DT and physical system allows the present and historical data analysis by human experts assisted by AI algorithms to derive solutions to improve operations.

As a complicated meshed network system, the smart grid is essential for the foundation and development of the country while keeping its operation stable and safe [9]. The various simulation tools were widely used in industry from the middle of the 1980s with the increasing spread of workstations and personnel [10,11]. Nowadays, the requirements for the simulation tools in smart grids are not limited to the solution of the numerical solver for the design of engineering-related problems; however, they should provide a more detailed, multi-domain and multi-level simulation to extend the modelling and optimisation capability [12]. Except for it, more simulation is required for the hardware-in-the-loop (HIL) simulation. It can be used to test the hardware and make the simulation closer to the physical situation.

The emergence of a real-time simulator like OPAL-RT provides a possible solution for the HIL [13]. A connection between the virtual world and the physical hardware can be built based on the HIL and the available interfacing media in real-time simulation technologies. The foundation of DT is that the simulation and the physical agent can interact with each other in real time.

According to the definition in Refs. [12,13], the establishment of DT for smart grid involves the measurement, data processing and communication between physical space and virtual environment, parameter estimation, virtual modelling and application. The procedures of building a DT for the smart grid in the lifecycle is described below. The feature states of the physical entity are measured and extracted before the data processing. The massive volume of data from the measurement is reprocessed with data cleaning, conversion and filtering [14]. Afterwards, the reprocessed data are exchanged to virtual space through the communication tunnel, like C37.118 integrated into the hardware [15]. The virtual digital model based on the modelling theory aims to build a virtual representation of the physical system, which needs the parameterisation process with measurement from the physical system for adaptively updating the parameters in the virtual model to evolve into a DT. The DT of the smart grid mirrors the physical power grid, which depicts the same feature and behaviours in digital space [16]. The applications of DT could be realised more efficiently with the participation of artificial intelligent analytical algorithms. Except for providing a real-time representation of the states in the physical agent that enables an operational insight for monitoring, and optimisation of the operation process, DT offers a deep hindsight on the operational history of the asset, making it possible both to aggregate behaviour and performance over time, as well as foretime for understanding root causes and patterns of behaviour [16–18].

46.2 Definition of Digital Twin in Smart Grid

The original definition of DT was a set of virtual information for a full description of a potential or actual physical product from the micro atomic level to the macro geometrical level. At this optimal stage, any information obtained from the inspection of a physically manufactured product is available from its DT. With the development of simulation technology, this definition is developed as the DT is a computer-based model that simulates, emulates and mirrors a physical entity's life linked to its physical twin through a unique key [19]. The integration of AI technology enables DT as a living, intelligent and evolving model, being the virtual counterpart of a physical entity. It allows the lifecycle monitoring, predicting and optimising of the physical entity. The continuous prediction of future states enables simulating and testing innovative configurations and concepts to apply preventive maintenance.

DTs can be classified into two types in Ref. [4]: DT Prototype (DTP) and DT Instance (DTI). DTP depicts the prototypical physical artefact. It contains the information sets for describing the physical entity that duplicates the virtual model, but the link between the digital model and the physical entity doesn't exist in this type. The DTI describes a specific corresponding physical entity that an individual DT remains linked to throughout the life of that physical unit. This link enables continuous data exchange between the digital model and physical entity. With the operational states captured from actual sensor data and the recorded data described the past performance, DTI enables the monitoring of the past and actual operational states of a physical entity in real time. DT Aggregate (DTA) is the aggregation of all the DTIs. DTI focuses on the dependent entity, but DTA is a computing construct that has access to all DTs and queries them proactively. It enables continual analysis of sensor data and correlation of these data with unhappened failure scenarios to enable prognostics.

The definition of DTs in a smart grid, according to the DTI, is more understood as an accuracy aggregated simulation model, which is interacted with the physical power grid in real time to simulate, emulate and monitor the condition and performance of the whole system. Figure 46.1 shows an overview of DT's definition and its application in the smart grid. The initial phase begins from the physical power grid data collection stored in the phasor data concentrator (PDC) database [17]. The measured data stored in the PDC database participate in the loop simulation through communication protocol C37.118 [14]. The real-time simulator receives the data and transfers it to the simulation environment. The mirroring process is completed by continuously adapting the parameters of the DT to duplicate the operational state of the physical power grid dynamically. Afterwards, DT runs in parallel to monitor the condition of the physical power grid and then immediately warns the operator that the operational behaviour deviates from expected predictive simulation (see right underpart in Figure 46.1). Intelligent algorithms like machine learning identify the deviation feature data to implement the appropriate repairs and redesigns. The whole detection process is classified into two steps. At first, in the AI algorithm development phase, the developer constructs the appropriate DT, and according to exchanged data, DT is properly parameterised with the characteristic features from the physical power grid. The fault states are then embedded in DT to collect the failure prediction data, which may happen in the physical power grid. The second phase is the training process and validation. Machine learning is used to form a detection model by learning the fault data in the first phase. The sufficient training data can accurately forecast and identify the malfunction in the early phase of the actual operation state in the power grid and rapidly prepare the appropriate reactions like protective actions to prevent damage to the physical system. The application of DT in smart grids is not only limited to the case mentioned above; the further cases for DT in smart grids are explained in Section 46.3.

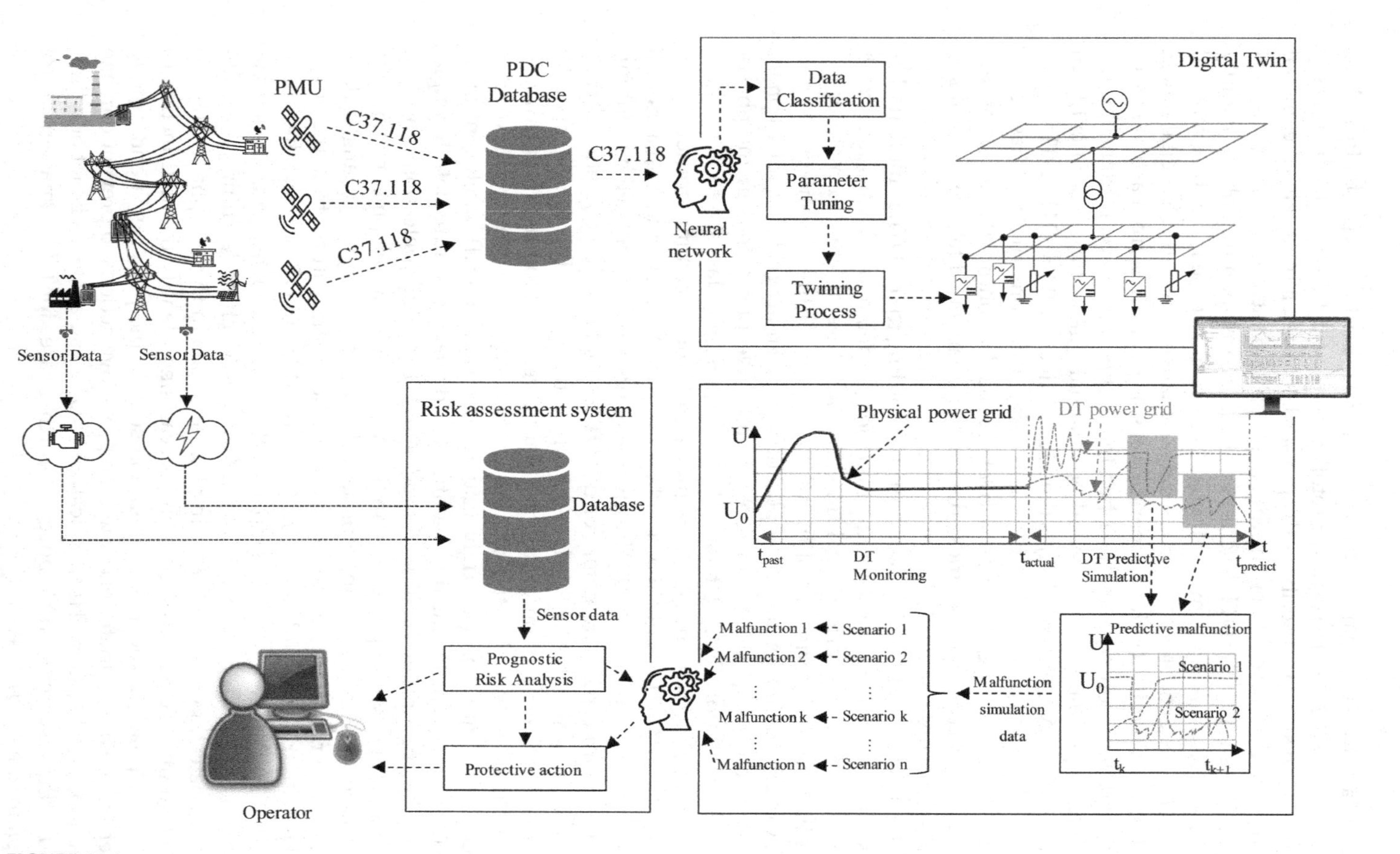

FIGURE 46.1
Digital twin in smart grid.

46.3 Modelling and Applications of Digital Twin in Smart Grid

With the use of DT in smart grid, plant operators can optimise immediate and transient control of the plant in terms of efficiency or performance. Furthermore, the operators can make informed decisions on components considering their lifespan, distribute loads over time and perform the proper maintenance tasks at the ideal time. The DT makes it possible to evaluate different scenarios in the smart grid and increase efficiency in decision-making [20,21]. DT can also demonstrate the actual and possible future system state, making it a solution for further improving the monitoring and control of the smart grids [22]. In this chapter, two DT models are presented, including their modelling and applications.

46.3.1 Artificial Neural Network-Based DT for State Estimation

The term state estimation describes estimating the current smart grid state (voltage magnitudes and angles) based on measurements. Fast and accurate state acquisition has a crucial role in monitoring and implementation of control design of the system. However, most states are not available or are too costly to measure. Therefore, the state estimation is required.

The purpose of constructing an artificial neural network (ANN)–based DT for state estimation is a reasonably accurate and effective estimation of unmeasurable states. Neural networks are known as fault-tolerant and able to recognise patterns as well as to prove the strong computing power, which is a primary advantage in the application of state estimation [7]. Process for the state estimation of an electrical network can be regarded as the determination of functions, which capture the relationship between the measured data and the network operating states. The functions are then to be constructed by using ANN.

46.3.1.1 ANN-Based DT

An ANN-based DT can be utilised in different applications. In this section, we are going to focus on ANN in smart grid state estimation. As mentioned, an ANN can be regarded as a function for mapping the input to output variables. For an ANN-based DT for state estimation, the well-trained ANNs are called the DT of the corresponding smart grid for its state estimation. For this DT, measurements from the actual power network are used as the input dataset, while the voltage amplitude and angle at every system bus are considered as the output dataset. Figure 46.2 shows, on the one hand, the configuration of the ANN-based DT, which consists of two ANNs. One ANN is trained to estimate the voltage amplitude at each bus in the smart grid, and the other ANN is for the corresponding voltage angle estimation. On the≈other hand,

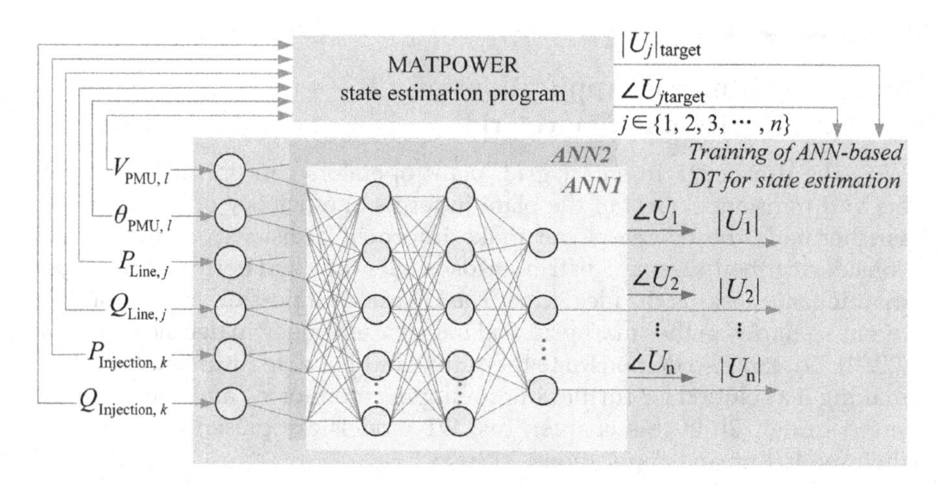

FIGURE 46.2
Training of the artificial neural network (ANN)–based digital twin and structure of the ANN.

the input and output sets for training the DT is also shown in this figure. Although training of two ANNs may increase the training effort, it results in the improved accuracy of this DT based on two ANNs.

Different measurement sets for the inputs are generated from the referenced smart grid. These measurements are used to perform a general state estimation programme in MATPOWER, which is a free software used by smart grid researchers. Afterwards, the estimated states of the smart grid corresponding to each measurement set are obtained. These estimated states are delivered into the training process as the target outputs of the ANNs-based DT.

The proposed methods are then applied to the IEEE 9-bus system. Two fully connected multilayer feedforward NNs have proved to be suitable for the state estimation problem of the smart grid. In this work, both ANNs have two hidden layers, and the number of neurons for both hidden layers are set for 8. Since ANN1 and ANN2 have 24 inputs for 12 measurements in the 9-bus system, and 9 outputs for the 9-bus, the architecture for both ANNs are 24-8-8-9. From the line power and power injection measurements, both active and reactive power are obtained. From the PMUs at bus 4, 6 and 8, their voltage amplitude and angle are acquired. From Figure 46.2, it is also clear that the ANN1 is used for voltage amplitude estimation, while the ANN2 is used for angle estimation.

To test the accuracy of the ANNs-based DT for state estimation, several test scenarios are conducted. The three loads that connect at bus 5, 7 and 9 in the system are changed from 10% to 90% with a step of 20%, respectively. To guarantee a converged state estimation result, the generated power from three generators should be correspondingly increased. The following table gives the maximum and average error for both voltage amplitude and angle

TABLE 46.1

Accuracy of the Artificial Neural Networks–Based Digital Twin for State Estimation

Voltage	Error in %	Bus 5	Bus 7	Bus 9
Amplitude	Maximum	0.22	0.18	0.23
	Average	0.08	0.06	0.07
Angle	Maximum	0.32	0.39	0.38
	Average	0.09	0.05	0.07

state estimation. Table 46.1 shows that the maximal error is at angle estimation of bus 7, which is smaller than 0.4%.

An ANNs-based DT for state estimation is presented in this section. The result shows that training with enormous measurement data in appropriate architectural setup ANNs, we can gain an accurate and reliable DT for fast state estimation. Compared to the DT in the last section, which is constructed based on the differential equations, the training effort in the ANNs-based DT is significantly reduced. However, the DT introduced in this section is remarkable for the state estimation. Another type of DT is presented in the next section, which is used to design its corresponding controller.

46.3.2 Data-Driven DT for Controller Design

As known, a DT can also be applied in designing a controller for smart grid devices. This section introduces a data-driven DT construction of a basic DC/DC boost converter, and its controller is designed accordingly based on this DT. Power converters are becoming critical in smart grids with the penetration of renewable energy in the electrical grid. A boost converter is required to increase the input voltage to the desired output voltage to connect a fuel cell or a photovoltaic into a power network. A DC/DC converter consists of an input dc voltage source V_{in}, an inductor L, a controlled switch S with switching period T, a diode, a capacitor C for filtering and a resistor R representing a load. The switching characteristics cause non-linearity in the converter model, leading to a complexity in the control design. The first step should be the construction of the converter model to derive a proper control technique. Due to the complexity of the interconnections between the components within the commercial DC/DC converter and the fact that some important parameters are kept unknown to protect the manufacturer's profits, its modelling is also considered a challenging task. Therefore, the local model network (LMN) technique for the converter's data-driven DT is introduced.

46.3.2.1 Local Model Networks

LMN is a notion for the integration of multiple local linear models (LLMs). It represents a globally complex non-linear model and provides a flexible framework for combining various model structures and learning algorithms. A typical algorithm for an LMN construction is called LOLIMOT (LLM tree) that is easy to implement and trains the LMN efficiently [23]. In LOLIMOT, the input space of the LMN is partitioned by a tree-structure algorithm, and the local models are interpolated by overlapping local basic functions [24]. Data-driven non-linear dynamic system construction consists of two steps, structure selection and parameter estimation. Using LOLIMOT, the amount of LLMs and the local parameters, as well as weight functions for each LLM, are optimised in the training of LMN [25].

LOLIMOT is an incremental or growing tree-construction algorithm that trains an LMN by adding one new LLM or a new rule in each iteration [26]. The worst performance LLM is split into two halves during the whole training process. Each is assigned to characterise the relationship between inputs and outputs within a specific range of operations. In this way, the model reflecting errors is minimised [27].

For each LLM_m, $m = 1,\ldots,M$ with i input variables x are expressed by [28]:

$$\hat{y}_m^{\text{LLM}} = w_{m,0} + w_{m,1}x_1 + w_{m,2}x_2 + \cdots + w_{m,i}x_i \tag{46.1}$$

Parameters for the LMNs are estimated separately, neglecting the interactions between the local models [29,30]. Efficient least square optimisation method WLS (Weighted Least Squares) is used for local models' parameter estimation. Many insights and methods in mature linear control theory could be transferred to non-linear areas by utilising LLMs [31].

In an LMN, the mth local model output $\hat{y}_m$ is calculated as the integration of the LLM with its corresponding weight function, which can be obtained by

$$\hat{y}_m = \Phi_m\left(\underline{x}\right)\hat{y}_m^{\text{LLM}} \tag{46.2}$$

where $\underline{x}$ contains i input vectors $\underline{x} = \begin{bmatrix} x_1 & x_2 & \cdots & x_i \end{bmatrix}^T$. Φ_m is the weight or validation function, which describes the operating ranges where the local models are valid, and contributes for each local model to the output [31] as well as determines the transition between neighbouring sub-models [32]. As for the calculation of Φ_m, an HBT (Hierarchical Binary Tree) structure based on the idea of decision tree [33] and motivated LOLIMOT algorithm are utilised. The structure of LMNs is similar to RBF (Radius Basis Function) neuron network, but the weight functions are radial and local models have constant values. The type of weight function is generally the normalised Gaussians [23], whose formula is shown in Eq. (46.3).

$$\Phi_m(\underline{x}) = \frac{\mu_m(\underline{x})}{\sum_{m=1}^{M} \mu_m(\underline{x})} \tag{46.3}$$

with

$$\mu_m(\underline{u}) = \exp\left(-\frac{1}{2}\left(\frac{(x_1 - c_{m1})^2}{\sigma_{m1}^2} + \cdots + \frac{(x_i - c_{mi})^2}{\sigma_{mi}^2}\right)\right)$$
$$= \exp\left(-\frac{1}{2}\frac{(x_1 - c_{m1})^2}{\sigma_{m1}^2}\right) \cdots\cdots \exp\left(-\frac{1}{2}\frac{(x_i - c_{mi})^2}{\sigma_{mi}^2}\right) \tag{46.4}$$

The centre coordinates c_{mi} and the standard deviations σ_{mi} determine the normalised Gaussian weight function $\Phi_m(\underline{x})$. These non-linear parameters present the partitioning of the input spaces. Based on the property of Gaussian function, centres c_{mi} determine the rectangles' centre while standard deviations σ_{mi} determine the extension. For higher-dimensional input spaces, the shape of the input spaces is transferred from rectangle to hyper-rectangles [28].

46.3.2.2 Data-Driven DT of DC/DC Boost Converter

A dataset containing sufficiently rich information about the converter behaviour should be generated and collected to construct a dynamically accurate model from the measurement data. The voltage controller introduced in the next section is designed for the output converter voltage to achieve the constant reference value without being affected by variations in input voltage and load. Therefore, the excitation signals for the generated dataset should be configured for the reference output voltage, input voltage and connected load. The three excitation signals are formed with amplitude-modulated pseudo-random binary signals (APRBS), covering all operating ranges of interest for the converter [28]. For the studied converter, with switching period T, its switch is closed for time $D * T$ and open for $(1-D) * T$, where D is the duty ratio for the converter at a steady state. The duty ratio can be then obtained with the following formula:

$$D = 1 - \frac{v_{\text{in}}}{v_{\text{out\_ref}}} \tag{46.5}$$

where v_{in} is the input voltage and $v_{\text{out\_ref}}$ represents the reference output voltage. The duty ratio D is calculated with the above formula while both excitation signals v_{in} and $v_{\text{out\_ref}}$ are set according to the APRBS signals. To imitate a series of changing duty ratios, 50 reference voltage levels and 10 input voltage levels are used to form the excitation signals. The minimum hold time for the excitation signal v_{in} is set to 1 second, while the time for $v_{\text{out\_ref}}$ is 0.2 second. As mentioned earlier, the changing load should also be considered as an APRBS signal in the data acquisition process. Its minimum hold time is set to 1 second. The three drafted excitation signals are then used as the converter's input to generate data of the output voltage $v_{out}(k)$ and the inductor current $i_L(k)$. Both data and the duty ratio $D(k)$ as well as their delayed signal $v_{out}(k-1), \ldots, v_{out}(k-n+1), i_L(k-1), \ldots, i_L(k-n+1), D(k-1), \ldots, D(k-n+1)$ that are extracted through a tapped delay line (TDL) are then collected and used to train the LMN-based converter DT. The output of the DT and its corresponding inputs can be represented with the formula (46.6).

$$\hat{v}_{out}(k+1) = f\left(\begin{array}{c} v_{out}(k), v_{out}(k-1), \ldots, v_{out}(k-n+1)), i_L(k), i_L(k-1), \\ \cdots, i_L(k-n+1), D(k), D(k-1), \cdots, D(k-n+1) \end{array}\right)$$
$$= \sum_{m=1}^{M} \Phi_m \times (\underline{\theta}_m \times \underline{x}(k)) + \sum_{m=1}^{M} \Phi_m (\theta_{m,3n} D(k)) \tag{46.6}$$

With $\underline{x}(k) = \left[1\ v_{out}(k) \cdots v_{out}(k-n+1)\ i_L(k) \cdots i_L(k-n+1) D(k-1) \cdots D(k-n+1)\right]^T$ and $\underline{\theta}_m = \left[\theta_{m,0} \quad \theta_{m,1} \quad \ldots \quad \theta_{m,3n-1}\right]$. Φ_m is the weight function of the mth local model and $\theta_{m,n}$ is the nth local parameter of the corresponding local model m. The larger the positive integer n that represents the order of the delayed signals, the smaller difference between the DT output and the measured converter output voltage. In other words, the more accurate DT can be obtained when more historical data are available. In the LMN-based DT, the converter dynamics for different operating ranges are modelled with different LLMs of the LMN. After getting the exact DT model, the designing of its controller proceeded.

46.3.2.3 Local Linear Control

Considering the dynamics of the constructed LMN-based DT model, the local linear controls (LLC) are designed for each corresponding LLM. An LLC is tasked with controlling the output voltage over a specific operating range. The control signal for a converter based on the averaged model is the duty ratio. Therefore, converter control design is based on calculating the proper duty ratio value. The error equation is solved in the precise tracking controller to establish the control input required for the following error value [34]:

$$v_{ref}(k) - v_{out}(k) = 0 \tag{46.7}$$

where $v_{ref}(k)$ represents the target voltage and $v_{out}(k)$ is the actual output voltage.

Considering the individual LLC over a particular operating range, the following formula for ith LLC can be obtained.

$$D_m(k) = \frac{v_{ref}(k+1) - \underline{\theta}_m \times \underline{x}(k)}{\theta_{m,3n}} \tag{46.8}$$

To cover all the possible operating ranges, the combined control signal for the converter can be obtained as following:

$$D(k) = \frac{\sum_{m=1}^{M} \Phi_m \theta_{m,3n} D_m(k)}{\sum_{m=1}^{M} \Phi_m \theta_{m,3n}}$$

$$= \frac{\sum_{m=1}^{M} \Phi_m v_{\text{ref}}(k+1) - \sum_{m=1}^{M} \Phi_m \times \left(\underline{\theta}_m \times \underline{x}(k)\right)}{\sum_{m=1}^{M} \Phi_m \theta_{m,3n}} \tag{46.9}$$

According to Eq. (46.3), we can get $\sum_{m=1}^{M} \Phi_m = 1$ for each operating point. Therefore, $D(k)$ is represented with (46.10), which can also be derived from combining Eqs. (46.6) and (46.7) [34].

$$D(k) = \frac{v_{\text{ref}}(k+1) - \sum_{m=1}^{M} \Phi_m \times \left(\underline{\theta}_m \times \underline{x}(k)\right)}{\sum_{m=1}^{M} \Phi_m \theta_{m,3n}} \tag{46.10}$$

According to the constructed LMN-based DT of the converter, its output voltage controller that controls the converter to achieve the constant reference value against input voltage and load variations is then designed and implemented in the converter (see Figure 46.3).

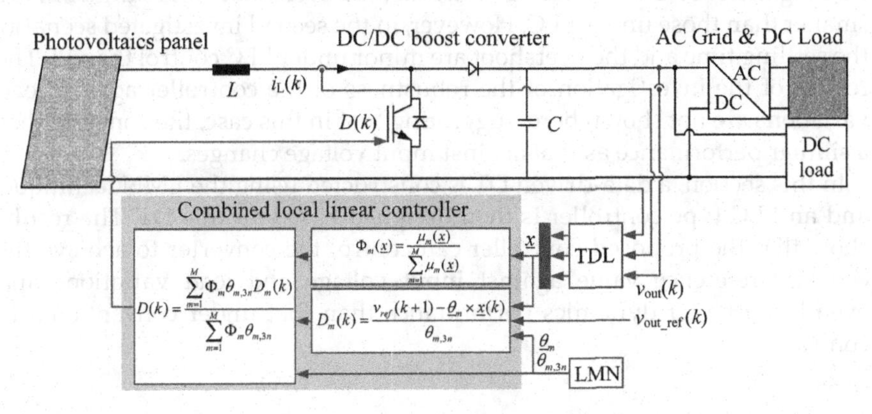

FIGURE 46.3
Designed local linear control based on digital twin of converter.

46.3.2.4 Simulation Results of LMN-Based DT and Control of LLC

The controller is designed based on exporting the proper duty cycle according to the current operating range. To this end, two steps are executed: construction of DC-DC converter DT model and design of the controller. The stand-alone DC-DC converter that utilises the average model is constructed using the LMN-based DT model. The converter parameters are set as follows: input voltage $V_{in}=24$ V, the inductance $L=0.25\,\text{mH}$, the capacitance $C=200\,\mu\text{F}$ and resistance $3\,\Omega$.

The LLM number of the LMN is selected as 6, while the order of the delayed signals is chosen as 4. To evaluate the accuracy of the DT model, the mean absolute percentage error (MAPE) is then used as the error criteria.

$$\text{MAPE} = \left(\frac{1}{N}\right) \cdot \sum_{k=1}^{N} \frac{\left|v_{\text{actual}}(k) - v_{\text{DT}}(k)\right|}{v_{\text{actual}}(k)} \tag{46.11}$$

where N is the number of test samples. For the operating range in this work, the MAPE = 0.0094%. According to this accurate LMN-based DT, the voltage controller LLC is then designed based on the structure and parameter of the DT. The reference voltage tracking and the robustness of the controller concerning input voltage variations and load changes are investigated using the proposed LLC in the converter. Moreover, the dynamic converter performance under LLC control is compared with the control of the PI controller, as to which parameters are determined with the Ziegler–Nichols control technique [35].

Compared to the dynamics under PI control, those under LLC show a better tracking performance in Figure 46.4a and an optimiser response against variations in the converter input voltage, as shown in Figure 46.4b. The settling time under LLC is shorter than that under PI control in each case. In the investigation of reference voltage tracking, the overshoot values under PI are smaller than those under LLC. However, in the second investigated scenario, the settling time and the overshoot are minor under LLC control than PI. The results of the investigation of the robustness of the controller against load variations are not shown here, suggesting that in this case, the converter has a similar performance as that against input voltage changes.

In this section, a data-driven DT is constructed using the LMN technique, and an LLC type controller is then designed based on this DT. The results show that the proposed controller can control the converter to achieve the constant reference value against input voltage and load variations and even has a better dynamics performance than that under conventional PI control.

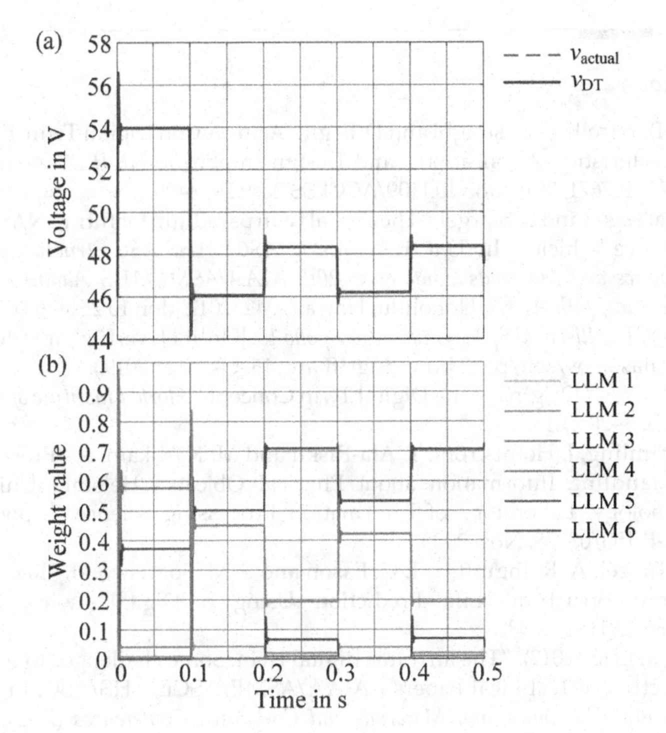

FIGURE 46.4
Output voltage response (a) for reference voltage tracking (b) to input voltage variations.

46.4 Conclusion

This chapter provides a comprehension of the DT in smart grid. A particular definition of the DT is presented from the perspective of power system researchers. It is understood more as an accurate aggregated simulation model that interacts with the physical grid in real time to simulate, emulate and monitor the condition and performance of the entire system. In order to give the reader a better picture of DT in the smart grid, we present a few applications and their compatible DT modelling techniques. A DT can be used in a smart grid, for example, to estimate the current state in a digital power plant and to design the appropriate controller in converters, whose amount recently increased significantly in the smart grid due to the energy transition. By training with numerous data collected from a running smart grid, the DT can model and track the state of the gird. As digitalisation progresses in all aspects of industry, DTs are gaining a greater focus in the development of smart grids. Researchers in power systems also need to continue to strengthen their research in this area so that the intelligence and digitisation of smart grids can proceed in a further stage.

References

1. B. R. Barricelli, E. Casiraghi and D. Fogli, "A Survey on Digital Twin: Definitions, Characteristics, Applications, and Design Implications", *IEEE Access*, Jg. 7, S. 167653–167671, 2019, doi: 10.1109/ACCESS.2019.2953499.
2. E. Glaessgen and D. Stargel, "The digital twin paradigm for future NASA and U.S. Air Force Vehicles" In *53rd AIAA/ASME/ASCE/AHS/ASC Structures, Structural Dynamics and Materials Conference; 20th AIAA/ASME/AHS Adaptive Structures Conference; 14th AIAA*, Honolulu, Hawaii, 04232012, doi: 10.2514/6.2012-1818.
3. NASA, *The Ill-Fated Space Odyssey of Apollo 13*. [Online]. Verfügbar unter: https://er.jsc.nasa.gov/seh/pg13.htm (Zugriff am: 16. Oktober 2019.).
4. M. Grieves, "Origins of the Digital Twin Concept", *Florida Institute of Technology*, Jg. 8, S. 3–20, 2016.
5. K. Främling, J. Holmström, T. Ala-Risku and M. Kärkkäinen, "Product Agents for Handling Information about Physical Objects", Helsinki University of Technology, Laboratory of Information Processing Science, Espoo, Finland TKO-B 153/03, 28. Nov. 2003.
6. E. J. Tuegel, A. R. Ingraffea, T. G. Eason and S. M. Spottswood, "Reengineering Aircraft Structural Life Prediction Using a Digital Twin", 2011, doi: 10.1155/2011/154798.
7. Tuegel, Eric. (2012). "The airframe digital twin: Some challenges to realization". Collection of Technical Papers - AIAA/ASME/ASCE/AHS/ASC. In *Structures, Structural Dynamics, and Materials and Co-located Conferences* (Zugriff am: 18. September 2020). 10.2514/6.2012-1812.
8. U.S. Air Force, *Global Horizons Final Report: United States Air Force Global Science and Technology Vision*. [Online]. Verfügbar unter: https://www.hsdl.org/?view&did=741377 (Zugriff am: 16. Oktober 2019.)
9. H. Pan, Z. Dou, Y. Cai, W. Li, X. Lei and D. Han, "Digital twin and its application in power system" In *2020 5th International Conference on Power and Renewable Energy (ICPRE)*, Shanghai, China, 9122020, S. 21–26, doi: 10.1109/ICPRE51194.2020.9233278.
10. G. L. Demidova et al., "Implementation of Digital Twins for Electrical Energy Conversion Systems in Selected Case Studies", *Proceedings of the Estonian Academy of Sciences*, Jg. 70, Nr. 1, S. 19, 2021, doi: 10.3176/proc.2021.1.03.
11. T. Orosz, "Evolution and Modern Approaches of the Power Transformer Cost Optimization Methods", *Periodica Polytechnica Electrical Engineering and Computer Science*, Jg. 63, Nr. 1, S. 37–50, 2019, doi: 10.3311/PPee.13000.
12. S. Boschert and R. Rosen, "Digital twin-The simulation aspect" In *Mechatronic Futures*, P. Hehenberger and D. Bradley, Hg., Cham: Springer International Publishing, 2016, S. 59–74, doi: 10.1007/978-3-319-32156-1_5.
13. M. D. Omar Faruque et al., "Real-Time Simulation Technologies for Power Systems Design, Testing, and Analysis", *IEEE Power Energy Technol. Syst. J.*, Jg. 2, Nr. 2, S. 63–73, 2015, doi: 10.1109/JPETS.2015.2427370.
14. S. Yun, J.-H. Park and W.-T. Kim, "Data-centric middleware based digital twin platform for dependable cyber-physical systems" In *2017 Ninth International Conference on Ubiquitous and Future Networks (ICUFN)*, Milan, 72017, S. 922–926, doi: 10.1109/ICUFN.2017.7993933.

15. *IEEE Standard for Synchrophasor Data Transfer for Power Systems,* in IEEE Std C37.118.2-2011 (Revision of IEEE Std C37.118-2005), pp.1–53, 28 Dec. 2011, doi: 10.1109/IEEESTD.2011.6111222.
16. S. Weyer, T. Meyer, M. Ohmer, D. Gorecky and D. Zühlke, "Future Modeling and Simulation of CPS-based Factories: An Example from the Automotive Industry", *IFAC-PapersOnLine,* Jg. 49, Nr. 31, S. 97–102, 2016, doi: 10.1016/j.ifacol.2016.12.168.
17. Q. Qi and F. Tao, "Digital Twin and Big Data towards Smart Manufacturing and Industry 4.0: 360 Degree Comparison", *IEEE Access,* Jg. 6, S. 3585–3593, 2018, doi: 10.1109/ACCESS.2018.2793265.
18. X. Song, T. Jiang, S. Schlegel and D. Westermann, "Parameter Tuning for Dynamic Digital Twins in Inverter-Dominated Distribution Grid", *IET Renewable Power Generation,* Jg. 14, Nr. 5, S. 811–821, 2020, doi: 10.1049/iet-rpg.2019.0163.
19. J. Ríos, J. Hernández, M. Oliva, and F. Mas, Hg., *Product Avatar as Digital Counterpart of a Physical Individual Product: Literature Review and Implications in an Aircraft,* 2015. ISPE International Conference on Concurrent Engineering (2015).
20. General Electric Company, *GE Digital Twin: Analytic Engine for the Digital Power Plant.*
21. Siemens, A. G. 2018, *Siemens Electrical Digital Twin: A Single Source of Truth to Unlock the Potential within a Modern Utility's Data Landscape.* [Online]. Verfügbar unter: siemens.com/electrical-digital-twin.
22. C. Brosinsky, D. Westermann and R. Krebs, "Recent and prospective developments in power system control centers: Adapting the digital twin technology for application in power system control centers" In *2018 IEEE International Energy Conference (ENERGYCON),* Limassol, 6/3/2018–
23. S. El Ferik and A. A. Adeniran, "Modeling and Identification of Nonlinear Systems: A Review of the Multimodel Approach-Part 2", *IEEE Transactions on Systems, Man, and Cybernetics: Systems,* Jg. 47, Nr. 7, S. 1160–1168, 2017, doi: 10.1109/tsmc.2016.2560129.
24. O. Nelles, "LOLIMOT - Lokale, lineare Modelle zur Identifikation nichtlinearer, dynamischer Systeme", *Automatisierungstechnik,* Jg. 45, Nr. 4, S. 163–174, 1997, doi: 10.1524/auto.1997.45.4.163.
25. O. Nelles and R. Isermann, "Basis function networks for interpolation of local linear models" In *35th IEEE Conference on Decision and Control,* Kobe, Japan, 11–13 Dec. 1996, S. 470–475, doi: 10.1109/CDC.1996.574356.
26. T. Fischer and O. Nelles, "Merging strategy for local model networks based on the Lolimot Algorithm" In *Artificial Neural Networks and Machine Learning -* ICANN 2014: *2014: 24th International Conference on Artificial Neural Networks, Hamburg, Germany,* 2014, S. 153–160.
27. D. Dovzan and I. Skrjanc, "Fuzzy Space Partitioning Based on Hyperplanes Defined by Eigenvectors for Takagi-Sugeno Fuzzy Model Identification", *IEEE Transactions on Industrial Electronics,* Jg. 67, Nr. 6, S. 5144–5153, 2020, doi: 10.1109/TIE.2019.2931243.
28. O. Nelles, *Nonlinear System Identification: From Classical Approaches to Neural Networks and Fuzzy Models.* Berlin, Heidelberg: Springer, 2001.
29. R. Murray-Smith, "Local learning in local model networks" In *4th International Conference on Artificial Neural Networks,* Cambridge, UK, 26–28 June 1995, S. 40–46, doi: 10.1049/cp:19950526.

30. O. Nelles, A. Fink and R. Isermann, "Local Linear Model Trees (LOLIMOT) Toolbox for Nonlinear System Identification", *IFAC Proceedings Volumes*, Jg. 33, Nr. 15, S. 845–850, 2000, doi: 10.1016/S1474-6670(17)39858-0.
31. O. Nelles, "Axes-oblique partitioning strategies for local model networks" In *2006 IEEE Conference on Computer Aided Control System Design, 2006 IEEE International Conference on Control Applications, 2006 IEEE International Symposium on Intelligent Control*, Munich, Germany, 10/4/2006–10/6/2006, S. 2378–2383, doi: 10.1109/CACSD-CCA-ISIC.2006.4777012.
32. R. Isermann and M. Münchhof, *Identification of Dynamic Systems: An Introduction with Applications / Rolf Isermann, Marco Münchhof*. Berlin: Springer, 2011.
33. A. D. Gordon, L. Breiman, J. H. Friedman, R. A. Olshen and C. J. Stone, "Classification and Regression Trees", *Biometrics*, Jg. 40, Nr. 3, S. 874, 1984, doi: 10.2307/2530946.
34. K. Rouzbehi, A. Miranian, J. M. Escaño, E. Rakhshani, N. Shariati and E. Pouresmaeil, "A Data-Driven Based Voltage Control Strategy for DC-DC Converters: Application to DC Microgrid", *Electronics*, Jg. 8, Nr. 5, S. 493, 2019, doi: 10.3390/electronics8050493.
35. G. Abbas, M. A. Samad, J. Gu, M. U. Asad and U. Farooq, "Set-point tracking of a DC-DC boost converter through optimized PID controllers" In *2016 IEEE Canadian Conference on Electrical and Computer Engineering (CCECE)*, Vancouver, BC, Canada, 2016, S. 1–5, doi: 10.1109/CCECE.2016.7726841.

47

Digital Twins in Graphene Technology

Elena F. Sheka
Institute of Physical Researches and Technology of the Peoples' Friendship University of Russia

47.1 Introduction

Digital twins (DTs) burst into our lives in 2002 from the podium of a Society of Manufacturing Engineers conference with the voice of Michael Greaves [1]. Although the term itself was proposed much earlier [2], it was Greaves who filled it with new content that constitutes the present-day DTs. Grieves proposed the DT as a conceptual model underlying the product lifecycle management. Eight years later, John Vickers suggested the DTs concept [3], which has been largely developed since. From that moment on, a wide march of this concept began in the industry, health, construction, business, education, social life, etc. This book will present the successes of this concept in various fields of human activity. In order not to duplicate other contributors to the book, I will limit myself to references to the most representative reviews [4–13], but a few.

Differing in nuances, a general presentation of the DTs concept concerns the trinity of the physical object, virtual/digital object, and the connection between them. The connection is provided by the data that flows from the physical object to the digital/virtual one and information that is available from the digital/virtual object to the physical environment. This concept, which is new for large massive fields of the human activity, has been widely exploited in academic researches since the first computer became available. Known as simulations or modeling, the concept implementation has provided a drastic development of the academic studies, related to the natural science, in particular.

Today it is impossible to imagine modern physics, chemistry (and same as them but prefixed with *bio-* and *geo-*) and material science without modeling. A huge leap in the development of computational programs and computing tools that has taken place over the past half century has led to the

DOI: 10.1201/9781003425724-56

fact that many previously predominantly empirical sciences have become virtual-empirical, while some of them predominantly virtual. Despite such rapid development, the relationship between the real object and its model, established as a subordinate for the model, has not changed until recently. As it turned out, this circumstance significantly limits the further development of science in the case of its predominantly virtual nature. And here the DTs concept in the above wording enters the scene. Indeed, physical and virtual objects form the basis of both the modeling and the DTs concept. The difference between these two approaches concerns the different sense, which is embedded in the understanding of the connection between these two objects. Using a grammatical term, it can be represented as a complex (principal–subordinate) sentence in the first case and compound (equal–equal) in the second case. In every language, substituting one sentence for another changes the meaning of the spoken speech. The same is true in the case of science. This chapter discusses precisely this change in semantic speech, which is introduced by the DTs concept into science, using the example of the virtual chemical physics of graphene [14].

Despite the widespread use of modeling in science, after a careful study of the available literature [12,15,16], I found only one reference concerning the DT concept [17], which precedes our works, however, in this case also related mainly not to polymer science but to polymer reaction engineering. However, the pioneer of the DTs concept, Greaves, in one of his interviews given in 2018, said [1]: "It does not have to be an all-or-nothing project. There is a wide range of information that I can collect and process with the twin. Digital twins could also be used in very specific, very limited scenarios". Graphenics has occurred to be one of these scenarios. Graphene is an outstanding academic project, strongly oriented on high-tech technology [18]. Heralded in 2012 with a list of promising applications on 850 pages and receiving an exclusive material for science financial support, Graphene Flagship faced insurmountable difficulties in fulfilling its obligations. As a result, the project, first divided into bad and good graphene [19], and then practically completely abandoned good one, referring to the dishonesty of graphene material manufacturers [20], turned out to be practically unfulfilled by the end. And one of the reasons for this plight is the virtual nature of graphenics, oversaturated with simulation results, which have contributed to the development of this field in the wrong direction. The real reasons for the failure were revealed by removing the models from the subordinate state and giving them the value of DTs equal to real objects. This chapter describes the result of such a transformation, which was the first application of the DTs concept in the academic materials science.

47.2 Graphene as a High-Tech Product

The name 'graphene' has a wide meaning covering a large set of different objects related to the modern graphenics [21]. The first concerns a honeycomb package of carbon atoms just forming bare graphene domains of different shape and size. Graphene crystal presents the next family member, which is simply a large domain, the linear dimensions of which exceed all significant size-limiting dimensions, starting from which such a domain can be described strictly in the language of solid-state physics [22]. In practice, empirical graphene crystal, obtained, say, by peeling away one layer of graphite with adhesive tape, is a one-atom thick sheet, linear size of which is usually over 1 μm. The domain edge atoms are valence not saturated thus possessing dangling bonds. In reality, these bonds are usually terminated by heteroatoms and chemical groups available in the object surrounding. In contrast, theoretical graphene crystal is a bare flat graphene domain generated by multiplication of an elementary cell consisting of two carbon atoms under the conditions of translational periodic invariance [23].

The next family member is presented with graphene molecules. As previously, graphene domains lay the foundation of the species. However, the domains linear size is less than that of the above-mentioned size-characteristic parameters, which distinguishes the molecular and crystalline communities. The molecular part is extremely large involving all possible derivatives of graphene domains. Since the latter is a topochemical polytarget species [24], the number of polyderivatives is practically endless. Nevertheless, the latter can be divided into two subclasses, which are structurally distinguished, namely, the subclass of sp^2 necklaced graphene molecules, edge atoms of whose domains are involved in chemical reactions only [25]. The second corresponds to sp^3 graphene molecules, for which not only edge atoms but also basal-plane atoms of the domains participate in the chemical modification as well [26,27].

The high practical appeal of using graphene in modern technology is based on the exceptional properties of its crystal [23] and molecules [14,28]. Readers interested in this issue are referred to the above monographs. However, the practical implementation of this intention faced difficulties from the initial steps. Thus, there was an understanding that a graphene molecule and a graphene crystal are completely different objects, which forced one to divide the relation to the graphene material into two parts that provide low performance (*lp*) and high performance (*hp*) technologies, spreading the implementation time of the corresponding technologies by decades [19]. The main problem concerned the production of graphene material. Thus, the adhesive-tape peeling of graphite cannot be seriously considered as the source of graphene mass production for *hp* technology. On the other hand, the production of molecular graphene for *lp* technologies, the need of which was realized in physical laboratories mainly, was correlated with the presence of the initial product in the form of natural graphite, which needs splitting into nanosized sheets. The best way to address the problem, suggested with chemists, is graphite oxidation [29] and hundreds of chemical laboratories around the world are engaged in the development of the most efficient and low-cost methods for the mass production of graphene oxide (GO) (see reviews [30–38] but a few). GO is a sp^3 graphene species. While analyzing the molecule application, which has been rapidly developing all this time, it was found that not GO but necklaced reduced graphene oxide (rGO) was much more effective and interesting in many cases. The finding opened the road for the reduction of the previously synthesized GO transferring it to the wished rGO. And again, hundreds of chemical laboratories were engrossed in the development of efficient and low-cost graphene product known as rGO [39]. A full technological chain looks like the following:

nanostructured Graphite ↔ Graphene oxide ↔ reduced Graphene oxide

Scheme 47.1

Here sign ↔ means that the connected two processes are reversible. At the same time, once stimulated by the rapid development of graphenics, old sp^2 amorphous carbons experienced the writing of a new history. Therefore, during the formation process of Shuntu carbon deposits millions of years ago, Scheme 47.1 was implemented in the Nature[40]. The same applies to coals and anthraxolites [41], carbon shells of silica [42], and other minerals. On the other hand, for the last century, the industry has managed to develop technologies for the tonnage production of synthetic sp^2 amorphous carbons, such as carbon black, activated carbon, carbon rubber fillers, etc. [43]. Research has found that rGO with different necklace chemical contents, as well as the shape and size of graphene domains, are the basis for all of these carbon natural [44]. Moreover, the humanity, without realizing it, is familiar with the body from the first bonfire, from the first shovel of coal thrown into the furnace, from the first filler of wheel tires, from the first… and this listing can be continued indefinitely. The ash of a burned fire, deposits of natural coal, synthetic black carbon, and, finally, a laboratory product of the chemical reduction of oxidized nanoscale graphite – this is a short list of a unique material, which is known as sp^2 amorphous carbon, the basic structural unit (BSU) of which is a polyatomic necklaced graphene molecule, which should be attributed to rGO. Thus, returning to the beginning of history, the simplest method of obtaining a massive amount of rGO by burning food waste is being revived in the hands of materials scientists [45]. In contrast to the rGO, the GO does not exist in nature. And, chemical synthesis is the only source of this product.

Graphene molecules do not exist individually and, once of either natural or synthetic origin, are fully incorporated in carbon amorphous solids. The solids have a complicated multilevel structure [40,41], which is schematically presented in panel A of Figure 47.1. The general picture presented in the

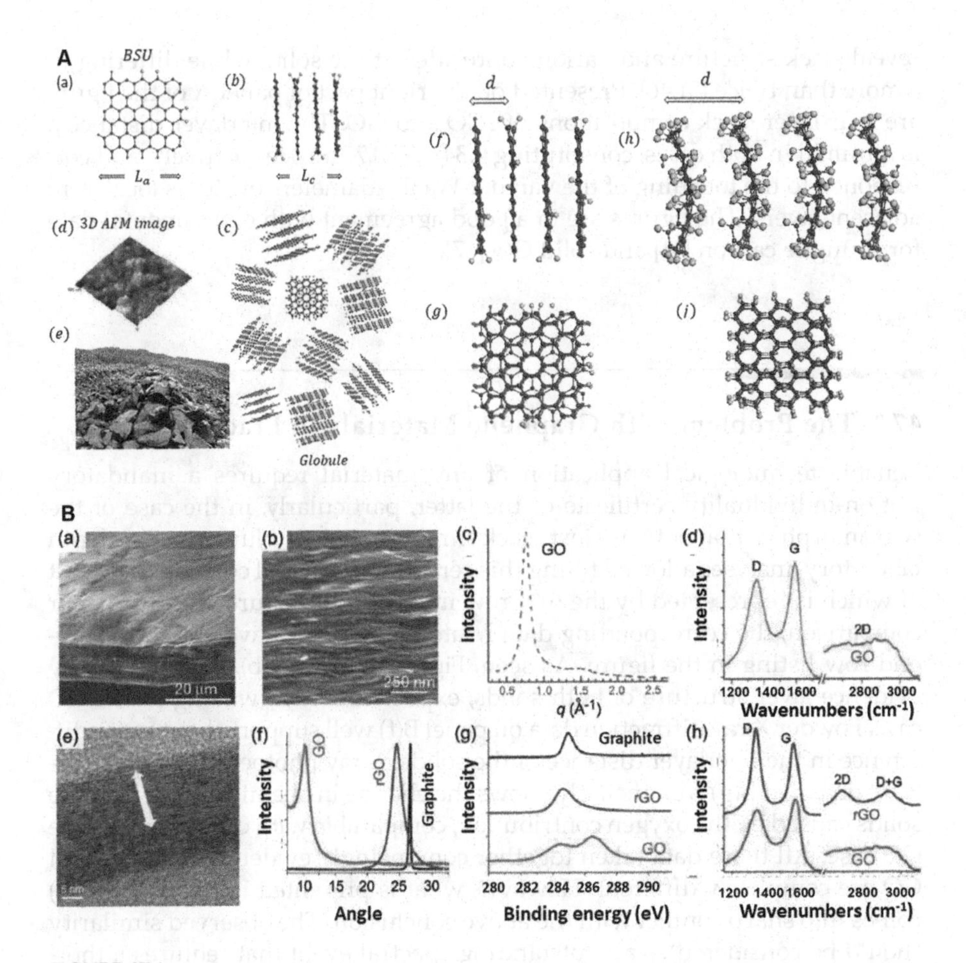

FIGURE 47.1
A. (a and b) Individual and four-layer stack of rGO BSU. (c) rGO stack globule. (d) General AFM image of shungite carbon. (e) Karelian shungite carbon deposit. (f and g) Side and top views of rGO stack, respectively. (h and i) Side and top views of Go stack, respectively. B. (a) Low- and (b) high-resolution SEM side-view images of a 10-mm-thick GO paper. (c) X-ray diffraction pattern of the GO paper sample. (d) Raman spectrum of a typical GO paper. (e) Cross-section TEM images of a stack of rGO platelets. (f) Powder XRD patterns of graphite, GO, and rGO. (g) XPS characterization of rGO platelets. (h) Raman spectra of rGO and the GO reference sample.

left side of the panel is typical to sp^2 amorphous rGOs. As seen, individual BSUs are stacked. Thus formed stacks agglomerate producing a solid body of highly variable porosity thus separating restricted regions of lower porosity (globules) in the whole of the body. Three main structural characteristics of these solids are the linear size of their BSUs, L_a, the stack thickness, L_c, and the distance between BSU layers in stacks, *d*. Amorphous GO retains the general motive of rGO. sp^3 Graphene molecules present its BSUs, L_a of which is comparable with that of rGO. Agglomerating, these BSUs clearly reveal stack structure and various porosities of the solid, while differing in *d* more than twice [31,46]. Presented on the right part of panel A of the figure are four-layer stack compositions of rGO and GO. The interlayer distance *d* is minimal in both cases, constituting 0.34 and 0.72 nm, respectively, and corresponds to the touching of the van der Waals diameters of atoms located in adjacent layers. The figures are in a good agreement with experimental data for shungite carbon [41] and solid GO [47].

47.3 The Problem with Graphene Materials in Practice

Reliable technological application of any material requires a mandatory test-on-individuality certificate of the latter, particularly, in the case of the two amorphics that both are just black similar powders. Numerous in-depth laboratory analyses allowed filling this certificate with real content, the result of which is represented by the first row in panel B of Figure 47.1 [31,35]. For comparison, the corresponding data related to rGO are involved in the second row listing in the figure. As seen, Figure 47.1B(a), B(b), and B(c) clearly evidence layer structure of both solids, expectedly more waving in the GO case. Powder X-ray diffraction data on panel B(f) well support about twice difference in the interlayer distance of the solids, X-ray photoelectron spectroscopy on panel B(g) convincingly shows the change in chemical content of the solids caused by the oxygen contribution, comparable with carbon one, in the GO case. All these data taken together convincingly evidence that rGO and GO are completely different. However, what is presented in Figure 47.1B(h) comes into sharp conflict with the above conclusion. The observed similarity should be considered as an outstanding spectral event that requires a thorough consideration.

The situation is significantly complicated by the fact that greatly characteristic D-G-bands Raman doublet is largely exploited in graphenics as express-analysis justifying the belonging of the studied material to the family of graphene materials [48]. While not disputing that the GO belongs to graphene materials, we nevertheless do not consider the observed similarity of the spectra to be a mere coincidence. Moreover, this circumstance leads to a number of undesirable results concerning practical application of materials. A technologist, who is not able to distinguish rGO and GO by express Raman spectra analysis cannot guarantee the fulfillment of at least two mandatory requests: (1) the absence of a reversible transformation rGO ↔ GO in due course of production and storage of the final product and (2) the absence of toxicity of the final product. This is vital in the use of rGO or GO in the manufacture of different medicals [49–51] and vaccines [52]. Suffice it to recall that rGO BSUs are stable, or sleeping, radicals [53], which can revive their activity at any moment under the action of surrounding, while GO is completely deactivated. It is also of great importance in the case of, for example, the use of carbon-based hydrogels in the production of lithium batteries [54]. It is clear that rGO $\leftrightarrow$ GO processes greatly influence the efficiency of devices and their validity term.

To be able to fight and predict the possible negative consequences of the rGO and GO practical applications, we must answer the following question: "What do we still not know about rGO and GO, which would explain reasons for the identity of the Raman spectra of two substances, which are completely different both in structure and in chemical content?" As occurred, the DT concept allows getting the answer to this question [25,55,56].

47.4 General Grounds of the Digital Twins Design

First used recently [57], the DT concept has revealed a high efficiency to solve intricately complicated problems. It was natural to apply to it looking for reasons of the identity of Raman spectra of rGO and GO and finding a way to discover this mystery. We look at the DT concept in Scheme 47.2.

Digital twins $\rightarrow$ Virtual device $\rightarrow$ IT product.

Scheme 47.2

Here, DTs are molecular models under study, virtual device is a carrier of a selected software, IT product covers a large set of computational results related to the DTs under different actions in the light of the soft explored. The quality of IT product highly depends on how broadly and deeply the designed DTs cover all the knowledge concerning the object under consideration and how adequate is the virtual device to handle the peculiarities of the object under study. The first requirement can be evidently met by a large set of the relevant models, in the current case up to a few hundreds, each consisting of N atoms (up to 264). As for the virtual device, it should not contradict with the object nature and perform quantum-chemical calculations providing the establishing of equilibrium structure of the designed DTs and obtaining their spectra of IR absorption and Raman scattering related to $3N$-6 vibrational modes. Semiempirical quantum-chemical programs cope with such a volume of cumbersome calculations. However, the radical nature of most rGO and/or GO DTs, which turns them into open-shell electronic systems, forced to abandon density functional theory (DFT) and MD programs and pay attention only to programs based on the unrestricted Hartree-Fock (UHF) approximation. In the current study, the virtual device HF Specrodyn [58] was used. A particular spin-density algorithm [59] was used to design DTs under study.

47.5 Vibrational Spectra of rGOs and GOs in Light of the Digital Twins Concept

Exclusiveness of the situation with Raman spectra of rGO and GO poses a few questions to be answered, namely:

- Is the spectra identity a chemical-content effect?
- Is the spectra identity a structural effect?

It is clear that the answers to these questions affect the deep essence of both oxides, each of which is a broad class of substances, and that they cannot be obtained empirically. In contrast, it is obvious that a wide chemical and structural modification of DTs makes it possible to find a reliable answer. To simplify further discussions, we classify rGO and GO as either polychrome or monochrome objects. The polychromicity concerns the variety of heteroatom content beyond the carbon one, which is typical to real products. The monochromicity will be used to limit the products' heteroatomic content to mono-atomic one or to special chemical groups. So, in accordance with Scheme 47.2, DTs represent a wide range of models of both the individual BSUs of the rGO and GO and their stacks of different composition and different sizes. The digital device is a quantum-chemical semiempirical software that implements the Hartree-Fock (HF) approximation in both versions concerning the UHF and restricted RHF (RHF) approximations. The IT product is a fully optimized DT structure on the one hand, and virtual one-phonon spectra of IR absorption and Raman scattering, on the other.

47.5.1 rGO in Light of the Digital Twins Concept

Two sets of designed DTs were considered. The first involves rGO DTs, based on the same graphene domain (5,5) NGr but differing with the heteroatom necklace content. The necklaces are both monochrome, consisting of either hydrogen or oxygen atoms, as well as of hydroxylic or carboxylic units only, and polychrome when the content is mixed. Figure 47.2 presents a collection of IR and Raman spectra characteristic for the set. Throughout the chapter, the virtual spectra are presented by stick-bars convoluted with Gaussian bandwidth of 10 cm^{-1}. Intensities are reported in arbitrary units, normalized per maximum values within each spectrum. Since the number of vibrational modes composing the spectra under consideration is too large, the excessive fine structure, statistically suppressed in practice, is covered by trend lines averaged over 50 next steps of 0.003 ÷ 0.010 cm^{-1} each. The figure is placed over a background composed of the relevant equilibrium DTs, the detailed study of which is discussed elsewhere [25,55]. A comparative analysis of

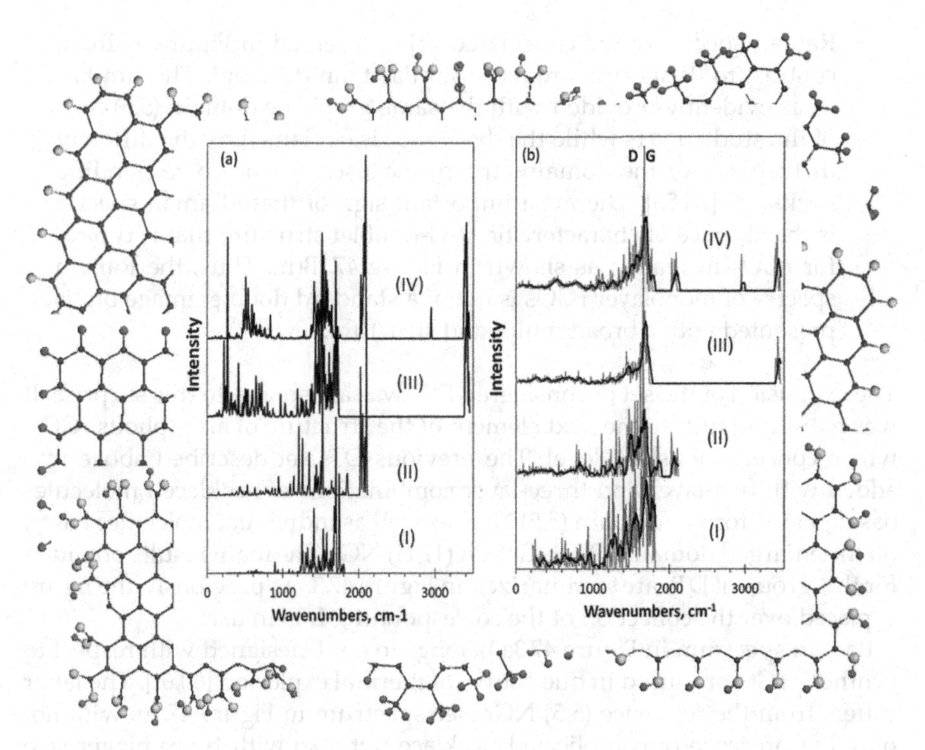

FIGURE 47.2
Virtual IR absorption (a) and Raman scattering (b) spectra of necklaced graphene molecules based on the domain (5,5) NGr, presenting graphene hydride $C_{66}H_{22}$ (I), and three reduced graphene oxides $C_{66}O_{22}$ (II), $C_{66}(OH)_{22}$ (III), and $C_{66}(COOH)_{18}$.(IV). UHF AM1 calculations.

calculated data obtained completes the digital cycle. As a result, the following conclusions important for the DT analytics of rGO were obtained.

- The carbon atoms of graphene domains of the rGOs and heteroatoms of their necklaces contribute to the optical vibrational spectra quite differently: the vibrations of the former are mainly responsible for Raman spectra, while the IR spectra are assigned to the vibrations of the necklaces heteroatoms.
- Virtual IR spectra in Figure 47.2a present well distinguished spectral signatures of the monochrome rGOs related to the contribution of atomic hydrogens and oxygens as well as hydroxyls and carboxyls to the relevant rGOs necklaces.
- Virtual IR spectra of polychrome rGOs retain superpositional character of spectral signatures, although not providing a quantitative analysis of the necklace content, which is connected with the domain-stimulated influence of heteroatom additives on each other.

- Raman spectra of the considered DTs presented in Figure 47.2b, in contrast to IR spectra, are more similar than different. The similarity is evidently provided with the same graphene domain (5,5) NGr of the studied DTs while the difference is explained by the different disturbance of the domain structure caused by the corresponding necklaces [25,55]. The most important sign of these Raman spectra is the absence of characteristic D-G-doublet structure that is typical for rGOs in reality, as shown in Figure 47.1B(h). Thus, the Raman spectra of monolayer rGOs is not of a standard doublet image but is presented with a broad multiband structure.

The expansion of the set of considered DTs was a response to this surprise. It was natural to turn to the next element of the structure of amorphous rGOs, which concerns its stack level. The previous DTs set described above was added with two-layer and three-layer combinations of necklaced molecules based on the former domain (5,5) NGr, as well as individual molecules based on the enlarged domains (9,9) NGr and (11,11) NGr. The main results obtained for this group of DTs are summarized in Figure 47.3. As previously, the figure is placed over the collection of the corresponding DTs in use.

Raman spectrum in Figure 47.3a belongs to a DT designed with respect to synthetic rGO produced in due course of thermal explosion [60,61]. The latter differs from the reference (5,5) NGr (see spectrum in Figure 47.2b) with not only the presence of complicated necklace but also with twice bigger size of the domain. As seen in the figure, this DT Raman spectrum drastically changes presenting the transformation of a broad multiband structure of the reference DT toward a single-band one. Such a transformation of Raman spectra is repeated for DTs involving bare and hydrogenated domains (9,9) NGr as well and becomes more pronounced in the case of bare domain (11,11) NGr [25]. The matter is connected with a typical size-effect concerning the Raman spectra of substances based on covalent bonds when the substance nature is transformed from molecularly amorphic to crystalline one [62,63]. Actually, when the linear dimension of the domains is achieving and/or exceeds $L_{ph} \sim 15\,\text{nm}$, which is a free path of the high-frequency optical phonon G of the graphene crystal [64], the domain dynamics acquires the characteristic features of crystalline behavior and Raman spectra tend to adopt a single-band appearance in favor of G band corresponding to asymmetrical sp^2C-C stretchings, the limit case of which is presented with the narrow G phonon band in graphene crystal.

In contrast to one-layer molecules, two-layer hydrogenated (5,5) NGr is characterized with clearly seen generation of D-G-band doublet in Figure 47.3c. This change in the spectrum shape becomes more pronounced when the two-layer DT is replaced with a three-layer one. Therefore, the performed DT analysis has revealed that the characteristic standard D-G-band-doublet structure of the Raman spectra of rGOs is a consequence of its stacking structure. A thorough analysis has shown [25, 55] that the feature results from one

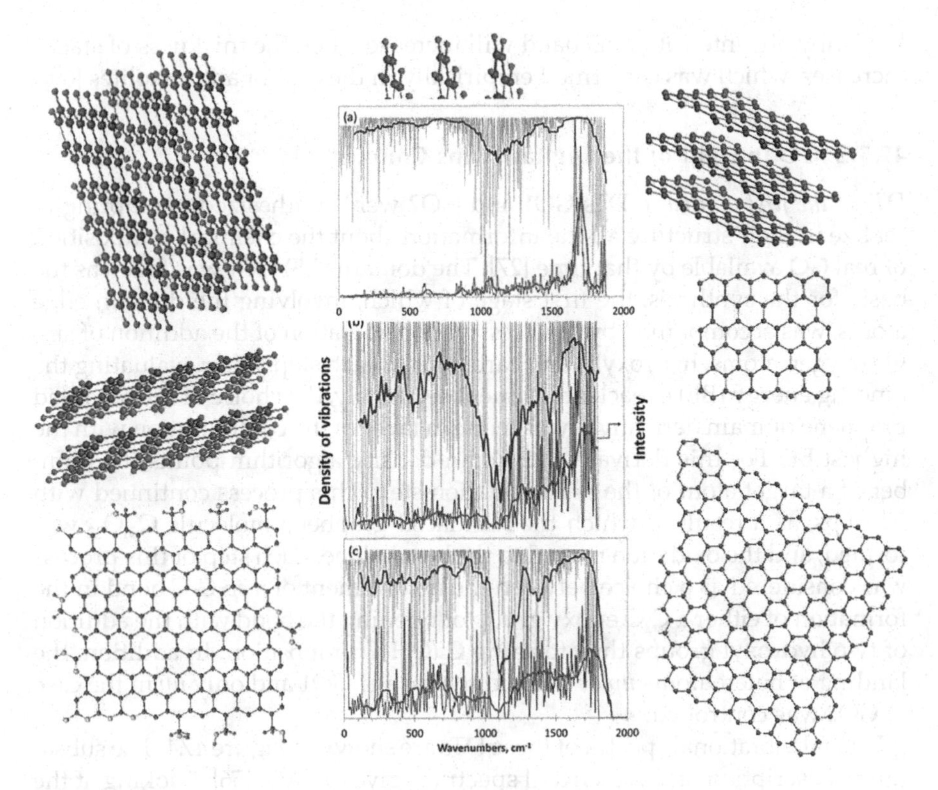

FIGURE 47.3
Raman spectra of DTs with different size of graphene domains and of different domain packing. (a and b) One-layer BSU based on domain (9,9) NGr ($C_{192}O_{19}H_{44}$) and bare domain (5,5) NGr (C_{66}), respectively; (c). Two layers of hydrogenated domains (5, 5) NGr ($C_{132}H_{44}$). Raman spectra (graph on the bottom) and density of vibrations (graph on the top) are accompanied with 50-points trend lines. UHF AM1 calculations.

more unique characteristic of carbon. It concerns the fact that the distance between adjacent layers in graphite-like structures is equal to the van der Waals diameter of a carbon atom and constitutes ≈0.335 nm. The distance is much larger than a standard size of a single sp^3C-C covalent bond of 0.153 nm, so that unpaired electrons of carbon atoms in the graphene domain basal planes of neighboring layers cannot be coupled while atoms themselves on their periphery contact each other. Meanwhile, the vibration of carbon atoms is not limited in space. Therefore, the special out of plane phonons appearing in multi-layer graphene crystals [65-68] support each other's out of plane displacement, thereby stimulating the emergence of dynamic sp3C-C bonds between domain layers [25]. The corresponding out-of-plane vibrations lead to the emergence of a broad band in the spectrum of the stack of nanosized graphene domains, which is in the place of a quite narrow D band of multilayer graphene crystal [65]. Therefore, the doublet of the D–G bands of sp^2 amorphous carbons is a sign marking the stack structure of the substance.

Evidently, the intensity of D band will increase when the thickness of stacks increases, which was confirmed empirically in the case of anthraxolites [69].

47.5.2 GO in Light of the Digital Twins Concept

DTs of the first approach: DTs GO1 and GO2 were synthesized attempting to realize in their structure all the information about the chemical composition of real GO available by that time [27]. The domain (5,5) NGr was taken as the basis for the synthesis, the first stage of which, involving the domain edge atoms, was accompanied by a successive consideration of the addition of single oxygen atoms, hydroxyls, and carboxyls at each step. After evaluating the binding energy (BE) of each attachment separately, the choice of the obtained graphene domain derivatives was made in favor of the configuration with the highest BE. For this derivative, the spin-density algorithm points the number of a target atom of the next oxidation step. This process continued with 22 steps, as a result of which the necklaced graphene molecule $C_{66}O_{22}$ was formed, and the oxidation moved to the basal plane. Each step of this process was considered as a choice between the involvement of a sp^2C-C bond in the formation of either a C_2O epoxy group or opening the bond with the addition of two hydroxyl groups thus forming $C_2(OH)_2$ composition. In addition, the landing of heteroatoms *up* or *down* in the case of GO1 and only *up* in the case of GO2 was controlled.

Virtual vibrational spectra of these DTs are shown in Figure 47.4. The subsequent description of these virtual spectra is given in Ref. [56]. Looking at the Raman spectra of DTs GO1 and GO2 in Figure 47.4b, one immediately notices a triplet of intense bands at ~1,760 cm^{-1} (**I**), ~1,900 cm^{-1}(**II**), and ~2,100 cm^{-1} (**III**). The bands should be attributed to the ν sp^3**C–C**, ν sp^2**C–C**, and ν sp^3**C = O** stretchings, respectively. The latter two assignings are non-alternative while the first one is ambiguous. The ν sp^3**C=O** origin of band **III** makes it possible to estimate the unavoidable shift of virtual frequencies in this region with respect to the experimental ones [70], which is a blue one and constitutes 200–300 cm^{-1}. As seen in the figure, applying this shift to experimental spectra of a real sample leads to an obvious agreement between the calculated and experimental spectra concerning both IR absorption and Raman scattering. In the latter case, the characteristic empirical doublet of D-G bands conveniently covers virtual bands **I** and **II**, making it possible to consider virtual **I–II** bands as wished DTs of the empirical D-G ones. Naturally, this is not about exact replicas but about agreement in the main elements of the structure interpretation.

Despite the apparent success in obtaining a doublet of **I–II** bands in the virtual Raman spectra of GO1 and GO2, the main question of the identity of the **I–II** and D-G doublets still remains. As occurred, a design of a particular set of DTs made it possible to obtain the following desired answer. The standard form of Raman spectra of GOs has a chemical origin. The D and G

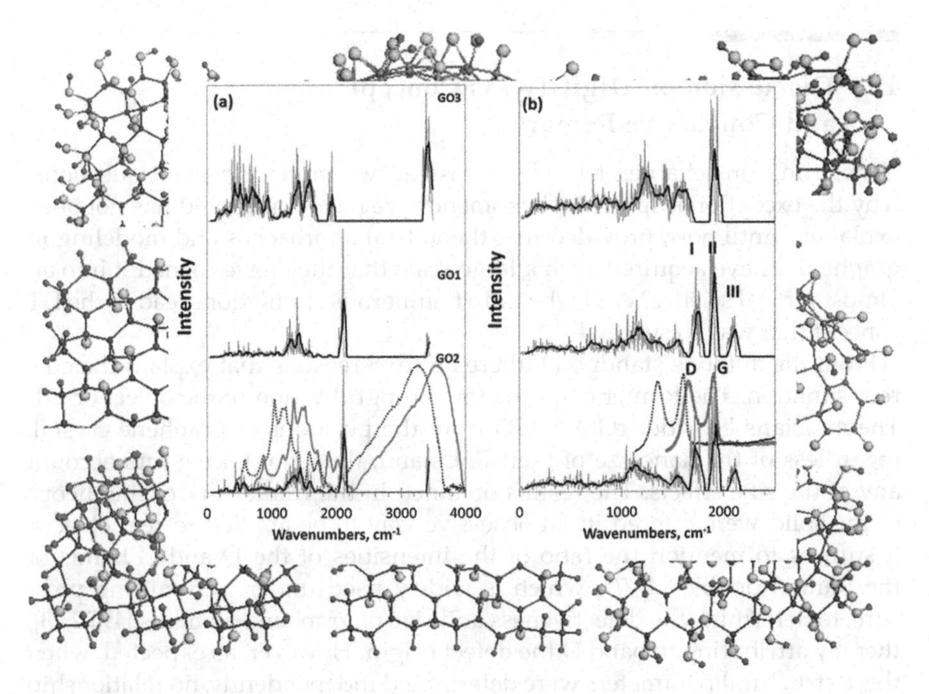

FIGURE 47.4
Virtual one-phonon IR absorption (a) and Raman scattering (b) spectra of DTs GO1, GO2, and GO3. Spectra plottings are accompanied with trend lines, corresponding to 50-point linear filtration. RHF AM1 calculations. Dotted plottings present original and the parallel plotting shifted on 300 cm^{-1} experimental spectra of GO produced by AkKo Lab Company [71], respectively.

bands are due to stretching vibrations of sp^3C–C and sp^2C–C bonds, respectively. In this case, sp^3 bonds constitute the main structural array of the GO graphene carcass, while sp^2 bonds are the result of stopping the oxidation of basal-plane carbon atoms. Accordingly, the main conclusion concerning the identity of Raman spectra of rGO and GO is the following. The identity does take the place and is due to the sp^3 and sp^2 character of D and G bands, respectively. These high-frequency stretchings correspond to the main pools of the bodies' covalent bonds and are manifested as D and G bonds of GO and rGO, respectively. Their sp^2 (G) and sp^3 (D) companions are of different origin, once presenting high-frequency stretchings of unoxidized sp^2C–C bonds caused by stopping oxidation of the graphene domain basal plane, thus marking incomplete oxidation of GO as a characteristic feature, and of dynamical sp^3C–C bonds resulted from the rGO stack structure. Over 400 DTs, taking together and subjected to the digital device HF Spectrodyn, have allowed clarifying this exclusive spectra case [56].

47.6 Discussion on Digit Twins Concept and Conclusive Remarks

Concluding this chapter, let's try to answer two more important questions: why the exceptional spectral phenomenon we have considered has not been explained until now, provided that theoretical approaches and modeling in graphenics have acquired such a large scale that they have turned it into an almost virtual science? And why didn't numerous simulations lead to the DT concept that was so successful?

From the author's standpoint, there are two reasons that explain the current situation. The former concerns the wrongful vision of the object or DT. Theoreticians consider rGO or GO from the positions of graphene crystal, regardless of the nanosize of their BSUs and, thus, not taking into account any of the size effects. The results obtained in the framework of the theory of the solid were offered in an obsessive way to be applied to real objects. It suffices to mention the ratio of the intensities of the D and G bands of the Raman doublet, I_D/I_G, which is widely used under unremitting pressure, to determine the defectiveness and size of graphene domains [49,72,73], thereby attributing to band D the defect origin. However, as expected, when these structural parameters were determined independently, no relationship between them and the intensity ratio was found [61]. The second reason concerns the identity of the Raman spectra of rGO and GO considered in the chapter. Widely accepted is the identity of the graphene domains of rGO and GO. The existence of more than 50% oxygen in the latter case was practically ignored. No sp^2-to-sp^3 transformation of the covalent bonds of the solids was considered. No conclusion was made to couple the G band existence with the predetermined incomplete oxidation of the graphene domain in the basal plane and the preservation of its inseparable structure. As for band D, as usual, it was considered as the sign of the graphene domain defectiveness.

Quantum chemists tried to keep the vision of the nanoscale object. However, the calculation programs used in the overwhelming majority of cases provided for the use of solid-state algorithms for calculating crystals with periodic boundary conditions, and therefore the proposed models were reduced to unfolded supercells. Both physicists and quantum chemists, in an overwhelming majority, did not take into account the radical nature of the graphene domain and the features associated with it. This led to improper choice of the digital device in Scheme 47.2.

The polyatomic nature of the models and/or the corresponding supercells required the use of fast and efficient calculation techniques. Only two of the most powerful semiempirical methods were considered, those based on the HF approximation and DFT. It so happened that the development of these methods was accompanied by intense competition, as a result of which the fans of the DFT method won (see a detailed discussion of the problem in

Ref. [74] and references therein) and virtual graphenics is predominantly a DFT one. However, the DFT approach, which is quite effective in the case of closed-shell molecules, turns out to be helpless when considering radicals with open shells. Concerning the problem, the DFT mainstream agreed to forget about the radical properties of graphene, thus opening up a wide road for the DFTzation of virtual graphenics. But the use of an inadequate virtual device in Scheme 47.2 either inevitably leads to incorrect results or does not allow you to see the essence of the process in question. Thus, the DFT Spectrodyn, if used, does not allow for controlled virtual synthesis of DTs of the required quality and has to be restricted with hand-made paintings only. It significantly distorts the consideration of out-of-plane displacements of carbon atoms of neighboring layers' domains, and does not allow detailed monitoring of the sp^2-to-sp^3 transformation of the covalent C–C bonds. As a result, the mystery of the identity of the Raman spectra of rGO and GO for the DFT device remains closed forever. In contrast, a conscious and controlled design of DTs, an adequate virtual device, and a wide comparative analysis of the obtained IT results, indeed, make it possible to sort out intricate problems, to reveal such hidden features as the ban on 100% oxidation of a simple graphene domain, provided in advance by nature.

The proposed analysis of the vibrational spectra of rGO and GO is based on the important concept of the radical nature of bare graphene domains and rGO. Many spears have been broken in discussions that cast doubt on the radical nature of graphene materials. The author would like to hope that the above described study will allow rejecting the last remaining doubts and will force researchers to work with materials familiar to them in a new way.

References

1. Waffenschmidt, S. Digital twin in the Industry 4.0: interview with a pioneer. t-systems.com›de/en/newsroom/best-practice/ issue 03, 2018.
2. Gelernter, D. H. *Mirror Worlds: or the Day Software Puts the Universe in a Shoebox-How It Will Happen and What It Will Mean*. Oxford; New York: Oxford University Press, 1991. ISBN 978-0195079067. OCLC 23868481.
3. Piascik, R., et al. (2010). Technology Area 12: Materials, Structures, Mechanical Systems, and Manufacturing Road Map. NASA Office of Chief Technologist.
4. Escorsa, E. (2018). *Digital Twin: A Glimpse at the Main Patented Developments*. Accessed on August 29, 2019. [Online]. https://www.i_claims.com/news/view/blog-posts/digital-.
5. *Prepare for the Impact of Digital Twins* by Gartner. https://www.gartner.com/smarterwithgartner/prepare-for-the-impact-of-digital-twins/.
6. Digital Twin and Big Data towards Smart Manufacturing. https://ieeexplore.ieee.org/document/8258937/.

7. *What Is Digital Twin Technology and Why Is It So Important.* Forbes. https://www.forbes.com/consent/?toURL=https://www.forbes.com/sites/bernardmarr/2017/03/06/what-is-digital-twin-technology-and-why-is-it-so-important/.
8. Digital Twins in Health Care: Ethical Implications of an Emerging Engineering Paradigm. https://doi.org/10.3389/fgene.2018.00031.
9. Finding Meaning, Application for the Much-Discussed "Digital Twin". https://doi.org/10.2118/0618-0026-jpt.
10. *Industry 4.0 and the Digital Twin* by Deloitte. https://www2.deloitte.com/insights/us/en/focus/industry-4-0/digital-twin-technology-smart-factory.html.
11. *Twins with Potential* by Siemens. https://www.siemens.com/customer-magazine/en/home/industry/digitalization-in-machine-building/the-digital-twin.html.
12. Rasheed, A., San, O., Kvamsdal, T. Digital twins: Values, challenges and enablers from a modeling perspective. *IEEE Access* 2020. https://doi.org/10.1109/ACCESS.2020.2970143.
13. Hartmann, D., Van der Auweraer, H. Digital twins. In: Cruz, M., Parés, C., Quintela, P. (eds) *Progress in Industrial Mathematics: Success Stories.* SEMA SIMAI Springer Series, vol 5. Springer, Cham, 2021. https://doi.org/10.1007/978-3-030-61844-5_1.
14. Sheka, E. F. *Spin Chemical Physics of Graphene.* Pan Stanford, Singapore, 2018.
15. Digital twins Research Papers. Accessed on 14.07.2022.
16. Dilmegani, C. Digital twins in 2022: What it is, Why it matters & Top Use Cases. research.aimultiple.com, updated on June 14, 2022.
17. Lazzari, S., Lischewski, A., Orlov, Y., Deglmann, P., Daiss, A., Schreiner, E., Vale, H. Toward a digital polymer reaction engineering. *Adv. Chem. Eng.* 2020, 56, 187–230.
18. Ferrari, A. C., et al. Science and technology roadmap for graphene, related two-dimensional crystals, and hybrid systems. *Nanoscale* 2015, 7, 4598.
19. Novoselov, K. S., Fal'ko, V. I., Colombo, L., Gellert, P. R., Schwab, M. G., Kim, K.. A roadmap for graphene. *Nature* 2012, 490, 192.
20. Kauling, A. P., Seefeldt, A. T., Pisoni, D. P., Pradeep, R. C., Bentini, R., Oliveira, R. V., Novoselov, K. S., Castro Neto, A. H. The worldwide graphene flake production. *Adv. Mater.* 2018, 30, 1803784.
21. Berger, M. All you need to know. Accessed on July 24, 2022. nanowerk.com›what_is_graphene.php.
22. Andrievski, R. A. Size-dependent effects in properties of nanostructured materials. *Rev. Adv. Mat. Sci.* 2006, 21, 107–133.
23. Shafraniuk, S. E. *Graphene: Fundamentals, Devices, Applications.* Pan Stanford: Singapore, 2015.
24. Sheka, E. F., Popova, V. A., Popova, N. A. Topological mechanochemistry of graphene. In: M. Hotokka et al. (eds) *Advances in Quantum Methods and Applications in Chemistry, Physics, and Biology,* Progress in Theoretical Chemistry and Physics 27, 2013, Springer International Publishing. pp. 285–302.
25. Sheka, E. F. Virtual vibrational spectrometry of stable radicals-necklaced graphene molecules. *Nanomaterials* 2022, 12, 597.
26. Sheka, E. F., Popova, N. A. Odd-electron molecular theory of the graphene hydrogenation. *J. Mol. Model.* 2012, 18, 3751–3768.

27. Sheka, E., Popova, N. Molecular theory of graphene oxide. *Phys. Chem. Chem. Phys.* 2013, 15, 13304–13332.
28. Sheka, E. F. *Fullerenes: Nanochemistry, Nanomagnetism, Nanomedicine, Nanophotonics.* CRC Press, Taylor&Francis Group, Boca Raton, FL, 2011.
29. Brodie, B. C. On the atomic weight of graphite. *Philos. Trans. R. Soc. London* 1859, 149, 249–259.
30. Yang, D., Velamakanni, A., Bozoklu, G., Park, S., Stoller, M., Piner, R. D., Stankovich, S., Jung, I., Field, D. A., Ventrice Jr, C. A., Ruoff, R. S. Chemical analysis of graphene oxide films after heat and chemical treatments by X-ray photoelectron and micro-Raman spectroscopy. *Carbon* 2009, 47, 145–152.
31. Moon, I. K., Lee, J., Ruoff, R. S., Lee, H. Reduced graphene oxide by chemical graphitization. *Nat. Commun.* 2010, 1, 73. https://doi.org/10.1038/ncomms1067.
32. Dreyer, D. R., Park, S., Bielawski, C. W., Ruoff, S. The chemistry of graphene oxide. *Chem. Soc. Rev.* 2010, 39, 228–240.
33. Chen, D., Feng, H., Li, J. Graphene oxide: Preparation, functionalization, and electrochemical applications. *Chem. Rev.* 2012, 112, 6027–6053.
34. Zhou, S., Bongiorno, A. Origin of the chemical and kinetic stability of graphene oxide. *Sci. Rep.* 2013, 3, 2484.
35. Wu, J. B., Lin, M. L., Cong, X., Liu, H. N., Tan, P. H. Raman spectroscopy of graphene-based materials and its applications in related devices. *Chem. Soc. Rev.* 2018, 47, 1822–1873.
36. Raidongia, K., Tan, A. T. L., Huan, J. Graphene oxide: Some new insights into an old material. *Carbon Nanotubes Graphene.* https://doi.org/10.1016/B978-0-08-098 232-8.00014-0341-375.
37. Backes, C., Abdelkader, A. M., Alonso, C., Andrieux-Ledier, A., Arenal, R., et al. Production and processing of graphene and related materials. *2D Mater.* 2020, 7, 022001.
38. Dideikin, A. T., Vul, A. Y. Graphene oxide and derivatives: The place in graphene family. *Front. Phys.* 2019, 6, 149.
39. Chua, C. K., Pumera, M. Chemical reduction of graphene oxide: a synthetic chemistry viewpoint. *Chem. Soc. Rev.* 2014, 43, 291–312.
40. Sheka, E. F., Rozhkova, N. N. Shungite as the natural pantry of nanoscale reduced graphene oxide. *Int. J. Smart Nano Mat.* 2014, 5, 1–16.
41. Golubev, Y. A., Rozhkova, N. N., Kabachkov, E. N., Shul'ga, Y. M., Natkaniec-Hołderna, K., Natkaniec, I., Antonets, I. V., Makeev, B. A., Popova, N. A., Popova, V. A., Sheka, E. F. sp^2 Amorphous carbons in view of multianalytical consideration: normal, expected and new. *J. Non-Cryst. Solids* 2019, 524, 119608.
42. Sadovnichii, R. M., Rozhkov, S. S., Rozhkova, N. N. The use of shungite processing products in nanotechnology: geological and mineralogical justification. *Smart Nanocomp.* 2016, 7, 111–119.
43. Harris, P. J. F. New perspectives on the structure of graphitic carbons. *Crit. Rev. Solid State Mater. Sci.* 2005, 30, 235–253.
44. Sheka, E. F., Golubev, Y. A., Popova, N. A. Amorphous state of sp^2 solid carbons. *FNCN* 2021, 29, 107–113.
45. Luong, D. X., Bets, K. V., Algozeeb, W. A., Stanford, M. G., Kittrell, C., Chen, W., Salvatierra, R. V., Ren, M., McHugh, E. A., Advincula, P. A., Wang, Z., Bhatt, M., Guo, H., Mancevski, V. R., Shahsavari, R., Yakobson, B. I., Tour, J. M. Gram-scale bottom-up flash graphene synthesis. *Nature* 2020, 577, 647–651.

46. Buchsteiner, A., Lerf, A., Pieper, J. Water dynamics in graphite oxide investigated with neutron scattering. *J. Phys. Chem. B* 2006, 110, 22328–22338.
47. Lerf, A., Buchsteiner, A., Pieper, J., Schoettl, S., Dekany, I., Szabo, T., Boehm, H. P. Hydration behavior and dynamics of water molecules in graphite oxide. *J. Phys. Chem. Sol.* 2006, 67, 1106–1110.
48. Ferrari, A. C.; Robertson, J. Raman spectroscopy of amorphous, nanostructured, diamond-like carbon, and nanodiamond. *Philos. Trans. R. Soc. A: Math. Phys. Eng. Sci.* 2004, 362, 2477–2512.
49. Savicheva, A., Tapilskaya, N., Spasibova, E., Gzgzyan, A., Kogan, I., Shalepo, K., Vorobiev, S., Kirichek, R., Pirmagomedov, R., Rybin, M. and Glushakov, R. Secure application of graphene in medicine. *Gynecol. Endocrin.* 2020, 36(sup1), 48–52.
50. Redzepi, S., Mulic, D., Dedic, M. Synthesis of graphene-based biosensors and its application in medicine and pharmacy - A review. *Medicon Med. Sci.* 2022, 2(1), 35–45.
51. Rhazouani, A., Gamrani, H., El Achaby, M., Aziz, K., Gebrati, L., Uddin, M. S., Aziz, F. Synthesis and toxicity of graphene oxide nanoparticles: A literature review of in vitro and in vivo studies. *BioMed Res. Int.* 2021, Article ID 5518999. vol. 2021, 19 pages, 2021. https://doi.org/10.1155/2021/5518999
52. Campra, P. Detection of graphene in COVID19 vaccines by micro-Raman spectroscopy. https://www.researchgate.net/publication/355979001, 2021.
53. Sheka, E. F. sp^2 Carbon stable radicals. *C Journ.Carb. Res.* 2021, 7, 31.
54. Babaeva, A. A., Zobova, M. E., Kornilov, D. Y., Tkachev, S. V., Terukov, E. I., Levitskii, V. S. Temperature dependence of electrical resistance of graphene oxide. *High Temp.* 2019, 57, 198–202.
55. Sheka, E. F., Popova, N. A. Virtual vibrational analytics of reduced graphene oxide. *Int. J. Mol. Sci.* 2022, 23, 6978.
56. Sheka, E. F. Digital Twins solve the mystery of Raman spectra of parental and reduced graphene oxides. *Nanomaterials* 2022, 12, 4209.
57. Sheka, E. F., Popova, N.A. Virtual vibrational spectrometer for sp^2 carbon clusters and dimers of fullerene C60. *FNCN* 2022, 30, 777–786.
58. Sheka, E. F., Popova, N. A., Popova, V. A. Virtual spectrometer for sp^2 carbon clusters. 1. Polycyclic benzenoid-fused hydrocarbons. *FNCN* 2021, 29, 703–715.
59. Sheka, E. F. Molecular theory of graphene chemical modification. In: Aliofkhazraei, M., Ali, N., Miln, W. I., Ozkan C. S., Mitura, S., and Gervasoni, J. (eds). *Graphene Science Handbook: Mechanical and Chemical Properties.* CRC Press, Taylor and Francis Group, Boca Raton, FL, 2016, pp. 312–338.
60. Sheka, E. F., Natkaniec, I., Ipatova, E. U., Golubev, Y. A., Kabachkov, E. N., Popova, V. A. Heteroatom necklaces of sp^2 amorphous carbons. XPS supported INS and DRIFT spectroscopy. *FNCN* 2020, 28, 1010–1029.
61. Sheka, E. F., Golubev, Y. A., Popova, N. A. Graphene domain signature of Raman spectra of sp^2 amorphous carbons. *Nanomaterials,* 2020, 10, 2021.
62. Dolganov, V. K., Kroo, N., Rosta, L., Sheka, E. F., Szabort, J. Multimode polymorphism of solid MBBA. *Mol. Cryst. Liq. Cryst.* 1985, 127, 187–194.
63. Sheka, E. F. Spectroscopy of amorphous substances with molecular structure. *Sov. Phys. Usp.* 1990, 33, 147–166.
64. Peelaers, H., Hernandez-Nieves, A. D., Leenaerts, O., Partoens, B. and Peeters, F. M. Vibrational properties of graphene fluoride and graphene. *Appl. Phys. Lett.* 2011, 98, 051914.

65. Park, J. S., Reina, A., Saito, R., Kongc, J., Dresselhaus, G., Dresselhaus, M. S. G′ band Raman spectra of single, double and triple layer graphene. *Carbon* 2009, 47, 1303–1310.
66. Cong, C., Yu, T., Saito, R., Dresselhaus, G. F., Dresselhaus, M. S. Second-order overtone and combination Raman modes of graphene layers in the range of 1690–2150 cm−1. *ACS Nano* 2011, 5. 1600–1605.
67. Sato, K., Park, J.S., Saito, R., Cong, C., Yu, T., Lui, C. H., Heinz, T. F., Dresselhaus, G., Dresselhaus, M. S. Raman spectra of out-of-plane phonons in bilayer graphene. *Phys. Rev. B* 2011, 84, 035419.
68. Rao, R., Podila, R., Tsuchikawa, R., Katoch, J., Tishler, D., Rao, A. M., Ishigami, M. Effects of layer stacking on the combination Raman modes in graphene. *ACS Nano* 2011, 5, 1594.
69. Golubev, Y. A., Sheka, E. F. Peculiarities of the molecular character of the vibrational spectra of amorphous sp^2 carbon: IR absorption and Raman scattering. *14th Intrnational Conference Carbon: Fundamental Problems of Material Science and Technology*. Moscow, Troitzk, 2022, pp. 59–60 (in Russian).
70. Dewar, M. J. S., Ford, G. P., McKee, M. L., Rzepa, H. S., Thiel, W., Yamaguchi, Y. Semiempirical calculations of molecular vibrational frequencies: The MNDO method. *J. Mol. Struct.* 1978, 43, 135–138.
71. AkKo Lab Company. Available online: www.akkolab.ru (accessed on 31 July 2022).
72. Tuinstra, F., Koenig, J. L. Raman spectrum of graphite. *J. Chem. Phys.* 1970, 53, 1126.
73. Pimenta, M. A., Dresselhaus, G., Dresselhaus, M. S., Cançado, L. G., Jorio, A., Saito, R. Studying disorder in graphite-based systems by Raman spectroscopy. *Phys. Chem. Chem. Phys.* 2007, 9, 1276–1290.
74. Sheka, E. F., Popova, N. A., Popova V. A. Physics and chemistry of graphene. Emergentness, magnetism, mechanophysics and mechanochemistry. *Phys. Usp.* 2018, 61, 645–691.

48

Applications of Triboelectric Nanogenerator in Digital Twin Technology

Jiayue Zhang
Tsinghua University

Jie Wang
Chinese Academy of Sciences
University of Chinese Academy of Sciences

48.1 Working Mechanism of TENG

Triboelectric nanogenerator (TNEG), first demonstrated by Zhong-Lin Wang's group in 2012, is a kind of energy harvesting device that can convert mechanical energy into electrical energy due to contact electrification and electrostatic induction (Fan, Tian, and Wang 2012). Contact electrification, also known as triboelectrification, is about a phenomenon that an object becomes electrically charged when it contacts or comes within close proximity of another object, which has been known for more than 2,600 years (Wang 2021). Although the physical mechanism behind this ancient and common phenomenon is still controversial and being studied, the charges generated by contact electrification can be utilized to generate electricity thanks to electrostatic induction. It is generally believed that when two objects contact, some charges are transferred from one object to another, and the transferred charges may be electrons, ions, free radicals, or other charged substances (Baytekin, Baytekin, and Grzybowski 2012, Baytekin et al. 2013, Wang and Wang 2019, Lin et al. 2019). The charges on the dielectric surface can drive the electrons in the electrodes to flow to balance the electric potential difference created by the change of the physical position, which is contributed by electrostatic induction. Based on such a principle, four fundamental working modes of TENGs have been proposed and elaborated, as illustrated in Figure 48.1 (Wang 2014).

Contact-separation mode is the simplest mode, whose general structure is two dielectric films, deposited with electrodes on the backward surfaces, facing each other, as shown in Figure 48.1a (Zhu et al. 2012, Wang, Lin, and

 DOI: 10.1201/9781003425724-57

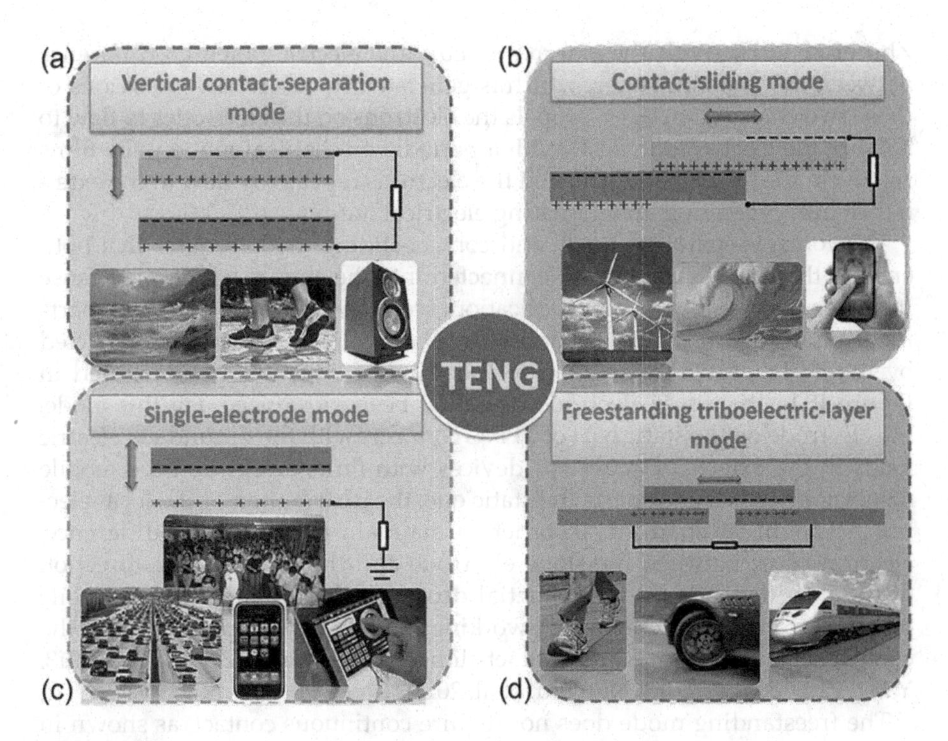

FIGURE 48.1
(a) contact-separation mode; (b) contact-sliding mode; (c) single-electrode mode; (d) freestanding mode. (Reprinted with permission from Wang, Zhong-Lin 2014. "Triboelectric nanogenerators as new energy technology and self-powered sensors–Principles, problems and perspectives." Faraday discussions 176:447–458.)

Wang 2012). A mechanical contact between two dielectric films introduces opposite charges on their surfaces. After the contact, when the two surfaces are separated by a small air gap, a potential difference is generated between the upper and lower electrodes. If the upper and lower electrodes are electrically connected by an external load, the potential difference would drive electrons to flow between these two electrodes. This process forms an opposite potential drop to balance the electrostatic field. Once the two surfaces approach each other and the gap is closed, the potential difference generated by the surface charges disappears, and the electrons in the electrodes thus flow back (Niu, Wang, et al. 2013). This process occurs repeatedly with repeated back-and-forth motions, and electrons repeatedly flow between the two plates due to electrostatic induction, outputting an alternating electrical signal.

As demonstrated in Figure 48.1b, contact-sliding mode is similar to contact-separation mode. Two dielectric films, deposited with electrodes, normally contact each other, and a relative slide in the tangential direction to the surfaces generates opposite charges on the two surfaces (Wang et al. 2013,

Zhu et al. 2013). With the external circuit, the electric potential difference between the two electrodes, which is generated by the relative positions of these two charged surfaces, propels the electrons on the electrodes to flow to balance the electrostatic field. With a periodic mechanical motion, the films relatively slide back and forth, and the electrons repeatedly flow in the external circuit, generating an alternating electrical output.

The contact-separation mode and contact-sliding mode require that both ends of the relative motion are connected into the circuit, which may cause inconvenience under some application scenarios, because the two parts involved in contact separation or relative sliding are difficult to be connected by wires. To provide a solution to this kind of situation, as illustrated in Figure 48.1c, the single-electrode mode has been introduced. For this mode, the electrode on the static part of TENG connects with the ground via electric loads in the external circuit. For devices with finite size, when the mobile object approaches or departs the static one, the distribution of the local electric field will be changed. In order to maintain this potential difference, electrons flow between the static electrode and the ground. The direction of the flow depends on the potential drop, which is reliant on the mobile object movement direction. This working mode can be applied to both the contact-separation mode and contact-sliding mode (Yang, Zhou, et al. 2013, Yang, Zhang, et al. 2013, Niu, Liu, et al. 2014).

The freestanding mode does not require continuous contact, as shown in Figure 48.1d. The charges generated by contact electrification can remain on the object for a long time, and how long the surface charges can be retained is determined by the properties of the materials. For our devices, the dielectric films generally could retain the charges for hours or even longer. Within this period of time, since the surface charge has been saturated due to contact electrified before, it is unnecessary to replenish it. Hence, there is no need for constant contact throughout the whole working period. The basic configuration of the freestanding mode is a pair of symmetric electrodes attached below a dielectric layer, and the two electrodes are connected by an external circuit with loads (Wang et al. 2014). If the size of these two electrodes and their spacing is of the same order of magnitude as the size of the moving object, the motion toward or away from electrodes generates an asymmetric charge distribution, causing the electrons flowing between the two electrodes to balance the electric potential difference. With a periodic motion of the object, this kind of TENG can output an alternative electrical signal without physical contact between the moving part and the static part, which can directly reduce or even estimate the wear of materials in contact-sliding mode and extend the durability.

The four modes described above are the most fundamental configurations and mechanisms of TENG. For diverse application scenarios, TENGs are adjusted to satisfy the corresponding requirements. For example, according to the actual physical motion, the specific motion for the contact-sliding mode can be planar motion, cylindrical rotation, or disc rotation (Jing et al.

2014, Lin, Wang, et al. 2013). As long as the device configurations are reasonably combined with the mechanical motion, tiny physical changes in the environment can be sensed. Moreover, TENG array integrating multiple nanogenerator units can be used to improve the signal-to-noise ratio of the output or increase the dimensions of the sensing signals (Niu, Liu, et al. 2013, Niu, Wang, et al. 2014).

48.2 Self-Powered Sensors Based on TENG

By collecting and converting mechanical energy into electrical energy, the TENG can either use the generated electrical energy to power itself and/or other sensors, or use the electrical signal to detect the physical changes. This kind of self-powered sensor has a wide application. Here, according to the working mechanisms of the TENG-based sensors, the sensors are classified and introduced by physical sensors and chemical sensors.

48.2.1 Self-Powered Physical Sensors Based on TENG

The basic principle of TENG-based self-powered physical sensors is to use the output voltage or current of a TENG device to indicate the relative position change between the two tribolayers, caused by mechanical triggers. The acoustic wave is a kind of vibration that exists everywhere in our daily life. Generally, the working mechanism of TENG-based acoustic sensors is: the sound wave reaches the vibrating sheet, causing the sheet to vibrate, and the vibrating sheet forms a contact-separation mode TENG with the stators. This kind of sensor has a high sensitivity of 51 mV Pa^{-1} with a fast response time of less than 6 ms, as well as a low detection limit of 2.5 Pa over a wide frequency range from 0.1 Hz to 3.2 kHz (Yang et al. 2015). By further designing the inner boundary architecture, the broadband response of this kind of sensor can be tuned from 100 to 5,000 Hz precisely and the sensitivity is up to 110 mV dB^{-1} (Guo et al. 2018). The TENG-based magnetic sensors generally utilized the interaction between magnetic field, force, and triboelectricity to find the relation between the magnitude of the voltage signal output and the magnetic field intensity. The detection sensitivity of this kind of sensor can be 0.0363 ± 0.0004 ln(mV) G^{-1} of magnetic field change and 0.0497 ± 0.0006 ln(mV) $(G\ s^{-1})^{-1}$ of the rate of the magnetic field change (Yang et al. 2012). And its response time can be about 0.13 second, while the reset time is about 0.34 second. The detection resolution of this kind of sensor is about 3 G and can work under low frequencies (<0.4 Hz), which is suitable for environmental monitoring, mineral exploration, and defense technology. When the TENG device serves as a self-powered force sensor or touch sensor, it has a high sensitivity of 0.31 kPa^{-1}, an ultra-fast response time of less than 5 ms, a long-term stability of 30,000 cycles, and a detection limit as low as 2.1 Pa (Lin, Xie, et al. 2013). The small-scale of TENG devices makes it possible to build sensing arrays, enabling it to be quite a broad application (Chen et al. 2015).

48.2.2 Self-Powered Chemical Sensors Based on TENG

The working principle of TENG-based self-powered chemical sensor is that changes in the chemical composition or environmental variables will change the chemical potential of the friction interface, which will further change the surface charge density generated by the contact of the two tribolayers, thereby affecting the output voltage and current of the TENG. For example, with a carefully designed structure, TENG-based pH sensor can be achieved (Wu et al. 2016). It has a wonderful performance in the range of pH values of 2 ~ 7. With a similar method, ion concentration detection based on the TENG can also be achieved for mercury and other heavy metal ions (Lin, Zhu, et al. 2013, Li et al. 2016). TENG can also serve as a UV sensor, which relies on photosensitive material. With a built-in photoresistor, when the range of the light intensity varies from 20 μW cm^{-2} to 7 mW cm^{-2}, both the logarithm of the output current and the output voltage have linear relationships with the incident light intensity. The response rate is as high as 280A W^{-1}, and it has excellent linearity in the intensity range of 20 μW cm^{-2} to 7 mW cm^{-2} (Lin et al. 2014). Theoretically, the dielectric constant of the dielectric material varies at different temperatures (Yadav et al. 2010), and the output voltage, output current, and transferred charge density of the TENG vary with temperature (Lu et al. 2017, Xu et al. 2018, Lin et al. 2019). This kind of self-powered temperature system has a wide application in abnormal temperature alarming, human health monitoring, and wearable sensing electronics.

48.3 TENG-Based Digital Twin Technology

The development of digital twin technology and Internet of Things (IoT) technology presents new challenges to the design and implementation of sensor networks. TENG-based sensors can continuously convert the mechanical energy in the surroundings into electrical energy while realizing the sensing function of various physical and chemical variables, which helps to solve the high-power consumption issues in sensor networks. Combining the self-powered TENG-based sensors with digital twin technology will play a significant role in advancing digitalization. Here, several typical application scenarios of digital twin technology employing TENG-based sensors are introduced.

48.3.1 3D Scanning System Based on Highly Sensitive TENG Contact Sensor

The digital twin can be regarded as a replica of an object in the real physical world in a virtual digital world. In this mapping process, 3D scanning technology is usually used to obtain point clouds of the objects (Figure 48.2a). In a contact scanning system, the probe moves according to a planned path and moves down at each planned point until a contact signal is received. The system records the 3D locations of the probe and processes the collected point cloud into a continuous surface (Figure 48.2b–d). This type of scanner has high precision and is less affected by the environment and the scanned object's features. However, based on the structure and mechanism of commonly used piezoresistive and piezoelectric sensors, contact 3D scanners possibly damage the surface of the object being measured. To achieve the goal of high accuracy and low damage at the same time, a highly sensitive sensor based on TENG is introduced into the contact scanning system (Zhang et al. 2022).

The TENG-based sensor was composed of a stator and a slider. The stator is a copper electrode, while the slider is a polytetrafluoroethylene (PTFE) film, whose surface charge density has already reached saturation, adhered on another copper electrode. The sensor serves as the probe for touching the scanned object. For each planned scanning position, the probe moves downwards. When the tip of the slider touches the surface of the object, the slider and the stator undergo a relative displacement, and thus the potential difference between them gradually increases. Compared with the commercial piezoelectric ceramics, the commercial strain gauge, and the contact-separation mode TENG pressure sensor, the contact-sliding mode TENG not only has excellent linearity between the input mechanical movement and the output voltage signals but also has a larger slope and a smaller detection limit. While being more sensitive, its sliding configuration also greatly reduces the surface damage of the scanned object. Besides, the utilization of the TENG-based sensor simplifies the circuit design and maintenance. Since this kind of sensor, designed based on the contact electrification and electrostatic induction effect, is self-powered, it does not consume additional power, which means it is not necessary to equip or replace batteries for it.

In practical application scenarios, the complex electromagnetic fields in the working environment would affect the sensing process of this sensor. This effect is mapped to the output electrical signals as noises and the baseline drift. To reduce the influence of noises, a threshold voltage has been set to distinguish the effective signal and the noises. When the potential difference between the slider and the stator is large enough to reach the threshold value, it is considered that the probe is in effective contact with the object, and the probe moves backward. The drift of the baseline would greatly influence the sensing results by radically changing the voltage value for each scanning position. The drift is irregular; it is difficult to predict it with mathematical

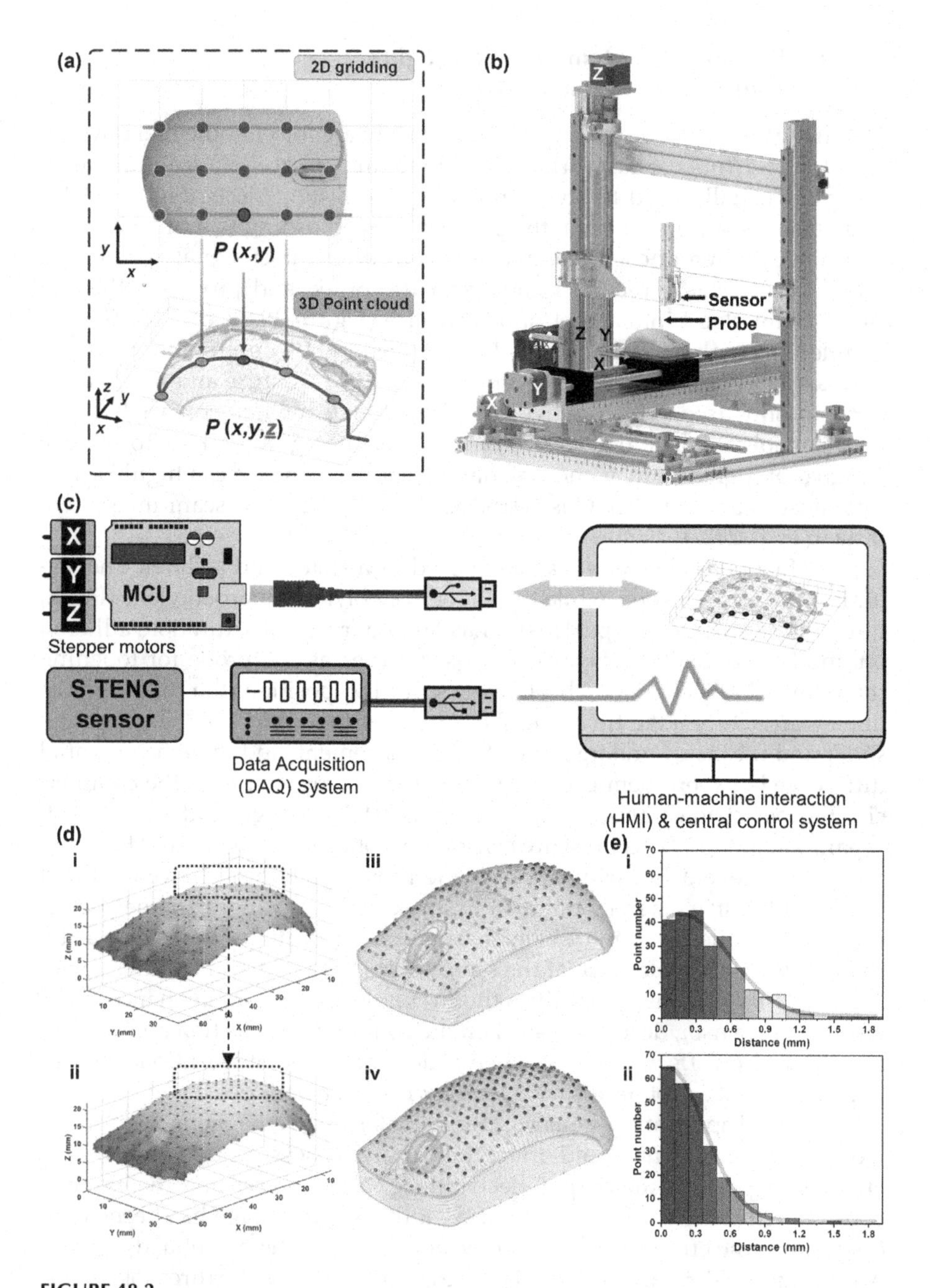

FIGURE 48.2
a. Schematic illustrations of the bionic b. Structure and mechanism of thetriboeletric auditory sensor (TAS). c. Calculated magnetic filed distribution in the coils and structure of the self-powered magnetic sensor. d. Schematic illustration of the flexible organic triboelectric transistor (FOTT). e. The structure of the triboelectric active sensor (TEAS). f. Structure of the intelligent keyboard.

models or machine learning (ML) methods. Therefore, a post-processing system has been proposed. For each planned scanning position, the process of the probe moving downward and upwards is reflected in the output electrical signals as a peak. Therefore, the start and end points of the peak signal have been found and linearly interpolated to determine the coordinates of the drifted baseline of the peak tip point, which is the valid contact point. Based on the linear relation between the displacement and the voltage, the actual displacement of the probe can be calculated after modifying the sensing voltage signals with the drifted baseline. This algorithm calculates the baseline drift for each sampling point with local peaks instead of absolute values, which makes it adaptive. The evaluation of this strategy is achieved by comparing the scanned point cloud and the digital model of the original sample. With this post-processing system, the accuracy increased by about 30% (Figure 48.2e).

For future applications, with the sensor and the post-processing strategy, this 3D scanning system can be mounted on mobile robots to detect and model in hazardous environments that humans cannot reach. By changing the hardware configuration and the number of sensors in the scanning system, more efficient 3D scanning can be achieved. This work demonstrates the potential of TENG-based sensors in surface scanning and sensing signal acquisition, contributing to the development of digital twin technology.

48.3.2 Smart Socks Based on TENG Tactile Sensors

With the development of wearable electronic devices, it is increasingly expected that they can play a role in human healthcare and medical applications. In order to monitor and detect the action of the human body, multiple sensors are usually placed on flexible wearable electronic devices, and the sensing information will be quickly exchanged with other devices through the wireless transmission network. TENG-based sensors can not only work with high sensitivity but also harvest energy from the surrounding environment and convert it into electrical energy, long-term sustainably supporting the operation of the wireless sensor network. Due to motion habits, foot activities are one promising source for collecting kinetic energy from the human body. In addition, a person's gait contains a variety of information, which can reflect health status. Under this background, a smart sock, employing a TENG-based self-powered sensor, supplemented by an AI algorithm, has been developed (Zhang et al. 2020). This sock applies digital modeling technology to achieve long-term physical state monitoring, gait recognition, and user identification.

The sensor placed at the bottom of the smart sock is a contact-separation mode TENG. A nitrile film and a silicone rubber film are used as two tribolayers. A conductive textile material is attached to the back of the two tribolayers, and an insulating textile material for encapsulating the entire device is attached outside it (Figure 48.3a). The pressure on a foot is in a large range of up to 200–300 kPa, so when making the TENG sensor, an millimeter-scale

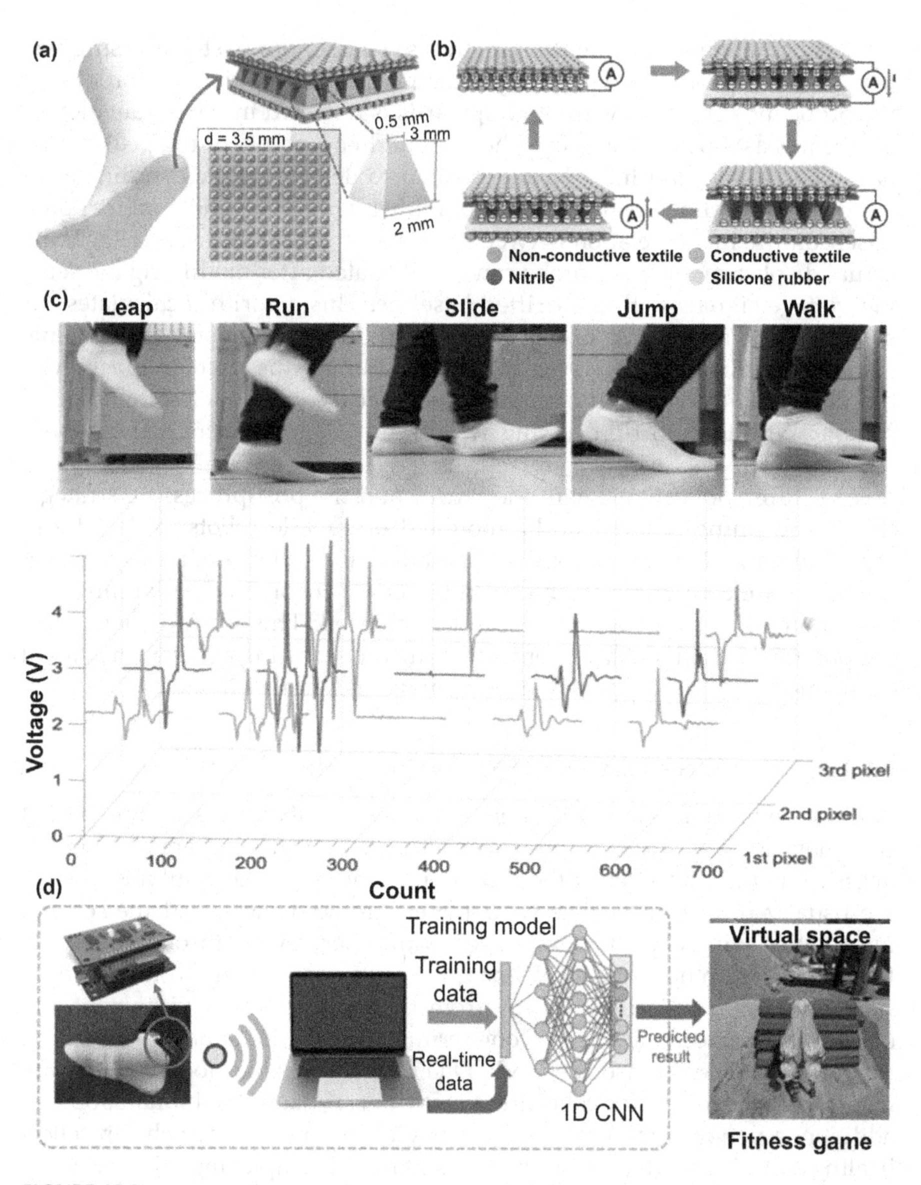

FIGURE 48.3
a. Structure design and working mechanism of the TENG-based heavy metal ion detector. b. Structure and fabrication process of the TENG-based UV sensor. c. The electrical signal curves of short-circuit current, open-voltage, and short-circuit transferred charges. d. The self-powered sensing system based on contact-separation mode TENG.

frustum structure has been made on the silicon rubber film to increase the pressure range from 72 kPa, which can be achieved with the TENG device without surface modification, to 200 kPa. Generally, a gait cycle contains four events, "heel contact", "toe contact", "heel leave", and "toe leave". Due to the working mechanism of the contact-separation mode TENG, when there is a contact action, there will be a positive peak mapped to the output voltage signals, while the negative peaks occur when the separation action happens (Figure 48.3b). For this human walking forward scenario, in a half gait, the two positive peaks indicate "heel contact" and "toe contact", while the two negative peaks indicate "heel leave" and "toe leave". With these features, the gait frequency, voltage peak magnitudes, and the time interval of every event can be extracted, and further used for the diagnoses of movement disorders and identity verifications by an AI algorithm. Thanks to TENG, in addition to the sensing function, this smart sock has promising electricity generation ability. Under the low-frequency motion of human walking, the electricity it generates can activate a Bluetooth module and send information about humidity and temperature to a smartphone, which can be utilized to monitor body temperature during exercise.

With a one-dimensional convolutional neural network (1D-CNN), a kind of ML method, the identification for the five participants reaches 96%. To further improve the accuracy, a sensor array has been built. This sensor array contains three TENG-based sensors at the locations that undertake the largest pressure. With this promotion, the identification accuracy increased to 100%. For more participants, the accuracy achieved was 93.54% with a dataset of 13 individuals. The smart socks can be employed to play a virtual reality (VR) fitness game. When a user leaps, runs, slides, jumps, and walks, the TENG-based sensor array will acquire corresponding electrical signals and the virtual character will be accordingly controlled with an intelligent algorithm (Figure 48.3c and d). Smart socks can not only identify the user's identity through gait analysis but also detect user abnormalities by monitoring their gaits. Besides, it can power other sensors by converting mechanical energy into electrical energy. When there are multiple users in the space, smart socks can be used to monitor the activities of each one. This feature can be used in scenarios such as movement monitoring of family members, especially the elderly and children, and smart classroom registration.

This work demonstrates systems for HMI entertainment, smart home, and smart workspace based on the smart sock. The wearable system can create a digital replica of a person in a virtual space by sensing the motion characteristics of the human body, namely, a digital twin of a human being. The TENG-based sensing network provides a low-cost and self-powered solution for digital humans. By designing a more comprehensive sensor network, this solution will further promote the development of the digitization of life and work in the future.

48.3.3 Smart Manipulator Based on TENG Tactile and Length Sensors

In 2019, the global pandemic of the coronavirus disease 2019 (COVID-19) forced many people to stay at home, which again pushed the technology that can support people's online life into a research boom. At present, online shopping has become a normal scenario in our daily life. The interactive online virtual store, based on digital twin technology, is bringing people an immersive experience. Through augmented reality (AR) or VR technology, people can learn more details about the product. With the development of IoT and 5G technology, it is possible to sense and provide feedback on the state of physical objects, including static states such as shape, size, color, and dynamic states such as position, transportation, and attitude. For this virtual shopping application, the digital twin technology and its affiliated sensing networks, which can show these features of the merchandise in real time, play a significant role. In order to better simulate the selection action of customers during offline shopping, a soft robotic manipulator has been developed. In order to match the properties of this manipulator, several TENG-based sensors and pyroelectric sensors are equipped on the manipulator, supplemented by ML methods for data processing. Based on this system, a virtual store has been built and provides customers with an immersive human–computer interaction environment (Sun et al. 2021).

As shown in Figure 48.4a and b, the soft robotic manipulator, manufactured by 3D printer, is composed of three pneumatic actuators. When the actuators are inflated, each of them would simultaneously deform, and the bending angle is linearly relative to the air pressure in a specific range. The length sensor based on TENG (L-TENG) in this system is a single-electrode contact-separation TENG. A gear covered by Ni-fabric conductive textile serves as the moving part. A PTFE film attached to the copper electrode serves as the fixed part. There is a strip fixed with the gear at one end, and fixed on the tip of the actuator at the other end. When the soft actuator is inflated, the gear will rotate driven by the strip. During the rotation, each tooth of the gear serves as the triolayer for the TENG. As the gear rotates, the electrical potential difference between the stator electrode and the ground periodically drives the charge flow between the stator electrode and the ground and generates output peaks. The contact-separation process of each tooth on the gear with PTFE film produces an output peak. Therefore, the deformation of the pneumatic finger can be calculated with the angle between every two teeth of the gear and the number of the output peak. The tactile sensor is based on an array of the contact-separation mode TENGs. This sensor is attached to a thermoplastic polyurethanes (TPU) substrate with a silicon rubber layer on it. Along the long edge of the pneumatic finger, three short electrodes are distributed with an interval, and a long electrode is attached along the long edge. From the experiments, the voltage ratio of the three short electrodes can tell the position of the contact, because the output voltage is inversely proportional to the distances between the contact point and the electrodes, while the voltage of the long electrode can tell the contact area, because the contact area will introduce different charges on the electrode, leading to the linearity between the contact area and the voltage magnitude. To better mimic the real-life shopping process, a pyroelectric-based temperature sensor has been also attached to the tip of the soft robotic manipulator.

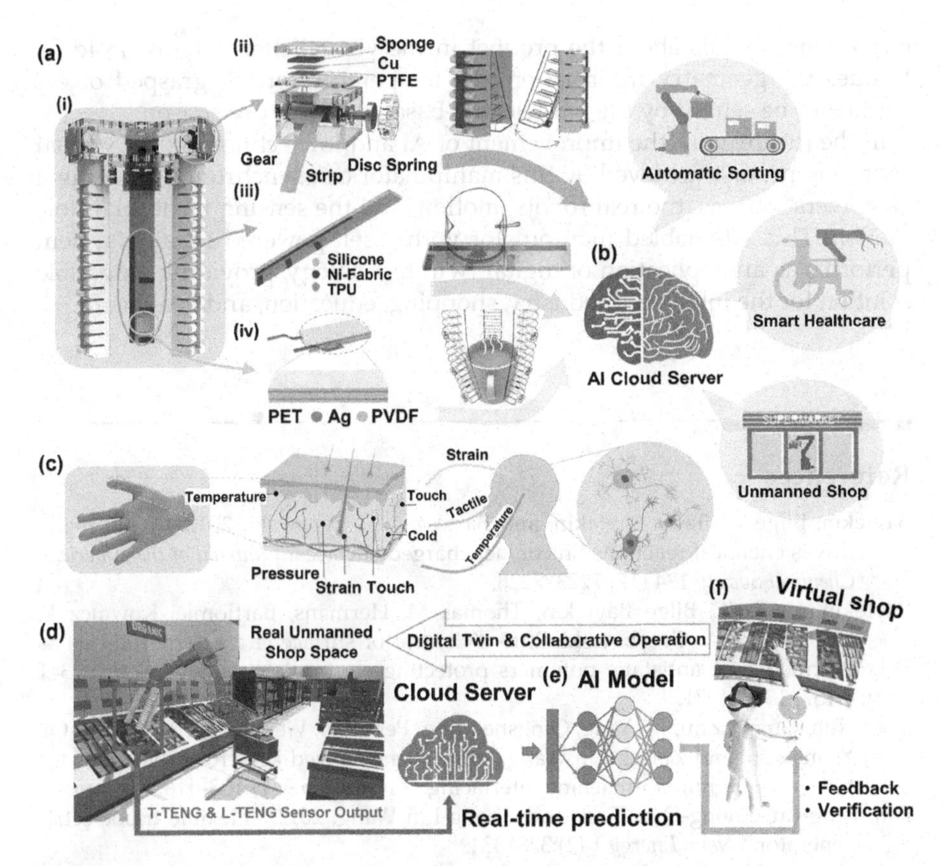

FIGURE 48.4
Working principle of TENG-based 3D scanning system. (a) Schematic diagram of diverse items in the real world and digital world. (b) The overall plan of 3D scanning. (c) The hardware setup of TENG-based 3D scanning system. (d) Electronic systems of TENG-based 3D scanning system. (Reprinted with permission from Zhang, Jiayue, Shaoxin Li, Zhihao Zhao, Yikui Gao, Di Liu, Jie Wang, and Zhong-Lin Wang. 2022. "Highly sensitive three-dimensional scanning triboelectric sensor for digital twin applications." Nano Energy 97:107198.)

For each finger of the pneumatic gripper, there are one L-TENG, consisting of one channel, and one T-TENG, consisting of four channels. There are a total of 15 channels for grasping data collection, including the contact area, contact position, and bending angle information. The excellent flexibility and compatibility with the soft robotic manipulator during the grasping process enable the TENG-based sensing network to feedback a 3D sensory space. Compared with a single sensor, this sensing network could reveal more information, such as the bending speed, collision vibration, contact sequence, and other characters with respect to time. With a three-layer 1D-CNN, this system can achieve perceiving accuracy as high as 96.1% even with varied grasping angles of the objects. Based on this smart manipulator, users can check more details about the product in the virtual shop (Figure 48.4d–f). Besides the geometry information, the temperature of the grasped object could also be sensed by the pyroelectric-based temperature sensor.

In the future, with the improvement of AI and IoT technology, the virtual shopping process achieved by this manipulator can synchronize the online customers' actions, the real robots' motion, and the sensing feedback information. This AI-enabled manipulator with a self-powered sensing system performs as an application of digital twin technology, providing a possible solution for the intelligent industry, shopping, education, and healthcare.

References

Baytekin, Bilge, H. Tarik Baytekin, and Bartosz A. Grzybowski. 2012. "What really drives chemical reactions on contact charged surfaces?" *Journal of the American Chemical Society* 134 (17):7223–7226.

Baytekin, H. Tarik, Bilge Baytekin, Thomas M. Hermans, Bartlomiej Kowalczyk, and Bartosz A. Grzybowski. 2013. "Control of surface charges by radicals as a principle of antistatic polymers protecting electronic circuitry." *Science* 341 (6152):1368–1371.

Chen, Jun, Guang Zhu, Jin Yang, Qingshen Jing, Peng Bai, Weiqing Yang, Xuewei Qi, Yuanjie Su, and Zhong-Lin Wang. 2015. "Personalized keystroke dynamics for self-powered human-machine interfacing." *ACS Nano* 9 (1):105–116.

Fan, Feng-Ru, Zhong-Qun Tian, and Zhong-Lin Wang. 2012. "Flexible triboelectric generator." *Nano Energy* 1 (2):328–334.

Guo, Hengyu, Xianjie Pu, Jie Chen, Yan Meng, Min-Hsin Yeh, Guanlin Liu, Qian Tang, Baodong Chen, Di Liu, and Song Qi. 2018. "A highly sensitive, self-powered triboelectric auditory sensor for social robotics and hearing aids." *Science robotics* 3 (20):eaat2516.

Jing, Qingshen, Guang Zhu, Peng Bai, Yannan Xie, Jun Chen, Ray P. S. Han, and Zhong-Lin Wang. 2014. "Case-encapsulated triboelectric nanogenerator for harvesting energy from reciprocating sliding motion." *ACS Nano* 8 (4):3836–3842.

Li, Zhaoling, Jun Chen, Hengyu Guo, Xing Fan, Zhen Wen, Min-Hsin Yeh, Chongwen Yu, Xia Cao, and Zhong-Lin Wang. 2016. "Triboelectrification-enabled self-powered detection and removal of heavy metal ions in wastewater." *Advanced Materials* 28 (15):2983–2991.

Lin, Long, Sihong Wang, Yannan Xie, Qingshen Jing, Simiao Niu, Youfan Hu, and Zhong-Lin Wang. 2013. "Segmentally structured disk triboelectric nanogenerator for harvesting rotational mechanical energy." *Nano Letters* 13 (6):2916–2923.

Lin, Long, Yannan Xie, Sihong Wang, Wenzhuo Wu, Simiao Niu, Xiaonan Wen, and Zhong-Lin Wang. 2013. "Triboelectric active sensor array for self-powered static and dynamic pressure detection and tactile imaging." *ACS Nano* 7 (9):8266–8274.

Lin, Shiquan, Liang Xu, Laipan Zhu, Xiangyu Chen, and Zhong-Lin Wang. 2019. "Electron transfer in nanoscale contact electrification: photon excitation effect." *Advanced Materials* 31 (27):1901418.

Lin, Zong-Hong, Gang Cheng, Ya Yang, Yu Sheng Zhou, Sangmin Lee, and Zhong Lin Wang. 2014. "Triboelectric nanogenerator as an active UV photodetector." *Advanced Functional Materials* 24 (19):2810–2816.

Lin, Zong-Hong, Guang Zhu, Yu Sheng Zhou, Ya Yang, Peng Bai, Jun Chen, and Zhong-Lin Wang. 2013. "A self-powered triboelectric nanosensor for mercury ion detection." *Angewandte Chemie* 125 (19):5169–5173.

Lu, Cun Xin, Chang Bao Han, Guang Qin Gu, Jian Chen, Zhi Wei Yang, Tao Jiang, Chuan He, and Zhong Lin Wang. 2017. "Temperature effect on performance of triboelectric nanogenerator." *Advanced Engineering Materials* 19 (12):1700275.

Niu, Simiao, Ying Liu, Sihong Wang, Long Lin, Yu-Sheng Zhou, Youfan Hu, and Zhong-Lin Wang. 2013. "Theory of sliding-mode triboelectric nanogenerators." *Advanced Materials* 25 (43):6184–6193.

Niu, Simiao, Ying Liu, Sihong Wang, Long Lin, Yu-Sheng Zhou, Youfan Hu, and Zhong-Lin Wang. 2014. "Theoretical investigation and structural optimization of single-electrode triboelectric nanogenerators." *Advanced Functional Materials* 24 (22):3332–3340.

Niu, Simiao, Sihong Wang, Long Lin, Ying Liu, Yu-Sheng Zhou, Youfan Hu, and Zhong-Lin Wang. 2013. "Theoretical study of contact-mode triboelectric nanogenerators as an effective power source." *Energy & Environmental Science* 6 (12):3576–3583.

Niu, Simiao, Sihong Wang, Ying Liu, Yu Sheng Zhou, Long Lin, Youfan Hu, Ken C. Pradel, and Zhong Lin Wang. 2014. "A theoretical study of grating structured triboelectric nanogenerators." *Energy & Environmental Science* 7 (7):2339–2349.

Sun, Zhongda, Minglu Zhu, Zixuan Zhang, Zhaocong Chen, Qiongfeng Shi, Xuechuan Shan, Raye Chen Hua Yeow, and Chengkuo Lee. 2021. "Artificial Intelligence of Things (AIoT) enabled virtual shop applications using self-powered sensor enhanced soft robotic manipulator." *Advanced Science* 8 (14):2100230.

Wang, Sihong, Long Lin, and Zhong-Lin Wang. 2012. "Nanoscale triboelectric-effect-enabled energy conversion for sustainably powering portable electronics." *Nano Letters* 12 (12):6339–6346.

Wang, Sihong, Long Lin, Yannan Xie, Qingshen Jing, Simiao Niu, and Zhong-Lin Wang. 2013. "Sliding-triboelectric nanogenerators based on in-plane charge-separation mechanism." *Nano Letters* 13 (5):2226–2233.

Wang, Sihong, Yannan Xie, Simiao Niu, Long Lin, and Zhong-Lin Wang. 2014. "Freestanding triboelectric-layer-based nanogenerators for harvesting energy from a moving object or human motion in contact and non-contact modes." *Advanced Materials* 26 (18):2818–2824.

Wang, Zhong-Lin 2014. "Triboelectric nanogenerators as new energy technology and self-powered sensors-Principles, problems and perspectives." *Faraday Discussions* 176:447–458.

Wang, Zhong-Lin 2021. "From contact electrification to triboelectric nanogenerators." *Reports on Progress in Physics* 84 (9):096502.

Wang, Zhong-Lin, and Aurelia Chi Wang. 2019. "On the origin of contact-electrification." *Materials Today* 30:34–51.

Wu, Ying, Yuanjie Su, Junjie Bai, Guang Zhu, Xiaoyun Zhang, Zhanolin Li, Yi Xiang, and Jingliang Shi. 2016. "A self-powered triboelectric nanosensor for PH detection." *Journal of Nanomaterials* 2016. vol. 2016, Article ID 5121572, 6 pages, 2016. https://doi.org/10.1155/2016/5121572

Xu, Cheng, Yunlong Zi, Aurelia Chi Wang, Haiyang Zou, Yejing Dai, Xu He, Peihong Wang, Yi-Cheng Wang, Peizhong Feng, and Dawei Li. 2018. "On the electron-transfer mechanism in the contact-electrification effect." *Advanced Materials* 30 (15):1706790.

Yadav, Vikram S., Devendra K. Sahu, Yashpal Singh, Mahendra Kumar, and Devi Charan Dhubkarya. 2010. "Frequency and temperature dependence of dielectric properties of pure poly vinylidene fluoride (PVDF) thin films." *AIP Conference Proceedings*. Hong Kong, (China), 17–19 March 2010

Yang, Jin, Jun Chen, Yuanjie Su, Qingshen Jing, Zhaoling Li, Fang Yi, Xiaonan Wen, Zhaona Wang, and Zhong-Lin Wang. 2015. "Eardrum-inspired active sensors for self-powered cardiovascular system characterization and throat-attached anti-interference voice recognition." *Advanced Materials* 27 (8):1316–1326.

Yang, Ya, Long Lin, Yue Zhang, Qingshen Jing, Te-Chien Hou, and Zhong-Lin Wang. 2012. "Self-powered magnetic sensor based on a triboelectric nanogenerator." *ACS Nano* 6 (11):10378–10383.

Yang, Ya, Hulin Zhang, Jun Chen, Qingshen Jing, Yu-Sheng Zhou, Xiaonan Wen, and Zhong-Lin Wang. 2013. "Single-electrode-based sliding triboelectric nanogenerator for self-powered displacement vector sensor system." *Acs Nano* 7 (8):7342–7351.

Yang, Ya, Yu-Sheng Zhou, Hulin Zhang, Ying Liu, Sangmin Lee, and Zhong-Lin Wang. 2013. "A single-electrode based triboelectric nanogenerator as self-powered tracking system." *Advanced Materials* 25 (45):6594–6601.

Zhang, Jiayue, Shaoxin Li, Zhihao Zhao, Yikui Gao, Di Liu, Jie Wang, and Zhong-Lin Wang. 2022. "Highly sensitive three-dimensional scanning triboelectric sensor for digital twin applications." *Nano Energy* 97:107198.

Zhang, Zixuan, Tianyiyi He, Minglu Zhu, Zhongda Sun, Qiongfeng Shi, Jianxiong Zhu, Bowei Dong, Mehmet Rasit Yuce, and Chengkuo Lee. 2020. "Deep learning-enabled triboelectric smart socks for IoT-based gait analysis and VR applications." *NPJ Flexible Electronics* 4 (1):1–12.

Zhu, Guang, Jun Chen, Ying Liu, Peng Bai, Yu-Sheng Zhou, Qingshen Jing, Caofeng Pan, and Zhong-Lin Wang. 2013. "Linear-grating triboelectric generator based on sliding electrification." *Nano Letters* 13 (5):2282–2289.

Zhu, Guang, Caofeng Pan, Wenxi Guo, Chih-Yen Chen, Yusheng Zhou, Ruomeng Yu, and Zhong-Lin Wang. 2012. "Triboelectric-generator-driven pulse electrodeposition for micropatterning." *Nano Letters* 12 (9):4960–4965.

Part 10

Digital Twins in Medicine and Life

49

Digital Twins in the Pharmaceutical Industry

João Afonso Ménagé Santos
Hovione Farmaciência S.A.
Universidade de Lisboa

João Miguel da Costa Sousa and Susana Margarida da Silva Vieira
Universidade de Lisboa

André Filipe Simões Ferreira
Hovione Farmaciência S.A.

49.1 Introduction

The concept of the digital twin (DT) is far-reaching in today's industrial paradigm, with the number of articles on the topic having doubled every year for almost 20 years, through google scholar search [12]. As a highly dynamic and tightly regulated industry, the pharmaceutical industry can benefit from the advantages offered by these systems in several stages of product development, manufacturing, and management, and several articles have been published on the academic and industrial literature.

The presence of DTs in pharmaceutical manufacturing is reviewed in Ref. [5]. The authors focused their review on the usage of DTs in pharma for smart manufacturing, with a focus on process analytical technology (PAT) and data collection, process modelling, and data integration platforms. Potential applications of DTs in pharma are presented. Several challenges and gaps are addressed in each of these components, namely:

- *PAT and data collection*: measurement accuracy in low-dose drug products; maintenance of calibration models; volume of the collected data.
- *Process modelling*: computational cost for complex models – requiring high-performance computing and preventing real-time usage; model maintenance, as models developed for the pharmaceutical industry are often static.

DOI: 10.1201/9781003425724-59

- *Data integration*: information communication between the physical components and the models and vice-versa – most applications only transfer data from the physical component to the virtual one; information security – paramount in the pharmaceutical industry.

It is concluded in Ref. [5] that while work has been done in all components of the DT, the process is still underdeveloped in the pharmaceutical industry, when compared with other sectors.

Further exploration of publications on the topic using Web of Science with the search query '*(digital twin*) AND (pharmaceutical OR pharma)*' provided additional insights on the topic. Forty-one articles were matched by the query. Out of these articles, eight were deemed out of scope – either did not feature DTs at length or had no prominent relation to the pharmaceutical industry. The remaining publications can be classified as follows:

- *Pharmaceutical manufacturing and processes*: Twenty publications regarded the usage of DTs for the manufacture of pharmaceutical products or associated with pharmaceutical processes. Most of these publications utilize the concept of the DT as part of the Quality-by-Design (QbD) framework, demanded by the regulatory authorities in the industry; on these, the DTs are mostly used to model the pharmaceutical processes that take place during the manufacturing [3,9,10,11,19,21].
- Other articles of the same classification deal with conceptually defining pharmaceutical DTs for manufacturing and its potential use [15]; DTs for automating the quality assurance of advanced therapy medicinal products [17]; DTs applied to the actual process of solutions preparation [6].
- *Supply chain & management*: ten articles were classified as using the DT concept applied to the pharmaceutical supply chain (SC) or as a supporting tool for managerial decisions.
- Articles classified as management-support researches were found to be either associated with the manufacturing of the pharmaceutical products, quality control (QC), automation, or expansion. Some examples of articles within these categories feature the usage of DTs to identify the endpoint of a pharmaceutical process so as to reduce energy expenditure [2]; using DTs with PATs with the objective of process automation, in order to achieve climate neutrality [18]; using DTs for QC resource planning and scheduling [13]; using a DT to assess and evaluate the introduction of automation in a QC laboratory [7].
- *Review*: besides the aforementioned review of DTs in pharmaceutical and biopharmaceutical manufacturing [5], two additional reviews

were matched by the search query [1,22]. Both of these articles are focused on DTs applied to manufacturing biopharmaceutical products.

This chapter goes into detail on the usage of DTs applied to pharmaceutical SCs and management. An analysis into pharmaceutical SC DTs is initially presented followed by pharmaceutical management DTs. These systems are defined, applications are presented, and their future is discussed. Finally, expected research directions are presented for the pharmaceutical DTs as a whole, along with overarching conclusions.

49.2 Supply Chain Digital Twins

SCs occupy a crucial role within enterprises. If an SC is unable to react to unpredictable situations, manufacturing operations may have to be halted, leading to deadlines not being met and eventually hefty costs impact the SC. On the other hand, a SC well prepared for unpredictable events may require the storage of large amounts of raw materials, leading to unnecessarily high holding costs. An ideal SC should contain only the necessary stocks for a small time frame, being able to predict when shortages may be likely and restock its materials accordingly; on the downstream, an ideal SC is even easier to define – the finished products should be delivered just in time, i.e., not before or after the agreed upon delivery date.

Pharmaceutical SCs have a few particularities. First of all, pharmaceutical raw materials are often perishable products, requiring strict adhesion to delivery schedules. In addition, some pharmaceutical raw materials need to be kept at cold temperatures, frequently freezing temperatures. This requires a specialized type of transportation, capable of fulfilling these temperature requirements. Another substantial particularity of pharmaceutical SCs is the tight control by regulatory entities, which is unlike any other industry. This supervision requires a strict QC over the purchased raw materials and an even stricter QC and quality assurance over the finished product, before being dispatched to the downstream SC.

The need for the optimal SC and the constraints that current pharmaceutical SCs face nowadays makes them ideal candidates for the adoption of digital approaches to improve their effectiveness. The application of DTs for the SC is a relatively new topic, but with many possibilities.

One definition that is open to debate is the term "supply chain digital twin". This concept is often named "digital supply chain twin". Some publications use both terms synonymously. Nevertheless, in this chapter, the two terms are considered to have separate meanings. The digital SC twin corresponds to the DT of a digital SC, i.e., the process of delivery of digital media such as songs or books. This concept is out of scope in the present chapter, meaning that only the supply chain digital twin (SCDT) term is used – a DT of the SC of a company.

Typical DTs use mostly sensor data collected from the real asset to construct its historical database and to supply it with live information. However, as the DT becomes less operational and more managerial, less operational data are required and more supervisory up to management data are used. This alludes to the five-layer automation pyramid, hierarchically organized from the base level of sensors and actuators (operational), the second level of control systems, the third level of supervisory control, the fourth level of planning, and the fifth level of management. SCDTs will mostly resort to the management level, utilizing the data from enterprise resource planners (ERPs). A DT which controls the SC requires data, e.g., shipment dates, location of the supplier, or quantity dispatched, all of which are contained in the ERP. However, operational data can also be used, if the application justifies it. Using the previously mentioned example of raw material transportation DT, the trucks used for transportation could be fitted with GPS data constantly supplying coordinates; RFID tags could be used for precise departure and arrival confirmations.

This concept can be applied to the SC as whole or to a single part of the SC. Some examples of where the DT can be applied within the SCs of companies are shown below:

- *Resilient supplier selection*: using a model of the material requirements on the internal SC and the suppliers' performance, on the upstream SC, it is possible to generate predictions of which suppliers will perform the best for the requirements of a company. This can ensure that for each material sourcing occasion, the option with the best-suited price, quality and leadtime can be selected, in order to attain a resilient SC performance. This was researched in Ref. [4], without being specifically applied to the pharmaceutical SC.
- *Internal supply chain*: while the SC is typically seen as a whole, it can be useful to segment it into its individual components. Perhaps, the best component for the adoption of digital strategies is the internal SC, since its data and models are more easily accessible as they are part of the company. The internal SC comprehends all the internal transportation of raw material among warehouse and for manufacturing (upstream internal SC) as well as transporting the finished products and coproducts for storage (downstream internal SC). While controlling this, internal SC is more accessible than the external SC, it is crucial that the process is correctly done, as it impacts the decision for raw material sourcing from the suppliers and raw material allocation for manufacturing. Different manufacturing processes may compete for the same raw materials which have to be taken into account.

- Managing the internal SC can easily be conjoined with management decision-making. For instance, allocating raw materials for multiple manufacturing procedures at similar times has to take into account equipment vacancy, and manufacturing and support areas (QC, warehousing) personnel availability. This can easily become a scheduling and capacity allocation problem.
- An application of a DT to the internal SC of a pharmaceutical company was done in Ref. [16]. The DT features a rough cut capacity planning simulation in order to estimate personnel monthly capacity requirements given the planned production orders on a scope of 2 years; the simulation takes into account raw material requirements and key equipment scheduling. A graphical user interface is also presented, where the users can not only interact with the simulation model but also visualize historical and live information.

The concept of the pharmaceutical SCDT is still a novel topic of research, without many articles published so far, and with limited presence in the industry. However, it is a topic with immense potential and it is expected to increase its importance in the coming years. In Ref. [8], modelling approaches for SC resilience for vaccines are analysed in the context of the COVID-19 pandemic. The authors mention DTs as tools that "will help supply chain managers to better quantify efficiency/resilience tradeoffs across all associated networks/domains and support optimal system performance post disruption". While the COVID-19 vaccine SC worked exceptionally well even considering all the constraints found at the global SC, digital strategies can improve its performance and visibility to stakeholders.

While scientific research often evolves in unpredictable directions, the topic of the pharmaceutical SCDT can be expected to evolve in a few directions:

- *Transparency*: improve visibility within the SC, in terms of suppliers, clients, and transportation of materials upstream and downstream of the SC.
- *Transportation estimation*: accurately predicting the lead times of suppliers and the duration of transportation can have major benefits. For one, safety stocks can be reduced as the uncertainty in the SC is better modelled and accounted for.
- *Sustainable supply chain*: the topic of sustainable or green SCs is one of the most researched in SC management nowadays. In the pharmaceutical industry, the adoption of SCDTs with a focus on sustainability can bring substantial reductions on the overall ecological footprint of the industry, e.g., by optimizing transportation routes or evaluating the efficiency of competing raw material suppliers.

49.3 Digital Twins for Management

Managing pharmaceutical companies is a complex process involving multiple stakeholders at different levels of the corporate ladder. All the specific characteristics of the industry make it paramount for decision-makers to have the most information possible, and that becomes easier to attain with each passing year, as the amount of data collected and the computational capacity to process it increases immensely. However, this is not sufficient nowadays for most industries, and even less for the pharmaceutical industry. Besides having large amounts of accurate information about past and present, it is also necessary to obtain future estimates. Knowing beforehand can mean that corrective action takes place before a critical failure happens; the company is quick to position itself on a new market segment, benefiting from the first-movers advantage; or that scale-up is adequate and the company is not overly or insufficiently ambitious when expanding.

While many predictive models have been presented in the academic literature and applied in the industry, DTs provide their estimations in a data-rich environment and through an online process, i.e., the models and their estimates are updated in real time. Furthermore, the DT supplies the obtained estimates along with processed historical and current information, making the decision-making process more data-driven.

An important distinction that strategic-level DTs have, when compared with more operational-level ones, is the fact that automatic updates from the model to the real-world system are often not possible. As per the definition of DT, a digital model of the real-world system must receive data from it, and must change the systems parameters according to its results, in a type of feedback loop. This makes sense on operational DTs, e.g., automatically change the maintenance schedule of a machine based on sensor readings that the model receives from it. On the other hand, DTs at the level of strategy should not be automatic and ultimately require analysis and decision by a stakeholder, e.g., increasing production capacity by purchasing an additional piece of multi-million dollar equipment. These types of DTs are graphically represented in Figure 49.1. This is possible since strategic decision-making is not as time sensitive as operational decisions.

Similar to SCDTs, DTs for management (hereby referred to as management DTs) tend to mainly resort to management data (mostly obtained from the ERP). There are applications that surely may resort to sensor-based data or other sources lower than the top category of the automation pyramid, but the core part of the data used should come from management-level sources.

On the topic of pharmaceutical management DTs, several application examples can be analysed.

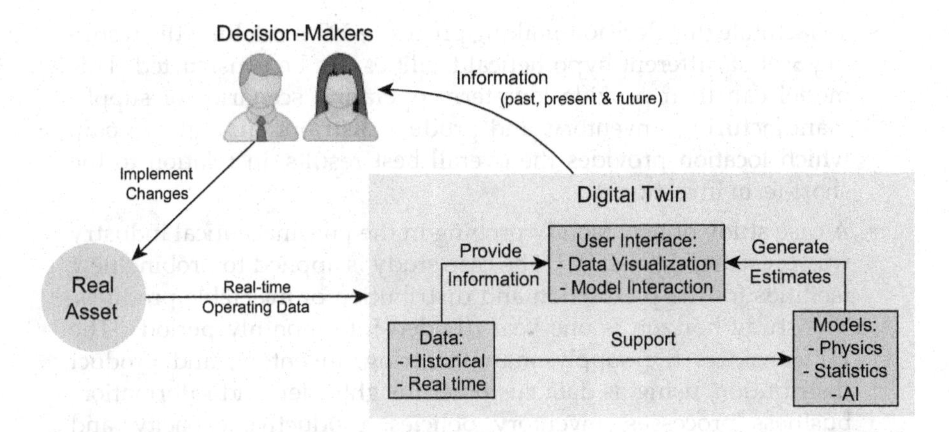

FIGURE 49.1
Digital twin (DT) schematic for strategic decision-making (Adapted from Ref. [16]). When decision-making requires human accountability, the automatic updates provided by the DT require a stakeholder as an intermediary.

- *Energy consumption optimization*: reducing the consumption of energy is desirable on a plethora of fronts: not only does it reduce costs for the company but it also reduces its ecological footprint, an increasingly important task for companies in the 21st century. Adequately modelling the equipment used for pharmaceutical manufacturing and using sensors for accurate and continuous data collection can help the processes become more efficient, by, e.g., reducing the process temperature or duration.
- While this application of DT may be considered a manufacturing application, the goal of energy consumption minimization and consequent cost reduction is more of a managerial issue. However, as these DTs are at a more operational level, the update to the parameters of the model may be done automatically within the DT feedback loop, in contrast to other more strategic applications of DTs. Furthermore, the data source for this type of DT is more sensor-based, given its operational level.
- In Ref. [2], a DT is proposed for the determination of the end point of processes in an industrial drying machine. Accurately knowing when a process reaches its end can guarantee that a process is correctly finalized and that it does not continue unnecessarily.
- *New facility probing*: investing on new facilities is an extremely expensive and delicate procedure. Not only is it necessary to acquire the grounds for construction but also the construction process itself, infrastructure, equipment, approval by the regulatory entities, and whether or not the location is practical for the workers has to be taken into account.

- To facilitate this decision-making process, a DT modelling the resulting SCs of different hypothetical facilities can be constructed. This model can then provide long-term operating scenarios of supply, manufacturing, inventory, and product distribution, and compare which location provides the overall best results, in relation to the short-term investment.
- A case study of new facility probing in the pharmaceutical industry was presented in Ref. [14]. The case study is applied to probing new facilities for the production and distribution of injectable products. The study horizon is one year divided into monthly periods. The DT considers the supply, manufacturing, inventory, and product distribution, using as data customer insights, demand information, business processes, inventory policies, productive capacity, and the location of the available facilities. Thus, DT provides predictive analytics in the form of key performance indicators, presented in a Power BI dashboard.
- *Planning & scheduling*: planning and scheduling are highly influential topics in most industries, and the pharmaceutical industry is no exception. While planning is a long-term task, scheduling is a short-term one, as it is more of an operational-level activity, typically done at a time horizon of just a few weeks [13]. Every area of a pharmaceutical company requires planning and scheduling. In terms of scheduling, examples of activities constantly scheduled include manufacturing operations, QC analyses, or warehouse movements. Planning is also applied to manufacturing operations, determining on a time horizon of a few months which products to manufacture in the upcoming months.
- Research on planning and scheduling is extensive, both in the industry as a whole and specifically in the pharmaceutical industry. However, the widespread use of DTs brought new opportunities for solving the problem in a resilient, comprehensive, and transparent way.
  - **Scheduling**: the main benefits of using a scheduling DT are its ability to create activity schedules, based on historical and live data, and constantly adapting them as unexpected occurrences take place. This means that they may operate at a time frame of weeks to generate, e.g., production schedules, but they can also be operational, able to quickly generate alternatives when the original schedules are not met. Furthermore, as the date of the produced schedule becomes closer, the schedule may be adapted considering changes in previous ones. Changes provided by the DT that do not require equipment or workforce changes, and simply correspond to small adjustments to the duration or start

of an operation, on a range of minutes to few hours, can be automatically applied without human approval.

A scheduling DT can use at its core the most common or state-of-the-art scheduling algorithms found in the literature – the difference from a typical implementation is the feedback loop where it is contained and its constant and automatic improvement, given the data-rich environment where it is contained.

Several implementations of scheduling DTs on the pharmaceutical industry were found on the industry. In Ref. [13], a discrete event simulation model of a QC laboratory is used to evaluate its capacity to meet predicted demand. Structured and free-for-all resource allocation policies for a new facility are compared, with the authors concluding that the free-for-all policy, where all QC groups share the equipment, is more beneficial. This work combines the short-term scheduling of QC tasks with the long-term strategic planning of the resource allocation policy, where changes require time and often costs.

In Ref. [7], the authors research the impact of automation on a QC laboratory. A DT is presented to evaluate the throughput of the laboratory after various degrees of automation have been implemented, and the results are compared against the as-is workflows. Scheduling is a key aspect of the research as the tasks performed by the QC laboratory analysts are comprised of regions of equipment and analyst effort, and regions of purely equipment effort. This means that the analysts can be scheduled to switch to other tasks when a sample is being processed only by the equipment. This process becomes complex as many analysts work in parallel on many different samples or pieces of equipment.

- **Planning**: planning is the strategic part of the planning and scheduling problem. Its time scope can vary quite substantially, from medium-term planning, e.g., creating production schedules, to long-term planning, e.g., business planning and managing production capability, to proper strategic time-frame planning, e.g., expanding infrastructure.

  On the medium-term planning, DTs can be used to manage raw material sourcing and suggest possible products that can be alternatively produced if a supplier fails to fulfil its delivery dates, avoiding or at least reducing shortage costs. Generally speaking, production plans are fixed in this time frame, but some alterations may be viable and beneficial.

  One possible application of DTs on the long-term planning is product selection. Companies often have a large amount of products that are able to produce; it is within the long-term planning that the products have to be selected, with these selections taking effect in the medium-term planning. DTs can be used to propose sets of products based on a wide array of parameters – production capacity, or expected demand, for example. Different combinations of products may have different consequences: some combinations of products may be easy to intertwine in the schedule part; some products may require high manufacturing but low QC effort, which could be useful to pair with products with opposite requirements. The possibilities for improvement during this stage are monumental, and using a comprehensive and predictive model of the production capabilities of a company, associated with demand forecast, in a DT architecture, can help companies generate the best production plan possible. In Ref. [16], a component of long-term planning was presented in the proposed DT. The monthly capacity for both the production workforce and support areas was calculated for the planned and user-defined production orders in a time horizon of 2 years. This estimated capacity was then compared with the available capacity. Decision-makers could use this simulation to investigate whether any production or support area requires additional capacity to accommodate a desired production plan, and if increasing that capacity is financially justifiable.

Finally, strategic time-frame planning, considered to be on a horizon of more than 2 years, often reaching 5 or 10 years, focuses on corporate strategy. The previously presented example of the application of a DT for new facility probing is also applicable in this category. Another example of application of DTs at the strategic time-frame planning can include new market segment identification.

DTs used for aiding the management of pharmaceutical companies comprise a wide array of applications. These may range from short-term decision-making, such as scheduling, to medium-term, such as production planning, to long-term, such as strategic decision-making. These types of DTs are possibly some of the most consequential, as managerial decisions often have a large impact in their companies. The power of the DT architecture can support this decision-making process.

Currently, there are not many publications on management-support DTs, and the field is expected to substantially change in the coming years. This tendency is even starker in the pharmaceutical industry, with only a small number of applications. However, given the current trends and developed work, it can be expected for research in the field of pharmaceutical management DTs to evolve in a few directions:

- Shop floor visibility and resilience: the problem of visibility is an issue faced by virtually all industries and within all aspects of corporations. Thanks to the widespread adoption of sensing technologies, data storage, and information reporting, this issue has been gradually corrected. Developing models of the pharmaceutical manufacturing operations within a DT architecture, where historical information regarding, e.g., previous durations, batch sizes, or unplanned stops, along with real-time sensor and enterprise data, can greatly improve visibility. This can be achieved through dashboards of running operations and their predicted end. At the same time, these models are able to generate alternative schedules in case of unexpected events, as well as estimating the probability of failure, given the past records.
- Large-scale planning: planning at the medium to long term can be considered one of the most essential indicators of future success. Most companies nowadays tend to already adopt strategies to improve this type of planning, but the concept of the DT is able to provide an integrated solution encompassing all the determining factors for success in the product choices that are made. Besides demand forecasting and manufacturing duration estimation, a DT to tackle this problem can include, e.g., bulk product discounts or frequent client discounts.
- A DT containing all this information would be capable of generating a production plan in the long term, and constantly adapting it according to changes to the market and company, until reaching the threshold of medium-term planning. Such plans would also provide when, to whom, and what quantities of raw materials to order for each product, where to store raw materials and final products, and when to increase workforce and equipment.
- Strategic outlooks: besides applications of DTs in the short and long term, proper strategic planning can also benefit from its capabilities. In contrast to long-term planning, one integrated tool for all things strategy is unfeasible and unnecessary. Instead, multiple DTs can be applied to different parts of strategic planning. The aforementioned work done in Ref. [14] for new facility probing is an excellent example of how the technology can be used. Capacity planning can also benefit from a dedicated DT – on instances where infrastructure needs to be installed, but when facility sourcing is not required or has already been done. The DT would be set to evaluate what equipment to install given long-term market trends and company strengths, while avoiding unnecessary quick expansion which could be costly and even hinder the capacity for the company to grow.
- Another area where strategic DTs could be useful, especially in the pharmaceutical industry, is research and development. It is well known

that the full pipeline of drug development is a lengthy and expensive process, frequently taking 10–12 years from basic research to an approved drug. A DT could be useful to estimate the economic feasibility of starting a new drug discovery process and to propose solutions based on the market, such as to start the process, postpone a couple of years, or outsource the procedure (or part of the procedure). A common practice that big pharmaceutical corporations tend to follow is to outsource the optimization of process development to pharmaceutical contract and development manufacturing organizations (CDMOs), which are encumbered with detecting alternatives in the production pipeline to expedite the process, without compromising the safety of the drug. This decision-making process could be supported by a DT estimating the costs of the different approaches followed.

49.4 Discussion

The model that the DTs contain can be one of various types, each with its own requirements, particularities, and benefits:

- *Static models*: models used for inference, which generate a prediction after being supplied an input. This could range from a model as simple as a linear regression to a more complex neural network. Given that these models are static, a periodic retraining is required, for it to update its prediction based on new data.
- On such models, inference is extremely fast, but the training can take a large amount of time, depending on the size of the model and dataset used to train it.
- *Simulation models*: models that run a modelled scenario for a long time and investigate the long-term effects of the selected parameters, or that evaluate the most likely outcome of a stochastic process. The most common algorithms are discrete event simulation and Monte Carlo simulation.
- These models are frequently used for process optimization and to investigate process bottlenecks.
- *Optimization algorithms*: consist of minimizing or maximizing one or more objective functions by changing some parameters of the problem. The algorithms range from simple linear programming to more complex metaheuristics, which can take a long time to obtain results and for that reason are often not adequate for operational-level DTs. However, for planning, SC, or strategic DTs, these algorithms can often be the best choice.

- Simulation-based optimization is also a possibility in DTs, combining the benefits of long-term prediction of simulation models, with the capacity to obtain the optimal parameters of optimization algorithms. This type of optimization can become extremely lengthy, but the results can be very helpful.

The pharmaceutical industry is an extremely complex one. For one, the pharmaceutical processes are often some of the most challenging ones and in need of extremely qualified workers; secondly, the product development pipeline is one of the longest and most expensive out of all industries; finally, the industry is very well regulated by independent regulatory agencies, such as USA's Food and Drug Administration or the European Medicines Agency (EMA). These circumstances create an industry which is slow to adapt to new digital technologies. In Ref. [20], the authors double down on this slowness attributing *on the one hand, to regulatory requirements that each process must be thoroughly tested and not negatively affect product quality and, on the other hand, to the very atypical nature of pharmaceutical manufacturing itself when compared to other industries.*

The application of the DT in the pharmaceutical industry is no exception to this slowness in digital adaptation. It is an underdeveloped research area, with only 33 articles found on the topic. Of the 21 publications cited in this chapter, seven were published in 2022, seven in 2021, six in 2020, and one in 2019. The novelty of the application of DTs in the pharmaceutical industry is clear. This underdevelopment of the research topic has a positive consequence – the topic is still very open for exploration.

49.5 Concluding Remarks

SCDTs in the pharmaceutical industry are expected to evolve toward transparency, prediction, and sustainability. Management DTs in the pharmaceutical industry, on the other hand, are expected to have a focus on the three types of planning – short-term planning and scheduling, long-term planning, and strategic planning. Both types of DTs are expected to become increasingly complex and reliable, verging on becoming autonomous. For accountability reasons, these types of DTs cannot be fully autonomous and require a decision-maker to process the updates. If information is adequately supplied, and estimations are abundant and accurate, the work of the decision-maker becomes more reliable.

This type of DT requires accountability from human stakeholders when changing parameters of the real asset. Fortunately, given the managerial degree that SC and strategic DTs have, parameter change is not required to be quick and to occur frequently, which would result in these systems becoming unhelpful. The more strategic a DT is, the larger the decision window.

References

1. Deenesh K. Babi, Jan Griesbach, Stephen Hunt, Francis Insaidoo, David Roush, Robert Todd, Arne Staby, John Welsh, and Felix Wittkopp. Opportunities and challenges for model utilization in the biopharmaceutical industry: current versus future state. *Current Opinion in Chemical Engineering*, 36:100813, 2022.
2. R. Barriga, M. Romero, D. Nettleton, and H. Hassan. Advanced data modeling for industrial drying machine energy optimization. *The Journal of Supercomputing*, 78(15):16820–16840, 2022.
3. Áron Kristóf Beke, Martin Gyürkés, Zsombor Kristof Nagy, György Marosi, and Attila Farkas. Digital twin of low dosage continuous powder blending-artificial neural networks and residence time distribution models. *European Journal of Pharmaceutics and Biopharmaceutics*, 169:64–77, 2021.
4. Ian M. Cavalcante, Enzo M. Frazzon, Fernando A. Forcellini, and Dmitry Ivanov. A supervised machine learning approach to data-driven simulation of resilient supplier selection in digital manufacturing. *International Journal of Information Management*, 49:86–97, 2019.
5. Yingjie Chen, Ou Yang, Chaitanya Sampat, Pooja Bhalode, Rohit Ramachandran, and Marianthi Ierapetritou. Digital twins in pharmaceutical and biopharmaceutical manufacturing: a literature review. *Processes*, 8(9):1088, 2020.
6. Tiago Coito, Paulo Faria, Miguel S. E. Martins, Bernardo Firme, Susana M. Vieira, João Figueiredo, and João M. C. Sousa. Digital twin of a flexible manufacturing system for solutions preparation. *Automation*, 3(1):153–175, 2022.
7. Tiago Coito, Miguel S. E. Martins, Bernardo Firme, João Figueiredo, Susana M. Vieira, and João M. C. Sousa. Assessing the impact of automation in pharmaceutical quality control labs using a digital twin. *Journal of Manufacturing Systems*, 62:270–285, 2022.
8. Maureen S. Golan, Benjamin D. Trump, Jeffrey C. Cegan, and Igor. Linkov. Supply chain resilience for vaccines: review of modeling approaches in the context of the covid-19 pandemic. *Industrial Management & Data Systems*, 121(7):1723–1748, 2021.
9. Heribert Helgers, Alina Hengelbrock, Axel Schmidt, and Jochen Strube. Digital twins for continuous mRNA production. *Processes*, 9(11):1967, 2021.
10. Heribert Helgers, Alina Hengelbrock, Axel Schmidt, Florian Lukas Vetter, Alex Juckers, and Jochen Strube. Digital twins for scFv production in Escherichia coli. *Processes*, 10(5):809, 2022.
11. Alina Hengelbrock, Heribert Helgers, Axel Schmidt, Florian Lukas Vetter, Alex Juckers, Jamila Franca Rosengarten, J¨orn Stitz, and Jochen Strube. Digital twin for HIV-Gag VLP production in HEK293 cells. *Processes*, 10(5):866, 2022.
12. Mengnan Liu, Shuiliang Fang, Huiyue Dong, and Cunzhi Xu. Review of digital twin about concepts, technologies, and industrial applications. *Journal of Manufacturing Systems*, 58:346–361, 2021.
13. Miguel R. Lopes, Andrea Costigliola, Rui Pinto, Susana Vieira, and Joao M. C. Sousa. Pharmaceutical quality control laboratory digital twin-a novel governance model for resource planning and scheduling. *International Journal of Production Research*, 58(21):6553–6567, 2020.
14. Jose Antonio Marmolejo-Saucedo. Design and development of digital twins: a case study in supply chains. *Mobile Networks and Applications*, 25(6):2141–2160, 2020.

15. Rui Portela, Christos Varsakelis, Anne Richelle, Nikolaos Giannelos, Julia Pence, Sandrine Dessoy, and Moritz von Stosch. When is an in silico representation a digital twin? A biopharmaceutical industry approach to the digital twin concept. In Herwig, Christoph, Ralf Pörtner, and Johannes Möller, eds. *Digital Twins:tools and concepts for smart biomanufacturing.* Cham: Springer International Publishing, 2021, pages 35–55, 2020.
16. João A. M. Santos, Miguel R. Lopes, Joaquim L. Viegas, Susana M. Vieira, and João M. C. Sousa. Internal supply chain digital twin of a pharmaceutical company. *IFAC-PapersOnLine*, 53(2):10797–10802, 2020.
17. Andreas Schmidt, Joshua Frey, Daniel Hillen, Jessica Horbelt, Markus Schandar, Daniel Schneider, and Ioannis Sorokos. A framework for automated quality assurance and documentation for pharma 4.0. In *International Conference on Computer Safety, Reliability, and Security*, SAFECOMP 2021, York, UK, September 8–10, pages 226–239. Springer, 2021.
18. Axel Schmidt, Dirk Köster, and Jochen Strube. Climate neutrality concepts for the german chemical-pharmaceutical industry. *Processes*, 10(3):467, 2022.
19. Kushal Sinha, Eric Murphy, Prashant Kumar, Kirsten A. Springer, Raimundo Ho, and Nandkishor K. Nere. A novel computational approach coupled with machine learning to predict the extent of agglomeration in particulate processes. *AAPS PharmSciTech*, 23(1):1–16, 2022.
20. Jannik Spindler, Thomas Kec, and Thomas Ley. Lead-time and risk reduction assessment of a sterile drug product manufacturing line using simulation. *Computers & Chemical Engineering*, 152:107401, 2021.
21. Botond Szilagyi, Ayse Eren, Justin L. Quon, Charles D. Papageorgiou, and Zoltan K. Nagy. Digital design of the crystallization of an active pharmaceutical ingredient using a population balance model with a novel size dependent growth rate expression. From development of a digital twin to in silico optimization and experimental validation. *Crystal Growth & Design*, 22(1):497–512, 2021.
22. Steffen Zobel-Roos, Axel Schmidt, Lukas Uhlenbrock, Reinhard Ditz, Dirk Köster, and Jochen Strube. Digital twins in biomanufacturing. In *Digital Twins*, pages 181–262, 2020.

50

Human Body Digital Twins: Technologies and Applications

Chenyu Tang
University of Cambridge

Yanning Dai, Jiaqi Wang, and Shuo Gao
Beihang University

50.1 Introduction

The concept of a digital twin (DT), which refers to the establishment of a virtual representation of a practical object, was first officially proposed by NASA in 2010 to address the working issues of flight engines, which play a significant role in Industry 4.0 and relevant fields, where the real-time monitoring of complex systems and the prediction of their future status in diverse conditions are required [1]. For example, in the manufacturing industry, DTs affect the entire product lifecycle management (PLM). Conventional approaches for PLM are extraordinarily time-consuming and complicated in terms of their design, manufacturing, service, and operations. However, DTs can extract digital information and overlay physical products with virtual models [2]. In this way, companies are capable of managing their products digitally throughout the overall product lifecycle. Furthermore, DTs are applied to urban planning in the form of interactive platforms [3]. With the captured real-time three-dimensional (3D) and four-dimensional (4D) spatial data, DTs establish digital models for urban environments [4,5].

Inspired by the successful utilization of DT technology in industries, biomedical researchers are focused on building DT models for the human body. This includes the digitization of the musculoskeletal nervous system, which can not only provide medical staff with patients' vital physiological information (e.g., blood glucose and respiration rate) in real time but can also report patients' potential body changes when certain drugs are delivered. Nevertheless, it is challenging for current mainstream technologies to achieve this goal for the following two reasons. First, the human body can be regarded as a complex system whose status involves the comprehensive

 DOI: 10.1201/9781003425724-60

interaction between different tissues, organs, and subsystems [6,7]. Medical tests usually examine one or several types of human body information, which may satisfy the diagnosis of certain diseases; however, they are insufficient to adequately represent the working mechanisms of human bodies. Further, most of the tests are discrete in the time domain, indicating that the continuous changing trend of indexes is not available. Second, it is difficult to establish relationships between different types of human body information. For example, the principle responsible for the control by the neuronal system on the musculoskeletal system is still vague, and only limited body movements can be explained [6].

During the last decade, there have been rapid developments in the fields of material science and fabrication technologies, and there has been an exponential increase in the number of reports on fundamental theories and applications. Advanced electronic devices exhibit desirable attributes such as small volume, biocompatibility, and low power consumption. Diverse multifunctional wearables and implantables have been demonstrated to monitor multidimensional human body information [8], which are essential, but they have been omitted in conventional measurements, indicating the potential to provide must-have elements for modeling human bodies. In addition, emerging artificial intelligence (AI) methods have proven to be useful in the determination of nonlinear relationships between multimodal data and in finding hidden links for historical and real-time information, based on which the future body status could be potentially forecasted [9]. Some state-of-the-art works have employed recurrent neural networks (RNNs) to predict users' motion strategies in a new environment after partially recording their daily activities [10–12], and reinforcement learning (RL) has now been applied to simulate the metabolism of drugs in an individual [13–15].

Although human body DTs have not yet been developed, many fundamental building blocks are now being prepared and are under construction. In the following sections, we first briefly review state-of-the-art wearable and implantable electronics and AI methods, which can monitor and analyze human body signals at all times and places, and then we provide different perspectives on new and advanced experience and services that are enabled by human body DTs.

50.2 Wearables and Implant-Based Human Body Signal Retrieving

The journey toward understanding the highly complex human body system can be traced to the Han dynasty of China, when the Huangdi Neijing was composed. Since then, there have been many attempts globally to create a foundation for modern medical science. Although there have been tremendous achievements, and many of them are now extensively used in biomedical research, a comprehensive recognition of the mystery of the human body is still underway, indicating the difficulty in establishing corresponding DTs. As briefly explained above, advanced electronics and machine learning algorithms have paved the way for the study of the human body. In this section, new material-based wearables and implants for retrieving multidimensional human information are explained.

50.2.1 Wearables

Advanced wearables with novel materials and MEMS technology provide higher detection accuracy and richer information dimensions [16–18]. They are discussed in terms of locomotion, physical signs, and external body fluid detection.

Locomotion analysis using wearables has seen significant achievements. In Ref. [19], a multisensory fusion-based technique achieved 96.53% accuracy for classifying five locomotion modes using insole force-sensing systems and an EMG-based intention detection system. In Ref. [20], a fusion of IMU, goniometer, and EMG data resulted in low error rates for steady-state and transitions, validating the feasibility of wearable sensor fusion in locomotion classification. J. Yoon et al. [21] developed a flexible skin-type sensing device to measure human metacarpophalangeal joint flexion with a capacitive strain sensor and discrete assembly of data acquisition, transmission, and processing modules.

However, using multiple devices increases complexity and burden on users. Solutions include reducing sensor types and numbers while increasing device performance and using smarter algorithms [22,23]. S. Liu et al. [22] designed an advanced motion tracking device that achieved high dynamic motion detection accuracy for boxing and kicking actions using a single device. X. Yi et al. [23] presented a real-time human motion tracking method using sparse inertial sensors, combining a neural kinematics estimator and a physics-aware motion optimizer for higher accuracy and realism.

Integrating sensing materials into clothing can reduce the physical load of multiple sensors while maintaining freedom of motion. Textile-based electronics show potential in applications like sensing [24–26], display [27], and energy harvesting [28–30]. For locomotion detection, textiles are used as resistive-type strain sensing materials, with examples in Refs. [27] and [31] using large-area textile touch sensors and graphene-based strain sensors, respectively. To enhance sensitivity, Z. Liu et al. [26] introduced microstructures along textile fibers.

Physical sign detection, such as heart rate and respiration rate, can provide a deeper understanding of body status [32–38]. Devices like the organic optoelectronic device in Ref. [38] and the epidermal electronic system (EES) technology in Ref. [39] offer high sensing performance and better user experience.

Wearable devices usually monitor surface signals, but Wang et al. [40] developed an ultrasonic device to obtain in-depth information, satisfying monitoring requirements for different body locations. This led to multimodal studies correlating motion information and physical signs.

Wearable chemical sensors analyze sweat for health information [41–49]. Examples include moisture-impermeable gloves [43], flexible Schottky barrier-based CCDs [44], wristbands for glucose, lactate and electrolyte measurements [45], and various other sweat sensors [46–48]. Some devices harvest energy from sweat, like the battery-free perspiration-powered electronic skin (PPES) in Ref. [49], which also responds to muscle contraction, indicating potential for locomotion classification.

50.2.2 Implants

It is either challenging or impossible to precisely monitor the internal conditions of the human body (such as human skin and fat) using conventional technologies, not only to protect organs but also to prevent external instruments from probing the internal environment. Therefore, mainstream technologies employ indirect means, such as computed tomography (CT) and ultrasound, to analyze the desired in vivo objectives, failing to obtain accurate and long-term detection. In contrast, novel implantable devices can make direct contact with internal targets, retrieving the desired information from the source. In this section, we explain how traditionally "impossible tasks" can be accomplished. The use of implantable devices to monitor the three most important body components, namely, the skeleton, visceral organs, and body fluid, which mainly provide functions for body support, powering life activities, and regular. L. Cai et al. [50] developed an implantable musculoskeletal bio-interface for monitoring bone health, promoting bone rehabilitation, and analyzing gait, with successful in vivo monitoring in rat femurs. A. D. Mickle et al. [51] designed a bio-optoelectronic implantable device for bladder state monitoring and bladder function modulation using soft strain gauges and μ-ILEDs, showing effective sensing and modulation of bladder activities in rats. K. Sim et al. [52] presented a soft epicardial bioelectronic patch for ECG monitoring, cardiac strain, temperature sensing, electrical pacing, and thermal ablation, which was experimentally verified using a beating porcine heart. Williams et al. [53] introduced an optical nanosensor made of an antibody-functionalized CNT complex to detect ovarian cancer biomarkers, successfully measuring human epididymis protein 4 (HE4) in murine models. R. Li et al. [54] demonstrated a wireless, flexible, and implantable sensor for nitric oxide (NO) detection, providing real-time NO monitoring in both in vitro and in vivo experiments. C. M. Boutry et al. [55] presented an implantable pressure sensor for pulse rate monitoring to assess artery patency, offering robust and real-time measurement in artificial artery models and rats. C. Liu et al. [56] proposed an implantable optoelectrochemical microprobe for real-time optogenetic interference and dopamine monitoring, demonstrating reliable performance in mice's ventral tegmental area (VTA). The following discusses its impact on the internal environment.

50.3 Machine Learning in Prediction

AI, which is a concept originally used in science fiction, is now becoming a ubiquitous technology in daily life. During the procedure of creating human body DTs, AI is crucial for abstracting hidden information from the complex and abundant data retrieved by wearable and implantable sensors, and finding relationships between phenomena and inputting information, although sometimes in a non-logical manner (i.e., black-box issue) [57]. In addition, AI can build a virtue representative of a practical object and apply physical laws to predict the performance under different conditions, which has been broadly used in finite-element analysis (FEA) for complicated mechanical structures, for example, engines [58]. In this section, we focus on the milestones in using AI to understand human motion and health status. We start with purely big data-based research and then discuss studies that combine big data and physical models.

Big data-based machine learning algorithms can generally be divided into two categories: traditional methods or deep learning methods. The former extracts features first and then uses random forest (RF), support vector machine (SVM), and artificial neural networks (ANNs) to implement classifications and regressions. In contrast, the latter provides predictions by using supervised learning methods, such as convolutional neural networks (CNNs) and RNNs, and unsupervised learning methods, such as K-nearest neighbor (KNN) and RL, which directly process the raw data in order to find the connections. Both machine learning categories have been used to address complex biomedical issues [59–61]. Among these, medicine effect prediction and locomotion prediction are explained here as examples because they are believed to be challenging and crucial tasks.

Many medicines are on the nanometer scale [62,63], but nanoparticles (NPs) induce an immune response when they spread in the human body. Conventional expensive and time-consuming animal experiments can filter only one or several specific characteristics of NPs. Moreover, simulation-based analyses cannot accurately mimic the immune response. In this context, researchers have attempted to use machine learning to predict effects of medicine, such as the tree-based RF feature importance and interaction network analysis framework (TBRFA) [64] and RL to model and verify the best delivery method for intravenous fluids and vasopressors [65]. Body locomotion monitoring is challenging due to the number of sensors and data processing required. Deep learning techniques like ANN models [22] and RNN-based networks [66] have been developed for efficient full-body motion estimation. Combining the body's musculoskeletal nerve model with big data methods can improve prediction accuracy and discover new body working mechanisms [67,68]. This combination has also been used in researching mental and neurological diseases, such as Parkinson's disease, addiction, and schizophrenia [69,70].

In addition to supervised and unsupervised learning, self-supervised deep learning methods have been used to understand and model the human body, like predicting 3D human poses in video clips [71] and analyzing various healthcare data, such as MRI scans, human brain scans, and pathology images [72]. The growing field of human DTs is expected to offer more momentum for self-supervised learning in reconstructing human body models.

Among diverse self-supervised learning frameworks, variants derived from transformer have distinguished advantages in the task of reconstructing the human model. The information of the human body is multimodal, while transformer architecture can extract the features of various digital quantities into the same semantic space, which has been utilized to unify the computer vision (CV) and natural language processing (NLP) in the deep learning area [73–75]. In addition, when deep learning researchers continue to deepen the depth of transformers and increase the amount of data, model saturation has not yet appeared (performance improvement slows down). For example, in GPT-3, the number of parameters has exceeded 100 billion, but no sign of saturation has been observed [76] (Figure 50.1).

In the issue of human health analysis, transformer-based self-monitoring learning focusing on some specific modality has already emerged. In Ref. [77], S. Park et al. presented a framework named distillation for self-supervision and self-train learning (DISTL), whose baseline model is vision transformer (ViT), for chest X-ray diagnosis through knowledge distillation, as shown in Figure 50.1a. The DISTL was validated in the diagnosis of tuberculosis, pneumothorax, and COVID-19 from three hospitals' public datasets and showcased considerable accuracy as well as robustness. In the diagnostic tasks of the three diseases, the calculated area under the receiver operating characteristics curves (AUCs) increased steadily as the unlabeled samples accumulated (from 0.948 to 0.974, from 0.832 to 0.913, and from 0.828 to 0.966), which even overwhelmed the performance of supervised learning with an equal amount of data. In addition, in Ref. [78], L. Rasmy et al. proposed Med-BERT, which adapts the BERT framework originally developed for the text domain to the structured electronic health record (EHR) domains, as shown in Figure 50.1b, and pre-trained this contextualized embedding model on a structured EHR dataset of 28,490,650 patients from Cerner Health Facts database. The model was evaluated on three phenotyped cohorts from two databases (two cohorts from Cerner and one cohort from Truven) in two tasks (Diabetes Heart Failure, DHF and Pancreatic Cancer, PaCa). The

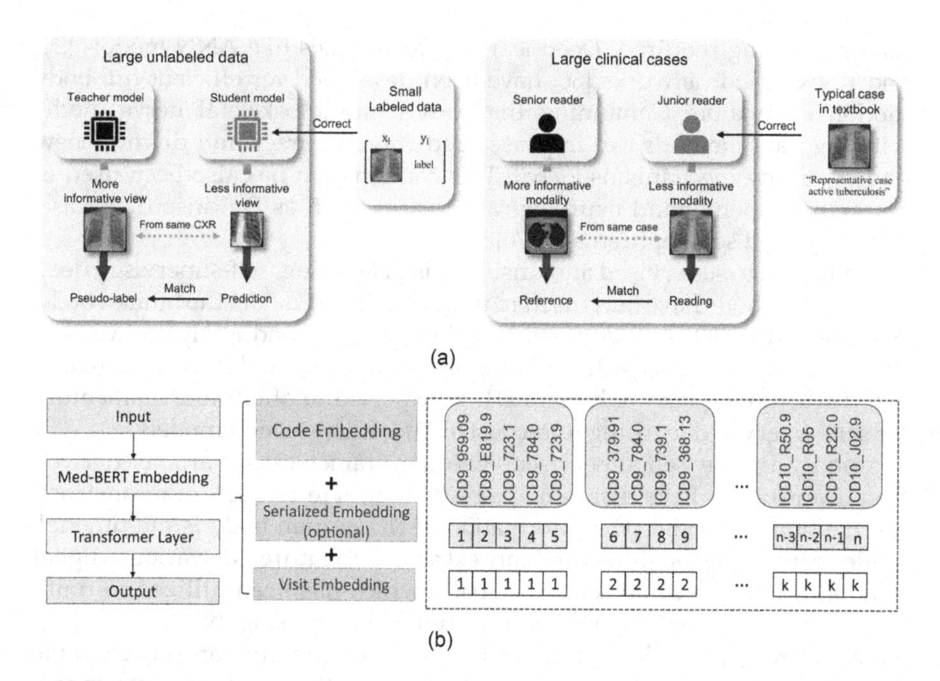

FIGURE 50.1
(a) The analogy between the artificial intelligence (AI) model training with the proposed DISTL method and human reader training [77]. (b) The structure of the proposed Med-BERT for disease prediction [78].

average AUC values and standard deviations for the three cohorts were 85.39 (0.05), 8.23 (0.29), and 80.57 (0.21), respectively, which increased by 1.21%–6.14% compared with previous works. Extensive experiments demonstrated that Med-BERT has the capacity to help boost the prediction performance of baseline deep learning models on different sizes of training samples and can obtain promising results. In particular, Med-BERT based on self-supervised learning enables training powerful deep learning predictive models with limited training sets.

The two examples demonstrated above correspond to the two application scenarios of self-supervised learning in the human body prediction models, the inputs are images and the inputs are sequences, respectively, which are the major forms of human body information. Both of them have shown commendable performance when the amount of labels is limited, which is instrumental in overcoming one of the most significant obstacles encountered in human modeling: the contradiction between massive data and few labels. Therefore, we can reasonably extrapolate that the transformers are expected to be conducive to the establishment of larger, more generalized, and more accurate human models by aggregating diverse modalities of the human body, potentially empowering the flourishing of the human body DTs.

50.4 Future of Human Body Virtual Representatives

With the above-explained advanced electronics and AI algorithms, rich and long-term human body data can be obtained and used to build human body DTs (as conceptually demonstrated in Figure 50.2) to address existing global challenges in the health and biomedical domains and to introduce novel experiences when interacting with the external world.

For the former, DTs are believed to play a major role in early stage disease detection and proper rehabilitation training design. The implementation of the first can enable effective and early intervention in diseases, limiting the damage as much as possible and reducing the risk of death and disability. However, the signs of diseases at early stages are not obvious and cannot be recognized by patients or even regular physical examinations [79,80]. With respect to the latter, DTs are essential in conveying users' intentions into cyberspace and reporting users' physical and mental conditions to prevent emergent issues.

With human body DTs, the invisible body's internal environment becomes transparent, and even tiny index changes can be recorded in real time by implantable devices installed surrounding the source [81,82]. For example, as introduced previously, the presence of NO and other desired chemicals, which cannot be detected by conventional techniques such as colorimetric measurement, can now be sensed. This information is processed by AI to calculate the potential for developing severe conditions. If poor results are obtained, DTs can be used to predict the impact of different medicine portions on the lesion site and other body components, which will help doctors to develop optimized medicine treatment plans.

In the area of rehabilitation, the training strategy has been proven to be the key factor correlated to rehabilitation progress. For example, in the United

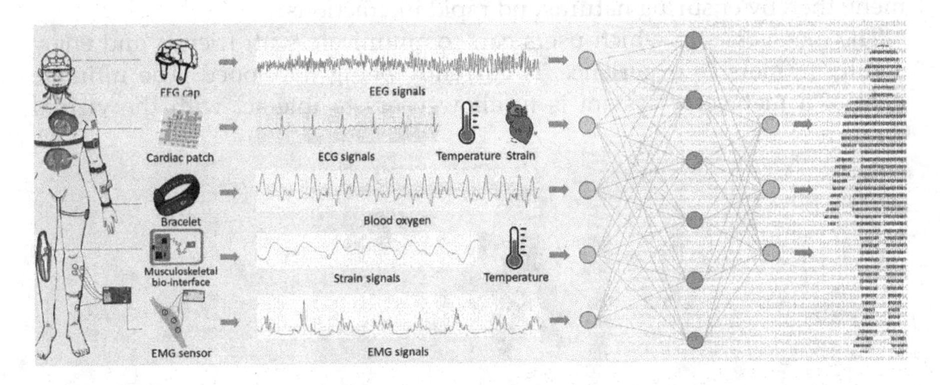

FIGURE 50.2
Conceptual depiction of the construction of the human body digital twin model using wearable and implantable sensors and machine learning algorithms.

States and Germany, approximately 47% of stroke patients can return to work and regain self-care abilities after undergoing well-designed physical training [83]. However, this number is approximately 25% in developing countries [84]. The main reason is that the stroke unit method, which is widely employed in developed countries in the design of proper physical training [85], requires rich medical resources that are not available in developing regions. In addition, physical training is irreversible, indicating that poor rehabilitation progress caused by improper training cannot be corrected. In the future, doctors can input different training strategies into patients' DT models and determine the best one. DT techniques may even generate better training strategies than existing techniques. DT has the potential to significantly accelerate the rehabilitation process and improve the training effect, relieving physical and mental pain for patients globally (Figure 50.3).

As the social aging problem intensifies, there is an increasing demand for intelligent assistive devices that return independence to users over their lifetimes. In this context, advanced human–robot interaction (HRI) techniques involving safe, adaptive, and intimate interactions are of utmost importance [86]. Human assist devices typically utilize two forms of interaction: human augmentation, such as wearable prosthetics or muscle strength-enhancing devices [87], and a stand-alone service robot that helps to handle delicate tasks such as preparing food items in the kitchen [88]. Soft robots are particularly suitable for these two aims owing to their salient body compliance, sensing sensitivity, and the large number of degrees of freedom (DoFs) [89,90]. With the data captured from the human body and soft robot during the interaction, a human DT model can provide robots with more comprehensive user physiological and psychological data, enabling a better understanding of human intentions and emotions. Furthermore, a DT model can enhance the HRI efficiency by predicting user behavior. For example, in a cooperative load-carrying mission, the DT model can predict the changes in human applied force in advance rather than responding after the measurement, thereby ensuring natural and rapid interactions.

The metaverse, in which users can communicate with friends and enjoy a fully immersive experience, is currently being developed. The ultimate goal of a metaverse system is to allow users to interact with the virtual

FIGURE 50.3
Conceptual depiction of the use of DTs to predict the best physical exercise strategy for rehabilitation training, and the impact of drugs on different in vivo systems.

environment as they would with a real one [91]. Here, DTs are required from at least two perspectives. First, highly digitized human bodies can provide essential information that can be used to precisely influence the digital environment. For example, the fingers block the sun and form a shadow on the ground; wind caused by running blows the paper off. All of these details will be used to create diverse extended reality (XR) scenarios and, finally, offer users some semblance of reality. In the meantime, long-term immersive experiences can make users addicted, which may result in physical and mental issues. Hence, DTs allow the system to determine whether a user should quit an entertainment activity. Alternatively, it can even predict how a user's body condition will change before extensive content is displayed.

50.5 Conclusion

The pace of today's technology development is exponential rather than linear. With the continuous maturity of basic science and engineering technologies, the development of human body DTs may be more rapid than expected. The history and current status of human body research have proven that each small step of deeper understanding can contribute to significant advances in the quality of life. The perspective provided in this article partially unveils the potential benefits of human body DTs, and the authors strongly believe that their development will positively impact society.

References

1. B. Piascik, J. Vickers, D. Lowry, S. Scotti, J. Stewart, A. Calomino, *Materials, Structures, Mechanical Systems, and Manufacturing Roadmap,* National Academies Press, Washington, DC, 2012.
2. F. Tao, H. Zhang, A. Liu, A. Y. Nee. Digital Twin in Industry: State-of-the-Art. *IEEE Trans. Industr. Inform.* 2018, 15, 2405.
3. Q. Lu, A. K. Parlikad, P. Woodall, G. D. Ranasinghe, X. Xie, Z. Liang, E. Konstantinou, J. Heaton,J. Schooling. Developing a digital twin at building and city levels: Case study of West Cambridge campus. *J. Manage. Eng.* 2020, 36, 05020004.
4. T. Ruohomäki, E. Airaksinen, P. Huuska, O. Kesäniemi, M. Martikka, J. Suomisto. In *2018 International Conference on Intelligent Systems (IS),* IEEE, Funchal, 2018.
5. A. Francisco, N. Mohammadi, J. E. Taylor. Smart city digital twin–enabled energy management: Toward real-time urban building energy benchmarking. *J. Manage. Eng.* 2020, 36, 04019045.

6. J. N. Kerkman, A. Daffertshofer, L. L. Gollo, M. Breakspear, T. W. Boonstra. Network structure of the human musculoskeletal system shapes neural interactions on multiple time scales. *Sci. Adv.* 2018, 4, eaat0497.
7. H. Gray, C. M. Goss. Anatomy of the human body. *Am. J. Phys. Med. Rehabil.* 1974, 53, 293.
8. G. H. Lee, H. Moon, H. Kim, G. H. Lee, W. Kwon, S. Yoo, D. Myung, S. H. Yun, Z. Bao, S. K. Hahn. Multifunctional materials for implantable and wearable photonic healthcare devices. *Nat. Rev. Mater.* 2020, 5, 149.
9. M. I. Jordan, T. M. Mitchell. Machine learning: Trends, perspectives, and prospects. *Science.* 2015, 349, 255.
10. J. Butepage, M. J. Black, D. Kragic, H. Kjellstrom. Deep Representation Learning for Human Motion Prediction and Classification. In *Proceedings of IEEE Conference on Computer Vision and Pattern Recognition (CVPR),* IEEE, Honolulu, 2017.
11. J. Martinez, M. J. Black, J. Romero. On Human Motion Prediction Using Recurrent Neural Networks. In Proceedings of IEEE Conference on Computer Vision and Pattern Recognition (CVPR), IEEE, Honolulu, 2017.
12. H. Wang, J. Dong, B. Cheng, J. Feng. A position-velocity recurrent encoder-decoder for human motion prediction. *IEEE Trans. Image Process.* 2021, 30, 6096.
13. B. Ribba, S. Dudal, T. Lavé, R. W. Peck. Model-informed artificial intelligence: Reinforcement learning for precision dosing. *Clin. Pharmacol. Ther.* 2020, 107, 853.
14. A. E. Gaweda, M. K. Muezzinoglu, G. R. Aronoff, A. A. Jacobs, J. M. Zurada, M. E. Brier. Individualization of pharmacological anemia management using reinforcement learning. *Neural Netw.* 2005, 18, 826.
15. S. Lee, J. Kim, S. W. Park, S.-M. Jin, S. M. Park. Toward a fully automated artificial pancreas system using a bioinspired reinforcement learning design: In silico validation. *IEEE J. Biomed. Health. Inf.* 2020, 25, 536.
16. W. A. D. M. Jayathilaka, K. Qi, Y. Qin, A. Chinnappan, W. S. García, C. Baskar, H. Wang, J. He, S. Cui, S. W. Thomas. Significance of nanomaterials in wearables: A review on wearable actuators and sensors. *Adv. Mater.* 2019, 31, 1805921.
17. C. Tang, et al. WMNN: Wearables-Based Multi-Column Neural Network for human activity recognition. *IEEE J. Biomed. Health Inform.* 2022, 27(1), 339–350.
18. M. Chu, et al. Multisensory fusion, haptic, and visual feedback teleoperation system under IoT framework. *IEEE Internet of Things J.* 2022, 9(20), 19717–19727.
19. Y. Zhao, J. Wang, Y. Zhang, H. Liu, Z. A. Chen, Y. Lu, Y. Dai, L. Xu, S. Gao. Flexible and Wearable EMG and PSD Sensors Enabled Locomotion Mode Recognition for IoHT-Based In-Home Rehabilitation. *IEEE Sens. J.* 2021:26311–26319. doi: 10.1109/JSEN.2021.3058429
20. J. Camargo, W. Flanagan, N. Csomay-Shanklin, B. Kanwar, A. Young. A machine learning strategy for locomotion classification and parameter estimation using fusion of wearable sensors. *IEEE. Trans. Biomed. Eng.* 2021, 68, 1569.
21. J. Yoon, Y. Joo, E. Oh, B. Lee, D. Kim, S. Lee, T. Kim, J. Byun, Y. Hong. Soft modular electronic blocks (SMEBs): A strategy for tailored wearable health-monitoring systems. *Adv. Sci.* 2019, 6, 1801682.
22. S. Liu, J. Zhang, Y. Zhang, R. Zhu, A wearable motion capture device able to detect dynamic motion of human limbs. *Nat. Commun.* 2020, 11, 1.

23. Yi, X., Zhou, Y., Habermann, M., Shimada, S., Golyanik, V., Theobalt, C., & Xu, F. (2022). Physical Inertial Poser (PIP): Physics-aware real-time human motion tracking from sparse inertial sensors. In *Proceedings of the IEEE/CVF Conference on Computer Vision and Pattern Recognition* (pp. 13167–13178).
24. W. Fan, Q. He, K. Meng, X. Tan, Z. Zhou, G. Zhang, J. Yang, Z. L. Wang. Machine-knitted washable sensor array textile for precise epidermal physiological signal monitoring. *Sci. Adv.* 2020, 6, eaay2840.
25. M. Liu, X. Pu, C. Jiang, T. Liu, X. Huang, L. Chen, C. Du, J. Sun, W. Hu, Z. L. Wang. Large-area all-textile pressure sensors for monitoring human motion and physiological signals. *Adv. Mater.* 2017, 29, 1703700.
26. Z. Liu, D. Qi, G. Hu, H. Wang, Y. Jiang, G. Chen, Y. Luo, X. J. Loh, B. Liedberg, X. Chen. Surface strain redistribution on structured microfibers to enhance sensitivity of fiber-shaped stretchable strain sensors. *Adv. Mater.* 2018, 30, 1704229.
27. X. Shi, Y. Zuo, P. Zhai, J. Shen, Y. Yang, Z. Gao, M. Liao, J. Wu, J. Wang, X. Xu, Q. Tong, B. Zhang, B. Wang, X. Sun, L. Zhang, Q. Pei, D. Jin, P. Chen, H. Peng. Large-area display textiles integrated with functional systems. *Nature.* 2021, 591, 240.
28. J. Chen, Y. Huang, N. Zhang, H. Zou, R. Liu, C. Tao, X. Fan, Z. L. Wang. Micro-cable structured textile for simultaneously harvesting solar and mechanical energy. *Nat. Energy.* 2016, 1, 16138.
29. J. Xiong, P. Cui, X. Chen, J. Wang, K. Parida, M. F. Lin, P. S. Lee. Skin-touch-actuated textile-based triboelectric nanogenerator with black phosphorus for durable biomechanical energy harvesting. *Nat. Commun.* 2018, 9, 4280.
30. H. Souri, H. Banerjee, A. Jusufi, N. Radacsi, A. A. Stokes, I. Park, M. Sitti, M. Amjadi. Wearable and stretchable strain sensors: Materials, sensing mechanisms, and applications. *Adv. Intell. Syst.* 2020, 2, 2000039.
31. Z. Yang, Y. Pang, X. Han, Y. Yang, J. Ling, M. Jian, Y. Zhang, Y. Yang, T. Ren. Graphene textile strain sensor with negative resistance variation for human motion detection. *ACS Nano.* 2018, 12, 9134.
32. T. Kim, C. Park, E. P. Samuel, S. An, A. Aldalbahi, F. Alotaibi, A. L. Yarin, S. S. Yoon. Supersonically sprayed washable, wearable, stretchable, hydrophobic, and antibacterial rGO/AgNW fabric for multifunctional sensors and supercapacitors. *ACS Appl. Mater. Interfaces.* 2021, 13, 10013.
33. H. Lee, E. Kim, Y. Lee, H. Kim, J. Lee, M. Kim, H. J. Yoo, S. Yoo. Toward all-day wearable health monitoring: An ultralow-power, reflective organic pulse oximetry sensing patch. *Sci. Adv.* 2018, 4, eaas9530.
34. A. Petritz, E. Karner-Petritz, T. Uemura, P. Schäffner, T. Araki, B. Stadlober, T. Sekitani. Imperceptible energy harvesting device and biomedical sensor based on ultraflexible ferroelectric transducers and organic diodes. *Nat. Commun.* 2021, 12, 2399.
35. E. O. Polat, G. Mercier, I. Nikitskiy, E. Puma, T. Galan, S. Gupta, M. Montagut, J. J. Piqueras, M. Bouwens, T. Durduran. Flexible graphene photodetectors for wearable fitness monitoring. *Sci. Adv.* 2019, 5, eaaw7846.
36. Y. Lee, J. W. Chung, G. H. Lee, H. Kang, J. Y. Kim, C. Bae, H. Yoo, S. Jeong, H. Cho, S. G. Kang. Standalone real-time health monitoring patch based on a stretchable organic optoelectronic system. *Sci. Adv.* 2021, 7, eabg9180.
37. Z. Zhen, Z. Li, X. Zhao, Y. Zhong, L. Zhang, Q. Chen, T. Yang, H. Zhu. Formation of uniform water microdroplets on wrinkled graphene for ultrafast humidity sensing. *Small.* 2018, 14, 1703848.

38. C. M. Lochner, Y. Khan, A. Pierre, A. C. Arias. All-organic optoelectronic sensor for pulse oximetry. *Nat. Commun.* 2014, 5, 5745.
39. D. H. Kim, N. Lu, R. Ma, Y. S. Kim, R. H. Kim, S. Wang, J. Wu, S. M. Won, H. Taoa, H. Islam, K. J. Yu, T. I. Kim, R. Chowdhury, M. Ying, L. Xu, M. Li, H. J. Chung, H. Keum, M. Mccormick, P. Liu, Y. W. Zhang, F. G. Omenetto, Y. Huang, T. Coleman, J. A. Rogers. Epidermal electronics. *Science.* 2011, 333, 838.
40. C. Wang, X. Li, H. Hu, L. Zhang, Z. Huang, M. Lin, Z. Zhang, Z. Yin, B. Huang, H. Gong. Monitoring of the central blood pressure waveform via a conformal ultrasonic device. *Nat. Biomed. Eng.* 2018, 2, 687.
41. M. Bariya, H. Y. Y. Nyein, A. Javey. Wearable sweat sensors. *Nat. Electron.* 2018, 1, 160.
42. Y. Yang, Y. Song, X. Bo, J. Min, O. S. Pak, L. Zhu, M. Wang, J. Tu, A. Kogan, H. Zhang. A laser-engraved wearable sensor for sensitive detection of uric acid and tyrosine in sweat. *Nat. Biotechnol.* 2020, 38, 217.
43. M. Bariya, L. Li, R. Ghattamaneni, C. H. Ahn, H. Y. Y. Nyein, L. C. Tai, A. Javey. Glove-based sensors for multimodal monitoring of natural sweat. *Sci. Adv.* 2020, 6, eabb8308.
44. S. Nakata, M. Shiomi, Y. Fujita, T. Arie, S. Akita, K. Takei. A wearable pH sensor with high sensitivity based on a flexible charge-coupled device. *Nat. Electron.* 2018, 1, 596.
45. W. Gao, S. Emaminejad, H. Y. Y. Nyein, S. Challa, K. Chen, A. Peck, H. M. Fahad, H. Ota, H. Shiraki, D. Kiriya, D. H. Lien, G. A. Brooks, R. W. Davis, A. Javey. Fully integrated wearable sensor arrays for multiplexed in situ perspiration analysis. *Nature.* 2016, 529, 509.
46. H. Lee, C. Song, Y. S. Hong, M. S. Kim, H. R. Cho, T. Kang, K. Shin, S. H. Choi, T. Hyeon, D. H. Kim. Wearable/disposable sweat-based glucose monitoring device with multistage transdermal drug delivery module. *Sci. Adv.* 2017, 3, 1601314.
47. W. He, C. Wang, H. Wang, M. Jian, W. Lu, X. Liang, X. Zhang, F. Yang, Y. Zhang. Integrated textile sensor patch for real-time and multiplex sweat analysis. *Sci. Adv.* 2019, 5, aax0649.
48. A. Koh, D. Kang, Y. Xue, S. Lee, R. M. Pielak, J. Kim, T. Hwang, S. Min, A. Banks, P. Bastien, M. C. Manco, L. Wang, K. R. Ammann, K. I. Jang, P. Won, S. Han, R. Ghaffari, U. Paik, M. J. Slepian, G. Balooch, Y. Huang, J. A. Rogers. A soft, wearable microfluidic device for the capture, storage, and colorimetric sensing of sweat. *Sci. Transl. Med.* 2016, 8, 366ra165.
49. Y. Yu, J. Nassar, C. Xu, J. Min, Y. Yang, A. Dai, R. Doshi, A. Huang, Y. Song, R. Gehlhar. Biofuel-powered soft electronic skin with multiplexed and wireless sensing for human-machine interfaces. *Sci. Robot.* 2020, 5, eaaz7946.
50. L. Cai, A. Burton, D. A. Gonzales, K. A. Kasper, A. Azami, R. Peralta, M. Johnson, J. A. Bakall, E. Barron Villalobos, E. C. Ross. Osseosurface electronics—thin, wireless, battery-free and multimodal musculoskeletal biointerfaces. *Nat. Commun.* 2021, 12, 6707.
51. A. D. Mickle, S. M. Won, K. N. Noh, J. Yoon, K. W. Meacham, Y. Xue, L. A. McIlvried, B. A. Copits, V. K. Samineni, K. E. Crawford. A wireless closed-loop system for optogenetic peripheral neuromodulation. *Nature.* 2019, 565, 361.

52. K. Sim, F. Ershad, Y. Zhang, P. Yang, H. Shim, Z. Rao, Y. Lu, A. Thukral, A. Elgalad, Y. Xi. An epicardial bioelectronic patch made from soft rubbery materials and capable of spatiotemporal mapping of electrophysiological activity. *Nat. Electron.* 2020, 3, 775.
53. R. M. Williams, C. Lee, T. V. Galassi, J. D. Harvey, R. Leicher, M. Sirenko, M. A. Dorso, J. Shah, N. Olvera, F. Dao. Noninvasive ovarian cancer biomarker detection via an optical nanosensor implant. *Sci. Adv.* 2018, 4, eaaq1090.
54. R. Li, H. Qi, Y. Ma, Y. Deng, S. Liu, Y. Jie, J. Jing, J. He, X. Zhang, L. Wheatley. A flexible and physically transient electrochemical sensor for real-time wireless nitric oxide monitoring. *Nat. Commun.* 2020, 11, 3207.
55. C. M. Boutry, L. Beker, Y. Kaizawa, C. Vassos, H. Tran, A. C. Hinckley, R. Pfattner, S. Niu, J. Li, J. Claverie. Biodegradable and flexible arterial-pulse sensor for the wireless monitoring of blood flow. *Nat. Biomed. Eng.* 2019, 3, 47.
56. C. Liu, Y. Zhao, X. Cai, Y. Xie, T. Wang, D. Cheng, L. Li, R. Li, Y. Deng, H. Ding. A wireless, implantable optoelectrochemical probe for optogenetic stimulation and dopamine detection. *Microsyst. Nanoeng.* 2020, 6, 64.
57. C. Rudin. Stop explaining black box machine learning models for high stakes decisions and use interpretable models instead. *Nat. Mach. Intell.* 2019, 1, 206.
58. X. Liang, M. Z. Ali, H. Zhang. Induction motors fault diagnosis using finite element method: A review. *IEEE Trans. Ind. Appl.* 2019, 56, 1205.
59. A. Gupta, S. Savarese, S. Ganguli,L. F. Fei. Embodied intelligence via learning and evolution. *Nat. Commun.* 2021, 12, 5721.
60. P. Schneider, W. P. Walters, A. T. Plowright, N. Sieroka, J. Listgarten, R. A. Goodnow, J. Fisher, J. M. Jansen, J. S. Duca,T. S. Rush. Rethinking drug design in the artificial intelligence era. *Nat. Rev. Drug Discov.* 2020, 19, 353.
61. Y. Huang, X. Sun, H. Jiang, S. Yu, C. Robins, M. J. Armstrong, R. Li, Z. Mei, X. Shi, E. S. Gerasimov. A machine learning approach to brain epigenetic analysis reveals kinases associated with Alzheimer's disease. *Nat. Commun.* 2021, 12, 4472.
62. Q. Sun, M. Barz, B. G. De Geest, M. Diken, W. E. Hennink, F. Kiessling, T. Lammers,Y. Shi. Nanomedicine and macroscale materials in immuno-oncology. *Chem. Soc. Rev.* 2019, 48, 351.
63. J. P. Ioannidis, B. Y. Kim, A. Trounson. How to design preclinical studies in nanomedicine and cell therapy to maximize the prospects of clinical translation. *Nat. Biomed. Eng.* 2018, 2, 797.
64. F. Yu, C. Wei, P. Deng, T. Peng, X. Hu. Deep exploration of random forest model boosts the interpretability of machine learning studies of complicated immune responses and lung burden of nanoparticles. *Sci. Adv.* 2021, 7, eabf4130.
65. M. Komorowski, L. A. Celi, O. Badawi, A. C. Gordon, A. A. Faisal. The artificial intelligence clinician learns optimal treatment strategies for sepsis in intensive care. *Nat. Med.* 2018, 24, 1716.
66. D. Kim,J. Kwon, S. Han, Y. L. Park, S. Jo. Deep full-body motion network for a soft wearable motion sensing suit. *IEEE Trans. Mechatron.* 2019, 24, 56.
67. S. Lee, M. Park, K. Lee, J. Lee. Scalable muscle-actuated human simulation and control. *ACM Trans. Graph.* 2019, 38(4), 1–13.
68. S. Park, H. Ryu, S. Lee, S. Lee, J. Lee. Learning predict-and-simulate policies from unorganized human motion data. *ACM Trans. Graph.* 2019, 38(6), 1–11.
69. T. V. Maia, M. J. Frank. From reinforcement learning models to psychiatric and neurological disorders. *Nat. Neurosci.* 2011, 14, 154.

70. A. Dezfouli, P. Piray, M. M. Keramati, H. Ekhtiari, C. Lucas, A. Mokri. A neurocomputational model for cocaine addiction. *Neural Comput.* 2009, 21, 2869.
71. K. Wang, L. Lin, C. Jiang, C. Qian, P. Wei. 3D human pose machines with self-supervised learning. *IEEE Trans. Pattern Anal. Mach. Intell.* 2019, 42(5), 1069–1082.
72. A. Chowdhury, J. Rosenthal, J. Waring, R. Umeton. (2021, September). Applying self-supervised learning to medicine: Review of the state of the art and medical implementations. In *Informatics* (Vol. 8, No. 3, p. 59). MDPI. https://doi.org/10.3390/informatics8030059
73. A. Dosovitskiy, L. Beyer, A. Kolesnikov, D. Weissenborn, X. Zhai, T. Unterthiner, … N. Houlsby. (2020). An image is worth 16x16 words: Transformers for image recognition at scale. arXiv preprint arXiv:2010.11929.
74. J. Devlin, M. W. Chang, K. Lee, K. Toutanova. (2018). Bert: Pre-training of deep bidirectional transformers for language understanding. arXiv preprint arXiv:1810.04805.
75. Y. H. H., Tsai, S. Bai, P. P. Liang, J. Z. Kolter, L. P. Morency, R. Salakhutdinov. (2019, July). Multimodal transformer for unaligned multimodal language sequences. In *Proceedings of the Conference. Association for Computational Linguistics. Meeting* (Vol. 2019, p. 6558). NIH Public Access.
76. T. Brown, B. Mann, N. Ryder, M. Subbiah, J. D. Kaplan, P. Dhariwal, … D. Amodei. Language models are few-shot learners. *Adv. Neural Inf. Process. Syst.* 2020, 33, 1877–1901.
77. S. Park, G. Kim, Y. Oh, J. B. Seo, S. M. Lee, J. H. Kim, … J. C. Ye. Self-evolving vision transformer for chest X-ray diagnosis through knowledge distillation. *Nat. Commun.*. 2022, 13(1), 1–11.
78. L. Rasmy, Y. Xiang, Z. Xie, C. Tao, D. Zhi. Med-BERT: Pretrained contextualized embeddings on large-scale structured electronic health records for disease prediction. *NPJ Digit. Med.* 2021, 4(1), 1–13.
79. P. R. Srinivas, B. S. Kramer, S. Srivastava. Trends in biomarker research for cancer detection. *Lancet Oncol.* 2001, 2, 698.
80. L. Clare. Managing threats to self: Awareness in early stage Alzheimer's disease. *Soc. Sci. Med.* 2003, 57, 1017.
81. R. Laubenbacher, J. P. Sluka, J. A. Glazier. Using digital twins in viral infection. *Science* 2021, 371, 1105.
82. S. A. Niederer, M. S. Sacks, M. Girolami, K. Willcox. Scaling digital twins from the artisanal to the industrial. *Nat. Comput. Sci.* 2021, 1, 313.
83. K. H. Culler, Y. C. Wang, K. Byers, R. Trierweiler. Barriers and facilitators of return to work for individuals with strokes: Perspectives of the stroke survivor, vocational specialist, and employer. *Top. Stroke Rehabil.* 2011, 18, 325.
84. J. P. Bettger, Z. Li, Y. Xian, L. Liu, X. Zhao, H. Li, C. Wang, C. Wang, X. Meng, A. Wang. Assessment and provision of rehabilitation among patients hospitalized with acute ischemic stroke in China: Findings from the China National Stroke Registry II. *Int. J. Stroke.* 2017, 12, 254.
85. P. Langhorne, M. J. O'Donnell, S. L. Chin, H. Zhang, D. Xavier, A. Avezum, N. Mathur, M. Turner, M. J. MacLeod, P. L. Jaramillo. Practice patterns and outcomes after stroke across countries at different economic levels (INTERSTROKE): An international observational study. *The Lancet.* 2018, 391, 2019.

86. P. Polygerinos, N. Correll, S. A. Morin, B. Mosadegh, C. D. Onal, K. Petersen, M. Cianchetti, M. T. Tolley, R. F. Shepherd. Soft robotics: Review of fluid-driven intrinsically soft devices; manufacturing, sensing, control, and applications in human-robot interaction. *Adv. Eng. Mater.* 2017, 19, 1700016.
87. H. Zhao, K. O. Brien, S. Li, R. F. Shepherd. Optoelectronically innervated soft prosthetic hand via stretchable optical waveguides. *Sci. Robot.* 2016, 1, Eaai7529.
88. D. Rus, M. T. Tolley. Design, fabrication and control of soft robots. *Nature.* 2015, 521, 467.
89. J. Byun, Y. Lee, J. Yoon, B. Lee, E. Oh, S. Chung, T. Lee, K. J. Cho, J. Kim, Y. Hong. Electronic skins for soft, compact, reversible assembly of wirelessly activated fully soft robots. *Sci. Robot.* 2018, 3, eaas9020.
90. T. Kim, S. Lee, T. Hong, G. Shin, T. Kim, Y. L. Park. Heterogeneous sensing in a multifunctional soft sensor for human-robot interfaces. *Sci. Robot.* 2020, 5, eabc6878.
91. H. Duan, J. Li, S. Fan, Z. Lin, X. Wu, W. Cai. Metaverse for Social Good: A University Campus Prototype. In *Proceedings on 29th ACM International Conference on Multimedia*, ACM, New York, 2021.

51

Digital Twins for Proactive and Personalized Healthcare – Challenges and Opportunities

Sai Phanindra Venkatapurapu, Marianne T. DeWitt, Marcelo Behar, and Paul M. D'Alessandro
Pricewaterhouse Coopers LLP

51.1 Introduction

With the explosion of data in healthcare and the advancements in science and technology, the dream of fully personalized healthcare is ever so close to becoming a reality. Personalized healthcare is a coordinated, strategic medical model to tailor care to individual patients using concepts of systems biology and personalized medicine tools (Nardini et al. 2021). Personalized medicine encompasses different approaches that often seek to exploit differences in genetic, physiological, anatomical, lifestyle, and environmental factors, to deliver treatments tailored to individual patients (Harutyunyan et al. 2018; Visvikis-Siest et al. 2020). According to an FDA report ("Paving the Way for Personalized Medicine: FDA's Role in a New Era of Medical Product Development." 2013), medications are ineffective in about 38%–75% of the patients, depending on the condition, the highest being in cancer patients. The lack of efficacy can be attributed to a complex interplay of factors such as genetics and environment, as well as other factors such as poor adherence and inadequate or inappropriate dosing regimens. The adoption of personalized approaches to medicine has resulted in the development and approval of therapies such as Herceptin® for breast cancer (Issa 2007). Expanding the concept of personalized medicine into broader applications across therapeutic areas and building tools for predictive, preventive, and participatory care will lead us into the future of personalized healthcare (Nardini et al. 2021).

Digital twins are increasingly being explored as tools for personalized healthcare (Kamel Boulos and Zhang 2021). Conceptually, digital twins are virtual representations of a physical entity (an object or a system), which can be used to gain more insights into and ultimately better control over the object/system itself. The two essential requirements for creating a digital

DOI: 10.1201/9781003425724-61

twin are (1) data about the physical entity and (2) a model of the entity that represents the parts within the system and their relationships to each other. Digital twins enable data to be transmitted between the physical and virtual worlds (El Saddik 2018). Data from a physical object create its digital twin, and in turn, forecasts using the digital twin can provide guidance to the physical object. Beyond forecasting, the twin also enables a critical exploration of the sensitivities or "pressure points" of the physical system itself. In systems dynamics language, these pressure points or "architectural control points" often prove to be critical in diagnosing and preventing unwanted or suboptimal maturation within the system. This bi-directional relationship (Fuller et al. 2020) between a digital twin and the physical object makes the digital twin a continuously evolving and informative replica of the object it represents.

The concept of a digital twin, originating from engineering disciplines, has disrupted many industries (Alsdurf 2021). For more than 30 years, engineers have been using 3D computer-aided design (CAD) models and process simulations in manufacturing (Pettey 2017). Most of the digital twin concepts that exist today are for Internet of Things (IoT) devices. Sensors in IoT devices generate a large amount of real-time data which could be used to build a digital twin of the device itself. Digital twins of machines/devices thus created could be used to monitor the real-time status of the device, and its location, and predict future failures by detecting anomalies in data collected, among other functions (You et al. 2022; Dinter, Tekinerdogan, and Catal 2022). Digital twins created using IoT devices and physics-based simulation models are currently being used in many advanced engineering disciplines to attain new levels of efficiency and innovation, ranging from automobile safety, semiconductor manufacturing optimization (Kusiak 2017; Lu et al. 2020), stealth aircraft design (Glaessgen and Stargel 2012), and nuclear power production (Kochunas and Huan 2021). Another growing area of digital twin application in recent years has been urban planning and development. Smart cities powered by digital twins (Triantafilou 2021) can help planners explore and evaluate proposed nuances to their design; traffic flow implications, safety concerns, power distribution, and other urban challenges.

More recently, companies have started using IoT sensors to build smart homes that leverage sensor data for monitoring the health of the residents (Macomber and Akiko Kanno 2022). Non-wearable or passive sensor-based technologies, smart beds, and smart chairs that can measure physiological data and instantly notify relevant care providers are driving innovation in home-based remote health monitoring, especially for elderly healthcare (Majumder et al. 2017). Home-based remote health monitoring is, in fact, one of the many potential applications of digital twin technology in the health industry. In this article, we explore the landscape of digital twins in healthcare today, outline existing challenges, and suggest future directions for realizing the goal of personalized healthcare.

51.2 Role of Digital Twins in Healthcare

In contrast to other industries, the application of digital twin technology is in its infancy in healthcare. One of the main reasons is the complexity of human biology compared to other physical objects. In the article "Can a biologist fix a radio? – Or, what I learned while studying apoptosis" (Lazebnik 2002), Yuri Lazebnik compares a radio to a signal transduction network in biology. Both systems under consideration are complex, but the radio parts are well understood and behave deterministically. An engineer trained in electronics will be able to test and identify which parts of the radio are malfunctioning and repair them to make the radio functional. In contrast, the level of understanding of how physiological systems function and especially "malfunction" is, in general, not as complete as it is for the radio. In addition, unlike radios, there is variability among the patients. Inter-individual variability and stochasticity intrinsic to many biological functions create additional challenges to digital twin applications seeking to capture the onset and progression of diseases, or how a person may respond to pharmacological treatments or other interventions.

Despite these obstacles, digital twins are making their way into healthcare; the same heterogeneity and stochasticity that complicate digital twin development also make the job of healthcare providers extremely challenging. For example, due to inconsistent clinical manifestations, and unpredictable treatment outcomes arising out of variability and stochasticity, physicians treating IBD patients face challenges in creating effective treatment plans for their patients (Fiocchi 2018). As evidenced in the FDA report referenced above ("Paving the Way for Personalized Medicine: FDA's Role in a New Era of Medical Product Development." 2013), treatments approved for certain diseases are not effective in a significant majority of the patients for whom they are prescribed. Physicians have to determine the cause of a patient's non-response to a particular drug and have to modify the treatment plan, even changing to another therapy whenever necessary (Dalal and Cohen 2015). This causes delays in initiating the right treatment for a patient. Delayed initiation of certain biologic treatments has been shown to impact the remission rates and long-term surgery rates in Crohn's disease patients (Frei et al. 2019; Mantzaris et al. 2021; Dulai et al. 2020). Predictive tools that account for inter-individual variability could provide better future health risk assessment for timely interventions, leading to better outcomes for patients (Gubatan et al. 2021; Cushing and Higgins 2021).

If the promise of personalized healthcare is to be realized fully, it is imperative to build tools to help patients manage their health (predictive and personalized). Healthcare providers must also be engaged to drive the adoption of predictive, personalized, and participatory (e.g., shared decision-making) tools to manage and ensure adherence to the treatment plan for their patients. Lastly, payors, employers, pharmaceutical companies, and governments will be critical to creating the right incentives for fostering a proactive approach to healthcare (preventative, predictive). Digital twins form an integral piece of this 4P medicine (predictive, preventative, personalized, participatory) approach for personalized healthcare (Nardini et al. 2021; Flores et al. 2013).

51.3 Current Applications of Digital Twins in Healthcare and a Commentary on the Enabling Technologies

Primary applications of digital twins in healthcare range from diagnostics to prognostics and as descriptive or prescriptive tools. Similar to the use cases in manufacturing and other industries such as monitoring the status of a device and predicting failures before they occur, digital twins can be used for monitoring the health, identifying potential issues, and rectifying them promptly, and scenario testing for identifying the best treatment for the individual and providing personalized treatment and lifestyle recommendations. Other applications could include risk assessment of populations (Venkatapurapu et al. 2018), educational tools for patients to improve adherence, digital companions for therapies, and beyond.

Digital twins in healthcare can be categorized based on the physical object they mimic and the enabling modeling technology behind the creation of the twin (Table 51.1). The physical object could be a specific organ such as the eye, heart, or lungs; systems such as circulatory or musculoskeletal; an individual patient or a population of patients. Several research groups and medical device firms are developing "digital twins" of various organs that mimic their physical and functional properties to aid physicians in deciding treatment approaches and predicting the proceedings of surgical interventions (Derycke et al. 2020; Aubert et al. 2021; Coorey et al. 2022). The Living Heart Project aims to create personalized digital human heart models for physicians and surgeons to virtually analyze their patients' health and plan therapies and surgeries (Baillargeon et al. 2014; Peirlinck et al. 2021). Physicians are using digital twins of gastrointestinal tracts loaded into a virtual reality platform, for planning surgeries to remove the guesswork and eliminate the need for time-consuming exploratory incisions (Campbell 2017). Computational models of knees facilitate the design and pre-clinical testing of efficient knee implants (Shu et al. 2021). Several groups have embarked upon a more ambitious task of creating digital twins of a human being (D'alessandro et al. 2019; *Business Wire* 2021). Like the digital twins of IoT devices and organs, the objective is to create a digital representation of individuals and use the representation to make decisions related to individual and population health.

The application of digital twins also depends on the enabling modeling technology. As discussed earlier, one of the two key requirements for

TABLE 51.1

Types of Digital Twin Technologies in Healthcare and Their Applications

Type of Digital Twin	Enabling Technology	Examples/Applications
Digital twins of organs	3D modeling using imaging	1. **NUREA** – SaMD, Decision support for surgeons (*Nurea*) 2. **PrediSurge** – Predict surgical intervention proceedings (*PrediSurge*) 3. **Philips Dynamic HeartModel** – Assessing cardiac function using 3D echo in clinical practice (*Philips*)
Digital twins of patients	Mechanistic simulation modeling	1. **PwC Bodylogical (*PwC*)** ***Payors, Providers, Governments*:** Predicting future health status of individuals and populations (Venkatapurapu et al. 2018; Sarkar et al. 2018) ***Pharma Med Affairs/Commerical:*** Physician decision support tools, Digital transformation for MedAffairs, Rx+ (Venkatapurapu et al. 2022) ***Pharma R&D:*** Early clinical asset prioritization, virtual clinical trials ***Wellness:*** Empowering health coaches, Corporate wellness
Digital twins of patients	Machine learning models	1. **Siemens AI-Pathway Companion** – Clinical decision support for oncology, cardiology, and infectious diseases (Henkel et al. 2022) 2. **Unlearn.ai** – Digital twin–based control arms for RCTs (*Business Wire* 2022)
Digital twins of patients	3D modeling using imaging	1. **Q-bio** – Whole-body scanner–based digital twin (*Business Wire* 2021)

creating a digital twin is a model of the physical object. But not all digital twins are created equal (Table 51.1). To create digital twins of organs for assisting surgeons, 3D modeling techniques are used to represent the anatomy of the organ. A 3D virtual model of the organ is constructed from a series of radiological or ultrasound images and is customized to the anatomy of a specific patient's organ. Digital twins created in this manner can be used to drive treatment decisions without requiring patients to undergo repeated imaging (*Nurea; PrediSurge; Philips*). Advanced imaging technologies like a whole-body scanner are increasingly being used to create digital twins of patients for clinical decision support (*Business Wire* 2021).

Besides 3D modeling of whole-body imaging data, digital twins of individuals can also be created using a mechanistic simulation model of human physiology. A mechanistic physiology simulation model is built by expressing the mechanistic interactions between relevant physiological components in the form of a system of ordinary differential equations (ODEs) (Sarkar et al. 2018; D'alessandro et al. 2019; Venkatapurapu et al. 2022). Typically, mechanistic simulation models strive to integrate relevant interactions that occur on multiple spatial and temporal scales to understand and model the behavior of the biological system (Dada and Mendes 2011). The simulation model with its default parameter set conceptually represents a prototypical human. Digital twins enhance the simulation model based on default parameters to capture the inter-individual variability (Chen et al. 2018; Venkatapurapu et al. 2022). By changing specific parameters, the simulation model can be calibrated to the data from an individual patient. This calibrated model represents the digital twin of the real patient and can be used to either run *in silico* experiments or gain insights into the future trajectory of the real patient. For example, digital twins can be used to gain a deeper understanding of the physiological factors driving the onset and progression of chronic diseases in individual patients (Sarkar et al. 2018). The twins can then be used to identify patients or even populations at risk for certain diseases (Venkatapurapu et al. 2018), computationally test the impact of certain treatments or interventions, and inform the design of clinical trials. Besides clinical research and clinical practice, mechanistic simulation model-based digital twin technology is also found to be useful in transforming medical affairs, and health economics and outcomes research (HEOR). A digital twin–powered Crohn's disease simulator is being used in Japan by medical science liaisons (MSLs) to have engaging scientific conversations with expert physicians ("Crohn's Disease Simulator for MSL-KOL Engagement") and for tools to help physicians involve patients in decisions about treatment plans ("Crohn's Disease Digital Twin Simulator for Shared Decision-Making"). Visiting Nurse Association of Texas has used PwC's Bodylogical® digital twin platform to quantify the health impact of their nutritional program for seniors (Matthew et al. 2019).

Another common approach to creating digital twins of patients involves machine learning (ML) as the enabling modeling technology. Data from a large number of patients are fed into a ML algorithm to build models based on the patterns that exist in the data. ML algorithms are used for classification or regression problems depending on whether the learning approach used is unsupervised or supervised. The algorithm can be used to create personalized digital models of patients for applications such as clinical decision support and generating control patient arms for more efficient randomized control trials (*Business Wire* 2022).

Unlike mechanistic modeling, ML approaches do not require a higher degree of domain knowledge. However, the black-box nature of ML algorithms presents challenges in interpreting the predictions in a clinical setting and identifying intervention areas to resolve a predicted issue. There is now an emphasis on using explainable AI to address the interpretability issue (Rao and Mane 2019).

Another challenge with ML algorithms is that they can be limited to making predictions that are related to the data the ML model has seen. For example, an ML model trained to predict the response of one type of biologic in patients cannot be used to predict the response for novel therapy. On the other hand, mechanistic models can generalize the "learning" from a patient's response to one treatment and extrapolate likely responsiveness to different regimens or drugs. This is possible because the response to treatment can be used to constrain the already existing causal relationships in the mechanistic model.

ML algorithms require large amounts of data to learn the relationships between inputs and outputs and make accurate predictions. This presents challenges in developing digital twin–based platforms for diseases where data are limited. In contrast, mechanistic models are capable of handling smaller datasets as they use scientific understanding of the system to build causal relationships between inputs and outputs, instead of re-learning the relationships from the data (Baker et al. 2018). However, if a sufficient level of mechanistic knowledge is not available about a system, an ML algorithm would be better suited for the problem than a mechanistic model. To address the shortcomings in both ML and mechanistic modeling approaches, researchers are proposing the integration of both modeling paradigms (Baker et al. 2018). The hybrid mechanistic-ML models eliminate the need for "re-learning" causal relationships from large datasets by incorporating mechanistic knowledge about the system and using the ML approach to fill in the knowledge gaps through data-driven modeling (Venkatapurapu et al. 2022). This hybrid modeling paradigm will result in generalizable, interpretable models that overcome data and domain knowledge limitations and enable the creation of powerful digital twin applications in healthcare.

51.4 Challenges in Using Digital Twins in Healthcare

Though the use of digital twin technology holds great promise in transforming healthcare, a wide range of ethical, technical, and philosophical challenges need to be addressed to realize the power of digital twins. Researchers in the field have been discussing the ethical implications of digital twins in healthcare (Braun 2021). At a societal level, the usage of digital twins in healthcare has the potential to deliver significant benefits but if the technology is not accessible for all, it could become a driver for inequality (Bruynseels, Santoni de Sio, and van den Hoven 2018). In this article, we focus on the technical and philosophical challenges related to the adoption of digital twins in healthcare. Among the technical challenges, we shed light on the complexity of models and scalability of digital twin creation. We also discuss a key philosophical challenge in the adoption of digital twins – managing expectations for successful adoption and realizing the full potential of digital twins in healthcare.

51.4.1 Complexity of Models

One of the often-discussed challenges in modeling is selecting the right model for the problem being solved. Model selection (Forster 2000; Myung 2000) determines the generalizability of the model, and hence becomes an important step in the digital twin creation process. Highly complex models can provide a good fit to the training data but may result in overfitting and poor performance when applied to the test dataset. This applies to both ML and mechanistic modeling techniques. Besides potential overfitting, as the model complexity increases, its interpretability decreases. This is especially true for ML models. The aim of model selection should be to achieve generalizability and interpretability of the model that is enabling digital twin creation.

51.4.2 Scalability of the Digital Twin Creation Process

The function of a digital twin depends on how close the digital twin is to the actual physical object, i.e., fidelity. High-fidelity digital twins offer more realistic detection and testing environments for the physical object, thus providing tremendous advantages in the functioning and maintenance of real-world objects. Oftentimes, the data required to create high-fidelity digital twins do not exist and even if they do, they may not be directly measurable outside of a specialized research setting. Mechanistic physiology models overcome this challenge as the inherent structure of the model built-in using the existing scientific knowledge reduces the burden of data requirement. However, one of the major challenges in creating high-fidelity digital twins using a mechanistic model is the computational complexity involved. The simulation model conceptually represents a generic human being. To create a digital twin of a specific individual, the simulation model needs to be customized to fit data specific to that individual, a process referred to as calibration or parameter estimation. The fidelity of the digital twin to its real-world human counterpart depends not only on the quality of the data but also on the calibration. The process of calibrating a simulation model to an individual's data thus becomes a critical step in the creation of a digital twin.

Calibration of a model to a specific patient requires extensive exploration of its parameter space. This requires hundreds if not thousands or millions of simulations to find a near-optimal solution in a high-dimensional parameter space (Wang 1997). Depending on the computing power, this process can take from several minutes to hours on a typical personal computer, or in the case of large models could be altogether unfeasible outside high-performance computing environments. This computational complexity presents two challenges for digital twin applications – scalability and execution in real time. To address the scalability challenge, we have developed a scalable infrastructure (Venkatapurapu et al.) using cloud computing to create high-fidelity digital twins in a semi-automatic manner. This methodology can be used in applications that require high-fidelity digital twins but do not require their creation to happen in real time. For applications that need real-time digital twin creation, but the fidelity requirement is not extremely high, for example, a physician or patient education tool, the process of parameter exploration can be replaced by matching from a virtual patient library (VPL). A VPL is a collection of model instances, each with some variation in parameters, that has been pre-simulated and key outputs, typically observable variables, are stored in a database (Venkatapurapu et al. 2022). In this approach, extensive simulations are replaced by a database query to identify model parameters consistent with the observed state of a patient. The quality of the match, and hence the fidelity of the digital twin, depends on the quality of the VPL. The VPL can be continuously updated whenever a patient with a poor match is discovered.

51.4.3 Expectations from a Digital Twin – Toward Realizing Their Full Potential

Beyond complexity, another major challenge for digital twins in healthcare is identifying the most purposeful and best application as well as managing expectations of those using digital twin–based technologies. A recap of the historical application of digital twins in Formula One illustrates this challenge. While initially the promise of models was thought to be able to predict the lap or even entire race finishing time of a car, it was soon realized that the twins were best served to inform sensitivity analyses. The other cars, environmental conditions, and even the performance of the drivers influenced performance prediction. However, when the focus was instead turned to how best to optimize fuel consumption, tire management, and even to inform drivers on when best to apply acceleration and braking, the true potential of the twins was realized. Similarly, in health, while longevity or discrete biomarker predictions far into the future might be envisioned, it is the potential to inform the sensitivity of systemic effects to inputs like a drug, diet, exercise, or sleep that create the incredible promise. Pivoting the users of digital twins from a singular mindset focused on prediction to instead an invaluable mechanism for steerage along the optimal path is often a challenge of both education and application.

51.5 Conclusion and Future Directions

How we drive today is much smarter than how we live. Today, cars are overloaded with sensors for fuel, tire pressure, exhaust gas, video, infrared, and rear-view cameras. These sensors acquire data related to vehicle conditions, road conditions, and traffic conditions to help the drivers be aware of the vehicle's health, stay focused on their path and see ahead of the curve, and with GPS, they can do so, literally. What if we could apply that same approach to humans and provide the ability to see ahead of the curve for an

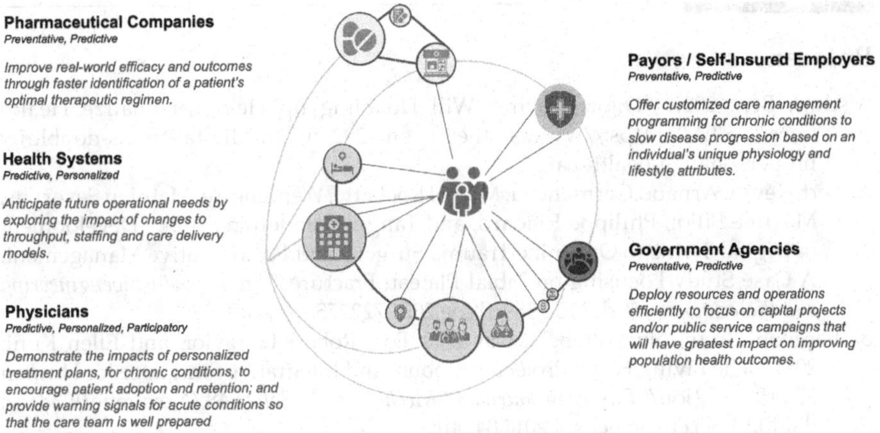

Pharmaceutical Companies
Preventative, Predictive

Improve real-world efficacy and outcomes through faster identification of a patient's optimal therapeutic regimen.

Health Systems
Predictive, Personalized

Anticipate future operational needs by exploring the impact of changes to throughput, staffing and care delivery models.

Physicians
Predictive, Personalized, Participatory

Demonstrate the impacts of personalized treatment plans, for chronic conditions, to encourage patient adoption and retention; and provide warning signals for acute conditions so that the care team is well prepared

Payors / Self-Insured Employers
Preventative, Predictive

Offer customized care management programming for chronic conditions to slow disease progression based on an individual's unique physiology and lifestyle attributes.

Government Agencies
Preventative, Predictive

Deploy resources and operations efficiently to focus on capital projects and/or public service campaigns that will have greatest impact on improving population health outcomes.

FIGURE 51.1
Digital twins for a patient-centric future of healthcare.

individual's health? Digital twins are an integral part of building a similar future for healthcare (Figure 51.1).

The advancement of digital twin technologies could help break down the current silos within healthcare by creating a line of sight to expected outcomes along the value chain. By integrating the data from wearable and non-wearable sensors along with simulation models of human physiology, one can create high-fidelity digital twins of individuals. The insights from digital twins could help patients in providing early warning signs, akin to a check engine light for the body , similar to a warning indicator in a car that alerts you to engine issues, obtaining an accelerated diagnosis, and more effective treatments enabled by stakeholders at all stages of the patient healthcare journey. Pharmaceutical companies would benefit greatly from faster identification of a patient's optimal therapeutic regimen to improve real-world efficacy. Physicians would have the tools to develop highly personalized treatment regimens for managing chronic conditions that clearly demonstrate expected patient outcomes at an individual level, supported by payor or employer-sponsored programs. Health systems would be able to flex care delivery models in anticipation of the rising demand. Government agencies would have the necessary insights to deploy resources and scale operations efficiently and design public health campaigns that result in improved population health outcomes and reduce overall healthcare costs. We envision a digital twin–powered patient-centric future in which individuals are aware of their health, stay on track with the help of timely nudges from their digital twins and a proactive health ecosystem, identify problems before they arise, and have full control of their life, thus shifting the focus of healthcare from treatment to prevention (Flores et al. 2013).

References

Alsdurf, Ben. 2021. "Digital Twins: Will Doubling up Help Personalize Health Care?" *STAT.* https://www.statnews.com/2021/08/10/digital-twins-doubling-up-personalize-health-care/.

Aubert, Kévin, Arnaud Germaneau, Michel Rochette, Wenfeng Ye, Mathieu Severyns, Maxime Billot, Philippe Rigoard, and Tanguy Vendeuvre. 2021. "Development of Digital Twins to Optimize Trauma Surgery and Postoperative Management. A Case Study Focusing on Tibial Plateau Fracture." *Frontiers in Bioengineering and Biotechnology* 9. doi:10.3389/fbioe.2021.722275.

Baillargeon, Brian, Nuno Rebelo, David D. Fox, Robert L. Taylor, and Ellen Kuhl. 2014. "The Living Heart Project: A Robust and Integrative Simulator for Human Heart Function." *European Journal of Mechanics. A, Solids* 48 (November): 38–47. doi:10.1016/j.euromechsol.2014.04.001.

Baker, Ruth E., Jose-Maria Peña, Jayaratnam Jayamohan, and Antoine Jérusalem. 2018. "Mechanistic Models versus Machine Learning, a Fight Worth Fighting for the Biological Community?" *Biology Letters* 14 (5). doi:10.1098/rsbl.2017.0660.

Braun, Matthias. 2021. "Represent Me: Please! towards an Ethics of Digital Twins in Medicine." *Journal of Medical Ethics* 47 (6): 394–400. doi:10.1136/medethics-2020-106134.

Bruynseels, Koen, Filippo Santoni de Sio, and Jeroen van den Hoven. 2018. "Digital Twins in Health Care: Ethical Implications of an Emerging Engineering Paradigm." *Frontiers in Genetics* 9. doi:10.3389/fgene.2018.00031.

Business Wire. 2021. "Q Bio Gemini," April 29. https://www.businesswire.com/news/home/20210429005437/en/Q-Bio-Announces-First-Clinical-%E2%80%9CDigital-Twin%E2%80%9D-Platform-and-Novel-Whole-Body-Scanner-and-Major-Investment-From-Kaiser-Foundation-Hospitals.

Business Wire. 2022. "Unlearn - TwinRCT," September 28. https://www.businesswire.com/news/home/20220928005391/en/European-Medicines-Agency-Qualifies-Unlearn%E2%80%99s-AI-powered-Method-for-Running-Smaller-Faster-Clinical-Trials.

Campbell, Sarah. 2017. "The Quantified Patient Checks In: Larry Smarr?S Experiments in Self-Tracking for Health." *IEEE Pulse* 8 (4): 4–10. doi:10.1109/MPUL.2017.2701739.

Chen, Julia H., Momoko Fukasawa, Qian Chen, Samuel P. Burns, Kei Kumar, Sr. Nirengi Shinsuke, Kaoru Takahashi, et al. 2018. "Diabetes Prevention Using a Simulation Model That Explains Individual Variability in Response to Diet Change." *Diabetes* 67 (Supplement_1). doi:10.2337/db18-1892-P.

Coorey, Genevieve, Gemma A. Figtree, David F. Fletcher, Victoria J. Snelson, Stephen Thomas Vernon, David Winlaw, Stuart M. Grieve, et al. 2022. "The Health Digital Twin to Tackle Cardiovascular Disease-a Review of an Emerging Interdisciplinary Field." *NPJ Digital Medicine* 5 (1): 126. doi:10.1038/s41746-022-00640-7.

"Crohn's Disease Digital Twin Simulator for Shared Decision-Making",2021, https://www.pwc.com/jp/en/press-room/takeda-project210518.html.

"Crohn's Disease Simulator for MSL-KOL Engagement", 2019, https://www.pwc.com/jp/en/press-room/takeda-project190912.html.

Cushing, Kelly, and Peter D. R. Higgins. 2021. "Management of Crohn Disease: A Review." *JAMA* 325 (1): 69–80. doi:10.1001/jama.2020.18936.

Dada, Joseph O., and Pedro Mendes. 2011. "Multi-Scale Modelling and Simulation in Systems Biology." *Integrative Biology* 3 (2): 86–96. doi:10.1039/c0ib00075b.

Dalal, Sushila R., and Russell D. Cohen. 2015. "What to Do When Biologic Agents Are Not Working in Inflammatory Bowel Disease Patients." *Gastroenterology & Hepatology* 11 (10): 657–665.

D'alessandro, Paul M., Mark Paich, Samuel Pierce Burns, Joydeep Sarkar, Gaurav Dwivedi, and Colleen Chelini. 2019. "System and Method for Physiological Health Simulation." Google Patents.

Derycke, Lucie, Jean Sénémaud, David Perrin, Stephane Avril, Pascal Desgranges, Jean-Noel Albertini, Frederic Cochennec, and Stephan Haulon. 2020. "Patient Specific Computer Modelling for Automated Sizing of Fenestrated Stent Grafts." *European Journal of Vascular and Endovascular Surgery* 59 (2): 237–246. doi:10.1016/j.ejvs.2019.10.009.

Dinter, Raymon van, Bedir Tekinerdogan, and Cagatay Catal. 2022. "Predictive Maintenance Using Digital Twins: A Systematic Literature Review." *Information and Software Technology* 151: 107008. doi:10.1016/j.infsof.2022.107008.

Dulai, Parambir S., Laurent Peyrin-Biroulet, Dirk Demuth, Karen Lasch, Kristen A. Hahn, Dirk Lindner, Haridarshan Patel, and Vipul Jairath. 2020. "Early Intervention with Vedolizumab and Longer-Term Surgery Rates in Crohn's Disease: Post Hoc Analysis of the GEMINI Phase 3 and Long-Term Safety Programmes." *Journal of Crohn's and Colitis* 15 (2): 195–202. doi:10.1093/ecco-jcc/jjaa153.

El Saddik, Abdulmotaleb. 2018. "Digital Twins: The Convergence of Multimedia Technologies." *IEEE MultiMedia* 25 (2): 87–92. doi:10.1109/MMUL.2018.023121167.

Fiocchi, Claudio. 2018. "Inflammatory Bowel Disease: Complexity and Variability Need Integration." *Frontiers in Medicine* 5: 75. doi:10.3389/fmed.2018.00075.

Flores, Mauricio, Gustavo Glusman, Kristin Brogaard, Nathan D. Price, and Leroy Hood. 2013. "P4 Medicine: How Systems Medicine Will Transform the Healthcare Sector and Society." *Personalized Medicine* 10 (6): 565–576. doi:10.2217/pme.13.57.

Forster, Malcolm R. 2000. "Key Concepts in Model Selection: Performance and Generalizability." *Journal of Mathematical Psychology* 44 (1): 205–231. doi:10.1006/jmps.1999.1284.

Frei, Roy, Nicolas Fournier, Jonas Zeitz, Michael Scharl, Bernhard Morell, Thomas Greuter, Philipp Schreiner, et al. 2019. "Early Initiation of Anti-TNF Is Associated with Favourable Long-Term Outcome in Crohn's Disease: 10-Year-Follow-up Data from the Swiss IBD Cohort Study." *Journal of Crohn's & Colitis* 13 (10): 1292–1301. doi:10.1093/ecco-jcc/jjz057.

Fuller, Aidan, Zhong Fan, Charles Day, and Chris Barlow. 2020. "Digital Twin: Enabling Technologies, Challenges and Open Research." *IEEE Access* 8: 108952–108971. doi:10.1109/ACCESS.2020.2998358.

Glaessgen, Edward, and David Stargel. 2012. "The Digital Twin Paradigm for Future NASA and U.S. Air Force Vehicles." In *53rd AIAA/ASME/ASCE/AHS/ASC Structures, Structural Dynamics and Materials Conference*. doi:10.2514/6.2012-1818.

Gubatan, John, Steven Levitte, Akshar Patel, Tatiana Balabanis, Mike T. Wei, and Sidhartha R. Sinha. 2021. "Artificial Intelligence Applications in Inflammatory Bowel Disease: Emerging Technologies and Future Directions." *World Journal of Gastroenterology* 27 (17): 1920–1935. doi:10.3748/wjg.v27.i17.1920.

Harutyunyan, Misak, Yunjie Huang, Kyu-Shik Mun, Fanmuyi Yang, Kavisha Arora, and Anjaparavanda P. Naren. 2018. "Personalized Medicine in CF: From Modulator Development to Therapy for Cystic Fibrosis Patients with Rare CFTR Mutations." *American Journal of Physiology. Lung Cellular and Molecular Physiology* 314 (4): L529–L543. doi:10.1152/ajplung.00465.2017.

Henkel, Maurice, Tobias Horn, Francois Leboutte, Pawel Trotsenko, Sarah Gina Dugas, Sarah Ursula Sutter, Georg Ficht, et al. 2022. "Initial Experience with AI Pathway Companion: Evaluation of Dashboard-Enhanced Clinical Decision Making in Prostate Cancer Screening." *PloS One* 17 (7): e0271183. doi:10.1371/journal.pone.0271183.

Issa, Amalia M. 2007. "Personalized Medicine and the Practice of Medicine in the 21st Century." *McGill Journal of Medicine: MJM : An International Forum for the Advancement of Medical Sciences by Students* 10 (1): 53–57.

Kamel Boulos, Maged N Kamel Boulos, and Peng Zhang. 2021. "Digital Twins: From Personalised Medicine to Precision Public Health." *Journal of Personalized Medicine* 11 (8). doi:10.3390/jpm11080745.

Kochunas, Brendan, and Xun Huan. 2021. "Digital Twin Concepts with Uncertainty for Nuclear Power Applications." *Energies* 14 (14). doi:10.3390/en14144235.

Kusiak, Andrew. 2017. "Smart Manufacturing Must Embrace Big Data." *Nature* 544 (7648): 23–25. doi:10.1038/544023a.

Lazebnik, Yuri. 2002. "Can a Biologist Fix a Radio?--Or, What I Learned While Studying Apoptosis." *Cancer Cell* 2 (3): 179–182. doi:10.1016/s1535-6108(02)00133-2.

Lu, Yuqian, Chao Liu, Kevin I.-Kai Wang, Huiyue Huang, and Xun Xu. 2020. "Digital Twin-Driven Smart Manufacturing: Connotation, Reference Model, Applications and Research Issues." *Robotics and Computer-Integrated Manufacturing* 61: 101837. doi:10.1016/j.rcim.2019.101837.

Macomber, John D., and Akiko Kanno. 2022. *Sekisui House and the In-Home Early Detection Platform*. Case 222-070. Harvard Business School.

Majumder, Sumit, Emad Aghayi, Moein Noferesti, Hamidreza Memarzadeh-Tehran, Tapas Mondal, Zhibo Pang, and M. Jamal Deen. 2017. "Smart Homes for Elderly Healthcare-Recent Advances and Research Challenges." *Sensors* 17 (11). doi:10.3390/s17112496.

Mantzaris, Gerassimos J., Christos Zeglinas, Angeliki Theodoropoulou, Ioannis Koutroubakis, Eleni Orfanoudaki, Konstantinos Katsanos, Dimitrios Christodoulou, et al. 2021. "The Effect of Early vs Delayed Initiation of Adalimumab on Remission Rates in Patients with Crohn's Disease with Poor Prognostic Factors: The MODIFY Study." *Crohn's & Colitis 360* 3 (4). doi:10.1093/crocol/otab064.

Matthew, C.Z.V., C.M. Chelini, G. Dwivedi, P.M. D'Alessandro, K. Krause, C. Culak, and S.P. Burns. 2019. "PMU94 Quantifying the Health Impact and Return on Investment of a Home Meal Delivery Service Using Simulation Modeling." *Value in Health* 22: S266. doi:10.1016/j.jval.2019.04.1255.

Myung, In Jae. 2000. "The Importance of Complexity in Model Selection." *Journal of Mathematical Psychology* 44 (1): 190–204. doi:10.1006/jmps.1999.1283.

Nardini, Christine, Venet Osmani, Paola G. Cormio, Andrea Frosini, Mauro Turrini, Christos Lionis, Thomas Neumuth, Wolfgang Ballensiefen, Elio Borgonovi, and Gianni D'Errico. 2021. "The Evolution of Personalized Healthcare and the Pivotal Role of European Regions in Its Implementation." *Personalized Medicine* 18 (3): 283–294. doi:10.2217/pme-2020-0115.

Nurea. "Praevaorta : An Unprecedented SAMD Solution for Vascular Diseases Quantification." https://www.nurea-soft.com/praevaorta-software/.

"Paving the Way for Personalized Medicine: FDA's Role in a New Era of Medical Product Development." 2013. Silver Spring: US Food and Drug Administration. https://www.fdanews.com/ext/resources/files/10/10-28-13-Personalized-Medicine.pdf.

Peirlinck, M., F. Sahli Costabal, J. Yao, J. M. Guccione, S. Tripathy, Y. Wang, D. Ozturk, et al. 2021. "Precision Medicine in Human Heart Modeling." *Biomechanics and Modeling in Mechanobiology* 20 (3): 803–831. doi:10.1007/s10237-021-01421-z.

Pettey, Christy. 2017. *Prepare for the Impact of Digital Twins*. Gartner. https://www.gartner.com/smarterwithgartner/prepare-for-the-impact-of-digital-twins.

Philips. "Philips Dynamic HeartModel." https://www.usa.philips.com/healthcare/resources/feature-detail/ultrasound-heartmodel.

PrediSurge. "PrediSurge -3D Numerical Simulation." https://www.predisurge.com/3d-numerical-simulation/.

PwC. "Bodylogical(r)." https://www.pwc.com/jp/en/industries/introducing-bodylogical.html.

Rao, Dattaraj Jagdish, and Shraddha Mane. 2019. "Digital Twin Approach to Clinical DSS with Explainable AI." *CoRR* abs/1910.13520. https://arxiv.org/abs/1910.13520.

Sarkar, Joydeep, Gaurav Dwivedi, Qian Chen, Iris E. Sheu, Mark Paich, Colleen M. Chelini, Paul M. D'Alessandro, and Samuel P. Burns. 2018. "A Long-Term Mechanistic Computational Model of Physiological Factors Driving the Onset of Type 2 Diabetes in an Individual." *PLoS One* 13 (2): 1–37. doi:10.1371/journal.pone.0192472.

Shu, Liming, Jiang Yao, Ko Yamamoto, Takashi Sato, and Naohiko Sugita. 2021. "In Vivo Kinematical Validated Knee Model for Preclinical Testing of Total Knee Replacement." *Computers in Biology and Medicine* 132: 104311. doi:10.1016/j.compbiomed.2021.104311.

Triantafilou, Josh. 2021. "The Role of Digital Twins in Smart Cities." *ESRI Canada*. https://resources.esri.ca/news-and-updates/the-role-digital-twins-in-smart-cities.

Venkatapurapu, Sai Phanindra, Ryuichi Iwakiri, Eri Udagawa, Nikhil Patidar, Zhen Qi, Ryoko Takayama, Kei Kumar, et al. 2022. "A Computational Platform Integrating a Mechanistic Model of Crohn's Disease for Predicting Temporal Progression of Mucosal Damage and Healing." *Advances in Therapy* 39 (7): 3225–3247. doi:10.1007/s12325-022-02144-y.

Venkatapurapu, Sai Phanindra, Mrinal K. Mandal, Jerome P. Offner, Rakesh V. C. Kapila, Gaurav Dwivedi, Qian Chen, Julia H. Chen, Samuel P. Burns, and Paul M. D'Alessandro. 2021. "Creating Digital Twins at Scale." https://patents.google.com/patent/US20210357556A1/en.

Venkatapurapu, Sai Phanindra, Chezev Matthew, Abhinav Aggarwal, and Gaurav Dwivedi. 2018. "Integrating Lifestyle Factors in a Quantitative Systems Pharmacology (Qsp) Model of Human Metabolism to Predict Long Term Cardiovascular Disease Risk." *Journal of Pharmacokinetics and Pharmacodynamics* 45: S88–S89.

Visvikis-Siest, Sophie, Danai Theodoridou, Maria-Spyridoula Kontoe, Satish Kumar, and Michael Marschler. 2020. "Milestones in Personalized Medicine: From the Ancient Time to Nowadays-the Provocation of COVID-19." *Frontiers in Genetics* 11. doi:10.3389/fgene.2020.569175.

Wang, Q. J. 1997. "Using Genetic Algorithms to Optimise Model Parameters." *Environmental Modelling & Software* 12 (1): 27–34. doi:10.1016/S1364-8152(96)00030-8.
You, Yingchao, Chong Chen, Fu Hu, Ying Liu, and Ze Ji. 2022. "Advances of Digital Twins for Predictive Maintenance." *Procedia Computer Science* 200: 1471–1480. doi:10.1016/j.procs.2022.01.348.

Printed in the United States
by Baker & Taylor Publisher Services

Printed in the United States
by Baker & Taylor Publisher Services